J. Belzner

Dr Johannes Belzner
Inst Organische Chem
Tammannstr 2
37077 GOTTINGEN,
Germany

American Chemical Society

D1751572

Organosilicon Chemistry II

Edited by N. Auner and J. Weis

VCH

© VCH Verlagsgesellschaft mbH. D-69451 Weinheim (Federal Republic of Germany), 1996

Distribution:
VCH, P.O. Box 10 11 61, D-69451 Weinheim (Federal Republic of Germany)
Switzerland: VCH, P.O. Box, CH-4020 Basel (Switzerland)
United Kingdom and Ireland: VCH (UK) Ltd., 8 Wellington Court, Cambridge CB1 1HZ (England)
USA and Canada: VCH, 220 East 23rd Street, New York, NY 10010-4606 (USA)
Japan: VCH, Eikow Building, 10-9 Hongo 1-chome, Bunkyo-ku, Tokyo 113 (Japan)

ISBN 3-527-29254-3

Organosilicon Chemistry II

From Molecules to Materials

Edited by Norbert Auner and Johann Weis

VCH Weinheim · New York ·
Basel · Cambridge · Tokyo

Prof. Dr. N. Auner
Institut für Chemie
Fachinstitut für Anorganische
und Allgemeine Chemie
Humboldt-Universität zu Berlin
Hessische Straße 1-2
D-10115 Berlin
Germany

Dr. J. Weis
Wacker-Chemie GmbH
Geschäftsbereich S
Werk Burghausen
Johannes-Heß-Straße 24
D-84489 Burghausen
Germany

This book was carefully produced. Nevertheless, authors, editors and publisher do not warrant the information contained therein to be free of errors. Readers are advised to keep in mind that statements, data, illustrations, procedural details or other items may inadvertently be inaccurate.

Published jointly by
VCH Verlagsgesellschaft mbH, Weinheim (Federal Republic of Germany)
VCH Publishers, Inc., New York, NY (USA)

Editorial Director: Dr. Thomas Mager
Production Manager: Dipl.-Ing. (FH) Hans Jörg Maier

Library of Congress Card No. applied for.

A catalogue record for this book is available from the British Library.

Deutsche Bibliothek Cataloguing-in-Publication Data:
Organosilicon chemistry : from molecules to materials / ed. by Norbert Auner ; Johann Weis. - Weinheim ; New York ; Basel ; Cambridge ; Tokyo : VCH.
  NE: Auner, Norbert [Hrsg.]
2 (1996)
  ISBN 3-527-29254-3

© VCH Verlagsgesellschaft mbH. D-69451 Weinheim (Federal Republic of Germany), 1996

Printed on acid-free and chlorine-free paper.

All rights reserved (including those of translation in other languages). No part of this book may be reproduced in any form - by photoprinting, microfilm, or any other means - nor transmitted or translated into machine language without written permission from the publishers. Registered names, trademarks, etc. used in this book, even when not specifically marked as such, are not to be considered unprotected by law.
Printing: betz-druck gmbh, D-64291 Darmstadt. Bookbinding: Wilh. Osswald + Co. GmbH + Co. KG, D-67433 Neustadt
Printed in the Federal Republic of Germany.

Preface

The "I. Münchner Silicontage" (First Munich Silicon Days) was jointly organized by the Gesellschaft Deutscher Chemiker and Wacker-Chemie GmbH on the occasion of the 50th anniversary of the Müller-Rochow synthesis. This meeting, held in August 1992, was met with so much enthusiasm by the German "Organosilicon Community" that the organizers were encouraged to hold a consecutive meeting in 1994.

During the "I. Münchner Silicontage" the "Wacker-Silicon-Preis" was awarded to the two pioneers of silicone chemistry - Prof. Dr. Richard Müller and Prof. Dr. Eugene Rochow. In the course of the "II. Münchner Silicontage" this award was conferred on Prof. Dr. Edwin Hengge, Technische Universität Graz, for his fundamental work on linear, cyclic, and oligomeric polysilanes.

In addition, two world renowned organosilicon chemists, Prof. Dr. Hubert Schmidbaur, Technische Universität München, and Prof. Dr. Nils Wiberg, Ludwig-Maximilians-Universität München, celebrated their 60th birthdays in 1994, thereby providing a special feature to the "II. Münchner Silicontage".

As in the first meeting it was possible to invite well-known scientists from Germany and abroad to give lectures. In addition, there was again a remarkable poster session. The scientific yield was once more so rich that the organizers felt obliged to collect the lectures and poster contributions in a second volume of "Organosilicon Chemistry - From Molecules to Materials".

This book is divided into four chapters:

I   Tetravalent Organosilicon Compounds: Chemistry and Structure
II Organosubstituted Silicon with Unusual Coordination Numbers
III Organosilicon Metal Compounds: Their Use in Organosilicon Synthesis, Coordination Chemistry, and Catalysis
IV Silicon Polymers: Formation and Application

We would like to thank the numerous authors for their enthusiastic cooperation, which made this overview of current organosilicon chemistry possible.

In this volume we have considered the constructive criticism of some reviewers of Volume I and have allowed more space, especially for the poster presentations. Consequently, this volume gained more weight - we think this is appropriate to the importance of organosilicon chemistry in research and application!

Im August 1995   *Prof. Dr. Norbert Auner, Dr. Johann Weis*

Acknowledgement

The highly motivated co-workers who made the attractive layout of this book possible are, in alphabetical order: Michael Backer, Brigitte Goetze, Martin Grasmann, Dr. Bernhard Herrschaft, Dr. Mathias Kersten, Dr. Stefanie König-Thieme, Karsten Korth, Dr. Thomas Müller, Björn Schade, Gabriele Sommer, Hans-Uwe Steinberger, Anja Wegner and Dr. Wolfgang Ziche. An outstanding task was performed by Claus-Rüdiger Heikenwälder, who coordinated the editorial work.

We thank them all for this formidable commitment!

Prof. Dr. Norbert Auner
Humboldt-Universität
zu Berlin

Dr. Johann Weis
Wacker-Chemie GmbH
Burghausen

Contents

I   Tetravalent Organosilicon Compounds: Chemistry and Structure

Introduction 1
N. Auner, J. Weis

Small Organosilicon Molecules – Synthesis, Stereochemistry, and Reactivity 3
H. Schmidbaur

Properties of Si-functionalized (Fluoromethyl)silanes 19
H. Beckers, H. Bürger, P. Moritz

Silicides as Catalysts for Hydrodehalogenation of Silicon Tetrachloride 31
H. Walter, G. Roewer, K. Bohmhammel, F. Vogt, W. Mörke

Silacyclobutenes – Synthesis and Reactivity 41
M. Backer, M. Grasmann, W. Ziche, N. Auner, C. Wagner, E. Herdtweck, W. Hiller, M. Heckel

Allyl Cleavage of 2-Silanorbornenes – A Facile Synthesis of Cyclopentenyl Compounds of 49
Organochlorosilanes and of 2-Silanorbornanes
H.-U. Steinberger, N. Auner

Base-catalyzed Hydrogenation of Chlorosilanes by Organotin Hydrides 55
U. Pätzold, G. Roewer

The Catalytic Hydrogenation of Methylchlorodisilanes with Tri-*n*butylstannane 63
U. Herzog, G. Roewer, K. Herzog

Disproportionation of Tetrachlorodimethyldisilane – NMR Spectroscopic Identification of 69
the Primary Products
E. Brendler, K. Leo, B. Thomas, R. Richter, G. Roewer, H. Krämer

Cyclotrisilane and Silacyclopropane – An Equilibrium 75
J. Belzner, H. Ihmels

Synthesis and Spectroscopy of Phenylated and Halogenated Trisilanes and Disilanes 81
W. Köll, K. Hassler

Investigations Concerning the Electrochemical Formaion of Di- and Polysilanes 89
S. Graschy, Ch. Jammegg, E. Hengge

Syntheses, ^{29}Si-NMR-Spectra, and Vibrational Spectra of Methylated Trisilanes 95
K. Schenzel, K. Hassler

New Ways to Difunctional Cyclosilanes 101
A. Spielberger, P. Gspaltl, E. Hengge

New Polycyclic Si-Ring Systems 105
P. Gspaltl, A. Spielberger, E. Hengge

Preparation and Reactions of Permethylated Cyclooligosilane Alkali Metal Derivatives 109
F. Uhlig, P. Gspaltl, E. Pinter, E. Hengge

Synthesis, Reactivity, and Spectroscopy of Phenylated Cyclotetrasilanes and Cyclopentasilanes 113
U. Pöschl, H. Siegl, K. Hassler

Oligosilanylsulfanes: New Functional Derivatives of Higher Silicon Hydrides 121
H. Stüger, P. Lassacher

Synthesis and Reactions of Iminosilanes 127
D. Großkopf, U. Klingebiel, J. Niesmann

1,3-Diazabutadienes and Amidinates as Ligands to Silicon Centers 133
H. H. Karsch, F. Bienlein

Specific Insertion of O=C=S into the N-Si Bond of an Extremely Electron-rich Molecule as 141
Evident from the X-Ray Crystal Structure

A. Ehlend, H.-D. Hausen, W. Kaim

Carbene and Carbenoid Chemistry of Silyldiazoacetic Esters: 149
The Silyl Group as a Substituent and a Functional Group

G. Maas

The Function of the Trimethylsilyl Substituent in the Syntheses of Low-valent Phosphorus and 161
Arsenic Containing Compounds

*G. Becker, G. Ditten, K. Hübler, U. Hübler, K. Merz, M. Niemeyer,
N. Seidler, M. Westerhausen, Z. Zheng*

Silicon and Phosphinomethanides: Novel Heteroelement Substituted Methanes and Ylides 187

H. H. Karsch, R. Richter

Supersilyl Phosphorus Compounds 195

N. Wiberg, A. Wörner, H. Nöth, K. Karaghiosoff

Silicon-Phosphorus, -Arsenic, -Antimony, and -Bismuth Cages: Syntheses and Structures 203

K. Hassler

Synthesis, Reactivity, and Molecular Structure of Diphosphadisilacyclobutanes, ... 209
Bis(stannyl)silyl-, and Bis(stannyl)germylphosphines

M. Waltz, M. Nieger, E. Niecke

On the Reaction of Silanols with Alcohols 215

I. Kohlheim, D. Lange, H. Kelling

4-Silyloxybenzothiopyrylium-Salts: A New Tool for the Stereoselective Annulation of 219
S-Heterocycles

U. Beifuss, H. Gehm, M. Tietze

Diastereoselective Synthesis of Annulated Quinolones by Transformations with 225
4-Silyloxyquinolinium-Salts

S. Ledderhose, U. Beifuss

Biological Recognition of Enantiomeric Silanes and Germanes: Syntheses and Antimuscarinic 231
Properties of the Enantiomers of the Si/Ge Analogues Cyclohexyl(hydroxymethyl)phenyl-
(2-piperidinoethyl)silane and -Germane and Their Methiodides

D. Reichel, R. Tacke, P. G. Jones, G. Lambrecht, J. Gross, E. Mutschler, M. Waelbroeck

Biotransformation as a Preparative Method for the Synthesis of Optically Active Silanes, ...237
Germanes, and Digermanes: Studies on the (R)-Selective Microbial Reduction of
MePh(Me$_3$C)ElC(O)Me (El = Si, Ge), MePh(Me$_3$Ge)GeC(O)Me, and MePh(Me$_3$Si)GeC(O)Me
Using Resting Cells of *Saccharomyces Cerevisiae* (DHW S-3) as Biocatalyst

S. A. Wagner, S. Brakmann, R. Tacke

Aldehyde-functionalized Silanes – New Compounds to Improve the Immobilization of 243
Biomolecules

J. Grobe, C. Brüning, M. Wessels

II Organosubstituted Silicon with Unusual Coordination Numbers

Introduction 249
N. Auner, J. Weis

Silylenes, Stable and Unstable 251
M. Denk, R. West, R. Hayashi, Y. Apeloig, R. Pauncz, M. Karni

Silylenes and Multiple Bonds to Silicon: Synergism between Theory and Experiment 263
Y. Apeloig, M. Karni, T. Müller

Heterosubstituted Silylenes: Cycloadditions with Heterodienes and -trienes 289
J. Heinicke, B. Gehrhus, S. Meinel

Decomposition Products of Methyltrichlorosilane 295
S. Leistner, S. Baumann, G. Marx

C$_2$H$_2$Si Isomers: Generation by Pulsed Flash Pyrolysis and Matrix-Spectroscopic Identification 303
G. Maier, H. P. Reisenauer, H. Pacl

Silylene and Disilene Additions to the Double Bonds of Alkenes, 1,3-Dienes, and 309
Hetero-1,3-dienes

E. Kroke, P. Will, M. Weidenbruch

Generation of Silylenes and Silaethenes by Dehalogenation of Chlorosilanes 317

J. Grobe, T. Schierholt

Triple Bonds to Silicon: Substituted Silanitriles Versus Silaisonitriles – A Theoretical Study 321

K. Albrecht, Y. Apeloig

The Nature of Organosilicon Cations and Their Interactions 329

C. Maerker, J. Kapp, P. von Ragué Schleyer

NMR Spectroscopic Investigation of the β-Silyl Effect in Carbocations 361

H.-U. Siehl, B. Müller, M. Fuß, Y. Tsuji

Underloaded and Overloaded Unsaturated Silicon Compounds 367

N. Wiberg

The Synthesis of Transient Silenes using the Principle of the Peterson Reaction 389

C. Krempner, H. Reinke, H. Oehme

Cycloaddition Reactions of 1,1-Dichloro-2-neopentylsilene with Pentafulvenes 399

C.-R. Heikenwälder, N. Auner

Silaneimines 405

N. Wiberg, H.-W. Lerner

Recent Explorations ot the Chemistry of Pentacoordinate Silicon 411

A. R. Bassindale, S. G. Glynn, J. Jiang, D. J. Parker, R. Turtle, P. G. Taylor, S. S. D. Brown

Syntheses, Structures, and Properties of Molecular λ^5Si-Silicates Containing Bidentate 427
1,2-Diolato(2−) Ligands Derived from α-Hydroxycarboxylic Acids, Acetohydroximic Acid,
and Oxalic Acid: New Results in the Chemistry of Pentacoordinate Silicon

R. Tacke, O. Dannappel, M. Mühleisen

Syntheses and Solution-State NMR Studies of Zwitterionic Spirocyclic λ^5Si-Organosilicates 447
Containing Two Identical Unsymmetrically Substituted 1,2-Benzenediolato(2–) Ligands

M. Mühleisen, R. Tacke

Intramolecular Ligand Exchange of Zwitterionic Spirocyclic Bis[1,2-benzenediolato(2–)]- 453
organosilicates: Ab Initio Studies of the Bis[1,2-benzenediolato(2–)]hydridosilicate(1–) Ion

O. Dannappel, R. Tacke

Highly Coordinated Silicon Compounds – Hydrazino Groups as Intramolecular Donors 459

J. Belzner, D. Schär

III Organosilicon Metal Compounds: Their Use in Organosilicon Synthesis, Coordination Chemistry, and Catalysis

Introduction 467

N. Auner, J. Weis

Silicon Frameworks and Electronic Structures of Novel Solid Silicides 469

R. Nesper, A. Currao, S. Wengert

Alkali Metal Derivatives of Tris(trimethylsilyl)silane – Syntheses and 493
Molecular Structures

K. W. Klinkhammer, G. Becker, W. Schwarz

Mono-, Bis-, Tris-, and Tetrakis(lithiomethyl)silanes: 499
New Building Blocks for Organosilicon Compounds

C. Strohmann, S. Lüdtke

Neutral and Anionic Lithiumsilylamides as Precursors of Inorganic Ring Systems 505

I. Hemme, U. Klingebiel

Lithium*t*butylamidosilanes – Syntheses and Structures 511

G. Becker, S. Abele, J. Dautel, G. Motz, W. Schwarz

Dilithiated Oligosilanes: Synthesis, Structure, and Reactivity 519

J. Belzner, U. Dehnert, D. Stalke

Synthesis of Gallium Hypersilyl Derivatives 525

R. Frey, G. Linti, K. Polborn, H. Schwenk

New Organometallic Silicon-Chalcogen Compounds 531

K. Merzweiler, U. Linder

Alternative Ligands XXXIII: Heterobimetallic Donor-Acceptor Interactions in Si/Ni-Cages: 541
Metallosilatranes

J. Grobe, H.-H. Niemeyer, R. Wehmschulte

Reactivity of π-Coordinated Chlorocyclopentadienylsilanes 547

S. Schubert, J. Hofmann, K. Sünkel

CO-, Isonitrile-, and Phosphine-induced Silyl Migration Reactions in Heterobimetallic 553
Fe–Pd and Fe–Pt Complexes

P. Braunstein, M. Knorr

Novel Diazomethylsilyl Substituted Fischer Carbene Complexes 565

D. Mayer, G. Maas

1-Metalla-2-sila-1,3-diene Compounds 569

M. Weinmann, H. Lang, O. Walter, M. Büchner

Novel Metallo-Silanols, -Silanediols, and -Silanetriols of the Iron and Chromium Group: 575
Generation, Structural Characterization, and Transformation to Metallo-Siloxanes

W. Malisch, S. Möller, R. Lankat, J. Reising, W. Schmitzer, O. Fey

Oligosilanyl-Tungsten Compounds – Precursors for Tungsten-Silicide CVD? 585

A. Zechmann, E. Hengge

IV Silicon Polymers: Formation and Application

Introduction 589
N. Auner, J. Weis

Silyl Modified Surfaces – New Answers to Old Problems 591
J. Grobe

Alkenyl Ethoxysilanes for the Synthesis of Silylester Functionalized Copolymers and 609
Surface Modification
J. Greb, J. Grobe

Silicone Surfactants – Development of Hydrolytically Stable Wetting Agents 613
K.-D. Klein, W. Knott, G. Koerner

Siloxanes Containing Vinyl Groups 619
S. Kupfer, I. Jansen, K. Rühlmann

Investigations of Chemical Heterogeneity in Polydimethylsiloxanes using SFC, HPLC, and 625
MALDI-MS
U. Just, R. P. Krüger

Vulcanization Kinetics of Addition-cured Silicone Rubber 633
D. Wrobel

Photo-crosslinked Polysiloxanes – Properties and Applications 649
K. Rose

Cationic Photo-crosslinking of α, ω-terminated Disiloxanes 655
A. Kunze, U. Müller, Ch. Decker, J. Weis, Ch. Herzig

Azo- and Triazene Modified Organosilicones as Polymeric Initiators for Graft Copolymers 659
R. Kollefrath, B. Voit, O. Nuyken, J. Dauth, B. Deubzer, T. Hierstetter, J. Weis

Core Shell Structures Based on Polyorgano Silicone Micronetworks Prepared in Microemulsion 665
F. Baumann, M. Schmidt, J. Weis, B. Deubzer, M. Geck, J. Dauth

Precrosslinked Poly(organosiloxane) Particles 673
M. Geck, B. Deubzer, J. Weis, G. Pepperl

Silsesquioxanes of Mixed Functionality – Octa[(3-chloropropyl)-$n$propyl-silsesquioxanes] and 685
Octa[(3-mercaptopropyl)-$n$propyl-silsesquioxanes] as Models of Organomodified Silica Surfaces
B. J. Hendan, H. C. Marsmann

New Functionalized Silsesquioxanes by Substitutions and Cage Rearrangement of 691
Octa[(3-chloropropyl)-silsesquioxane]
H. C. Marsmann, U. Dittmar, E. Rikowski

Investigations on the Thermal Behavior of Silicon Resins by Thermoanalysis, ^{29}Si NMR, and 697
IR Spectroscopy
H. Jancke, D. Schultze, H. Geissler

Convenient Approach to Novel Organosilicon Polymers 703
W. Uhlig

Silicon Carbide and Carbonitride Precursors via the Silicon-Silicon Bond Formation: 709
Chemical and Electrochemical New Perspectives
M. Bordeau, C. Biran, F. Spirau, J.-P. Pillot, M. Birot, J. Dunoguès

Synthesis of Spinnable Poly(silanes/-carbosilanes) and Their Conversion into SiC Fibers 719
R. Richter, G. Roewer, H.-P. Martin, E. Brendler, H. Krämer, E. Müller

Polymeric Silylcarbodiimides – Novel Route to Si–C–N Ceramics 725
A. Kienzle, K. Wurm, J. Bill, F. Aldinger, R. Riedel

New Modified Polycarbosilanes 733
W. Habel, W. Haeusler, A. Oelschläger, P. Sartori

Heteropolysiloxanes by Sol-Gel Techniques: Composite Materials with Interesting Properties 737
H. K. Schmidt

Fumed Silica – Production, Properties, and Applications 761
H. Barthel, L. Rösch, J. Weis

Poly(dimethylsiloxane) Chains at a Silica Surface 779
V. M. Litvinov

2,5-Bis(*t*butyl)-2,5-diaza-1-germa-cyclopentane – A New Precursor for Amorphous 815
Germanium Films
J. Prokop, R. Merica, F. Glatz, S. Veprek, F.-R. Klingan, W. A. Herrmann

Getting Light from Silicon: From Organosilanes to Light Emitting Nanocrystalline Silicon 821
S. Veprek

A Novel Liquid Xenon IR Cell Constructed from a Silicon Single Crystal 837
M. Tacke, P. Sparrer, R. Teuber, H.-J. Stadter, F. Schuster

Author Index 841

Subject Index 847

List of Contributors

Prof. Dr. Yitzhak Apeloig
Department of Chemistry
Technion – Israel Institute of Technology
Technion City
32000 Haifa
Israel

Dr. Herbert Barthel
Wacker-Chemie GmbH
Geschäftsbereich S
Werk Burghausen
Johannes-Heß-Str. 24
D-84489 Burghausen
Germany

Prof. Dr. Gerd Becker
Institut für Anorganische Chemie
Universität Stuttgart
Pfaffenwaldring 55
D-70550 Stuttgart
Germany

Dr. Helmut Beckers
Anorganische Chemie
Bergische Universität –
Gesamthochschule Wuppertal
Gaußstr. 20
D-42097 Wuppertal
Germany

Dr. Uwe Beifuss
Institut für Organische Chemie
Georg-August-Universität Göttingen
Tammannstr. 2
D-37077 Göttingen
Germany

Dr. Johannes Belzner
Institut für Organische Chemie
Georg-August-Universität Göttingen
Tammannstr. 2
D-37077 Göttingen
Germany

Prof. Dr. Pierre Braunstein
Laboratoire de Chimie de Coordination
Associé au CNRS (URA 0416)
Université Louis Pasteur
4 Rue Blaise Pascal
F-67070 Strasbourg Cedex
France

Dr. Erica Brendler
Institut für Analytische Chemie
Technische Universität Bergakademie Freiberg
Leipziger Str. 29
D-09599 Freiberg
Germany

Prof. Dr. Jacques Dunoguès
Laboratoire de Chimie Organique et
Organométallique
Université Bordeaux
Université Bordeaux 1
F-33405 Talence
France

Dr. Michael Geck
Wacker-Chemie GmbH
Geschäftsbereich S
Werk Burghausen
Johannes-Heß-Str. 24
D-84489 Burghausen
Germany

Prof. Dr. Joseph Grobe
Anorganisch-Chemisches Institut
Westfälische Wilhelms-Universität
Wilhelm-Klemm-Str. 8
D-48149 Münster
Germany

Priv. Doz. Dr. Karl Hassler
Institut für Anorganische Chemie
Technische Universität Graz
Stremayrgasse 16
A-8010 Graz
Austria

Prof. Dr. Joachim Heinicke
Institut für Anorganische Chemie
Ernst-Moritz-Arndt-Universität Greifswald
Soldtmannstr. 16
D-17487 Greifswald
Germany

Prof. Dr. Edwin Hengge
Institut für Anorganische Chemie
Technische Universität Graz
Stremayrgasse 16
A-8010 Graz
Austria

Dr. H. Jancke
Bundesanstalt für Materialforschung
und -prüfung (BAM)
Unter den Eichen 87
D-12489 Berlin
Germany

Dr. U. Just
Bundesinstitut für Materialforschung
und -prüfung(BAM)
Unter den Eichen 87
D-12205 Berlin
Germany

Prof. Dr. Wolfgang Kaim
Institut für Anorganische Chemie
Universität Stuttgart
Pfaffenwaldring 55
D-70550 Stuttgart
Germany

Prof. Dr. Hans Heinz Karsch
Anorganisches-Chemisches Institut
Technische Universität München
Lichtenbergstr. 4
D-85747 Garching
Germany

Prof. Dr. Hans Kelling
Fachbereich Chemie
Abteilung Anorganische Chemie
Universität Rostock
Buchbinderstr. 9
D-18051 Rostock
Germany

Dr. Andreas Kienzle
Institut für Werkstoffwissenschaft
Pulvermetallurgisches Laboratorium (PML)
Max-Planck-Institut für Metallforschung
Heisenbergstr. 5
D-70569 Stuttgart
Germany

Dr. K.-D. Klein
Th. Goldschmidt AG
Goldschmidtstr. 100
D-45127 Essen
Germany

Prof. Dr. Uwe Klingebiel
Institut für Anorganische Chemie
Georg-August-Universität Göttingen
Tammannstr. 4
D-37077 Göttingen
Germany

Dr. Karl Wilhelm Klinkhammer
Institut für Anorganische Chemie
Universität Stuttgart
Pfaffenwaldring 55
D-70550 Stuttgart
Germany

Dr. Michael Knorr
Anorganische Chemie
Universität des Saarlandes
Postfach 151150
D-66041 Saarbrücken
Germany

Priv.-Doz. Dr. Heinrich Lang
Anorganisch-Chemisches Institut
Ruprecht-Karls-Universität Heidelberg
Im Neuenheimer Feld 270
D-69120 Heidelberg
Germany

Dr. Susanne Leistner
Fachbereich Chemie
Lehrstuhl Physikalische Chemie
Technische Universität Chemnitz-Zwickau
Straße der Nationen 62
D-09111 Chemnitz
Germany

Dr. Gerald Linti
Institut für Anorganische Chemie
Ludwig-Maximilians-Universität München
Meiserstr. 1
D-80333 München
Germany

Prof. Dr. V. M. Litvinov
DSM Research B.V.
PAC–MC
P.O.Box 18
6160 MD
The Netherlands

Prof. Dr. Gerhard Maas
Fachbereich Chemie
Universität Kaiserslautern
Erwin-Schrödinger-Strasse
D-67663 Kaiserslautern
Germany

Prof. Dr. Günther Maier
Institut für Organische Chemie
Justus-Liebig-Universität Gießen
Heinrich-Buff-Ring 58
D-35392 Gießen
Germany

Prof. Dr. Wolfgang Malisch
Institut für Anorganische Chemie
Bayerische Julius-Maximilians-Universität
Würzburg
Am Hubland
97074 Würzburg
Germany

Prof. Dr. Heinrich Christian Marsmann
Anorganische und Analytische Chemie
Universität – Gesamthochschule – Paderborn
Warburger Str. 100
D-33098 Paderborn
Germany

Prof. Dr. Kurt Merzweiler
Institut für Anorganische Chemie
Martin-Luther-Universität Halle-Wittenberg
Weinbergweg 16
D-06120 Halle (Saale)
Germany

Dr. U. Müller
Institut für Organische Chemie
Martin-Luther-Universität Halle-Wittenberg
Geusaer Strasse
D-06217 Merseburg
Germany

Prof. Dr. Reinhard Nesper
Laboratorium für Anorganische Chemie
Eidgenössische Technische Hochschule Zürich
ETH-Zentrum
CH-8092 Zürich
Switzerland

Prof. Dr. Edgar Niecke
Institut für Anorganische Chemie
Rheinische Friedrich-Wilhelms-Universität Bonn
Gerhard-Domagk-Str. 1
D-53121 Bonn
Germany

Prof. Dr. Oskar Nuyken
Institut für Technische Chemie
Lehrstuhl Makromolekulare Chemie
Technische Universität München
Lichtenbergstr. 4
D-85747 Garching
Germany

Prof. Dr. Hartmut Oehme
Fachbereich Chemie
Abteilung Anorganische Chemie
Universität Rostock
Buchbinderstr. 9
D-18051 Rostock
Germany

Prof. Dr. Gerhard Roewer
Institut für Anorganische Chemie
Technische Universität Bergakademie Freiberg
Leipziger Str. 29
D-09599 Freiberg
Germany

Dr. Klaus Rose
Fraunhofer-Institut für Silicatforschung
Neunerplatz 2
D-97082 Würzburg
Germany

Prof. Dr. Klaus Rühlmann
Institut für Anorganische Chemie
Technische Universität Dresden
Mommsenstr. 13
D-01062 Dresden
Germany

Prof. Dr. Peter Sartori
Anorganische Chemie
Gerhard Mercator Universität –
Gesamthochschule – Duisburg
Lotharstr. 1
D-47048 Duisburg
Germany

Prof. Dr. Paul von Ragué Schleyer
Institut für Organische Chemie
Friedrich-Alexander Universität
Erlangen-Nürnberg
Henkestr. 42
D-91052 Erlangen
Germany

Prof. Dr. Hubert Schmidbaur
Anorganisch-Chemisches Institut
Technische Universität München
Lichtenbergstr. 4
D-85747 Garching
Germany

Prof. Dr. Helmut K. Schmidt
Institut für Neue Materialien gem. GmbH
Im Stadtwald, Geb. 43 A
D-66123 Saarbrücken
Germany

Prof. Dr. Manfred Schmidt
Makromolekulare Chemie
Universität Bayreuth
Universitätsstr. 30
D-95440 Bayreuth
Germany

Prof. Dr. Hans-Ullrich Siehl
Institut für Organische Chemie
Eberhard-Karls-Universität Tübingen
Auf der Morgenstelle 18
D-72076 Tübingen
Germany

Dr. H.-J. Stadter
Wacker-Chemitronic
Werk Burghausen
Johannes-Heß-Str. 24
D-84489 Burghausen
Germany

Dr. C. Strohmann
Anorganische Chemie
Universität des Saarlandes
Postfach 1150
D-66041 Saarbrücken
Germany

Dr. Harald Stüger
Institut für Anorganische Chemie
Technische Universität Graz
Stremayrgasse 16
A-8010 Graz
Austria

Priv. Doz. Dr. Karlheinz Sünkel
Institut für Anorganische Chemie
Ludwig-Maximilians-Universität München
Meiserstr. 1
D-80333 München
Germany

Prof. Dr. Manfred Weidenbruch
Fachbereich Chemie
Carl-von-Ossietzky-Universität Oldenburg
Carl-von-Ossietzky-Str. 9–11
D-26111 Oldenburg
Germany

Dr. Matthias Tacke
Institut für Anorganische Chemie
Universität Fridericiana zu Karlsruhe (TH)
Engesserstrasse, Geb. 30.45
D-76128 Karlsruhe
Germany

Prof. Dr. Robert West
Department of Chemistry
University of Wisconsin
Madison
Wisconsin 53706
USA

Prof. Dr. Reinhold Tacke
Institut für Anorganische Chemie
Universität Fridericiana zu Karlsruhe (TH)
Engesserstrasse, Geb. 30.45
D-76128 Karlsruhe
Germany

Prof. Dr. Nils Wiberg
Institut für Anorganische Chemie
Ludwig-Maximilians-Universität München
Meiserstr. 1
D-80333 München
Germany

Dr. Wolfram Uhlig
Laboratorium für Anorganische Chemie
Eidgenössische Technische Hochschule Zürich
ETH-Zentrum
CH-8092 Zürich
Switzerland

Dr. Dieter Wrobel
Bayer AG
D-51368 Leverkusen
Germany

Prof. Dr. Stan Veprek
Institut für Chemie der
Informationsaufzeichnung
Technische Universität München
Lichtenbergstr. 4
D-85747 Garching
Germany

Editors:

Prof. Dr. Norbert Auner
Institut für Chemie
Fachinstitut für Anorganische und Allgemeine
Chemie
Humboldt-Universität zu Berlin
Hessische Str. 1–2
D-10115 Berlin
Germany

Dr. Johann Weis
Wacker-Chemie GmbH
Geschäftsbereich S
Werk Burghausen
Johannes-Heß-Str. 24
D-84489 Burghausen
Germany

Tetravalent Organosilicon Compounds: Chemistry and Structure

Norbert Auner, Johann Weis

Despite the great interest in low- and hyper-valent silicon compounds four coordinated, tetravalent silicon covers the largest part of silicon compounds in nature and in industrial applications. The chemistry of such species still yields new and extraordinary results.

The priority in this chapter lies in the description of small organosubstituted silanes and of small heterocycles that might be precursors for the production of new materials for high tech applications (e. g. polycrystalline silicon for photovoltaic purposes, thin ceramic coatings from CVD processes).

The synthetic potential of silicon substituents in organic and organometallic chemistry has by far not been fully exploited, which is evidenced by the numerous contributions in this chapter. This is examplified for the description of the silyl group as a substituent and as a functional group in carbene and carbenoid chemistry, for the function of the trimethylsilyl substituent in the synthesis of low-valent compounds containing elements of group 15 and for the influence of a supersilyl ligand to a phosphorous center.

There is a steadily increasing interest in the synthesis of new silaheterocyclic systems not only with respect to the potential formation of new polymers but also for a deeper insight into the structural and electronic features of silicon in those cycles.

This chapter deals with the chemistry of silacyclopropanes, -cyclobutenes, and -norbornenes, of cyclosilanes and polycyclic derivatives, of cyclic iminosilanes, and of cages containing silicon in the coordination sphere of group 15 elements. In general very often a salt elimination route is used for the synthesis of silaheterocycles. For that reason the editors had to decide whether the contributions of this chapter dealing with the use of metals in the synthesis of those compounds had to be included into chapter I or III. The guideline for the decision was to collect all articles in chapter I, which focus on the chemistry of the resulting silicon species. All papers which emphasize on the metal compounds were integrated into chapter III.

References:

[1] R. West, in: *The Chemistry of Organic Silicon Compounds* (Eds.: S. Patai, Z. Rappoport), Wiley, Chichester, **1989**, p. 1207.

[2] *Studies in Organic Chemistry, Catalyzed Direct Reactions of Silicon* (Eds.: K. M. Lewis, D. G. Rethwisch), Elsevier, Amsterdam, **1993**.

[3] D. W. H. Rankin, in: *Organosilicon Chemistry – From Molecules to Materials* (Eds.: N. Auner, J. Weis), VCH, Weinheim, **1994**, p. 3.

[4] H. Bock, J. Meuret, Ch. Näther, K. Ruppert, in: *Organosilicon Chemistry – From Molecules to Materials* (Eds.: N. Auner, J. Weis), VCH, Weinheim, **1994**, p. 11.

[5] W. Uhlig, in: *Organosilicon Chemistry – From Molecules to Materials* (Eds.: N. Auner, J. Weis), VCH, Weinheim, **1994**, p. 21.

[6] U. Klingebiel, in: *Organosilicon Chemistry – From Molecules to Materials* (Eds.: N. Auner, J. Weis), VCH, Weinheim, **1994**, p. 51.

Small Organosilicon Molecules – Synthesis, Stereochemistry, and Reactivity

Hubert Schmidbaur

Anorganisch-Chemisches Institut

Technische Universität München

Lichtenbergstraße 4, D-85747 Garching, Germany

Summary: Systematic studies have been continued in the chemistry of simple silylated hydrocarbons of the alkane, alkene, alkyne, and arene series. Polysilylated species with SiH_3 groups have been the main synthetic target. Through their low molecular mass, high volatility, and specific thermal, plasma or photochemical fragmentation behavior, these species are promising precursors for the production of tailored materials based on the ternary system Si–C–H or even the binary system Si–C. The prototypes studied previously included the poly(silyl)methanes $(H_3Si)_nCH_{4-n}$ -ethenes $(H_3Si)_nC_2H_{4-n}$ and -ethynes $H_3SiC\equiv CX$ (X = H, Hal). This chemistry has now been extended to poly(silyl)arenes from the benzene and naphthalene series: The Merker-Scott type reaction of the appropriate halohydrocarbon with aryldihydrohalosilanes and magnesium in THF affords suitable arylated precursors, which are readily converted into the hydrogenated products through regioselective Si-aryl cleavage by strong acid, followed by hydride reduction. Crystal structures of the arylated precursors and the simple silyl compounds have been determined for selected examples. Nitrogen- and/or oxygen-containing molecules (quaternary systems such as silyl-amines, silyl-hydrazines, (hetero)siloxanes, and silyl-hydroxylamines) have also been considered and some new prototypes investigated, i.e., small molecules from the Si/N/C/H, Si/O/C/H, and Si/N/O/C/H systems, respectively.

Introduction

SiH-functional poly(silylated) aromatic hydrocarbons are important starting materials for the preparation of arene-bridged polysilanes through thermal or catalytic dehydrogenative coupling reactions.

Depending on the substitution pattern of the monomers, different kinds of cross-linking can be anticipated, which can lead to small oligomers, but also to rings, chains, or polydimensional networks [1]. This applies to standard benzene derivatives, as well as to condensed arenes with more sophisticated molecular structures. These products are interesting in their own right owing to their physical and optophysical properties, but also as precursors for refractory materials prepared in pyrolysis or plasma processes [1].

While poly(silylated) arenes with triorganosilyl groups have been extensively studied, simple poly(silyl)arenes bearing solely hydrogen at the silicon substituents were largely unknown until some very recent investigations, which have shown the way to at least a few of the possible isomeric poly(silyl)benzenes [2, 3]. In these initial studies it has been recognized that the methods established for the preparation of one isomer may not be applicable to the other. Of the three di(silyl)benzenes, e.g., only the *meta*- and *para*-substituted isomers are readily available through standard methods [3], while the *ortho*-isomer is difficult to prepare owing to steric hindrance in the more crowded intermediates. Therefore new methods, or new variations of established procedures, had to be developed to make the missing species accessible.

In previous work in this laboratory [4–9] the silylation of halohydrocarbons with aryl(dihydro)halosilanes and magnesium in a highly-polar ether solvent like THF proved to be very successful in both the alkane and alkene/alkyne series. A recent extension into the arene series also turned out to be a promising endeavor [10–12]: The investigations aiming at the first synthesis of hexa(silyl)benzene have shown that complete silylation of benzene is feasible if suitably substituted aryl groups are chosen as protecting and leaving groups (Scheme 1).

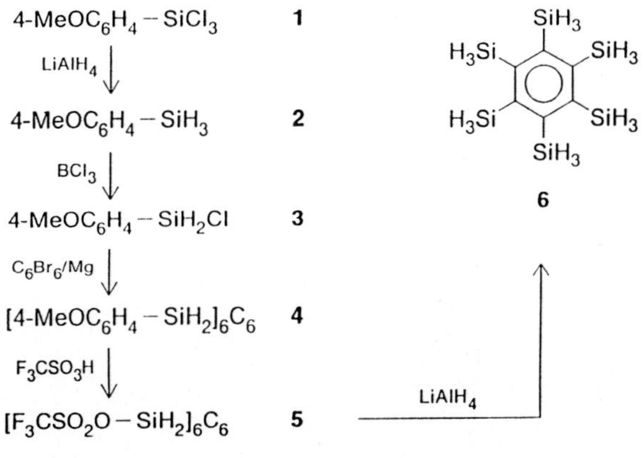

$$
\begin{array}{l}
\text{4-MeOC}_6\text{H}_4-\text{SiCl}_3 \quad \mathbf{1} \\
\quad \downarrow \text{LiAlH}_4 \\
\text{4-MeOC}_6\text{H}_4-\text{SiH}_3 \quad \mathbf{2} \\
\quad \downarrow \text{BCl}_3 \\
\text{4-MeOC}_6\text{H}_4-\text{SiH}_2\text{Cl} \quad \mathbf{3} \\
\quad \downarrow \text{C}_6\text{Br}_6/\text{Mg} \\
[\text{4-MeOC}_6\text{H}_4-\text{SiH}_2]_6\text{C}_6 \quad \mathbf{4} \\
\quad \downarrow \text{F}_3\text{CSO}_3\text{H} \\
[\text{F}_3\text{CSO}_2\text{O}-\text{SiH}_2]_6\text{C}_6 \quad \mathbf{5} \\
\quad \xrightarrow{\text{LiAlH}_4} \quad \mathbf{6} \; \text{C}_6(\text{SiH}_3)_6
\end{array}
$$

Scheme 1. Five-step synthesis of hexa(silyl)benzene [11]

This concept has now been applied to the preparation of some of the selectively substituted benzenes and naphthalenes, where standard procedures failed to give the expected products [3]. The partially arylated intermediates are very useful model compounds for the elucidation of structural data. In most cases they can be obtained as single crystals which allow the application of standard X-ray diffraction techniques. For the low-melting and volatile final products similar data can only be collected from electron-diffraction studies [5, 13].

Following similar conceptual lines, a series of silazanes, siloxanes, silylhydrazines and silylhydroxylamines have also been included into the current studies, thus building connections between the Si–C–H and the related Si–N–H, Si–O–H and Si–N–O–H systems. Again the arylated intermediates served as crystalline images of the liquid and volatile basic members of the three series.

Selected Poly(silyl)benzenes and -naphthalenes

O-Di(silyl)benzene is available through a three-step procedure as presented in Scheme 2. Treatment of o-dibromobenzene with chloro(p-tolyl)silane and magnesium in THF affords 1,2-bis[(p-tolyl)-silyl]benzene in 60 % yield. The regioselective cleavage of both p-tolyl groups by two equivalents of trifluoromethylsulfonic acid gives the corresponding bis-triflate (quantitative conversion), which can be reduced to the target compound with LiAlH$_4$ in 87 % yield. The product is obtained as a colorless, distillable liquid [14].

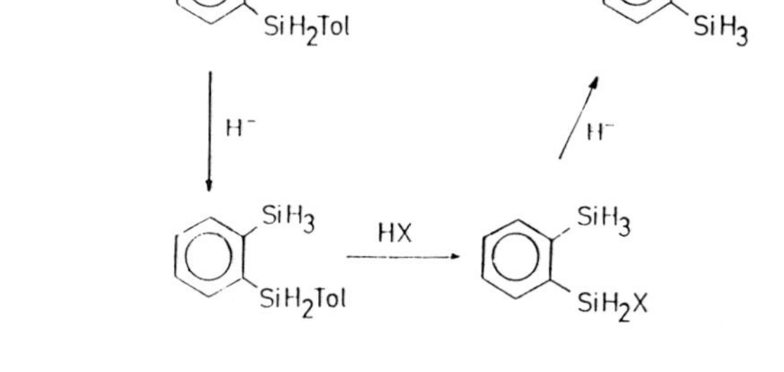

Scheme 2. Synthesis of 1,2-di(silyl)benzene and its mono(aryl) derivatives [14]

If only one equivalent of triflic acid is employed, the mono-triflate is generated (again in quantitative regioselective conversion). This product gives the mono-arylated intermediate after treatment with LiAlH$_4$. Cleavage of the remaining p-tolyl groups yields the mono-triflate, which can again be converted into 1,2-di(silyl)benzene by LiAlH$_4$ (Scheme 2) [14].

1,2-Di(silyl)benzene gives the di(chlorosilyl)benzene reagent upon treatment with boron trichloride (Scheme 3). Contrary to the parent compound, this derivative is readily solvolyzed and gives cyclic siloxanes, silthianes, silazanes and silylphosphines on contact with water, hydrogensulfide, ammonia, amines, or phosphines, respectively.

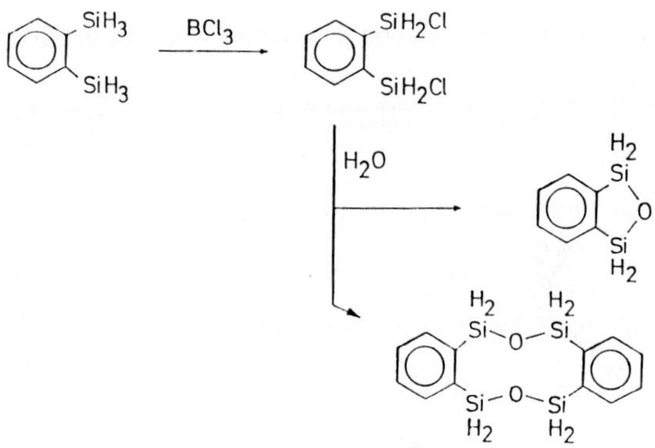

Scheme 3. Functionalization of 1,2-di(silyl)benzene [14]

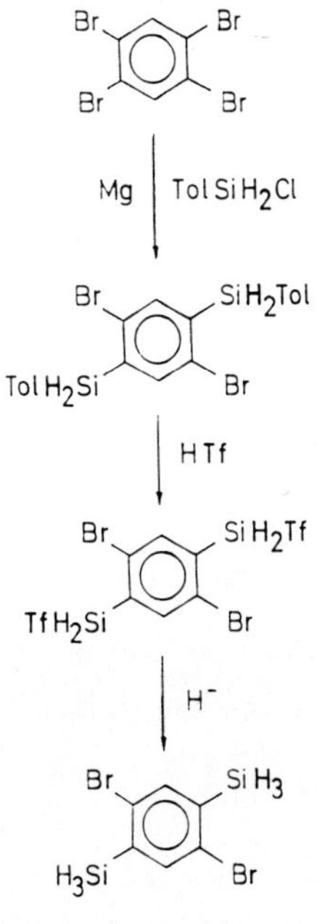

Scheme 4. Silylation of 1,2,4,5-tetrabromobenzene [14]

1,3- and 1,4-dibromobenzene have also been converted into their bis[(p-tolyl)silyl] derivatives following the same procedures, but yields were found to vary with the substitution pattern (*meta:* 35 %, *para:* 70 %). The regioselectivity of the (partial) aryl cleavage is also clearly dependent on neighbouring group effects. For more distant groups the discrimination between mono- and disubstitution is less effective and mixtures of products are obtained.

Complete silylation of the related 1,2,4,5-tetrabromobenzene proved to be more difficult (Scheme 4) [14]. Even with an excess of reagents and under more forcing conditions and longer reaction times only the disubstituted products could be obtained. 1,4-Dibromo-2,5-di(silyl)benzene is readily available from this procedure in high yield. It is a stable crystalline solid, the structure of which has been determined (Fig. 1).

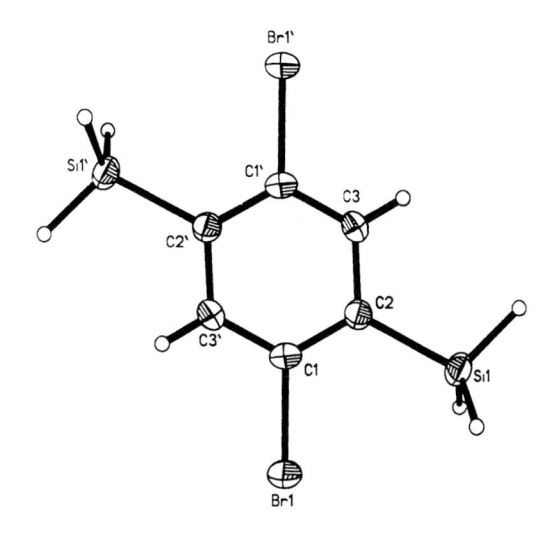

Fig. 1. Molecular structure of 1,4-dibromo-2,5-di(silyl)benzene [14]

The molecular geometry shows no unusual features. Steric crowding in this molecule and in its di(p-tolyl) precursor (Fig. 2) can therefore be ruled out as a reason for the failure of the attempts to accomplish complete silylation. Electronic effects are more likely to retard tri- and tetra-substitution.

In the naphthalene series, the synthesis of the peri-substituted 1,8-di(silyl)naphthalene represents the greatest synthetic challenge. From earlier studies [15] it appears that even two small *methyl* groups in the *peri*-positions are experiencing significant steric strain, which is even more severe as the substituents get bigger. Previous examples the structures of which have been determined are bis(trimethylsilyl), -germyl, and -stannyl derivatives [16–18]. The analogous boryl and amino compounds are known as hydride and proton sponges, respectively, referring to the excellent chelating properties originating from the proximity of the functional groups [19, 20].

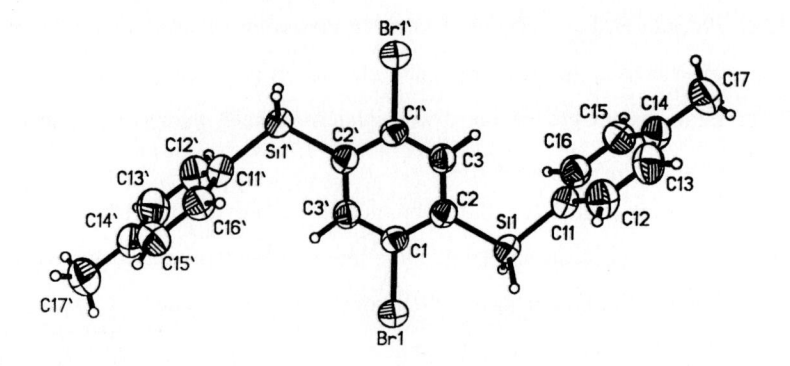

Fig. 2. Molecular structure of 1,4-dibromo-2,5-di(*p*-tolylsilyl)benzene [14]

The target molecule has been synthesized according to the reaction sequence illustrated in Scheme 5.

Scheme 5. Synthesis of 1,8-di(silyl)naphthalene [18]

It should be noted that the *p*-anisyl substituent has to be employed instead of the *p*-tolyl group in order to reach complete regioselectivity of the acid cleavage (Scheme 6).

Scheme 6. Regioselective cleavage of *p*-anisyl groups from peri-substituted naphthalenes

The structures of 1,8-di(silyl)naphthalene and its mono- and di(*p*-anisyl) derivatives have been determined and are shown in Fig. 3–5. While the naphthalene part of the molecules appears to be largely undistorted, the two silyl groups are clearly bent away from each other in the molecular plane in order to avoid closer repulsive contacts. This steric crowding enhances the chemical reactivity of the molecule and makes the compound a versatile starting material for numerous derivatives. For substitution control the conversion into the symmetrical dichlorosilane is possible using boron trichloride (Scheme 5).

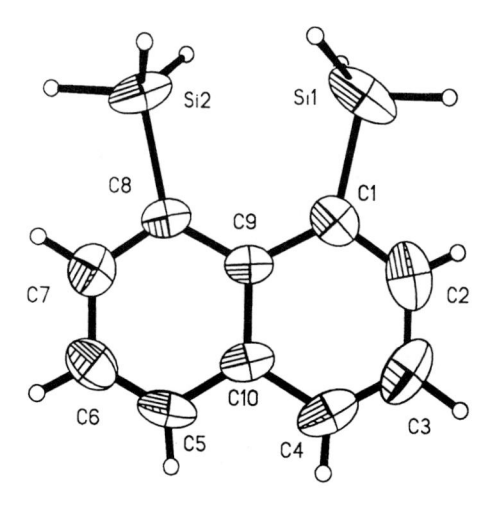

Fig. 3. Molecular structure of 1,8-di(silyl)naphthalene [18]

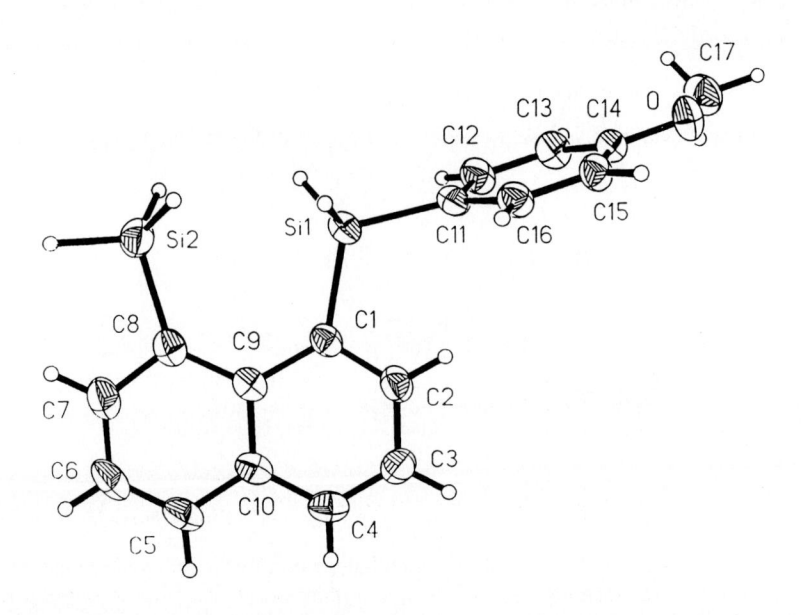

Fig. 4. Molecular Structure of 1-silyl-8-(*p*-anisylsilyl)-naphthalene

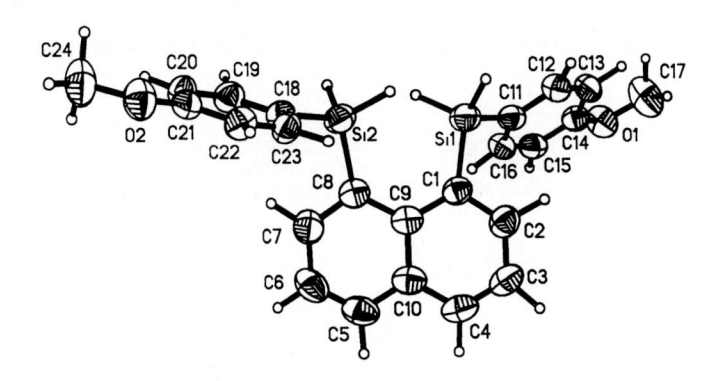

Fig. 5. Molecular structure of 1,8-bis(*p*-anisylsilyl)naphthalene [18]

Polysilylated Amines, Hydrazines, and Hydroxylamines

Since the first structure determination of tri(silyl)amine, which has shown the molecule to have a planar configuration at nitrogen [21], the unexpected structural features of silylated amines have attracted considerable interest. The planar structure is associated with a low nitrogen basicity and with poor donor properties of the silylated amines. Pertinent conclusions had to be drawn from rather limited structural information, however, since most of the silyl-amines are difficult to crystallize. The partially phenylated silylamines obtained as precursors for small hydrogen-rich silazane molecules (Introduction) offered a chance to provide a larger variety of structural data. An example in case is tri(phenylsilyl)amine [22], the tri*sila* analogue of tribenzylamine. It is immediately obvious from Fig. 6, that the carbon compound has a pyramidal NC_3 skeleton, whereas the silicon anlogue has a planar NSi_3 unit.

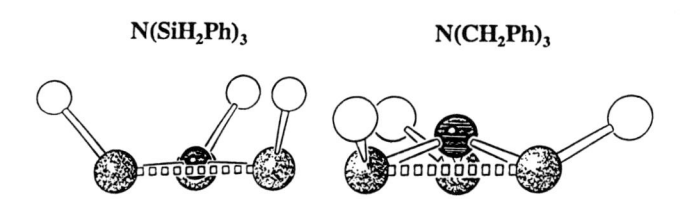

$N(SiH_2Ph)_3$ $N(CH_2Ph)_3$

Fig. 6. Planar and pyramidal skeleton of tri(phenylsilyl)amine and tri(benzyl)amine [22]

A series of other structures recently determined has confirmed the rule, that in standard cases already one silicon substituent is enough to induce planarization at nitrogen (Fig. 7, 8) [23–25].

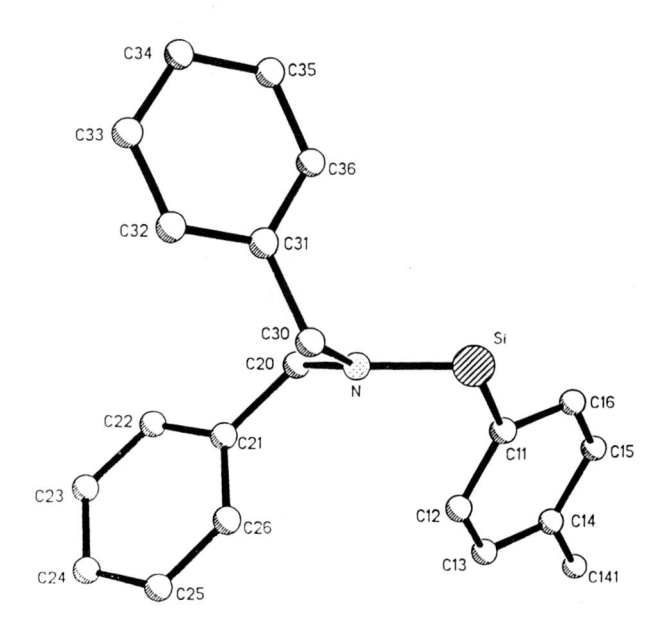

Fig. 7. Molecular structure of dibenzyl-(*p*-tolylsilyl)amine [23–25]

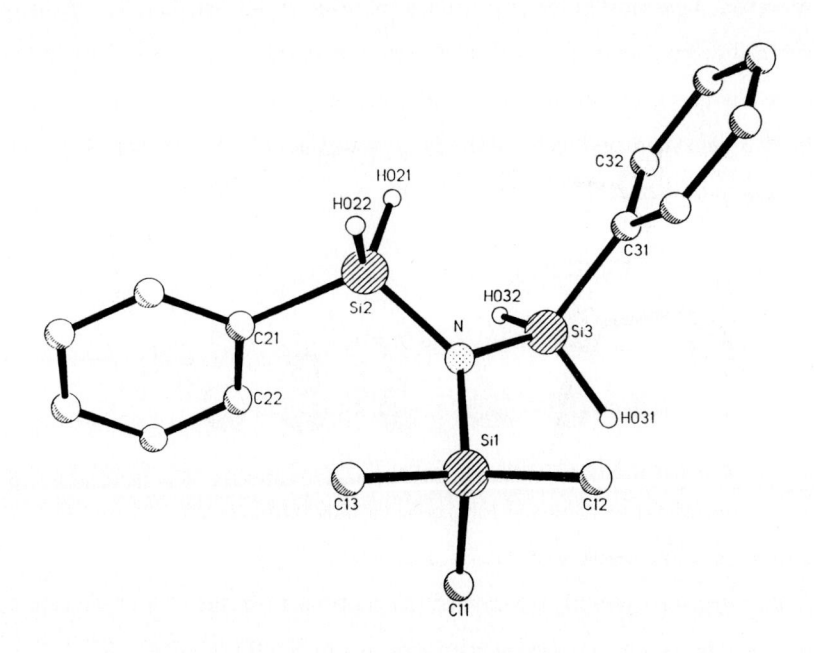

Fig. 8. Molecular structure of bis(phenylsilyl)(trimethylsilyl)amine [23–25]

This rule appears to be also valid for silylated hydrazines, an example of which is presented in Fig. 9 [26].

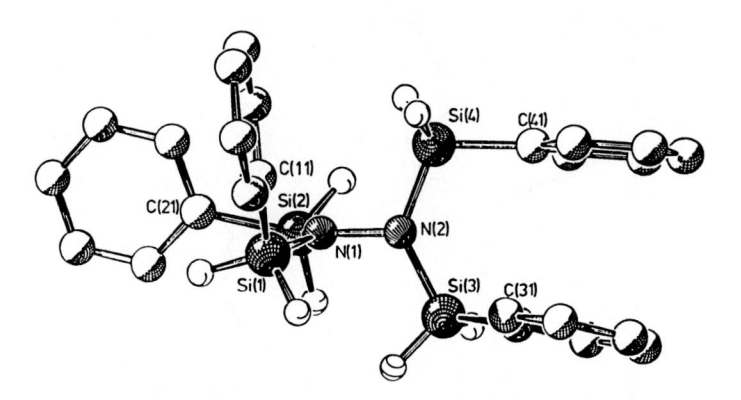

Fig. 9. Molecular structure of tetrakis(phenylsilyl)hydrazine [27]

Even after integration of the N_2Si_4 moiety into ring systems, no significant deviation from planarity at both nitrogen atoms is observed (Fig. 10, 11) [27].

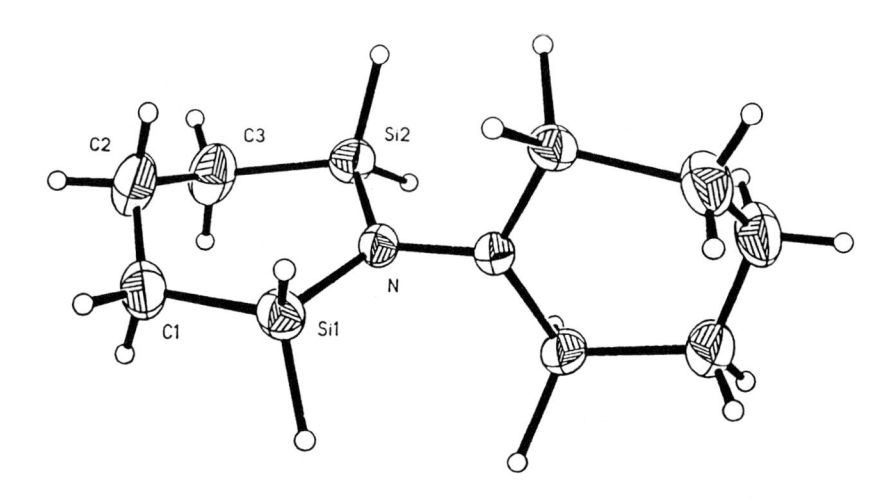

Fig. 10. Molecular structure of a bicyclic tetrasilylated hydrazine [(CH$_2$)$_3$(SiH$_2$)$_2$N]$_2$ [27]

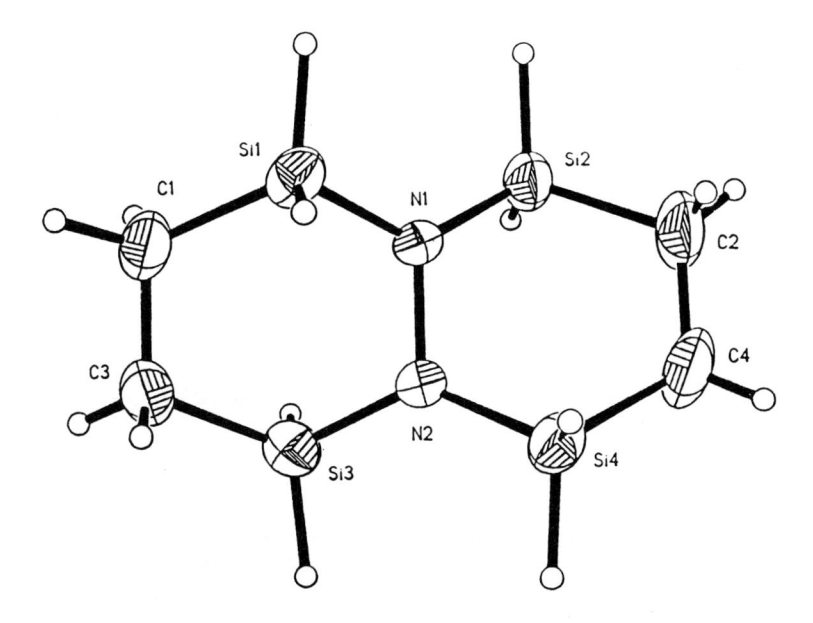

Fig. 11. Molecular structure of a bicyclic tetrasilylated hydrazine [(CH$_2$)$_2$(SiH$_2$)$_2$]$_2$N$_2$ [27]

These small bicyclic molecules are very stable, show high volatility, and are ideal precursors for the deposition of "plasma nitride" from the vapour phase. In this respect the materials are valuable alternatives for the (disilyl)amines successfully introduced previously for this purpose [28, 29].

In substituted hydrazine molecules with both alkylated and silylated nitrogen atoms, the products of reaction with Lewis acids clearly show the reduced basicity of the nitrogen atoms bound to silicon. In the BH_3 adducts only the nonsilylated amino groups undergo complexation [30].

It is only for related hydroxylamines, that pyramidalization at silylated nitrogen atoms is encountered. The molecular structure presented in Fig. 12 for a prototype of this series features a nonplanar nitrogen center, thus representing one of the very few exceptions to the rule [31].

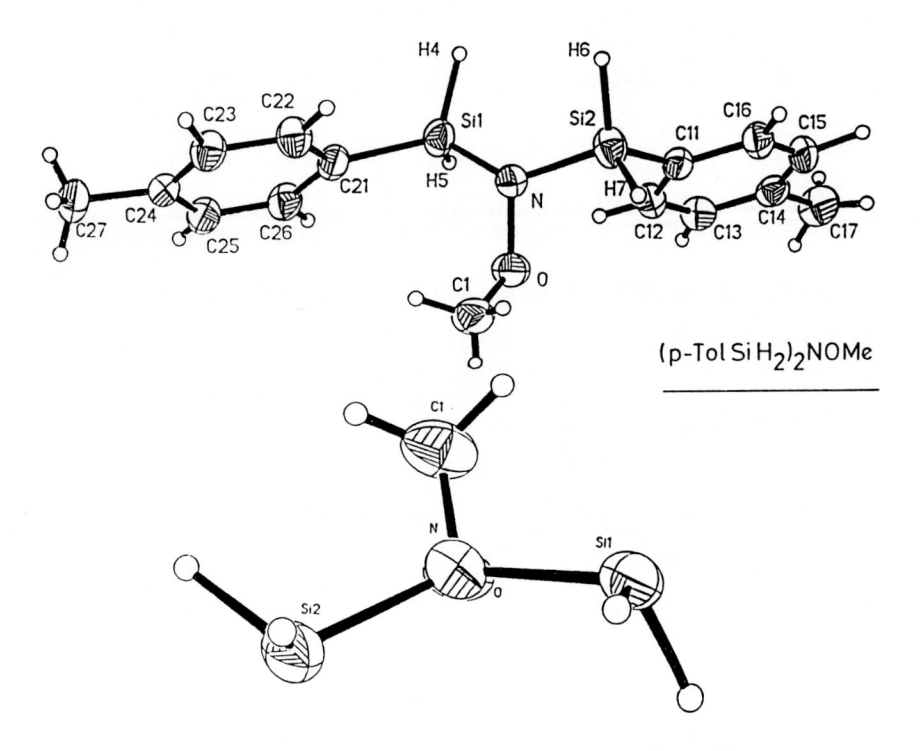

Fig. 12. Molecular structure (and its heavy atom detail) of *O*-methyl-*N*-bis(*p*-tolylsilyl)hydroxylamine [31]

The preparation of the silylated amines, hydrazines, and hydroxylamines follows conventional synthetic methods established previously for species with fully or partially alkylated silyl groups. The low molecular masses and the absence of substituent branching lead to new physical properties, however, which make the simple hydrogen-rich molecules attractive for vapour phase technology.

Siloxanes and Heterosiloxanes

New open-chain and cyclic siloxanes are available through controlled substitution of functional groups or hydrogen in the polysilylated arenes presented above. Similar results have recently been obtained also with functionalized poly(silyl)alkanes, however, which include species with silica and silicon carbide units in the same molecule. Scheme 7 gives the synthesis of the molecule $(H_3Si)_3CSiH_2OSiH_2C(SiH_3)_3$ with a SiOSi linkage between two CSi_4 units [32].

$$p\text{-TolH}_2\text{SiC}(\text{SiH}_3)_3 \xrightarrow{CF_3SO_3H} \text{TfH}_2\text{SiC}(\text{SiH}_3)_3$$

$$\text{TfH}_2\text{SiC}(\text{SiH}_3)_3 \xrightarrow[-\text{Et}_3\text{NHTf}]{H_2O/\ Et_3N} H_3Si-\underset{\underset{SiH_3}{|}}{\overset{\overset{SiH_3}{|}}{C}}-SiH_2-O-SiH_2-\underset{\underset{SiH_3}{|}}{\overset{\overset{SiH_3}{|}}{C}}-SiH_3$$

Scheme 7. Synthesis of Si,Si'-bis[tri(silyl)methyl]disiloxane [32]

Pyrolysis and plasma decomposition leads to silica/silicon carbide phases, the properties of which are strongly dependent on the process conditions.

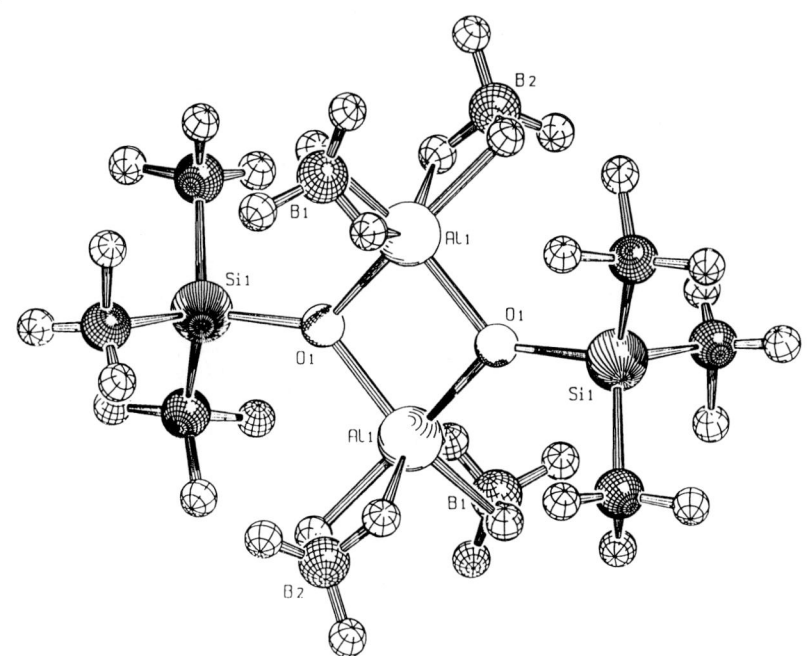

Fig. 13. Molecular structure of $[Me_3SiOAl(BH_4)_2]_2$, dimeric trimethylsilyloxy-aluminum-bis(tetrahydroborate) [33]

Hydrogen-rich *alumo*siloxanes were a little investigated class of compounds as compared to their well established alkylated analogues. Recent studies have provided a much larger spectrum of stoichiometries, structures and properties [33, 34] (Fig. 13).

Only a few specialities can be mentioned here, with hydride and borohydride functions present at silicon or at aluminum atoms [33], or with unsaturated alkene or alkyne side chains for addition reactions or cross-linking [34] (Fig.14). The scope of this chemistry has not yet been explored, and more detailed studies are currently under way.

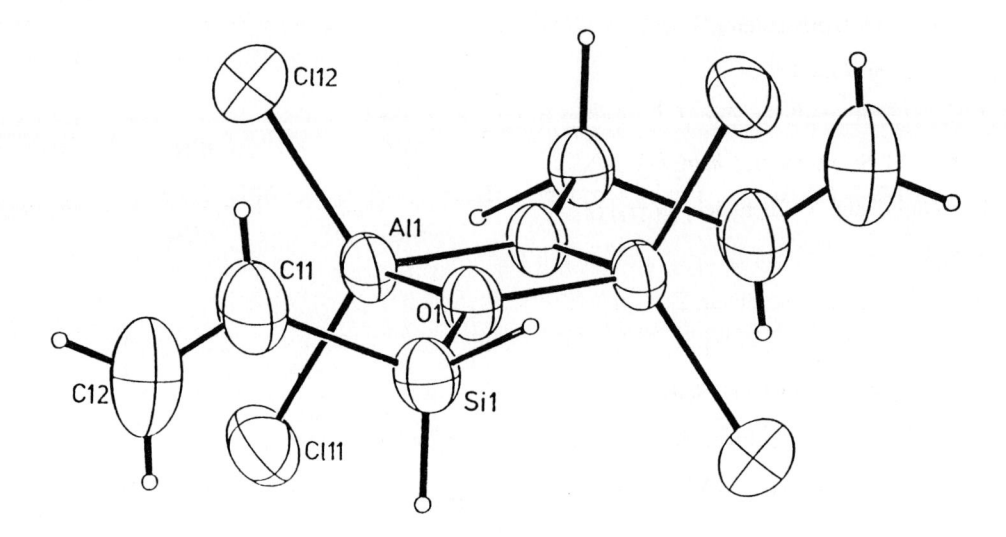

Fig. 14. Molecular structure of dimeric vinylsilyloxy-aluminum dichloride [(CH_2=$CHSiH_2$)$OAlCl_2$]$_2$ [34]

Acknowledgement: The work summarized in this account has been supported by the *Deutsche Forschungsgemeinschaft*, by the *Fonds der Chemischen Industrie*, by the *Bundesministerium für Forschung und Technologie*, by the *European Community*, and – through the donation of chemicals – by *Wacker-Chemie GmbH*. The author gratefully acknowledges the fruitful collaboration with his students and colleagues, whose names are given in the list of references.

References:

[1] J. Ohshita, D. Kanaya, M. Ishikawa, T. Koike, *Macromolecules* **1991**, *24*, 2106; T. J. Barton, S. Ijadi-Maghsoodi, Y. Pang, *Macromolecules* **1991**, *24*, 1257; T. Iwahara, S. Hayase, R. West, *Macromolecules* **1990**, *23*, 1298.

[2] D. R. Anderson, J. M. Holovka, *J. Chem. Soc.* **1965**, 2269.

[3] H.-G. Woo, J. F. Walzer, T. D. Tilley, *Macromolecules* **1991**, *24*, 6863.

[4] R. Hager, O. Steigelmann, G. Müller, H. Schmidbaur, *Chem. Ber.* **1989**, *122*, 2115.

[5] R. Hager, O. Steigelmann, G. Müller, H. Schmidbaur, H. E. Robertson, D. W. H. Rankin, *Angew. Chem.* **1990**, *102*, 204.

[6] J. Zech, H. Schmidbaur, *Eur. J. Solid State Inorg. Chem.* **1992**, *29*, 5.

[7] J. Zech, H. Schmidbaur, *Chem. Ber.* **1990**, *123*, 2087.

[8] H. Schmidbaur, J. Ebenhöch, G. Müller, *Z. Naturforsch.* **1986**, *41B*, 142; *Z. Naturforsch.* **1988**, *43B*, 49.

[9] S. Bommers, H. Schmidbaur, *Z. Naturforsch.* **1994**, *49B*, 337.

[10] C. Rüdinger, P. Bissinger, H. Beruda, H. Schmidbaur, *Organometallics* **1992**, *11*, 2867.

[11] C. Rüdinger, H. Beruda, H. Schmidbaur, *Chem. Ber.* **1992**, *125*, 1401.

[12] H. Schmidbaur, C. Rüdinger, R. Schröck, *Phosphorus, Sulfur, and Silicon*, in press.

[13] H. Schmidbaur, J. Zech, D. W. H. Rankin, H. E. Robertson, *Chem. Ber.* **1991**, *124*, 1953.

[14] R. Schröck, A. Sladek, H. Schmidbaur, *Z. Naturforsch.* **1994**, *49B*, 1036.

[15] D. Bright, I. E. Maxwell, J. de Boer, *J. Chem. Soc., Perkin Trans.* **1973**, *2*, 2101.

[16] J. Handal, J. G. White, R. W. Franck, U. H. Yuh, N. L. Allinger, *J. Am. Chem. Soc.* **1977**, *99*, 3345; J. F. Blount, F. Cozzi, J. R. Damewood, L. D. Iroff, U. Sjöstrand, K. Mislow, *J. Am. Chem. Soc.* **1980**, *102*, 99.

[17] D. Seyferth, S. C. Vick, *J. Organomet. Chem.* **1977**, *141*, 1973; R. J. Wrocynski, M. W. Baum, D. Kost, K. Mislow, S. Vick, D. Seyferth, *J. Organomet. Chem.* **1979**, *170*, C29; F. Cozzi, U. Sjöstrand, K. Mislow, *J. Organomet. Chem.* **1979**, *174*, C1.

[18] R. Soorlyakumaran, P. Boudjouk, *Organometallics* **1982**, *1*, 218.

[19] H. A. Staab, T. Saupe, *Angew. Chem.* **1988**, *100*, 895.

[20] H. E. Katz, *J. Am. Chem. Soc.* **1985**, *107*, 1420.

[21] K. Hedberg, *J. Am. Chem. Soc.* **1955**, *77*, 6491.

[22] N. W. Mitzel, A. Schier, H. Schmidbaur, *Chem. Ber.* **1992**, *125*, 2711.

[23] N. W. Mitzel, J. Riede, A. Schier, M. Paul, H. Schmidbaur, *Chem. Ber.* **1993**, *126*, 2027.

[24] N. W. Mitzel, A. Schier, H. Beruda, H. Schmidbaur, *Chem. Ber.* **1992**, *125*, 1053.

[25] N. W. Mitzel, K. Angermaier, H. Schmidbaur, *Chem. Ber.* **1994**, *127*, 841.

[26] N. W. Mitzel, P. Bissinger, H. Schmidbaur, *Chem. Ber.* **1993**, *126*, 345.

[27] N. W. Mitzel, P. Bissinger, J. Riede, K.-H. Dreihäupl, H. Schmidbaur, *Organometallics* **1993**, *12*, 413.

[28] H. Schuh, T. Schlosser, P. Bissinger, H. Schmidbaur, *Z. Anorg. Allg. Chem.* **1993**, *619*, 1347.

[29] T. Schlosser, A. Sladek, W. Hiller, H. Schmidbaur, *Z. Naturforsch.*, in press.

[30] N. W. Mitzel, K. Angermaier, A. Schier, H. Schmidbaur, *Inorg. Chem.*, in press.

[31] N. W. Mitzel, H. Schmidbaur, *Z. Anorg. Allg. Chem.* **1994**, *620*, 1087; N. W. Mitzel, K. Angermaier, H. Schmidbaur, *Organometallics* **1994**, *13*, 1762; N. W. Mitzel, M. Hofmann, E. Waterstradt, P. v. R. Schleyer, H. Schmidbaur, *J. Chem. Soc., Dalton Trans.* **1994**, 2503.

[32] S. Bommers, H. Schmidbaur, *Chem. Ber.* **1994**, *127*, 1359.

[33] P. Bissinger, M. Paul, J. Riede, H. Schmidbaur, *Chem. Ber.* **1993**, *126*, 2579.

[34] P. Bissinger, P. Mikulcik, J. Riede, A. Schier, H. Schmidbaur, *J. Organomet. Chem.* **1993**, *446*, 37.

Properties of Si-functionalized (Fluoromethyl)silanes

H. Beckers*, H. Bürger, P. Moritz

Anorganische Chemie

Bergische Universität – Gesamthochschule Wuppertal

Gaußstr. 20, D-42097 Wuppertal, Germany

1 Routes to Si-functionalized (Fluoromethyl)silanes

Since the direct synthesis of alkylchlorosilanes has enabled the industrial development of silicones, α-fluoroalkyl derivatives were also given some attention because perfluorosilicones presumably could combine the outstanding properties of both conventional silicones and perfluorocarbon polymers [1].

However, the work of Haszeldine [2] soon showed that perfluoroalkyl silicon compounds have relatively low thermal stability and are readily amenable to decomposition by nucleophilic or electrophilic reagents. While early reports on the synthesis of (trifluoromethyl)silanes [1] could not be confirmed, Ruppert's versatile route to Me_3SiCF_3 ($TMSCF_3$), published some ten years ago [3], brought about a breakthrough for this chemistry. There are now different pathways leading to tetraorganosilanes which carry at least one fluoromethyl group [4]. Likewise, $TMSCF_3$ has become a commercially available, efficient trifluoromethylation agent in organic chemistry [5].

We have been interested in the synthesis and the chemical properties of Si-functionalized (fluoromethyl)silanes of the general formula $F_nH_{3-n}CSiX_3$, (**A**) with at least one X group being halogen or hydrogen. The thermal and IR laser-induced decomposition of the silanes ($X = H$) have also been investigated because these are promising precursors for chemical vapour deposition of fluorinated SiC(F, H)-films, which reveal some interesting photovoltaic properties [6]. Since standard synthetic procedures for introducing fluorine by either nucleophilic fluorination at the α-carbon attached to silicon [7], or treatment of Si halides by (fluoromethyl)metal compounds [8], have failed, special routes to the target molecules **A** had to be developed [9–13] (Scheme 1).

Scheme 1. Routes to Si-functionalized (Fluoromethyl)silanes (**A**)

F_3CSiCl_3 has become easily accessible by trifluoromethylation of $SiCl_4$ with Ruppert's reagent $CF_3Br/P[N(C_2H_5)_2]_3$ [9, 10]. The first functionalized silanes bearing F_2HC groups have recently been synthesized by selective insertion of CF_2 into SiH bonds [10]. The perfluorinated derivative F_3CSiF_3, first obtained from CF_3I and SiF_2 by Sharp et al. [11], was shown to be a particularly convenient source of CF_2. FH_2CSiH_3 was obtained by dibromo-fluoromethylation of $SiCl_4$ with the reagent combination $CFBr_3$/TDME (TDME = $[(CH_3)_2 N]_2 C=C[(CH_3)_2 N]_2$) and subsequent hydrogenation of the intermediate FBr_2CSiCl_3 with nBu_3SnH in an overall yield $\geq$ 30 % [12].

2 F_3CSiCl_3, a Versatile Precursor for F_3CSi Derivatives

F_3CSiCl_3 has been shown to be a versatile precursor for inorganic and organometallic F_3CSi derivatives [9] owing to the unexpected resistance of its F_3CSi moiety towards nucleophilic substitution at Si. Some of its reactions are displayed in Scheme 2.

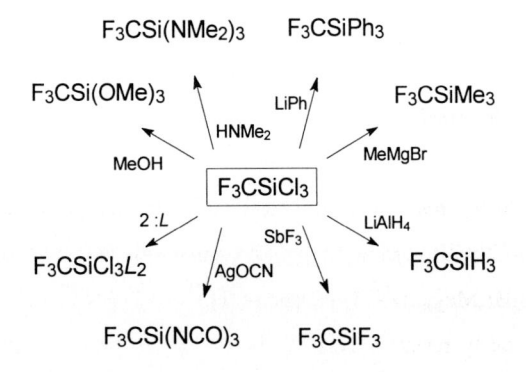

Scheme 2. Reactivity of F_3CSiCl_3

According to the position of Si–Cl bonds in the well-known conversion series [14], F_3CSiCl_3 reacts with AgOCN to yield $F_3CSi(NCO)_3$, and conversion of F_3CSiCl_3 into F_3CSiF_3 was achieved by treatment with SbF_3. Substitution of the Si–Cl bond by several nucleophiles, like $LiAlH_4$, MeOH, $HNMe_2$, LiPh, and MeMgBr, afforded the corresponding F_3CSiR_3 derivatives (R = H, OMe, NMe_2, Ph, Me) in moderate to high yield without rupture of the C–Si bond. In contrast to $MeSiCl_3$, F_3CSiCl_3 forms stable 1:1 and 1:2 complexes with dipyridyl and pyridine, respectively. However, the C–Si bond was cleaved by aqueous alkali ($\rightarrow HCF_3 + SiO_2^- \cdot nH_2O$), or at temperatures ≥ 150 °C.

3 From Si-H Derivatives to (Monofluoromethyl)silanes

The silicon atoms of FH_2CSiH_3 and its methylated derivatives $FH_2CSiH_nMe_{3-n}$, n = 1, 2 [12], can be functionalized as displayed in Scheme 3.

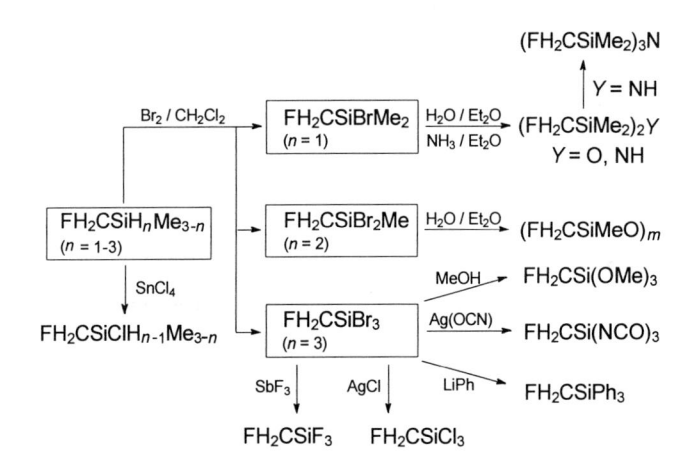

Scheme 3. Reactivity of FH_2CSi derivatives

Thus, treatment with $SnCl_4$ results in selective chlorination of only one Si–H function of $FH_2CSiH_nMe_{3-n}$ to yield $FH_2CSiClH_{n-1}Me_{3-n}$, n = 1–3, while bromination of the former with Br_2 at room temperature affords $FH_2CSiBr_nMe_{3-n}$, n = 1–3, nearly quantitatively. The FH_2CSi moiety of the novel products was shown to be fairly resistant against hydrolysis as well as attack by several nucleophilic reagents. On the other hand, the reaction of $FH_2CSiBrMe_2$ with H_2O or NH_3, dissolved in Et_2O, affords the air stable disiloxane and disilazane $(FH_2CSiMe_2)_2Y$, Y = O and NH. Lithiation of the latter with nBuLi and further treatment of the intermediary amide $(FH_2CSiMe_2)_2NLi$ with one equivalent $FH_2CSiBrMe_2$

gave $(FH_2CSiMe_2)_3N$ in an overall yield ≥ 50 %. Oligocyclosiloxanes are also accessible by hydrolysis of FH_2CSiBr_2Me with aqueous Et_2O. Although cyclosiloxanes with different configurations are formed by this route, NMR spectra of the reaction products confirmed that hydrolysis of the Si–Br bond took place without any cleavage of the C–Si link. The mass spectrum of a crystalline siloxane with b.p. 57–60 °C at ≈ 0.001 mbar, which had been isolated from the reaction products in ≈ 10 % yield, was consistent with a tetrameric cyclosiloxane [15].

Several derivatives FH_2CSiR_3, where R = F, Cl, OMe, NCO, or Ph, have been synthesized by substitution of the Br atoms of FH_2CSiBr_3 by standard procedures. Thus, the reactivity of FH_2CSi compounds (Scheme 3) closely resembles that of corresponding CH_3Si derivatives.

4 Structural Investigations of (Fluoromethyl)silanes

There are several structural investigations of (fluoromethyl)silanes by microwave, electron diffraction and single crystal X-Ray diffraction methods [16–22]. Some results are shown in Table 1.

X	r (SiC) [pm]	r (SiX) [pm]	r (CF) [pm]	∠ (FCF) [°]	∠ (XSiX) [°]
FH_2CSiH_3					
H[a]	189.29 (75)	149.02 (26) 147.34 (9)	140.42 (71)	[b]	[b]
F_3CSiX_3					
H[b]	189.97 (71)	146.00 (7)	136.00 (29)	105.41 (34)	111.69 (3)
H[d]	192.3 (3)	148.2 (5)	134.8 (1)	106.7 (5)	110.3 (10)
F[e]	191.0 (5)	155.7 (2)	134.2 (2)	107.6 (3)	108.9 (3)
C_6H_5	193.3 (4)[f]	185.8 (3)[f]	137.0 (3)[f, g]	105.2 (3)[f, g]	112.5 (6)[f, g]

Table 1. Geometries of FH_2CSiH_3 and F_3CSiX_3 compounds:
[a] r_0 values, see text; r (CH) = 108.98(18) [pm] [21]
[b] Angles in [°]: (FCSi) = 109.56(18), (H_aSiC) = 109.33(14), (H_sSiC) = 107.71(18), (H_aSiH_s) = 110.04(12), (H_aSiH_a) = 110.34(24), ((HSiH) derived from dependent parameters); H_s is the hydrogen atom in the heavy atom plane and H_a is for the hydrogen atoms out of plane
[c] r_0 values [20]
[d] r_{0a} values [16]
[e] r_g-values [17]
[f] Average values from X-ray data [19]
[g] Values are corrected for thermal motion

The main features are rather long C–Si (r_0, [pm] = 190.0(7) (F_3CSiH_3) [20], 189.3(7) (FH_2CSiH_3) [21], 186.86(2) (H_3CSiH_3) [23]), and C–F (r_0, [pm] = 136.0(3) (F_3CSiH_3), 140.4(7) (FH_2CSiH_3)) bonds. While the Si–C bond is elongated by successive fluorination at carbon, the C–F bond is shortened. The long C–Si bond in (fluoromethyl)silanes is associated with a low barrier of internal rotation about this bond [20–22]. It was concluded from the microwave spectra of FH_2CSiH_3 [21] that the SiH_3 group has lost its threefold symmetry. The in-plane SiH_s bond length *trans* to the C–F bond (r_0, [pm] = 149.02(26)) is significantly longer than the out-of-plane SiH_a bonds (r_0, [pm] = 147.34(9)), and the pseudo C_3 axis of the silyl group is tilted by 1.08 ° towards the out of plane hydrogen atoms of the CH_2F group. The rather long and more polar bonds to the antiperiplanar substituents F and H possibly facilitate a unimolecular rearrangement of FH_2CSiH_3 to H_3CSiFH_2 via a brigded transition state, which indeed has been observed for FH_2CSiH_3 in the gas phase at elevated temperature.

The monohydrides FH_2CSiHR_2 exhibit at least two conformations in which the SiH bond is in either *gauche* or *anti* position to the CF bond. In the gas phase IR spectra different Si–H stretching vibrations were assigned to these conformers with R = Me at 2153 and 2134 [cm^{-1}] and R = Et at 2140, 2128, and 2118 [cm^{-1}] [12].

The rather long and more polar C–F bond in (monofluoromethyl)silanes induces intermolecular interactions in the liquid phase which are evident from physical and spectroscopic properties. Thus, a substantial liquid phase (Raman) to gas phase (IR) shift of the C–F stretching vibration wavenumber of about 15–23 cm^{-1} has been observed. Furthermore, some FH_2CSi derivatives are less volatile than their F_3CSi counterparts. For example, FH_2CSiF_3 has a higher boiling point (+1.5 °C) than F_3CSiF_3 (–61.4 °C) or H_3CSiF_3 (–30.2 °C) and an unusually high entropy of vaporization (ΔS_v = 106 KJ mol^{-1}K^{-1}) [12]. The reactivity of the more polar C–F bond of FH_2CSi compounds should be enhanced by fluoride acceptors.

This might account for the failure to synthesize FH_2CSi derivatives by treatment of (monochloromethyl)silanes with fluorinating agents like AgF, KF, or SbF_5, which led to a breakdown of the (chloromethyl)silicon moiety to give the corresponding fluorosilanes [7]. Additional indications of the reactivity of the C–F bond have been obtained by studying the thermal decomposition of these compounds.

5 Thermolysis of (Fluoromethyl)silanes

While most of the mentioned (fluoromethyl)silanes are stable compounds at ambient temperature, the partially Si-chlorinated silanes F_2HCSiH_2Cl and $FClHCSiH_nCl_{3-n}$, n = 1, 2, are dangerously shock-sensitive. In line with intermolecular interaction in the liquid phase of these compounds is a large phase shift of 20 cm^{-1} of the SiCl stretching vibration of FH_2CSiH_2Cl from the gas phase IR to the liquid phase Raman spectra. Such interactions might well be responsible for the quantitative decomposition of the fluoride F_2HCSiH_2F at 25 °C within 5 days [10]. Its boiling point, 45 °C, is relatively high and above that of F_2HCSiH_2Cl (b.p.: 27 °C).

In order to get more insight into the factors that determine thermal stability, which is crucial for the safe manipulation of (fluoromethyl)silanes, we have investigated the conventional thermal [24] and laser-induced decomposition [25, 26] of the parent silanes $F_nH_{3-n}CSiH_3$. These compounds are stable gases at ambient temperature, and they may be conveniently studied by IR and multinuclear NMR spectroscopy. Therefore these appear to be particularly suited for gas phase investigations.

The decrease of thermal stability of (fluoromethyl)silanes in comparison to the non-fluorinated methylsilanes is evident from their decomposition temperature under static conditions. This drops significantly in the series H_3CSiH_3 ($\geq$ 340 °C), F_3CSiH_3, and F_2HCSiH_3 ($\geq$ 180 °C) to FH_2CSiH_3 ($\geq$ 120 °C). While H_3CSiH_3 decomposes under these conditions primarily by geminal elimination of molecular H_2 from the silicon moiety and secondarily by a slow surface-catalysed reaction yielding H• and CH_3SiH_2• radicals [27], the significantly lower decomposition temperatures of (fluoromethyl)silanes require different low energy decomposition pathways.

6 Decomposition Pathways of (Fluoromethyl)silicon Compounds

Although the C–Si bond of (fluoromethyl)silanes is rather long and supposedly weak, its homolytic cleavage is apparently not energetically favoured. However, the fragmentation pattern of the molecular ion of (fluoromethyl)silanes, after electron impact ionization in the mass spectrometer, is dominated by ions emerging from homolytic C–Si bond fission. While (trifluoromethyl)phosphorous or sulfur hydrides tend to decompose via vicinal HF elimination [28], we were unable to direct the decomposition of F_3CSiH_3 to HF and F_2CSiH_2. Neither the dynamic thermolysis of F_3CSiH_3 in the presence of fluoride acceptors like KF, which accelerate its decomposition at lower temperature, nor its laser-induced decomposition by multiphoton excitation of the Si–H bending or the C–F stretching vibration, or simultaneous excitation of both the Si–H and the C–F vibrational modes, promoted any processes involving vicinal HF elimination. The different polarity of Si–H and P–H bonds may account for this selectivity.

Thermolysis of both F_3CSiH_3 and F_2HCSiH_3 in the presence of efficient carbene trapping agents have shown that they decompose predominantly by α-fluorine shift to give fluorosilane and the fluorocarbenes CF_2 and CHF, respectively [24]. Secondary reactions at elevated temperature in the absence of any carbene trap eventually result in numerous volatile fluoro-, (fluoromethyl)-, and (methyl)fluorosilanes via insertion of initially formed fluorocarbenes into SiH bonds of fluorosilanes, and some dismutation of the SiH_nF_{3-n} moieties. The decomposition temperature of F_3CSiX_3 is significantly lowered as the electronegativity of X increases. While the parent silane ($X =$ H) decomposes at temperatures ≥ 180 °C, the perfluorinated derivative ($X =$ F) has been shown to eliminate difluorocarbene already at temperatures below 100 °C, which makes it a convenient CF_2 source. The lowering of the decomposition temperature of $(F_3C)_nSiX_{3-n}$ upon increase of the number n of F_3C groups attached to Si should also be noted. We found that bis(trifluoromethyl)silane $(F_3C)_2SiX_2$ decomposes at temperatures about 100 °C below those of $(F_3C)SiX_3$, $X =$ H and Cl. Of the tris(trifluoromethyl)silanes, only those stabilized by bulky dialkylamino groups attached to silicon could be isolated at ambient temperature. $(F_3C)_2SiCl_2$, which is a possible precursor of perfluorinated silicones, decomposes at $\approx$ 45 °C by CF_2 elimination. Therefrom one can conclude that perfluorinated silicones themselves would not be stable at ambient temperature.

Unexpectedly, FH_2CSiH_3 is thermally less stable than both F_3CSiH_3 and F_2HCSiH_3. The analysis of the decomposition products revealed a clean rearrangement of FH_2CSiH_3 to H_3CSiH_2F, with some SiF/SiH dismutation occurring additionally at $\geq$ 120 °C. Since neither CF_2 nor CH_2 could be detected by means of carbene trapping agents, this gas phase isomerization is apparently a unimolecular dyotropic rearrangement. On the other hand, $FH_2CSiHClMe$ and FH_2CSiBr_2Me, dissolved in C_6D_6, rearranged completely and selectively within several months already at 25 °C to yield $ClH_2CSiHFMe$ and $BrH_2CSiBrFMe$, respectively [12]. Thus, migration of Cl or Br atoms from Si to C is faster than that of H or CH_3 moieties.

7 Laser-Induced Decomposition of (Fluoromethyl)silanes

The CO_2 laser-induced multiphoton decomposition of silanes, known to be a really homogeneous reaction, was utilized for the chemical vapour deposition of fluorine containing SiC films from the parent (fluoromethyl)silanes [25, 26]. In contrast to work with H_3CSiH_3, irradiation of (fluoromethyl)silanes with a single unfocused CO_2 laser pulse at fluence of ≤ 0.9 Jcm^{-2} tuned to absorption bands of either the SiH bending or the CF stretching vibrations results in an explosive reaction. This is accompanied by an intense chemoluminescence when the sample pressure exceeds a certain limit in the range of 0.1–6.7 kPa (Fig. 1).

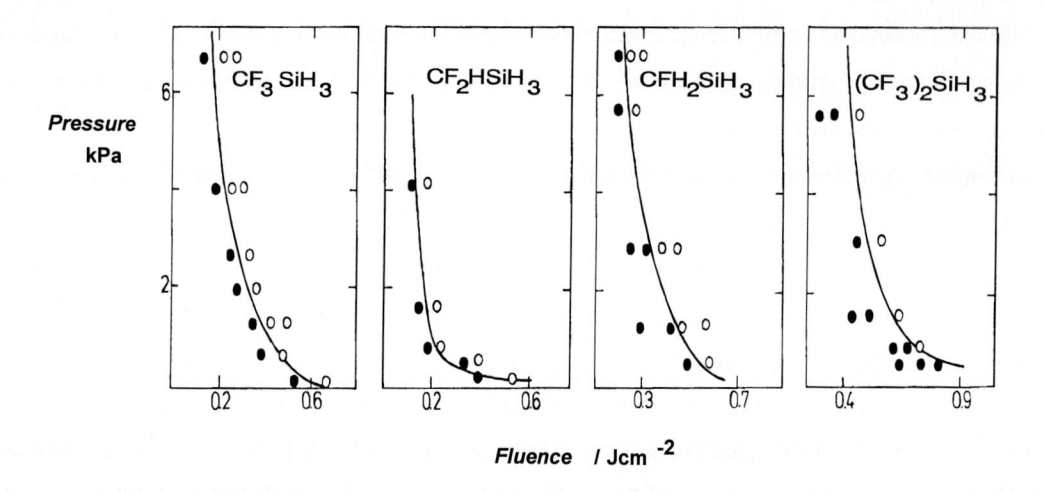

Fig. 1. Explosion limit for the IR-multiphoton decomposition of (fluoromethyl)silanes, irradiating wavelengths in the given order are 916.6, 1025.3, 927.0, and 922.9 cm^{-1}; empty and full circles relate to explosive and no reaction, respectively (reprinted from reference [25])

The explosion limit dropped to lower laser fluence upon increasing the pressure of the (fluoromethyl)silanes. Below the explosion limit, chemical changes of the (fluoromethyl)silanes could only be detected after irradiation with as many as 10^3 laser pulses. The reaction products revealed an almost quantitative reduction of the CF bond and afforded the gaseous fluorosilanes SiF_4 and SiF_3H, the hydrocarbons CH_4 and C_2H_2, and different solid Si/C/F/H materials.

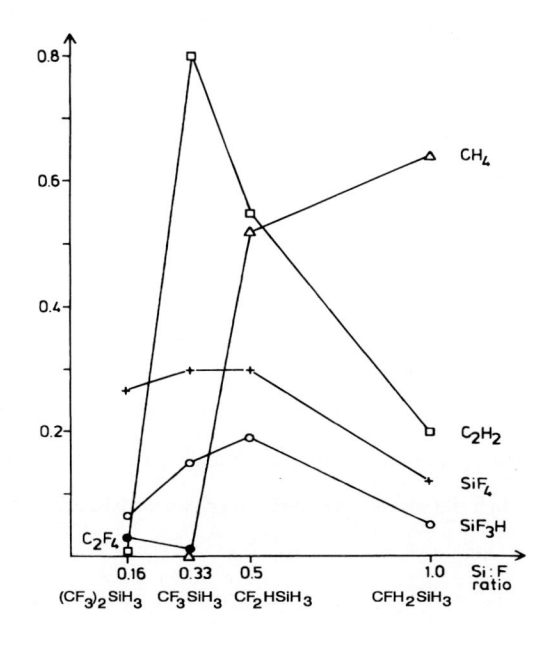

Fig. 2. Mole fraction of gaseous products formed by the IR-multiphoton decomposition of (fluoro-methyl)silanes (reprinted from reference [25])

It was supposed that, because of a rapid intramolecular energy pooling, infrared laser multiphoton decomposition at low laser fluence usually proceeds via the energetically most favourable reaction channel [26]. From the material balance of gaseous decomposition products (Fig. 2) it was inferred that there are common primary steps for both the thermal and the explosive laser-induced decomposition of (fluoromethyl)silanes.

Nevertheless, in the gaseous products of the explosive decomposition of F_3CSiH_3, ethyne and some traces of other carbon-containing compounds were found which had not been detected among the products of the conventional thermolysis.

In contrast to this, methane was, in addition to ethyne, the main gaseous product of the explosive decomposition of FH_2CSiH_3, while in the intermediate case of F_2HCSiH_3 both hydrocarbons were found in roughly the same amount. CH_4 was inferred to be eliminated according to (1) from precursors with a CH_3SiH moiety which have been obtained in the conventional thermolysis by a dyotropic rearrangement of FH_2CSi compounds.

$$H_3C-SiH \longrightarrow H_4C \ + \ |Si{<} \qquad (1)$$

$$H_2Si{<} \longrightarrow H_2 \ + \ |Si{<} \qquad (2)$$

$$2 \ FC| \longrightarrow FC=CF \qquad (3)$$

CH_4 elimination was also observed as the primary decomposition pathway of CH_3SiH_3 under shock-tube conditions [29]. C_2H_2 was supposed to be generated via the sequence of low energy reactions Eq. 3–6, in which fluorovinylsilane and ethynylsilane intermediates are formed by initial reactions of fluorocarbenes and silylenes.

$$FC=CF \ + \ |Si{<} \longrightarrow FC=CSiF \longrightarrow -C{\equiv}C- \ + \ F_2Si{<} \qquad (4)$$

$$FC{\equiv}C- \ + \ |Si{<} \longleftarrow FC=C- \overset{\diagdown \ \diagup}{\underset{\diagup \ \diagdown}{Si}} \longrightarrow -C{\equiv}CSiF \qquad (5)$$

$$-\text{C}\equiv\text{CSiH} \rightleftarrows \text{HC}=\overset{\overset{\diagdown\diagup}{\text{Si}}}{\underset{\diagup\diagdown}{\text{C}}}- \rightleftarrows \text{HC}\equiv\text{C}- + |\text{Si}\!\!<\quad\quad(6)$$

Ethynylsilanes were indeed identified among the products of the explosive decomposition of both F_2HCSiH_3 and F_3CSiH_3. This reaction sequence ensures complete CF–SiH exchange, with generation of C_2H_2 by irreversible Si–F bond formation.

The different decomposition pathways of the various (fluoromethyl)silanes do not only determine the composition of the gas phase products but also the composition and the texture of the deposited Si/C/H/F materials. They were shown to consist of carbon and a blend of silicon-carbon based frameworks containing some Si–H and C–H bonds by XPS, SEM, IR, and Raman spectroscopy [26].

References:

[1] J. H. Simons, R. D. Dunlap, *US* 2 651 651,**1953**; N. Haszeldine, *Nature* **1951**, *168*, 1028; N. Haszeldine, *Angew. Chem.* **1954**, *66*, 693.

[2] R. E. Banks, R. N. Haszeldine, *Adv. Inorg. Chem. Radiochem.* **1961**, *3*, 337; N. Haszeldine, C. Parkinson, P. J. Robinson, W. J. Williams, *J. Chem. Soc. Perkin II* **1979**, 954; and references therein.

[3] I. Ruppert, K. Schlich, W. Volbach, *Tetrahedron Lett.* **1984**, *25*, 2195.

[4] D. Seyferth, S. P. Hopper, *J. Organomet. Chem.* **1973**, *51*, 77; Chvalovsky, J. Vcelák, L. Roman, *Coll. Czech. Chem. Commun.* **1976**, *41*, 2708; Fuchikami, I. Ojima, *J. Organomet. Chem.* **1981**, *212*, 145; Josten, I. Ruppert, *J. Organomet. Chem.* **1987**, *329*, 313; Broicher, D. Geffken, *J. Organomet. Chem.* **1990**, *381*, 315; Pawelke, *J. Fluorine Chem.* **1989**, *42*, 429; G. Pawelke, *J. Fluorine Chem.* **1991**, *54*, 60.

[5] G. K. S. Prakash, R. Krishnamurti, G. A. Olah, *J. Am. Chem. Soc.* **1989**, *111*, 393; J.-P. Bosmans, *Janssen Chimica Acta* **1992**, *10*, 22.

[6] A. Morimoto, T. Miura, M. Kumeda, T. Shimizu, *Japan. J. Appl. Phys.* **1983**, *22*, 908; Uesugi, H. Ihara, H. Matsumura, *Japan. J. Appl. Phys.* **1985**, *24*, 1263; Matsumura, T. Uesugi, H. Ihara, *Japan. J. Appl. Phys.* **1985**, *24*, L24; Saito, *Philosophical Magazine* **1987**, *B55*, 615.

[7] J. M. Bellama, A. G. McDiarmid, *J. Organomet. Chem.* **1969**, *18*, 275; J. Hairston, D. H. O'Brien, *J. Organomet. Chem.* **1970**, *23*, C41; F. Cunico, B. B. Chou, *J. Organomet. Chem.* **1978**, *154*, C45; Damrauer, V. E. Yost, S. E. Danahey, B. K. O'Connell, *Organometallics* **1985**, *4*, 1779.

[8] J. A. Morrison, L. L. Gerchman, R. Eujen, R. J. Lagow, *J. Fluorine Chem.* **1977**, *10*, 333; H. Lange, D. Naumann, *J. Fluorine Chem.* **1985**, *27*, 115.

[9] H. Beckers, H. Bürger, P. Bursch, I. Ruppert, *J. Organomet. Chem.* **1986**, *316*, 41.

[10] H. Bürger, R. Eujen, P. Moritz, *J. Organomet. Chem.* **1991**, *401*, 249.

[11] K. G. Sharp, T. D. Coyle, *J. Fluorine Chem.* **1971/72**, *1*, 249.

[12] H. Bürger, P. Moritz, *Organometallics* **1993**, *12*, 4930.

[13] H. Beckers, H. Bürger, R. Eujen, *J. Fluorine Chem.* **1985**, *27*, 461; H. Beckers, H. Bürger, R. Eujen, *Z. Anorg. Allg. Chem.* **1988**, *563*, 38; see also R. Josten, I. Ruppert, [4].

[14] E. A. V. Ebsworth, *Volatile Silicon Compounds*, Pergamon Press, Oxford, **1963**.

[15] H. Bürger, P. Moritz, unpublished results.

[16] H. Beckers, H. Bürger, R. Eujen, B. Rempfer, H. Oberhammer, *J. Mol. Struct.* **1986**, *140*, 281.

[17] B. Rempfer, G. Pfafferott, H. Oberhammer, H. Beckers, H. Bürger, R. Eujen, J. E. Boggs, *Rev. Chim. Miner.* **1986**, *23*, 551.

[18] J. G. Smith, *J. Mol. Spectrosc.* **1988**, *128*, 288.

[19] H. Beckers, D. J. Brauer, H. Bürger, C. J. Wilke, *J. Organomet. Chem.* **1988**, *356*, 31.

[20] J. R. Durig, G. Attia, P. Groner, H. Beckers, H. Bürger, *J. Chem. Phys.* **1988**, *88*, 545.

[21] J. R. Durig, H. Nanaie, H. Bürger, P. Moritz, to be published.

[22] D. R. Lide, Jr., D. R. Johnson, K. G. Sharp, T. D. Coyle, *J. Chem. Phys.* **1972**, *57*, 3699.

[23] J. L. Duncan, J. L. Harvie, D. C. McKean, *J. Mol. Struct.* **1986**, *145*, 225.

[24] H. Beckers, H. Bürger, *J. Organomet. Chem.* **1990**, *385*, 207; H. Bürger, P. Moritz, *J. Organomet. Chem.* **1992**, *427*, 293.

[25] J. Pola, Z. Bastl, J. Tláskal, H. Beckers, H. Bürger, P. Moritz, *Organometallics* **1993**, *12*, 171.

[26] J. Pola, H. Beckers, H. Bürger, *Chem. Phys. Lett.* **1991**, *178*, 192; Pola, Z. Bastl, J. Tláskal, H. Beckers, H. Bürger, P. Moritz, P. Weiss, M. Sigrist, *Appl. Organomet. Chem.* **1993**, *7*, 381; C. Sigüenza, L. Diaz, P. F. Gonzáles-Diaz, H. Beckers, Bürger, J. Pola, *Spectrochim. Acta A* **1994**, *50*, 1207.

[27] P. S. Neudorfl, E. M. Lown, I. Safarik, A. Jodhahn, O. P. Strausz, *J. Am. Chem. Soc.* **1987**, *109*, 5780; and references therein.

[28] H. W. Kroto, J. F. Nixon, N. P. C. Simmons, N. P. C. Westwood, *J. Am. Chem. Soc.* **1978**, *100*, 446.

[29] B. A. Sawrey, H. E. O'Neal, M. A. Ring, D. Coffey, Jr., *Int. J. Chem. Kinetics* **1984**, *16*, 7.

Silicides as Catalysts for Hydrodehalogenation of Silicon Tetrachloride

H. Walter, G. Roewer*, K. Bohmhammel

Institut für Anorganische Chemie/Institut für Physikalische Chemie
Technische Universität Bergakademie Freiberg
Leipziger Str. 29, D-09599 Freiberg, Germany

F. Vogt, W. Mörke

Institut für Anorganische Chemie
Martin-Luther-Universität Halle-Wittenberg
Weinbergweg 16, D-06120 Halle (Saale), Germany

Introduction

The catalytical hydrodehalogenation of $SiCl_4$, from the "Siemens process", to give $HSiCl_3$ is a fundamental scientific and technological problem. In this paper, a method to form metal silicides in situ as efficient catalysts will be described, which make it possible to carry out the hydrodehalogenation under mild conditions maintaining the high purity of the products. Starting from the dominating reactions of:

$$x\ SiCl_4 + 2x\ H_2 + y\ M \longrightarrow M_ySi_x + 4x\ HCl$$

Eq. 1. Silicide generation

$$M_ySi_x + 3\ HCl \longrightarrow M_ySi_{x-1} + HSiCl_3 + H_2$$

Eq. 2. Silicide destruction

$$SiCl_4 + H_2 \rightleftharpoons HSiCl_3 + HCl$$

Eq. 3. $HSiCl_3$ formation

The generation of all silicide phases in dependence on temperature can be thermodynamically explained by means of equilibrium calculations. Taking into account our experimental results for the $H_2/SiCl_4/Ni$ system, a picture of the mechanism of hydrodehalogenation is developed which combines thermodynamic and kinetic results.

Catalytic Hydrogenation

The proposed concept combines the advantages of some known processes without their disadvantages. The catalyst itself is created in a flow reactor from dispersed pure metals in contact with an ultragrade $SiCl_4/H_2$ mixture. This prevents the inclusion of electronically active elements and compounds. The produced silicides show excellent activity and selectivity with respect to $HSiCl_3$ formation. The obtained yield was nearly constant over a long residence time range.

Mechanism of Catalytic Hydrodechlorination

The obtained results on the reaction mechanism can be summarized as follows: The metal silicides form cluster structures which represent electron buffer systems. They can be oxidized or reduced easily by surface reactions. The adsorption of $SiCl_4$ molecules at the cluster surface is immediately followed by an electron transfer from the cluster to the silicon atom of $SiCl_4$, the cluster is oxidized. As a result of such a process a silylene species is formed at the surface of the catalyst. Chloride ions act as counter ions to the positive cluster, supporting the redox step (Eq. 4).

$$(Ni_2Si)_n + SiCl_4 \rightleftharpoons [(Ni_2Si)_n SiCl_2]Cl_2$$

Eq. 4.

In the next step, an H_2 molecule transfers its electrons to the cluster, reducing it again. Two hydrochloride molecules are produced and located near the silylene species (Eq. 5).

$$[(Ni_2Si)_nSiCl_2]Cl_2 + H_2 \rightleftharpoons (Ni_2Si)_nSiCl_2 + 2\ HCl$$

Eq. 5.

The silylene species undergoes an oxidative addition. The newly formed $HSiCl_3$ is desorbed from the catalysts surface (Eq. 6).

$$(Ni_2Si)_nSiCl_2 + HCl \rightleftharpoons (Ni_2Si)_n + HSiCl_3$$

Eq. 6.

Activation of the Catalyst

The catalyst has to be created before the hydrodehalogenation reaction starts because disperse metals are inactive. An induction period is always observed when the $H_2/SiCl_4$ mixture is passed over the metal powders. This is caused by the gradual formation process of the silicides which are stable under the specific reaction conditions. In the Ni/Si system at 600 °C this is the Ni_2Si phase. Only after this phase has been formed at least on the particle surface (identified by X-ray diffraction), the silylene group concentration appears to be high enough for the oxidative addition of HCl to take place. Then the fast reduction to $Si^{\pm 0}$ (silicides) is essentially completed. From this moment, trichlorosilane can be detected in the reaction gas. Its content gradually increases at the same rate as phases with a higher silicon content are formed by solid state reactions inside the catalyst grains (Eq. 7, 8).

$$Ni_2Si + Ni \longrightarrow Ni_3Si$$

Eq. 7.

$$Ni_2Si + Ni_3Si \longrightarrow Ni_5Si_2$$

Eq. 8.

Variation of the Reaction Volume Stream

Unlike the uncatalyzed hydrodehalogenation reaction, the yield of this catalytical conversion of $SiCl_4$ into $HSiCl_3$ is almost independent of the reaction volume stream and the residence time (0.5 or 0.025 s) (Fig. 1).

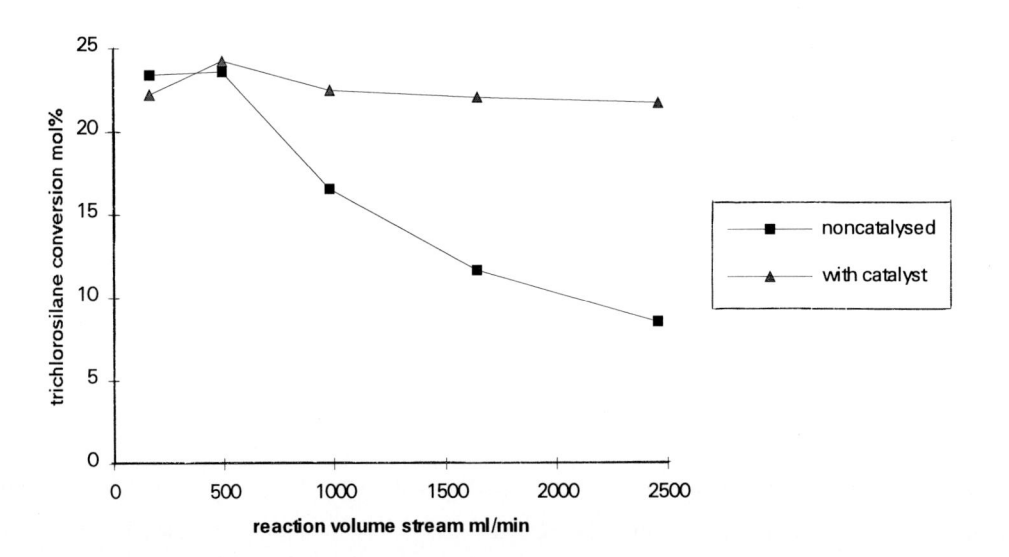

Fig. 1. Dependence of conversion of the reaction volume stream at 900 °C

This result leads to the conclusion that in accordance with the proposed reaction mechanism, the oxidative addition of HCl to the silylene species is the step limiting the rate.

HCl Pulse

Trichlorosilane formation is finally caused by the reaction of HCl and a silylene species. A completely formed catalyst which was cooled to room temperature under reaction gas, flushed with H_2 and treated again with a $H_2/SiCl_4$ mixture at the previous reaction temperature, is inactive. The reaction can be restarted on the silylene species-covered surface by a HCl pulse. The conversion reaches the original level.

FMR Investigations

Based on the above ideas on the reaction mechanism, it is expected that the silylene species at higher temperatures display an increased electron acceptor ability compared to the silicide cluster. Evidence for that is provided by investigations of catalyst samples using ferro-magnetic resonance (FMR).

$$(Ni_2Si)_n SiCl_2 \xrightarrow{> 350\ °C} [(Ni_2Si)_n Si]Cl_2$$

diamagnetism — electron deficit in the cluster (magnetism)

Eq. 9.

A completely formed catalyst does not emit any signals in the FMR spectrum. In contrast, samples which were thermally stressed at temperatures above 350 °C under inert gas produce a strong FMR signal. However, the crystalline phase remains unchanged. A redox reaction between catalyst and chemisorbed silylene species provides a possible explanation for this phenomenon. The Ni_2Si cluster causes the further reduction of the silicon of the silylene group by donation of electrons. As a result of this electron transfer, the cluster should contain unpaired electrons, which are the source of the obtained FMR signal (Eq. 9).

Conjugate Phase Transitions

In accordance with our thermodynamic calculations we have observed the generation of the NiSi phase at temperatures above 700 °C. The incorporated silicon atoms originate from the gas phase. After the generation period the trichlorosilane conversion increases to the level determined by the equilibrium calculations. Taking into account the proposed mechanism of hydrodehalogenation, this leads to the same consequences as in the case of Ni_2Si silicide catalyst, but the selectivity is less. Traces of H_2SiCl_2 are found. These are attributed to noncatalyzed subreactions in addition to $HSiCl_3$ (Eq. 10, 11).

$$SiCl_4 + H_2 \rightleftharpoons SiCl_2 + 2 HCl$$

Eq. 10.

$$SiCl_2 + H_2 \rightleftharpoons H_2SiCl_2$$

Eq. 11.

The generation of the silicide phases is reversible. According to thermodynamic calculations, the NiSi phase was changed to the Ni_2Si phase below 700 °C in a $SiCl_4/H_2$ gas phase.

$$\begin{bmatrix}
(NiSi)_n + SiCl_4 & \rightleftharpoons & [(NiSi)_nSiCl_2]Cl_2 \\
[(NiSi)_nSiCl_2]Cl_2 + H_2 & \rightleftharpoons & (NiSi)_nSiCl_2 + 2 HCl \\
(NiSi)_nSiCl_2 + HCl & \rightleftharpoons & (NiSi)_n + HSiCl_3 \\
(NiSi)_n + 2 HCl & \longrightarrow & (Ni_nSi_{n-1})SiCl_2 + H_2 \\
(Ni_nSi_{n-1})SiCl_2 + HCl & \longrightarrow & Ni_nSi_{n-1} + HSiCl_3
\end{bmatrix}_{0.5 n}^{3}$$

Eq. 12. - Eq. 16.

During this conversion, an excess of Si was generated. The segregation of silicon causes enrichment of silylenes on the surface of the catalyst (Eq. 16). This effect was connected with a higher yield of $HSiCl_3$ in the reaction gas atmosphere (Eq. 12 – Eq. 16).

When elemental silicon was added to the NiSi phase, this phase remained stable, at 600 °C until the silicon was consumed in the reactions, shown by Eq. 17 and Eq. 18.

Eq. 17.
$$2\ NiSi + SiCl_4 + 2\ HCl \longrightarrow Ni_2Si + 2\ HSiCl_3$$

Eq. 18.
$$Ni_2Si + Si \longrightarrow 2\ NiSi$$

This means, that silicon does not react directly, but via intermediate NiSi, Ni_2Si and other phases.

Use of b-Group Elements

Silicides of other b-group elements, such as Cu, Fe, Pd, Pt, Mo, and Re basically react in the same way as nickel silicides. They are catalytically active as soon as stable silicides are formed at preselected temperatures. In case of the Cu/Si system, this is the Cu_5Si phase. The results obtained for the Cu_5Si phase are comparable with those obtained with the Ni_2Si catalyst. When elemental silicon was added to the Cu_5Si compound, the Cu_3Si phase was dominating at 600 °C. In any case, Cu_3Si remained stable, until the excess silicon was consumed forming $HSiCl_3$ (compare with Eq. 12–18).

Organochlorosilanes and Müller-Rochow Synthesis

The described ideas about the mechanism can also be used to explain results of the hydrodehalogenation of organochlorsilanes, for instance in the case of CH_3SiCl_3. During its hydrodehalogenation the byproducts $HSiCl_3$ and CH_4 always appear in addition to the main product CH_3SiHCl_2. According to the thermodynamic calculations, however CH_4 and $HSiCl_3$ should be the preferred products. The proposed mechanism provides a good interpretation of the dominating formation of dichloromethylsilane (Scheme 1).

Scheme 1 demonstrates that different products result in dependence on the type of connection between the CH_3SiCl_3 molecule and the silicide cluster on the catalyst surface. The preferred connection over two chlorine atoms is plausible and seems to be the reason for the kinetically controlled generation of dichloromethylsilane.

The proposed silylene mechanism gives an explanation for the high selectivity of $(CH_3)_2SiCl_2$ formation in the "Direct Synthesis" of methylchlorosilanes (Müller-Rochow process). Via an oxidative addition of CH_3Cl to methylsilylenes on the surface of a Cu/Si catalyst, $(CH_3)_3SiCl_2$ is produced in a kinetically controlled process (Scheme 2).

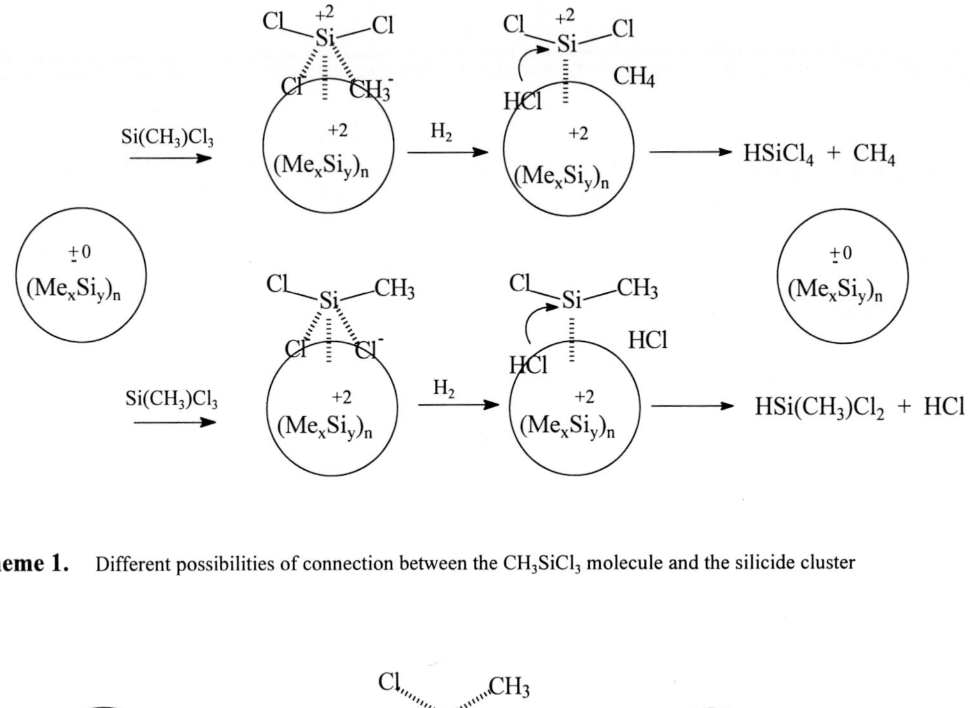

Scheme 1. Different possibilities of connection between the CH_3SiCl_3 molecule and the silicide cluster

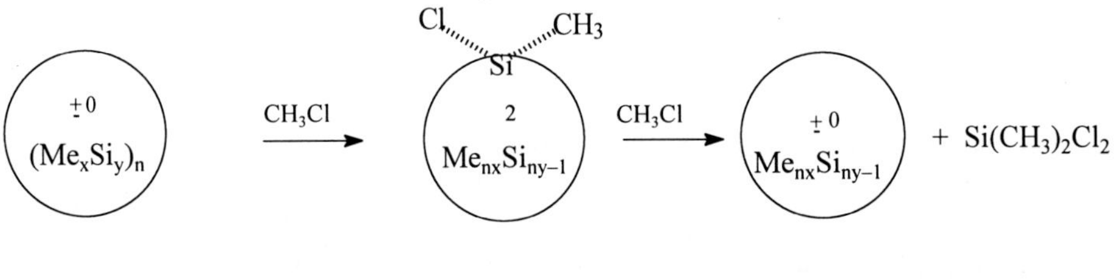

Scheme 2. Oxidative addition of CH_3Cl by "Müller-Rochow process"

The intermediate Cu_3Si phase should be only one of some catalytical active silicides under these conditions. The mechanism, which suggests the intermediate formation of surface fixed silylenes, requires that the silicide clusters alternately release and accept silicon atoms (see hydrodehalogenation).

The intermediate Cu_3Si phase should be only one of some catalytical active silicides under these conditions. The mechanism, which suggests the intermediate formation of surface fixed silylenes, requires that the silicide clusters alternately release and accept silicon atoms (see hydrodehalogenation).

References:

[1] M. E. Schlesinger, *Chem. Rev.* **1990**, *90*, 4.
[2] W. M. Ingle, M. S. Peffley, *J. Electrochem. Soc.: Solid-State Science and Technology* **1985**, 1236.
[3] M. P. Clarke, I. M. T. Davidson, *J. Organomet. Chem.* **1991**, *408*, 149.

Silacyclobutenes – Synthesis and Reactivity

Michael Backer, Martin Grasmann, Wolfgang Ziche, Norbert Auner*
Fachinstitut für Anorganische und Allgemeine Chemie
Humboldt-Universität zu Berlin
Hessische Str. 1-2, D-10115 Berlin, Germany

Carola Wagner, Eberhardt Herdtweck[1], Wolfgang Hiller[1], Maximilian Heckel[1]
Anorganisch-Chemisches Institut
Technische Universität München
Lichtenbergstr. 4, D-85747 Garching, Germany

Summary: $Cl_2Si=CHCH_2tBu$ is a versatile building block in organosilicon synthesis which reacts with a wide variety of multiple bond systems. The unique reactivity is shown by the predominance of [2+2] over [4+2] cycloaddition reactions. This unusual behaviour is caused by the two chlorine substituents at the silicon atom. This presentation shows the synthesis of silacyclobutenes from $Cl_2Si=CHCH_2tBu$ and disubstituted alkynes and some chemistry of these four membered ring compounds.

$Cl_2Si=CHCH_2tBu$ is easily synthesized by the reaction of trichlorovinylsilane with *t*butyllithium in nonpolar solvents [2–4] and reacts in a [2+2] cycloaddition reaction with disubstituted alkynes to give 1,1-dichloro-2,3-diorganyl-4-neopentyl-1-silacyclo-2-butenes [5].

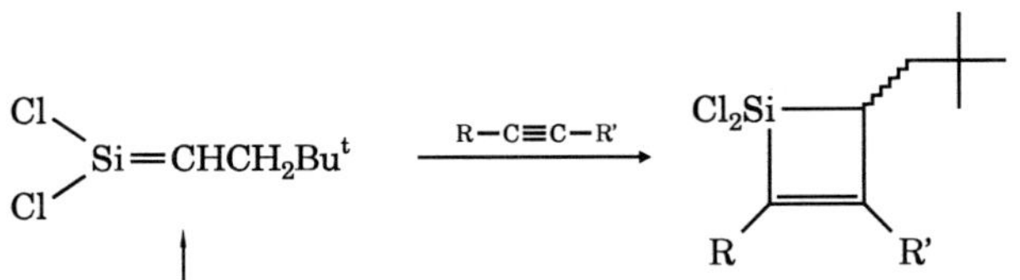

1 : R = R' = Ph
2 : R = Me$_3$Si, R' = Ph
3 : R = Me, R' = Ph

Scheme 1.

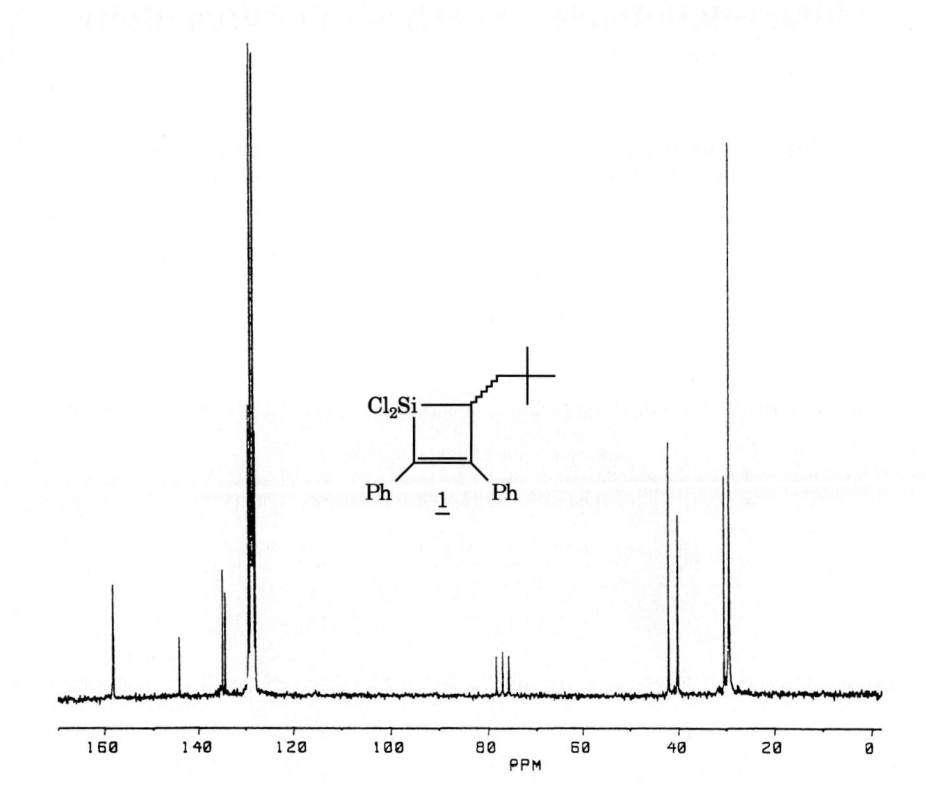

Fig. 1. 100 MHz ^{13}C NMR spectrum of **1**

Fig. 2. Crystal structure of **2**

The substitution pattern of the silaethenes, $R_2Si=CR'CH_2tBu$, may be changed; it is, however, mandatory that the substituents at the silicon atom are π-donors (R = Cl, R' = H: **A**; R = Me$_3$SiO, R' = H: **B**), π-acceptors can be bound to the carbon atom (R = Cl, R' = Ph: **C**). Best yields of silacyclobutenes are restricted to the use of 1,1-dichloro-2-neopentylsilaethene **A** [6].

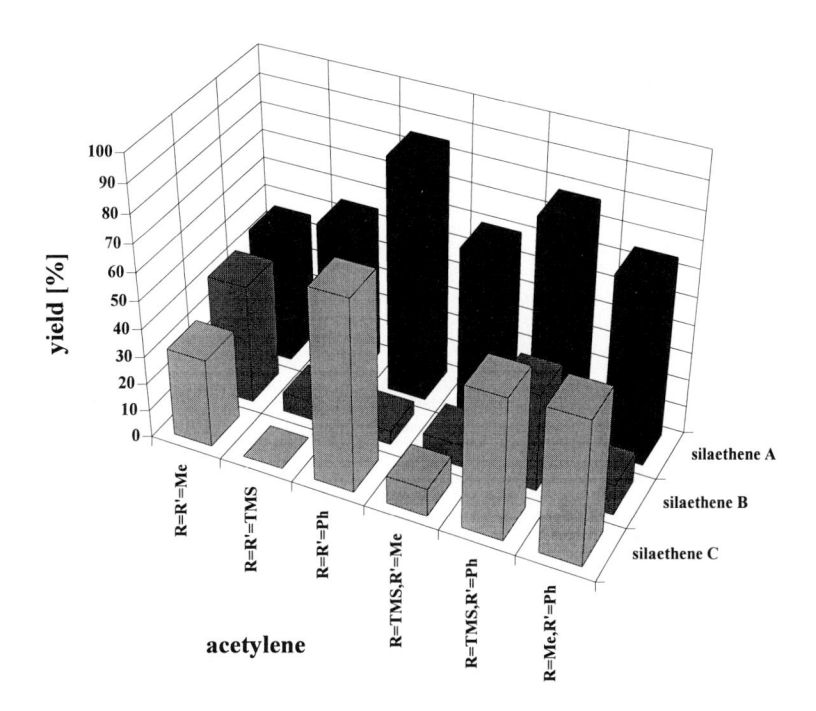

Fig. 3. Yield of silaethene as a function of substitution

The two chlorine substituents at the silicon allow facile derivatizations of the cycloadducts by treatment with methyllithium or lithiumalanate.

Scheme 2.

Fig. 4. Crystal structures of **4** and **5**

Transition metal derivatives are known to stabilize unusual molecules in their coordination sphere. First studies on silanediyl- and silyl complexes of iron carbonyls have been carried out [7–9].

Scheme 3.

$\underline{6}$: R = Ph, $\underline{7}$: R = Me

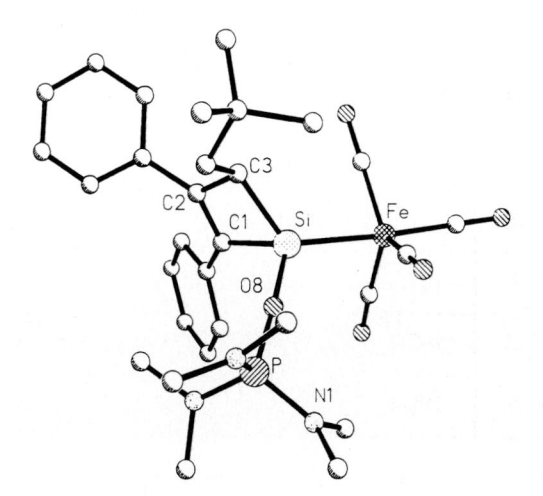

Fig. 5. Crystal structure of **6**

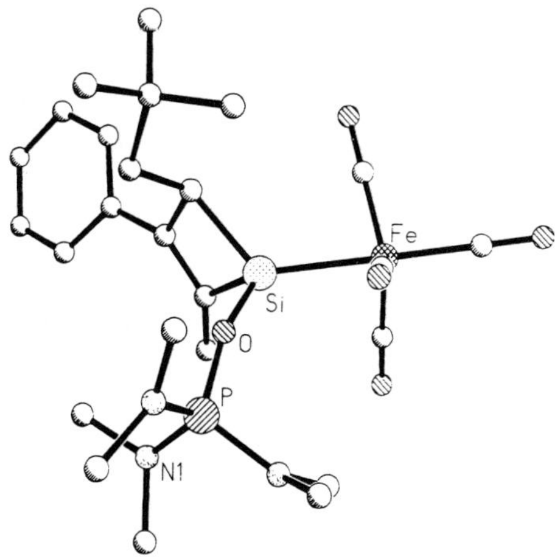

Fig. 6. Crystal structure of 7

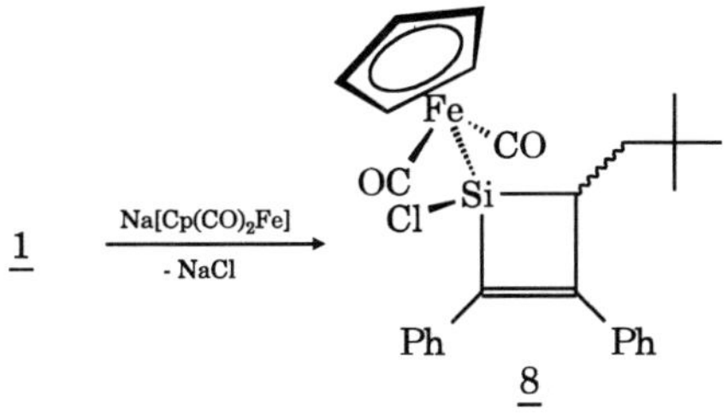

Scheme 4.

Compound	Si–C1	Si–C3	C1–C2	M=Si	∠ C3–Si–C1	Σ(endocyclic)
2	182.8	186.4	135.4		80.7	359.9
4	185.8	191.0	135.6		77.1	359.6
5	184.8	189.0	135.0		77.5	360.0
6	187.8	190.3	135.5	227.0	76.3	360.0
7	186.5	190.4	134.5	227.7	76.7	360.0

Table 1. Selected bond length [pm] and angles [°]

Photolysis of silacyclobutenes allows a straightforward approach to the generation of 1-sila-1,3-butadienes. Furthermore the wide variability of substituents at the silicon atom enables a study of their influence on the ring opening reactions. First results of the photolysis of a dimethylated silacyclobutene **2** demonstrate the synthetic potential of this class of compounds. On the one hand the intermediate silabutadiene reacts like a silaethene with methoxytrimethylsilane, on the other hand it shows "butadiene behaviour" in a [4+2] cycloaddition reaction with acetone [10–13].

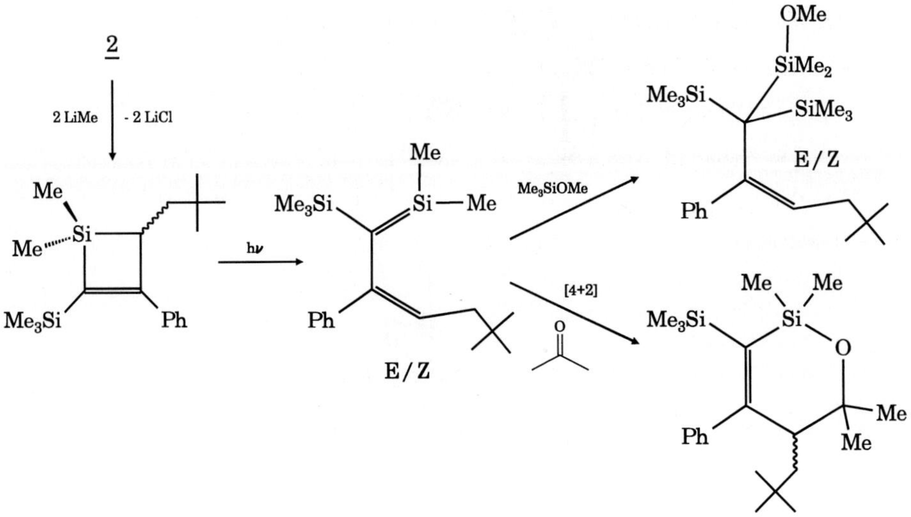

Scheme 5.

Acknowledgement: N. A. gratefully thanks the *Stiftung Volkswagenwerk* for financial support, *Dow Corning Ltd.* (Barry), *Wacker-Chemie GmbH*, and *Chemetall GmbH* for chlorosilanes and lithium alkyls. W. H. thanks *Hewlett Packard* for support.

References:

[1] Crystal structure determination.
[2] N. Auner, *J. Organomet. Chem.* **1988**, *353*, 275.
[3] N. Auner, in: *Organosilicon Chemistry – From Molecules to Materials* (Eds.: N. Auner; J. Weis), VCH, Weinheim, **1994**, p. 103.

[4] N. Auner, *J. Prakt. Chem.* **1995**, *337*, 79.
[5] N. Auner, C. Seidenschwarz, N. Sewald, E. Herdtweck, *Angew. Chem.* **1991**, *103*, 1172; *Angew. Chem., Int. Ed. Engl.* **1991**, *30*, 1151.
[6] N. Auner, C.-R. Heikenwälder, C. Wagner, *Organometallics* **1993**, *12*, 4135.
[7] T. D. Tilley, in: *The Chemistry of Organic Silicon Compounds* (Eds.: S. Patai, Z. Rappoport), Wiley, Chichester, **1989**, p. 1415.
[8] C. Zybill, *Nachr. Tech. Lab.* **1989**, *37*, 248.
[9] C. Wagner, N. Auner, E. Herdtweck, W. Hiller, M. Heckel, *Bull. Soc. Chim. Fr.*, in press.
[10] P. B. Valkovich, W. P. Weber, *Tetrahedron Lett.* **1975**, *26*, 2153.
[11] R. Okazaki, K. T. Kang, N. Inamoto, *Tetrahedron Lett.* **1981**, *22*, 235.
[12] R. T. Conlin, S. Zhang, M. Namavari, K. L. Bobbitt, M. J. Fink, *Organometallics* **1989**, *8*, 571.
[13] N. Auner, M. Backer, in preparation.

Allyl Cleavage of 2-Silanorbornenes – A Facile Synthesis of Cyclopentenyl Compounds of Organochlorosilanes and of 2-Silanorbornanes

Hans-Uwe Steinberger, Norbert Auner*

Fachinstitut für Anorganische und Allgemeine Chemie

Humboldt-Universität zu Berlin

Hessische Str. 1-2, D-10115 Berlin, Germany

Summary: 1,1-Dichloro-2-neopentylsilaethene, $Cl_2Si=CHCH_2tBu$, is a versatile building block in organosilicon chemistry. It is easily synthesized from trichlorovinylsilane and *t*butyllithium in nonpolar solvents and reacts with a wide variety of multiple bond systems to give mono- and bicyclic silicon compounds [1]. The easily accessible cycloadducts allow extensive investigations of their interesting chemistry [2]. This paper describes the simple synthesis of new cyclopentenyl compounds containing a silicon functionality in β-position to the five membered ring moiety.

Compounds **1** and **3** are prepared according to published procedures [3], and the silanorbornenes **2** and **4** can be synthesized from **1** in good yield [4].

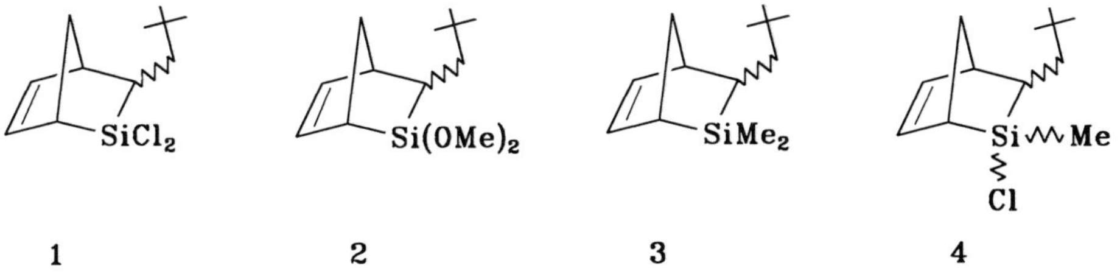

Fig. 1.

The triorganochlorosilanes **5**, **6**, and **7** are obtained by cleavage of the bicyclic precursors **2**, **3**, and **4** with HCl gas (23.5 g/100 g diethyl ethersolution at 22.5 °C) in excellent yield (Eq. 1). Notably no addition of HCl to the C=C double bond of the norbornene skeleton is observed.

Starting from **1** the compounds **8** and **9** are synthesized using CH_3OD and CD_3OD as reaction partners

[4]. The first step in the reaction course is the methanolysis of the Si–Cl bond that generates HCl which is capable to cleave the methoxylated bicyclic skeleton by a preliminary proton attack at the allylic position relative to the silicon.

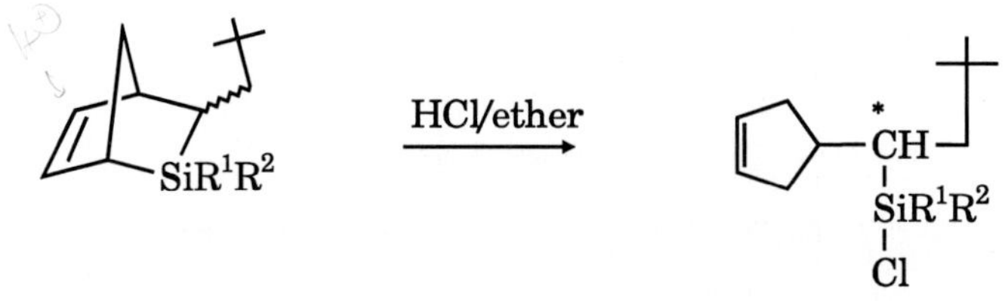

$R^1 = R^2 = OMe : 2$ $R^1 = R^2 = OMe : 5$
$R^1 = R^2 = Me : 3$ $R^1 = R^2 = Me : 6$
$R^1 = Me, R^2 = Cl : 4$ $R^1 = Me, R^2 = Cl : 7$

Eq. 1.

Then the bond between silicon and the bridgehead carbon breaks and the double bond shifts. The final step is an addition of a chlorine which then reacts with an excess of deuterated methanol to a methoxy group. This reaction route is best described analogous to a S_E-mechanism, which has already been published for other group 14 elements (Si, Ge, Sn) [5].

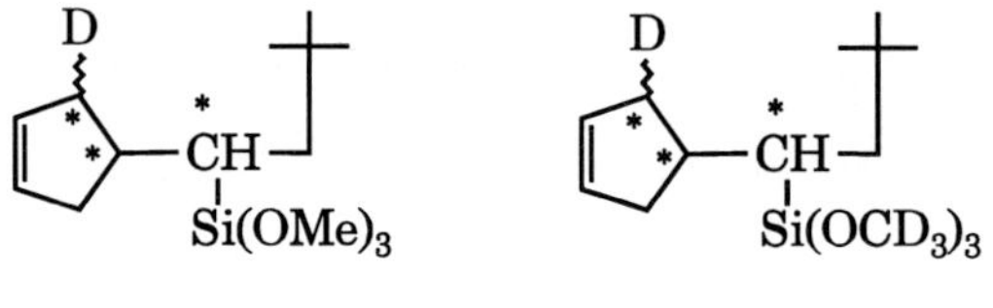

8   9

Fig. 2.

Compared to the organosubstituted silanorbornenes **2**, **3**, and **4** the silicon dichloro functionalized heterocycle **1** is characterized by a completely different reactivity:

In chloroform **1** reacts with HCl gas nearly quantitatively (> 91 %) to give the product of the HCl addition to the double bond **10**, which is a mixture of two diastereomers (the *exo/endo* ratio of the neopentyl group is the same as in **1** – 60:40). Exclusively the *exo*-6-chloro-compounds are formed and this proves the HCl addition to be stereo- and regiospecific. When an HCl/ether solution is used, the trichloroorganosilane **11** (resulting from an allylic cleavage) is formed along with the adducts **10** as an unseparable mixture (**10:11** = 66:34). In contrast the stronger acid CF_3SO_3H does not add to the double bond of **1** and the cleavage product **12** is selectively formed in high yield (Scheme 1).

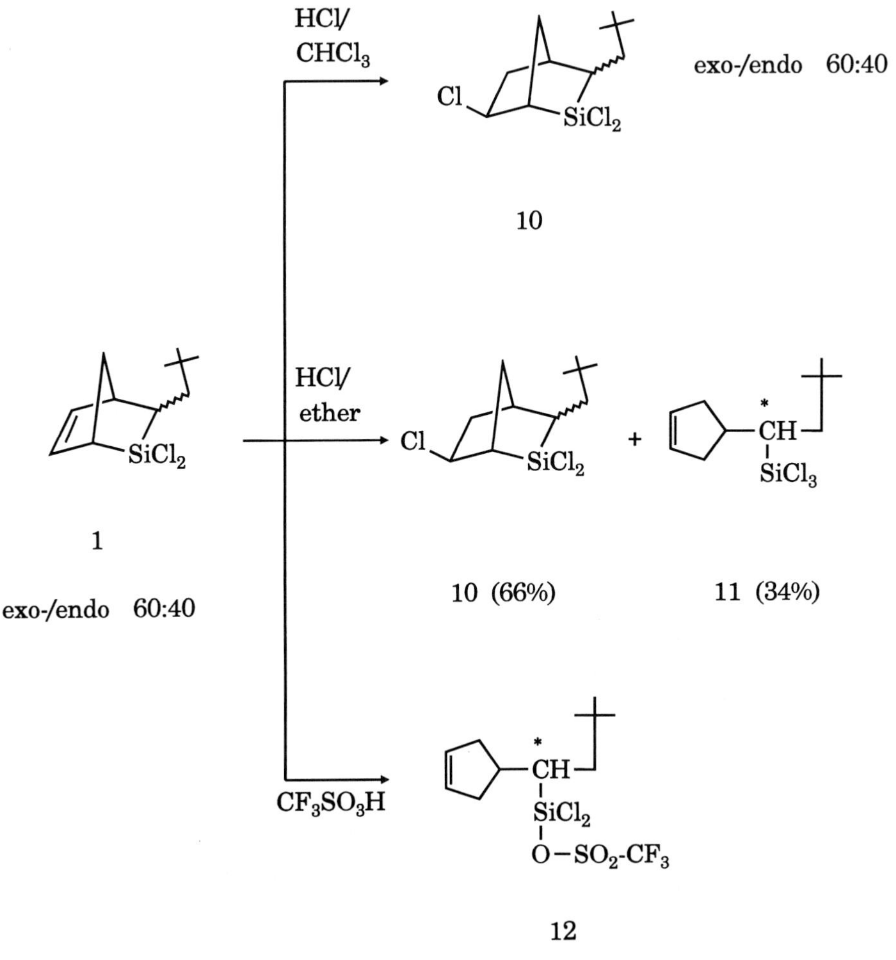

Scheme 1.

The chlorine substituents at silicon allow an easy derivatization of **5, 6,** and **7**, e.g., the transformation of **6** into the corresponding hydridosilanes **13** (80 % yield). These cyclopentenyl-diorganohydridosilane are useful precursors for the syntheses of saturated bicyclic heterocycles. Compound **14** was identified as a mixture of *exo/endo* diastereomers (80:20) (Scheme 2).

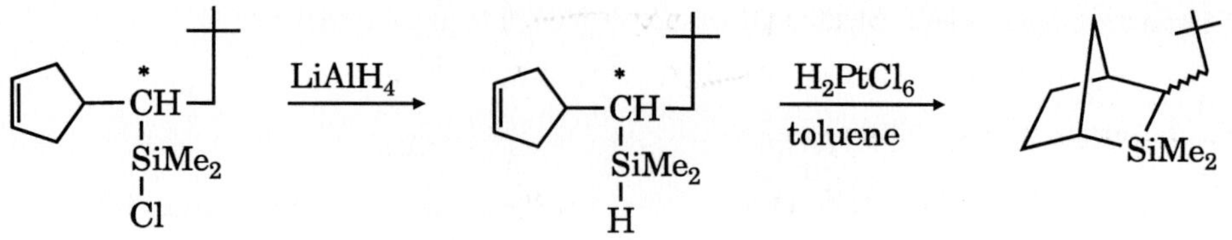

Scheme 2.

Furthermore, compounds **15** and **16** are examples of alkoxy substituted cyclopentenyl compounds which can be synthesized in nearly quantitative yield from **1** [4].

Fig. 3.

References:

[1] Review: N. Auner, in: *Organosilicon Chemistry – From Molecules to Materials* (Eds.: N. Auner, J. Weis), VCH, Weinheim, **1994**, p. 103.

[2] Thermal isomerisation reactions: N. Auner, C. Seidenschwarz, N. Sewald, E. Herdtweck, *Angew. Chem.* **1991**, *103*, 425; *Angew. Chem., Int. Ed. Engl.* **1991**, *30*, 444; N. Auner, C. Seidenschwarz, N. Sewald, *Organometallics* **1992**, *11*, 1137; N. Sewald, W. Ziche, A. Wolff, N. Auner, *Organometallics* **1993**, *12*, 4123; W. Ziche, C. Seidenschwarz, N. Auner, E. Herdtweck, N. Sewald, *Angew. Chem.* **1994**, *106*, 93; *Angew. Chem., Int. Ed. Engl.* **1994**, *33*, 77;

The reaction of Si-dichloro substituted silacyclobutenes and -butanes with metal fragments yield stable silylene complexes; no ring cleavage is observed: – Silacyclobutenes: N. Auner, C. Wagner, E. Herdtweck, M. Heckel, W. Hiller, *Bull. Soc. Chim. Fr.*, in press. – Silacyclobutanes: N. Auner, M. Grasmann, B. Herrschaft, P. Kiprof, *J. Organomet. Chem.*, in press;

Transformation of cycloadducts into polymers with and without retention of the ring moiety: N. Auner, B. Biebl, H.-U. Steinberger, in preparation.

[3] P. R. Jones, T. F. O. Lim, *J. Am. Chem. Soc.* **1980**, *102*, 4970; N. Auner, J. Grobe, *J. Organomet. Chem.* **1980**, *190*, 129.

[4] N. Auner, H.-U. Steinberger, *Z. Naturforsch.* **1994**, *49b*, 1743.

[5] D. Y. Young, W. Kitching, G. Wickham, *Tet. Lett.* **1983**, *24*, 5789.

Base-catalyzed Hydrogenation of Chlorosilanes by Organotin Hydrides

Uwe Pätzold, Gerhard Roewer*

Institut für Anorganische Chemie
Technische Universität Bergakademie Freiberg
Leipziger Str. 29, D-09599 Freiberg, Germany

Summary: Bu_3SnH is an effective reagent for partial conversion of Si–Cl into Si–H groups. The hydrogenation mechanism postulates the coordination of the catalyst or the solvent to silicon giving a hypervalent intermediate in the first step, followed by the attack of tributyltin hydride by a single electron transfer or a synchronous hydride transfer. This mechanism implies that the intermediate containing a hypervalent silicon atom reacts faster than the starting tetracoordinate silane.

Introduction

Chlorosilanes are readily reduced to silicon hydrides by lithium aluminium hydride in ethereal solvents. The use of $LiAlH_4$, however, has certain disadvantages caused by the formation of $AlCl_3$ which can result in the disproportionation of silicon hydrides and the polymerisation of vinyl compounds. If ionic hydrides are used, it will not be possible to obtain partially reduced species, even if a deficiency of the reducing agent is present. Organotin hydrides have selective, stepwise reducing properties towards a variety of halosilanes [1]. They are ineffective as reducing agents for alkyl-substituted halosilanes [2]. The synthetic applications of organosilicon compounds including catalysis by nucleophiles were extensively reviewed recently [3, 4]. The hydrogenation of chlorosilanes with tributyltin hydride involving nucleophilic activation of Si–Cl bonds will be presented in this paper. A mechanistic interpretation will be proposed and the role of the catalysts discussed.

Solvent Effects

The reaction course of hydrogenation and the composition of hydrogenated products strongly depend on polarity and donor strength of the used solvent. No hydrogenation reaction was observed in nonpolar or less polar solvents, such as hydrocarbons and aromatics. In these solvents the reaction starts by the addition of radical initiators, e.g., AIBN or dibenzoyl peroxide. The radical initiated reactions result in the formation of $SiHCl_3$. A similar course was observed in less polar ethereal solvents, such as 1,4-dioxane, anisole, and dibenzyl ether. Radical steps seem to play an important role for the hydrogenation in these solvents.

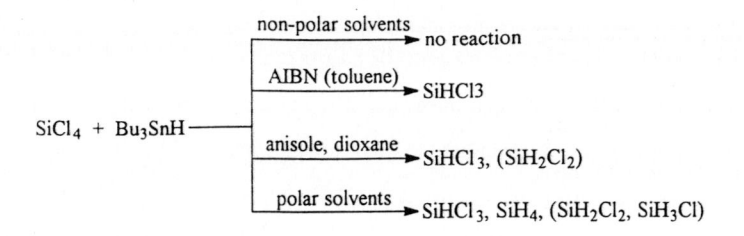

$$SiCl_4 + Bu_3SnH \begin{cases} \text{non-polar solvents} \rightarrow \text{no reaction} \\ \text{AIBN (toluene)} \rightarrow SiHCl_3 \\ \text{anisole, dioxane} \rightarrow SiHCl_3, (SiH_2Cl_2) \\ \text{polar solvents} \rightarrow SiHCl_3, SiH_4, (SiH_2Cl_2, SiH_3Cl) \end{cases}$$

Fig. 1. Reaction of Bu_3SnH with $SiCl_4$ in organic solvents

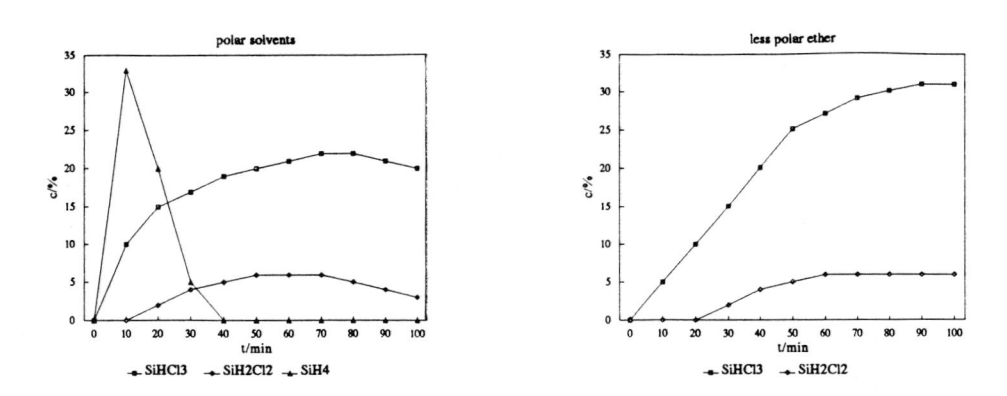

Fig. 2. Comparison of the hydrogenation course in ethereal solvents and in polar solvents

Hydrogenation in more polar solvents (THF, diglyme, nitriles, pyridine) proceeds via a different reaction pathway. This behavior resembles the hydrogenation of chlorosilanes by $LiAlH_4$ in THF, where reaction steps which include polar or ionic species can be expected. The use of polar solvents leads to the formation of a mixture of hydrogenation products. With increasing polarity and donor strength of the solvents SiH_4 becomes the major product. The Lewis base properties of the solvents lead to interactions with $SiCl_4$. These interactions result in the formation of solid adducts in solvents with DN > 30 (DN = donor number established by Gutmann [5]).

Lewis Bases as Catalysts for the Hydrogenation of Chlorosilanes

Experiments with catalysts were carried out in toluene. Quaternary ammonium or phosphonium halides are suitable as catalysts to get fully hydrogenated products, whereas halide anions (Cl^-, Br^-, I^-) are the effective catalytic species. Ammonium and phosphonium halides show the same effect. The reactivity is not changed significantly by the variation of organic groups. Different nitrogen and phosphorous ligands were investigated concerning their catalytic effects:

- Amines, nitrogen heterocycles and 1,3-diaza-1,4-dienes (DAD)
- Phosphanes and phosphorous acid esters

The reaction is accelerated using a series of amines and N-heterocycles. SiH_4 becomes the preferable product with increasing basicity of the ligands (Table 1).

Catalyst	Basicity (pK_a)	Hydrogenation Products (yield)
Pyrazine	0.61	$SiHCl_3$ (40 %), SiH_2Cl_2 (trace)
Pyridazine	1.55	$SiHCl_3$ (55 %), SiH_2Cl_2 (trace)
Pyrimidine	2.33	$SiHCl_3$ (58 %), SiH_2Cl_2 (15 %), SiH_3Cl (trace)
Quinoline	4.94	main products $SiHCl_3$ and SiH_4
Pyridine	5.25	main products $SiHCl_3$ and SiH_4
N-Methylimidazole	7.13	main product SiH_4

Table 1. Products of catalytic hydrogenation (in toluene) depending on catalyst basicity

A comparison of the catalytic effects of compounds containing different groups is difficult when accounting only for the basicity. For example, Ph_2NH and pyrazine (pK_a = 0.79 and 0.61, respectively) show different catalytic effects. Besides the basicity of the catalysts other electronic and steric effects also play an important role. The hydrogenation is also catalyzed by DAD (Fig. 3).

1.3-diaza-1.4-dienes

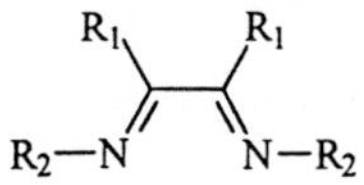

$R_2-N=CR_1-CR_1=N-R_2$

decreasing catalytic activity

R_1 = H ; R_2 = Cyh ; t-Bu ; n-Bu ; Ph

R_1 = Me ; R_2 = n-Pr ; Bz ; Ph

R_2 = Ph ; R_1 = H ; Me ; Ph

Fig. 3. Catalytic activity of diazadienes

In all cases, a mixture of hydrogenation products was obtained. The variation of organic groups (R^1, R^2) results in a change of the catalytic activity of DAD. Increasing electron density at the nitrogen atoms leads to an increase of the donor strength of DAD. Increasing catalytic activity has been found for the glyoxalic derivatives (R^1 = H) in the series R^2 = Ph < nBu < tBu < Cy. The replacement of group R^1 by methyl or phenyl leads to decreasing activity.

Tolman [6] has established the electronic parameter χ and the cone angle θ to characterize the collective electron donor-acceptor properties of phosphorus(III) ligands. The electronic parameter χ seems to be well appropriated to classify the donor compounds in the order of their catalytic activity. For example, use of PCyh$_3$ (χ = 1.4) as catalyst gives SiH$_4$ as the main product, but P(OPh)$_3$ (χ = 30.2) preferably leads to SiHCl$_3$.

Hydrogenation of Methylchlorosilanes

In addition to the basic strength of catalysts the acceptor strength of chlorosilanes is important for the hydrogenation course. With the increasing number of methyl groups (instead of Cl atoms) the silane reactivity decreases (Table 2). This tendency seems to be associated with the decreasing Lewis-acid strength of methyl chlorosilanes.

Only the strong bases HMPT and NMI catalyze the hydrogenation of trimethyl chlorosilane. Both catalysts NMI and P(NEt$_2$)$_3$ form solid adducts with Me$_3$SiCl, but no hydrogenation was observed in the case of P(NEt$_2$)$_3$.

Catalyst	MeSiCl₃	Me₂SiCl₂	Me₃SiCl
PPh$_3$	no reaction	no reaction	no reaction
TMEDA	MeSiHCl$_2$ (trace)	no reaction	no reaction
PCyh$_3$	all products	all products (low yield)	no reaction
P(NEt$_2$)$_3$	all products	all products	no reaction
Ph$_4$PBr	mainly MeSiH$_3$	mainly Me$_2$SiH$_2$ (low yield)	Me$_3$SiH (trace, after 80 min)
HMPT	mainly MeSiH$_3$	mainly Me$_2$SiH$_2$	Me$_3$SiH (fast)
NMI	mainly MeSiH$_3$	mainly Me$_2$SiH$_2$	Me$_3$SiH (fast)

Table 2. Hydrogenation of methyl chlorosilanes in toluene

Hydrogenation of Vinyl Trichlorosilane

Vinyl trichlorosilane can undergo a number of different reactions with tributyltin hydride. Apart from the expected conversion of Si–Cl into Si–H function, the hydrogenation of the vinyl group, addition of Bu$_3$SnH to the double bond or the polymerization of ViSiCl$_3$ may take place. All these reactions were not observed. ViSiCl$_3$ is hydrogenated more efficiently than the methyl chlorosilanes. The reaction with NMI, HMPT, P(NEt$_2$)$_3$, and PCyh$_3$ as catalysts yielded the major product ViSiH$_3$ and only small amounts of ViSiHCl$_2$ and ViSiH$_2$Cl. A higher yield of partially hydrogenated vinyl silanes was obtained when Ph$_4$PBr, phen, and bipy as catalysts in acetonitrile were used. Another kind of reaction took place by the addition of AIBN or PPh$_3$, involving the release of HCl. The obtained product has been not fully characterized yet, but NMR spectra point to ViCl$_2$Si–SnBu$_3$ (Eq. 1).

$$\text{ViSiCl}_3 + \text{Bu}_3\text{SnH} \longrightarrow \text{ViCl}_2\text{Si–SnBu}_3 + \text{HCl}$$

Eq. 1.

Mechanistic Considerations

Some suggestions about the hydrogenation mechanism of halosilanes by organotin hydrides involve a synchronous H–Cl exchange via a planar four-centered transition state between tin and silicon [2]. Such a reaction pathway seems to be less probable under such conditions when a small donor amount already catalyzes the reaction to give fully hydrogenated products. It is also difficult to associate the yield of fully hydrogenated silanes with radical reactions. The following steps are substantial for our suggestion of the mechanism:

1) The first step is the nucleophilic attack of the catalyst at the silicon atom, caused by the donor-acceptor interactions between the Lewis base and the silicon compound. The components may be associated very weakly. Here no interaction between $SiCl_4$ and PPh_3 was found by means of ^{31}P NMR. In some cases the interactions lead to solid adducts. In solution they dissociate into the original components. Complexes are formed in which the coordination at the silicon atom is increased to five or six.

    The Si–Cl distances at the octahedrally coordinated Si atom are lengthened compared with those at the tetrahedral silicon compounds. Hypervalent silicon compounds are more reactive than tetracoordinate ones and exhibit their own pattern of reactivity.

2) Although organometal hydrides are regarded both as potential hydride donors and as electron donors, little is known about the mechanism of the hydride- and electron-transfer. For some reduction reactions using Bu_3SnH an electron-transfer mechanism is assumed [7, 8]. Hydrogenation steps take place preferably on higher coordinated silicon species. A step involving an electron transfer from the organotin hydride to the hypervalent silicon intermediate could be assumed which gives a radical ion pair (Eq. 2).

    Such a suggestion describes a strongly polar transition state.

$$D-SiCl_4 + HSnBu_3 \longrightarrow [(D-SiCl_4)^{\cdot -},(H\dot{S}nBu_3)^{+}]_{solv}$$

Eq. 2.

3) The electron transfer should be irreversible, because of the fast cleavage of the stannyl cation Bu_3SnH^+. The resulting products undergo further reactions in a solvent cage (Eq. 3, 4).

Eq. 3.
$$[(D-SiCl_4)^-,(H^\bullet SnBu_3)^+]_{solv} \longrightarrow [(D-SiCl_4)^-,(SnBu_3)^+,{}^\bullet H]_{solv}$$

Eq. 4.
$$[(D-SiCl_4)^-,(SnBu_3)^+,{}^\bullet H]_{solv} \longrightarrow D-SiHCl_3 + ClSnBu_3$$

The electron transfer (ET) may take place via π^* orbitals of the ligands at the silicon atom (outer-sphere model). Inner-sphere ET could involve a chloro bridge between silicon and tin. It leads to a four-centered transition state in a special case.

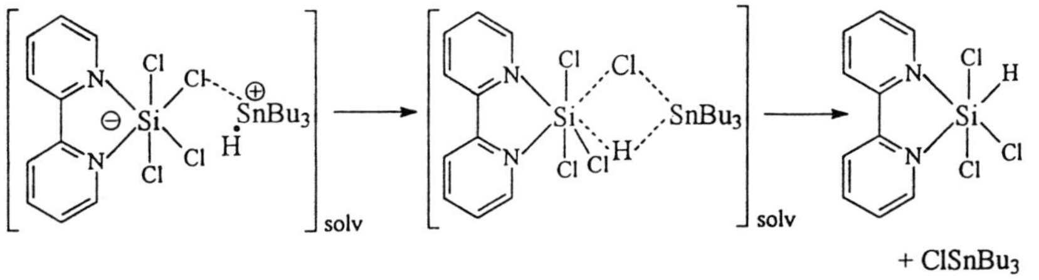

Fig. 4. Four-centered transition state

4) Successive hydrogenation steps occur only at the silicon atom coordinated by donor species. Because of the decreasing Lewis-acid strength in the series $SiCl_4$ down to SiH_4, the interactions between catalyst and hydrogenation products become more weakly. Strong Lewis-bases are able to coordinate also partially hydrogenated silanes, so that a fast sequence of hydrogenation steps at the origin donor-acceptor complex results in the formation of SiH_4. After releasing the catalyst by the formation of SiH_4 another $SiCl_4$ molecule is coordinated. The adducts of trichloro- or dichlorosilane and weaker bases dissociate more easily. In such a case the hydrogenation leads to $SiHCl_3$ as the major product.

A mechanism including a simple exchange of Cl^- and H^- ions between Bu_3SnH and $D-SiCl_4$ should be an alternative to the reduction pathway via an ET mechanism.

References:

[1] J. J. D'Errico, K. G. Sharp, *US* 4 798 713, **1989**.
[2] J. J. D'Errico, K. G. Sharp, *Inorg. Chem.* **1989**, *28*, 2177.
[3] C. Chuit, R. J. P. Corriu, C. Reye, J. C. Young, *Chem. Rev.* **1993**, *93*, 1371.
[4] G. G. Furin, O. A. Vyazankina, B. A. Gostevsky, N. S. Vyazankin, *Tetrahedron* **1988**, *44*, 2675.
[5] V. Gutmann, *Coord. Chem. Rev.* **1976**, *18*, 225.
[6] C. A. Tolman, *Chem. Rev.* **1977**, *77*, 313.
[7] R. J. Klingler, K. Mochida, J. K. Kochi, *J. Am. Chem. Soc.* **1979**, *101*, 6626.
[8] C. L. Wong, R. J. Klingler, J. K. Kochi, *Inorg. Chem.* **1980**, *19*, 423.

The Catalytic Hydrogenation of Methylchlorodisilanes with Tri-*n*butylstannane

U. Herzog, G. Roewer*, K. Herzog

Institut für Anorganische Chemie
Technische Universität Bergakademie Freiberg
Leipziger Str. 29, D-09599 Freiberg, Germany

Summary: Tri-*n*butylstannane is a valuable reagent for partial conversion of Si–Cl into Si–H functions. Now it is possible to obtain disilanes containing methyl, chlorine and hydrogen simultaneously by treatment of methylchlorodisilanes with Bu_3SnH in the presence of a catalyst.

Introduction

Hydrogenation of methylchlorodisilanes or Si_2Cl_6 with $LiAlH_4$ in diethylether leads only to the formation of the fully hydrogenated methyldisilanes [1] or disilane, respectively. The synthesis of partially hydrogenated methylchlorodisilanes was only possible by a stepwise substitution of hydrogen bonded to the silicon with AgCl [2], BCl_3 [3], HCl or by equilibration with a methylchlorodisilane [4]. Nine of 27 possible disilanes containing methyl, chlorine and hydrogen simultaneously are known. ^{29}Si NMR data have been reported in only four cases (SiX_2Me–SiX_2Me, X = H, Cl).

Experimental

Disilane (e.g., $SiCl_2Me$–$SiCl_2Me$) (4 mmole) was added to 10 mg of the catalyst, 0.75 g dried toluene and 2.4 g (8 mmole) Bu_3SnH (in the same case (see below) the molar ratio of Bu_3SnH:disilane exceeded 2:1) in a Schlenk tube. After standing at room temperature for two days, a disilane mixture in toluene could be trapped at 77 K. ^{29}Si NMR spectra of the disilane mixtures were recorded with a BRUKER MSL 300 NMR spectrometer.

Results and discussion

1 Hydrogenation of SiCl$_2$Me–SiCl$_2$Me (1)

The hydrogenation of **1** follows the general equation:

$$SiCl_2Me-SiCl_2Me + nBu_3SnH \longrightarrow SiH_iCl_{2-i}Me-SiH_jCl_{2-j}Me + nBu_3SnCl \qquad n = i + j$$

Eq. 1.

The formation of Bu$_3$SnCl was detected by ^{119}Sn NMR. The hydrogenation of **1** was carried out with several catalysts in order to compare the properties of the catalysts used with the obtained product spectra.

a) Without a catalyst no reaction took place.

b) Ph$_3$N led to the formation of small amounts of SiH$_2$Me–SiCl$_2$Me (**2**) and SiH$_2$Me–SiH$_2$Me (**3**), but the reaction was not completed within two days.

c) With benzildianile as catalyst a colorless crystalline powder was isolated. Elemental analysis, ^{1}H and ^{13}C NMR are in agreement with the assumed structure 2,2,7,7-tetrachloro-3,4,5,6-tetraphenyl-2,7-disila-3,6-diazaoctene. This addition product made the catalyst ineffective.

d) Ph$_3$P, (PhO)$_2$(iC$_3$H$_7$O)P, and CH$_3$CN yielded only **2** (ca. 60–70 mole%) and **3** as hydrogenation products. With a molar ratio of Bu$_3$SnH:**1** > 4:1 pure **3** could be isolated.

e) Phenanthroline and pyrazine yielded besides **2** and **3** small amounts (2.5 mole%) of SiHClMe–SiCl$_2$Me (**4**) and SiHClMe–SiH$_2$Me (**5**). No SiHClMe–SiHClMe (**6**) was detected.

f) 2,2'-Bipyridyl, pyridazine, (Ph$_3$P)$_2$NCl, Et$_4$NCl, Bu$_4$PBr, Bu$_4$PI, and Ph$_3$MePI yielded **2** (55–65 mole%), **3** (2–3 mole %), **4**, and **5** (10–20 mole% each) and small amounts (2–6 mole%) of **6**. With a molar ratio of Bu$_3$SnH:**1** = 1:1 or 3:1 the compounds **4** and **5**, respectively, could be obtained in over 30 % yield. As well small amounts of methylchloromonosilanes SiX$_3$Me, X = H, Cl (0–5 mole% Si) were detected. In the case of 2,2'-bipyridyl initially the yellow adduct Si$_2$Cl$_4$Me$_2$·bipy was formed. After a few minutes the reaction mixture became dark green (λ_{max} = 820 nm) and gave an intensive ESR signal.

g) With (NMe$_2$)$_3$P=N–P(O)(NMe$_2$)$_2$, pyridine and 1-methylimidazole, the content of monosilanes increased to 13, 34 and 71 mole% Si, respectively.

2 Comparison of the Properties of the Catalysts with the Product Spectra

The comparison of the basicity of the catalysts with the product spectra showed a close correlation (Table 1). Deviations occured on sterically more demanding catalysts, such as 1,10-phenanthroline and Ph_3P. In the case of weak bases as catalysts, only disilanes with $-SiH_2Me$ and $-SiCl_2Me$ groups were formed. Stronger bases yielded also disilanes with $-SiHClMe$ groups. The strongest bases used as catalysts yielded mainly monosilanes moieties SiX_3Me (X = H, Cl). The results are shown in Scheme 1.

Catalyst	Product Spectrum	$pK_S(BH^+)$ [°C]
1-Methylimidazole	g	7.13/25
Pyridine	g	5.25/25
Quinoline	f	4.92/25
2,2'-Bipyridyl	f	4.37/20
Pyridazine	f	2.39/25
1,10-Phenanthroline	e	4.84/25
Pyrazine	e	0.63/25
Ph_3P	d	2.73/25
$(PhO)_2(iPrO)P$	d	0.00/25
CH_3CN	d	< 0
Ph_3N	b	<< 0

Table 1. Comparison product spectra-base strength of the catalysts

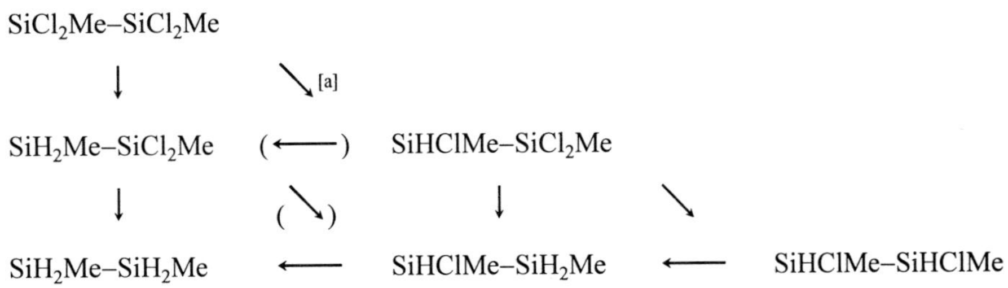

```
SiCl₂Me–SiCl₂Me
    ↓           ↘ [a]
SiH₂Me–SiCl₂Me  (←)  SiHClMe–SiCl₂Me
    ↓           (↘)       ↓              ↘
SiH₂Me–SiH₂Me    ←   SiHClMe–SiH₂Me   ←   SiHClMe–SiHClMe
```

Scheme 1. [a] only with strong bases as catalysts

With very strong bases the Si–Si bonds were cleaved:

$$SiCl_2Me–SiCl_2Me + B| \longrightarrow SiCl_3Me + B{\rightarrow}{:}SiClMe \; (\longrightarrow SiMe(SiCl_2Me)_3)$$

Eq. 2.

Three reaction mechanisms were considered: a radical chain mechanism, a hydride transfer mechanism and an electron transfer reaction from Bu_3SnH to the disilane followed by H· transfer. The first mechanism should lead to high yields of Bu_6Sn_2, but this was not observed. We thus assume the last mechanism, which is also in agreement with other investigated reactions of trialkylstannanes [5–7].

3 Hydrogenation of Other Disilanes

Besides the basicity of the catalyst, the acceptor strength of the disilane is important for the course of the hydrogenation. If the formation of a base adduct is the first step, in mixtures of several disilanes the disilane with the highest acceptor strength is hydrogenated first.

3.1 Hydrogenation of Si_2Cl_6

The hydrogenation of Si_2Cl_6 with bipy, or Bu_4PI as catalysts in the molar ratio $Bu_3SnH:Si_2Cl_6 = 2:1$ leads only to partially hydrogenated monosilanes, mainly $SiHCl_3$ and SiH_2Cl_2. With Ph_3P as catalyst, the partially hydrogenated disilanes $SiHCl_2–SiCl_3$ (18 mole% Si), $SiHCl_2–SiHCl_2$ (17 mole% Si), $SiH_3–SiCl_3$ (5 mole% Si), and $SiH_3–SiHCl_2$ (8 mole% Si) occured in addition to 50 mole% Si monosilanes, mainly $SiHCl_3$ and SiH_2Cl_2. The content of monosilanes could be lowered to 28 mole% Si with Ph_3N as catalyst. No disilanes with α-SiH_2Cl group were detected. This can be explained with the following reaction scheme:

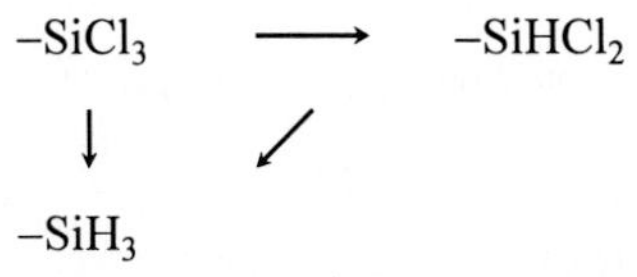

$$\begin{array}{ccc} -SiCl_3 & \longrightarrow & -SiHCl_2 \\ \downarrow & \nearrow & \\ -SiH_3 & & \end{array}$$

Scheme 2. The acceptor strength of –$SiHCl_2$ should be close to that of the –$SiCl_2Me$, i.e., that with Ph_3N and Ph_3P only the complete hydrogenated –SiH_3 group is formed.

3.2 Hydrogenation of $SiCl_2Me-SiClMe_2$ (7) besides $SiCl_2Me-SiCl_2Me$ (1)

To investigate the hydrogenation of **7**, a mixture of 45 mole% **7** and 55 mole% **1** was used. Hydrogenation with varying molar ratios Bu_3SnH:disilane = 1:1, 2:1, and 3:1 (Bu_4PI as catalyst) showed that **1** was hydrogenated at first. The hydrogenation of **7** started, when **1** was completely hydrogenated to **2** and **4**. The novel compounds $SiX_2Me-SiXMe_2$ (X = H, Cl) were detected by ^{29}Si NMR spectroscopy. All ^{29}Si chemical shifts and coupling constants $^1J_{SiH}$ were determined. The major products formed were $SiH_2Me-SiClMe_2$ (up to 45 mole% of the total $SiX_2Me-SiXMe_2$) and $SiCl_2Me-SiHMe_2$ (up to 30 mole%).

3.3 Hydrogenation of $SiClMe_2-SiClMe_2$ (8) and $SiCl_2Me-SiMe_3$ (9)

Hydrogenation of **8** with Bu_4PI, bipy and 1-methylimidazole lead to the formation of $SiHMe_2-SiClMe_2$ (up to 60 mole% in a molar ratio Bu_3SnH:**8** = 1:1) and $SiHMe_2-SiHMe_2$. Both compounds were detected by ^1H and ^{29}Si NMR spectroscopy.

Hydrogenation of $SiCl_2Me-SiMe_3$ with Bu_4PI (molar ratio Bu_3SnH:**9** = 1:1) leads only to the formation of $SiH_2Me-SiMe_3$ beside of **9**, no $SiHClMe-SiMe_3$ (**10**) was obtained. In the case of hydrogenation with 1-methylimidazole as catalyst **10** occured in small amounts (4 mole% of the initial **9**) but in the product mixture also $SiClMe_3$ (about 30 mole% of the initial **9**) and some $SiHMe_3$ were present, due to the cleavage of the Si–Si bond by this strong base. Besides these cleavage products with three methyl substituents no volatile monosilanes with one methyl substituent occured. Because of the lower acceptor strength of **9** compared with **1** stronger bases had to be used to obtain products with –SiHClMe groups and to cleave the Si–Si bond.

4 NMR data of the Novel Disilanes

Compound	δ_A [ppm]	$|^1J_{SiH}|$ [Hz]	δ_B [ppm]	$|^1J_{SiH}|$ [Hz]
Si^{A}HMe$_2$–Si^{B}ClMe$_2$	–38.73	186	22.90	–
SiHMe$_2$–SiHMe$_2$	–39.02	172	–	–
Si^{A}HClMe–Si^{B}Me$_3$	2.07	199	–17.17	–
Si^{A}H$_2$Me–Si^{B}Me$_3$	–65.61	177	–17.89	–
Si^{A}HClMe–Si^{B}ClMe$_2$	–3.74	218	17.24	–
Si^{A}H$_2$Me–Si^{B}ClMe$_2$	–64.62	190	22.56	–
Si^{A}Cl$_2$Me–Si^{B}HMe$_2$	33.77	–	–34.98	190
Si^{A}HClMe–Si^{B}HMe$_2$	1.83	218	–38.16	180
Si^{A}H$_2$Me–Si^{B}HMe$_2$	–65.84	179	–39.46	179

Table 2. ^{29}Si chemical shifts and coupling constants $^1J_{SiH}$ of the novel disilanes

References:

[1] A. D. Craig, A. D MacDiarmid, *J. Inorg. Nucl. Chem.* **1962**, *24*, 161.

[2] A. J. Vanderwielen, M. A. Ring, *Inorg. Chem.* **1972**, *11*, 246.

[3] J. E. Drake, N. Goddard, *J. Chem. Soc.* **1970**, 2587.

[4] H. Schmölzer, E. Hengge, *J. Organomet. Chem.* **1984**, *260*, 31.

[5] J. K. Kochi, *Angew. Chem.* **1988**, *100*, 1331.

[6] R. J. Klingler, K. Mochida, J. K. Kochi, *J. Am. Chem. Soc.* **1979**, *101*, 6626.

[7] J. Chojnowski, W. Fortuniak, W. Stanczyk, *J. Am. Chem. Soc.* **1987**, *109*, 7776.

Disproportionation of Tetrachlorodimethyldisilane – NMR-Spectroscopic Identification of the Primary Products

Erica Brendler*, Katrin Leo, Berthold Thomas, Robin Richter,
Gerhard Roewer, Hans Krämer

Institut für Analytische Chemie
Technische Universität Bergakademie Freiberg
Leipziger Str. 29, D-09599 Freiberg, Germany

Summary: The primary oligomers formed during the disproportionation of 1,1,2,2-tetrachlorodimethyldisilane have been identified and characterized by ^{29}Si and ^{13}C NMR spectroscopy. After initial formation of 1,1,2,3,3-pentachlorotrimethyltrisilane, a branched tetrasilane – tris(dichloromethylsilyl)methylsilane – is formed, demonstrating the preference of branched structures over linear ones. ^{29}Si, ^{13}C, and ^{1}H chemical shifts for both silanes have been determined. Knowing the chemical shifts of the oligomers it is possible to assign the signal groups in the polymer spectra to certain structural units and therewith to calculate branching and polymerization degrees for the finally obtained polysilanes.

Today, polysilanes (PS) and polycarbosilanes (PCS) as precursors for SiC ceramics and fibers can be obtained by various routes. A very promising way is the disproportionation of chloromethyldisilanes giving chloromethylmonosilanes and polysilanes. The thus formed polymers differ in their properties depending on the reaction pathway (homogeneous or heterogeneous), the catalyst employed and temperature regime [1, 2]. To influence the reaction towards the desired product it is essential to investigate the reaction mechanism and therefore the oligomers formed during the first steps of the disproportionation process. The existing proposals for mechanisms are contradictory and postulate either linear or branched oligosilanes [3–7].

It was the aim of this work to identify the primary oligomers formed in the reaction mixture and to characterize them by ^{29}Si, ^{13}C, and ^{1}H NMR spectroscopy. The investigations were carried out using samples of the monosilanes and oligosilanes obtained during the disproportionation of a disilane mixture (72 % 1,1,2,2-tetrachlorodimethyldisilane (TCDMS), 20 % 1,1,2-trichlorodimethyldisilane, and 4 %

1,2-dichlorodimethyldisilane) or of pure TCDMS. Spectra were recorded with a Bruker MSL 300 using gated decoupling and adding of the relaxation reagent [Cr(acac)$_3$].

Starting with TCDMS, two different trisilanes and the corresponding monosilanes can be formed in a first reaction step:

$$2\ [Cl_2CH_3Si]_2 \longrightarrow Cl_2CH_3Si-SiCH_3Cl-SiCH_3Cl_2 + CH_3SiCl_3$$
$$\mathbf{1}$$
$$\longrightarrow Cl_2CH_3Si-SiCl_2-SiCH_3Cl_2 + (CH_3)_2SiCl_2$$
$$\mathbf{2}$$

Eq. 1.

Our NMR investigations proved that no dichlorodimethylsilane (DMS) is formed during the disproportionation of pure TCDMS, hence it follows that reactions giving DMS as a by-product can be excluded from further considerations.

If oligosilane **1** reacts further with TCDMS two tetrasilanes can be formed:

$$(\mathbf{1}) + TCDMS \longrightarrow CH_3SiCl_3 + Cl_2CH_3Si-SiCH_3Cl-SiCH_3Cl-SiCH_3Cl_2$$
$$\mathbf{3}$$
$$\longrightarrow CH_3SiCl_3 + (Cl_2CH_3Si)_3SiCH_3$$
$$\mathbf{4}$$

Eq. 2.

The decisive question whether linear or branched silanes are formed preferentially is thus answered; the properties of the final polysilane depend on this difference. Tetrasilanes **3** and **4** can continue reacting with TCDMS to give a rapidly increasing variety of products.

^{29}Si and ^{13}C NMR spectra taken immediately after the start of the TCDMS-disproportionation showed two resonances with an intensity ratio of 2:1 besides the signal for TCDMS, which could be assigned to trisilane **1**. The exact chemical shifts are listed in Table 1. As can be seen in Fig. 1 ^{29}Si NMR spectra of oligosilane mixtures exhibit one very strong resonance at 30.73 ppm. The intensity of this signal correlates to that of a resonance at –63.51 ppm. The ^1J(SiSi) coupling constant has a value of 86.4 Hz for both resonances, which confirms the connection of the involved silicon atoms. The satellites of the highfield signal have a remarkable intensity of 7 % indicating a tertiary SiSi$_3$ group. Therefore the resonances were assigned to the branched tetrasilane **4**.

Compound		δ ^{29}Si [ppm]	δ ^{13}C [ppm]	δ ^{1}H [ppm]	^{1}J(SiSi) [Hz]
(Cl$_2$CH$_3$Si$_A$)$_2$Si$_B$CH$_3$Cl	A	23.77	6.58	0.988	111.77
	B	−4.93	−2.6	0.82	
(Cl$_2$CH$_3$Si$_A$)$_3$Si$_B$CH$_3$	A	30.73	9.08	1.038	86.4
	B	−63.51	−12.54	0.575	

Table 1. NMR-data of the newly characterized oligosilanes

The ^{13}C NMR spectra confirm these conclusions. It is interesting to note that the corresponding methoxy compound was isolated by Atwell *et al.* [3] after disproportionation of tetramethoxy-dimethyldisilane. The formation of the branched silane was considered as a proof for a silylene reaction mechanism which is also discussed for our system.

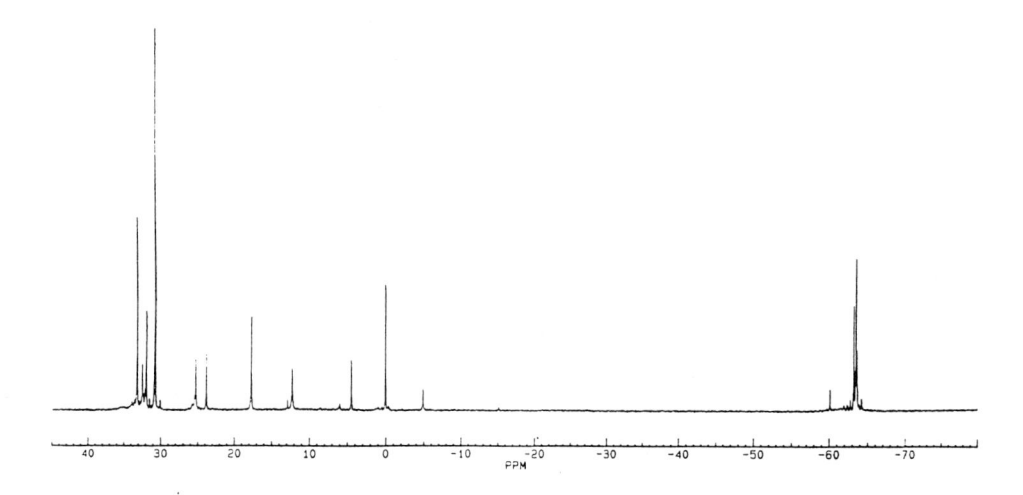

Fig. 1. ^{29}Si NMR spectrum of a oligosilane mixture obtained by disproportionation of TCDMS

In further reaction steps the pentasilane (SiCH$_3$Cl$_2$)$_2$SiCH$_3$–SiCH$_3$Cl–SiCH$_3$Cl$_2$ and hexasilane [(Cl$_2$CH$_3$Si)$_2$SiCH$_3$]$_2$ are formed with high probability. The corresponding ^{29}Si and ^{13}C NMR signals could already be assigned and will be published soon.

The results show that branched silanes are the preferred products determining the properties of the final polysilanes. The disproportionation of the disilane mixture gives similar results, whereas the nonsymmetric trichlorodimethyldisilane reacts under abstraction of DMS. Comparing the oligosilane spectrum with those of a polymer three signal groups can be distinguished and assigned to certain structural groups owing to the knowledge of the formed oligosilanes, which is demonstrated by the ^{29}Si NMR spectrum in Fig. 2.

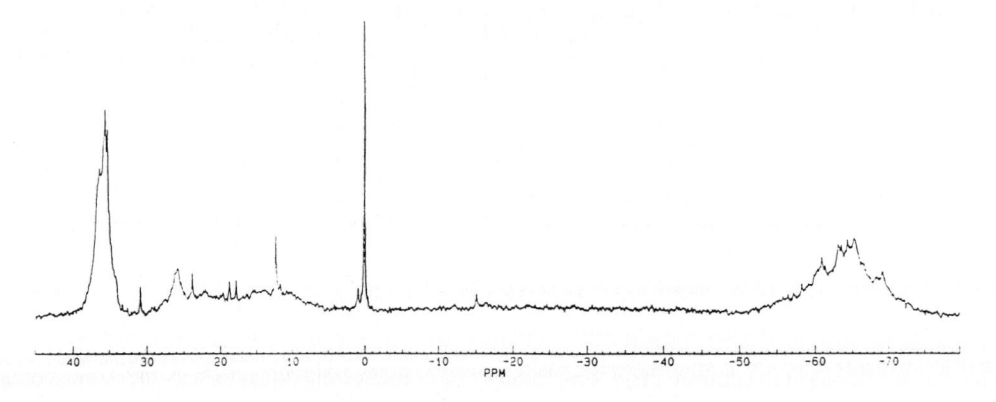

Fig. 2. ^{29}Si NMR spectrum of a polysilane obtained by disproportionation of TCDMS

δ ^{29}Si [ppm]	δ ^{13}C [ppm]	Assignment
30...40	9...12	–SiCH$_3$Cl$_2$-endgroups at tertiary Si$_3$SiCH$_3$-units
–5...30	8...–5	linear units –SiCH$_3$Cl– and –SiCH$_3$Cl$_2$– endgroups connected to them
–60...–70	–5...–12	tertiary Si$_3$SiCH$_3$ units

Table 2. Assignment of the polysilane resonances to structural groups, resulting in a general formula for the polymer: Cl$_2$CH$_3$Si–[(CH$_3$)SiR]$_x$–SiCH$_3$Cl$_2$; R = Cl, SiCH$_3$R$_2$

It has to be clarified whether linear and branched units alternate or if the latter are dominating. Because of the assignment of the ^{29}Si and ^{13}C NMR resonances it is now possible to estimate the branching and polymerization degree of the polymers. For that purpose the ^{29}Si NMR spectra are more suitable due to their higher structural sensitivity.

References:

[1] R. Calas, J. Dunoguès, G. Deleris, N. Duffaut, *J. Organomet. Chem.* **1968**, *5*, 237.
[2] R. F. Trandell, G. Urry, *J. Inorg. Nucl. Chem.* **1978**, *9/5*, 1305.
[3] W. H. Atwell, D. R.Weyenberg, *J. Organomet. Chem.* **1966**, *5*, 594.
[4] E. Hengge, W. Kalchauer, *Monatsh. Chemie* **1990**, *121*, 793.
[5] R. Lehnert, M. Höppner, H. Kelling, *Z. Anorg. Allg. Chem.* **1990**, *591*, 209.
[6] R. Richter, G. Roewer, K. Leo, B. Thomas, *Freiberger Forschungsh.* **1993**, A 832.
[7] E. Brendler, G. Roewer, D 4304256.2, 12.02.**1993**.

Cyclotrisilane and Silacyclopropane – An Equilibrium

Johannes Belzner*, Heiko Ihmels
Institut für Organische Chemie
Georg-August-Universität Göttingen
Tammannstr. 2, D-37077 Göttingen, Germany

Summary: New silacyclopropanes were synthesized quantitatively under mild thermal conditions by reaction of olefins with cyclotrisilane (cyclo-$(Ar_2Si)_3$, Ar = $Me_2NCH_2C_6H_4$) **1**, which transfers all of its three silylene subunits to terminal and strained internal olefins. Thermolysis of silacyclopropanes **3a** und **3b** indicated these compounds to be in a thermal equilibrium with cyclotrisilane **1** and the corresponding olefin. Silaindane **13** was synthesized by reaction of **1** with styrene via initially formed 2-phenyl-1-silacyclopropane **3d**. Reaction of **1** with conjugated dienes such as 2,3-dimethyl-1,3-butadiene, 1,3-cyclohexadiene or anthracene resulted in the formation of the expected 1,4-cycloaddition products in high yield.

Since the first synthesis of silacyclopropanes (siliranes) by Seyferth and Lambert [1] above all the addition of thermally or photolytically generated silylenes to olefins [2] has been shown to be a useful route for the synthesis of silacyclopropanes, although isomerization of the silacyclopropanes is often observed under these conditions [3]. Here we show that cyclotrisilane **1** [4] transfers all of its three silylene subunits **2** to a variety of olefins and dienes under mild thermal conditions [5].

Spectroscopically pure silacyclopropane **3a** was formed quantitatively, when **1** was heated at 40 °C for 12 h with an excess of 1-pentene in C_6D_6. Especially the characteristic ^{29}Si NMR shift values [6] (**3a**: δ = –76.6 ppm) proved the proposed cyclic structure. Isolation of analytically pure **3a** from the reaction mixture was impossible due to its inherent thermal instability. Heating a concentrated solution of **3a** in C_6D_6 for 5.5 h at 40 °C resulted in a 1:1:3 mixture of **3a**, **1**, and 1-pentene. An analogous partial retro reaction to **1** and an olefin was also shown by silacyclopropane **3b**, which was similarly synthesized by reaction of **1** with excess 1-hexene.

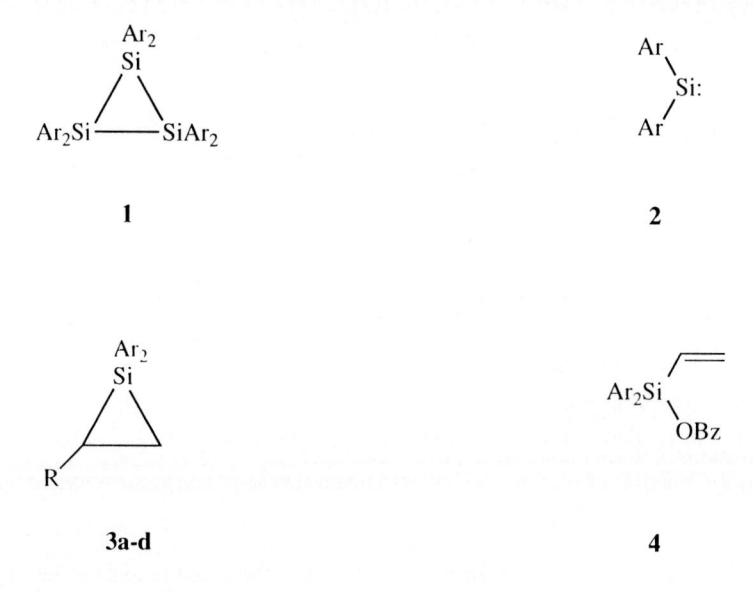

Fig. 1. Ar = 2-(Me$_2$NCH$_2$)C$_6$H$_4$; **3a**: R = nPr; **3b**: R = nBu; **3c**: R = OBz; **3d**: R = Ph

A mechanism for this unprecedented equilibrium between silacyclopropanes **3a** or **3b** and cyclotrisilane **1** is represented in Scheme 1.

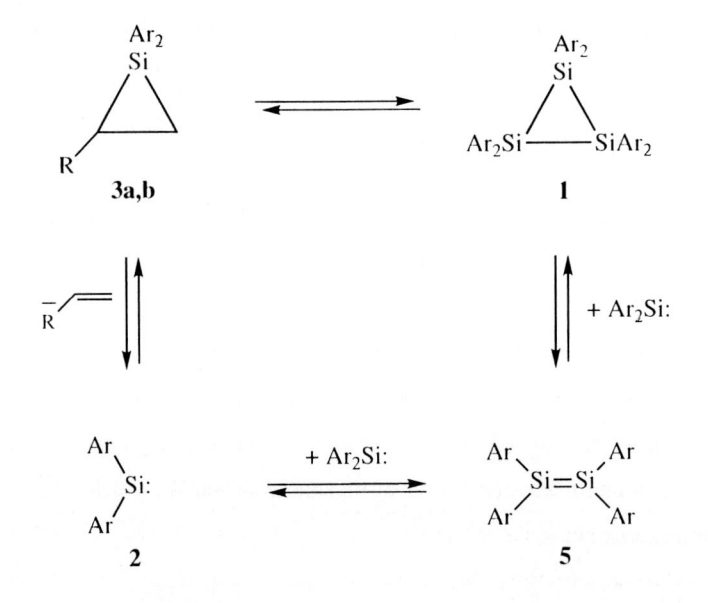

Scheme 1.

The reaction is initiated by extrusion of a silylene **2** from **3a** or **3b**, thus paralleling the well known equilibrium between hexamethylsilacyclopropane and dimethylsilylene [7]. In the absence of a silylene trapping reagent [8] dimerization of **2** to disilene **5** takes place. Addition of a third silylene to the Si=Si double bond eventually yields cyclotrisilane **1** [9]. The reversibility of the cyclotrisilane formation from **3a** and **3b** provides evidence, that the reverse reaction of **1** with olefins includes free silylenes **2** as reactive species as well.

It is tempting to assume, that the facile formation of silylene **2** from cyclotrisilane **1** is due to the effective stabilization of **2** by intramolecular coordination of the dimethylamino group to the silicon centre [10], which should lower the activation energy of a dissociation process from **1** to **2**. Reaction of **1** with benzylvinylether resulted in a complex reaction mixture, from which 12 % of vinylsilane **4** was isolated; **4** is presumably formed by rearrangement of the unstable oxy-substituted silacyclopropane **3c**.

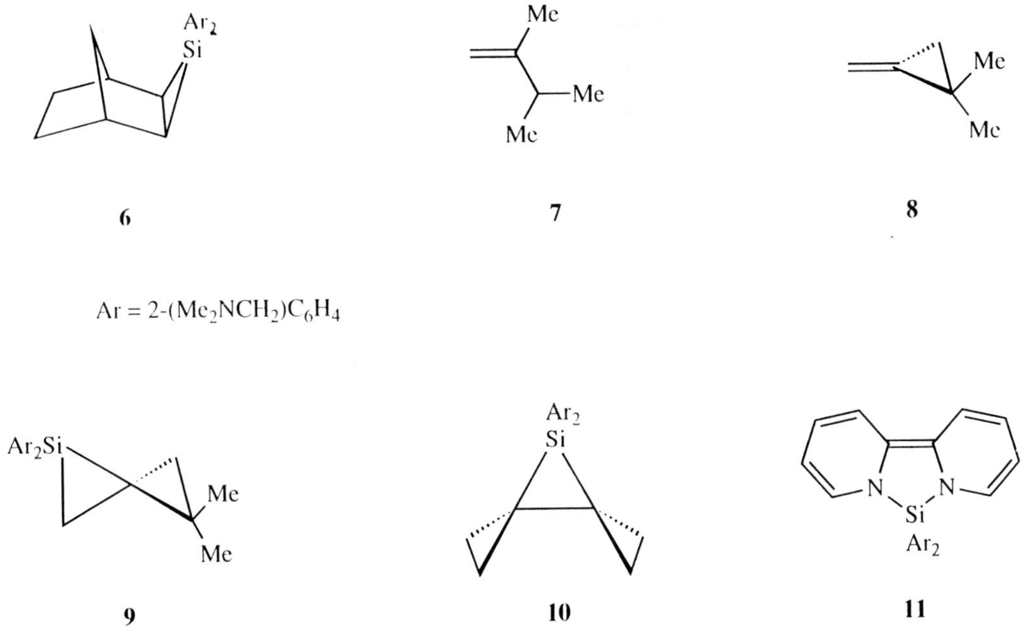

Ar = 2-(Me_2NCH_2)C_6H_4

Fig. 2.

Compound **1** did not react with unstrained internal olefins such as tetramethylethylene, *trans*-3-hexene, *trans*-stilbene, cyclooctene, cyclohexene, or cyclopentene. But imposing strain to the olefinic moiety resulted in a clean silylene transfer to the double bond: Norbornene formed with **1** the tricyclic silacyclopropane **6**. Whereas **2** did not add to the double bond of **7**, methylene cyclopropane **8** could be transformed into spiro[2.2]pentane **9** by reaction with **1**. Addition of **2** to bicyclopropylidene allowed the convenient synthesis of dispiro[2.0.2.1]heptane **10** in a quantitative manner.

In contrast to **3a** and **3b**, **6** as well as **9** and **10** were stable at room temperature. Only at temperatures above 110 °C, **6** transferred its silylene unit to 2,2'-bipyridyl to give the known [4a] tricyclic system **11** [11].

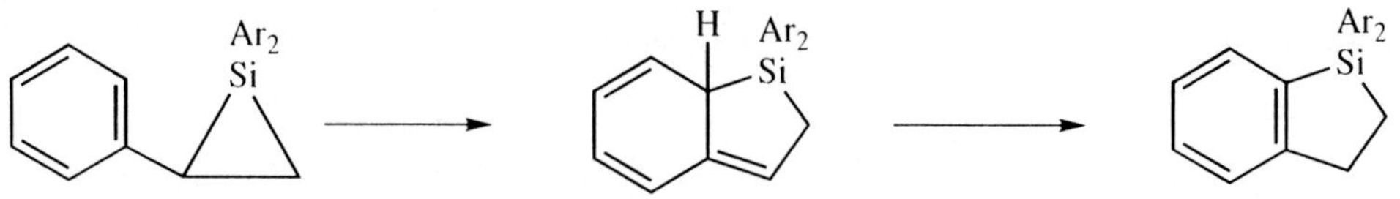

    3d   **12**   **13**

Scheme 2.

Stirring **1** for 12 h at 40 °C with excess styrene led quantitatively to silaindane **13** [12]. The silacyclopropane **3d** was identified as an intermediate in this reaction by its ^{29}Si NMR shift (δ = –82.5 ppm) [6]. Thus, **13** appears to be formed by initial formation of **3d**, which rearranges to intermediate **12**. Rearomatization eventually yields **13** (Scheme 2). This pathway resembles the well known mechanism of the reaction of silylenes with conjugated olefins via initial formation of vinylsilacyclopropanes [3].

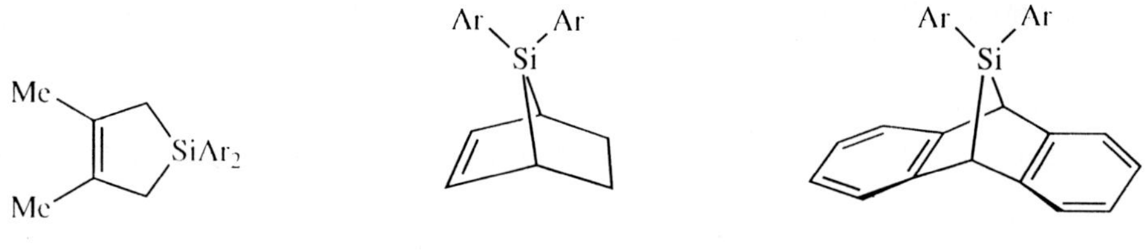

    14   **15**   **16**

Fig. 3.   Ar = 2-(Me$_2$NCH$_2$)C$_6$H$_4$

When **1** was stirred with three equiv. 2,3-dimethyl-1,3-butadiene at 40 °C for 12 h, silacyclopentene **14** was formed quantitatively [10a]. In a similar manner, 7-silanorbornene **15** [12] and dibenzo-norbornadiene **16** were formed without any side products by reaction of **1** with 1,3-cyclohexadiene or anthracene, respectively.

Acknowledgment: This work was supported by the *Deutsche Forschungsgemeinschaft* (financial support, fellowship to J. B.), *Fonds der Chemischen Industrie* (financial support), and *Friedrich-Ebert-Stiftung* (fellowship to H. I.). We thank Dr. R. Herbst-Irmer, Dr. R. O. Gould, and cand.-chem. B. O. Kneisel for the determination of X-ray structures, and Dr. V. Belov, Dipl.-Chem. S. Bräse, and Dipl.-Chem. T. Späth for samples of **8** and bicyclopropylidene.

References:

[1] R. L. Lambert, Jr., D. Seyferth, *J. Am. Chem. Soc.* **1972**, *94*, 9246.

[2] V. J. Tortorelli, M. Jones, Jr., S. H. Wu, Z. H. Li, *Organometallics* **1983**, *2*, 759; W. Ando, M. Fujita, H. Yoshida, A. Sekiguchi, *J. Am. Chem. Soc.* **1988**, *110*, 3310; D. H. Pae, M. Xiao, M. Y. Chiang, P. P. Gaspar, *J. Am. Chem. Soc.* **1991**, *113*, 1281; P. Boudjouk, E. Black, R. Kumarathasan, *Organometallics* **1991**, *10*, 2095.

[3] M. Weidenbruch, E. Kroke, H. Marsmann, S. Pohl, W. Saak, *J. Chem. Soc., Chem. Comm.* **1994**, 1233; P. P. Gaspar, D. Lei, *Organometallics* **1986**, *5*, 1276; M. P. Clarke, I. M. T. Davidson, *J. Chem. Soc., Chem. Commun.* **1988**, 241; S. Zhang, R. T. Conlin, *J. Am. Chem. Soc.* **1991**, *113*, 4272.

[4] a) J. Belzner, *J. Organomet. Chem.* **1992**, *430*, C51.
b) J. Belzner, H. Ihmels, *Tetrahedron Lett.* **1993**, 6541.

[5] J. Belzner, H. Ihmels, *Organometallics*, accepted for publication.

[6] E. A. Williams, in: *The Chemistry of Organic Silicon Compounds* (Eds.: S. Patai, Z. Rappoport), Wiley, Chichester **1989**, 511.

[7] D. Seyferth, D. C. Annarelli, S. C. Vick, *J. Organomet. Chem.* **1984**, *272*, 123; and references therein.

[8] Compounds **2** can be intercepted by alkynes, yielding known [4b] silacyclopropenes.

[9] G. R. Gillette, G. Noren, R. West, *Organometallics* **1990**, *9*, 2925.

[10] a) R. Corriu, G. Lanneau, C. Priou, F. Soulairol, N. Auner, R. Probst, R. Conlin, C. Tan, *J. Organomet. Chem.* **1994**, *466*, 55.
b) R. T. Conlin, D. Laakso, P. Marshall, *Organometallics* **1994**, *13*, 838.
c) Y. Apeloig, in: *Heteroatom Chemistry* (Ed.: E. Block), VCH, Weinheim, **1990**, p. 27.

[11] See also: M. Weidenbruch, A. Lesch, H. Marsmann, *J. Organomet. Chem.* **1990**, *385*, C47.

[12] The structures of **13** and **15** could be further confirmed by X-ray analysis.

Synthesis and Spectroscopy of Phenylated and Halogenated Trisilanes and Disilanes

Wolfgang Köll, Karl Hassler*

Institut für Anorganische Chemie

Technische Universität Graz

Stremayrgasse 16, A-8010 Graz, Austria

Summary: The introduction of *p*-tolyl groups into phenylated trisilanes allows the preparation of asymmetrically substituted phenylated trisilanes which cannot be prepared from Ph_8Si_3 or $(Ph_3Si)_2SiH_2$ and HX (X = Cl, Br, I) in good yield. These trisilanes can be transformed into iodocompounds. The presented reaction pathways very likely allow the preparation of further phenylated trisilanes $H_nSi_3Ph_{8-n}$ (n = 0–7), which in subsequent steps may be transformed into the halogenated compounds $H_nSi_3X_{8-n}$ (X = Cl, Br, I). Triflation followed by reaction with LiX (X = F, Cl, Br, I) also allows the selective synthesis of mixed halodisilanes.

Introduction

Until recently only a very small number of the possible 29 isomers of phenylated trisilanes $H_nSi_3Ph_{8-n}$ (n = 0–7) have been synthesized. Octaphenyltrisilane and 1,1,1,3,3,3-hexaphenyltrisilane can be prepared by salt elimination from $KSiPh_3$ and Cl_2SiPh_2 or I_2SiH_2 respectively [1, 2].

1,1,1,2,3,3,3-heptaphenyltrisilane is prepared from $KSiPh_3$ and Cl_2HSiPh and 1,1,1,2,2,3,3-heptaphenyltrisilane by the reaction of $KSiPh_3$ with $(ClPh_2Si)_2$ followed by reduction with $LiAlH_4$ [3, 4]. One method for synthesizing other phenylated trisilanes is the partial substitution of phenyl groups by halogen with HX (X = Cl, Br, I) of the trisilanes mentioned above and subsequent reduction with $LiAlH_4$. These procedures require long reaction times and only symmetrically substituted phenylated trisilanes can be synthesized in good yield.

A method for the preparation of asymmetrically substituted phenylated trisilanes is the reaction of triflic acid (TfOH) with arylated trisilanes containing phenyl and *p*-tolyl groups. Due to the higher reactivity of p-tolyl groups towards triflic acid Si–p-Tol bonds can be cleaved selectively without any cleavage of the Si–Ph bonds. Further advantageous is the higher solubility of the p-tolyl compounds. The triflate compounds can either be reduced with LiAlH$_4$ to form Si–H bonds or they can be transformed into halosilanes using LiX (X = F, Cl, Br, I). The same types of reactions can also be performed using arylated disilanes. The reaction of a triflate compound with LiF followed by further triflation and reaction with LiCl allows the selective preparation of mixed halogenated disilanes, which so far have been prepared by equilibration reactions in most cases. These always result in mixtures of several isomers.

Using these reactions, triflation of an arylated disilane or trisilane followed by reaction with LiAlH$_4$ or LiX (X = F, Cl), allows the preparation of a wide range of compounds.

Synthesis

The phenylated trisilanes HPh$_2$Si–SiPh$_2$–SiPh$_2$H, H$_2$PhSi–SiPh$_2$–SiPhH$_2$, H$_2$PhSi–SiPhH–SiPhH$_2$, and H$_2$PhSi–SiH$_2$–SiPhH$_2$ are prepared by reduction of the corresponding halogenated phenylated trisilanes with LiAlH$_4$ [5, 6]. These are synthesized from Ph$_8$Si$_3$ or (Ph$_3$Si)$_2$SiH$_2$ and liquid HCl, HBr, or HI.

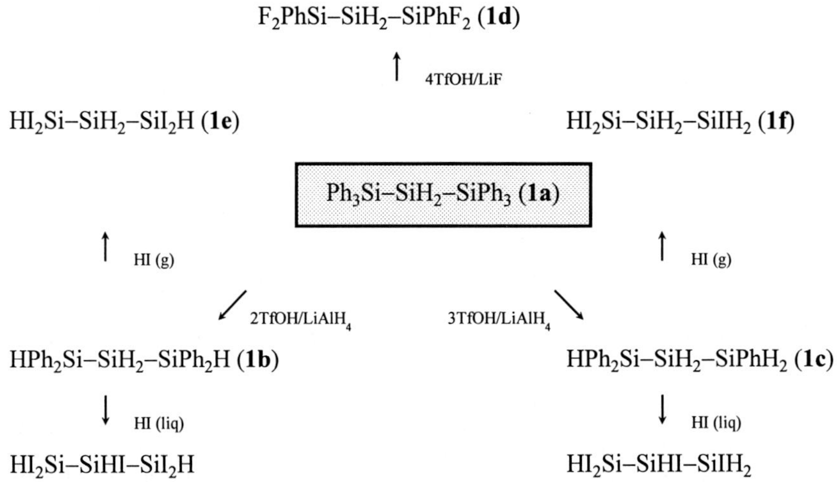

$$F_2PhSi-SiH_2-SiPhF_2 \ (\mathbf{1d})$$
$$\uparrow \text{4TfOH/LiF}$$

HI$_2$Si–SiH$_2$–SiI$_2$H (**1e**)   HI$_2$Si–SiH$_2$–SiIH$_2$ (**1f**)

$$\boxed{Ph_3Si-SiH_2-SiPh_3 \ (\mathbf{1a})}$$

↑ HI (g)   ↗ 2TfOH/LiAlH$_4$ 3TfOH/LiAlH$_4$ ↘   ↑ HI (g)

HPh$_2$Si–SiH$_2$–SiPh$_2$H (**1b**)   HPh$_2$Si–SiH$_2$–SiPhH$_2$ (**1c**)

↓ HI (liq)   ↓ HI (liq)

HI$_2$Si–SiHI–SiI$_2$H   HI$_2$Si–SiHI–SiIH$_2$

Scheme 1. Reactions of 1,1,1,3,3,3-hexaphenyltrisilane (**1a**)

When 1,1,1,3,3,3-hexaphenyltrisilane (**1a**) is used as a starting material in the reaction with two or three equivalents of TfOH respectively, the trisilanes **1b** and **1c** can be isolated after reduction with LiAlH$_4$. Compound **1a** with four equivalents of TfOH and subsequent reaction with LiF yields **1d** (see

Scheme 1). The phenylated trisilanes can easily be transformed into iodotrisilanes by reaction with gaseous HI and catalytic amounts of AlI_3 (**1e** and **1f**) [7]. When liquid HI is used and reaction times of several days are applied additional iodination of the central SiH_2-group of **1b** and **1c** are observed besides the expected Ph/I exchange.

All Si–p-Tol groups of p-Tol$_3$Si–SiPh$_2$–SiPh$_3$ (**2a**) can be cleaved selectively with triflic acid without any Si–Ph cleavage (Scheme 2). Several new trisilanes $Si_3F_nH_mAr_{8-n-m}$ (n, m = 0–3; Ar = Ph, p-Tol) have been prepared starting with the reaction of **2a** with one equivalent of TfOH followed by reaction with LiF to give **2b** and reduction of the resulting fluorotrisilane with LiAlH$_4$ (**2c**).

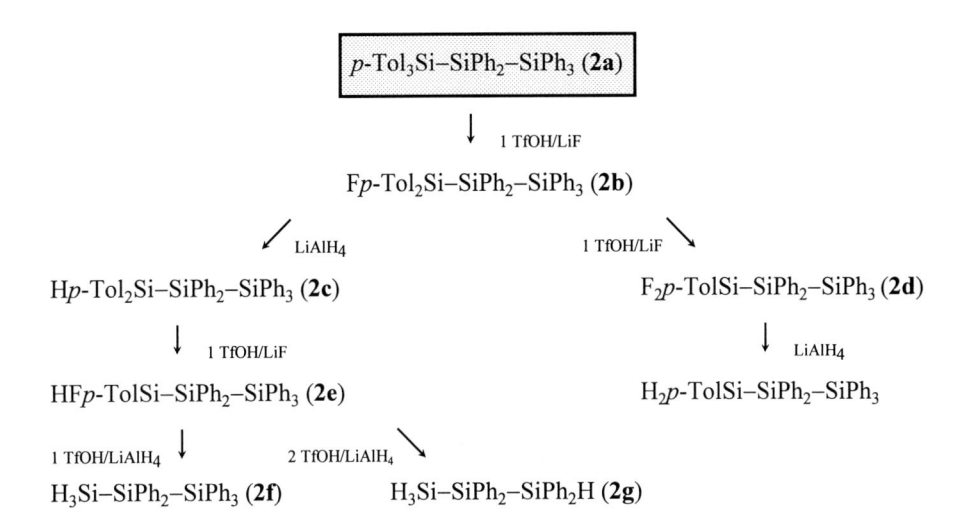

p-Tol$_3$Si–SiPh$_2$–SiPh$_3$ (**2a**)
↓ 1 TfOH/LiF
Fp-Tol$_2$Si–SiPh$_2$–SiPh$_3$ (**2b**)

↙ LiAlH4   ↘ 1 TfOH/LiF

Hp-Tol$_2$Si–SiPh$_2$–SiPh$_3$ (**2c**)   F$_2p$-TolSi–SiPh$_2$–SiPh$_3$ (**2d**)

↓ 1 TfOH/LiF   ↓ LiAlH4

HFp-TolSi–SiPh$_2$–SiPh$_3$ (**2e**)   H$_2p$-TolSi–SiPh$_2$–SiPh$_3$

↙ 1 TfOH/LiAlH4   ↘ 2 TfOH/LiAlH4

H$_3$Si–SiPh$_2$–SiPh$_3$ (**2f**)   H$_3$Si–SiPh$_2$–SiPh$_2$H (**2g**)

Scheme 2.   Reactions of p-Tol$_3$Si–SiPh$_2$–SiPh$_3$ (**2a**)

Further triflation of **2b** and **2c** followed by reaction with LiF results in **2d** and **2e**. **2e** can be transformed into the pentaphenyltrisilane **2f** and the tetraphenyltrisilane **2g**. The yield for each single step of this reaction sequence is about 90 % or higher so that an overall yield for the preparation of 1,1,1,2,2-pentaphenyltrisilane of 60 % can be achieved.

The selective preparation of some mixed halodisilanes has been performed starting from p-TolPh$_2$Si–SiPh$_2p$-Tol (**3a**) or Si_2Ph_6. The reaction of **3a** with four equivalents of TfOH followed by treatment with LiF gives **3b**, which is transformed to **3c** by reaction with liquid HI (Scheme 3).

Scheme 3

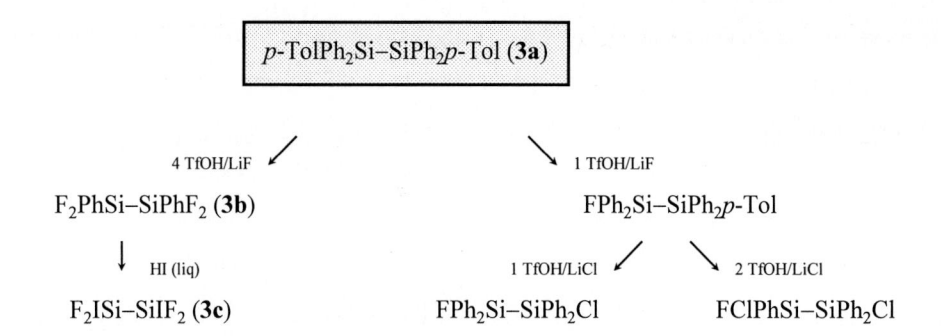

```
                    p-TolPh₂Si–SiPh₂p-Tol (3a)
                    /                        \
            4 TfOH/LiF                       1 TfOH/LiF
       F₂PhSi–SiPhF₂ (3b)              FPh₂Si–SiPh₂p-Tol
            ↓ HI (liq)                   /              \
       F₂ISi–SiIF₂ (3c)        1 TfOH/LiCl            2 TfOH/LiCl
                           FPh₂Si–SiPh₂Cl         FClPhSi–SiPh₂Cl
```

Scheme 3. Reactions of p-TolPh₂Si–SiPh₂p-Tol (3a)

NMR-Spectra

Tables 1 to 3 show the ^{29}Si and ^{19}F NMR data of all the synthesized compounds (300 MHz, C$_6$D$_6$, 20 °C, TMS respectively CFCl$_3$). In most cases coupling constants ^2J (Si, H) can only be resolved when the corresponding Si-atom is not connected to an aromatic group. Otherwise coupling with aromatic protons ^3J (Si, C, C, H) causes broadening of the signals.

Generally, ^1J (Si, H) increases with the electronegativity of the other substituents at silicon. Especially the presence of a CF$_3$O$_2$SO-group has great influence on ^1J (Si, H). A comparison of ^1J (Si, H) in (Ph$_2$TfSi)$_2$SiH$_2$ (J = 196.4 Hz), (Ph$_2$HSi)$_2$SiH$_2$ (J = 188.0 Hz) and (Ph$_3$Si)$_2$SiH$_2$ (J = 173.6 Hz) clearly shows, that the group electronegativity of the SiPh$_2$X group (X = Tf, H, Ph) decreases in the order X = Tf > H > Ph. Higher triflation (for Tf$_2$PhSi–SiH$_2$–SiPh$_2$Tf J = 208.0 Hz and for (Tf$_2$PhSi)$_2$SiH$_2$, J = 210.5 Hz) shows even greater values of ^1J (Si, H), which indicates a further increase in the group electronegativity.

^{19}F NMR spectra of the following trisilanes have been measured:

- Fp-Tol$_2$Si–SiPh$_2$–SiPh$_3$: $\delta = -172.9$ ppm
- HFp-TolSi–SiPh$_2$–SiPh$_3$: $\delta = -188.8$ ppm, 41.8 (^2J (H, F))
- (F$_2$PhSi)$_2$SiH$_2$: $\delta = -126.4$ ppm

Compound	δ (Si)	δ (*Si)	δ (**Si)	J [Hz]	
p-Tol$_3$Si–*SiPh$_2$–**SiPh$_3$	−19.4	−42.4	−18.5		
Tfp-Tol$_2$Si–*SiPh$_2$–**SiPh$_3$	+17.2	−44.1	−19.4		
Fp-Tol$_2$Si–*SiPh$_2$–**SiPh$_3$	+10.1	−45.4	−19.0	316.8 (d, ^1J(Si, F))	27.7 (d, ^2J(*Si, F))
Hp-Tol$_2$Si–*SiPh$_2$–**SiPh$_3$	−31.1	−42.3	−19.2	188.0 (d, ^1J(Si, H))	
FTfp-TolSi–*SiPh$_2$–**SiPh$_3$	−1.8	−45.5	−20.0	371.5 (d, ^1J(Si, F))	20.0 (d, ^2J(*Si, F))
HTfp-TolSi–*SiPh$_2$–**SiPh$_3$	+10.4	−43.9	−19.8	217.2 (d, ^1J(Si, H))	6.0 (t, ^3J(Si, H))
F$_2$p-TolSi–*SiPh$_2$–**SiPh$_3$	−8.2	−47.1	−20.0	355.2 (t, ^1J(Si, F))	27.7 (t, ^2J(*Si, F))
HFp-TolSi–*SiPh$_2$–**SiPh$_3$	+9.8	−46.6	−20.2	320.1 (d, ^1J(Si, F))	211.4 (d, ^1J(Si, H))
					5.8 (t, ^3J(Si, H))
H$_2$p-TolSi–*SiPh$_2$–**SiPh$_3$	−57.9	−43.1	−20.5	186.3 (t, ^1J(Si, H))	
HFTfSi–*SiPh$_2$–**SiPh$_3$	−2.8	−46.6	−21.2	376.3 (d, ^1J(Si, F))	255.8 (d, ^1J(Si, H))
					15.0 (d, ^2J(*Si, F))
H$_3$Si–*SiPh$_2$–**SiPh$_3$	−94.8	−42.2	−20.1	191.7 (q, ^1J(Si, H))	
H$_3$Si–*SiPh$_2$–**SiPh$_2$H	−96.7	−42.2	−31.9	192.7 (q, ^1J(Si, H))	188.4 (d, ^1J(**Si, H))

Table 1. ^{29}Si NMR shifts (ppm) and coupling constants (Hz) of the trisilanes prepared from p-Tol$_3$Si–SiPh$_2$–SiPh$_3$ (Tf = OSO$_2$CF$_3$)

Compound	δ (Si)	δ (*Si)	δ (F)	J [Hz]	
p-TolPh$_2$Si-SiPh$_2$p-Tol	−24.3				
TfPh$_2$Si-*SiPh$_2$p-Tol	+12.4	−25.3			
FPh$_2$Si-*SiPh$_2$p-Tol	+7.1	−26.7	−178.2	317.8 (d, ^1J(Si, F))	27.1 (d, ^2J(*Si, F))
FTfPhSi-*SiPh$_2$Tf	+0.4	+3.8		314.3 (d, ^1J(Si, F))	37.1 (d, ^2J(*Si, F))
FPh$_2$Si-*SiPh$_2$Cl	+0.6	−6.6	−179.1	315.3 (d, ^1J(Si, F))	35.2 (d, ^2J(*Si, F))
				5.1 (t, ^3J(Si, H))	5.8 (t, ^3J(*Si, H))
FClPhSi-*SiPh$_2$Cl	−0.1	−8.9	−143.6	365.3 (d, ^1J(Si, F))	35.9 (d, ^2J(*Si, F))
				6.5 (t, ^3J(Si, H))	6.5 (t, ^3J(*Si, H))
IF$_2$Si-SiF$_2$I	−77.1			351.4 (t, ^1J(Si, F))	72.1 (t, ^2J(Si, F))

Table 2. ^{29}Si and ^{19}F NMR-shifts (ppm) and coupling constants (Hz) of the disilanes prepared from p-TolPh$_2$Si–SiPh$_2$p-Tol

Compound	δ (Si)	δ (*Si)	δ (**Si)	J [Hz]	
Ph$_3$Si–*SiH$_2$–SiPh$_3$	−15.3	−105.2			173.6 (t, ^1J(*Si, H))
Tf$_2$PhSi–*SiH$_2$–SiPhTf$_2$	−4.9	−100.2		6.4 (t, ^3J(Si, H))	210.5 (t, ^1J(*Si, H))
F$_2$PhSi–*SiH$_2$–SiPhF$_2$	−5.0	−122.3		344.1 (t, ^1J(Si, F))	198.1 (t, ^1J(*Si, H))
				3.6 (t, ^3J(Si, H))	33.9 (p, ^2J(Si, F))
TfPh$_2$Si–*SiH$_2$–SiPh$_2$Tf	+16.3	−106.2			196.4 (t, ^1J(*Si, H))
Tf$_2$PhSi–*SiH$_2$–**SiPh$_2$Tf	0.0	−103.4	+11.9		208.0 (t, ^1J(*Si, H))
HPh$_2$Si–*SiH$_2$–SiPh$_2$H	−31.3	−108.6		193.4 (d, ^1J(Si, H))	181.3 (t, ^1J(*Si, H))
					10.0 (t, ^2J(*Si, H))
H$_2$PhSi–*SiH$_2$–**SiPh$_2$H	−60.2	−109.4	−31.2	194.5 (t, ^1J(Si, H))	185.6 (t, ^1J(*Si, H))
					193.4 (d, ^1J(*Si, H))

Compound	δ (Si)	δ (*Si)	δ (**Si)	J [Hz]	
H$_2$PhSi–*SiH$_2$–SiPhH$_2$	–60.1	–110.6		196.9 (t, ^1J(Si, H))	188.0 (t, ^1J(*Si, H))
					7.3 (p, ^2J(*Si, H))
H$_2$PhSi–*SiPhH–SiPhH$_2$	–58.7	–68.1		193.3 (t, ^1J(Si, H))	185.5 (t, ^1J(*Si, H))
H$_2$PhSi–*SiPh$_2$–SiPhH$_2$	–58.2	–41.7		191.0 (t, ^1J(Si, H))	
HPh$_2$Si–*SiPh$_2$–SiPh$_2$H	–31.9	–42.4		188.9 (d, ^1J(Si, H))	
HI$_2$Si–*SiIH–**SiIH$_2$	–92.5	–83.9	–81.8	267.8 (d, ^1J(Si, H))	232.3 (d, ^1J(*Si, H))
				14.1 (d, ^2J(Si, H))	19.7 (d, ^2J(*Si, H))
					13.8 (t, ^2J(*Si, H))
					241.0 (t, ^1J(**Si, H)
					12.4 (d, ^2J(**Si, H)
HI$_2$Si–*SiH$_2$–SiI$_2$H	–89.7	–82.7		264.7 (d, ^1J(Si, H))	214.5 (d, ^1J(*Si, H))
				6.9 (t, ^2J(Si, H))	17.4 (d, ^1J(*Si, H))
HI$_2$Si–*SiIH–SiI$_2$H	–94.9	–82.8		270.0 (d, ^1J(Si, H))	236.7 (d, ^1J(*Si, H))
				14.5 (d, ^2J(Si, H))	20.4 (t, ^2J(*Si, H))
H$_2$ISi–*SiH$_2$–SiIH$_2$	–77.6	–99.0		230.6 (t, ^1J(Si, H))	205.3 (d, ^1J(*Si, H))
					11.1 (q, ^2J(*Si, H))
H$_2$ISi–*SiHI–SiIH$_2$	–80.7	–88.3		236.4 (t, ^1J(Si, H))	227.5 (d, ^1J(*Si, H))
				10.3 (d, ^2J(Si, H))	13.7 (p, ^2J(*Si, H))
I$_3$Si–*SiH$_2$–SiI$_3$	–165.8	–96.5		10.7 (t, ^2J(Si, H))	220.1 (d, ^1J(*Si, H))
I$_3$Si–*SiI$_2$–SiI$_3$	–145.6	–165.7			

Table 3. ^{29}Si NMR shifts (ppm) and coupling constants (Hz) of the trisilanes prepared from Ph$_3$Si–SiH$_2$–SiPh$_3$ and Ph$_8$Si$_3$

Acknowledgement: The authors thank the *Fonds zur Förderung der wissenschaftlichen Forschung* (Wien) for financial support.

References:

[1]   H. Gilman, T. C. Wu, H. A. Hartzfeld, G. A. Guter, A. G. Smith, J. J. Goodman, S. H. Eidt, *J. Am. Chem. Soc.* **1952**, *74*, 561.
[2]   K. Hassler, *Monatsh. Chem.* **1988**, *119*, 1051.
[3]   E. Hengge, F. K. Mitter, *Z. Anorg. Allg. Chem.* **1985**, *529*, 22.
[4]   H. Gilman, H. J. S. Winkler, *J. Org. Chem.* **1962**, *27*, 254.
[5]   K. Hassler, U. Katzenbeisser, *J. Organomet. Chem.* **1991**, *421*, 151.
[6]   K. Hassler, U. Katzenbeisser, B. Reiter, *J. Organomet. Chem.*, in press.
[7]   K. Hassler, U. Katzenbeisser, *J. Organomet. Chem.*, in press.

Investigations Concerning the Electrochemical Formation of Di- and Polysilanes

Susanne Graschy, Christa Jammegg, Edwin Hengge*

Institut für Anorganische Chemie

Technische Universität Graz

Stremayrgasse 16, A-8010 Graz, Austria

Summary: The electroreductive coupling of chlorosilanes is a relatively new method for the formation of Si–Si bonds [1]. Adopting the reaction conditions published by Dunoguès and coworkers [2], we developed a special kind of hydrogen anode (Eq. 1) and succeeded in the electrolysis of several chlorosilanes (Eq. 2a, 2b) [3, 4].

$$\text{anode:} \quad 2\ Cl^- + H_2 \longrightarrow 2\ HCl + 2e^- \quad (1)$$

$$\text{cathode:} \quad 2\ R_3SiCl + 2\ e^- \longrightarrow R_3SiSiR_3 + 2\ Cl^- \quad (2a)$$

$$n\ R_2SiCl_2 + 2n\ e^- \longrightarrow -(R_2Si)_n- + 2n\ Cl^- \quad (2b)$$

To test the applicability of this method and to understand the arising problems, several basic investigations were effected.

Solvents

For electrolysis with the hydrogen anode we use a mixture of THF and HMPA (hexamethylphosphoric triamide) with 0.02 M Et_4NBF_4 as solvent/electrolyte system. Trying to find a substitute for HMPA, we tested several aprotic solvents, known to replace HMPA very effectively in organic synthesis (see Table 1).

Except THF alone, all of the chosen mixtures showed a sufficient conductivity. But when using them for electrolysis of trimethylchlorosilane, either the solvent or the supporting electrolyte is not stable, thus leading to the formation of hexamethyldisiloxane instead of hexamethyldisilane. With TMEDA Me_3SiCl reacts to an insoluble, white complex that can not be electrolyzed.

Solvent	Supporting electrolyte [0.02 M]	Specific conductivity [μScm^{-1}]
DMF	Et_4NBF_4	241.90
THF/NMP[a]	Et_4NBF_4	64.19
THF/HMPA	Et_4NBF_4	41.89
THF/DMPU[b]	Et_4NBF_4	34.46
DME	Bu_4NClO_4	4.79
THF/TMEDA	LiCl	1.90
THF	$BzlBu_3NBr$	0.11

Table 1. Specific conductivity of solvent/electrolyte systems at 20 °C; [a] N-Methylpyrrolidone, [b] N,N'-Dimethylpropyleneurea

Overpotential

In the galvanostatic electrolysis the potentials of the electrodes adapt automatically. To get an idea of the order of the emerging overpotential, we recorded the cathodic potential in relation to the current flow (Fig. 1) [5]. The declination of the exponential form indicates mixed control regime, i.e., the current is controlled by mass and electron transfer steps.

For the phenylchlorosilanes, the trend of lower reduction potentials (Table 2) is veryfied in the galvanostatic electrolysis, though additional overpotentials emerge. For Ph_2SiCl_2 we measured a cathodic potential of –28 V (vs SCE) at a current density of 0.3 mA/cm^2 [5]. This effect above all is responsible for the low current yield in comparison with the methylchlorosilanes.

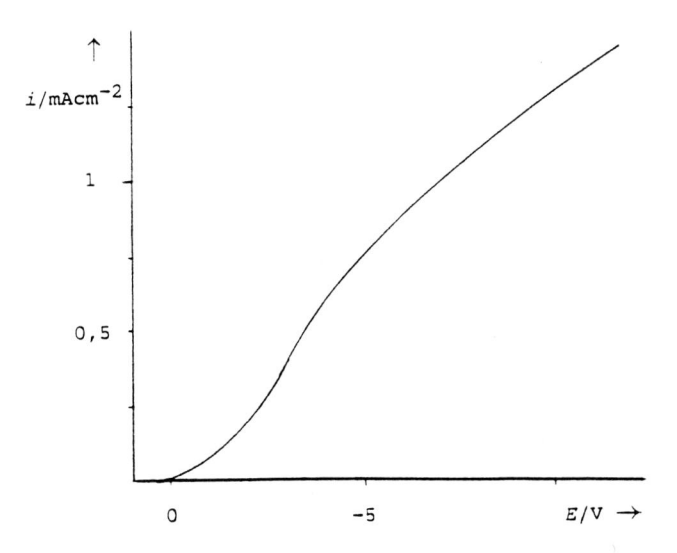

Fig. 1. Relationship between current density i and cathodic potential E for the reduction of Me_3SiCl [5]

Cyclic Voltammetry

The reduction properties of chloro-, alkoxy-, and hydrosilanes were investigated by cyclic voltammetry.

Silane	E_{red} [V]	Silane	E_{red} [V]
Me_3SiCl	−0.1	Ph_3SiCl	−0.2
Me_2SiCl_2	0.5/−0.7	Ph_2SiCl_2	−0.2/−0.9
$ClMe_2SiSiMe_2Cl$	−0.6	Ph_3SiH	−0.8
Me_3SiOMe	−0.8	Ph_2SiH_2	−0.9

Table 2. Reduction potentials E_{red} vs SCE (scan rate 0.1 V/s)

Concerning the chlorosilanes, the methyl substituted ones showed in each case higher reduction potentials than the corresponding phenylsilanes. This is in accordance with the results of the galvanostatic electrolysis. The investigation of Me_2SiCl_2 led us to the assumption of the following reduction mechanism:

$$Me_2SiCl_2 + e^- \longrightarrow [Me_2SiCl_2]^- \quad (3)$$
$$[Me_2SiCl_2]^- \longrightarrow ClMe_2Si + Cl^- \quad (4)$$
$$2\, ClMe_2Si \longrightarrow ClMe_2Si\text{–}SiMe_2Cl \quad (5)$$
$$ClMe_2Si\text{–}SiMe_2Cl + e^- \longrightarrow [ClMe_2Si\text{–}SiMe_2Cl]^- \quad (6)$$

After acceptance of an electron, chloride is eliminated. The so formed chlorodimethylsilyl radical combines with another one to give 1,2-dichlorotetramethyldisilane, which accepts an electron again. The two reduction peaks in the cyclic voltammogram are related to the first (Eq. 3) and the second (Eq. 6) electron acceptance.

The results of the cyclovoltammetric measurements of 1,2-dichlorotetramethyldisilane, an intermediate in this mechanism, corroborate our assumptions. There is only one reduction peak, corresponding to the second peak of dichlorodimethylsilane (see Fig. 2).

Methoxy- and hydrosilanes are reduced at lower values than chlorosilanes. Additionally, the reduction seems to proceed very slowly, because the measured current density is low in comparison with the chlorosilanes.

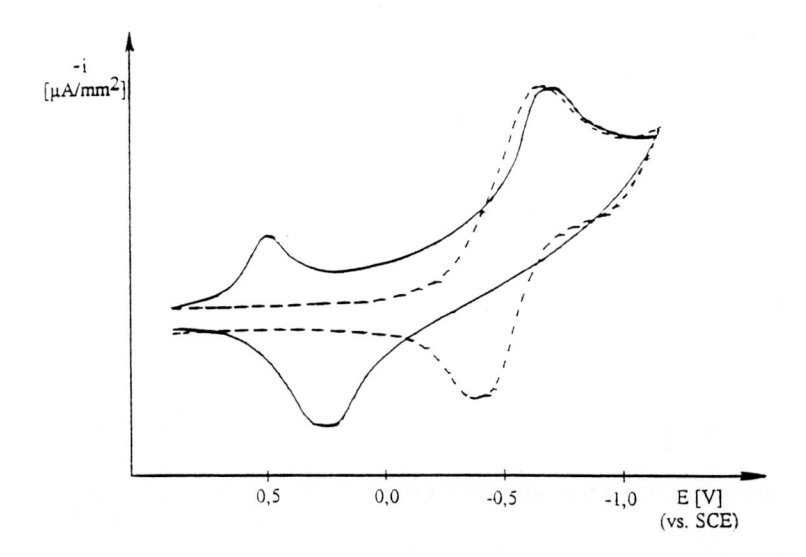

Fig. 2. Cyclic voltammograms of Me_2SiCl_2 (—) and $ClMe_2SiSiMe_2Cl$ (---)

Acknowledgement: We wish to thank the *Wacker-Chemie GmbH* for support of this study.

References:

[1] E. Hengge, G. Litscher, *Angew. Chem.* **1976**, *88*, 414; *Angew. Chem., Int. Ed. Engl.* **1976**, *15*, 370.
[2] C. Biran, M. Bordeau, P. Pons, M. P. Leger, J. Dunoguès, *J.Organomet.Chem.* **1990**, *382*, C17.
[3] Ch. Jammegg, S. Graschy, E. Hengge, *Organomet.* **1994**, *13*, 2397.
[4] E. Hengge, Ch. Jammegg, W. Kalchauer, *German Patent Application*, **1993**.
[5] All values refer to the saturated calomel electrode (SCE). No IR drop compensation was effected.

Syntheses, ^{29}Si NMR Spectra, and Vibrational Spectra of Methylated Trisilanes

Karla Schenzel

Institut für Analytik und Umweltchemie
Martin-Luther-Universität Halle-Wittenberg
Weinbergweg 16, D-06126 Halle (Saale), Germany

*Karl Hassler**

Institut für Anorganische Chemie
Technische Universität Graz
Stremayrgasse 16, A-8010 Graz, Austria

Introduction

Methylated trisilanes $Me_nSi_3X_{8-n}$ (X = H, F, Cl, Br, I, Ph, OMe) can serve as model compounds for the study of correlations between structures and ^{29}Si chemical shifts, ^{29}Si^{29}Si coupling constants, bond strengths and bond lengths, as determined with vibrational spectroscopy and electron diffraction or for the study of rotational isomerism of Si–Si bonds. A report of these ongoing activities is given in this work.

Preparations

Several methods, all starting from appropriate phenylated trisilanes [1–3] were used to prepare the title compounds:

a) $Ph_nSi_3Me_{8-n}$ + HX/AlX_3 $\longrightarrow$ $X_nSi_3Me_{8-n}$ X = Cl, Br, I

b) $Ph_nSi_3Me_{8-n}$ + $n\,CF_3SO_3H$ $\longrightarrow$ $(CF_3SO_3)_nSi_3Me_{8-n}$ + $n\,C_6H_6$

c) $(CF_3SO_3)_nSi_3Me_{8-n}$ + $LiAlH_4$ $\longrightarrow$ $H_nSi_3Me_{8-n}$
   $X_nSi_3Me_{8-n}$

d) $(CF_3SO_3)_nSi_3Me_{8-n}$ + LiF $\longrightarrow$ $F_nSi_3Me_{8-n}$
   $Br_nSi_3Me_{8-n}$

e) $X_nSi_3Me_{8-n}$ + MeOH $\longrightarrow$ $(MeO)_nSi_3Me_{8-n}$

f) with CF_3SO_3H followed by a reduction with $LiAlH_4$ or reaction with LiF, it is possible to synthesize partially dearylated trisilanes like $(PhX_2Si)_2SiMe_2$ or $(Ph_2XSi)_2SiMe_2$ and $(XH_2Si)_2SiMe_2$ or $(X_2HSi)_2SiMe_2$ (X = F, Cl, Br, I) [4].

^{29}Si NMR Spectra

Table 1 summarizes the ^{29}Si chemical shifts and $^{29}Si^{29}Si$ one and two bond coupling constants of all trisilanes synthesized so far. The coupling constants were measured using INADEQUATE pulse techniques.

The ^{29}Si chemical shift data of Table 1 can be used to elucidate trends that may be helpful for the prediction and interpretation of ^{29}Si NMR spectra of unknown Si-compounds. One can, for instance, think of the analysis of spectra of complex reaction mixtures so often encountered in silicon chemistry. Our aim is to develop a model based on mutual interaction terms [5] to describe the influence of neighboring Si atoms and their substituents on ^{29}Si chemical shifts. As can be seen from Table 1, Si atoms that are two bonds away do not influence ^{29}Si shifts of $SiMe_3$ groups significantly (δ = 16 ± 1 ppm for $X_nMe_{3-n}SiSiMe_2SiMe_3$) but this is no longer the case for X_3Si groups (X = halogen). For instance δ ($SiBr_3$) values for $Br_3SiSiMe_2SiMe_3$ and $Br_3SiSiMe_3SiBr_3$ differ by 13.5 ppm. The influence of the substituents on neighboring Si atoms can be quite large, as can be seen from $\delta(SiMe_2)$ of $(F_3Si)_2SiMe_2$ (δ = –60,2 ppm) and $(Br_3Si)_2SiMe_2$ (δ = –19,4 ppm). $^{29}Si^{29}Si$ one bond and two bond coupling constants ($^1J, ^2J$) are equally helpful for the same purposes. Generally $^1J(SiSi)$ increases with increasing electronegativities of the substituents, but there is no linear dependence on the sum of the electronegativities [6].

	X	δ(Si)	δ(*Si)	δ(**Si)	¹J(Si*Si)	¹J(*Si**Si)	²J(Si**Si)
XMe₂Si*SiMe₂**SiMe₃	H	−37.1	−48.1	−16.0			
	F	37.5	−50.3	−16.5			
	Cl	26.3	−45.8	−16.3			
	Ph	−18.6	−48.2	−15.8			
X₂MeSi*SiMe₂**SiMe₃	H	−64.6	−47.6	−16.0	68.3	75.3	7.9
	F	17.4	−52.0	−16.7	99.4	75.1	9.4
	Cl	37.8	−39.9	−16.2	83.1	75.6	11.2
	Br	26.9	−36.8	−16.2	74.6	75.4	11.8
	I	−22.4	−34.7	−16.0	64.4	74.7	12.2
	OMe	−5.0	−52.6	−16.6	106.0	74.2	5.9
	Ph	−18.9	−47.9	−15.6	73.5	72.7	7.6
XMeSi(*SiMe₃)₂	H	−72.7	−15.0		71.1		
	F	36.3	−19.3		79.6		
	Cl	9.9	−15.0		74.5		
	Br	2.4	−15.2		72.1		
	I	−22.8	−15.4		69.0		
	Ph	−46.2	−15.8		72.4		
X₂Si(*SiMe₃)₂	H	−104.6	−14.0		69.3		
	F	31.3	−21.3		86.9		
	Cl	34.7	−11.3		75.6		
	Br	23.0	−10.2		70.1		
	I	−28.6	−10.8		62.4		
	OMe	24.8	−17.1		86.1		
	Ph	−38.7	−16.0		71.3		
(XMe₂Si)₂*SiMe₂	H	−36.8	−47.5		72.3		
	F	36.9	−53.7		85.9		
	Cl	25.0	−43.9		81.5		

	X	δ(Si)	δ(*Si)	δ(**Si)	$^1J(Si^*Si)$	$^1J(^*Si^{**}Si)$	$^2J(Si^{**}Si)$
(XMe$_2$Si)$_2$*SiMe$_2$	Br	19.6	−42.6		78.7		
	I	−1.8	−41.6		74.6		
	OMe	18.7	−53.7		85.1		
	Ph	−18.5	−47.9		72.7		
(XMe$_2$Si)$_2$*SiMeX	H	−40.2	−76.4		70.4		
	F	30.2	24.2		96.8		
	Cl	19.8	−0.3		89.3		
	Br	11.9	−11.0		84.4		
	I	−12.7	−38.9		76.8		
	OMe	14.0	+6.5		94.9		
	Ph	−18.6	−46.4		72.1		
(X$_3$Si)$_2$*SiMe$_2$	F	−59.7	−60.2				
	Cl	13.0	−25.7		121.1		
	Br	−16.5	−19.4		105.7		
	I	−135.2	−33.9		82.2		
	OMe	−61.5	−40.2		143.7		
	Ph	−16.6	−45.7		74.4		
(XMe$_2$Si)$_2$*SiX$_2$	Cl	17.9	14.1		98.1		
	Br	9.6	−3.0		89.8		
	Ph	−19.1	−40.0		71.4		
X$_3$Si*SiMe$_2$**SiMe$_3$	Cl	21.1	−30.9	−15.9			
	Br	−3.0	−23.8	−15.5			
	Ph	−15.3	−47.5	−16.5	73.8	73.0	7.3

Table 1. ^{29}Si chemical shifts [ppm, TMS] and ^{29}Si^{29}Si coupling constants of methylated trisilanes X$_n$Si$_3$Me$_{8-n}$

Vibrational Spectra and Rotational Isomerism

With the help of a local symmetry force field for the methyl groups, the vibrational spectra of all trisilanes were assigned using normal coordinate analyses. As can be seen from Table 2, the calculated SiSi-force constants f(SiSi) increase with increasing sum of the electronegativities of the substituents. There seems to be a rather linear relationship between ^1J(SiSi) and f(SiSi) (Fig. 1), as is also observed for ^1J(CC) and f(CC) [7]. One of our goals therefore is to prepare Si compounds with either very large or very small ($J < 0$) coupling constants or force constants.

Molecule	f [Nm^{-1}]	Molecule	f [Nm^{-1}]
Si_3Me_8	150	$Me_2Si(SiCl_3)_2$	200
$F_2Si(SiMe_3)_2$	163	$Me_2Si(SiBr_3)_2$	190
$MeFSi(SiMe_2F)_2$	170	$Me_2Si(SiOMe_3)_2$	200

Table 2. Selected SiSi valence force constants of methylated trisilanes

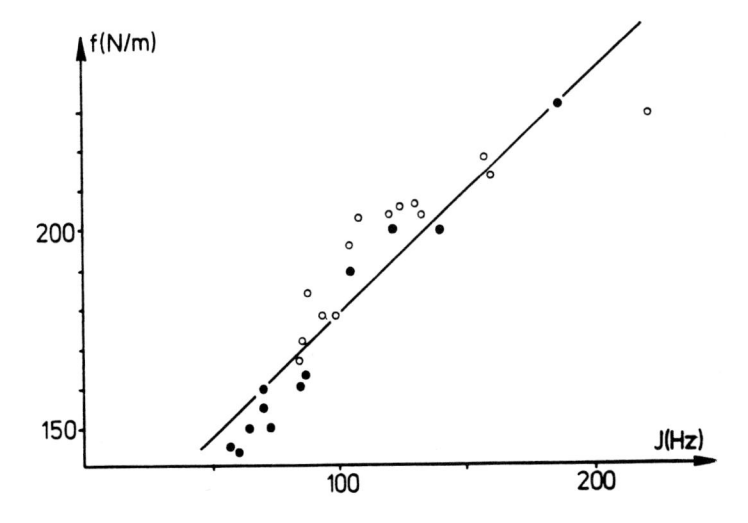

Fig. 1. Correlation between J(SiSi) and f(SiSi) for a large number of silicon compounds, including trisilanes; f(SiSi) calculated by NCA (●), f(SiSi) estimated by interpolation (O)

Many of the trisilanes exist as mixtures of rotational rotamers, as can be proved with variable temperature Raman spectroscopy. As an example Fig. 2 shows the v(SiCl) region of the Raman spectrum of $ClMe_2SiSiMe_2SiMe_3$, which exists as a mixture of *anti*- and *gauche*- rotamers at room temperature.

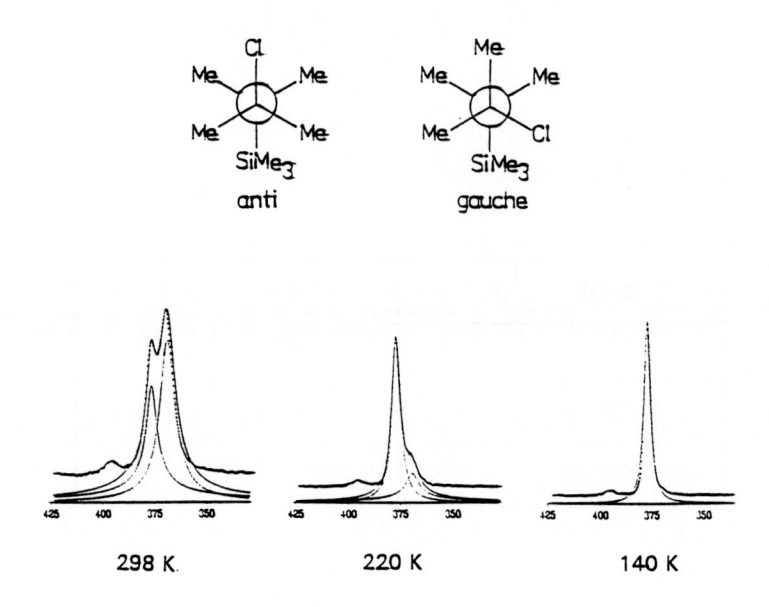

Fig. 2. Rotational isomerism and temperature dependence of v(SiCl) of $ClMe_2SiSiMe_2SiMe_3$; the more stable isomer is *anti*

Acknowledgement: One of the authors (K. S.) wishes to express her thanks to the *Deutsche Akademie der Naturforscher Leopoldina* for financial support.

References:

[1]   K. Hassler, G. Bauer, *Spectrochim. Acta* **1987**, *43A*, 1325.
[2]   K. Hassler, R. Neuböck, *Spectrochim. Acta* **1993**, *49A*, 95.
[3]   K. Schenzel, K. Hassler, *Spectrochim. Acta* **1994**, *50A*, 127.
[4]   K. Schenzel, K. Hassler, to be published.
[5]   M. Vongehr, H. C. Marsmann, *Z. Naturforsch.* **1976**, *31b*, 1423.
[6]   K. Hassler, G. Bauer, *J. Organomet. Chem.* **1993**, *460*, 149.
[7]   K. Kamienska-Trela, *Spectrochim. Acta* **1979**, *36A*, 239.

New Ways to Difunctional Cyclosilanes

Andreas Spielberger, Peter Gspaltl, Edwin Hengge*
Institut für Anorganische Chemie
Technische Universität Graz
Stremayrgasse 16, A-8010 Graz, Austria

Summary: Mixtures of structural isomers of permethylated dichlorocyclohexasilanes and disubstituted cyclopentasilanes can easily be separated after hydrolysis. The syntheses and a crystal structure are discussed.

Introduction

Difunctional cyclosilanes, which are useful monomers for the synthesis of oligocyclic and cage like silanes, are obtained after chlorination of dodecamethylcyclohexasilane with $SbCl_5$ in CCl_4 [1, 2] or with $Me_3SiCl/AlCl_3$ [3], respectively.

Syntheses

Scheme 1 shows the hydrolysis and subsequent separation of 1,3- and 1,4-dichloro-decamethylcyclohexasilane, which are produced by chlorination with $SbCl_5$. The dichloro isomers can easily be regained by chlorination of the products with acetylchloride. Under acidic conditions the hydrolyzed products give oligomeric and polymeric structures, which are subject of the ongoing investigations.

Scheme 1. ● = SiMe$_{2-n}$ (n = 0, 1)

The isomeric mixture of 1-chloro-2-(chlorodimethylsilyl)octamethyl-cyclopentasilane and 1-chloro-3-(chlorodimethylsilyl)octamethylcyclopentasilane, which are formed in the reaction with Me$_3$SiCl/AlCl$_3$, behaves quite similarly. In this case decamethyl-2-oxa-1,3,4,5,6,7-hexasilanorbornane is formed (Fig. 1).

Fig. 1. ● = SiMe$_{2-n}$ (n = 0, 1)

X-Ray Structure

A single crystal containing a mixture of decamethyl-7-oxa-1,2,3,4,5,6-hexasilanorbornane and 1,4-dihydroxydecamethylcyclohexasilane in a 2:1 ratio was used to determine the molecular structure. Some interesting data of decamethyl-7-oxa-1,2,3,4,5,6-hexasilanorbornane are shown in Fig. 2 and listed in Table 1.

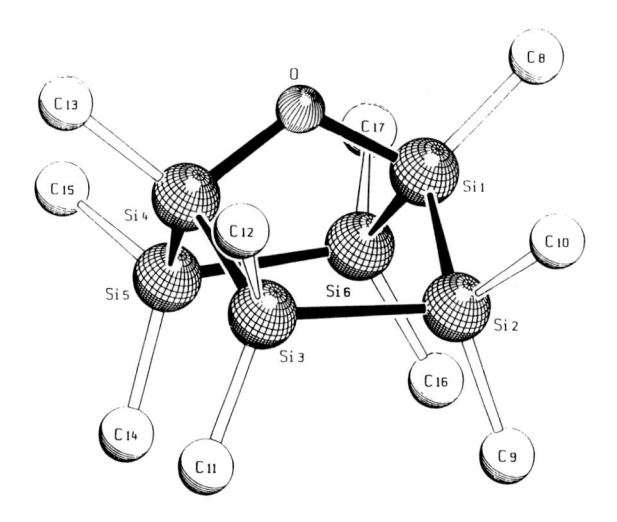

Fig. 2.

Bonds	Angle [°]	Atoms	Distance [nm]
Si2–Si1–Si6	116.81(0.05)	Si1–Si2	237.3(0.1)
Si1–Si2–Si3	96.13(0.05)	Si1–O	169.7(0.2)
Si1–O–Si4	116.38(0.11)	Si1–C8	186.7(0.3)
Si2–Si1–C8	114.15(0.12)	Si2–Si3	235.9(0.2)
C9–Si2–C10	110.12(0.24)	Si2–C9	188.4(0.4)

Table 1. Characteristic bond lengths and angles of decamethyl-7-oxa-1,2,3,4,5,6-hexasilanorbornane

Acknowledgement: The authors are grateful to the *Fonds zur Förderung der wissenschaftlichen Forschung* (Wien) for the financial support of this work.

References:

[1] W. Wojnowski, B. Dreczewski, *Angew. Chem.* **1985**, *97*, 978; *Angew. Chem., Int. Ed. Engl.* **1985**, *24*, 992.

[2] E. Hengge, M. Eibl, *J. Organomet. Chem.* **1992**, *428*, 335.

[3] T. J. Drahnak, R. West, J. C. Calabrese, *J. Organomet. Chem.* **1980**, *198*, 55.

New Polycyclic Si-Ring Systems

Peter Gspaltl, Andreas Spielberger, Edwin Hengge*
Institut für Anorganische Chemie
Technische Universität Graz
Stremayrgasse 16, A-8010 Graz, Austria

Summary: 1,4-Di(undecamethylcyclohexasilanyl)benzene has been synthesized and its X-ray structure determined. The new cyclohexasilanes Ph–(*cyclo*-Si_6Me_{10})–X (X = CF_3SO_3, H, Cl, Br) have been prepared. The coupling reaction between Ph–(*cyclo*-Si_6Me_{10})–H and (*t*Bu)$_2$Hg resulted in the bicyclic Si-system (Ph–*cyclo*-Si_6Me_{10})$_2$Hg, which is easily converted to (Ph–*cyclo*-Si_6Me_{10})$_2$ by UV-irradiation.

Introduction

Methylated polycyclic Si-ring systems, bearing phenyl substituents, in particular 1,4-di(cylosilanyl)benzenes, should be able to form stable radical anions. This offers the possibility to investigate the electronic situation by ESR. Now we are able to report on the syntheses of the first representatives of this new class of compounds.

Syntheses

'One-pot *in situ*' Grignard reaction [1] of monohalocyclosilanes (*cyclo*-$Me_{11}Si_6$–X; X = F, Cl, Br) with *p*-dibromobenzene and Mg affords 1,4-di(undecamethylcyclohexasilanyl)benzene in very poor yields depending on X. Bi(undecamethycyclohexasilanyl), the separation of which requires chromatographic methods, is formed as the main product by transmetallation of the halosilane. Formation of 1,4-di(undecamethylcyclohexasilanyl)benzene in suitable yields without any transmetallation products therefore is achieved by treating chloroundecamethylcyclohexasilane with BrMg–C_6H_4–MgBr in THF under sonication [2] (Scheme 1).

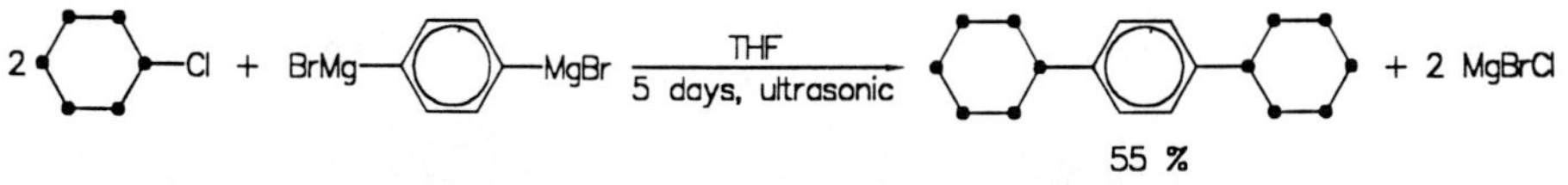

Scheme 1. • = SiMe$_{2-n}$ (n = 0, 1)

By treatment of 1,4-diphenyldecamethylcyclohexasilane with CF$_3$SO$_3$H one phenyl group is split off selectively. The reaction sequence outlined in Scheme 2 demonstrates the further steps to form new polycyclic Si-ring systems.

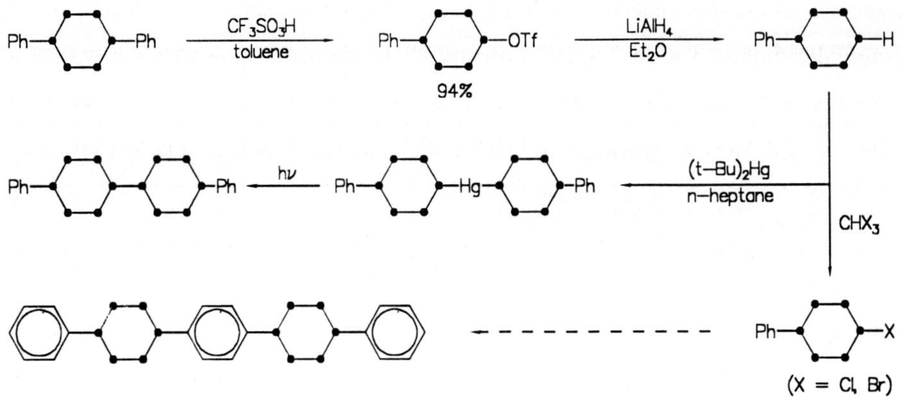

Scheme 2.

X-Ray Structure of 1,4-Di(undecamethylcyclohexasilanyl)benzene

The X-ray study (Fig. 1) revealed a triclinic unit cell containing three molecules, with $a = 10.088$ (12), $b = 10.216$ (10), $c = 41.507$ (54) Å, $\alpha = 83.48$ (9) °, $\beta = 86.77$ (8) °, $\gamma = 61.28$ (7) °. The benzene ring occupies the axial sites of the two cyclohexasilane chairs, whereas the two phenyl rings in 1,4-diphenyldecamethylcyclohexasilane are in equatorial position [3]. The shortest C–C distance between two CH$_3$-groups bonded to different rings of the same molecule is 3.827 (0.014) Å. Si–Si distances exhibit rather usual values between 2.350 (6) and 2.330 (6) Å. Interesting distances and angles are given in Table 1.

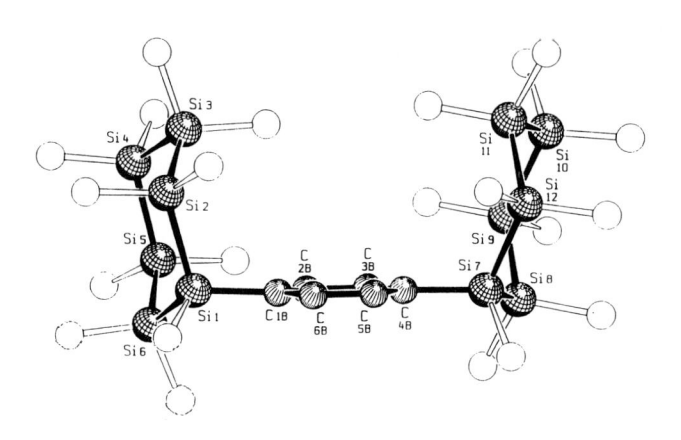

Fig. 1. X-ray structure of 1,4-Di(undecamethylcyclohexasilanyl)benzene

Atoms	Distance [Å]	Bonds	Angle [°]
Si(3)–Si(11)	7.390 (0.006)		
Si(5)–Si(9)	7.914 (0.006)	Si(6)–Si(1)–CB(1)	111.26 (0.43)
Si(1)–CB(1)	1.874 (0.011)	Si(8)–Si(7)–CB(4)	110.98 (0.40)
Si(7)–CB(4)	1.876 (0.011)	Si(2)–Si(1)–Si(7)–Si(12)	17.20 (0.23)
CB(1)–CB(2)	1.371 (0.015)	CB(6)–CB(1)–CB(2)	115.27 (1.01)
CB(2)–CB(3)	1.371 (0.014)	CB(1)–CB(2)–CB(3)	122.94 (1.06)
CB(3)–CB(4)	1.404 (0.014)	CB(2)–CB(3)–CB(4)	122.08 (1.02)
CB(4)–CB(5)	1.387 (0.015)	CB(3)–CB(4)–CB(5)	114.43 (0.97)
CB(5)–CB(6)	1.371 (0.014)	CB(4)–CB(5)–CB(6)	122.51 (1.04)
CB(6)–CB(1)	1.380 (0.015)	CB(5)–CB(6)–CB(1)	122.73 (1.06)

Table 1. Interatomic distances and angles

References:

[1] H. Bock, J. Meuret, K. Ruppert, *J. Organomet. Chem.* **1993**, *446*, 113.

[2] P. Gspaltl, A. Spielberger, E. Hengge, in preparation.

[3] K. Kumar, M. H. Litt, R. K. Chada, J. E. Drake, *Can. J. Chem.* **1987**, *65*, 437.

Preparation and Reactions of Permethylated Cyclooligosilane Alkali Metal Derivatives

F. Uhlig, P. Gspaltl, E. Pinter, E. Hengge*

Institut für Anorganische Chemie
Technische Universität Graz
Stremayrgasse 16, A-8010 Graz, Austria

Introduction

The highly reactive potassium undecamethylcyclohexasilane is a suitable starting material for the synthesis of several cyclic silanes. Potassium undecamethylcyclohexasilane **1** has been synthesized by two methods. Both methods have some disadvantages, either the use of dangerous, highly toxic solvents (HMPA) [1a] or reagents (Hg) [1b] and/or sequence of steps with a low yield. On the basis of these facts we tried to develop a simple new way to synthesize potassium undecamethylcyclohexasilane.

Synthesis

Potassium undecamethylcyclohexasilane is formed in the reaction of the permethylated cyclohexasilane with potassium *t*butanolate in glyme solvents at room temperature over a period of two weeks. This provides a simple access to **1**.

The observed reaction products can be explained by a complicated ring-chain-ring equilibrium (Scheme 1). The first step is an Si–Si bond cleavage followed by a methyl group shift with trimethyl(*t*butoxy) silane as a byproduct. The resulting potassium nonamethylcyclopentasilane reacts slowly to give **1**. The byproduct of this step is a polymeric silane which contains H, C, Si, and K. The theoretical yield is nearly 80 % of the starting cyclosilane. Practically, the reaction of **1**, e.g., with propylchloride, yield 65–75 % propylundecamethylcyclohexasilane.

Sodium and lithium *t*butanolates yields only small amounts (5–10 %) of the alkali metal undecamethylcyclohexasilane.

Conversions of decamethylcyclopentasilane and trimethylsilylnonamethylcyclopentasilane with potassium *t*butanolate lead only to **1** in the described way. Potassium nonamethylcyclopentasilane is only an intermediate which cannot be isolated.

Scheme 1. ^{29}Si NMR data of **1** (δ [ppm]): Si$_a$ = –109.1; Si$_b$ = –34.2; Si$_c$ = –40.9; Si$_d$ = –42.1; SiSi coupling constants are given in Hz: $^1J_{Si_a-Si_b}$ = n.f.; $^1J_{Si_b-Si_c}$ = 45.2; $^1J_{Si_c-Si_d}$ = 56.9; $^2J_{Si_a-Si_c}$ = 3.2; $^2J_{Si_b-Si_d}$ = 10.5

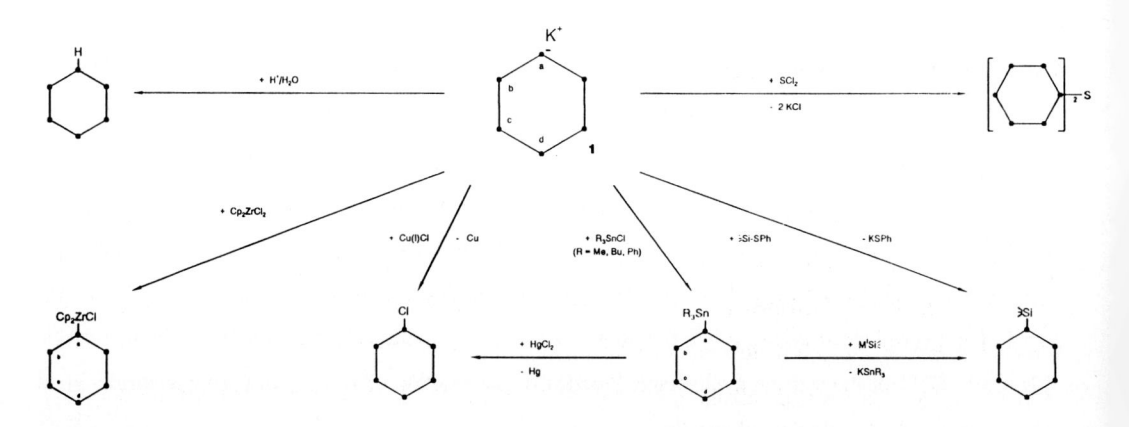

Scheme 2.

Synthesis of Group IV – Undecamethylcyclohexasilanederivatives: **1** reacts with zirconocenedichloride at −78 °C to give a dark red colored product (mp: 188–190 °C), which is the first group IV derivative of a cyclic silane. A second substitution step does not occur even in the presence of two equivalents of potassium undecamethylcyclohexasilane.

^{29}Si NMR (δ [ppm]):

$Si_a = -22.3$; $Si_b = -31.5$; $Si_c = -39.7$; $Si_d = -42.7$

SiSi coupling constants are given in Hz:

$^1J_{Si_a-Si_b} = 40.5$; $^1J_{Si_b-Si_c} = 57$; $^1J_{Si_c-Si_d} = 61$; $^2J_{Si_a-Si_c} = 2.0$; $^2J_{Si_b-Si_d} = 10.5$

The product was further characterized by ^{1}H, ^{13}C, IR, and mass spectroscopy.

Chloroundecamethylcyclohexasilane: Conversion of triorganostannylundecamethylcyclohexasilane with mercurydichloride gave in a simple way chloroundecamethylcyclohexasilane. The yield is nearly quantitative. In the case of the trimethylstannyl-derivative the purity of the product is very high. Copper(I)chloride reacts directly with **1** to the chloroundecamethylcyclohexasilane with bis(undecamethylcyclohexasilane) as the only byproduct (5–15 %).

Triorganostannylderivatives: **1** reacts with various triorganostannylchlorides to the new triorganostannylundecamethylcyclohexasilanes.

Formation of Silicon-Silicon Bonds

Si–Si bonds are formed when **1** is reacted with phenylthiooligosilanes. Alkali metal thiophenolates, which are byproducts in the reaction can be easily removed. Also reaction of triethyl- or tributylstannylundecamethylcyclohexasilane with alkali metal silanes leads to the formation of Si–Si bonds. While the reaction of phenylthiooligosilanes proceeds at room temperature the conversion of the tin compound requires temperatures below 0 °C to avoid decomposition of the product.

In contrast to the related alkali metal halide elimination reactions of organohalooligosilanes, transmetallations are never observed in these reactions. Examples are described in [2].

Acknowledgement: The authors thank the *Austrian Science Foundation* (Wien) for the financial support of this work.

References:

[1] A. L. Allred, R. T. Smart, D. A. van Beek, *Organometallics* **1992**, *11*, 4225; E. Hengge, P. K. Jenkner, *J. Organomet. Chem.* **1986**, *314*, 1; E. Hengge, P. K. Jenkner, P. Gspaltl, A. Spielberger, *Z. Anorg. Allg. Chem.* **1988**, *560*, 27.

[2] F. Uhlig, B. Stadelmann, A. Zechmann, P. Lassacher, H. Stüger, E. Hengge, *Phosphorus & Sulfur,* in press.

[3] M. Ishikawa, M. Kumada, *J. Chem. Soc., Chem. Commun.* **1969**, 567.

Synthesis, Reactivity, and Spectroscopy of Phenylated Cyclotetrasilanes and Cyclopentasilanes

Ulrich Pöschl, Harald Siegl, Karl Hassler*

Institut für Anorganische Chemie

Technische Universität Graz

Stremayrgasse 16, A-8010 Graz, Austria

Summary: Phenylated cyclopentasilanes $Si_5Ph_{10-n}X_n$ (X = H, F, Cl, Br, I; n = 1, 2) have been prepared by dearylation of Si_5Ph_{10} with triflic acid. Disubstitution gives high yields of the *trans*-1,3-isomer. The crystal structure of *trans*-1,3-difluorooctaphenylcyclopentasilane has been determined. Triflation of the novel educt (*p*-TolSiPh)$_4$ for the first time allows the selective monofunctionalization of a perarylated cyclotetrasilane. The synthesized compounds have been characterized by ^{29}Si, ^{19}F, and ^1H NMR spectroscopy (including Si–Si coupling constants) and will be used for the preparation of mixed halogenated cyclosilanes as well as for further NMR studies.

Introduction

The perphenylated cyclosilanes Si_4Ph_8, Si_5Ph_{10}, and Si_6Ph_{12} synthesized by F. S. Kipping in 1921 were the first cyclosilanes to be known at all [1]. Until a few years ago the reaction with hydrogen halides under pressure or with aluminum halides as catalysts was the only way to functionalize perarylated cyclosilanes. This kind of dearylation only allows the preparation of $(SiCl_2)_m$, $(SiBr_2)_m$, and $(SiI_2)_m$ with m = 4, 5, 6 as well as the preparation of $Si_5Ph_5I_5$. Otherwise the reaction products are complex mixtures of partially halogenated cyclosilanes with different degrees of substitution. Selective mono- and difunctionalisation of Si_5Ph_{10} by use of triflic acid, CF_3SO_3H (TfOH), was reported by Uhlig in 1989 and 1992 [2–4]. K. Matyjaszewski studied the reaction of Si_4Ph_8 with triflic acid but could not achieve selective monofunctionalisation [5].

For the synthesis and characterization of phenylated cyclosilanes with the formulae $Si_5Ph_{10-n}X_n$ and $Si_4Ph_{8-n}X_n$ (X = H, F, Cl, Br, I; n = 1, 2) we had three main incentives. As far as the twofold substituted cycles are concerned, we were interested in the synthesis and characterization of different constitutional and steric isomers. Furthermore we are going to use the arylcyclosilanes for the preparation of mixed halogenated cyclosilanes (e.g., $Si_5Cl_8F_2$ or Si_4Br_7F) as well as for the preparation of polycyclic silanes. Finally, we are planning to do detailed NMR studies like relaxation measurements and determination of coalescence temperatures and activation energies for the pseudorotation of cyclopentasilanes and the ring flip of cyclotetrasilanes.

Syntheses, Results and Discussion

All cyclopentasilanes investigated so far have been prepared from deca-phenylcyclopentasilane by dearylation with triflic acid (triflation) and subsequent reaction of the triflate compounds with lithium halides, potassium halides or lithium aluminum hydride. The reaction sequences are symbolized below (Scheme 1, phenyl groups are not shown).

Scheme 1.

The use of one equivalent of TfOH leads to Si_5Ph_9X (X = OTf, H, F, Cl, Br, I) in yields of more than 90 %. Spectroscopic data are given in Table 2. The use of 2 equivalents of TfOH leads to $Si_5Ph_8X_2$ and allows the isolation of the respective *trans*-1,3-isomers as crystalline substances in yields of up to 70 %. Fig. 1 shows the crystal structure of *trans*-1,3-difluorooctaphenylcyclopentasilane. The most significant bond lengths are given in Table 1. Complete results of the X-ray diffraction analysis as well as experimental details will be published in a subsequent paper. Triflation and fluorination of Si_5Ph_9F leads to the formation of all possible constitutional and steric isomers of $Si_5Ph_8F_2$ in practically equal amounts. Only 1,1-difluorooctaphenylcyclopentasilane was not detected. Among others this fact indicates that the high electronegativity of triflate groups is not alone decisive for the high stereoselectivity of dearylation by triflic acid.

Bond	Bond length [Å]	Bond	Bond length [Å]
F(1)–Si(1)	1.628 ± 0.003	Si(2)–C(7)	1.879 ± 0.005
F(2)–Si(3)	1.634 ± 0.003	Si(2)–C(13)	1.886 ± 0.005
Si(1)–Si(2)	2.377 ± 0.002	Si(3)–C(19)	1.872 ± 0.005
Si(2)–Si(3)	2.386 ± 0.002	Si(4)–C(25)	1.892 ± 0.005
Si(3)–Si(4)	2.363 ± 0.002	Si(4)–C(31)	1.881 ± 0.005
Si(4)–Si(5)	2.371 ± 0.002	Si(5)–C(37)	1.885 ± 0.005
Si(5)–Si(1)	2.372 ± 0.002	Si(5)–C(43)	1.885 ± 0.005
Si(1)–C(1)	1.873 ± 0.005		

Table 1. Bond lengths of *trans*-1,3-Si$_5$Ph$_8$F$_2$

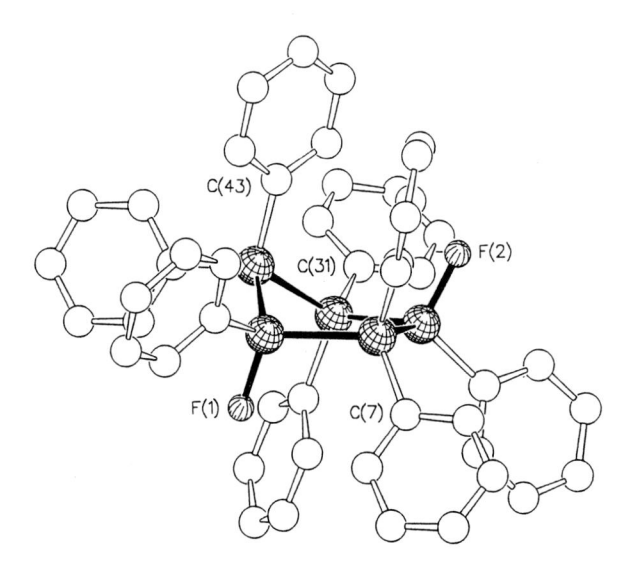

Fig. 1. X-ray structure of *trans*-1,3-Si$_5$Ph$_8$F$_2$ (hydrogen atoms are not shown)

Despite manifold variations of the reaction conditions we could not achieve selective monofunctionalization of Si$_4$Ph$_8$ by use of TfOH. In order to increase solubility and reactivity of the four-membered ring, we synthesized the mixed phenyl and *p*-tolyl substituted cyclotetrasilane (*p*-TolSiPh)$_4$. Due to this modification of the educt, which retains all the chemical properties required for the reactions and investigations further intended, selective monosubstitution of a perarylated cyclotetrasilane was rendered possible for the first time. Compared to phenyl groups, the Si–C bond in p-tolyl substituted silanes is well known to be significantly more reactive towards protonic acids as dearylating reagents [6]. Accordingly, the dearylation of (p-TolSiPh)$_4$ by use of one equivalent TfOH and subsequent reactions described above lead to Si$_4$Ph$_4$ p-Tol$_3$X in yields of over 95 %. The educt (p-TolSiPh)$_4$, prepared from p-TolPhSiCl$_2$ by ring closure with lithium, only gives rise to one sharp ^{29}Si NMR signal (–22.1 ppm referred to TMS, half band width 3 Hz), although it consists of four diastereomers in equal amounts (Scheme 2, phenyl groups are not shown, p-tolyl groups are indicated by simple vertical lines).

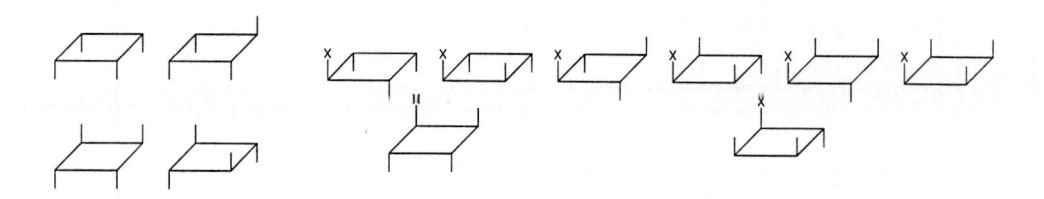

Scheme 2.   **Scheme 3.**

The products Si$_4$Ph$_4$ p-Tol$_3$X consist of 6 diastereomers and 2 enantiomers (Scheme 3). Accordingly, the corresponding NMR spectra show multiplet structures consisting of 6 sometimes overlapping lines with characteristic intensity ratios. Table 4 gives averaged values for the chemical shift. In particular the three groups of ^{29}Si signals of the triflate compound Si$_4$Ph$_4$ p-Tol$_3$OTf and the ^{19}F spectrum of Si$_4$Ph$_4$p-Tol$_3$F show six clearly separate lines exactly in the required intensity ratio 1:2:1:1:2:1 (Fig. 2).

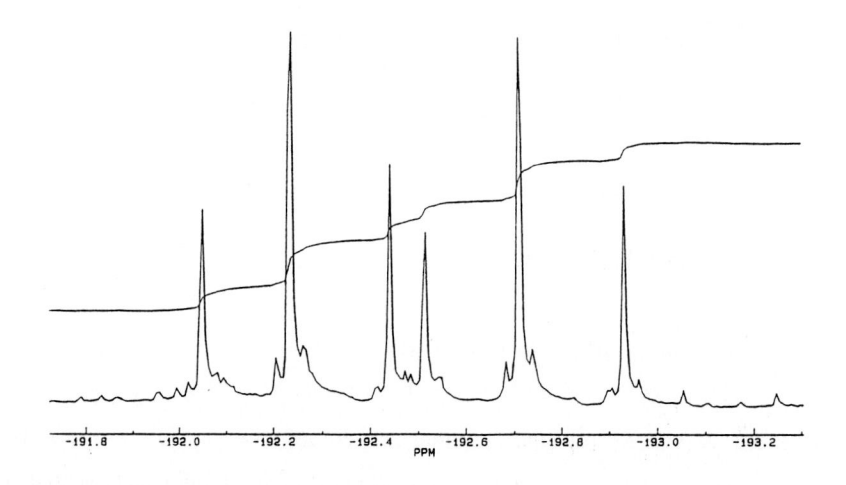

Fig. 2. ^{19}F spectrum of Si$_4$Ph$_4$p-Tol$_3$F

NMR Experiments

The NMR measurements were performed in C_6D_6, partly mixed with toluene in order to increase solubilities. Chemical shifts are referred to TMS or $CFCl_3$. All experiments were carried out on a Bruker 300 MSL spectrometer. ^{29}Si spectra were recorded at 59.6 MHz using INEPT and INEPT-INADEQUATE pulse sequences among others. ^{19}F Spectra were recorded at 282.4 MHz and ^1H spectra at 300.1 MHz.

X	$^{29}Si(1)$		$^{29}Si(2, 5)$		$^{29}Si(3, 4)$		$^1H/^{19}F$
	δ [ppm]	J [Hz]	δ [ppm]	J [Hz]	δ [ppm]	J [Hz]	δ [ppm]
OTf	33.3		−39.0		−35.0		
F	27.4	^1J(Si, F) 341.1	−41.3	^2J(Si, F) 15.6	−34.9		191.9
		^1J(Si, Si) 66.0		^1J(Si, Si) 66.0		^2J(Si, Si) 13.1	
		^2J(Si, Si) 13.1		^2J(Si, Si) 9.5		^2J(Si, Si) 9.5	
Cl	6.4		−37.1		−35.4		
Br	−5.5	^2J(Si, Si) 13.7	−36.8		−35.1	^2J(Si, Si) 13.7	
				^2J(Si, Si) 10.1		^2J(Si, Si) 10.1	
I	−29.4	^1J(Si, Si) 55.8	−37.0	^1J(Si, Si) 55.8	−35.0	^2J(Si, Si) 14.0	
		^2J(Si, Si) 14.0		^2J(Si, Si) 10.2		^2J(Si, Si) 10.2	
H	−60.7	^1J(Si, H) 173.5	−35.3	^1J(Si, Si) 60.0	−33.7	^1J(Si, Si) 60.0	5.69
				^2J(Si, Si) 10.6		^2J(Si, Si) 10.6	

Table 2. NMR data of Si_5Ph_8X (C_6D_6, 20°C, TMS or $CFCl_3$)

X	^{29}Si(1,3) δ [ppm]	^{29}Si(1,3) J [Hz]	^{29}Si(2) δ [ppm]	^{29}Si(2) J [Hz]	^{29}Si(4, 5) δ [ppm]	^{29}Si(4, 5) J [Hz]	^{1}H/^{19}F δ [ppm]
OTf	28.9		−40.8		−39.0		
F	23.8	^1J(Si, F) 342.1	−46.6	^2J(Si, F) 15.7	−40.5	^2J(Si, F) 17.5	195.4
		^3J(Si, F) 3.5		^1J(Si, F) 65.3		^1J(Si, F) 66.6	
		^1J(Si, Si) 66.1		^2J(Si, Si) 14.1		^2J(Si, Si) 14.1	
		^2J(Si, Si) 10.3				^2J(Si, Si) 10.3	
Cl	−0.5	^1J(Si, Si) 62.2	−38.2	^1J(Si, Si) 62.5	−37.2	^1J(Si, Si) 61.5	
		^2J(Si, Si) 11.9				^2J(Si, Si) 11.9	
				^2J(Si, Si) 11.5		^2J(Si, Si) 11.5	
Br	−7.8	^2J(Si, Si) 12.2	−37.9		−37.4	^2J(Si, Si) 12.2	
				^2J(Si, Si) 10.6		^2J(Si, Si) 10.6	
I	−31.2	^1J(Si, Si) 56.8	−38.4	^1J(Si, Si) 56.9	−38.1	^1J(Si, Si) 56.6	
		^2J(Si, Si) 12.8				^2J(Si, Si) 12.8	
				^2J(Si, Si) 9.8		^2J(Si, Si) 9.8	
H	−61.2	^1J(Si, H) 176.1	−36.1	^2J(Si, Si) 8.4	−34.0	^2J(Si, Si) 8.4	5.48

Table 3. NMR data of *trans*-1,3-Si$_5$Ph$_8$X$_2$ (C$_6$D$_6$, 20°C, TMS or CFCl$_3$)

X	^{29}Si(1) δ [ppm]	^{29}Si(1) J [Hz]	^{29}Si(2, 4) δ [ppm]	^{29}Si(2, 4) J [Hz]	^{29}Si(3) δ [ppm]	^{29}Si(3) J [Hz]	^{1}H/^{19}F δ [ppm]
OTf	33.6		−21.3		−30.3		
F	33.3	^{1}J(Si, F) 369.3 ^{1}J(Si, Si) 63.5	−20.7	^{2}J(Si, F) 16.6 ^{1}J(Si, Si) 63.5	−29.5		192.3
Cl	10.8		−19.3		−24.2		
Br	2.4		−20.5		−23.7		
I	−21.6		−22.1		−23.9		
H	−54.4	^{1}J(Si, H) 170.2	−22.4		−23.2		

Table 4. NMR data of Si$_4$Ph$_4$ p-Tol$_3$X (C$_6$D$_6$, 20°C, TMS or CFCl$_3$)

Acknowledgement: The authors thank the *Fonds zur Förderung der wissenschaftlichen Forschung* (Wien) for financial support.

References:

[1] F. S. Kipping, H. E. Sands, *J. Chem. Soc.* **1921**, *119*, 830.
[2] W. Uhlig, A. Tzschach, *J. Organomet. Chem.* **1989**, *378*, C1.
[3] W. Uhlig, C. Tretner, *J. Organomet. Chem.* **1992**, *436*, C1.
[4] W. Uhlig, *Chem. Ber.* **1992**, *125*, 47.
[5] K. Matyjaszewski, *Organometallics* **1992**, *11*, 3257.
[6] G. Pollhammer, *Ph. D. Thesis*, Technische Universität Graz, **1986**.

Oligosilanylsulfanes: New Functional Derivatives of Higher Silicon Hydrides

Harald Stüger*, Paul Lassacher

Institut für Anorganische Chemie

Technische Universität Graz

Stremayrgasse 16, A-8010 Graz, Austria

Summary: α,ω-Bis(organylthio)-derivatives of linear oligosilanes RS–SiH$_2$(SiH$_2$)$_n$SiH$_2$–SR (n = 0, 1, 2; R = Me, Ph, 2-Naphthyl, SiPh$_3$) are formed in the reaction of the corresponding α,ω-dibromooligosilanes with suitable alkali metal thiolates or with HSSiPh$_3$/Et$_3$N without destruction of the oligosilane skeleton. Product distribution and yields, however, are strongly influenced by the nucleophilic strength of the sulfur nucleophile. The structures proposed for the oligosilanylsulfanes thus prepared are proved by ^{29}Si, ^{1}H NMR, and MS investigations.

Introduction

Just three silyl compounds, bis(disilanyl)sulfane, disilanylsilyl sulfane, and the heterocycle 1,4-dithiacyclohexasilane, are known which contain the –H$_2$Si–H$_2$Si–S– unit. They can be prepared from H$_3$SiSiH$_2$I and Hg$_2$S [1] or by redistribution reactions of appropriate disilanyl chlorides and H$_3$SiSSiH$_3$ [2]. Recently it has been shown that the reaction of halofunctional higher silicon hydrides with amines or alkali metal amides can be taken for the synthesis of oligosilanylamines [3]. Using a similar approach now we succeeded in elaborating a more systematic access to *S*-functional derivatives of higher silicon hydrides by nucleophilic displacement of the halo-substituents in suitable halooligosilane precursors.

Synthesis

When mild reaction conditions are applied (–40 °C, hydrocarbon solvent), only the X-groups in α,ω-dihalooligosilanes $XSiH_2(SiH_2)_nSiH_2X$ (X = Cl, Br; n = 0, 1, 2) are substituted by suitable sulfur nucleophiles. The oligosilane skeleton remains intact and the corresponding α,ω-bis(organylthio)-derivatives are obtained in excellent yields.

The course of the reaction, however, is strongly governed by the nucleophilic strength of the sulfur component. Thus, α,ω-dibromooligosilanes and $HSSiPh_3/Et_3N$ react to give the corresponding α,ω-bis(organylthio)-derivatives, whereas the use of the stronger nucleophilic system $HSPh/Et_3N$ causes extended Si–Si and Si–H bond scission (reactions **A** and **B** in Scheme 1). The reaction products form a complex mixture of various silylsulfanes, which cannot be separated. No reaction at all takes place between simple sulfanes and α,ω-dibromooligosilanes under the conditions applied.

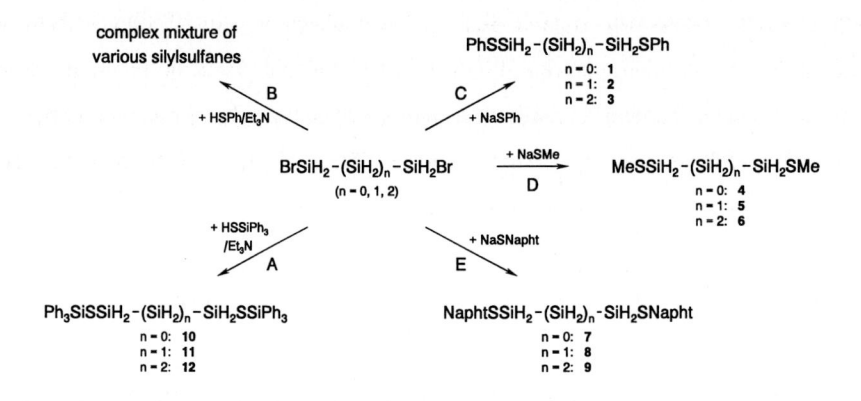

Scheme 1. Reactions of α,ω-dibromooligosilanes with selected sulfur nucleophiles

When alkali metal thiolates are employed, the auxiliary base Et_3N, which is likely to catalyze Si–Si bond cleavage and skeletal rearrangement reactions in organosilicon chemistry [4], can be omitted, and the desired oligosilanylsulfanes again are obtained in excellent yields (reactions **C**, **D** and **E** in Scheme 1). In this case, the course of the reaction is remarkably insensitive towards the nature of the R-groups on sulfur. Neither with simple alkyl groups nor with bulky substituents like 2-naphthyl significant side reactions occur.

Properties

All substances have been characterized by common spectroscopic techniques like MS, IR, ^{1}H, and ^{29}Si NMR and by elemental analysis. As can be expected, ^{29}Si NMR spectroscopy turned out to be particularly useful to prove the proposed structures of the oligosilanylsulfanes **1–12**. Selected ^{29}Si parameters are depicted in Table 1.

R	n	Compound	$\delta\ ^{29}$Si(1)	$\delta\ ^{29}$Si(2)	^1J(Si(1), H)	^1J(Si(2), H)
Phenyl	0	1	−45.7	–	216	–
	1	2	−43.5	−107.1	216	198
	2	3	−42.1	−109.1	215	197
Methyl	0	4	−45.5	–	210	–
	1	5	−42.1	−109.1	209	193
	2	6	−41.7	−111.1	210	192
2–Naphthyl	0	7	−45.6	–	216	–
	1	8	−43.7	−106.3	216	196
	2	9	−42.2	−108.5	214	198
Ph$_3$Si	0	10	−52.8	–	213	–
	1	11	−50.6	−99.8	216	196
	2	12	−50.2	−101.9	215	197

Table 1. ^{29}Si chemical shifts δ and coupling constants ^1J(Si–H) of α,ω-bis(organylthio)oligosilanes in C$_6$D$_6$ solution vs external TMS; RSSSi(1)H$_2$–(Si(2)H$_2$)$_n$–Si(1)H$_2$SR

Due to the large number of hydrogen atoms in the molecules rather complex splitting patterns are obtained because of ^{29}Si^{1}H couplings, which allow an unequivocal structure assignment in each case. As an illustrative example the coupled ^{29}Si NMR spectrum of 1,3-bis(triphenylsilylthio)trisilane is shown in Fig. 1.

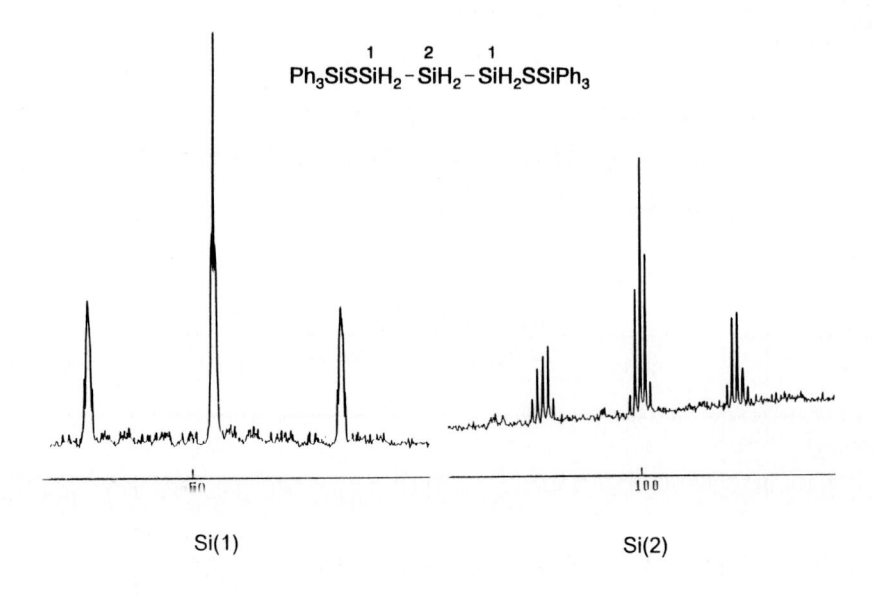

Fig. 1. Coupled ^{29}Si NMR spectrum of 1,3-bis(triphenylsilylthio)trisilane in C$_6$D$_6$ solution vs; TMS in the range –40 to –120 ppm

The broad, poorly resolved Ph$_3$Si signal appears outside the measuring range in Fig. 1 near 1 ppm. The residual lines can easily be assigned to Si(1) and Si(2). Both signals are split into triplets by couplings to the directly bonded hydrogens. The Si(1) resonance lines are further split into triplets by long range couplings to the hydrogen atoms on Si(2). The hyperfine structure of the Si(2) triplet exhibits quintett splitting of each line due to ^{29}Si^{1}H couplings to the 4 magnetically equivalent hydrogen substituents on the adjacent silicon atoms.

General properties of oligosilanylsulfanes selectively may be altered by variation of the R-groups on sulfur. The methyl derivatives, for instance, are colorless liquids, which can be distilled without decomposition, while the 2-naphthyl substituted compounds are non volatile crystalline solids. If R = Ph or Ph$_3$Si, the reaction products are obtained as viscous oils, which cannot be purified further because of their thermal and hydrolytic lability.

Acknowledgement: This work was supported by the *Fonds zur Förderung der wissenschaftlichen Forschung* (Wien). The authors are grateful to *Wacker-Chemie GmbH*, for the supply of organochlorosilane starting materials.

References:

[1] L. G. L. Ward, A. G. MacDiarmid, *J. Inorg. Nucl. Chem.* **1961**, *21*, 287.
[2] A. Haas, R. Süllentrup, C. Krüger, *Z. Anorg. Allg. Chem.* **1993**, *619*, 819.
[3] H. Stüger, P. Lassacher, *Monatsh. Chem.* **1994**, *125*, 615.
[4] W. Raml, E. Hengge, *Monatsh. Chem.* **1980**, *111*, 29.

Synthesis and Reactions of Iminosilanes

Dorothee Großkopf, Uwe Klingebiel*, Jörg Niesmann

Institut für Anorganische Chemie

Georg-August-Universität Göttingen

Tammannstr. 4, D-37077 Göttingen, Germany

Summary: Lithiated aminofluorosilanes can often be regarded as LiF adducts of iminosilanes on account of their structures [1]. They are suitable precursors to iminosilanes after fluorine-chlorine exchange followed by elimination of lithium chloride [1, 2]. This synthesis shows the following features: (a) the method is a general one in the sense that it can be applied to differently substituted aminosilanes; (b) the iminosilanes are obtained, either as THF adducts or as free imines in good yield; (c) the tedious and sometimes complicated synthesis of aminochlorosilanes as precursors is avoided, since the fluorine-chlorine exchange takes place in one step in the lithium derivative.

Lithium Derivatives of Aminofluorosilanes

Essentially three factors determine the structural make-up of the lithium derivatives [1]:

1) A common feature is that the Lewis acid lithium prefers to bind the Lewis base fluorine.
2) The structure is influenced by the presence or absence of solvent bases which can coordinate to the lithium.
3) The decreasing basicity of the nitrogen in the series of lithium derivatives with R = alkyl, silyl, aryl finally leads to rupture of the Li-N bond.

e.g., crystal structure of

$$\begin{array}{c} \text{CMe}_3 \quad\quad \text{CMe}_3 \\ | \quad\quad\quad | \\ \text{Me}_3\text{C} - \underset{\underset{\underset{(\text{THF})_3}{\text{Li}}}{|}}{\underset{|}{\text{Si}}} = \text{N} - \text{Si} - \text{Ph} \\ \quad\quad\quad\quad | \\ \quad\quad\quad\quad \text{CMe}_3 \end{array}$$

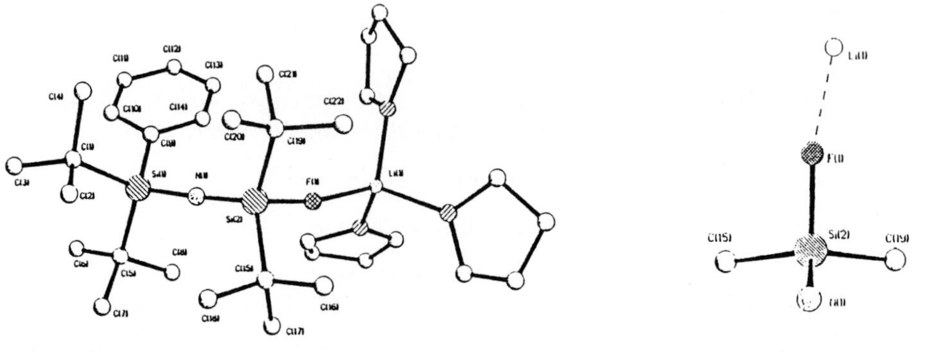

Fig 1. Si(1)–N(1): 165.2 pm   Si(1)–N(1)–Si(2): 176.3 °
Si(2)–N(1): 160.8 pm   Σ < C$_2$ Si(2) N(1): 344.4 °
Si(1)–F(1): 109.2 pm
Li(1)–F(1): 184.5 pm
δ^{29}Si [ppm]: −17.60, SiPh, d, $^3J_{SiF}$ = 6.3 Hz
            − 5.05, SiF, d, $^1J_{SiF}$ = 126.9 Hz

This molecule may best be considered as an LiF adduct of an iminosilane.

Synthesis of Iminosilanes

Lithiated aminochlorosilanes are obtained in reactions of lithiated aminofluorosilanes with Me$_3$SiCl by a fluorine-chlorine exchange. LiCl is easily eliminated, and monomeric or dimeric iminosilanes are formed.

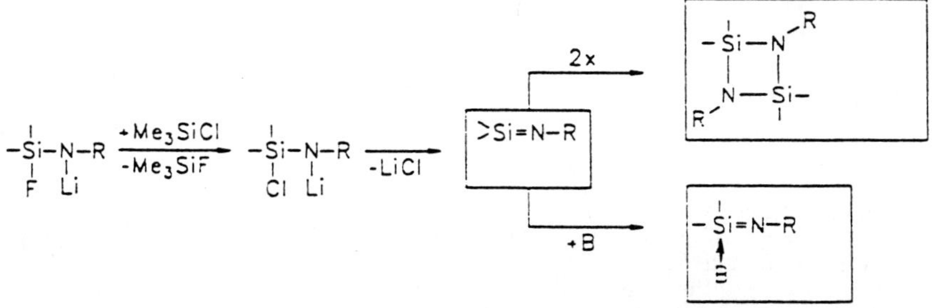

Scheme 1.

Our monomeric iminosilanes **1** and **2** are remarkably stable. They could be separated from LiCl by distillation, **1** [2] without losing THF and **2** [3] as the free iminosilane.

Fig. 2. 1: Me–Si(CMe₃)(CMe₃)–N=Si(CMe₃)(CMe₃)·THF

1: δ^{29}Si [ppm]: –14.57 SiMe
3.60 SiO
bp: 65 °C/0.01 mbar

Fig. 3. 2: Ph–Si(CMe₃)(CMe₃)–N=Si(CMe₃)₂

2: δ^{29}Si [ppm]: –13.33 SiPh
80.43 Si=N
bp: 108 °C/0.01 mbar

In the ^{13}C NMR spectrum of **1**, the C(2)–O-signal of the THF molecule appears at 73.78 ppm – a downfield shift of nearly 6 ppm compared with free THF. This reflects the strong Lewis acid character of the three-coordinated silicon atom.

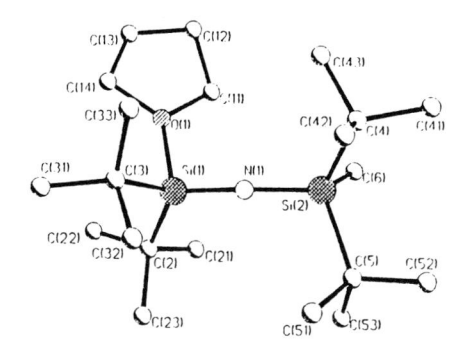

Fig. 4. Crystal structure of **1**:
Si(1)–N1): 159.6 pm
Si(2)–N(1): 166.1 pm
Si(1)–O(1): 190.2 pm
Si(1)–N(1)–Si(2): 174.3 °
Σ < C₂ Si(1) N(1): 347.7

Reactions of Iminosilanes

These iminosilanes react with H-acid compounds like water and ethanol by insertion into the O–H bond, with acetone and 2,3-dimethyl-1,3-butadiene to give alkenes, and with methacrolein, ethylvinylether, silyl and aryl azides to give cycloadducts.

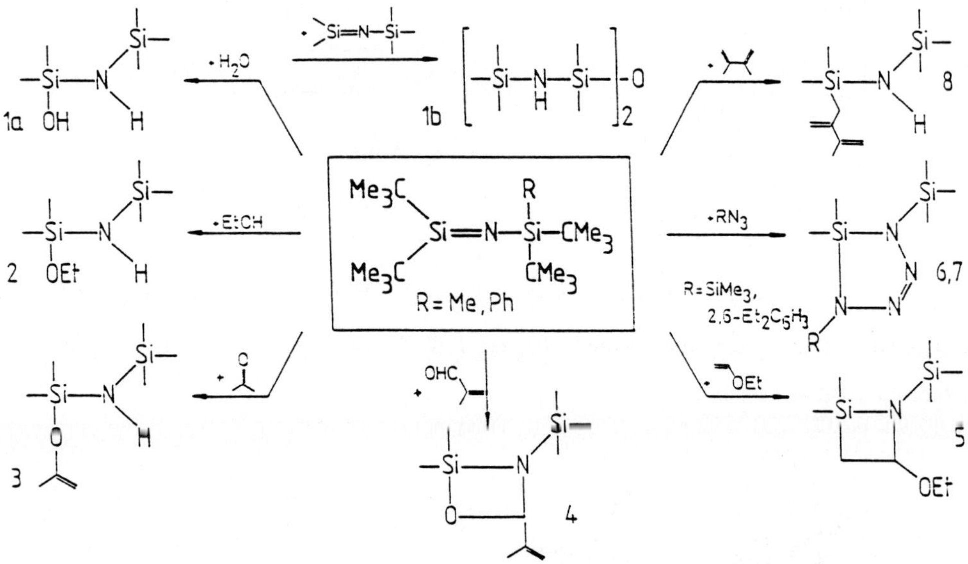

Scheme 2.

	1a	1b	2	3
δ^{29}Si [ppm]	−3.27 SiO	−10.10 SiO	−7.99 SiO	−8.39 SiO
	0.25 SiPh	−1.06 SiPh	0.60 SiPh	0.93 SiPh

	4	5	6	7	8
δ^{29}Si [ppm]	−6.30 SiPh	−3.82 SiPh	2.36 SiMe	−0.42	−0.78 SiPh
	27.82 SiO	25.88 SiC	8.62 11.26	1.80	5.47 SiC

Table 1.

Fig. 5. Crystal structure of **5**
N(1)–Si(1): 178.0 pm
N(1)–Si(2): 175.2 pm
Si(1)–C(1): 187.9 pm
N(1)–C(2): 147.3 pm
Si(1)–N(1)–Si(2): 144.6 °
C(1)–Si(1)–N(1): 79.5 °

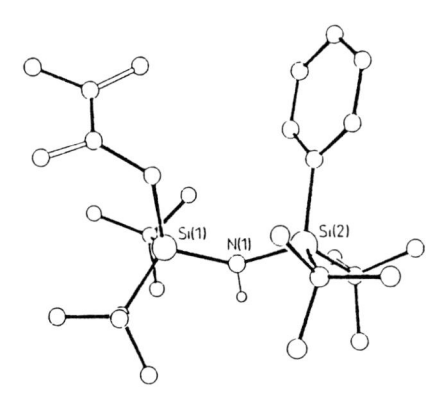

Fig. 6. Crystal structure of **7**
N(1)–Si(1): 178.9 pm
N(1)–Si(2): 179.8 pm
N(2)–N(3): 124.8 pm
Si(1)–N(1)–Si(2): 149.3 °
N(1)–Si(1)–N(4): 87.7 °

Fig. 7. Crystal structure of **8**
Si(1)–N(1): 174.9 pm
N(1)–Si(2): 174.7 pm
Si(1)–C(6): 190.8 pm
Si(2)–N(1)–Si(1): 147.15 °

Acknowledgement: We thank the *Deutsche Forschungsgemeinschaft* and the *Fonds der Chemischen Industrie* for financial support.

References:

[1] S. Walter, U. Klingebiel, *Coord. Chem. Rev.* **1994**, *130*, 481.
[2] S. Walter, U. Klingebiel, D. Schmidt-Bäse, *J. Organomet. Chem.* **1991**, *412*, 319.
[3] D. Großkopf, L. Marcus, U. Klingebiel, M. Noltemeyer, to be published.

1,3-Diazabutadienes and Amidinates as Ligands to Silicon Centers

Hans Heinz Karsch*, Fritz Bienlein

Anorganisches-Chemisches Institut

Technische Universität München

Lichtenbergstr. 4, D-85747 Garching, Germany

Summary: A novel strategy for the access to dichlorosilylene $SiCl_2$ and its incorporation into silaheterocycles has been worked out successfully by performing the Benkeser reaction in the presence of dienes and heterodienes. For example, a 1,3-diazabutadiene is used to prepare the five membered heterocycle $t\overline{BuN-C(Ph)=N-C(CF_3)_2-\dot{S}iCl_2}$, the molecular structure of which has been determined by X-ray diffraction. It turned out to be isomorphous to the germanium analog, obtained from $GeCl_2$ and the respective 1,3-diazabutadiene. The silacycle reacts with water to give the first silanol with pentacoordinated silicon and an amidinate ligand, thus demonstrating the "Umpolung" of the 1,3-diazabutadiene on complexation. The X-ray structure determination of the silanol reveals very short ligand to silicon bonds. Obviously, amidinate ligands are ideal for the stabilization of high coordination numbers at silicon. Nevertheless, an amidinate, prepared from PhCN and LiNHtBu in reaction with $SiCl_4$ (2:1) does not give a bisamidinate silicon complex with hexacoordinated silicon, but rather leads to the novel amidine derivative [H(tBu)N−C(Ph)=N]$_2$SiCl$_2$ (X-ray). In order to extend the amidinate ligand series to mixed N/P systems, [Ph−C(PSiMe$_3$) (NSiMe$_3$)]Li was synthesized from PhCN and LiP(SiMe$_3$)$_2$. Its reaction with Me$_3$SiCl gives a novel access to a phosphaalkene, for example (Me$_3$Si)P=C(Ph)[N(SiMe$_3$)$_2$].

Introduction

Silylenes are electron deficient, in most cases transient species. Only a few stable derivatives are known so far [1]. For syntheses employing silylenes, they have to be generated in situ. For this purpose alkyl- and aryl-silylenes can be prepared, e.g., from 7-silanorbornadiene compounds by thermal or photochemical treatment [2, 3]. Dimethylsilylene can be generated by heating of hexamethylsilirane [4] or photolysis of dodecamethylcyclohexasilane [5]. Another elegant route to SiMe$_2$ is the reduction of bis-[bis(P,P´(bisdimethylphosphino)(trimethylsilyl)methanido]dimethylsilane **1** with Ni(COD)$_2$ [6] (Eq. 1).

Eq. 1. Reaction of **1** with Ni(COD)$_2$

With the exception of this latter reaction type, which is also applicable to the generation of SiCl$_2$, dichlorsilylene can be generated only under drastic conditions [7], not allowing following reactions to be well controlled.

Results

It is well established that silicochloroform is easily deprotonated by amines to give the highly nucleophilic anion SiCl$_3^-$ [8] (Eq. 2).

$$HSiCl_3 + NR_3 \longrightarrow (HNR_3)(SiCl_3)$$

Eq. 2.

Bis(trifluoromethyl) substituted 1,3-diazabutadienes are highly electrophilic on the one hand and on the other hand, they add readily to carbene analogues via [4, 1] cycloaddition to give five membered heterocycles [9]. Regarding these facts, we treated a mixture of HSiCl$_3$/DBU with 1-*t*butyl-2-phenyl-4,4-bistrifluoromethyl-1,3-diazabuta-1,3-diene **2** to prepare the silacycle **3** [10] (Eq. 3).

Eq. 3.

Compound **3** is isomorphous to the corresponding germacycle **4**, which is obtained from **2** and $GeCl_2$. Thus, at least with electrophilic heterodienes, the $HSiCl_3$/DBU system may serve as a silylene synthon.

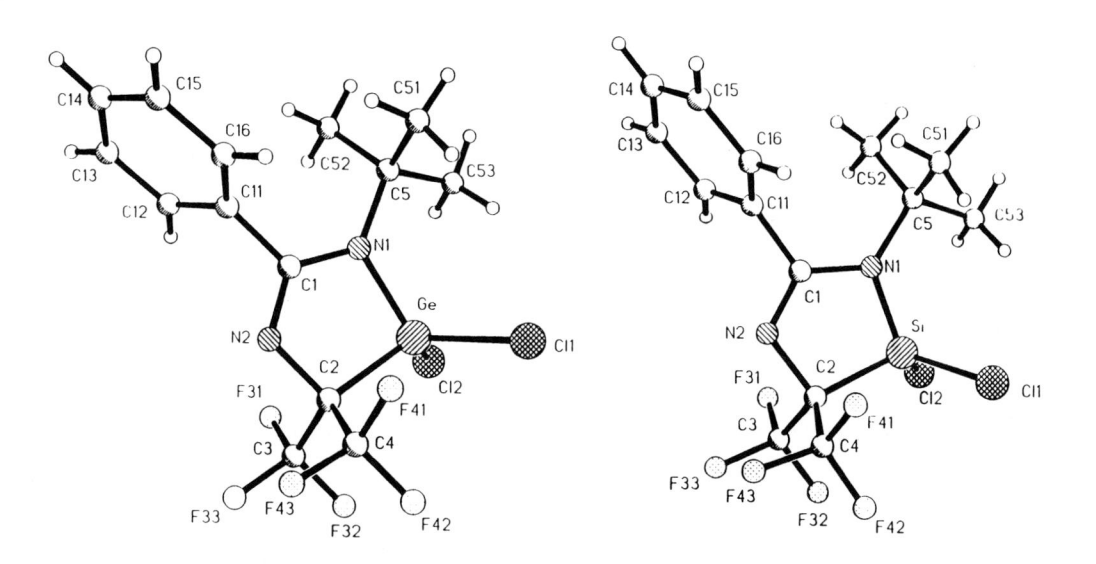

Fig. 1. Molecular structure of **3** and **4**

The five membered ring skeletons of **3** and **4** are perfectly planar. The bond lengths of C(1)–N(2) are comparable to double bonds C=N, the distances C(1)–N(1) correspond to slightly shortened single bonds. The planes of the phenylsubstituents are nearly orthogonal to those of the heterocycles.

The addition of water to **3** reveals the nucleophilic nature of the bis(trifluoromethyl) substituted carbon in **3**, in contrast to its reactivity in the parent diazabutadiene **2** ("Umpolung") (Eq. 4).

Eq. 4.

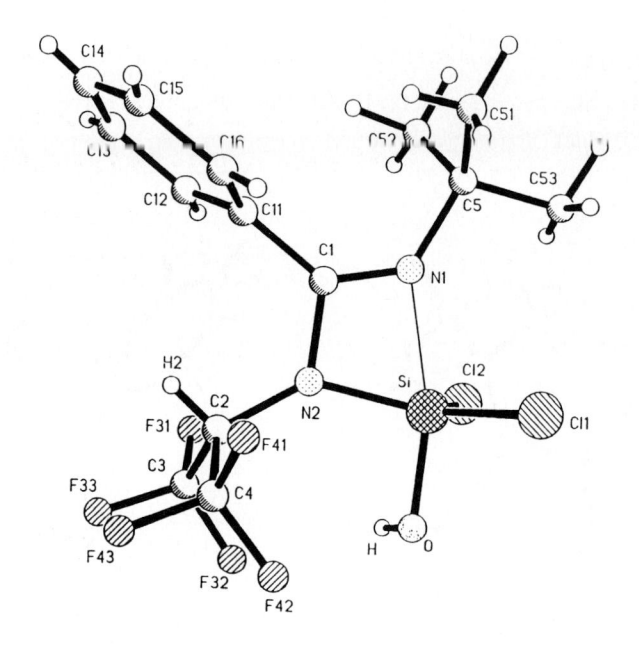

Fig. 2. Molecular structure of **5**

A single X-ray structure determination reveals **5** (Fig. 2) to be the first silanol with pentacoordinated silicon and the first stable dichlorosilanol derivative. Compound **5** is monomeric in the solid state and there is no tendency to the formation of H-bridges or elimination of HCl. The silicon center adopts a distorted trigonal bipyramidal arrangement. The four membered ring (SiNCN) is planar. All silicon to element bonds are relatively short, compared with other compounds with pentacoordinated silicon [11]. In particular, the Si–O bond is very short (1.61 Å), the Si–N bonds are at the lower limit for hypervalent silicon compounds (1.79/1.97 Å), suggesting that amidinate ligands may stabilize silicon in high coordination states.

As an approach to unsymmetrical amidinate ligands, we reacted benzonitrile with LiNH(*t*Bu). The ligand H(*t*Bu)N–C(Ph)=NLi **6**, reacts with $SiCl_4$ (2:1) to give compound **7** (Eq. 5).

$$2 \text{ (}^t\text{Bu)HN-C(Ph)=NLi} + \text{SiCl}_4 \xrightarrow{-2 \text{ LiCl}} \text{Cl}_2\text{Si[N=C(Ph)-NH(}^t\text{Bu)]}_2$$

$$ \mathbf{6} \mathbf{7}$$

Eq. 5.

As shown by X-ray structure determination (Fig. 3), **7** does not contain a hexacoordinated silicon atom stabilized by amidinate ligands but rather may be regarded as an amidine derivative with a tetracoordinated silicon atom.

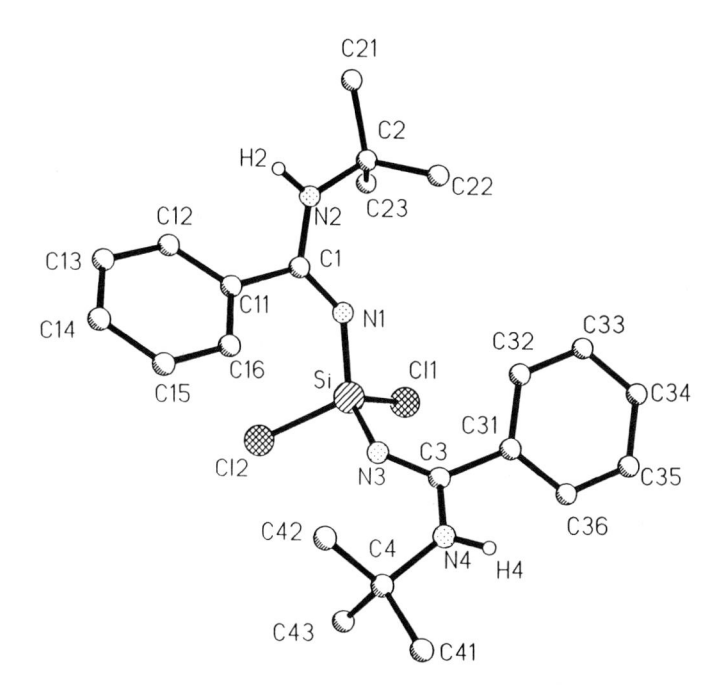

Fig. 3. Molecular structure of **7**

The bond lengths C(1)–N(1) (1.287 Å) and C(3)–N(3) (1.291 Å) correspond to standard C=N double bonds, the distances C(3)–N(4) (1.346 Å) and C(1)–N(2) (1.348 Å) indicate shortened single bonds. The other bond lengths and angles are within the expected range.

In order to extend the series of 1,3-diheteroallyl anions and following the strategy of Sanger [12] we synthesized the novel ligand [(Ph)C(NSiMe$_3$)(PSiMe$_3$)]Li **8** from PhCN and LiP(SiMe$_3$)$_2$ (Eq. 6).

$$PhCN + LiP(SiMe_3)_2 \longrightarrow [(Ph)C(NSiMe_3)(PSiMe_3)]Li$$
$$\mathbf{8}$$

Eq. 6.

The reaction of **8** with Me$_3$SiCl leads to the formation of E/Z isomers (1:1) of the phosphaalkene **9** (Eq. 7), thus affording a novel access to phosphaalkenes.

$$[(Ph)C(NSiMe_3)(PSiMe_3)]Li + Me_3SiCl \longrightarrow \underset{Me_3Si \quad SiMe_3}{\underset{N}{Ph}}C=P\text{\large\textasciitilde}SiMe_3$$
$$\mathbf{8} \qquad\qquad\qquad\qquad\qquad\qquad\qquad\qquad \mathbf{9}$$

Eq. 7.

Conclusions

The system HSiCl$_3$/DBU may be used as a SiCl$_2$ synthon in the presence of electrophilic heterobutadienes. The resulting heterocycles are not only precursors for fluorine substituted imidazoles [13], but also may be transformed to amidinate complexes with silicon in a pentacoordinated state. Monophosphorous analogues to amidinates are synthons for silyl-substituted phosphaalkenes.

Further Perspectives

The silicochloroform/amine route may be extended to a wide range of (hetero)butadienes to give a convenient access to novel silaheterocycles. The "Umpolung" in these systems will be exploited in organic synthesis. Reduction of these heterocycles possibly will lead to novel silicon(II) species. Amidinate ligands seem to be ideal candidates to promote high coordination numbers in main group chemistry. Silanols with pentacoordinated silicon may serve as model system for the amine catalyzed polysiloxane synthesis. Their potential for the generation of a variety of unusual species, e. g., donor stabilized silanones, silaimines etc. will be tested as well in the near future.

References:

[1] P. Jutzi, D. Kanne, C. Krüger, *Angew. Chem.* **1986**, *98*, 163; *Angew. Chem., Int. Ed. Engl.* **1986**, *25*, 164; D. B. Poranik, M. J. Fink, *J. Am. Chem. Soc.* **1989**, *111*, 595; H. H. Karsch, U. Keller, S. Gamper, G. Müller, *Angew. Chem.* **1990**, *102*, 297; *Angew. Chem., Int. Ed. Engl.* **1990**, *29*, 295; M. Denk, R. Lenon, R. Hayashi, R. West, A.-V. Belyakov, H.-P. Verne, A. Haaland, M. Wagner, N. Metzler, *J. Am. Chem. Soc.* **1994**, *116*, 2691.

[2] H. Gilman, S. G. Cottis, W. H. Atwell, *J. Am. Chem. Soc.* **1964**, *86*, 1596.

[3] H. Gilman, S. G. Cottis, W. A. Atwell, *J. Am. Chem. Soc.* **1964**, *86*, 5584.

[4] D. Seyferth, D. C. Anarelli, D. Duncan, *Organometallics* **1982**, *1*, 1288.

[5] G. Levin, P. K. Das, C. L. Lee, *Organometallics* **1988**, *7*, 1288.

[6] H. H. Karsch, in: *Organosilicon Chemistry – From Molekules to Materials* (Eds.: N. Auner, J. Weis), VCH, Weinheim, **1994**, p. 95.

[7] W. A. Atwell, D. R. Weyenberg, *Angew. Chem.* **1985**, *81*, 485; *Angew. Chem., Int. Ed. Engl.* **1985**, *24*, 488; P. P. Gaspar, in: *Reactive Intermediates, Vol 2* (Eds.: M. Jones Jr., R. A. Moss), Wiley, Chichester, **1981**, p. 335; G. Rabe, J. Michl, *Chem. Rev.* **1985**, *85*, 419.

[8] R. A. Benkeser, *Acc. Chem. Res.* **1974**, 94.

[9] K. Burger, U. Waßmuth, S. Penzinger, *J. Fluorine Chem.* **1982**, *20*, 813; H. H. Karsch, F. Bienlein, M. Paul, K. Burger, *Z. Naturforsch.*, to be submitted; K. Burger, K. Geith, N. Sewald, *J. Fluorine Chem.* **1990**, *46*, 105.

[10] H. H. Karsch, F. Bienlein, A. Sladek, M. Heckel, K. Burger, *J. Am. Chem. Soc.*, to be submitted.

[11] C. Chrit, R. J. P. Corriu, C. Raye, J. C. Young, *Chem Rev.* **1993**, *93*, 1371.

[12] A. R. Sanger, *Inorg. Nucl. Chem. Letters* **1973**, *9*, 351.

[13] K. Burger, K. Geith, N. Sewald, *J. Fluorine Chem.* **1990**, *46*, 105.

Specific Insertion of O=C=S into the N–Si Bond of an Extremely Electron-rich Molecule as Evident from the X-Ray Crystal Structure

A. Ehlend, H.-D. Hausen, W. Kaim*

Institut für Anorganische Chemie

Universität Stuttgart

Pfaffenwaldring 55, D-70550 Stuttgart, Germany

Summary: 1,4-Bis(trimethylsilyl)-1,4-dihydropyrazine **1**, a very electron rich silylamine with eight conjugated π electrons in the planar six-membered ring, reacts with the unsymmetrical heterocumulene O=C=S to yield an O-silyl substituted thiourethane as the only mono-insertion product. Si-O affinity and the stronger π-acceptor character of C=S vs C=O are probably responsible for the formation of the isomer with the N–C(=S)–OSiMe$_3$ function. The π-acceptor effect of C=S causes full planarity at the heterocyclic nitrogen atom whereas the N–SiMe$_3$ group at the other end of the molecule shows a small pyramidalization. The heterocycle proper bends into a slight boat conformation.

Introduction

1,4-Bis(trimethylsilyl)-1,4-dihydropyrazine **1** and related species are unique in that they are thermally quite stable molecules which contain a planar six-membered heterocycle with eight cyclically conjugated π electrons [1, 2]. In addition to the "hetero-antiaromatic" structure [3] these molecules are distinguished by being extremely electron rich as evident from ionization potentials in the gas phase (ca. 6 eV) or oxidation potentials in solution (ca. –0.4 V vs SCE) [2, 3]. It is not surprising, therefore, that the typical reactivity of **1** and derivatives under non-hydrolyzing conditions involves (outer-sphere) electron transfer to acceptors such as TCNQ, TCNE [4], O$_2$, or C$_{60}$ [5].

A second type of reaction is the exchange of the SiMe$_3$ substituents by SiPh$_3$ or BMes$_2$ which requires elevated temperatures and an ion-stabilizing solvent such as acetonitrile [5, 6]. Obviously, this is a polar reaction which does not proceed via electron transfer.

A third kind of reactivity is related to the silyl substitution mechanism in that it involves an intermediate N–Si bond cleavage. Insertion of heterocumulenes such as CO_2 is common for N–M bonds [7], including the $-NSiR_3$ function in **1** or other molecules [8, 9]. We now report the result of the reaction between **1** and the unsymmetrical carbonyl sulfide, O=C=S.

Results and Discussion

Insertion of unsymmetrical heterocumulenes such as O=C=X (X = S, NR), into an N–Si bond can produce two different constitutional isomers [8]. In the case of compound **1** and X = NR, R = alkyl or aryl, there is rapid insertion into *both* N–Si functions and the resulting situation is extremely complex with respect to the configurational (O,N-silylation) and conformational isomerism [5, 8, 10]. In contrast, O=C=S inserts rapidly into *only one* N–Si bond of **1**, there is no evidence for any second insertion even after several days. The formation of small amounts of pyrazine as oxidation product even under strictly anaerobic conditions suggests some electron transfer reactivity. Spectroscopic [10] and especially the structural evidence given below show that of the two conceivable constitutional isomers [11] of the insertion reaction only the O–Si bonded species **2** with intact thiocarbonyl function is formed.

Crystals of **2** for structure analysis were obtained by slow cooling of an *n*hexane solution. The results of the X-ray structure analysis are presented in Tables 1–3 [12].

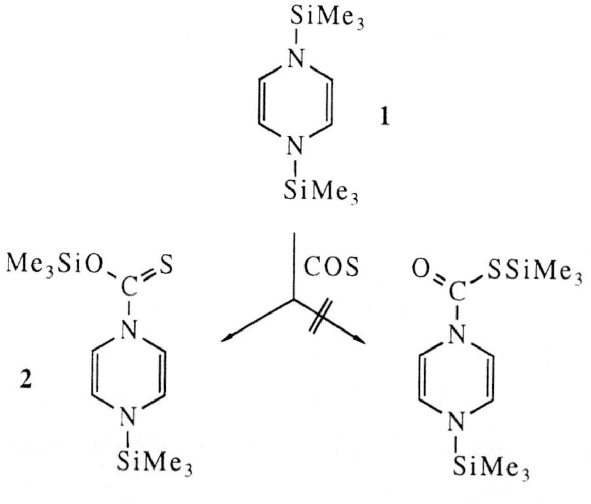

Scheme 1.

Atom	x	y	z	U(eq)[a]
Si(1)	1131(1)	6008(1)	1816(1)	305(2)
C(11)	739(3)	4512(2)	1531(2)	405(6)
C(12)	−214(3)	7309(2)	858(2)	395(6)
C(13)	4008(3)	6059(2)	1601(2)	493(7)
N(1)	−24(2)	6140(1)	3314(1)	325(4)
C(1)	−2014(3)	5815(2)	3845(1)	318(5)
C(2)	−3224(3)	6199(2)	4831(2)	323(5)
N(2)	−2570(2)	7007(1)	5412(1)	273(4)
C(3)	−425(3)	7199(2)	4980(2)	315(5)
C(4)	708(3)	6800(2)	3988(2)	334(5)
C(5)	−3903(3)	7562(1)	6289(1)	284(4)
S	−6737(1)	7383(1)	6803(1)	415(2)
O	−2939(2)	8271(1)	6718(1)	323(4)
Si(2)	−4031(1)	9437(1)	7511(1)	279(1)
C(21)	−5642(4)	10655(2)	6615(2)	514(7)
C(22)	−1630(3)	9944(2)	7682(2)	464(7)
C(23)	−5578(4)	8898(3)	8983(2)	580(9)

Table 1. Atomic coordinates (x10^4) and equivalent isotropic displacement coefficients (pm^2); [a] Equivalent isotropic U defined as one third of the trace of the orthogonalized U$_{ij}$ tensor

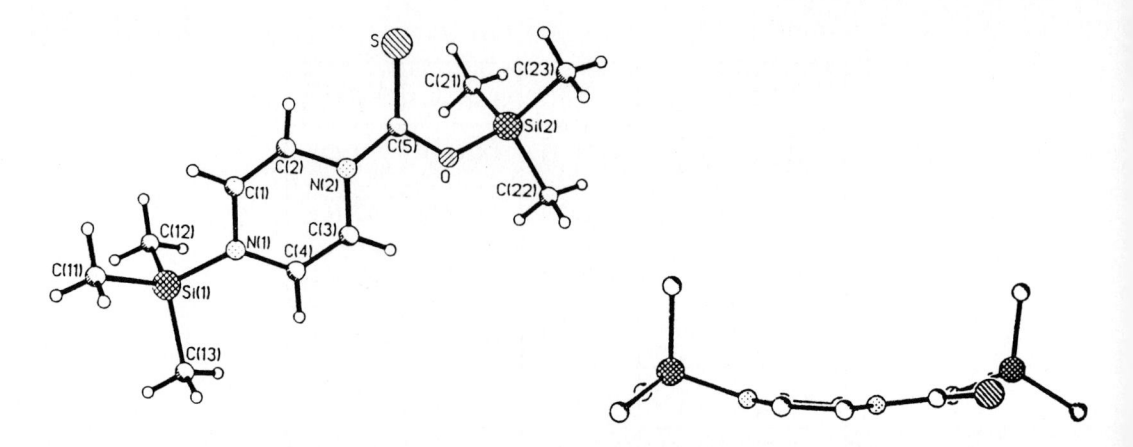

Fig. 1a: Molecular structure of 2; molecular structure and atomic numbering of compound 2

Fig. 1b: Molecular structure of 2; side view of the ring system

In the crystal, the molecules **2** adopt an almost planar arrangement of the non-methyl group atoms with a *syn* position of the O–Si bond to the thioamide function. There is an obvious deviation into the direction of a boat conformation (cf. below) which is in contrast to the very slight chair arrangement in **1** [1].

Bond A–B	Bond Length [pm]	Bond A–B	Bond Length [pm]
Si(1)–C(11)	185.2(2)	N(2)–C(5)	134.3(2)
Si(1)–C(12)	185.4(2)	C(3)–C(4)	132.7(2)
Si(1)–C(13)	185.4(2)	C(5)–S	167.1(2)
Si(1)–N(1)	176.4(1)	C(5)–O	134.2(2)
N(1)–C(1)	139.9(2)	O–Si(2)	168.5(1)
N(1)–C(4)	139.8(3)	Si(2)–C(21)	185.0(2)
C(1)–C(2)	133.0(2)	Si(2)–C(22)	184.7(3)
C(2)–N(2)	141.8(3)	Si(2)–C(23)	184.9(2)
N(2)–C(3)	142.7(2)		

Table 2. Bond lengths in **2**

Angle A–B–C	Angle [°]	Angle A–B–C	Angle [°]
C(11)–Si(1)–C(12)	110.5(1)	C(11)–Si(1)–C(13)	111.0(1)
C(12)–Si(1)–C(13)	111.1(1)	C(11)–Si(1)–N(1)	107.8(1)
C(12)–Si(1)–N(1)	108.8(1)	C(13)–Si(1)–N(1)	107.6(1)
Si(1)–N(1)–C(1)	121.7(1)	Si(1)–N(1)–C(4)	123.4(1)
C(1)–N(1)–C(4)	113.6(1)	N(1)–C(1)–C(2)	124.1(2)
C(1)–C(2)–N(2)	121.0(2)	C(2)–N(2)–C(3)	115.1(1)
C(2)–N(2)–C(5)	121.8(1)	C(3)–N(2)–C(5)	123.1(2)
N(2)–C(3)–C(4)	120.6(2)	N(1)–C(4)–C(3)	124.5(2)
N(2)–C(5)–S	124.6(1)	N(2)–C(5)–O	111.6(1)
S–C(5)–O	123.8(1)	C(5)–O–Si(2)	129.3(1)
O–Si(2)–C(21)	108.4(1)	O–Si(2)–C(22)	101.4(1)
C(21)–Si(2)–C(22)	111.5(1)	O–Si(2)–C(23)	112.0(1)
C(21)–Si(2)–C(23)	112.8(1)	C(22)–Si(2)–C(23)	110.1(1)

Table 3. Bond angles in **2**

All bond lengths and angles of compound **2** are within the usual range [1, 2], the distances between the ring members are compatible with the localized formulation of **2**. However, small but significant effects are visible on inspection of dihedral angles between planes of selected portions of the molecule, such as the following:

N(1)–C(1)–C(2)–N(2) (plane A)
N(2)–C(3)–C(4)–N(1) (plane B)
C(4)–N(1)–C(1)   (plane C)
C(2)–N(2)–C(3)   (plane D)
C(1)–C(2)–C(3)–C(4) (plane E)

The boat conformation of the six-membered ring in **2** may be quantified by the dihedral angle $\alpha(A/B) = 171.0$ °, indicating a small [1] but non-negligible deviation from planarity. The situation at the two different ends of the molecule is apparent from comparable "bow" angles $\beta(C/E) = 174.6$ ° and $\beta(D/E) = 171.3$ °. These values are both larger than the 178.7 ° determined for the starting material **1** [1]. There is also a small difference for the configuration at the nitrogen centers. The presence of N–SiMe$_3$ and N–C(=S)OSiMe$_3$ functions in the same molecule indicates by comparison that the unusually strong [12] thiocarbonyl π-acceptor group causes full planarity at N(2) ($\Sigma = 360$ °) in contrast to the non-π type acceptor SiMe$_3$ which leads to $\Sigma = 358.7$ ° (359.4 ° in **1** [1]). The particularly strong interaction between the thiocarbonyl π-acceptor and the very π-electron-rich diene-amine donor function of 1,4-dihydropyrazine [2–4] is also responsible for the rather short "single" bond N(2)–C(5) of 134.3 pm. The double bond C(5)–S, on the other hand, lies in the expected range.

It is interesting to note that the formation of the first C(=S)OSiMe$_3$ group precludes a facile second insertion of O=C=S into the other N–Si bond of the molecule **2**. In contrast to the behaviour of isocyanates [5, 8, 10] which always yield unsymmetrical bis(insertion) products with O- *and* N-silylation, the straightforward situation described here may be discussed with regard to the very strong π-electron withdrawing effect of the C=S group [13].

In summary, the insertion of O=C=S into the Si–N bond of reactive **1** proceeds to give only one apparently rather stable primary product which owes its stability to the matching of thiocarbonyl π-acceptor and dihydropyrazine π-donor capacities.

Acknowledgement: This work was supported by *Deutsche Forschungsgemeinschaft* and *Fonds der Chemischen Industrie*.

References:

[1] H.-D. Hausen, O. Mundt, W. Kaim, *J. Organomet. Chem.* **1985**, *296*, 321.

[2] J. Baumgarten, C. Bessenbacher, W. Kaim, T. Stahl, *J. Am. Chem. Soc.* **1989**, *111*, 2126; *J. Am. Chem. Soc.* **1989**, *111*, 5017.

[3] W. Kaim, *J. Am. Chem. Soc.* **1983**, *105*, 707.

[4] W. Kaim, *Angew. Chem.* **1984**, *96*, 609; *Angew. Chem., Int. Ed. Engl.* **1984**, *23*, 613.

[5] A. Lichtblau, *Ph.D. Thesis*, Universität Stuttgart, **1994**.

[6] A. Lichtblau, H.-D. Hausen, W. Schwarz, W. Kaim, *Inorg. Chem.* **1993**, *32*, 73.

[7] M.-F. Lappert, B. Prokai, *Adv. Organomet. Chem.* **1967**, *8*, 243.

[8] W. Kaim, A. Lichtblau, T. Stahl, E. Wissing, in: *Organosilicon Chemistry – From Molecules to Materials* (Eds.: N. Auner, J. Weis), VCH, Weinheim, **1994**, p. 41.

[9] W. Fink, *Chem. Ber.* **1964**, *97*, 1433.

[10] A. Lichtblau, A. Ehlend, W. Kaim, manuscript in preparation.

[11] H. Schumann, P. Jutzi, M. Schmidt, *Angew. Chem.* **1965**, *77*, 812; *Angew. Chem., Int. Ed. Engl.* **1965**, *4*, 869.

[12] Crystal data for **2**: $C_{11}H_{22}N_2OSSi_2$, M = 286.56 g/mol, triclinic ($P\bar{1}$), a = 655.0(1) pm, b = 1121.0(2) pm, c = 1171.9(2) pm; α = 78.86(1) °, β = 76.60(1) °, γ = 77.59(1) °. N = 808.1(2) x 10^6 pm^3, Z = 2, ρ_{calc} = 1.178 g cm^{-3}, $\mu(MoK_\alpha)$ = 0.327 mm^{-1}. 4723 Reflections (4708 independent, 3 ° $\leq 2\Theta \leq$ 60 °) were collected at –100 °C on a crystal of 0.4x0.3x0.1 mm, 3799 reflections with F > 4 σ(F) were used in the refinement (full matrix), R = 0.039, R_w = 0.049, GOF = 0.63. Difference Fourier syntheses gave the hydrogen positions of the 1,4-diazine ring which were used in the refinement with isotropic temperature factors. Hydrogen atoms of the methyl groups were calculated with ideal geometry and C–H distances of 96 pm and were refined using the riding model. Further information on the structure determination may be obtained from Fachinformationszentrum Karlsruhe GmbH, D-76344 Eggenstein-Leopoldshafen, Germany, on quoting the depository number CSD 58368, the names of the authors, and the journal citation.

[13] A. Modelli, D. Jones, S. Rossini, G. Distefano, *Tetrahedron* **1984**, *40*, 3257.

Carbene and Carbenoid Chemistry of Silyldiazoacetic Esters: The Silyl Group as a Substituent and a Functional Group

Gerhard Maas

Fachbereich Chemie

Universität Kaiserslautern

Erwin-Schrödinger-Str., D-67663 Kaiserslautern, Germany

Summary. Intra- and intermolecular carbene or carbenoid reactions resulting from the photochemical and Cu(I)-, Rh(II)-, or Ru(I)-catalyzed decomposition of α-diazo-α-silylacetic esters are described. Among the products reported are (alkoxysilyl)ketenes, silaheterocycles, 1-trialkylsilylcyclopropane-1-carboxylates, and products derived from transient carbonyl ylides.

Introduction

In contrast to the carbene and carbenoid chemistry of simple diazoacetic esters, that of α-silyl-α-diazoacetic esters has not yet been developed systematically [1]. Irradiation of ethyl diazo(trimethylsilyl)acetate in an alcohol affords products derived from O–H insertion of the carbene intermediate, Wolff rearrangement, and carbene→silene rearrangement [2]. In contrast, photolysis of ethyl diazo(pentamethyldisilanyl)acetate in an inert solvent yields exclusively a ketene derived from a carbene→silene→ketene rearrangement [3]. Photochemically generated ethoxycarbonyltrimethylsilylcarbene cyclopropanates alkenes and undergoes insertion into aliphatic C–H bonds [4]. Copper-catalyzed and photochemically induced cyclopropenation of an alkyne with methyl diazo(trimethylsilyl)acetate has also been reported [5].

In this contribution, it will be shown that a silyl group directly attached to the carbene center offers several more synthetic opportunities, mainly under the following aspects:

a) Silicon-attached substituents can interact with the reactive carbene (or carbenoid) center in the same molecule. In addition to the already known 1,2-(Si→C) migration leading to short-lived silaethenes, [3, 4] intramolecular reaction modes pave the way for the syntheses of various silaheterocycles.

b) Silyl groups can be employed as substituents with adjustable steric demand; this aspect may be useful for stereoselective reactions.

c) The π-acceptor character of trialkylsilyl groups influences the stability and reactivity of the diazo compound, the carbene (or carbenoid), and other carbene-derived reactive intermediates (e.g., carbonyl ylides).

Synthesis of α-Silyl-α-diazoacetic Esters

Simple silyldiazoacetic esters are obtained readily and in high yield by silylation of an alkyl diazoacetate with a silyl triflate ($R_3Si-OSO_2CF_3$, R_3Si = Me_3Si, Et_3Si, iPr_3Si, Ph_2tBuSi, $Me_3Si-SiMe_2$, etc.) in the presence of a tertiary amine [6–8]. Silicon-functionalized analogues are prepared in a two-step one-pot reaction by treating a silyl bis(triflate) first with one equivalent of a diazoester, then with another carbon-, oxygen-, or nitrogennucleophile. Some examples are given in Scheme 1 (**1** [9], **2** [10], **3** [10], **4** [11], **5** [12]).

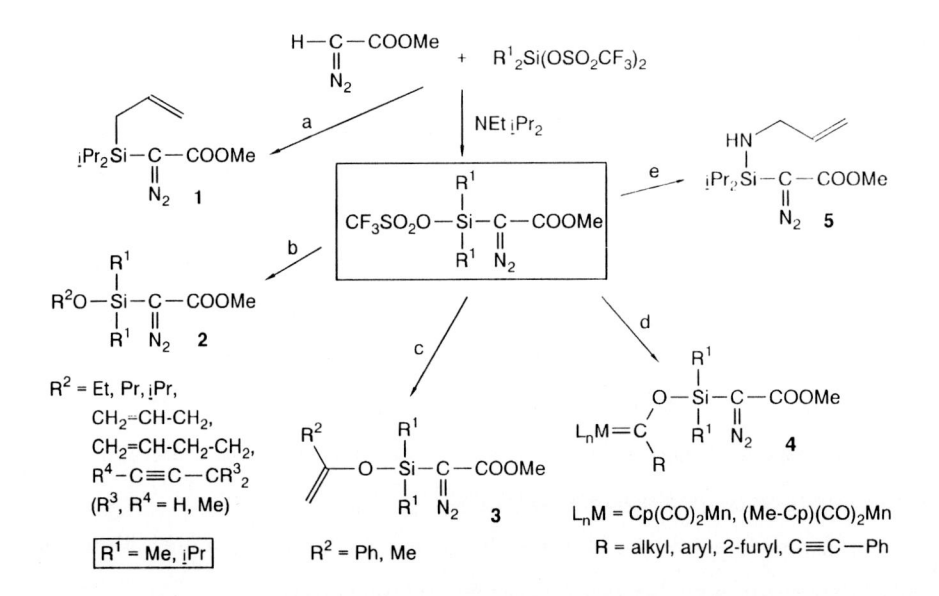

Scheme 1. Synthesis of silyl-functionalized silyldiazoacetic esters:
a: (CH_2=$CHCH_2$)MgCl is combined with the bis(triflate), then with diazoester and NEtiPr$_2$; b: R^2OH, NEtiPr$_2$; c: R^2–CO–CH$_3$, NEt$_3$; d: L_nM=C(R)O$^-$Li$^+$; e: allylamine, NEtiPr$_2$

For the synthesis of chloro- (azido-, isocyanato-, isothiocyanato-)silyldiazoacetic esters **6–9**, the same strategy has been used, but with chlorosilyl triflates as starting materials [12] (Scheme 2).

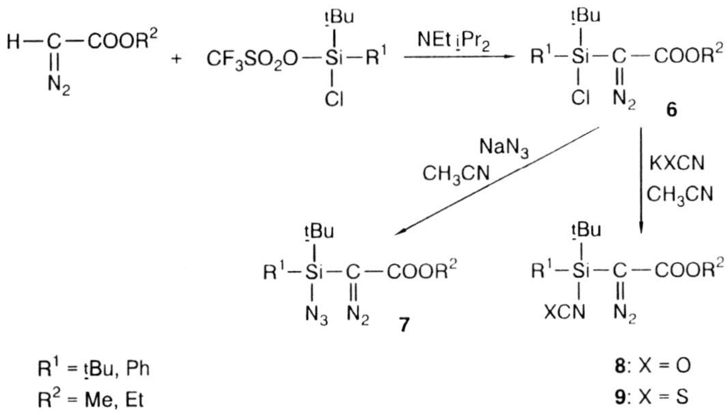

Scheme 2. Synthesis of chloro- (azido-, isocyanato-, isothiocyanato-)silyldiazoacetic esters

Intramolecular Carbene and Carbenoid Chemistry

The mode of nitrogen extrusion from a diazo compound – photochemically, thermally, or by transition-metal catalysis – is often crucial for the product pattern. For the last-mentioned method, believed to generate short-lived metal-carbene complexes rather than free carbenes, the nature of the metal can also influence the reaction course. A trialkylsilyl group attached to a diazo function reduces the nucleophilicity of the diazo carbon atom and causes steric shielding of this atom, especially if R_3Si is very voluminous. Therefore, only the most efficient decomposition catalysts (i.e. those that are highly electrophilic and/or have an easily accessible coordination site) are appropriate. In fact, it was found that copper(I) triflate (CuO_3SCF_3, CuOTf) decomposes even the bulkiest diazoesters, **10d** and **10e**, at room temperature, whereas rhodium(II) perfluorobutyrate [$Rh_2(OOCC_3F_7)_4$, $Rh_2(pfb)_4$], is active towards **10e**, but not towards **10d**. The less electrophilic catalysts $Rh_2(OAc)_4$ and $[Ru_2(CO)_4(OAc)_2]_n$ require higher temperatures, but even then they decompose only diazoesters bearing less bulky silyl groups. The products obtained from the copper- or rhodium-catalyzed decomposition reactions of diazoesters **10a–f** in toluene or tetrachloromethane are shown in Scheme 3 [7]. The sensitivity of the product pattern to the substrate, the catalyst, and even to the solvent is clearly obvious.

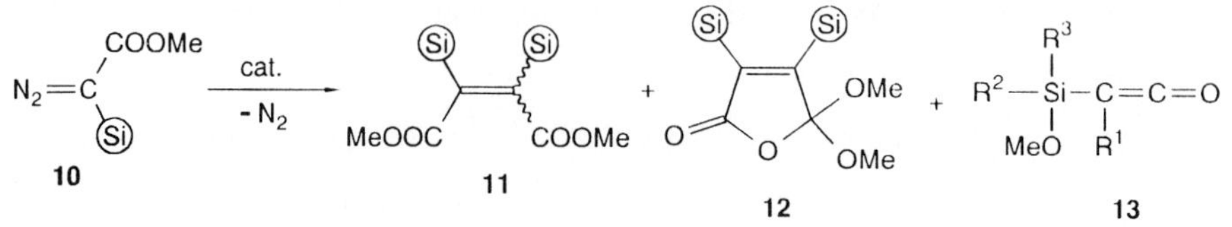

	Ⓢ = SiR¹R²R³			products (yield, %)			
10 - 13	R¹	R²	R³	CuO₃SCF₃, CCl₄, 20 °C	CuO₃SCF₃, toluene, 20 °C	Rh₂(OAc)₄, toluene, 110 °C	Rh₂(OOCC₃F₇)₄, toluene, 20 °C
a	Me	Me	Me	E-11 (95)	E-11 (51), Z-11 (8)	E-11 (19), Z-11 (7)	a
b	Et	Et	Et	E-11 (64)	E-11 (36), 12 (36)	E-11 (43)	13 (45)
c	Me	Me	tBu	E-11 (43) b	E-11 (40), 12 (40)	c	13 (18)
d	iPr	iPr	iPr		12 (31)	no reaction	no reaction
e	Ph	Ph	tBu		13 (23)		13 (50)
f	Me	Me	SiMe₃		13 (40)		13 (45)

a Mixture of products, including a ketene and PhCH₂CH(SiMe₃)COOMe. b The ketazine derived from **10** was also isolated (23% yield). c Unseparable mixture of products.

Scheme 3. Transition-metal-catalyzed decomposition of silyldiazoacetic esters

Intramolecular reaction pathways are involved in the formation of **12** and **13**. Thus, the Wolff rearrangement product, ketene **14,** is a likely intermediate en route to **12**. The doubly rearranged ketene **13f** has already been obtained upon photolysis of **10f** (EtO instead of MeO) [3], most likely via silene **15** that was observable under matrix isolation conditions [13]. Since photolysis of **10b** and **10d**, in contrast to that of **10f,** did not proceed cleanly and did not reproduce the result of the Rh₂(pfb)₄-catalyzed reaction, it can be concluded that the latter reaction does not involve a free carbene nor a silene. A feasible intermediate is the metal-carbene complex **16**, which may directly rearrange to ketene **13** as shown. It should also be noted that alkyl, Ph, and SiMe₃ groups undergo the 1,2-(Si→C) migration in the Rh₂(pfb)₄-catalyzed reaction, whereas in the CuOTf-catalyzed case, no alkyl migration is observed; a higher electron-deficiency at the rhodium-substituted carbon atom in **16** could account for this difference.

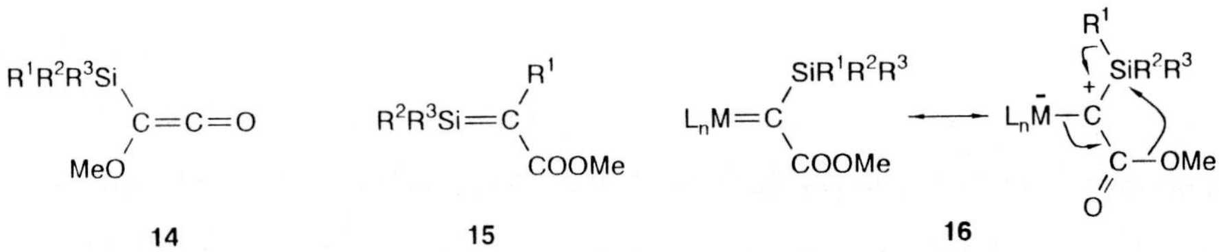

Fig. 1.

A variety of silaheterocycles can be prepared by intramolecular carbene or carbenoid pathways of silyldiazoacetic esters (Scheme 4). Thus, several Si–O-containing saturated or unsaturated five- and six-membered rings have been obtained from alkyloxysilyl-, alkenyloxysilyl-, and alkinyloxysilyl-substituted diazoesters **2** and **3**. Examples include C–H insertion, cyclopropanation, and cyclopropenation reactions [14]. Similarly, photolyis of diazoesters **6–9** (R^1 = *t*Bu) produced silacyclobutanecarboxylates by insertion of the carbene center into a C–H bond of the *t*Bu groups [12, 14]. However, when R^1 was a phenyl group, photolysis with 300 or 254 nm light in benzene solution did not afford an isolable monomeric product.

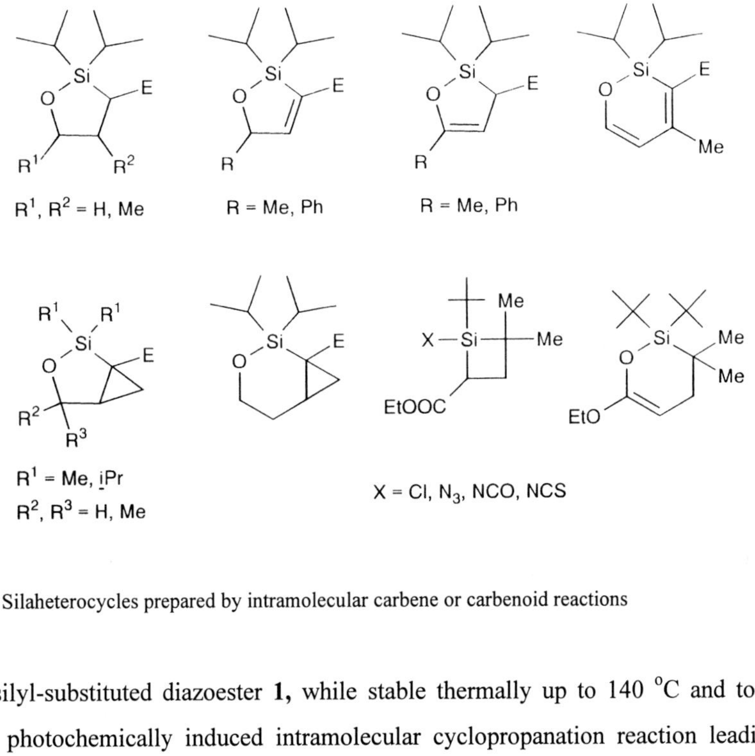

Scheme 4. Silaheterocycles prepared by intramolecular carbene or carbenoid reactions

The allylsilyl-substituted diazoester **1**, while stable thermally up to 140 °C and towards $Rh_2(pfb)_4$, undergoes a photochemically induced intramolecular cyclopropanation reaction leading to the rather strained 2-silabicyclo[2.1.0]butane **17** in surprisingly good yield [9].

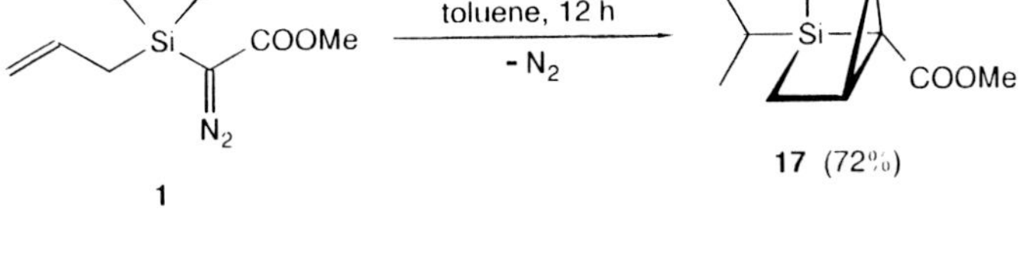

Scheme 5.

Cyclopropanation of Alkenes

When diazoester **10–SiMe₃** is decomposed with the catalysts mentioned above in neat styrene or 1-hexene, cyclopropanecarboxylates **18** are obtained as a mixture of diastereomers (Scheme 5) [15, 16]. In all cases, the stereoisomer in which the SiMe₃ group is *trans* to R^1 is formed preferentially. A consistent relationship between catalyst and diastereomer ratio cannot be seen, which is perhaps not surprising in view of the limited set of data and the different reaction temperatures. It should be remembered that cyclopropanations of a representative set of alkenes with ethyl diazoacetate have shown that a higher *trans*-diastereoselectivity in general results with copper rather than with rhodium catalysts [17]. The preference for the formation of (*E*)-**18** (R^1 = Ph) increases only modestly in the series SiMe₃ < SiEt₃ < Si*i*Pr₃. However, catalytic cyclopropanation of 1-hexene with **10–Si*i*Pr₃** fails with all catalysts, even with CuOTf. Presumably, the reduced accessibility of the diazo carbon atom does not allow **10–Si*i*Pr₃** to compete with the nucleophilic alkene for the coordination site(s) of the catalyst.

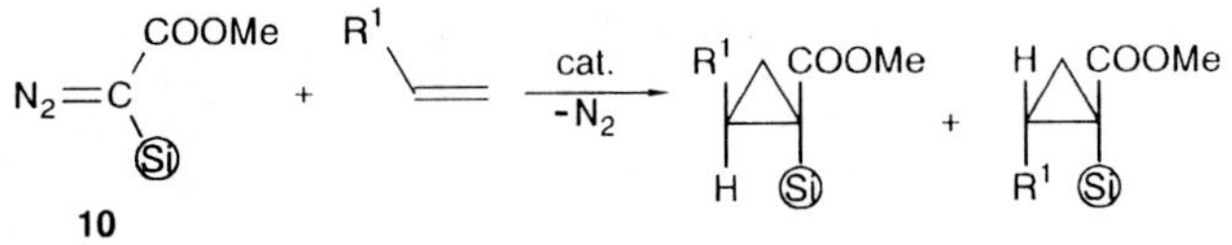

	R^1 = Ph			R^1 = Bu		
catalyst, mol-%	temp. (°C)	yield (%)	E/Z	catalyst, mol-%	temp. (°C)	yield (%)	E/Z
CuOTf, 7.3	20	82	3.2	5.6	20	53	2.9
Rh₂(OAc)₄, 6.0	50	48	3.4		63	no reaction	
Rh₂(pfb)₄, 2.5	70	53	1.2	4.5	20	47	3.2
[Ru₂(CO)₄(OAc)₂]ₙ, 3.4	70	67	1.8	3.2	63	89	3.2
hν	20	28	0.65		20	73	0.83

catalyst: CuOTf (5.6 - 8.2 mol-%), 20 °C

	R^1 = Ph		R^1 = Bu	
Si	yield (%)	E/Z	yield (%)	E/Z
SiMe₃	82	3.6	53	2.9
SiEt₃	61	4.2	52	3.0
Si*i*Pr₃	63	4.7	no reaction	

Scheme 6. Cyclopropanation of monosubstituted alkenes with methyl diazo(trialkylsilyl)acetates

In contrast to the metal-catalyzed reactions, photochemically generated methoxycarbonyl-trimethylsilylcarbene cyclopropanates styrene and 1-hexene with a slight Z-selectivity. While the preference for the sterically more congested cyclopropane is somewhat surprising, the rather stereounselective behavior of the carbene is in agreement with earlier observations by Schöllkopf et al. [4].

As anticipated, catalytic cyclopropanation of cyclohexene occurs with a much higher preference for the sterically less hindered diastereomer (*anti*-**20**) than in the case of the monosubstituted alkenes (Scheme 7). However, the reaction succeeds only with [Ru$_2$(CO)$_4$(OAc)$_2$]$_n$ as catalyst, whereas decomposition of **10a**–SiMe$_3$ with CuOTf leads to the carbene dimers **11**, and decomposition with Rh$_2$(pfb)$_4$ yields the apparent product of allylic C–H insertion (**19**) besides some ketazine.

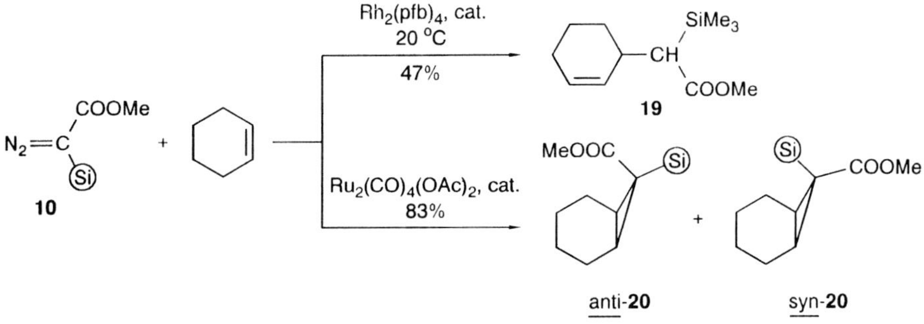

(Si)	yield, %	anti/syn
SiMe$_3$	54	7.0
SiEt$_3$	61	> 95% anti

Scheme 7. Metal-catalyzed decomposition of **10a** in the presence of cyclohexene

Carbonyl Ylide Chemistry

Carbonyl ylides, often generated from a carbene and an aldehyde or ketone, are in general reactive intermediates that can isomerize to stable products (oxiranes, enolethers) or can be trapped by [3+2] cycloaddition reactions [18]. As Scheme 8 shows, silyldiazoacetic esters can also be used to generate carbonyl ylides [19].

Thus, metal-catalyzed decomposition of **10a** in the presence of an equimolar amount of benzaldehyde generates the carbonyl ylide **21** which can be trapped with a suitable electron-poor dipolarophile to give tetrahydrofurans **23**, **24**, and dihydrofuran **25**, respectively. These heterocycles were obtained as sole products in 41–62 % yield when Rh$_2$(pfb)$_4$ or [Ru$_2$(CO)$_4$(OAc)$_2$]$_n$ were used as catalysts, whereas catalysis by CuOTf afforded mainly the oxirane **22** (38 %) besides some cycloaddition product (**23**: 11.5 %; **24**: 16 %).

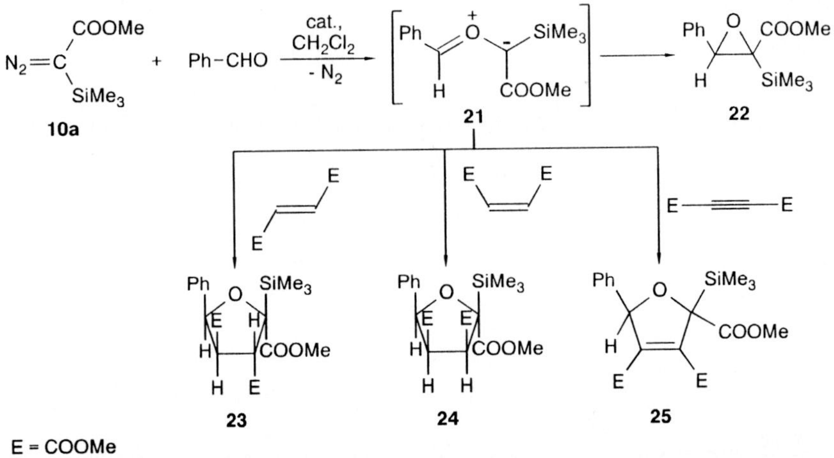

Scheme 8. Carbonyl ylide reactions derived from **10a**

In this three-component reactions, the SiMe$_3$ group has a multiple function. Due to its π-acceptor character, it suppresses the direct 1,3-dipolar-cycloaddition of the intact diazoester to the added dipolarophile (a competition reaction that is observed when methyl diazoacetate is employed) and it probably stabilizes the dipolar carbonyl ylide intermediate. Furthermore, the diastereospecific formation of **23**–**25** suggests that the SiMe$_3$ group occupies the *exo*-position in the W-shaped, planar carbonyl ylide **21** and that this configuration is intercepted in a stereospecific [3+2] cycloaddition reaction.

The tandem carbonyl ylide/cycloaddition reaction is also observed when crotonaldehyde or acetone is used instead of benzaldehyde (dimethyl fumarate as dipolarophile), whereas with cyclohexanone, an enol ether derived from the carbonyl ylide is isolated [19] (Scheme 9).

Rhodiumcatalyzed decomposition of **10a** in the presence of benzaldehyde and methyl cyanoformate does not afford the dihydrooxazole derived from carbonyl ylide **21**, but rather the interestingly functionalized 1,3-oxazole [19]. This product is likely to arise from cyclization of the nitrile ylide intermediate **26** [20].

Dioxolanone **33** is obtained when the unsaturated silyldiazoester **30** is decomposed by Rh$_2$(pfb)$_4$ in the presence of an aldehyde or of acetone (Scheme 11) [21]. The reaction sequence is likely to include formation and (probably reversible) 1,5-cyclization of carbonyl ylide **31**, and Cope rearrangement of the allylvinylether **32**. In analogy to carbonyl ylide **21**, the SiMe$_3$ should occupy the *exo*-position in **31**, thereby bringing the ester carbonyl in a geometry that is favorable to the cyclization step. Again, the choice of catalyst determines the product pattern, since CuOTf catalysis affords not only **33**, but also oxirane **22** and the intramolecular cyclopropanation product **34**.

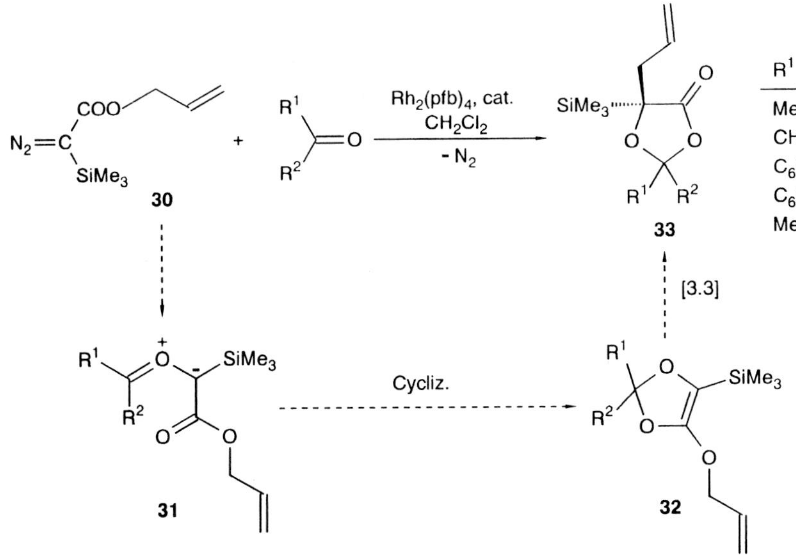

R¹	R²	yield, %
Me	H	48
CH=CHMe	H	56
C₆H₅	H	42
C₆H₄-4-OMe	H	51
Me	Me	39

Scheme 9. Nitrile ylide vs carbonyl ylide formation

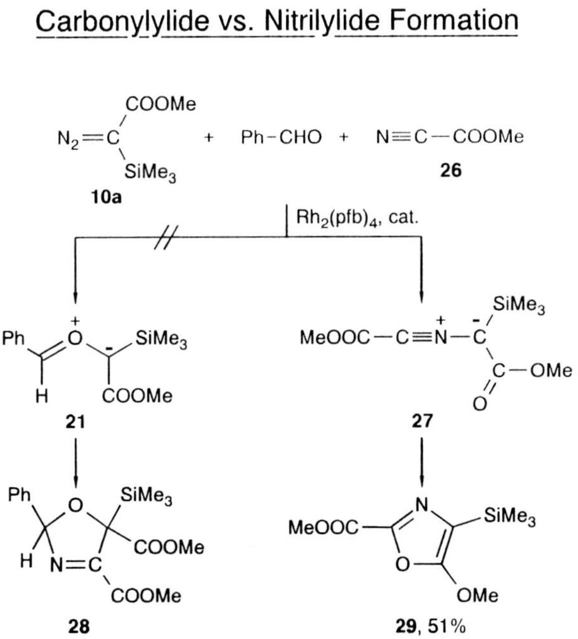

Carbonylylide vs. Nitrilylide Formation

Scheme 10. Transitionmetalcatalyzed decomposition of unsaturated diazoester **30** in the presence of carbonyl compounds

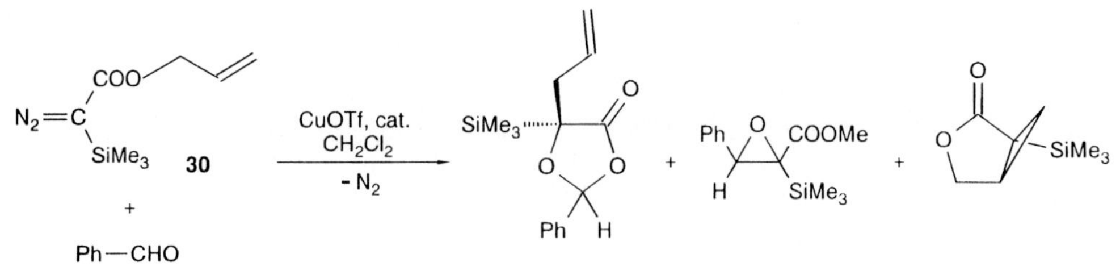

Scheme 11. Transitionmetalcatalyzed decomposition of unsaturated diazoester **30** in the presence of carbonyl compounds

Acknowledgement: I wish to thank the following coworkers for their patient and dedicated work: Mechthild Alt, Susanne Bender, Birgit Daucher, Monika Gimmy, Fred Krebs, Dieter Mayer, and Thorsten Werle. Furthermore, financial support by the *Deutsche Forschungsgemeinschaft*, the *Stiftung Volkswagenwerk*, and the *Fonds der Chemischen Industrie* is gratefully acknowledged.

References:

[1] Review on acyl-silyl-carbenes: H. Tomioka, in: *Methoden der organischen Chemie, Houben-Weyl*, Vol. E19b, Part 2 (Ed.: M. Regitz), Thieme, Stuttgart, **1989**, p. 1431.

[2] W. Ando, T. Hagiwara, T. Migita, *J. Am. Chem. Soc.* **1973**, *95*, 7518; W. Ando, A. Sekiguchi, T. Hagiwara, T. Migita, V. Chowdhry, F. H. Westheimer, S. L. Kammula, M. Green, M. Jones, Jr., *J. Am. Chem. Soc.* **1979**, *101*, 6393.

[3] W. Ando, A. Sekiguchi, T. Sato, *J. Am. Chem. Soc.* **1981**, *103*, 5573.

[4] U. Schöllkopf, N. Rieber, *Angew. Chem.* **1967**, *79*, 906; *Angew. Chem., Int. Ed. Engl.* **1967**, *6*, 884; U. Schöllkopf, D. Hoppe, N. Rieber, V. Jacobi, *Liebigs Ann. Chem.* **1969**, *730*, 1.

[5] G. Maier, D. Volz, J. Neudert, *Synthesis* **1992**, 561.

[6] T. Allspach, H. Gümbel, M. Regitz, *J. Organomet. Chem.* **1985**, *290*, 33.

[7] G. Maas, M. Gimmy, M. Alt, *Organometallics* **1992**, *11*, 3813.

[8] K. K. Laali, G. Maas, M. Gimmy, *J. Chem. Soc., Perkin Trans.* **1993**, 2,1387.

[9] B. Daucher, G. Maas, unpublished results.

[10] A. Fronda, F. Krebs, B. Daucher, T. Werle, G. Maas, *J. Organomet. Chem.* **1992**, *424*, 253.

[11] D. Mayer, G. Maas, in: *Organosilicon Chemistry II* (Eds. N. Auner, J. Weis), VCH, Weinheim, **1995**, p. 565.

[12] S. Bender, *Ph. D. Thesis*, Universität Kaiserslautern, **1994**.

[13] A. Sekiguchi, W. Ando, K. Honda, *Tetrahedron Lett.* **1985**, *26*, 2337.

[14] F. Krebs, S. Bender, B. Daucher, T. Werle, G. Maas, in: *Organosilicon Chemistry – From Molecules to Materials* (Eds.: N. Auner, J. Weis), VCH, Weinheim, **1994**, p. 57.

[15] G. Maas, M. Alt, D. Mayer, manuscript in preparation.

[16] G. Maas, T. Werle, M. Alt, D. Mayer, *Tetrahedron* **1993**, *49*, 881.

[17] M. P. Doyle, R. L. Dorow, W. E. Buhro, J. H. Griffin, W. H. Tamblyn, M. L. Trudell, *Organometallics* **1984**, *3*, 44.

[18] A. Padwa, S. Hornbuckle, *Chem. Rev.* **1991**, *91*, 263.

[19] M. Alt, G. Maas, *Tetrahedron* **1994**, *50*, 7435.

[20] In a similar reaction, such a nitrile ylide has been intercepted by a dipolarophile: T. Ibata, K. Fukushima, *Chem. Lett.* **1992**, *21*, 2196.

[21] M. Alt, G. Maas, *Chem. Ber.*, in press.

The Function of the Trimethylsilyl Substituent in the Syntheses of Low-valent Phosphorus and Arsenic Containing Compounds

G. Becker*, G. Ditten, K. Hübler, U. Hübler, K. Merz, M. Niemeyer,
N. Seidler, M. Westerhausen, Z. Zheng

Institut für Anorganische Chemie
Universität Stuttgart
Pfaffenwaldring 55, D–70550 Stuttgart, Germany

Summary: In recent years trimethylsilylphosphanes proved to be versatile starting compounds in the syntheses of phosphaalkenes and phosphaalkynes. The tris(trimethylsilyl) derivative, for example, reacts with acyl chlorides in a molar ratio of 1:1 give the corresponding [1-(trimethylsiloxy)-alkylidene]trimethylsilylphosphanes first. From a subsequent hexamethyldisiloxane elimination catalyzed by solid sodium hydroxide at 110–120 °C many phosphaalkynes have been obtained so far. In order to understand the underlying reaction mechanism, studies on the chemical behavior of [1-bis(1,2-dimethoxy-ethane-O,O')lithoxy-2,2-dimethylpropylidene]trimethylsilylphosphan prepared by lithiation of the related trimethylsiloxy derivative are being started.

Surprisingly, the reaction of diethyl carbonate and O,O'-diethyl thiocarbonate with lithium bis(trimethylsilyl)phosphanide has been found to give ethoxy trimethylsilane and the phosphaalkynes (dme)$_2$Li–O–C≡P and [(dme)$_3$Li]$^+$ [S–C≡P]$^-$ respectively. As known so far, lithoxymethylidynephosphane shows a great tendency to undergo [2+2] cycloaddition and reduction reactions. The NMR parameters of the [S–C≡P]$^-$-anion resemble much more the values of diorganylamino-phosphaalkynes than those of its oxygen homologue.

Through a rather complicated sequence of meanwhile fully understood reaction steps ethyl benzoate and lithium bis(trimethylsilyl)phosphanide form tris(1,2-dimethoxyethane-O,O')-lithium 3-phenyl-1,3-bis(trimethylsilyl)-1,2-diphosphapropenide and 3,5-diphenyl-1,2-bis(trimethylsilyl)-1,2,4-triphospholide. X-ray structure determinations on orange or green, metallically lustrous, crystals show the compounds to be ionic in the solid and to contain a 1,2-diphosphaallyl and a 2-phosphaallyl anion, respectively. Dark red tetrakis(tetrahydrofuran)lithium 3,5-bis(2,4,6-trimethylphenyl)-1,3-bis(trimethylsilyl)-1,2,4-triphosphapentadienide, isolated from a similar reaction with 2,4,6-trimethylbenzoyl chloride, corresponds to an intermediate which could not be detected in the above mentioned sequence of reaction steps. Furthermore, syntheses and structures of tris(1,2-dimethoxyethane-O,O')-lithium 3,5-di-$t$butyl-1,2,4-triphospholide and tetrakis(tetra-hydrofuran)lithium 2,3,4,5-tetraphenyl-phospholide are discussed.

In contrast to phosphaalkynes, nitriles show quite a different chemical reactivity towards lithium trimethylsilylphosphanides. Whereas with benzonitrile and one equivalent of the lithium phenyltrimethylsilyl compound 1-[(1,2-dimethoxyethane-O,O')lithium-trimethylsilylamido]benzylidenephosphane is formed, 1-(1,2-dimethoxyethane-O,O')lithium bis(trimethylsilyliminobenzoyl)phosphanide has been isolated from a similar reaction with lithium bis(trimethylsilyl)phosphanide in a molar ratio of 2:1. Solvent coordinate lithium is not bound to phosphorus, but to both the nitrogen atoms. Protonation gives the related bis(trimethylsilyliminobenzoyl)phosphane, which exists only as imino-enamine tautomer in the solid as well as in even very polar solvents.

Introduction

By pyrolysis of a mixture of hydrogen phosphide and silicon hydride, it was Fritz [1] who first succeeded in the preparation of a simple compound with an Si–P bond. About forty years ago, neither the author nor his colleagues anticipated that the silyl phosphane H_3Si-PH_2 would become the parent compound of a widely used class of starting materials. On the contrary, silylphosphanes remained peculiarities until a convenient synthesis of the tris(trimethylsilyl) derivative $(Me_3Si)_3P$ had been developed [2].

Nowadays this important compound can be easily prepared on a scale up to 200 g when an emulsion of white phosphorus in 1,2-dimethoxyethane is treated with liquid sodium-potassium alloy and excess chlorotrimethylsilane is added to the suspension of the hitherto scarcely characterized Zintl phases Na_3P and K_3P, respectively. With decreasing yield the tris(trimethylsilyl) derivatives of arsane [3], stibane [4], and bismuthane [5] have also been obtained in the same way (Eq. 1). Meanwhile, hesitation to handle dangerous sodium-potassium alloy or white phosphorus led to the development of similar methods to prepare the phosphane [6, 7].

$$E \xrightarrow{+\ 3\ Na\ /\ 3\ K} Na_3E\ /\ K_3E \xrightarrow[-\ 3\ NaCl\ /\ 3\ KCl]{+\ 3\ Me_3SiCl} (Me_3Si)_3E$$

$E = P_{white},\ As,\ Sb,\ Bi;\quad Me = CH_3$

Eq. 1.

With methyllithium in 1,2-dimethoxyethane tris(trimethylsilyl)phosphane and -arsane are quantitatively converted to their related lithium derivatives (Eq. 2). These highly sensitive compounds crystallize as dimers building up four-membered heterocycles with dme-coordinate lithium and bis(trimethylsilyl)phosphanyl [8] or -arsanyl fragments alternately arranged [9].

$$2\ (Me_3Si)_3E \xrightarrow[-\ 2\ Me_4Si;\ \langle dme \rangle]{+\ 1/2\ (LiMe)_4} (Me_3Si)_2E \underset{\underset{dme}{Li}}{\overset{\overset{dme}{Li}}{\diamondsuit}} E(SiMe_3)_2$$

$E = P,\ As;\ dme = 1,2\text{-dimethoxyethane}$

Eq. 2.

From Phosphaalkenes to Phosphaalkynes [47]

Syntheses and Structures: In subsequent reactions with acyl chlorides carefully freed from impurities such as hydrogen chloride, one trimethylsilyl group of tris(trimethylsilyl)phosphane and -arsane or the lithium atom of both the related bis(trimethylsilyl) compounds can be replaced to give the corresponding acyl derivatives. Keeping in mind the weakness of a Si–P or Si–As bond on the one hand and the strength of an Si–O bond on the other, the thermal lability of these compounds becomes immediately understandable. With a 1,3-shift of a trimethylsilyl substituent from phosphorus or arsenic to the oxygen of the carbonyl group they rearrange to [1-(trimethylsiloxy)alkylidene]trimethylsilylphosphanes [10] and -arsanes [11] (Eq. 3). The formation of a P=C double bond combined with a relatively high barrier of planar inversion at the phosphorus atom implicate the occurence of *E*- and *Z*-isomers which have been detected several times by NMR spectroscopic methods [12, 13].

$(Me_3Si)_3E$ $\xrightarrow[- Me_3SiCl]{+ R\text{-}CO\text{-}Cl}$

$1/2\ [(Me_3Si)_2E\text{-}Li(dme)]_2$ $\xrightarrow[- LiCl;\ - dme]{+ R\text{-}CO\text{-}Cl}$ $(Me_3Si)_2E\text{-}CO\text{-}R \xrightarrow{\Omega} Me_3Si{\sim}E{=}C{\diagup}^{O-SiMe_3}_{\diagdown R}$

$E = P, As$

Eq. 3.

The introduction of a P≡C triple bond into these molecules is easily achieved by an obvious elimination of hexamethyldisiloxane in the presence of small amounts of commercially available, unpurified sodium hydroxide [14]. Several years ago Regitz et al. [15] improved this method by avoiding a solvent, dropping the liquid alkylidenephosphane onto the hot (110–120 °C) solid catalyst, removing the products in vacuo and separating the phosphaalkyne from hexamethyldisiloxane by distillation (Eq. 4). Since its first preparation in 1981 the *t*butyl derivative $Me_3C\text{-}C{\equiv}P$ especially has turned out to be a very versatile starting compound for further syntheses [16–20]. Contrary to expectations, only one alkylidynearsane has been isolated so far [21]; therefore, this class of compounds will not be discussed further.

$Me_3Si{\sim}P{=}C{\diagup}^{O-SiMe_3}_{\diagdown R}$ $\xrightarrow[- Me_3Si\text{-}O\text{-}SiMe_3]{<NaOH>_{solid}}$ $P{\equiv}C\text{-}R$

Eq. 4.

If the greenish-yellow mixture of excess Z- and little E-isomeric [2,2-dimethyl-1-(trimethylsiloxy)propylidene]trimethylsilylphosphane obtained from the tris(trimethylsilyl) compound and 2,2-dimethylpropionyl chloride in a molar ratio of 1:1 (Eq. 3) is stored in a refrigerator before further use, it usually remains liquid even at −20 °C. However, several months ago colourless squares crystallized from that mixture by chance. Due to their low melting point of about −5 °C they had to be handled very carefully at −50 °C and transferred in sealed capillaries to the diffractometer. An X-ray structure determination showed the Z-isomer which has the trimethylsilyl and the trimethylsiloxy group on the same side of the P=C unit, to be present in the solid state (Fig. 1). Most of the molecular parameters such as a P=C distance of 169.4 pm and angles at the sp^2-hybridized carbon, which sum up to 359.9 ° are as expected; compared with standard values of 227 and 141 pm, P–Si (225.3) and C–O bond lengths (136.5 pm) are slightly shorter [22].

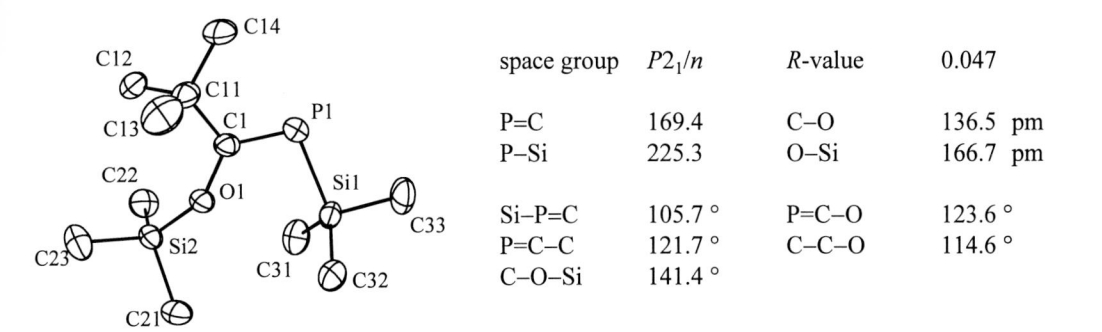

space group	$P2_1/n$	R-value	0.047
P=C	169.4	C–O	136.5 pm
P–Si	225.3	O–Si	166.7 pm
Si–P=C	105.7 °	P=C–O	123.6 °
P=C–C	121.7 °	C–C–O	114.6 °
C–O–Si	141.4 °		

Fig. 1. Molecular structure of Z-[2,2-Dimethyl-1-(trimethylsiloxy)propylidene]trimethylsilylphosphane; 50 %probability; hydrogen atoms omitted; measurement at -100 ± 3 °C

When the Z-isomer crystallizes at a relatively low temperature, the E-isomer is as a consequence enriched in the remaining greenish-yellow liquid. The NMR parameters of both species can now be determined very easily and compared to each other (Table 1). To begin with, the chemical shifts of the ^{31}P-, ^{13}C(P)-, and the ^{29}Si(P)- and the ^{29}Si(O)-nuclei differ considerably in part.

	Z	E		Z	E
$\delta\,^{31}$P	121.7	106.7	$^1J_{C=P}$	80.1	69.1
$\delta\,^{13}$C=P	225.6	229.3	$^1J_{Si-P}$ [a]	51.4	73.0
$\delta\,^{29}$Si–P	–4.9	–3.8	$^2J_{C-Si-P}$	9.6	7.5
$\delta\,^{13}$C–Si–P	1.5	0.8	$^2J_{C-C=P}$	24.5	12.3
$\delta\,^{29}$Si–O	13.7	19.2	$^3J_{Si\cdot P}$ [a]	2.8	5.0
$\delta\,^{13}$C–Si–O	2.6	2.5	$^3J_{P\cdot\cdot CH_3}$	13.1	2.3
$\delta\,^{13}$C(CH$_3$)$_3$	45.3	46.5	$^4J_{P\cdot\cdot CH_3}$	1.5	<0.5
$\delta\,^{13}$CH$_3$–C	30.6	30.6	$^1J_{C-Si(P)}$	48.9	[a]
			$^1J_{C-Si(O)}$	57.5	[b]

Table 1. Characteristic NMR parameters of Z- and E-isomeric [2,2-dimethyl-1-(trimethylsiloxy)-propylidene]trimethylsilylphosphane; chemical shift δ (ppm); coupling constants J (Hz); [D$_6$]-benzenesolution; [a] Unreproducible coupling constants in [12]; [b] undetermined

$$\begin{array}{cc} \text{Me}_3\text{Si}\diagdown\diagup\text{O-SiMe}_3 & \diagup\text{O-SiMe}_3 \\ \text{P=C} & \text{P=C} \\ \diagup\diagdown\text{CMe}_3 (Z) & \text{Me}_3\text{Si}\diagup\diagdown\text{CMe}_3 (E) \end{array}$$

As for the coupling constants the higher $^1J_{CP}$- and the lower $^1J_{SiP}$-parameter are both found in the Z-isomer. Remarkably, the coupling between phosphorus and a nucleus in a substituent at the sp^2-hybridized carbon is high when the non-bonding electron pair at phosphorus and that substituent are arranged *cis* and it is low when they are in a *trans* position. This phenomenon has already been observed in several other E/Z-isomeric phosphaalkenes [13, 23].

Reaction Mechanism: The mechanism of the sodium hydroxide-catalyzed elimination of hexamethyldisiloxane may easily be understood when the reaction is compared to the well-known Peterson olefination in organic chemistry [24]. Provided that an enolate anion is formed as an intermediate, either directly or *via* a preceeding hydrolysis of the O–Si bond with traces of water which are always present on the hot surface of the crude catalyst, trimethylsilanolate splits off readily and thus the P≡C triple bond is introduced into the molecule (Eq. 5). Subsequent attack of trimethylsilanolate at the trimethylsiloxy group of the starting compound results in a formation of hexamethyldisiloxane and the initial enolate anion so that the reaction circle is closed.

$$\text{Me}_3\text{Si}\diagdown_{}\diagup\text{O-SiMe}_3 \xrightarrow[-\text{Me}_3\text{SiOH}]{+\text{OH}^\ominus} \text{Me}_3\text{Si}\diagdown_{}\diagup\text{O}^\ominus \xrightarrow{-\text{Me}_3\text{SiO}^\ominus} \text{P}\equiv\text{C-R}$$

Eq. 5.

With respect to the mechanism just discussed, the statement of Cowley et al. [25] that merely Z-isomeric [2,2-dimethyl-1-(trimethylsiloxy)propylidene]trimethylsilylphosphane can eliminate hexamethyldisiloxane, does need further verification. From our point of view the E- and Z-isomer of the mesomeric enolate anion are readily interconverted by a rotation around the P–C bond of the keto form (Eq. 6) which is supposed to be an easily accessible transition state. At any rate, we were not able to confirm their results as the reaction of lithium bis(trimethylsilyl)phosphanide with 2,2-dimethylpropionyl chloride at −78 °C in cyclopentane solution gives exclusively the E-isomeric phosphaalkene, whereas at room temperature the Z-isomer prevails.

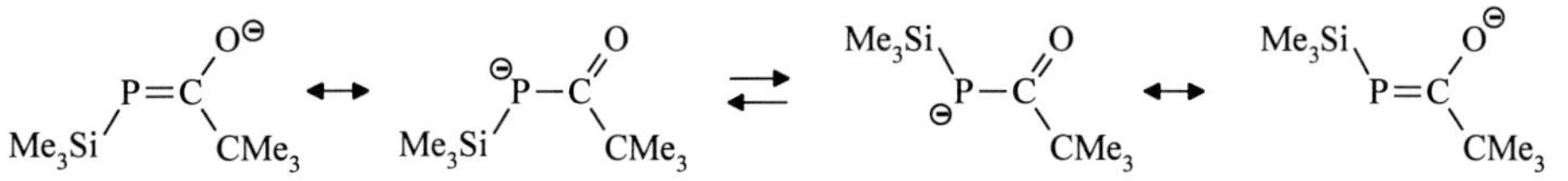

Eq. 6.

A better insight into the underlying mechanism of hexamethyldisiloxane elimination, however, may be gained from future investigations of the chemical reactivity of [1-bis(1,2-dimethoxyethane-O,O')lithoxy-2,2-dimethylpropylidene]trimethylsilylphosphane (Fig. 2) – a compound recently isolated from the reaction of the E/Z-isomeric 1-(trimethylsiloxy) derivative and *n*butyl lithium at –55 °C in a mixture of 1,2-dimethoxyethane and *n*hexane (Eq. 7) [26, 48].

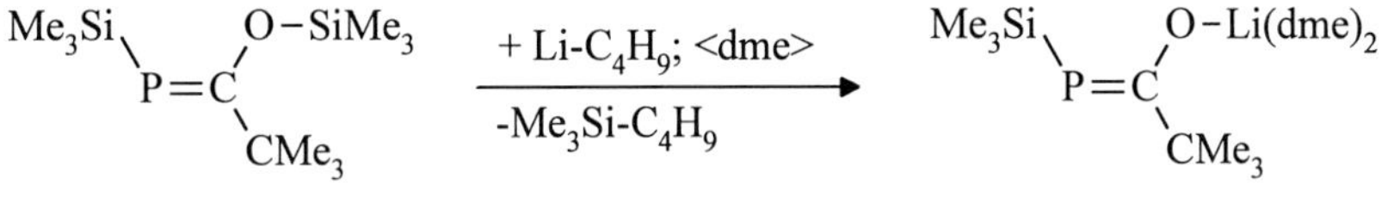

Eq. 7.

Due to its low thermal stability, only a rather broad singlet at –20.6 ppm was obtained from a ^{31}P NMR spectrum of a [D$_6$]-benzene/1,2-dimethoxyethane solution. However, colorless squares could be isolated when recrystallizing the compound from toluene. An X-ray structure determination shows the Z-isomer which has the trimethylsilyl and the bis(1,2-dimethoxyethane-O,O')lithoxy group at the same side of the P=C unit, to be present in the solid. P=C and C–O distances of 176.7 and 126.9 pm deviate significantly from standard values for a double or a single bond respectively; this indicates strong electronic interactions in the mesomeric heteroallyl anion. Studies with regard to an obvious formation of a phosphaalkyne are being started.

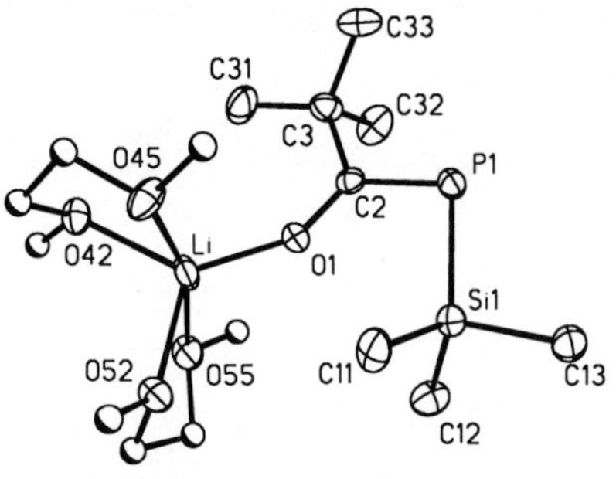

space group	C2/c	wR$_2$-value	0.194
P=C	176.7	C–O	126.9 pm
P–Si	222.4	O–Li$_{av.}$	184.0 pm
Si–P=C	98.8 °	P–C–O	125.4 °
P–C–C	118.6 °	C–C–O	116.0 °
C–O–Li	152.4 °		

Fig. 2. Molecular structure of [1-Bis(1,2-dimethoxyethan-O,O')lithoxy-2,2-dimethylpropyliden]trimethylsilylphosphane; 50 % probability for most of the heavier atoms; hydrogen atoms omitted; measurement at –100 ± 3 °C

Lithium Acylphosphanides: [1-Bis(1,2-dimethoxyethane-O,O')lithoxy-2,2-dimethylpropylidene]trimethylsilylphosphane may be considered as a remarkable representative of the up to now scarcely studied class of lithiated acylphosphanes. Since in accordance with Pearson's acid-base concept the hard lithium cation is bound to the hard oxygen anion leading to a phosphorus atom of coordination number two, as a consequence phosphanides of this type are phosphaalkenes. Compounds with a P–H group are easily prepared by reaction of (1,2-dimethoxyethane-O,O')lithium phosphanide with carboxylic esters or acyl chlorides at –50 °C in 1,2-dimethoxyethane solution (Eq. 8) [27].

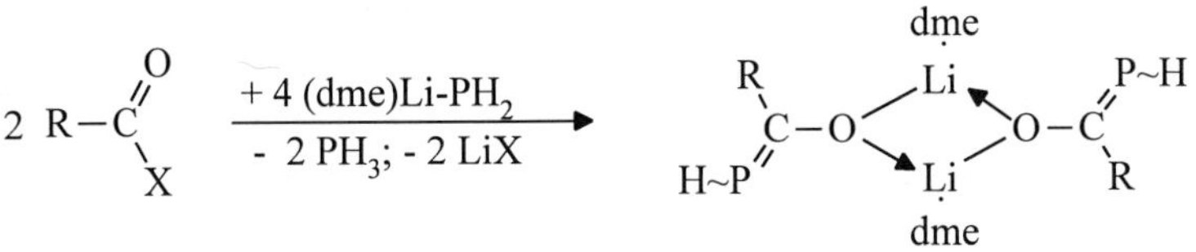

R = H or alkyl, X = OR'; R = aryl, X = Cl

Eq. 8.

With ethyl formate, for instance, a mixture of liquid *E*- and *Z*-isomeric (1,2-dimethoxyethane-O,O')-lithoxy-methylidenephosphane has been isolated and fully characterized by NMR methods [13]. Bond lengths and angles are, however, available from solid 1-(1,2-dimethoxyethane-O,O')lithoxy-1-(2,4,6-trimethylphenyl)methylidenephosphane prepared with 2,4,6-trimethylbenzoyl chloride and recrystallized from 1,2-dimethoxyethane. The compound has been found to be dimeric in the solid state forming a four-membered ring of solvent coordinate lithium and oxygen atoms alternately arranged. Considering the fact that hydrogen atoms may be located with a rather high degree of uncertainty only, the crystal studied by an X-ray structure analysis contains the Z-isomer (Fig. 3). As in the compound previously described the phosphaalkene shows a slightly elongated P–C double bond (172.3 pm) and a strongly shortened C–O single bond (128.7 pm). In a [D$_6$]-benzene solution of both the E- (δH–P 4.23 ppm; $^1J_{HP}$ 156.2 Hz; δ^{31}P 31.4 ppm) and Z-isomer (δH–P 4.59 ppm; $^1J_{HP}$ 135.4 Hz; δ^{31}P 16.0 ppm) can be detected [28].

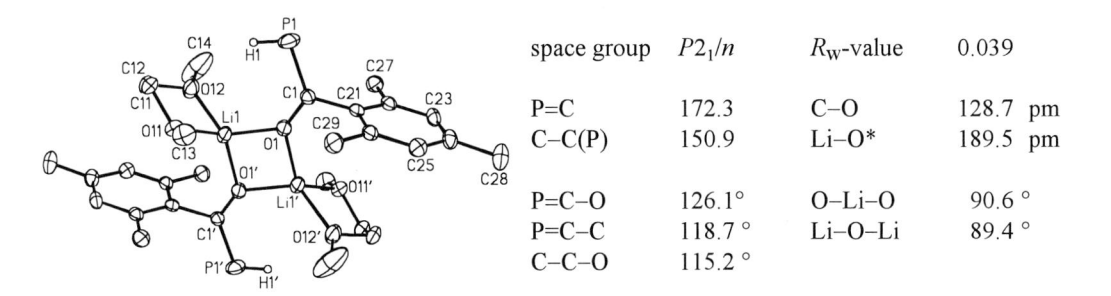

space group	$P2_1/n$	R_W-value	0.039
P=C	172.3	C–O	128.7 pm
C–C(P)	150.9	Li–O*	189.5 pm
P=C–O	126.1°	O–Li–O	90.6°
P=C–C	118.7°	Li–O–Li	89.4°
C–C–O	115.2°		

Fig. 3. Molecular structure of 1-(1,2-Dimethoxyethane-*O,O'*)lithoxy-1-(2,4,6-trimethylphenyl)methylidenephosphane; 40 % probability; measurement at −100 ± 3 °C;. with exception of the PH-groups hydrogen atoms have been omitted for clearness; the marked distance corresponds to the 2,4-dioxa-1,3-dilithietane ring

Heteroatom Substituted Phosphaalkynes

Lithoxymethylidynephosphane: The simple preparation of 1-(1,2-dimethoxyethane-*O,O'*)lithoxy-alkylidenephosphanes from carboxylic esters and (1,2-dimethoxyethane-*O,O'*)lithium phosphanide (Eq. 8) motivated us to investigate the analogous reaction with carbonic acid esters very carfully. Whereas with the educt (dme)Li–PH$_2$ several, difficult to separate by-products are formed, lithium bis(trimethylsilyl)phosphanide and diethyl carbonate react at 0 °C in 1,2-dimethoxyethane with elimination of two equivalents of ethoxy trimethylsilane to give the heteroatom substituted phosphaalkyne (dme)$_2$Li–O–C≡P in a nearly 80 % yield (Eq. 9) [13].

$$O=C(O-Et)(O-Et) + Me_3Si-P(-Li)-SiMe_3 \xrightarrow[-2\ Et-O-SiMe_3]{<dme>} (dme)_2Li-O-C\equiv P$$

Eq. 9.

From an 1,2-dimethoxyethane solution of the initially obtained red oily residue first, colorless to pale yellow, rapidly efflorescing crystals of bis(1,2-dimethoxyethane-*O,O'*)lithoxymethylidynephosphane precipitate within three days at −20 °C. In the solid, monomeric neutral complexes with an almost linear

P≡C–O–Li fragment and a lithium cation in a trigonal bipyramidal coordination sphere of five oxygen atoms, from two chelating 1,2-dimethoxyethane ligands and the P≡C–O⁻ anion are found (Fig. 4). Characteristic NMR parameters will be discussed later in context with those of the [P≡C–S]⁻-anion.

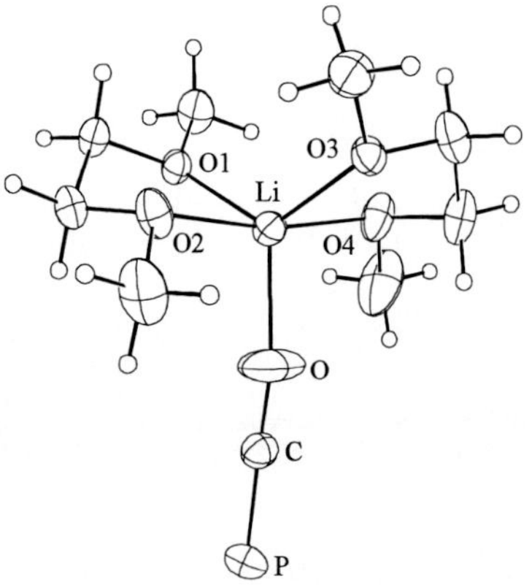

space group	$P2_1/n$
R-value	0.050
P≡C	155.5 pm
C–O	119.8 pm
(C)O–Li	187.8 pm
P–C–O	178.5 °
C–O–Li	170.7 °

Fig. 4. Structure of the phosphaalkyne P≡C–O–Li(dme)$_2$; 50 % probability; measurement at –100±3 °C

With respect to the reaction mechanism a nucleophilic attack of the bis(trimethylsilyl)phosphanide anion at the sp^2-hybridized carbon of diethyl carbonate, followed by an elimination of one molecule of ethoxy trimethylsilane and formation of a still undetected phosphaalkene is supposed. Splitting off a second molecule ethoxytrimethylsilane the P–C double bond of the intermediate is converted into a triple bond thereafter (Eq. 10).

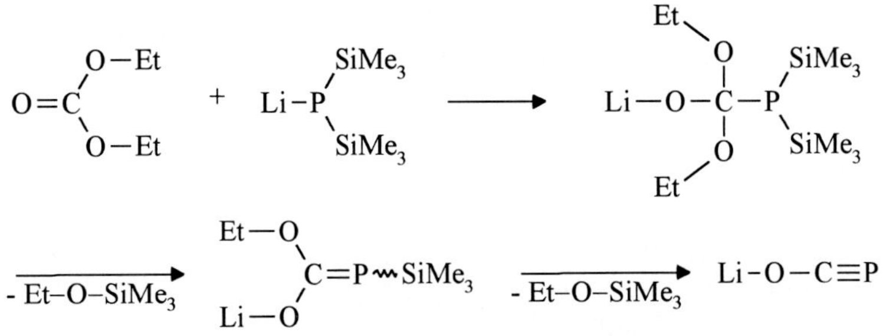

Eq. 10.

Only a few chemical properties of the phosphaalkyne P≡C–O–Li(dme)$_2$ have been studied in detail so far. Protonation of the anion probably leads to hydroxymethylidynephosphane, the separation of which from the solvent 1,2-dimethoxyethane and its characterization have not yet been achieved. Furthermore, lithoxymethylidynephosphane shows a great tendency to undergo [2+2] cycloaddition reactions; with ethyl phenylpropiolate in a molar ratio of 1:2 dimeric 2,6-diphenyl-4-(tetrahydrofuran)lithoxy-3,5-bis-(carbethoxy)-λ^3-phosphinine **A** is formed [29]. Oxidation with sulfur dioxide or iodine at −50 °C in 1,2-dimethoxyethane leads to a compound with a butterfly structure (**B**) which has been isolated in a very high yield.

A **B**

An X-ray structure determination {$Cmcm$; R_1= 0.061} of pale yellow crystals of compound **B** obtained from the sulfur dioxide reaction, shows a neutral complex of symmetry $mm2$ to be present in the solid. The anionic part of the molecule consists of two anellated 1,2-dihydro-5-oxo-1,2,4-triphosphol-3-olate rings which share the central P–P group (P–P 215.3; P–C 189.1; P∷C 178.4; C∷O 123.9 pm; C–P–P 98.4°; C–P–C 91.2°; C∷P∷C 98.7°). The lithium cations are coordinated square pyramidally (Li–O 193.5 to 209.1 pm), each binding a 1,2-dimethoxyethane and a tetrahydrofuran ligand in addition [30].

The [P≡C–S]¯ Anion: Very surprisingly, in a reaction with carbon bisulphide at −50 °C in 1,2-dimethoxy-ethane the oxygen of the [P≡C–O]¯ anion is exchanged for a sulphur atom to produce the till then unknown species [P≡C–S]¯, the phosphorus homologue of thiocyanate [N≡C–S]¯. Presumably a cyclic, but still undetected intermediate may be formed (Eq. 11); carbon oxysulfide also present in solution does not show any further reactivity.

Eq. 11.

By analogy to Eq. 9 the new phosphaalkyne may also be obtained from O,O'-diethyl thio-carbonate and lithium bis(trimethylsilyl)phosphanide dissolved in 1,2-dimethoxyethane. Both educts react at about 0 °C to give ethoxy trimethylsilane and the finally isolated compound tris(1,2-dimethoxyethane-O,O')-lithium 2-phosphaethynylsulphide (Eq. 12). Emphasis should be laid on its NMR parameters (δ^{31}P −121.3; δ^{13}C 190.8 ppm; $^1J_{CP}$ 18.2 Hz; [D$_8$]-THF solution) in that they correspond much more to the values of diisopropylaminophosphaalkyne (δ^{31}P −99.6; δ^{13}C 152.2 ppm; $^1J_{CP}$ 14.7 Hz [31]) than to those of bis(1,2-dimethoxyethane-O,O')lithoxymethylidynephosphane (δ^{31}P −384.2; δ^{13}C 166.6 ppm; $^1J_{CP}$ 41.5 Hz; [D$_8$]-THF solution [13]). IR absorptions at 1762 and 747 cm^{-1} is assigned to the $\tilde{\nu}$(P≡C) and $\tilde{\nu}$(C–S) stretching vibrations.

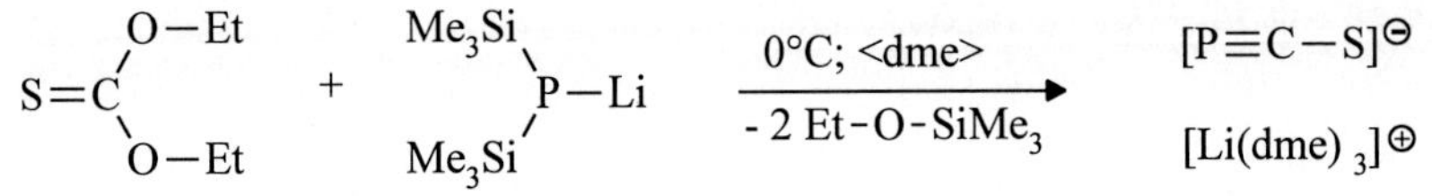

$$S=C\begin{smallmatrix}O-Et\\O-Et\end{smallmatrix} \quad + \quad \begin{smallmatrix}Me_3Si\\Me_3Si\end{smallmatrix}P-Li \quad \xrightarrow[-2\,Et-O-SiMe_3]{0°C;\,<dme>} \quad [P\equiv C-S]^{\ominus} \quad [Li(dme)_3]^{\oplus}$$

Eq. 12.

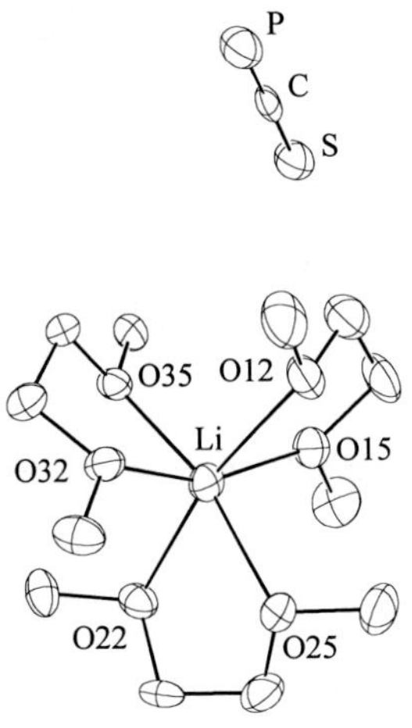

space group	$P2_1/c$	wR_1-value	0.064
P≡C	155.5°	C–S	162.0 pm
P≡C–S	178.9°	Li–O	206.4–220.3 pm

Fig. 5. Structure of the phosphaalkyne [(dme)$_3$Li]$^+$[P≡C–S]$^-$; 30 % probability; measurement at −100 ± 3 °C; the reproduction of cation and anion reflects their mutual arrangement in the solid.

The phosphaalkyne [(dme)$_3$Li]$^+$[P≡C–S]$^-$ may be stored at −25 °C for months without any perceptible decomposition; it melts at +16 °C to give a dark red oil that is difficult to recrystallize. An X-ray structure determination of very sensitive, pale yellow crystals isolated from a concentrated 1,2-dimethoxyethane solution reveals an ionic structure built up of lithium cations coordinated by three 1,2-dimethoxyethane ligands and linear [P–C≡S]$^-$ anions (Fig. 5) [32].

Five Membered Phosphorus Containing Heterocycles

The first step – a 1,2-diphosphaallyl anion [49]: In contrast to the syntheses of dimeric 1-(1,2-dimethoxyethane-*O,O'*)lithoxyalkylidenephosphanes from ethyl formate or 2,4,6-trimethylbenzoyl chloride and (1,2-dimethoxyethane-*O,O'*)lithium phosphanide already mentioned (Eq. 13), the reaction of ethyl benzoate with lithium bis(trimethylsilyl)phosphanide at about 0 °C in 1,2-dimethoxyethane allows an access to several highly remarkable compounds with low coordinate phosphorus atoms. To begin with a molar ratio of 1:3, very air-sensitive orange crystals of *E/E*-isomeric tris(1,2-dimethoxyethane-*O,O'*)lithium 3-phenyl-1,3-bis(trimethylsilyl)-1,2-diphosphapropenide have been isolated; however, the less than 20 % yield is highly unsatisfying. The formation of this 1,2-diphosphaallyl anion may be explained as follows (Scheme 1):

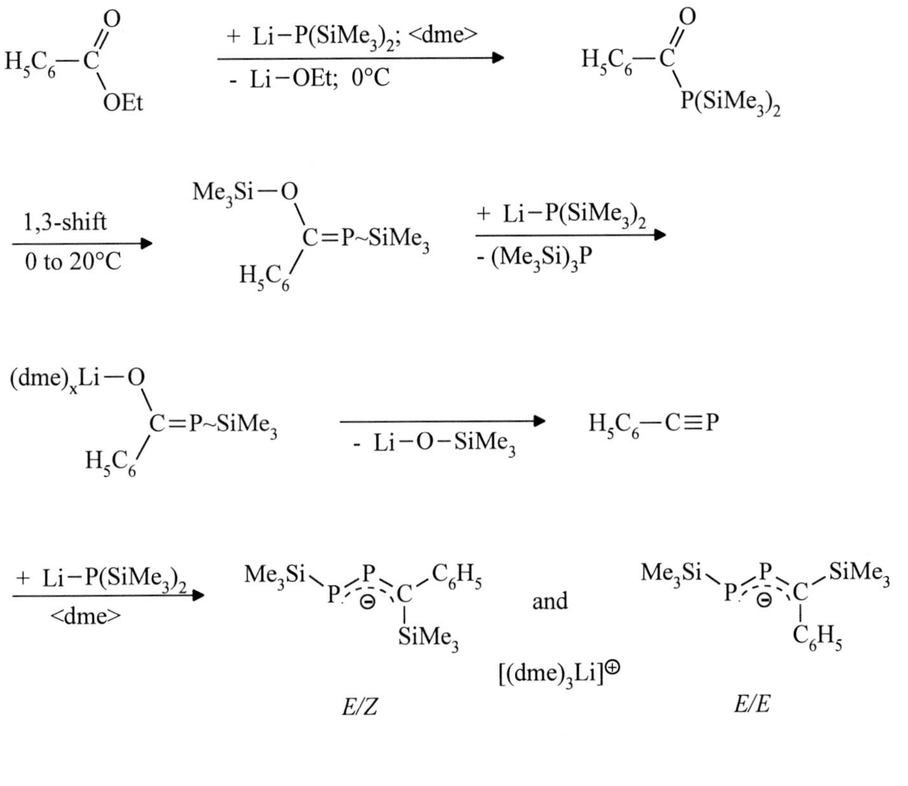

Scheme 1. Reaction steps in the preparation of *E/Z* tris(1,2-dimethoxyethane-*O,O'*)lithium 3-phenyl-1,3-bis(trimethylsilyl)-1,2-diphosphapropenide

Substitution of the ethoxy group transforms the ester first into thermally unstable benzoylbis(trimethylsilyl)phosphane, which at least partially rearranges on warming to room temperature to give the corresponding [1-(trimethylsiloxy)benzylidene]trimethylsilyl compound. In a subsequent reaction with excess lithium bis(trimethylsilyl)phosphanide, exchange of the trimethylsilyl substituent bound to the oxygen atom, for solvent coordinate lithium leads to the analogous lithoxy derivative. Compounds of this type, i.e., lithium acyltrimethylsilylphosphanides, have already been discussed earlier (Eq. 7); in contrast to the acylbis(trimethylsilyl)phosphane and its related phosphaalkene mentioned before, this intermediate can be detected in the NMR spectra of the reaction mixture (δ^{31}P −2.1 ppm, broad signal; δ(H$_3$C)$_3$Si 0.61 ppm; $^3J_{HP}$ 3.2 Hz). Elimination of lithium trimethylsilanolate then gives very reactive phenylmethylidynephosphane, which immediately is attacked nucleophilicly by an bis(trimethylsilyl)-phosphanide anion at the positively charged phosphorus of the P≡C triple bond. At last stabilization is achieved by a 1,3-shift of a trimethylsilyl substituent from phosphorus to carbon [35].

Recently, detailed ^{31}P NMR studies of the formation of the 1,2-diphosphaallyl anion were carried out in order to increase its unacceptably low yield. This led to the very surprising discovery that an unstable E/Z- (δ^{31}P–Si +41.4; δ^{31}P–C +454.2 ppm; $^1J_{PP}$ 460.6 Hz) as well as the finally isolated E/E-isomer (δ^{31}P–Si +8.0; δ^{31}P–C 411.1 ppm; $^1J_{PP}$ 461.2 Hz; 1,2-dimethoxyethane/[D$_6$]-benzene solution) are present in the reaction mixture. An X-ray structure determination shows the solid compound to be built up of discrete tris(1,2-dimethoxyethane-O,O′)lithium cations and E/E-isomeric 3-phenyl-1,3-bis(trimethylsilyl)-1,2-diphosphapropenide anions (Fig. 6).

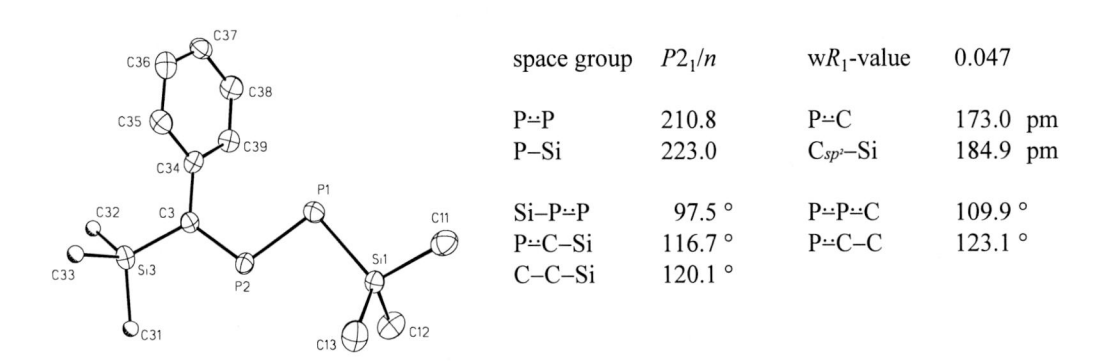

space group	P2$_1$/n	wR$_1$-value	0.047
P⋯P	210.8	P⋯C	173.0 pm
P–Si	223.0	C$_{sp^2}$–Si	184.9 pm
Si–P⋯P	97.5 °	P⋯P⋯C	109.9 °
P⋯C–Si	116.7 °	P⋯C–C	123.1 °
C–C–Si	120.1 °		

Fig. 6. Structure of the E/E-isomeric 3-phenyl-1,3-bis(trimethylsilyl)-1,2-diphosphapropenide anion; 30 % probability for most of the heavier atoms; hydrogen atoms omitted; measurement at −120±3 °C

With values of 210.8 and 173.0 pm both the lengths of the P⋯P and the P⋯C bond correspond to a bond order of 1.5. Similar to other phosphaalkenes the distance between two coordinate phosphorus and silicon of a trimethylsilyl group is found to be shortened with respect to the standard value (223.0 vs 227 pm) [35].

In order to understand the formation of compounds now being discussed one has to realize that tris(trimethylsilyl)phosphane originating in the desilylation of [1-(trimethylsiloxy)benzylidene]-trimethylsilylphosphane (Scheme 1) reacts slowly with lithium ethanolate to give lithium bis(trimethylsilyl)phosphanide again (Eq. 13). This compound must then be considered a continuous source for the phosphaalkyne $H_5C_6-C\equiv P$, provided that a sufficient amount of ethyl benzoate is present in solution.

$$(Me_3Si)_3P \xrightarrow[-Me_3Si-OEt]{+Li-OEt} Li-P(SiMe_3)_2$$

Eq. 13.

A 1,2,4-triphosphapentadienide anion from 2,4,6-trimethylbenzoyl chloride [36]: A subsequent nucleophilic attack of the negatively polarized sp^2-hybridized carbon of 3-phenyl-1,3-bis(trimethylsilyl)-1,2-diphosphapropenide at phosphorus, i.e., the positively polarized atom in the P≡C group of phenylmethylidynephosphane, and an attendant 1,3-shift of the trimethylsilyl group from carbon in position 3 to carbon in position 5 should result in the formation of the open chained 3,5-diphenyl-1,5-bis(trimethylsilyl)-1,2,4-triphosphapentadienide anion (Eq. 14).

Eq. 14.

Although this species could not be detected by ^{31}P NMR spectroscopy with ethyl benzoate as an educt, we succeeded in an isolation of analogous tetrakis(tetrahydrofuran)lithium 3,5-bis(2,4,6-trimethylphenyl)-1,5-bis(trimethylsilyl)-1,2,4-triphosphapentadienide treating a THF solution of lithium bis(trimethylsilyl)-phosphanide with 2,4,6-trimethylbenzoyl chloride in a molar ratio of 3:2. As a consequence of special reaction conditions, such as a temperature range near −60 °C, concentration, and addition rate applied, synthesis, purification and especially the originally high yield of 86 % have been difficult to reproduce. Generally the obtained product has been found to be more or less contaminated with bis(tetrahydrofuran)lithium bis(2,4,6-trimethylbenzoyl)phosphanide (Eq. 15).

Considering the underlying mechanism we assume that the reaction process is determined mainly by the chemical behaviour of [lithoxy(2,4,6-trimethylphenyl)methylidene]trimethylsilylphosphane formed in analogy to Scheme 1. Provided that this intermediate is allowed to split off lithium trimethylsilanolate, the

1,2,4-triphosphapentadienide will be obtained *via* further addition reactions of the phosphaalkyne produced, whereas substitution of the trimethylsilyl substituent for a second 2,4,6-trimethylbenzoyl group gives the lithium diacylphosphanide (Eq. 15).

Eq. 15.

Remarkably the NMR parameters of tetrakis(tetrahydrofuran)lithium 3,5-bis(2,4,6-trimethylphenyl)-1,5-bis(trimethylsilyl)-1,2,4-triphosphapentadienide depend strongly on the solvents used; therefore, in Table 2 values of a $[D_8]$-THF solution are given only.

As determined by an X-ray structure analysis of dark red, shining, air sensitive crystals obtained at −20 °C from a tetrahydrofuran/*n*pentane solution the solid is built up of discrete tetrakis(tetrahydrofuran)lithium cations and 3,5-bis(2,4,6-trimethylphenyl)-1,5-bis(trimethylsilyl)-1,2,4-triphosphapentadienide anions. A characteristic P–P distance of 206.0 pm and an average of 272 pm for three almost equally long P–C bonds indicate an electronically balanced system; the sequence of three phosphorus, two carbon, and two silicon atoms at both ends are arranged in a staggered, almost planar chain. Since the two 2,4,6-trimethylphenyl substituents in positions 3 and 5 adopt a *cis* constitution, their planes are nearly perpendicularly arranged with respect to that chain. Some further characteristic structure parameters are given in Fig. 7.

space group	$Pna2_1$	R_w-value	0.055

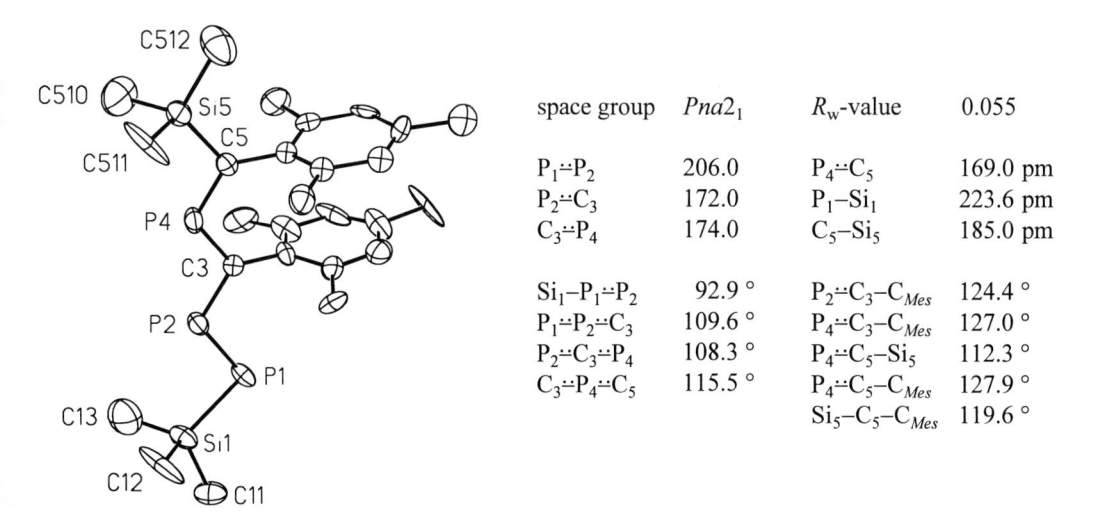

$P_1\dot{-}P_2$	206.0		$P_4\dot{-}C_5$	169.0 pm
$P_2\dot{-}C_3$	172.0		P_1-Si_1	223.6 pm
$C_3\dot{-}P_4$	174.0		C_5-Si_5	185.0 pm
$Si_1-P_1\dot{-}P_2$	92.9 °		$P_2\dot{-}C_3-C_{Mes}$	124.4 °
$P_1\dot{-}P_2\dot{-}C_3$	109.6 °		$P_4\dot{-}C_3-C_{Mes}$	127.0 °
$P_2\dot{-}C_3\dot{-}P_4$	108.3 °		$P_4\dot{-}C_5-Si_5$	112.3 °
$C_3\dot{-}P_4\dot{-}C_5$	115.5 °		$P_4\dot{-}C_5-C_{Mes}$	127.9 °
			$Si_5-C_5-C_{Mes}$	119.6 °

Fig. 7. Structure of the 3,5-bis(2,4,6-trimethylphenyl)-1,5-bis(trimethylsilyl)-1,2,4-triphosphapentadienide anion; 50 % probability; hydrogen atoms omitted; measurement at -100 ± 3 °C

$\delta^{31}P\{^1H\}$	P_1	126.5 (ddd)[a]	P_2	480.6 (ddd)	P_4	287.8 (ddd)
	$^1J_{PP}$	488.2	$^2J_{PP}$	349.1	$^3J_{PP}$	20.1
$\delta^{13}C\{^1H\}$	C_3	171.6 (dpt)[b]			C_5	[c]
	$^1J_{CP}$	70.7	$^2J_{PC}$	14.6		–
$\delta^{29}Si\{^1H\}$	Si_1	6.0 (dd)			Si_5	2.2 (d)
	$^1J_{SiP}$	61.1	$^2J_{SiP}$	26.7	$^2J_{SiP}$	49.5

Table 2. Characteristic NMR parameters of the 3,5-bis(2,4,6-trimethylphenyl)-1,5-bis(trimethyl-silyl)-1,2,4-triphosphapentadienide anion; chemical shift δ (ppm); coupling constants (Hz); [D$_8$]-THF solution; [a] Doublet of doublets of doublets; [b] doublet of pseudotriplets, assignment very uncertain; [c] undetermined due to impurities coming from a subsequent cyclization reaction
Numbering Scheme:

Mes = 2,4,6-Me$_3$H$_2$C$_6$

A 2-phosphaallyl anion as part of a five membered heterocycle [27, 35]: When in the already discussed reaction between ethyl benzoate and lithium bis(trimethylsilyl)phosphanide the molar ratio of the educts is changed finally from 1:3 to 1:1, dark green, metallically lustrous, air sensitive crystals of tris(1,2-dimethoxyethane-O,O')lithium 3,5-diphenyl-1,2-bis(trimethylsilyl)-1,2,4-triphospholide can be isolated from a just as dark green 1,2-dimethoxyethane solution in about 50 % yield. Obviously, the preceeding, but still undetected 3,5-diphenyl-1,5-bis(trimethylsilyl)-1,2,4-triphosphapentadienide undergoes a very fast cyclization reaction in that the negatively charged carbon at one end of the anion attacks the phosphorus at the other nucleophilicly. Simultaneously, one trimethylsilyl substituent is shifted from this carbon to the other phosphorus atom in position 2 and a 2-phosphaallyl anion is formed (Eq. 16).

Eq. 16.

The ^{31}P{^{1}H} NMR spectrum obtained from a [D$_8$]-toluene/1,2-dimethoxyethane solution shows a doublet at –87.6 and a triplet at 274.8 ppm with a $^2J_{PP}$-value of 33.0 Hz. Unfortunately, the sp^2-hybridized carbon within the heterocyclic P–C$\doteq$P fragment gives rise to a merely weak unstructured multiplet at 114.0 ppm. The solid is built up of discrete tris(1,2-dimethoxy-O,O')lithium cations and planar 1,2,4-triphospholide anions with the trimethylsilyl substituents in a *trans* position at the P–P group. Remarkably, the two arene rings and the five membered heterocycle lie almost in a plane; as a consequence, extended electronic interactions might occur and hence stabilize the anion. Characteristic bond lengths and angles do not differ on both sides of the heterocycle (Fig. 8).

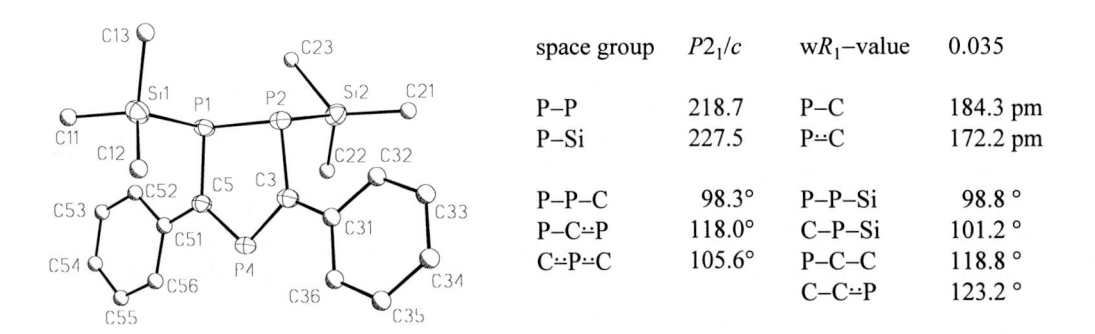

space group	$P2_1/c$	wR$_1$–value	0.035
P–P	218.7	P–C	184.3 pm
P–Si	227.5	P$\doteq$C	172.2 pm
P–P–C	98.3°	P–P–Si	98.8 °
P–C$\doteq$P	118.0°	C–P–Si	101.2 °
C$\doteq$P$\doteq$C	105.6°	P–C–C	118.8 °
		C–C$\doteq$P	123.2 °

Fig. 8. Structure of the 3,5-diphenyl-1,2-bis(trimethylsilyl)-1,2,4-triphospholide anion; 50 % probability for the silicon atoms and the atoms of the heterocycle; hydrogen atoms omitted; measurement at –120±3 °C; average values

A 1,2,4-triphospholide and a phospholide anion: Since in subsequent studies on the synthesis of a *t*butyl substituted 1,2-bis(trimethylsilyl)-1,2,4-triphospholide, the phosphaalkyne Me$_3$C–C≡P has turned out to be too reactive a starting compound, its precursor [2,2-dimethyl-1-(trimethylsiloxy)pro-pylidene]trimethylsilylphosphane was treated with half the molar amount of lithium bis(trimethylsilyl)-phosphanide at –50 °C in 1,2-dimethoxyethane. Probably all reactions steps will be run through as discussed before; finally, however, both the trimethylsilyl substituents are lost in a still unknown way. Accordingly, long, pale yellow needles of tris(1,2-dimethoxyethane-*O,O'*)lithium 3,5-di*t*butyl-1,2,4-triphospholide can be isolated at –30 °C from the reaction mixture in about 25 % yield. In the ^{31}P{^{1}H} NMR spectrum of the compound dissolved in a mixture of [D$_6$]-benzene and 1,2-dimethoxyethane a doublet and a triplet at 242.5 and 248.4 ppm, respectively, with a $^2J_{PP}$, 47.5 Hz are observed. An X-ray structure determination shows the solid to be built up of discrete tris(1,2-dimethoxyethane-*O,O'*)lithium cations and planar 1,2,4-triphospholide anions. Bond lengths and angles (Fig. 9) are as expected for a cyclic 6 π-electron system [27, 37]. In recent years, Nixon and coworkers [38] have been utilizing this compound widely in the syntheses of sandwich complexes of transition elements.

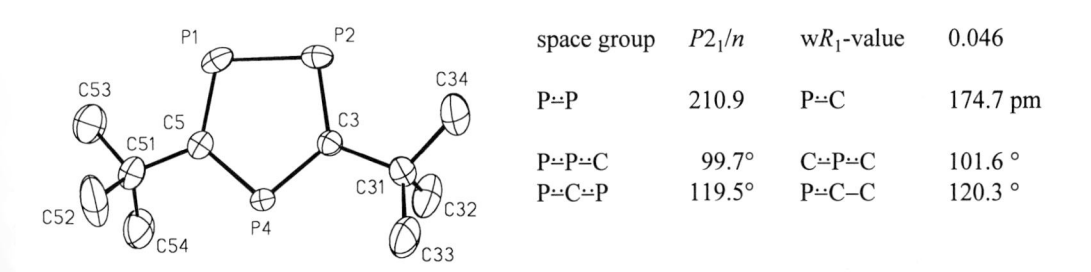

space group	$P2_1/n$	wR_1-value	0.046
P∺P	210.9	P∺C	174.7 pm
P∺P∺C	99.7°	C∺P∺C	101.6 °
P∺C∺P	119.5°	P∺C–C	120.3 °

Fig. 9. Structure of the 3,5-di*t*butyl-1,2,4-triphospholide anion; 50 % probability; hydrogen atoms omitted; measurement at –120±3 °C; average values

Surprisingly, the reaction of phosphaalkynes with lithium bis(trimethylsilyl)phosphanide may be transferred to alkynes. With diphenylethyne, for example, tetrakis(tetrahydrofuran)lithium 2,3,4,5-tetraphenylphospholide has been isolated in about 60 % yield from a THF solution (Eq. 17); again, the fate of both trimethylsilyl substituents remains unknown. The solid is built up of discrete tetrakis(tetrahydrofuran)lithium cations and planar phospholide anions (space group $P\bar{1}$; wR_2-value 0.173; P∺C 176; C∺C 140 to 143 pm); the phenyl groups are tilted out of the C$_4$P plane by angles of 47–65° [39].

$$H_5C_6-C\equiv C-C_6H_5$$
$$+$$
$$Li-P(SiMe_3)_2$$
$$+$$
$$H_5C_6-C\equiv C-C_6H_5$$

refluxing thf; 7 d
− 2 Me$_3$Si·

→ [phospholide anion with C$_6$H$_5$ groups] [(thf)$_4$Li]$^+$

Eq. 17.

Reactions with Nitriles

Since replacement of phosphorus by nitrogen changes the polarity of an E≡C triple bond considerably from $P^{\delta+}-C^{\delta-}$ to $N^{\delta-}-C^{\delta+}$, phosphaalkynes and nitriles are expected to show quite a different chemical reactivity towards lithium trimethylsilylphosphanides. However, only nitriles without a CH group in α-position have been found to be convertible to phosphaalkenes; accordingly from a 1,2-dimethoxyethane solution of acetonitrile and lithium dihydrogenphosphanide only the already known [40] trimer 6-amino-2,4-dimethylpyrimidine could be isolated in 44 % yield.

1-(Lithium-amido)benzylidenephosphanes [41]: In contrast to this observation benzonitrile reacts with an equivalent amount of lithium phenyltrimethylsilylphosphanide at −50 °C in 1,2-dimethoxyethane to give 1-[(1,2-dimethoxyethane-O,O′)lithium-trimethylsilylamido]benzylidenephenylphosphane (Eq. 18). Presumably the still undetected intermediate arising from a nucleophilic attack of the phosphanide anion at the *sp*-hybridized carbon of the nitrile function stabilizes rapidly by a 1,3-shift of the trimethylsilyl group from phosphorus to nitrogen. The P≐C group of the compound is characterized by the following NMR parameters: $\delta^{31}P$ 60.0; $\delta^{13}C$ 211.6 ppm; $^1J_{CP}$ 62.2 Hz; [D$_8$]-THF solution.

Eq. 18.

An X-ray structure analysis of red rodlets isolated from an 1,2-dimethoxyethane/*n*pentane solution at –5 °C shows the compound to crystallize as a Z-isomeric neutral complex (Fig. 10). P⸪C and C⸪N bond lengths of 176.9 and 132.3 pm respectively are in good agreement with the formation of an electronically stabilized mesomeric heteroallyl anion. Lithium primarily bound to nitrogen and two oxygen atoms of the 1,2-dimethoxyethane ligand increases its coordination number from 3 to 4 or probably 5 by interactions with the *ipso*-carbon of the phenyl substituent and phosphorus.

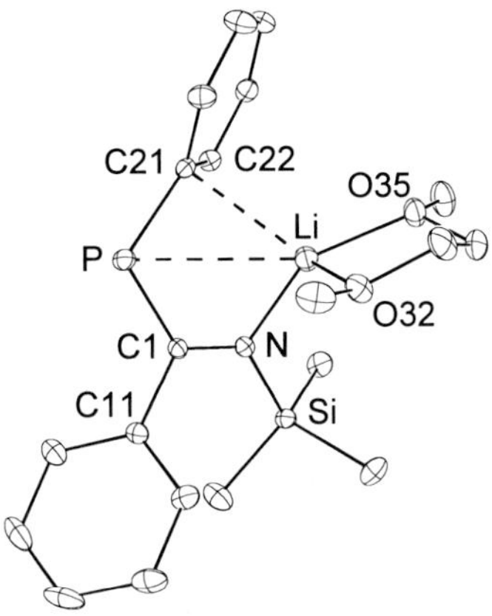

space group	Cc	R-value	0.042
P⸪C	176.9	N–Si	172.5 pm
P–C$_6$H$_5$	183.5	Li–N	197.0 pm
C⸪N	132.3	Li··P	92.4 pm
		Li··C$_6$H$_5$	240.4 pm
C–P–C	103.6 °	P–C–N	128.0 °
P–C(N)–C	111.0 °	N–C–C	120.7 °

Fig. 10. Structure of the neutral complex Z-1-[(1,2-dimethoxyethane-*O,O'*)lithium-trimethylsilylamido]benzylidene-phenylphosphane; 30 % probability; hydrogen atoms omitted; measurement at –100 ± 3 °C

Lithium bis(iminobenzoyl)phosphanides [41, 50]: From the reaction of lithium bis(trimethylsilyl)phosphanide with two equivalents of benzonitrile at –50 °C the 1,2-dimethoxyethane complex of lithium bis(trimethylsilyliminobenzoyl)phosphanide is obtained in about 70 % yield (Eq. 19). ^{31}P{^{1}H} and ^{13}C{^{1}H} NMR spectra taken from a [D$_8$]-THF solution show a singlet at +63.3 and a doublet at 208.3 ppm with $^1J_{CP}$ 83.4 Hz respectively, for the C⸪P⸪C fragment.

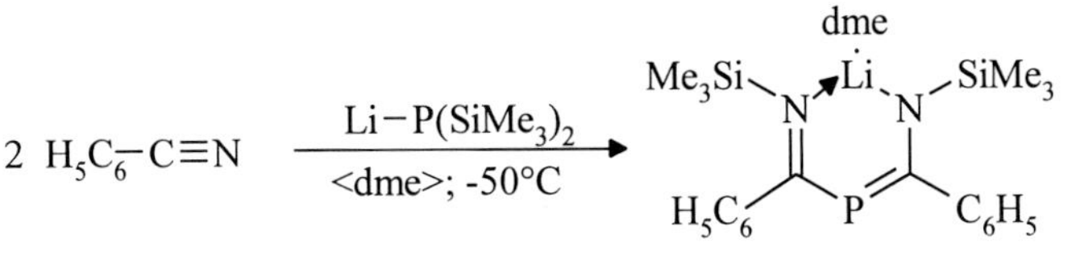

$$2\ H_5C_6-C\equiv N \xrightarrow[\text{<dme>; -50°C}]{Li-P(SiMe_3)_2} \underset{H_5C_6}{\overset{Me_3Si}{\underset{}{}}}\!\!\!\!\!\!\!\! \text{N}\overset{\text{dme}}{\underset{}{\text{Li}}}\text{N}\!\!\!\!\!\!\!\! \underset{C_6H_5}{\overset{SiMe_3}{\underset{P}{}}}$$

Eq. 19.

When, however, benzonitrile is treated in diethylether with only one equivalent of lithium bis-(trimethylsilyl)phosphanide, [1-(lithium-trimethylsilylamido)benzylidene]trimethylsilylphosphane, the product of the first reaction step, can be detected by its NMR spectra (δ^{31}**P** –11.7; δ^{13}**C** 222.6 ppm;

$^1J_{CP}$ 53.5 Hz). The intermediate is formed by a nucleophilic attack of the phosphanide anion, combined with an 1,3-shift of one trimethylsilyl group from phosphorus to nitrogen (Eq. 20); it rearranges within three days at −50 °C to give lithium bis(trimethylsilyliminobenzoyl)- and lithium bis(trimethylsilyl)-phosphanide. In 1,2-dimethoxyethane the reaction is too fast to observe the intermediate.

$$H_5C_6-C\equiv N \quad \xrightarrow[<et_2O>;\ -50°C]{+\ Li-P(SiMe_3)_2} \quad H_5C_6-C\begin{smallmatrix}N-SiMe_3\\ \|\\ P\cdots SiMe_3\end{smallmatrix}^{Li_{(solv.)}}$$

Eq. 20.

As (1,2-dimethoxyethane-O,O')lithium bis(trimethylsilyliminobenzoyl)phosphanide crystallizes from an 1,2-dimethoxyethane/*n*pentane solution within several days at −5 °C in small, but well shaped yellow plates, its structure (Fig. 11) could be determined by X-ray diffraction methods. The hard lithium cation is not bound to the soft phosphorus, but to the harder nitrogen atoms and increases its coordination number by an interaction with both the oxygen atoms of an 1,2-dimethoxyethane ligand. Bond lengths and angles within the electronically balanced chelating bis(trimethylsilyliminobenzoyl)phosphanide anion of the neutral complex are very similar to values determined from lithium diacylphosphanides [30, 44].

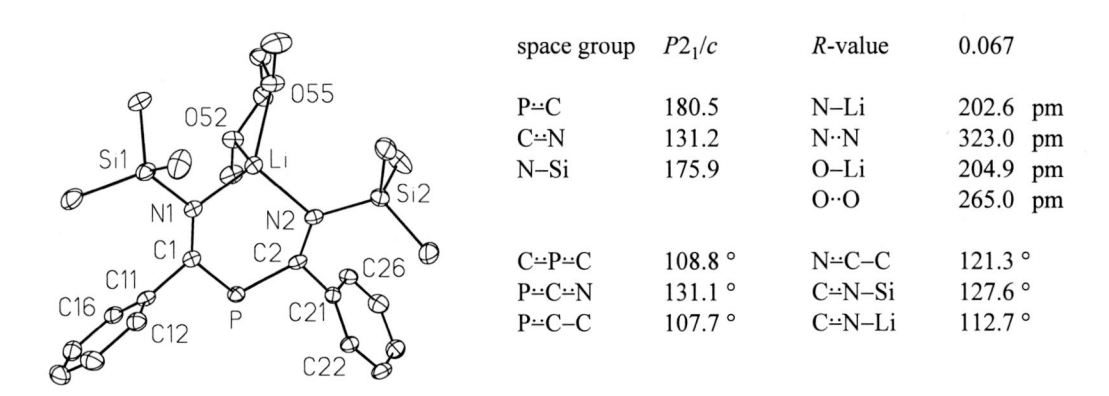

space group	$P2_1/c$		R-value	0.067
P∺C	180.5		N–Li	202.6 pm
C∺N	131.2		N··N	323.0 pm
N–Si	175.9		O–Li	204.9 pm
			O··O	265.0 pm
C∺P∺C	108.8 °		N∺C–C	121.3 °
P∺C∺N	131.1 °		C∺N–Si	127.6 °
P∺C–C	107.7 °		C∺N–Li	112.7 °

Fig. 11. Structure of the neutral complex 1-(1,2-dimethoxyethane-O,O')lithium bis(trimethylsilyliminobenzoyl) phosphanide; 30 % probability; hydrogen atoms omitted; measurement at −100 ± 3 °C; average values

Treating (1,2-dimethoxyethane-O,O')lithium bis(trimethylsilyliminobenzoyl)phosphanide with trifluoroacetic acid in 1,2-dimethoxyethane solution the cation can be exchanged for a proton to give bis(trimethylsilyliminobenzoyl)phosphane (Eq. 21). In contrast to diacylphosphanes which show a keto-enol equilibrium in solution [45], this compound has been found to exist only as an imino-enamine tautomer, even in highly polar solvents. Using [D$_8$]-THF as a solvent, for example, the following characteristic NMR parameters of the C⋕P⋕C fragment can be obtained: δ^{31}P 58.1; δ^{13}C 205.8 ppm; $^1J_{CP}$ 76.4 Hz; δ^1H–N 14.2 ppm).

Eq. 21.

As found by an X-ray structure determination of thin yellow to orange needles obtained at −5 °C from an acetonitrile/*n*pentane solution, the imino-enamine tautomer also exists in the solid. As the hydrogen atom is very asymmetrically bound to the two nitrogen atoms, P–C (183.6 vs 174.5 pm) and C–N bond lengths (128.0 vs 135.5 pm) differ substantially in both parts of the molecule (Fig. 12). Since in the enol tautomers of diacylphosphanes P–C and C–O bond lengths are much more equalized, several solid compounds show an O···H···O bridge which is symmetric, however, for crystallographic reasons [46].

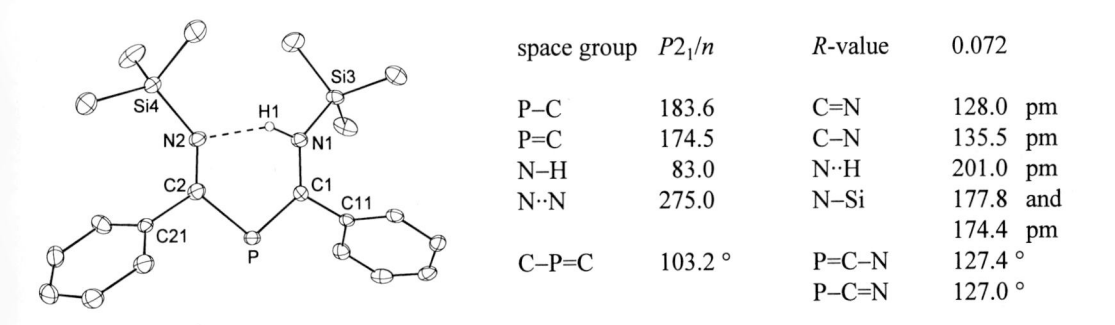

space group	$P2_1/n$	R-value	0.072
P–C	183.6	C=N	128.0 pm
P=C	174.5	C–N	135.5 pm
N–H	83.0	N··H	201.0 pm
N··N	275.0	N–Si	177.8 and 174.4 pm
C–P=C	103.2 °	P=C–N	127.4 °
		P–C=N	127.0 °

Fig. 12. Structure of the imino-enamine tautomer of bis(trimethylsilyliminobenzoyl)phosphane; 30 % probability; measurement at −100 ± 3 °C; with exception of the NH group hydrogen atoms have been omitted

Conclusion

This short review clearly demonstrates how the introduction and subsequent elimination or the 1,3-shift of trimethylsilyl substituents can be used in a manifold manner to synthesize new classes of low-valent phosphorus compounds. Since, at present the underlying principles and mechanisms are only partially understood, further investigations in this fascinating area will be needed in order to open new routes to even more unexpected species. As far as the recently accessible anionic phosphaalkenes and phosphaalkynes of this article are concerned, at the moment they are represented by a few examples only. Therefore, many more experiments have to be carried out in order to elucidate the chemical and physical properties of these compounds and to integrate them into the still enlarging field of phosphorus chemistry.

Acknowledgement: We thank *Deutsche Forschungsgemeinschaft*, *Fond der Chemischen Industrie*, and *Hoechst AG* for generous financial support, Frau G. Weckler and Frau Dr. S. Abele for typing the manuscript.

References:

[1] G. Fritz, *Z. Naturforsch.* **1953**, *8B*, 776.

[2] G. Becker, W. Hölderich, *Chem. Ber.* **1975**, *108*, 2484.

[3] G. Becker, G. Gutekunst, H. J. Wessely, *Z. Anorg. Allg. Chem.* **1980**, *462*, 113.

[4] G. Becker, H. Freudenblum, O. Mundt, M. Reti, M. Sachs, in: *Brauer, Handbook of Preparative Inorganic Chemistry* (Ed.: W. A. Herrmann), in press.

[5] G. Becker, M. Rößler, *Z. Naturforsch.* **1982**, *B37*, 91.

[6] W. Uhlig, A. Tzschach, *Z. Anorg. Allg. Chem.* **1989**, *576*, 281; E. Niecke, H. Westermann, *Synthesis* **1988**, 330.

[7] H. H. Karsch, F. Bienlein, T. Rupprich, F. Uhlig, E. Herrmann, M. Scheer, in: *Brauer, Handbook of Preparative Inorganic Chemistry* (Ed.: W. A. Herrmann), in press.

[8] G. Becker, H.-M. Hartmann, W. Schwarz, *Z. Anorg. Allg. Chem.* **1989**, *577*, 9.

[9] G. Becker, C. Witthauer, *Z. Anorg. Allg. Chem.* **1982**, *492*, 28.

[10] G. Becker, *Z. Anorg. Allg. Chem.* **1976**, *423*, 242; *Z. Anorg. Allg. Chem.* **1977**, *430*, 66.

[11] G. Becker, G. Gutekunst, *Z. Anorg. Allg. Chem.* **1980**, *470*, 131; *Z. Anorg. Allg. Chem* **1980**, *470*, 144.

[12] G. Becker, M. Rößler, W. Uhl, *Z. Anorg. Allg. Chem.* **1981**, *473*, 7.

[13] G. Becker, W. Schwarz, N. Seidler, M. Westerhausen, *Z. Anorg. Allg. Chem.* **1992**, *612*, 72; and references therein.

[14] G. Becker, G. Gresser, W. Uhl, *Z. Naturforsch.* **1981**, *B36*, 16.

[15] W. Rösch, U. Vogelbacher, T. Allspach, M. Regitz, *J. Organomet. Chem.* **1986**, *306*, 39.

[16] M. Regitz, P. Binger, *Angew. Chem.* **1988**, *100*, 1541; *Angew. Chem., Int. Ed. Engl.* **1988**, *27*, 1484.

[17] J. F. Nixon, *Chem. Rev.* **1988**, *88*, 1327.

[18] L. N. Markovski, V. D. Romanenko, *Tetrahedron* **1989**, *45*, 6019.

[19] M. Regitz, *Chem. Rev.* **1990**, *90*, 191.

[20] *Multiple Bonds and Low Coordination in Phosphorus Chemistry* (Eds.: M. Regitz, O. J. Scherer), Thieme, Stuttgart, **1990**.

[21] G. Märkl, H. Sejpka, *Angew. Chem.* **1986**, *98*, 286; *Angew. Chem., Int. Ed. Engl.* **1986**, *25*, 264; see, however: J.-C. Guillemin, L. Lassalle, P. Dréan, G. Wlodarczak, J. Demaison, *J. Am. Chem. Soc.* **1994**, *116*, 8930.

[22] G. Becker, K. Merz, W. Schwarz, unpublished results.

[23] S. J. Goede, F. Bickelhaupt, *Chem. Ber.* **1991**, *124*, 2677.

[24] D. J. Peterson, *J. Org. Chem.* **1968**, *33*, 780; L. Birkofer, O. Stuhl, *Top. Current Chem.* **1980**, *88*, 33.

[25] A. R. Barron, A. H. Cowley, S. W. Hall, *J. Chem. Soc. Chem. Commun.* **1987**, 980.

[26] G. Becker, G. Ditten, M. Niemeyer, W. Schwarz, unpublished results.

[27] G. Becker, W. Becker, R. Knebl, H. Schmidt, U. Weeber, M. Westerhausen, *Nova Acta Leopoldina, Neue Folge No. 264*, **1985**, *59*, 55.

[28] G. Becker, M. Westerhausen, unpublished results.

[29] G. Becker, N. Seidler, Z. Zheng, unpublished results.

[30] G. Becker, G. Heckmann, K. Hübler, W. Schwarz, *Z. Anorg. Allg. Chem.* **1995**, *621*, 34.

[31] J. Grobe, D. Le Van, B. Lüth, M. Hegemann, *Chem. Ber.* **1990**, *123*, 2317.

[32] G. Becker, K. Hübler, *Z. Anorg. Allg. Chem.* **1994**, *620*, 405.

[33] E. Niecke, M. Nieger, P. Wenderoth, *J. Am. Chem. Soc.* **1993**, *115*, 6989; *Angew. Chem.* **1994**, *106*, 362; *Angew. Chem., Int. Ed. Engl.* **1994**, *33*, 353; *Angew. Chem.* **1994**, *106*, 2045; *Angew. Chem., Int. Ed. Engl.* **1994**, *33*, 1953.

[34] A. C. Gaumont, X. Morise, J. M. Denis, *J. Org. Chem.* **1992**, *57*, 4292; see also: F. Mercier, C. Hugel-Le Goff, L. Ricard, F. Mathey, *J. Organomet. Chem.* **1990**, *389*, 389; and references therein.

[35] G. Becker, W. Becker, G. Ditten, unpublished results.

[36] G. Becker, M. Schmidt, unpublished results.

[37] G. Becker, H.-D. Hausen, U. Hildenbrand, U. Hübler, to be published.

[38] R. Bartsch, A. Gelessus, P. B. Hitchcock, J. F. Nixon, *J. Organomet. Chem.* **1992**, *430*, C10; and references therein.

[39] G. Becker, Z. Zheng, unpublished results.

[40] A. R. Ronzio, W. B. Cook, *Org. Synth. Coll.* **1955**, *3*, 71.

[41] G. Becker, U. Hübler, unpublished results.

[42] H. H. Karsch, F. Bienlein, in: *Organosilicon Chemistry II* (Eds.: N. Auner, J. Weis), VCH, Weinheim, **1995**, p. 133.

[43] P. B. Hitchcock, M. F. Lappert, D.-S. Liu, *J. Chem. Soc., Chem. Comm.* **1994**, 1699.

[44] G. Becker, K. Hübler, M. Niemeyer, N. Seidler, B. Thinus, *Z. Anorg. Allg. Chem.*, in press.

[45] G. Becker, M. Schmidt, W. Schwarz, M. Westerhausen, *Z. Anorg. Allg. Chem.* **1992**, *608*, 33; and references therein.

[46] G. Becker, W. Becker, M. Schmidt, W. Schwarz, M. Westerhausen, *Z. Anorg. Allg. Chem.* **1991**, *605*, 7.

[47] Since alkylidene- and alkylidynephosphanes show an alkene- or alkyne-like chemical reactivity, they are often called phosphaalkenes and phosphaalkynes.

[48] The homologous trimethylstannyl compound recently prepared according to Eq. 7 turned out to be much more.

[49] Phosphaallyl anions are also being studied in the research groups of, e.g., Niecke [33] and Denis [34].

[50] Similar results have been obtained by Karsch and coworkers [42]. The group of Lappert was able to isolate dimeric lithium bis(trimethylsilyliminobenzoyl)methanide [43].

Silicon and Phosphinomethanides: Novel Heteroelement Substituted Methanes and Ylides

Hans Heinz Karsch*, Roland Richter

Anorganisch-Chemisches Institut

Technische Universität München

Lichtenbergstr. 4, D-85747 Garching, Germany

Summary: $LiCH(PMe_2)_2$ reacts with $SiCl_4$ or with PhR_2SiCl (R = Me, Ph) via the carbanion to give silyl substituted phosphinomethanes. $HC(PMe_2)_2(SiMe_2Ph)$ is deprotonated by nBuLi to give trimeric $\{Li[C(PMe_2)_2(SiMe_2Ph)]\}_3$. $\{Li[C(PMe_2)(SiMe_3)_2]\}_2 \cdot$TMEDA **3** reacts with RR'SiCl$_2$ via both the carbanion, thus generating a tetraheteroatom substituted methane moiety and, in a second substitution step, via phosphorus, thus forming an ylidic moiety. This proposed reaction pathway can be verified by the reaction of $PhSiCl_3$ with a half equivalent of **3** where "C"-coordination of the ligand is obtained. Novel skeletal rearrangements and C–H activation reactions starting from $RSiCl_3$ (R = Ph, tBu, Me) under formation of five-membered heterocycles are reported.

Introduction

To promote the formation of the rather weak silicon-phosphorus bonds, we have introduced anionic monophosphinomethanide and diphosphinomethanide ligands of type **I** or **II** (Fig. 1), respectively, into the coordination sphere of a silicon center.

Depending on the substituents R, X, and Y, the reactivity of the ligand is tunable, e.g., silyl or phosphino C-substituents reduce the carbanion nucleophilicity. For instance, Me_3SiCl reacts with $LiCH_2PMe_2$ via Si–C bond formation **4** [1]. In contrast, Me_3SiCl reacts with the fully heteroelement substituted **2** under Si–P bond formation **5** [2].

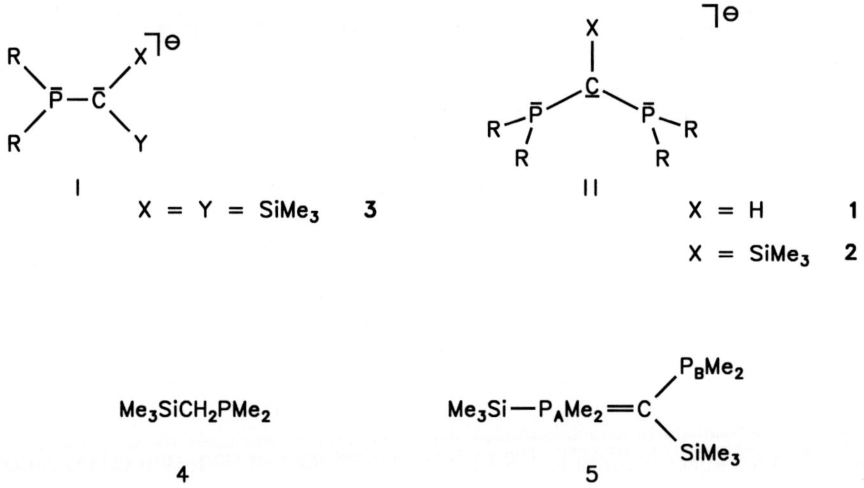

Fig. 1.

I X = Y = SiMe₃ **3**

II X = H **1**
    X = SiMe₃ **2**

Me₃SiCH₂PMe₂ **4**

Me₃Si—P_AMe₂=C(P_BMe₂)(SiMe₃) **5**

Furthermore, these ligands stabilize high phosphine coordination numbers at the silicon. Me$_2$SiCl$_2$ reacts with two equivalents of **2** to form a hypervalent silicon complex with a hexacoordinated silicon center **6** (Fig. 2) [3].

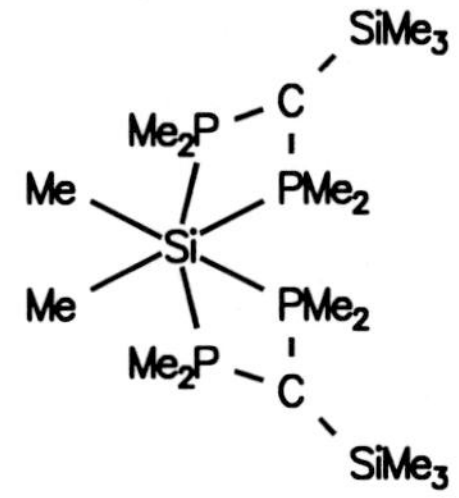

Fig. 2.

Structure **6**: hexacoordinate Si center bonded to two Me groups and two [C(SiMe₃)(PMe₂)₂] chelating ligands.

1 Interaction of Chlorosilanes with Li[CH(PMe₂)₂] (1)

Various "C-coordinated compounds" are obtained by the reaction of different chlorosilanes with **1** (Scheme 1) [4]. Compound **7** is available as a colorless liquid; **8** and **9** are colorless, crystalline solids, which are characterized by X-ray structure determinations (Fig. 3).

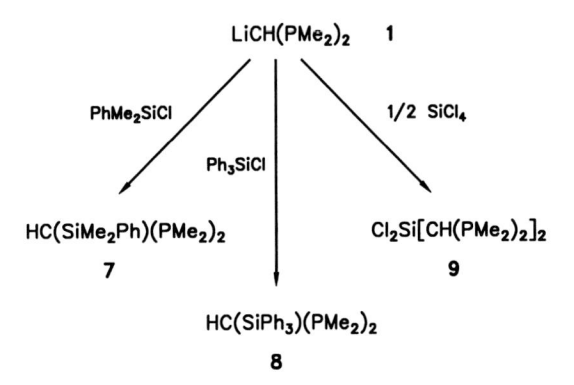

Scheme 1.

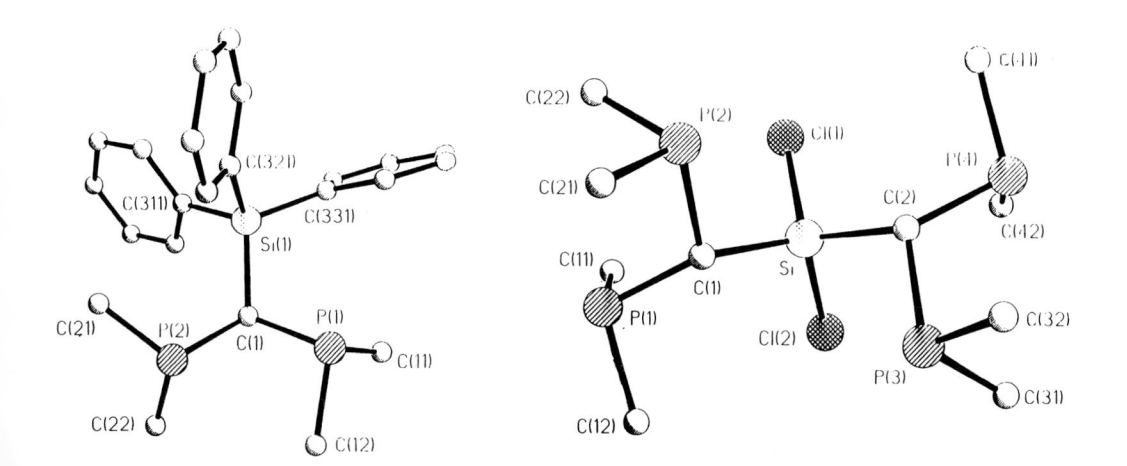

Fig. 3. Molecular structure of 8 and 9

A further deprotonation of 7 with nBuLi gives the first coligand-free diphosphinomethanide in a crystalline form [4] (Eq. 1). The molecular structure of the trimeric {Li[C(PMe$_2$)$_2$(SiMe$_2$Ph)]}$_3$ (10) shows that each Li atom is surrounded by two phosphorus and one carbon. Two of the three Li atoms show additional very weak lithium-phenyl contacts (Fig. 4).

$$(Me_2P)_2(PhMe_2Si)CH \ + \ ^nBuLi \ \xrightarrow{- BuH} \ 1/3 \ \{Li[C(PMe_2)_2(SiMe_2Ph)]\}_3$$

7   10

Eq. 1.

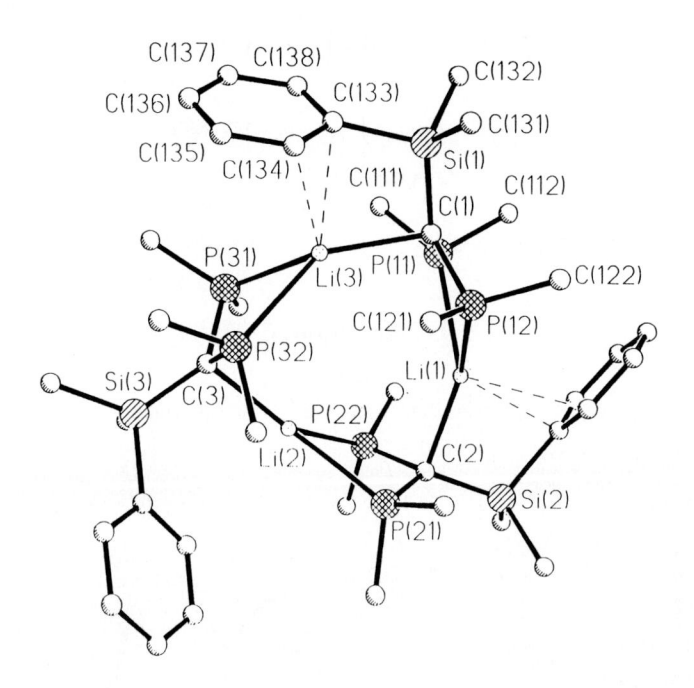

Fig. 4. Molecular structure of **10**

2 Interaction of Chlorosilanes with {Li[C(PMe$_2$)(SiMe$_3$)$_2$]}$_2$·TMEDA (3)

Me$_3$SiCl reacts with **3** forming the ylide **11** and the isomeric methane **12** (Eq. 2) [2].

$$\text{Me}_3\text{SiCl} + \mathbf{3} \longrightarrow \begin{cases} \text{Me}_3\text{Si–PMe}_2\text{=C(SiMe}_3)_2 \\ \mathbf{11} \\ \\ (\text{PMe}_2)\text{C(SiMe}_3)_3 \\ \mathbf{12} \end{cases}$$

Eq. 2.

According to Eq. 3, the di- or tetrafunctional chlorosilanes react with two equivalents of **3** in an analogous way: both "P"- and "C"-coordination at the silicon center is obtained [4, 5].

$$RR'SiCl_2 \;+\; \{Li[C(PMe_2)(SiMe_3)_2]\}_2 \cdot TMEDA$$

$$\mathbf{3}$$

$$\xrightarrow{-2\,LiCl} (Me_3Si)_2(Me_2P_B)C-SiRR'-P_AMe_2=C(SiMe_3)_2$$

$$R = R' = Me \quad \mathbf{13}$$
$$R = R' = Cl \quad \mathbf{14}$$
$$R = Ph;\; R' = Cl \quad \mathbf{15}$$

Eq. 3.

In the first substitution step **3** reacts under Si-C bond formation, generating the tetraheteroatom substituted methane moiety. For sterical reasons in the second substitution step only "P"-coordination can be achieved. The preference of "C"-coordination in the first step can be confirmed by the 1:1 reaction of PhSiCl$_3$ with **3** (Eq. 4).

$$PhSiCl_3 \;+\; \mathbf{3} \longrightarrow (PhSiCl_2)C(PMe_2)(SiMe_3)_2$$

$$\mathbf{16}$$

Eq. 4.

The obtained compounds (Me$_3$Si)$_3$(Me$_2$P)C–SiR$_2$PMe$_2$=C(SiMe$_3$)$_2$ (R = Me, **13** Cl) **14** are structurally characterized by X-ray structure determinations (Fig. 5).

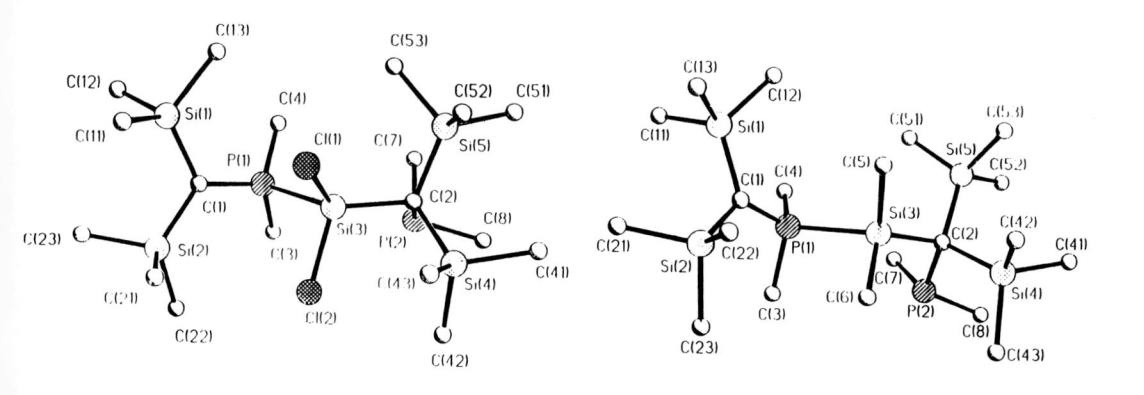

Fig. 5. Molecular structure of **13** and **14**

3 Interaction of RSiCl₃ with Three Equivalents of Li[C(PMe₂)₂(SiMe₃)]

A novel type of heterocycle can be obtained by the reaction of PhSiCl$_3$ with three equivalents of **2** [5]. The formation of **17** proceeds via a pentacoordinate, isolable intermediate and involves a P–P coupling reaction, probably via a π-complex, as proposed for a similar reaction of tBuSiCl$_3$ with **2** [6]. The second step is followed by a C,H-activation according to Scheme 2 [7]. The molecular structure of **17** is shown in Fig. 6.

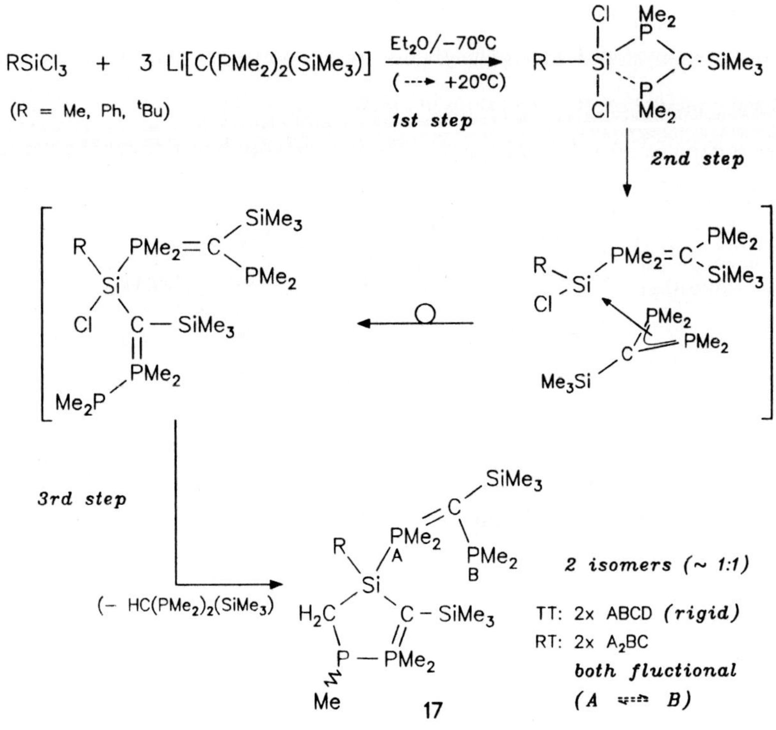

Scheme 2.

Conclusion

The C vs P reactivity of phosphinomethanides towards silicon centers may be turned by the substitution pattern of both, the phosphinomethanide and the chlorosilane. Novel heterocycles, skeleton rearrangements and high coordinations numbers may be achieved. **9** and **14** potentially be precursors for a specific type of unsaturated silicon compounds, adding to the growing class of compounds at the interface between organophosphorus and organosilicon chemistry.

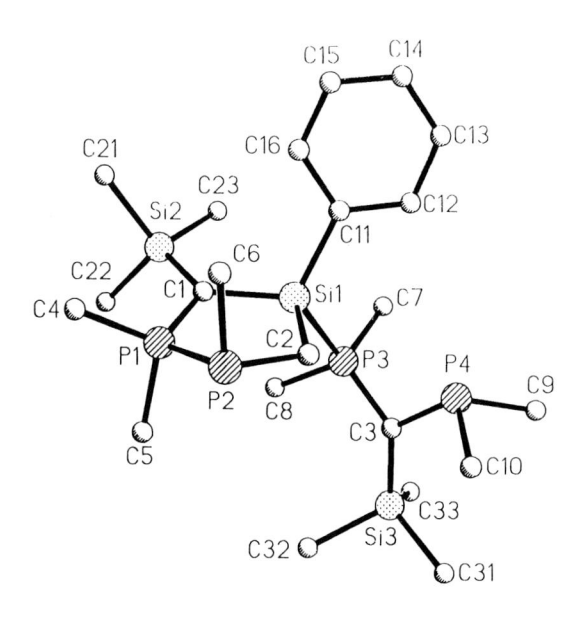

Fig. 6. Molecular structure of **17** in the crystal

References:

[1] H. H. Karsch, A. Appelt, *Z. Naturforsch.* **1983**, *38B*, 1399.
[2] H. H. Karsch, R. Richter, A. Schier, *Z. Naturforsch.* **1993**, *48B*, 1533.
[3] U. Keller, *Ph. D. Thesis*, Technische Universität München, **1992**.
[4] H. H. Karsch, R. Richter, B. Deubelly, A. Schier, M. Paul, M. Heckel, K. Angermeier, W. Hiller, *Z. Naturforsch.* in press.
[5] H. H. Karsch, R. Richter, unpublished.
[6] H. H. Karsch, in: *Organosilicon Chemistry – From Molecules to Materials* (Eds.: N. Auner, J. Weis), VCH, Weinheim, **1994**, p. 95.
[7] H. H. Karsch, *Russ. Chem. Bull.*, in press.

Supersilyl Phosphorus Compounds

Nils Wiberg*, Angelika Wörner, Heinrich Nöth[1], Konstantin Karaghiosof[2]

Institut für Anorganische Chemie
Ludwig-Maximilians-Universität München
Meiserstrasse 1, D-80333 München, Germany

1 Reaction of Supersilyl Sodium with White Phosphorus

The formation of the phosphorus compounds **1–4** depends on stoichiometry of the starting materials, on the reaction temperature and on the type of solvent.

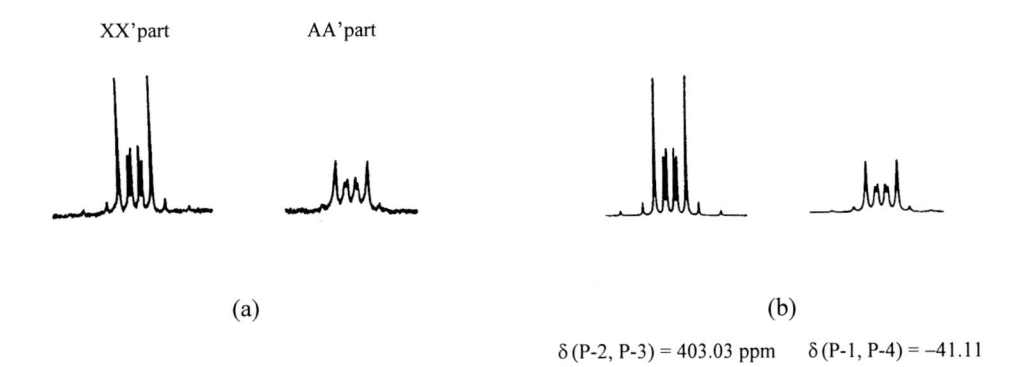

$$2\ t\text{Bu}_3\text{SiNa} + \text{P}_4 \xrightarrow{(\text{THF}, -78°\text{C})} (t\text{Bu}_3\text{Si})_2\text{P}_4\text{Na}_2 \quad \mathbf{1}$$

$$2\ t\text{Bu}_3\text{SiNa} + \text{P}_4 \xrightarrow[(\text{THF, r.t.})]{-\text{Na}_x\text{P}_y} (t\text{Bu}_3\text{Si})_2\text{P}_3\text{Na} \quad \mathbf{2}$$

$$2\ t\text{Bu}_3\text{SiNa} + \text{P}_4 \xrightarrow[(\text{THF, toluene, r.t.})]{-2\ \text{Na}_x\text{P}_y} 2\ (t\text{Bu}_3\text{Si})_3\text{P}_5\text{Na}_2 \quad \mathbf{3}$$

$$2\ t\text{Bu}_3\text{SiNa} + 2\ \text{P}_4 \xrightarrow[(\text{THF}, -78°\text{C})]{+2\ "\text{H}"} t\text{Bu}_3\text{SiP}_7\text{Na}_2 + t\text{Bu}_3\text{SiPH}_2 \quad \mathbf{4}$$

XX'part   AA'part

(a)   (b)

δ(P-2, P-3) = 403.03 ppm   δ(P-1, P-4) = –41.11

Fig. 1. Experimental (a) and simulated ^{31}P NMR spectrum (b) of **1**

1.1 ^{31}P NMR Data of 1

$t\text{Bu}_3\text{Si}-\text{P}(1)_A\text{Na}-\text{P}(2)_X=\text{P}(3)-\text{P}(4)-\text{Na}-\text{Si}t\text{Bu}_3$

AA'XX' spin system

coupling constants [Hz]

$^1J(2, 3) = -502.59$
$^1J(1, 2) = {}^1J(3, 4) = -432.26$
$^2J(1, 3) = {}^2J(2, 4) = -34.66$
$^3J(1, 4) = -184.06$

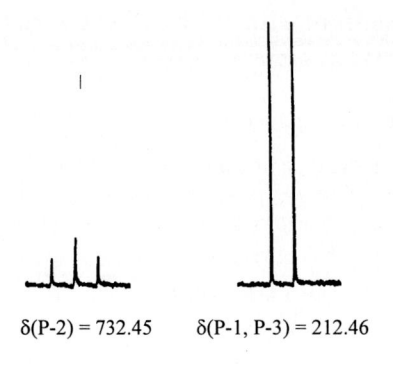

$\delta(\text{P-2}) = 732.45 \quad \delta(\text{P-1, P-3}) = 212.46$

Fig. 2. Experimental ^{31}P NMR spectrum of **2**

1.2 NMR Data of 2

$t\text{Bu}_3\text{Si} - \overset{A}{\underset{1}{\text{P}}} \overset{\overset{2}{\underset{X}{\text{P}}}}{\text{-}} \underset{\text{Na}^+}{\underset{3}{\text{P}}} - \text{Si}t\text{Bu}_3$

A$_2$X spin system

coupling constant [Hz]

$J(1, 2) = {}^1J(2, 3) = 552.6$

AA' part	R part	MM' part		XX' part

(a) (b)

(c) (d)

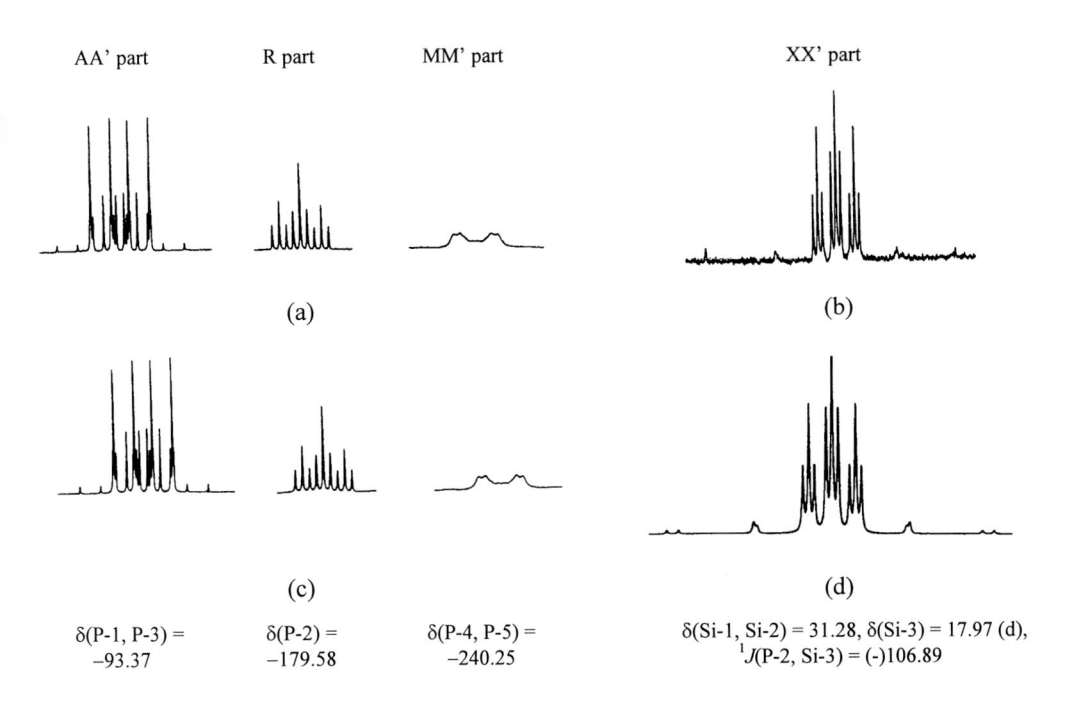

δ(P-1, P-3) =   δ(P-2) =   δ(P-4, P-5) =   δ(Si-1, Si-2) = 31.28, δ(Si-3) = 17.97 (d),
−93.37   −179.58   −240.25 ^1J(P-2, Si-3) = (−)106.89

Fig. 3. Experimental ^{31}P NMR (a), ^{29}Si NMR spectrum (b), and simulated ^{31}P NMR (c), ^{29}Si NMR spectrum (d) of 3

1.3 NMR Data of 3

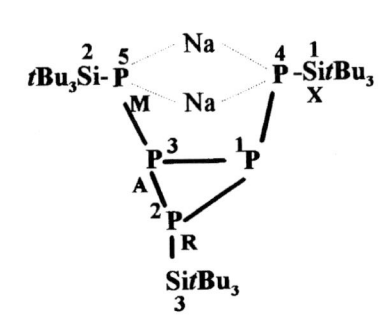

AA'RMM' spinsystem

coupling constants[Hz]

$^1J(1, 3) = -201.6(3)$
$^1J(1, 4) = {}^1J(3, 5) = -348.9(3)$
$^2J(1, 5) = {}^2J(3, 4) = -5.1(3)$
$^3J(4, 5) = 231.8(3)$
$^1J(1, 2) = {}^1J(2, 3) = -191.9(2)$
$^2J(2, 4) = {}^2J(2, 5) = 64.9(2)$
^1J(P-4, Si-1) = ^1J(P-5, Si-2) = −90.8(2)
^4J(P-4, Si-2) = ^4J(P-5, Si-1) = 21.4(2)
^2J(P-1, Si-1) = ^2J(P-3, Si-2) = −15.6(3)
^3J(P-3, Si-1) = ^3J(P-1, Si-2) = −2.1(3)
^3J(P-2, Si-1) = ^3J(P-2, Si-2) = 0.0

1.4 X-Ray Structure Analysis of 3

Crystal Data: Empirical formula: $C_{52}H_{113}Na_2O_4P_5Si_3$, color and habit: orange cube, crystal system: monoclinic, space group: $P2_1/n$, $Z = 4$. Unit cell dimensions: $a = 13.087(3)$ Å, $b = 23.773(9)$ Å, $c = 21.556(8)$ Å, $\beta = 100.78(3)°$.

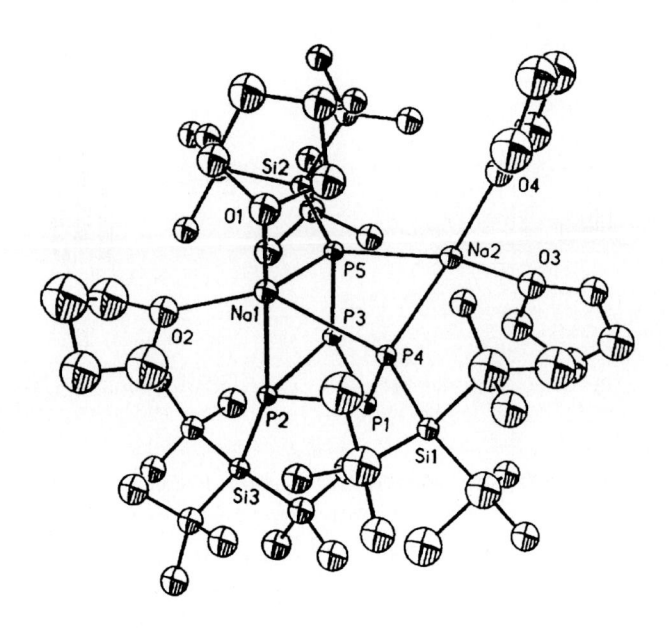

Fig. 4. Structure of $(tBu_3Si)_3P_5Na_2 \cdot 4$ THF

Bond lengths [Å]		Bond angles [°]		Bond angles [°]	
P(1)–P(2)	2.236(4)	P(2)–P(1)–P(3)	60.3(1)	Na(1)–P(4)–Na(2)	96.1(1)
P(1)–P(4)	2.179(4)	P(3)–P(1)–P(4)	106.3(1)	P(3)–P(5)–Na(1)	93.6(1)
P(3)–P(5)	2.183(3)	P(1)–P(3)–P(5)	105.8(1)	P(3)–P(5)–Na(2)	99.3(1)
P(1)–P(3)	2.244(4)	P(2)–P(1)–P(4)	106.2(1)	Na(1)–P(5)–Na(2)	95.6(1)
P(2)–P(3)	2.250(3)	P(1)–P(2)–P(3)	60.0(1)	P(2)–Na(1)–P(5)	72.3(1)
P(4)–Na(1)	2.879(4)	P(1)–P(3)–P(2)	59.7(1)	P(1)–P(2)–Si(3)	110.1(1)
P(5)–Na(2)	2.851(5)	P(2)–P(3)–P(5)	107.5(1)	Si(3)–P(2)–Na(1)	162.2(1)
P(2)–Na(1)	3.167(5)	P(1)–P(2)–Na(1)	85.8(1)	P(1)–P(4)–Si(1)	106.3(1)
P(5)–Na(1)	2.878(5)	P(2)–Na(1)–P(4)	71.3(1)	Si(1)–P(4)–Na(1)	129.1(1)
P(4)–Na(2)	2.826(5)	P(4)–Na(1)–P(5)	73.6(1)	Si(1)–P(4)–Na(2)	124.8(1)
P(2)–Si(3)	2.307(4)	P(4)–Na(2)–P(5)	74.8(1)	P(3)–P(5)–Si(2)	107.9(2)
P(5)–Si(2)	2.233(4)	P(3)–P(2)–Na(1)	85.0(1)	Si(2)–P(5)–Na(1)	125.0(1)
P(4)–Si(1)	2.239(4)	P(1)–P(4)–Na(1)	94.3(1)	Si(2)–P(5)–Na(2)	127.9(1)
		P(1)–P(4)–Na(2)	98.7(1)	P(3)–P(2)–Si(3)	109.7(1)

Table 1. Selected bond lengths and bond angles of **3**

2 Reaction of Supersilyl Disilane with White Phosphorus and Phosphorus Trichloride

The compounds **5** and **6** are obtained by reaction of $tBu_3Si\text{-}SitBu_3$ with white phosphorus in the presence of solvents, the products **7** and **8** by reaction of $tBu_3Si\text{-}SitBu_3$ with PCl_3 in the absence of a solvent.

$$6\ tBu_3Si\text{-}SitBu_3 + 7\ P_4 \xrightarrow{\text{(toluene, 110°C)}} 4\ (tBu_3Si)_3P_7\ [3]$$
$$\mathbf{5}$$

$$tBu_3Si\text{-}SitBu_3 + P_4 \xrightarrow{\text{(THF, 100°C)}} (tBu_3Si)_2P_4$$
$$\mathbf{6}$$

$$tBu_3Si\text{-}SitBu_3 + PCl_3 \xrightarrow[\text{(120°C)}]{-tBu_3SiCl} tBu_3SiPCl_2$$
$$\mathbf{7}$$

$$3\ tBu_3Si\text{-}SitBu_3 + 2\ PCl_3 \xrightarrow[\text{(120°C)}]{-4\ tBu_3SiCl} tBu_3SiPCl\text{-}PClSitBu_3$$
$$\mathbf{8}$$

2.1 ^{31}P NMR data of 6

$tBu_3Si - \overset{3}{P}\diagdown\overset{2}{\underset{A}{P}}\diagup\overset{4}{P} - SitBu_3$
$\diagdown\overset{1}{\underset{X}{P}}\diagup$

A_2X_2 spin system

coupling constant [Hz]

$^1J(1, 3) = 170.5$

$\delta(\text{P-3, P-4}) = -139.14(t);$

$\delta(\text{P-1, P-2}) = -334.40(t)$ ppm

2.2 NMR data of 7 and 8

tBu_3SiPCl_2 (**7**) $tBu_3SiPCl\text{--}PClSitBu_3$ (**8**)

$\delta\ ^{31}P = 211.9$ $\delta\ ^{31}P = 188.09$ ppm

$\delta\ ^{29}Si = 12.0(d)$ ppm, $^1J(\text{Si-P}) = 91.82$ Hz

References:

[1] X-ray structure analysis of **3**.
[2] Spectra simulation.
[3] N. Wiberg, H. Schuster, K. Karaghiosoff, I. Kovács, G. Baum, G. Fritz, D. Fenske, *Z. Anorg. Allg. Chem.* **1993**, *619*, 453.

Silicon-Phosphorus, -Arsenic, -Antimony, and -Bismuth Cages: Syntheses and Structures

Karl Hassler

Institut für Anorganische Chemie

Technische Universität Graz

Stremayrgasse 16, A-8010 Graz, Austria

Introduction

One method for the synthesis of cyclic silicon-phosphorus compounds is the reaction of alkaliphosphides with dichlorosilanes [1]. Using sodium/potassium phosphide, -arsenide, -antimonide and -bismutide Na_3E/K_3E (E = P, As, Sb and Bi) [2] with various di-, tri-, and tetrachloroorganosilanes, cages of the following types (Fig 1.) have been synthesized so far.

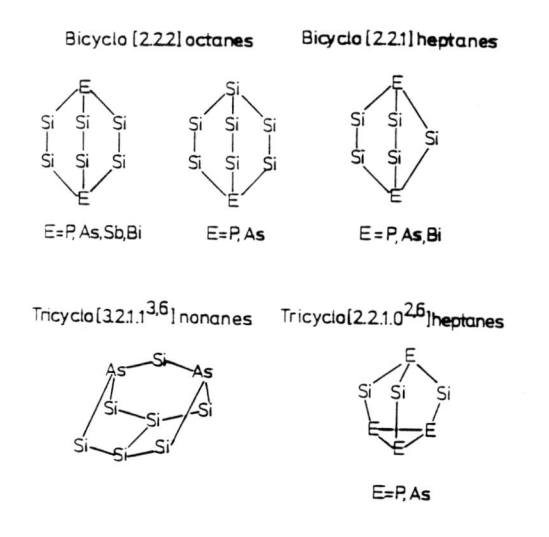

Fig. 1.

Bicyclo[2.2.2]octanes

1,2-Dichlorotetramethyldisilane reacts with Na_3E/K_3E (E = P, As, Sb, Bi), prepared from the elements and sodium-potassium alloy to form dodecamethyl-1,3-diphospha, -diarsa, -distiba, and -dibismutha 2,3,5,6,7,8-hexasilabicyclo[2.2.2]octane [3, 4] in yields up to 20 %. The thermal stability of the cages decreases in the order P > As > Sb > Bi. The bismuth cage decomposes within a few minutes at room temperature. Infrared and Raman spectra, combined with a normal coordinate analysis, are consistent with a twisted ESi_6E skeleton (symmetry D_3): In solution, the cages have symmetry D_{3h} on the NMR time scale. With B_2H_6, a cage with the structure $H_3BP(SiMe_2SiMe_2)_3PBH_3$ is formed.

The heptasilane $MeSi(SiMe_2SiMe_2Cl)_3$ [5], prepared by the following reaction sequence (Eq. 1–3) reacts with Na_3P/K_3P or Na_3As/K_3As to form tridecamethyl-1-phospha- and tridecamethyl-1-arsa-7-heptasilabicyclo[2.2.2]octanes (Scheme 1) in surprisingly high yields (P: 50 %, As: 40 %).

$$MeSi(SiMe_2Cl)_3 + 3LiSiMe_2Ph \longrightarrow MeSi(SiMe_2SiMe_2Ph)_3 \qquad (1)$$

$$MeSi(SiMe_2SiMe_2Ph)_3 + 3CF_3SO_3H \longrightarrow MeSi(SiMe_2SiMe_2OSO_2CF_3)_3 \qquad (2)$$

$$MeSi(SiMe_2SiMe_2OSO_2CF_3)_3 + 3LiCl \longrightarrow MeSi(SiMe_2SiMe_2Cl)_3 \qquad (3)$$

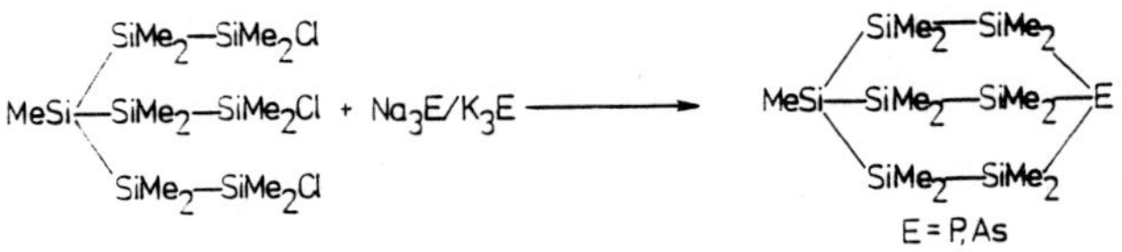

Scheme 1. Formation of tridecamethyl-1-phospha- and tridecamethyl-1-arsa-7-heptasilabicyclo[2.2.2]octanes

In the crystal, the $AsSi_7$ skeleton is twisted (symmetry D_3, see Fig. 2). The ^{13}C and ^1H NMR spectra of both cages are consistent with C_{3v} symmetry in the liquid state.

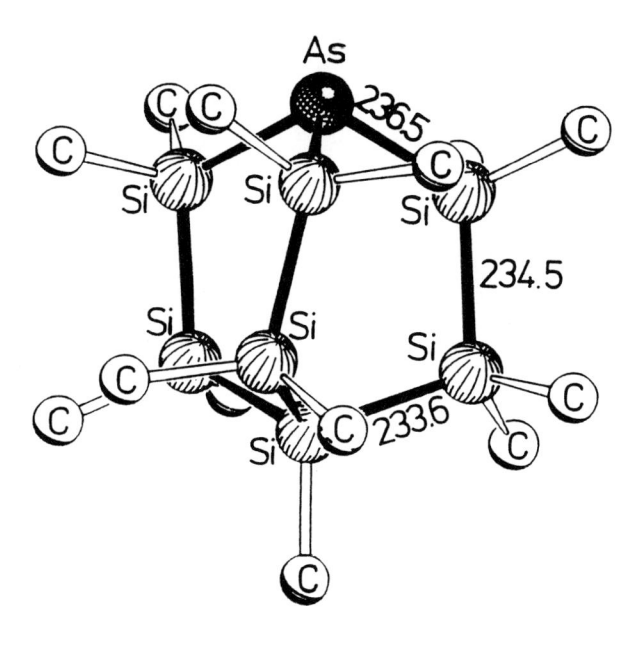

Fig. 2. Molecular structure of AsSi$_7$Me$_{13}$ in the crystal [6]; bond angles: AsSiSi: 115.9°, SiAsSi 101.7°

Bicyclo[2.2.1]heptanes

The reactions of Na$_3$E/K$_3$E with chlorosilanes are very complex. If DME is used as a solvent breaking of Si–Si bonds occurs quite easily, leading to new cages. From MeSi(SiMe$_2$Cl)$_3$ and Na$_3$P/K$_3$P or Na$_3$As/K$_3$As, one obtains decamethyl-1,3-diphospha- and decamethyl-1,3-diarsa-2,4,5,6,7-pentasila-bicyclo[2.2.1]heptane.

Scheme 2. Formation of decamethyl-1,3-diphospha- and decamethyl-1,3-diarsa-2,4,5,6,7-pentasilabicyclo[2.2.1]heptane

Besides the breaking of Si–Si bonds, the formation of new Si–Si bonds also occurs in these reactions. If Me$_2$SiCl$_2$ reacts with Na$_3$Bi/K$_3$Bi, decamethyl-1,3-dibismutha-2,4,5,6,7-pentasilabicyclo[2.2.1]heptane is formed. The structures of P$_2$(SiMe$_2$)$_5$ and Bi$_2$(SiMe$_2$)$_5$ are shown in Fig. 3. As$_2$(SiMe$_2$)$_5$ was identified from ^{29}Si NMR spectra, mass spectra and elemental analysis.

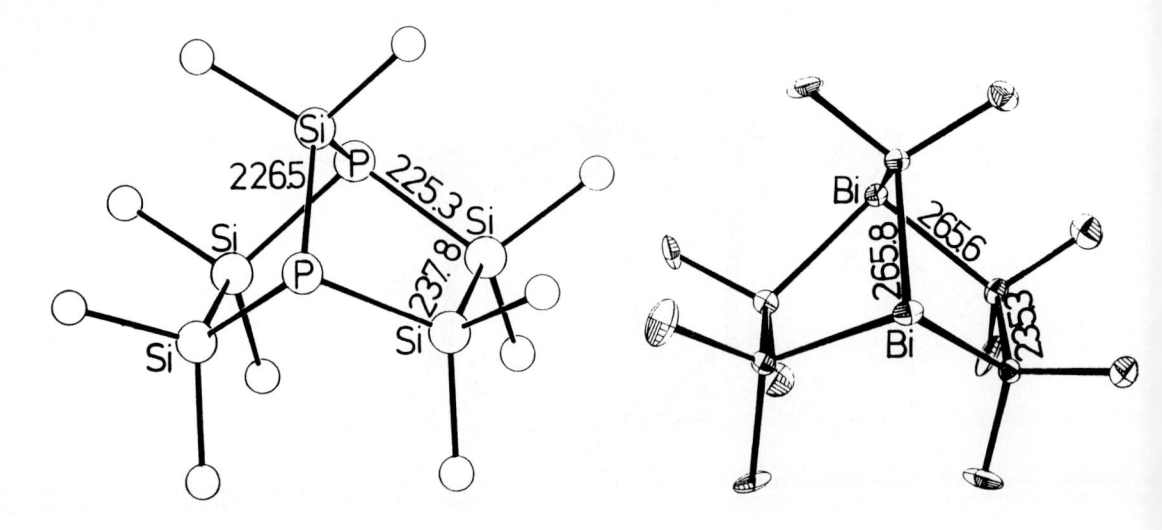

Fig. 3. Molecular structures of $P_2(SiMe_2)_5$ and $Bi_2(SiMe_2)_5$

Tricyclo[2.2.1.0^{2,6}]heptanes

With dichlorodimethylsilane and dichloromethylphenylsilane Na_3E/K_3E (E = As, P) form nortricyclane cages $As_4(SiMe_2)_3$ and $P_4(SiMePh)_3$ that contain three-membered E_3 rings. Depending on the arrangement of the phenyl groups $P_4(SiMePh)_3$ exists as a mixture of isomers (Fig. 4), that can be separated by fractional crystallization.

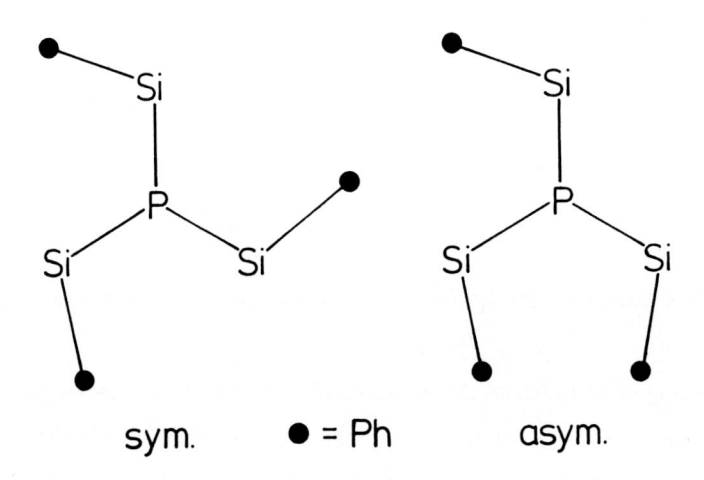

Fig. 4. Isomers of $P_4(SiMePh)_3$

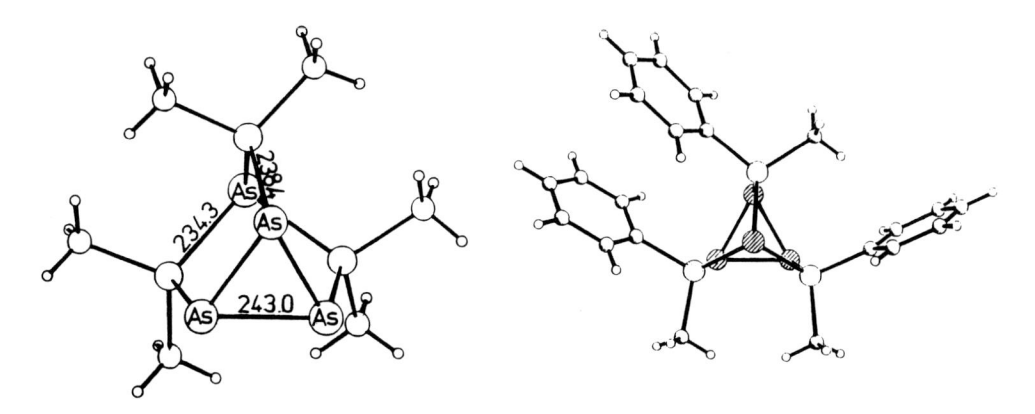

Fig. 5. Molecular structures of As$_4$(SiMe$_2$)$_3$ and asymmetrical P$_4$(SiMePh)$_3$

Tricyclo[3.2.1.1^{3,6}]nonanes

With Na$_3$As/K$_3$As the tetrachlorodecamethylhexasilane (Me$_2$ClSi)$_2$SiMeSiMe(SiMe$_2$Cl)$_2$ [7] forms dodecamethyl-1,3-diarsa-2,4,5,6,7,8,9-heptasilatricyclo[3.2.1.1^{3,6}]nonane. Its structure was confirmed by mass spectroscopy, elemental analysis, and ^{29}Si NMR spectroscopy (Fig. 6).

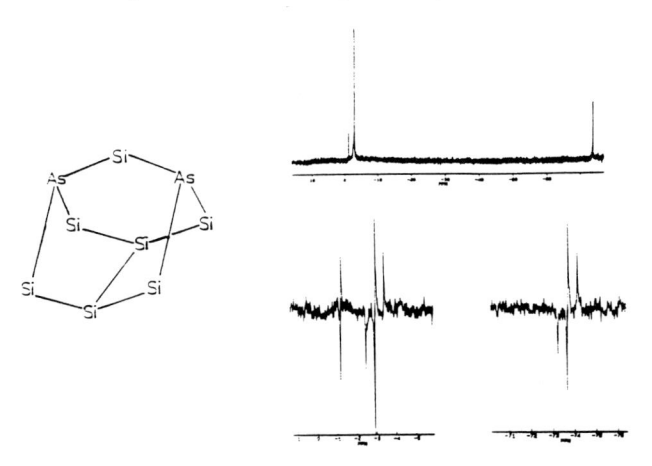

Fig. 6. ^{29}Si and ^{29}Si INADEQUATE NMR spectra of As$_2$Si$_7$Me$_{12}$

Acknowledgement: The author thanks the *Fonds zur Förderung der wissenschaftlichen Forschung* (Wien) for financial support.

References:

[1] G. Fritz, *Comments Inorg. Chem.* **1982**, *6*, 329; and references cited therein.

[2] G. Becker, W. Hölderich, *Chem. Ber.* **1975**, *108*, 2484; G. Becker, M. Rössler, *Z. Naturforsch.* **1982**, *37B*, 91.

[3] K. Hassler, *J. Organomet. Chem.* **1983**, *246*, C31.

[4] K. Hassler, S. Seidl, *J. Organomet. Chem.* **1988**, *347*, 27.

[5] K. Hassler, *Monatsh. Chem.* **1986**, *117*, 6134.

[6] K. Hassler, G. Kollegger, H. Siegl, to be published.

[7] K. Hassler, G. Kollegger, *J. Organomet. Chem.*, in print.

Synthesis, Reactivity, and Molecular Structure of Diphosphadisilacyclobutanes, Bis(stannyl)silyl-, and Bis(stannyl)germylphosphines

M. Waltz, M. Nieger, E. Niecke*

Institut für Anorganische Chemie

Rheinische Friedrich-Wilhelms-Universität Bonn

Gerhard-Domagk-Str. 1, D-53121 Bonn, Germany

Summary: $RECl_3$ **1a–1d** (E = Si, Ge; R = Cp*, Mes, Is) react with $Li[Al(PH_2)_4]$ via Si–P or Ge–P bond formation to give the triphosphinosilanes **2b**, **2c**, the trisphosphinogermane **2a**, or monocyclic $[Cp^*Si(PH)PH_2]_2$ (**3**) respectively. The reactions with RLi, $Hg(tBu)_2$, and the thermolysis of **3** are described, yielding the lithiated product **4**, the bicyclic $[Cp^*Si(PH)PH]_2$ (**5**) and binary SiP_2 (**6**). The chlorinated compounds **7a–7d** are obtained by reaction of $RECl_3$ with $P(SnMe_3)_3$. Further reaction of **7b–7d** leads to silyl substituted diphosphenes **8b–8d**. All compounds have been characterized by NMR and by X-ray analysis in the cases of **3**, **7a**, **7d**, and **8d**.

Introduction

In recent years, polycyclic P–Si and P–Ge clusters respectively have been obtained by reaction of organosilyltrihalides with threefold functionalized phosphines [1–3]. A decisive influence on the nature and the steric hindrance of the ligand R of the adequately substituted $RECl_3$ on the reaction can be observed.

Results and Discussion

Treatment of Cp*GeCl$_3$ (**1a**) or RSiCl$_3$ (**1b**, **1c**) with Li[Al(PH$_2$)$_4$] (LAP) leads to the triphosphinogermane **2a** or the triphosphinosilanes **2b**, **2c** respectively. In contrast, the formation of monocyclic [Cp*Si(PH)PH$_2$]$_2$ (**3**) was observed in the phosphination of Cp*SiCl$_3$ (Scheme 1).

A mixture of the three isomers **3α–3γ**, which differ in configuration of the endocyclic phosphorus atoms, was identified in solution. Their ^{31}P{^{1}H} NMR spectra at –50 °C show characteristic AMX$_2$ and A$_2$X$_2$ spin systems with highfield shifts typical for threefold coordinated silylphosphine phosphorus atoms (–160 to –240 ppm). Compound **3α** was investigated by X-ray analysis [4] (Fig. 1). The exocyclic P–Si bond lengths (2.24 Å) are a little shorter than the endocyclic P–Si bond lengths (2.27 Å). A deviation from ideal tetrahedral geometry can be observed for the silicon atoms with Si–C bond lengths (1.92 Å) which differ distinctly from standard values (1.86 Å) [5].

Scheme 1. a: E = Ge, R = Cp*; b: E = Si, R = Mes; c: E = Si, R = Is; d: E = Si, R = Cp*; (Cp* = Me$_5$C$_5$; Mes = 2,4,6-Me$_3$C$_6$H$_2$; Is = 2,4,6-iPr$_3$C$_6$H$_2$; Mes* = 2,4,6-tBu$_3$C$_6$H$_2$)

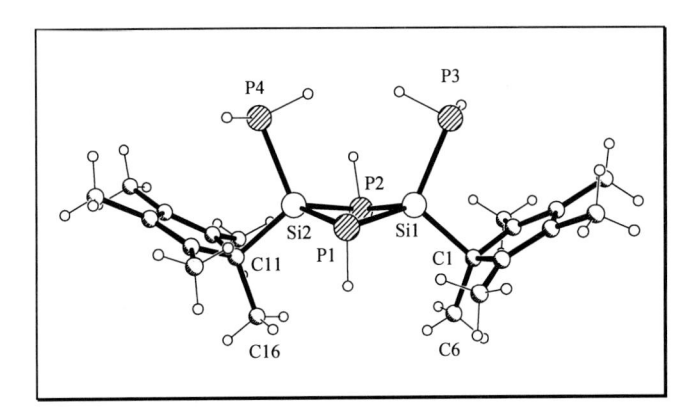

Fig. 1. X-ray structure of the disiladiphosphacyclobutane (3α)

The lithiation of 3α–3γ with one equivalent RLi yields the lithiated compounds 4α, 4β, which differ in the configuration of the endocyclic phosphorus atoms. Both isomers were characterized by their ^{31}P NMR spectra, which show characteristic AMX$_2$ spin systems with shifts in the typical range of lithium silylphosphides(–160 to –210 ppm).

A mixture of three isomers α–γ of the bicyclic compound [Cp*Si(PH)PH]$_2$ (5) could be obtained by oxidative P–P bond formation between the PH$_2$ substituents of the isomers 3α–3γ by means of one equivalent Hg(tBu)$_2$. The isomers 5α–5γ can be distinguished by their characteristic AMKX, AA'MM', and AA'XX' spin systems (δ ^{31}P = –130 to –220 ppm), and show characteristic values of $^1J_{PP}$ (55–170 Hz) and $^2J_{PP}$ (5–150 Hz), respectively in their ^{31}P{^{1}H} NMR spectra. The presence of P–H-units in 5α–5γ was identified by ^{1}H-detected 2D-^{31}P{^{1}H} NMR experiment.

The thermolysis of the compounds 3α–3γ at 700 °C leads to the binary compound SiP$_2$ 6 via liberation of two equivalents Cp*H and H$_2$. The reaction was investigated by _d_ifferential-_t_hermo-_a_nalysis (DTA) and _t_hermo _g_ravimetric (TG) measurements.

Other interesting compounds, which possibly could be precursors to phosphagermaclusters or phosphasilaclusters are the chlorinated germylstannylphosphine (7a) and the silylstannylphosphines 7b–7d, which were obtained by treatment of the corresponding RECl$_3$ 1a–1d with one equivalent P(SnMe$_3$)$_3$ (Scheme 2).

The Cp* substituted compounds 7a and 7d were characterized by X-ray analysis (Fig. 2). The P–Ge bond in 7a (2.27 Å) is shorter than reported values for a P–Ge single bond (2.30–2.36 Å) [6]. This may be caused by effects of hyperconjugation. The *Newman* projection along the P–Ge axis shows the *syn*-periplanar configuration of the tin and the chlorine atoms.

Scheme 2. a: E = Ge, R = Cp*; b: E = Si, R = Mes; c: E = Si, R = Is; d: E = Si, R = Cp*;
(Cp* = Me$_5$C$_5$; Mes = 2,4,6-Me$_3$C$_6$H$_2$; Is = 2,4,6-iPr$_3$C$_6$H$_2$; Mes* = 2,4,6-tBu$_3$C$_6$H$_2$)

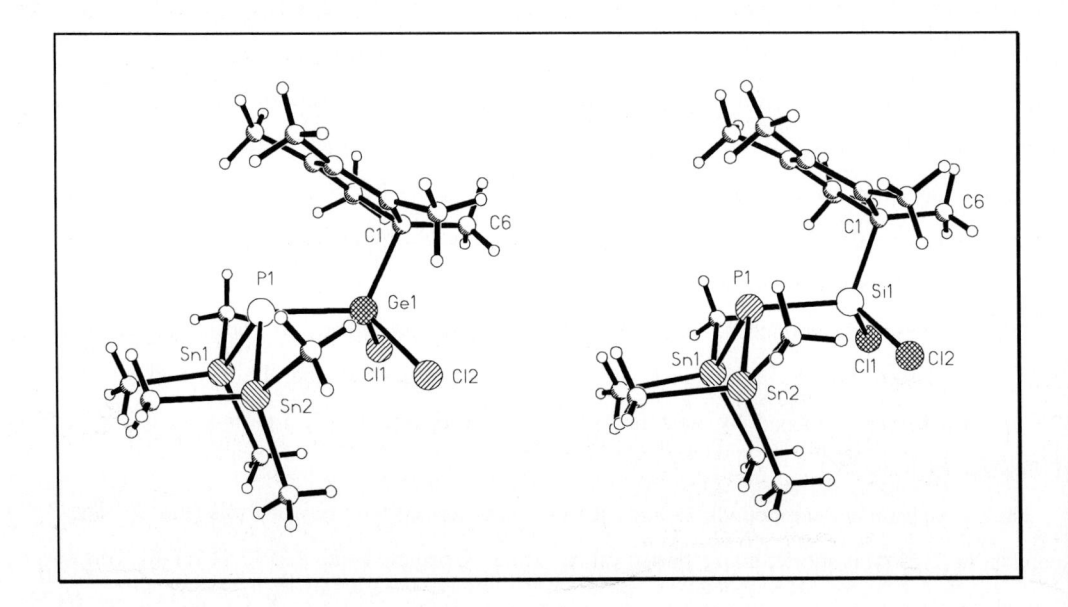

Fig. 2. X-ray structures of **7a** and **7d**

Similar structural properties as for **7a** can be observed for **7d**. The P–Si bond (2.20 Å) is distinctly shorter than the theoretically calculated value for a P–Si single bond (2.24–2.28 Å) [5]. Also the Sn, P, Si, and Cl atoms form two almost planar trapezoids with rather short Cl–Sn distances (3.63 Å, 3.70 Å).

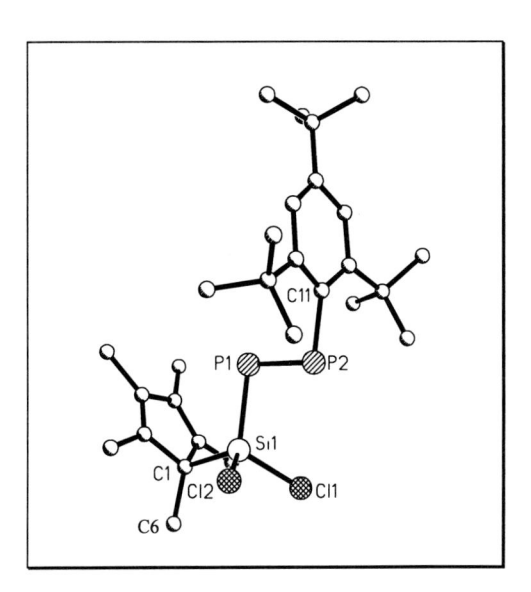

Fig. 3. X-ray structure of **8d**

Treatment of **7b–7d** with one equivalent Mes*PCl$_2$ yields the unsymmetrical silyl substituted diphosphenes **8b–8d**. Red crystals of compound **8d** were investigated by X-ray analysis (Fig. 3). The P=P bond length (2.03 Å) and the Si–P–P and P–P–C bond angles (95.7 °, 99.2 °) do not essentially differ from theoretically calculated values (2.00 Å, 96,5 °, 96.4 °) [7].

References:

[1] H. Schumann, H. Benda, *J. Organomet. Chem.* **1970**, *21*, 12.
[2] M. Baudler, G. Scholz, K. F. Tebbe, M. Feher, *Angew. Chem.* **1989**, *101*, 352; *Angew. Chem., Int. Ed. Engl.* **1989**, *28*, 339.
[3] M. Baudler, W. Oehlert, K. F. Tebbe, *Z. Anorg. Allg. Chem.* **1991**, *598/599*, 9.
[4] E. Niecke, M. Waltz, M. Nieger, *Z. Anorg. Allg. Chem.*, to be submitted.
[5] P. Rademacher, in: *Grösse und Gestalt von Molekülen, Strukturen organischer Moleküle, Physikalische Organische Chemie* (Ed.: M. Klessinger), VCH, Weinheim, **1987**.

[6] F. Bickelhaupt, in: *Multiple Bonds and Low Coordination in Phosphorus Chemistry* (Eds.: M. Regitz, O. J. Scherer), Thieme, Stuttgart, New York, **1990**.

[7] W. W. Schöller, in: *Multiple Bonds and Low Coordination in Phosphorus Chemistry* (Eds.: M. Regitz, O. J. Scherer), Thieme, Stuttgart, New York, **1990**.

On the Reaction of Silanols with Alcohols

Ines Kohlheim, Dieter Lange, Hans Kelling*

Fachbereich Chemie, Abteilung Anorganische Chemie

Universität Rostock

Buchbinderstr. 9, D-18051 Rostock, Germany

Summary: Compared with the frequently investigated alkoxysilane hydrolysis the reverse silanol alcoholysis has been investigated at model silanols of type XMe_2SiOH. The equilibrium constants are within the order of 0.1. The reaction rate of the acid-catalyzed silanol alcoholysis decreases with increasing electronegativity of the substituents X at Si, whereas the base-catalyzed reaction is accelerated. Various alcohols ROH affect the reaction rates preferably by the sterical influence of R. The acid-catalyzed silanol alcoholysis is slighly more affected by substituents X as the reverse alkoxysilane hydrolysis.

Introduction

Within the scope of systematic investigations on basic reactions in industrial silicone syntheses in the system – alkoxysilane, silanol, siloxane, alcohol, water, catalyst, solvent – after previous investigations on the equilibrium of the silanol alkoxysilane heterocondensation and the reverse siloxane cleavage with alcohol (Eq. 1) [1]

$$\equiv SiOH + \equiv SiOR \rightleftharpoons \equiv SiOSi\equiv + ROH$$

Eq. 1.

we are interested in the silanol alcoholysis / alkoxysilane hydrolysis equilibrium (Eq. 2).

$$\equiv SiOH + \equiv ROH \rightleftharpoons \equiv SiOR + H_2O$$

Eq. 2.

Whereas the alkoxysilane hydrolysis is already frequently investigated, there is only little information on the reverse silanol alcoholysis. Thus, we investigated the reaction of silanols XMe_2SiOH (for X see Table 1) with methanol and other alcohols with respect to equilibrium constants K and rate constants k_a and k_b of the acid and base-catalyzed reaction and their substituent dependence.

Results and Discussion

The equilibrium constants K corresponding to Eq. 3 have been determined in dioxane solution starting with definite concentrations of silanol or alkoxysilane, alcohol, and water and measuring the equilibrium concentrations of silanol and alkoxysilane by GC. The results given in Table 1 show that K is within the order of 0.1. This means that the silanol alcoholysis is, especially at an excess of alcohol, a reaction not to be neglected in the system mentioned above.

$$K = c(XMe_2SiOR) \cdot (H_2O) \,/\, c(XMe_2SiOH) \cdot c(ROH)$$

Eq. 3.

The K values decrease with increasing electronegativity of the substituents X. This effect is stronger for aryl- than for alkyl-substituents. The dependence of K on the various alcohols seems to be determined also by sterical effects.

The acid-catalyzed silanol alcoholysis, in dioxane with HCl as catalyst and a 20-fold excess of alcohol, has been investigated measuring the decrease of silanol concentration corresponding to Eq. 4.

$$-dc(XMe_2SiOH) \,/\, dt = k_a \cdot c(XMe_2SiOH) \cdot c(HCl)$$

Eq. 4.

The resultant rate constants k_a are shown in Table 1. The k_a values of the XMe_2SiOH methanolysis decrease with increasing electronegativity of X. In agreement with other nucleophilic substitutions at Si–O compounds [2] this should be determined by a protonation preequilibrium Eq. 5 generating the reactive form of the silanol for the ROH attack.

$$XMe_2SiOH + H^+(Solvent) \rightleftarrows XMe_2SiOH_2^+ + Solvent$$

Eq. 5.

X	R	K	k_a	k_b	k
Me	Me	0.110	28.47		75.3
Et	Me	0.125	28.67		78.3
nPr	Me	0.130	28.99		84.6
ClCH$_2$	Me	0.076	1.85		
Bzl	Me	0.110	7.46	0.020	48.2
p-CH$_3$O–Ph	Me	0.270	4.39		51.5
p-CH$_3$–Ph	Me	0.260	10.99	0.038	50.3
p-Cl–Ph	Me	0.062	3.38	0.080	
Ph	Me	0.130	6.18	0.058	48.6
Ph	Et	0.099	4.48		21.5
Ph	nPr	0.144	3.99		6.65
Ph	iPr	0.154	2.45		10.38
Ph	nBu	0.164	3.60		15.90
Ph	iBu	0.104	3.04		1.15
Ph	tBu	0.014			0.71
Ph	Ph	0.018			0.38

Table 1. Equilibrium constants K, rate constants k_a and k_b for the acid and base-catalyzed alcoholysis of XMe$_2$SiOH and rate constants k for the acid-catalyzed hydrolysis of XMe$_2$SiOR

Going from methanol to other alcohols ROH the decrease of k_a seems to be determined preferably by the steric effects of R. We also have investigated the acid-catalyzed hydrolysis of the corresponding alkoxysilanes XMe$_2$SiOR under similar conditions, except that an excess of water was used here. The k values corresponding to Eq. 6 (see Table 1) are absolutely higher than the k_a, as expected from the equilibrium constants. They also decrease with increasing electronegativity of X, but weaker as the k_a.

Eq. 6.
$$-dc(XMe_2SiOR)/dt = k \cdot c(XMe_2SiOR) \cdot c(HCl)$$

Finally the rate constants k_b of the base-catalyzed reaction of only four different silanols with methanol have been determined under similar conditions as for the acid-catalyzed reaction using *t*BuOK as catalyst. The k_b (see Table 1) corresponding to Eq. 7 increase with increasing electronegativity of X, which is in agreement with other base-catalyzed nucleophilic reactions [2] of Si–O compounds. The virtually lower rates compared with the acid-catalyzed reaction should be due to the very small effective concentration of the RO⁻ nucleophile, because inspite of the ROH excess preferably the silanolate is formed.

Eq. 7.
$$-dc(XMe_2SiOH)/dt = k_b \cdot c(XMe_2SiOH) \cdot c(tBuO\,K)$$

Acknowledgement: This work was supported by the *Fonds der Chemischen Industrie*.

References:

[1] H. Kelling, D. Lange, A. Surkus, in: *Organosilicon Chemistry – From Molecules to Materials*, (Eds.: N. Auner, J. Weis), VCH, Weinheim, **1994**, p. 65.

[2] J. Boe, *J. Organomet. Chem.* **1976**, *107*, 139.

4-Silyloxybenzothiopyrylium-Salts: A New Tool for the Stereoselective Annulation of S-Heterocycles

Uwe Beifuss*, Henning Gehm, Mario Tietze

Institut für Organische Chemie

Georg-August-Universität Göttingen

Tammannstr. 2, D-37077 Göttingen, Germany

Summary: Sequential intermolecular 1,2-addition/intramolecular 1,4-addition-reactions of 4-silyloxy-1-benzothiopyrylium-salts with 2-silyloxy-1,3-butadienes highly diastereoselectively give annulated thiochromanones.

Thioxanthenes are not only valuable synthetic intermediates [1], they also find considerable interest in pharmaceutical research [2] and as dyes for numerous applications [3]. Only few methods are known for the stereoselective synthesis of S-heterocycles [4]. Sequential transformations are a powerful strategy in organic synthesis frequently allowing the synthesis of complex molecules in a single synthetic operation [5]. It was anticipated that 4-silyloxy-1-benzothiopyrylium-salts **1** [6], which represent double-activated 4-thiochromanones, can be used for the stereoselective synthesis of annulated thiochromanones **3** by reaction with 2-silyloxy-1,3-butadienes **2** [7].

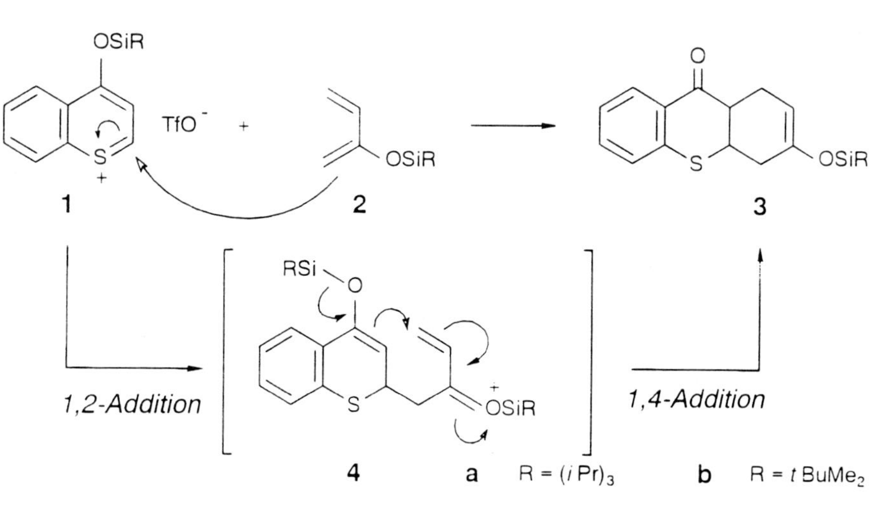

Scheme 1.

In the first step we expected an *intermolecular 1,2-addition* of the silylenolether functionality in **2** to the C=S$^+$ bond of the benzothiopyrylium-salt **1**, and in the second step we expected an *intramolecular 1,4-addition* of the silylenolether to the newly formed acceptor system in **4** to take place.

Following this concept we found that *cis*-thioxanthones **5** can be diastereoselectively synthesized by sequential reactions of 4-silyloxy-1-benzothiopyrylium-salts **1** with unsubstituted 2-silyloxy-1,3-butadienes **2**.

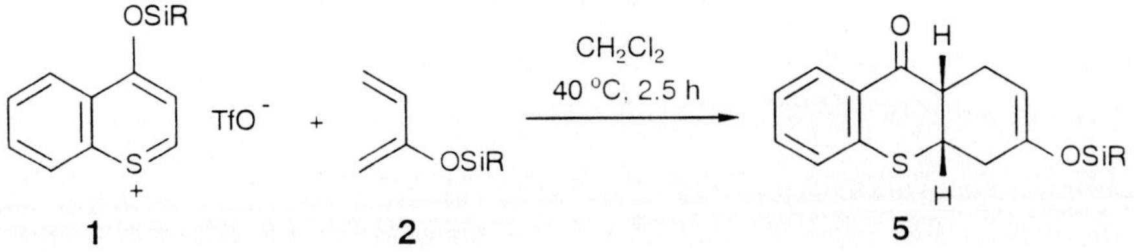

Eq. 1.

It is remarkable that both reactants, the 4-silyloxy-1-benzothiopyrylium-salts **1** as well as the 2-silyloxy-1,3-butadienes **2**, are formed *in situ* in one pot. First, the 4-silyloxy-1-benzothiopyrylium-salt **1** is obtained from the thiochromen-4-one **6** by treatment with a trialkylsilyltriflate [8] and then, the 2-silyloxy-1,3-butadiene **2** is formed by treatment of methyl vinyl ketone **7** with a trialkylsilyltriflate and 2,6-lutidine in the same flask.

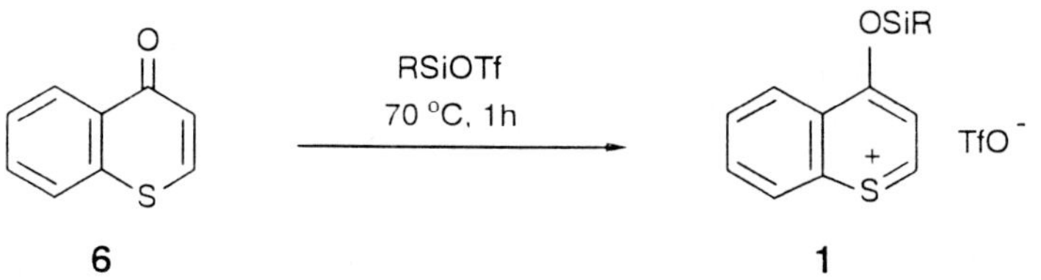

Eq. 2.

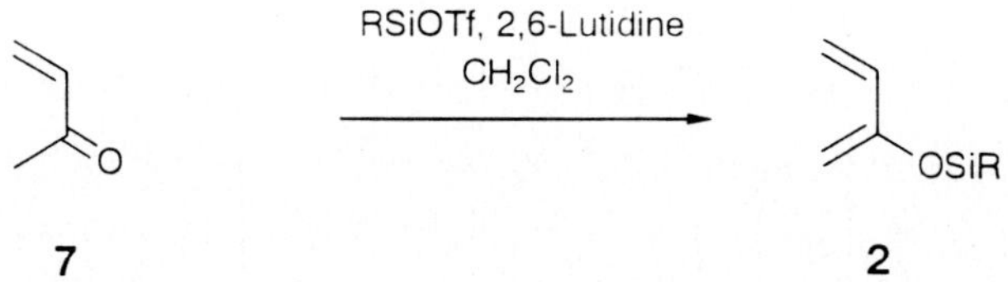

Eq. 3.

The sequential reaction of the 4-silyloxy-1-benzothiopyrylium-salt **1** with 4-phenyl-2-silyloxy-1,3-butadiene **8a** exclusively yields the all-*cis*-annulated thiochromanone **9a** with high yield.

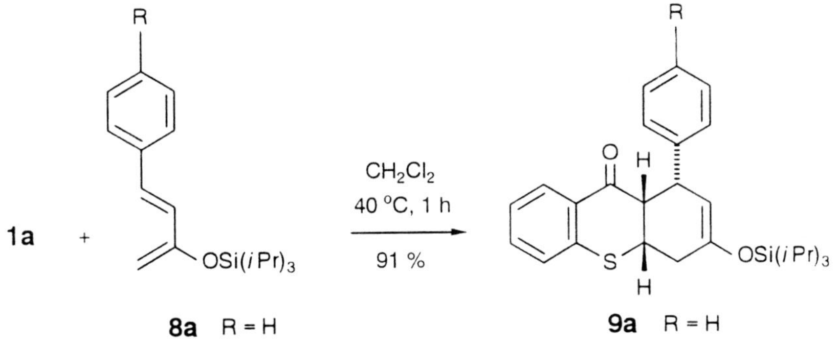

Eq. 4.

The stereochemical outcome of this new annulation process can be rationalized by assuming the transition state structure **10**. The *synclinal* arrangement of the respective hydrogen atoms exclusively gives the all-*cis*-annulation product **9a**.

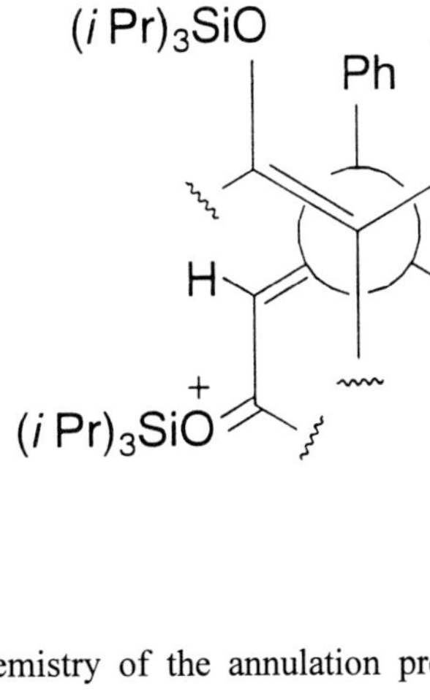

Fig. 1.

The all-*cis*-stereochemistry of the annulation product **9a** was unambiguously confirmed by X-ray crystal structure analysis [9]. Subsequently, the reactions of the 4-silyloxy-1-benzothiopyrylium-salt **1a** with a number of 4-substituted-2-silyloxy-1,3-butadienes and 3,4-disubstituted-2-silyloxy-1,3-butadienes (Table 1) were studied. In each case only one out of 4 diastereomers was isolated, namely the all-*cis*-stereoisomer. Yields are ranging from 68 to 91 %.

Product	R¹	R²	Reaction Time[h]	Yield [%]
9a	H–	SiiPr$_3$	1.0	91
9b	MeO–	SiiPr$_3$	2.0	79
9c	MeO$_2$C–	SiiPr$_3$	2.0	91
9d	NO$_2$–	SiiPr$_3$	1.5	80
11		SiiPr$_3$	2.5	75
12		SiiPr$_3$	2.5	68
13		SiiPr$_3$	3.0	87
14		SiiPr$_3$	3.0	87

Table 1. Diastereoselective sequential transformations of 4-silyloxy-1-benzothiopyrylium-salt **1a** with 4-substituted and 3,4-disubstituted 2-silyloxy-1,3-butadienes

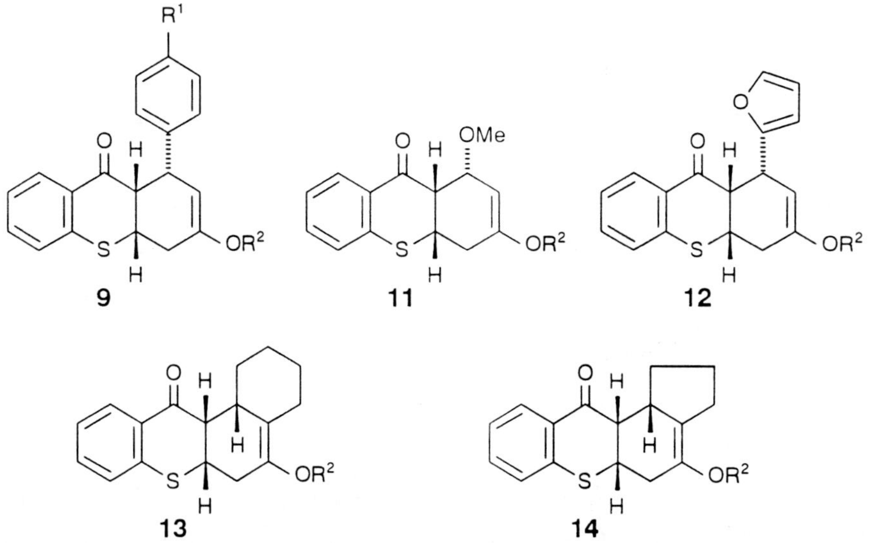

Fig. 2.

The reactions presented may either proceed as a stepwise sequential *intermolecular 1,2-addition/ intramolecular 1,4-addition process* or as a one-step *intermolecular Diels-Alder reaction*. In general, it is difficult to differentiate between these two alternatives. In our case, however, there is evidence in favour of the sequential two-step mechanism over the one-step Diels-Alder mechanism. In some reactions the formation of side products like **16**, which may be derived from the proposed intermediate **4** of the first step in the sequential reaction, is observed.

Eq. 5.

Moreover, the cyclization of the keto-silylenolether **17** [10] diastereoselectively yields the all-*cis*-annulation product **9a** exclusively. The outcome of this experiment clearly supports a sequential two-step mechanism for the reactions in question.

Eq. 6.

Acknowledgement: This work was supported by the *Fonds der Chemischen Industrie* and the *Georg-August-Universität Göttingen*. H. G. thanks the *Fonds der Chemischen Industrie* for a doctoral fellowship. We thank Dr. M. Noltemeyer and Mr. H. G. Schmidt for X-ray analyses. We are greatful to Professor L. F. Tietze for his generous support and his continuous interest in our work.

References:

[1] L. I. Belen'kii, in: *Chemistry of Organosulfur Compounds* (Ed.: L. I. Belen'kii), Ellis Horwood, Chichester, **1990**, p. 193.

[2] L. A. Damani, *Sulphur-Containing Drugs and Related Organic Compounds*, Ellis Horwood, Chichester, **1989**.

[3] K. H. Pfoertner, M. Voelker, *J. Chem. Soc. Perkin Trans.* **1991**, *2*, 527; K. H. Pfoertner, *J. Chem. Soc. Perkin Trans* **1991**, *2*, 523; W. Fischer, *Helv. Chim. Acta* **1991**, *74*, 1119; C. H. Chen, J. L. Fox (Eastman Kodak Co) Eur. Pat. Appl. EP 330,444, **1989**; *Chem. Abstr.* **1990**, *112*, 138909r.

[4] A. H. Ingall, in: *Comprehensive Heterocyclic Chemistry, Vol. 3* (Eds.: A. R. Katritzky, C. W. Rees), Pergamon, Oxford, **1984**, p. 885.

[5] L. F. Tietze, U. Beifuss, *Angew. Chem.* **1993**, *105*, 137; *Angew. Chem., Int. Ed. Engl.* **1993**, *32*, 131.

[6] I. Stahl, in: *Methoden der Organischen Chemie, Houben-Weyl, Vol. E7a, Part 1, Hetarene II* (Ed.: R. P. Kreher), Thieme, Stuttgart, **1991**, 205.

[7] F. Fringuelli, A. Taticchi, *Dienes in the Diels-Alder Reaction*, Wiley, Chichester, **1990**; P. Brownbridge, *Synthesis* **1983**, 1; *Synthesis* **1983**, 85.

[8] G. Simchen, in: *Advances in Silicon Chemistry, Vol. 1* (Ed.: G.L. Larson), JAI Press, Greenwich, **1991**, p. 189.

[9] Dr. M. Noltemeyer, H. G. Schmidt, Institut für Anorganische Chemie der Georg-August-Universität Göttingen, Göttingen, **1994**.

[10] The keto-silylenolether **17** was obtained by 2,6-lutidiniumtriflate mediated reaction of **6** with **8a**. After 50 % consumption of the 2-silyloxy-1,3-butadiene **8a** the reaction was worked up and besides 50 % of the starting material **8a** 19 % of **9a** and 16 % of **17** were isolated.

Diastereoselective Synthesis of Annulated Quinolones by Transformations with 4-Silyloxyquinolinium-Salts

Sabine Ledderhose, Uwe Beifuss*

Institut für Organische Chemie

Georg-August-Universität Göttingen

Tammannstr. 2, D-37077 Göttingen, Germany

Summary: A new and efficient method for the highly diastereoselective annulation of 4-quinolones using 4-silyloxyquinolinium-salts is presented. Reactions of the quinolinium-salts with several 2-silyloxy-1,3-butadienes give *cis-cis* fused acridones with yields of 62–99 %.

The stereoselective synthesis of *N*-heterocycles like quinolines [1] and acridines [2] is of great interest because many of these compounds can be used for the synthesis of alkaloids. In addition, compounds with the quinoline and acridine skeletons frequently exhibit pharmaceutically interesting properties [3].

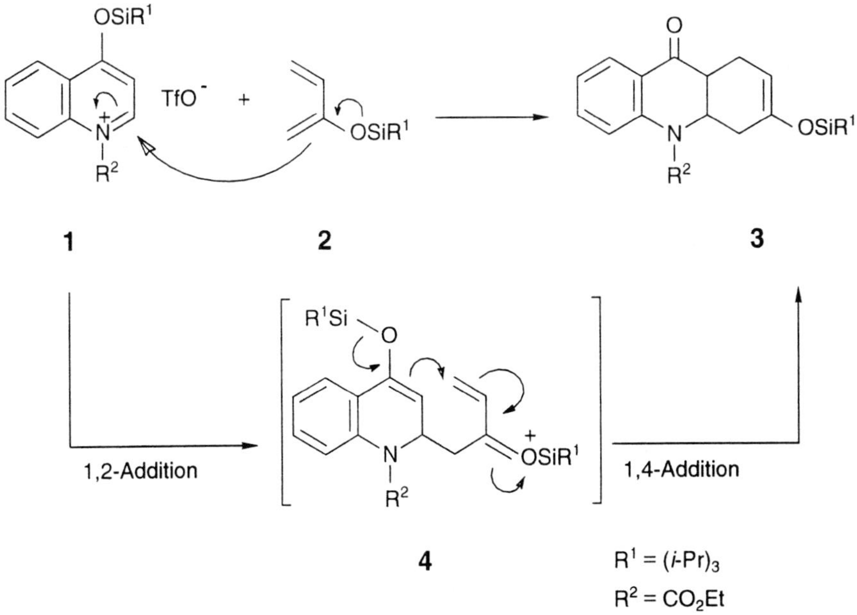

Scheme 1.

Annulations of pyridinium- and quinolinium-salts however have not been reported so far. From the addition reactions of *C*-nucleophiles like Grignard, organozinc and organotin reagents to pyridinium- and quinolinium-salts have been reported to be valuable methods for the construction of substituted dihydropyridine [4a–4d] and dihydroquinoline-derivatives [4d–4e] results of our studies in the field of benzothiopyrylium-salts [5] we concluded that the sequential intermolecular 1,2-addition/intramolecular 1,4-addition of 4-silyloxyquinolinium-salts **1** with 2-silyloxy-1,3-butadienes **2** might be useful for the diastereoselective preparation of annulated quinolones **3**.

As expected, the reaction of the 4-silyloxyquinolinium-salt **1**, which can be regarded as a double activated 4-quinolone, with the unsubstituted 2-silyloxy-1,3-butadiene **2** diastereoselectively gives the *cis*-fused acridone **5** with 86 % yield.

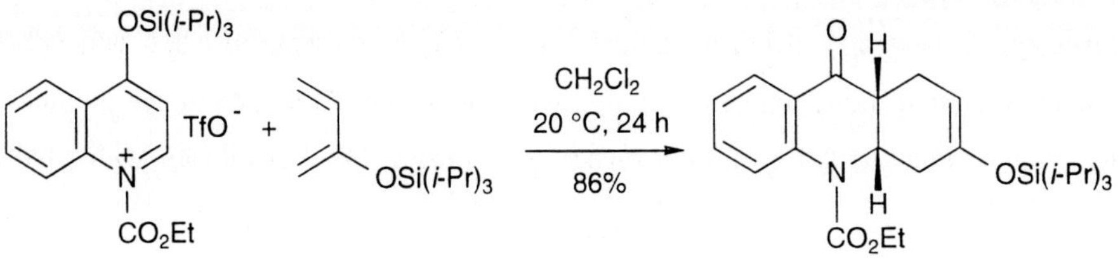

Eq. 1.

A great advantage of the new method is that both reactants, the double activated 4-silyloxyquinolinium-salt **1** and the 2-silyloxy-1,3-butadiene **2** [6], are obtained *in situ* from the corresponding 4-quinolone **6** and methyl vinyl ketone **7** by reaction with triisopropylsilyltrifluoromethanesulfonate (TIPSOTf) and TIPSOTf/2,6-lutidine, respectively.

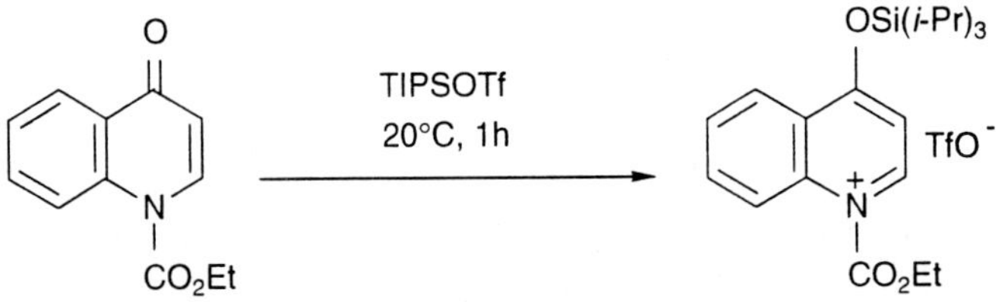

Eq. 2.

Eq. 3.

Reaction of the 4-silyloxyquinolinium-salt **1** with the 3,4-disubstituted 2-silyloxybutadiene **8** exclusively yields the *cis-cis* annulation product **9** with excellent yield. Again, both reactants are formed *in situ* in one flask. First, the 4-silyloxyquinolinium-salt **1** is obtained from the quinolone **6** by treatment with TIPSOTf and then subsequently the 2-silyloxybutadiene **8** is formed from 1-acetylcylopentene [7] by reaction with TIPSOTf and 2,6-lutidine.

Eq. 4.

Using this method a number of annulation reactions of the quinolinium-salt **1** with various 3,4-disubstituted 2-silyloxy-1,3-butadienes have been performed to give highly diastereoselectively the annulation products **9–13** in yields of 62–99 % (see Table 1). In each case the exclusive formation of the *cis-cis* diastereomer is observed.

The assignment of the relative configuration of the annulation products **9–13** is based on NMR spectroscopy. In the ^{1}H NMR spectrum of **9**, for example, 11a-H, which absorbs at $\delta = 2.58$ ppm, occurs as a triplet with a coupling constant $J = 4.5$ Hz ($^3J_{5a,11a} = 4.5$ Hz, $^3J_{11a,11b} = 4.5$ Hz; Fig. 2). This pattern clearly proves the *cis-cis* fusion of the annulation product. It is noteworthy that the success of the annulation reactions is crucially dependent on the double activation of the 4-quinolone **6** as 4-silyloxyquinolinium-salt **1**, because the thermal reaction of the 4-quinolone **6** with the triisopropylsilyl enol ether of 1-acetylcyclohexene gives no annulation product **11** at all.

Product	Reaction time [h]	Yield [%]
9	2	99
10	24	91
11	3	86
12	24	77
13	24	62

Table 1. Diastereoselective transformations of 4-silyloxyquinolinium-salt **1** with 3,4-disubstituted 2-silyloxy-1,3-butadienes

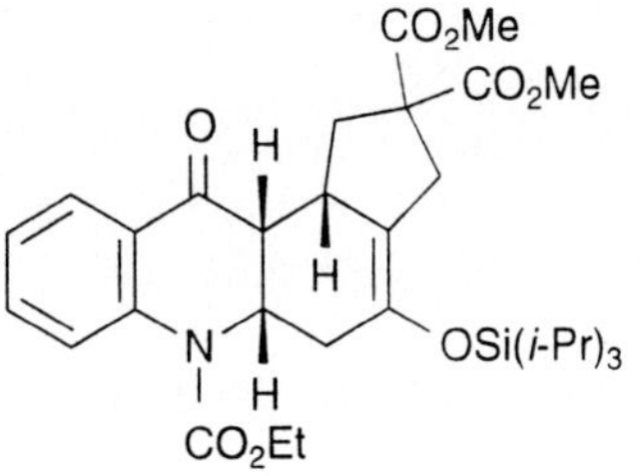

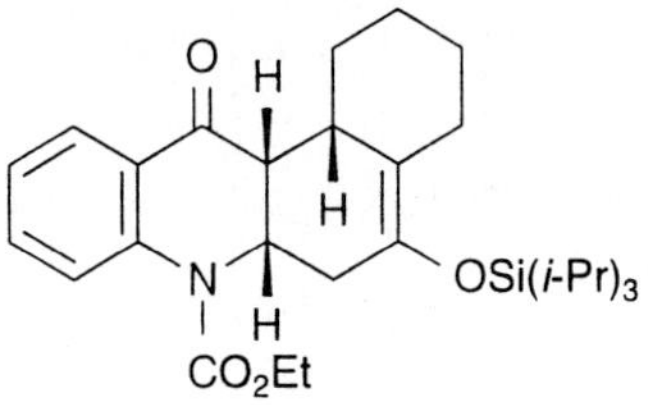

Fig. 1.

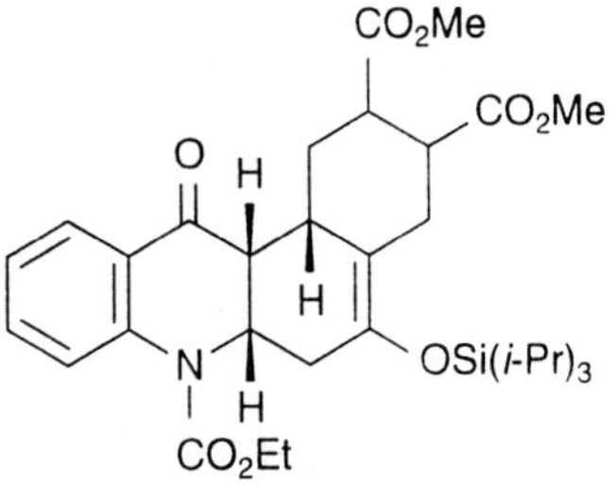

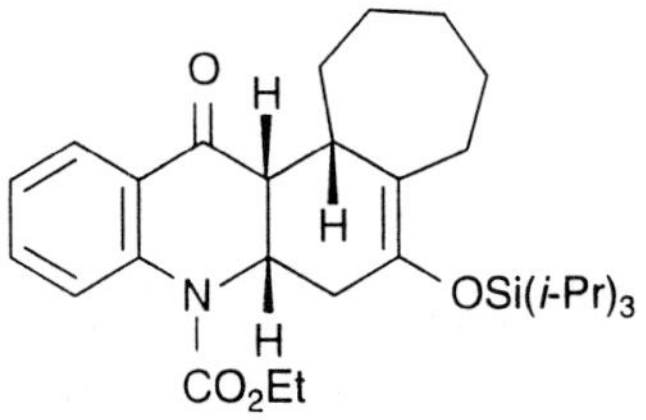

From a mechanistic point of view the annulations may either proceed as a stepwise, sequential intermolecular 1,2-addition/intramolecular 1,4-addition process or as a one-step intermolecular Diels-Alder reaction. Currently, we are performing experiments to differentiate between these mechanistic alternatives.

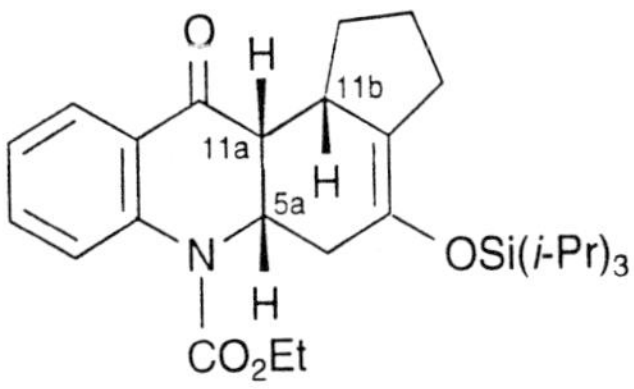

Fig. 2.

$^3J_{5a, 11a} = 4.5$ Hz

$^3J_{11a, 11b} = 4.5$ Hz

(structure **9**)

It is remarkable that the *cis-cis* annulated ketosilyl enol ethers can easily be transformed to the corresponding *cis-syn-trans* diketones by hydrolysis of the silyl enol ether functionality. For example, treatment of **13** with HF in acetonitrile [8], exclusively yields the diketone **14** in 88 % yield.

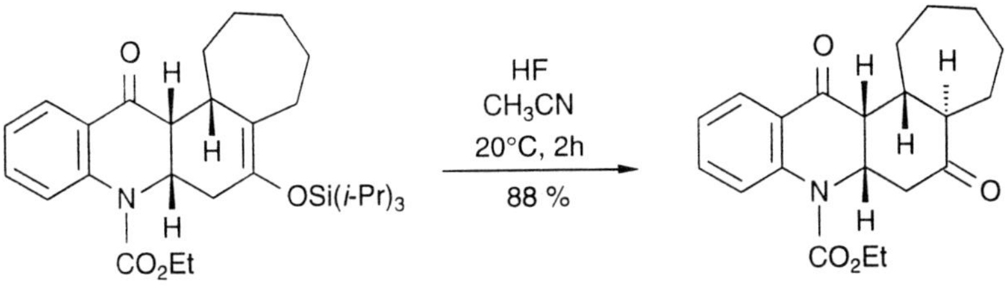

13 → (HF, CH$_3$CN, 20°C, 2h, 88 %) → **14**

Eq. 5.

Acknowledgement: This work was supported by the *Fonds der Chemischen Industrie*, the *Cusanuswerk* and the *Georg-August-Universität Göttingen*. S. L. thanks the *Cusanuswerk* for a doctoral fellowship. We thank Professor L. F. Tietze for his generous support and interest in our work.

References:

[1] C. D. Johnson, in: *Rodd's Chemistry of Carbon Compounds, Vol. 4, Part F* (Ed.: M. F. Ansell), Elsevier, Amsterdam, **1987**, p. 119.

[2] J. D. Hepworth, in: *Rodd's Chemistry of Carbon Compounds, Vol. 4, Part G* (Ed.: M. F. Ansell), Elsevier, Amsterdam, **1987**, p. 1.

[3]   W. A. Denny, in: *Chemistry of Antitumor Agents* (Ed.: D. E. V. Wilman), Blackie, Glasgow, **1990**, p. 1; M. F. Grundon, in: *The Alkaloids, Vol. 32* (Ed.: A. Brossi), Academic Press, New York, **1988**, p. 341; M. F. Grundon, in: *Alkaloids: Chemical and Biological Perspectives, Vol. 6* (Ed.: S. W. Pelletier), Wiley, Chichester, **1988**, p. 339; K. Gerzon, G. H. Svoboda, in: *The Alkaloids 21* (Ed.: A. Brossi), Academic Press, New York, **1983**, p. 1.

[4]   a) Y. Génisson. C. Marazano, B. C. Das, *J. Org. Chem.* **1993**, *58*, 2052.

b) D. L. Comins, D. H. LaMunyon, *J. Org. Chem.* **1992**, *57*, 5807.

c) D. L. Comins, S. O'Connor, *Tetrahedron Lett.* **1987**, *28*, 1843.

d) R. Yamaguchi, M. Moriyasu, M. Yoshioka, M. Kawanisi, *J. Org. Chem.* **1988**, *53*, 3507.

e) W. Bradley, S. Jeffrey, *J. Chem. Soc.* **1954**, 2770.

[5]   U. Beifuss, H. Gehm, M. Tietze, in: *Organosilicon Chemistry II* (Eds.: N. Auner, J. Weis), VCH, Weinheim, **1995**, p. 219.

[6]   F. Fringuelli, A. Taticchi, *Dienes in the Diels-Alder Reaction*, Wiley, Chichester, **1990**.

[7]   P. J. Casals, *Bull. Soc. Chim. France* **1963**, 253.

[8]   J. L. Masareñas, A. Mouriño, L. Castedo, *J. Org. Chem.* **1986**, *51*, 1269.

Biological Recognition of Enantiomeric Silanes and Germanes: Syntheses and Antimuscarinic Properties of the Enantiomers of the Si/Ge Analogues Cyclohexyl(hydroxymethyl)phenyl(2-piperidinoethyl)silane and -Germane and Their Methiodides

Dirk Reichel, Reinhold Tacke*

Institut für Anorganische Chemie

Universität Fridericiana zu Karlsruhe (TH)

Engesserstr., Geb. 30.45, D-76128 Karlsruhe, Germany

Peter G. Jones

Institut für Anorganische und Analytische Chemie

Technische Universität Carolo-Wilhelmina zu Braunschweig

Hagenring 30, D-38023 Braunschweig, Germany

Günter Lambrecht, Jan Gross, Ernst Mutschler

Pharmakologisches Institut für Naturwissenschaftler, Biozentrum Niederursel

Johann Wolfgang Goethe-Universität Frankfurt am Main

Marie-Curie-Straße 9, Geb. N 260, D-60439 Frankfurt, Germany

Magali Waelbroeck

Laboratoire de Chimie Biologique et de la Nutrition, Faculté de Médecine et de Pharmacie

Université Libre de Bruxelles

Route de Lennik 808, B-1070 Bruxelles, Belgium

Summary: The enantiomers of the Si/Ge analogues cyclohexyl(hydroxymethyl)phenyl-(2-piperidinoethyl)silane and -germane and their methiodides were synthesized and investigated with respect to their affinities at muscarinic M1, M2, and M3 receptors. The compounds displayed pronounced stereoselective antimuscarinic action, the (*R*)-enantiomers being more potent than the corresponding (*S*)-antipodes. The receptor affinities of the respective Si/Ge analogues were found to be very similar, indicating a strongly pronounced Si/Ge bioisosterism.

Introduction

As a part of our systematic studies on biologically active silicon [1–4] and germanium [2, 5] compounds, we synthesized the (R)- and (S)-enantiomers of cyclohexyl(hydroxymethyl)phenyl(2-piperidinoethyl)silane (**1a**), a derivative of the muscarinic antagonist sila-trihexyphenidyl (**3**) (Fig 1.) [6]. In addition, the antipodes of the quaternary ammonium derivative **2a** and the (R)- and (S)-enantiomers of the corresponding germanium analogues **1b** and **2b** were prepared (Fig. 1.). The antimuscarinic properties of the antipodes of **1a**, **1b**, **2a**, and **2b** were investigated in functional pharmacological experiments. These studies were carried out with a special emphasis on the comparison of the biological properties of the respective Si/Ge analogues (studies on silicon/germanium bioisosterism).

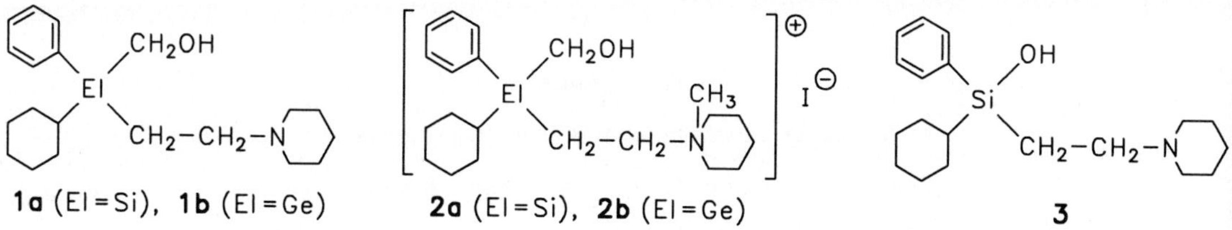

1a (El = Si), **1b** (El = Ge)   **2a** (El = Si), **2b** (El = Ge)   **3**

Fig. 1.

Results and Discussion

The preparation of the optically active title compounds is based on the synthesis of the racemic silane *rac*-**1a** and the racemic germane *rac*-**1b**, followed by their resolution into the respective (R)- and (S)-enantiomers and subsequent transformation of the latter compounds into the antipodes of the quaternary ammonium derivatives **2a** and **2b**.

The racemic compounds *rac*-**1a** and *rac*-**1b** were prepared by a four-step synthesis, starting from (chloromethyl)cyclohexyl(phenyl)vinylsilane and (chloromethyl)cyclohexyl(phenyl)vinylgermane, respectively (Scheme 1).

In the next step, the (R)- and (S)-enantiomers of **1a** and **1b** were obtained by resolution of *rac*-**1a** and *rac*-**1b**, using the antipodes of *O,O'*-di-*p*-toluoyltartaric acid as resolving agents (Scheme 2). Subsequent reaction of the antipodes of **1a** and **1b** with methyl iodide gave the corresponding *N*-methyl derivatives (R)-**2a**, (S)-**2a**, (R)-**2b**, and (S)-**2b** (Scheme 2).

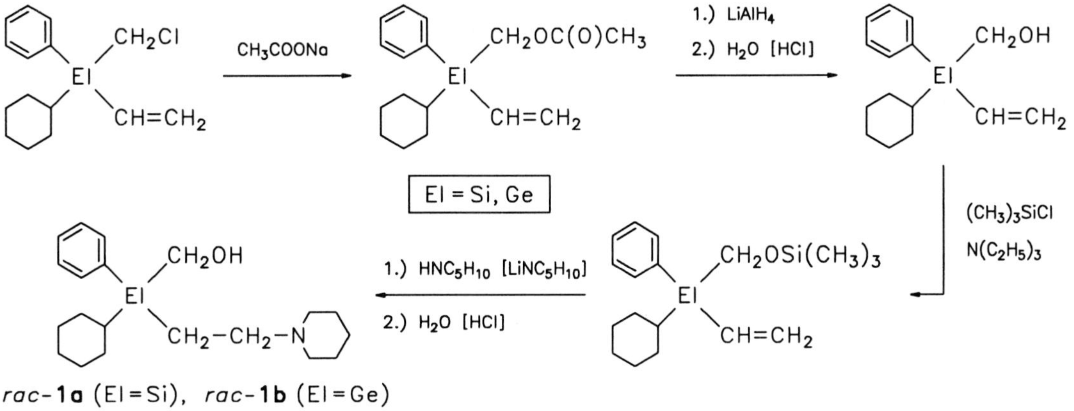

Scheme 1.

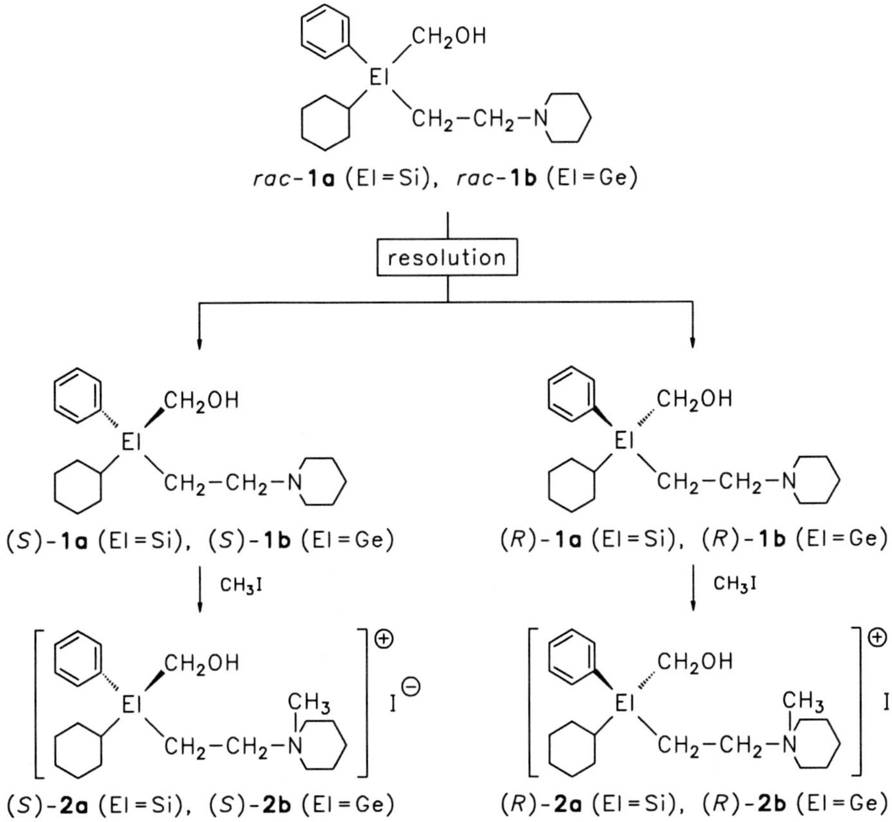

Scheme 2.

The (*R*)- and (*S*)-enantiomers of **1a**, **1b**, **2a**, and **2b** were isolated as colorless crystalline solids. Their identity was established by elemental analyses, NMR spectroscopic studies, and mass spectrometric investigations. In addition, the laevorotatory enantiomers of **2a** and **2b** were structurally characterized by single-crystal X-ray diffraction analyses. According to these studies, (–)-**2a** and (–)-**2b** (optical rotations measured for solutions in CHCl$_3$ at 546 nm) are the (*R*)-enantiomers (Fig. 2). As the *N*-methylation of **1a** (→ **2a**) and **1b** (→ **2b**) does not affect the absolute configuration at the Si or Ge atoms, the absolute configurations of the enantiomers of **1a** and **1b** could also be assigned.

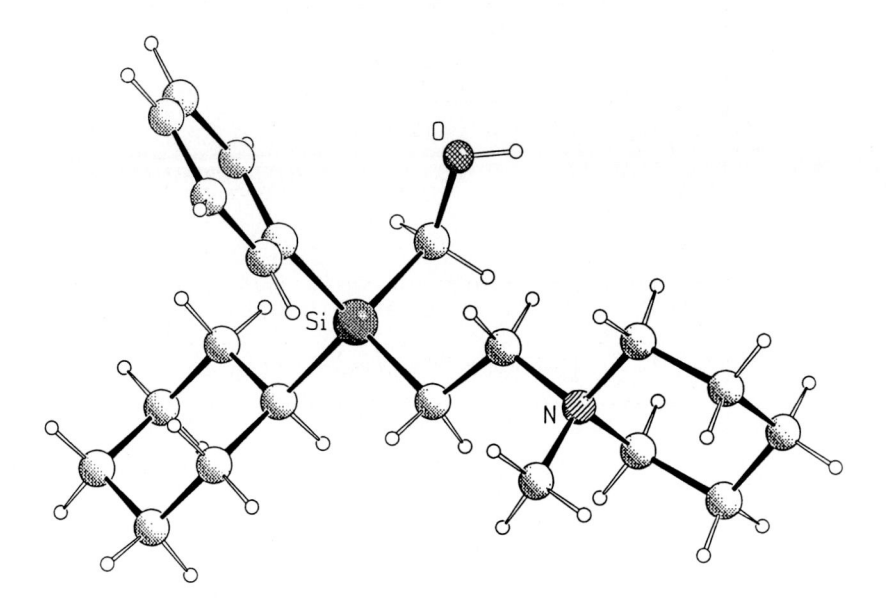

Fig. 2. Structure of the cation of (*R*)-**2a** in the crystal [(*R*)-**2b** is isostructural]

The enantiomeric purities of the antipodes of **1a** and **1b** were determined by ^{1}H NMR experiments using the chiral shift reagent (–)-2,2,2-trifluoro-1-(9-anthryl)ethanol. The (*R*)- and (*S*)-enantiomers of **1a** and **1b** could be clearly discriminated by NMR spectroscopy (characteristic resonance signals for the ElCH$_2$OH units; El = Si, Ge) and therefore quantitatively determined by integration of the respective resonance signals (400.1 MHz; composition of the samples: 30.2 µmol **1a** or **1b**, 152 µmol shift reagent, 0.5 mL CDCl$_3$). According to these studies, the enantiomeric purities of the resolved antipodes of **1a** and **1b** were ≥ 98 % *ee*. As the *N*-methylation of these compounds does not affect the absolute configurations at the Si or Ge atoms, the same enantiomeric purities can be assumed for the (*R*)- and (*S*)-enantiomers of **2a** and **2b**.

The antipodes of **1a**, **1b**, **2a**, and **2b** were studied with respect to their affinities at muscarinic M1 (rabbit vas deferens), M2 (guinea-pig atrium), and M3 receptors (guinea-pig ileum) using functional pharmacological tests. In these studies, 4-F-PyMcN$^+$ (M1 receptors) or arecaidine propargyl ester (M2 and

M3 receptors) were used as agonists (for methods, see [1]). All compounds behaved as simple competitive antagonists. Their receptor affinities (pA_2 values) are listed in Table 1. In addition, the pA_2 values of the enantiomers of **2a** and **2b** are depicted in Fig. 3.

	M1	M2	M3
(*R*)-**1a**	7.39±0.05	6.76±0.04	7.32±0.03
(*S*)-**1a**	6.53±0.04	6.26±0.03	6.15±0.02
(*R*)-**1b**	7.08±0.05	6.56±0.05	7.14±0.03
(*S*)-**1b**	6.24±0.08	6.23±0.05	6.12±0.04
(*R*)-**2a**	9.09±0.06	8.21±0.01	8.64±0.05
(*S*)-**2a**	7.74±0.04	7.56±0.04	7.36±0.03
(*R*)-**2b**	9.17±0.07	8.09±0.03	8.40±0.04
(*S*)-**2b**	7.76±0.05	7.47±0.04	7.15±0.04

Table 1. Affinities (pA_2 values[a]) of the (*R*)- and (*S*)-enantiomers of **1a**, **1b**, **2a**, and **2b** for muscarinic M1, M2, and M3 receptors; [a] $pA_2 = -\log K_D$ (K_D = dissociation constant of the drug-receptor complex)

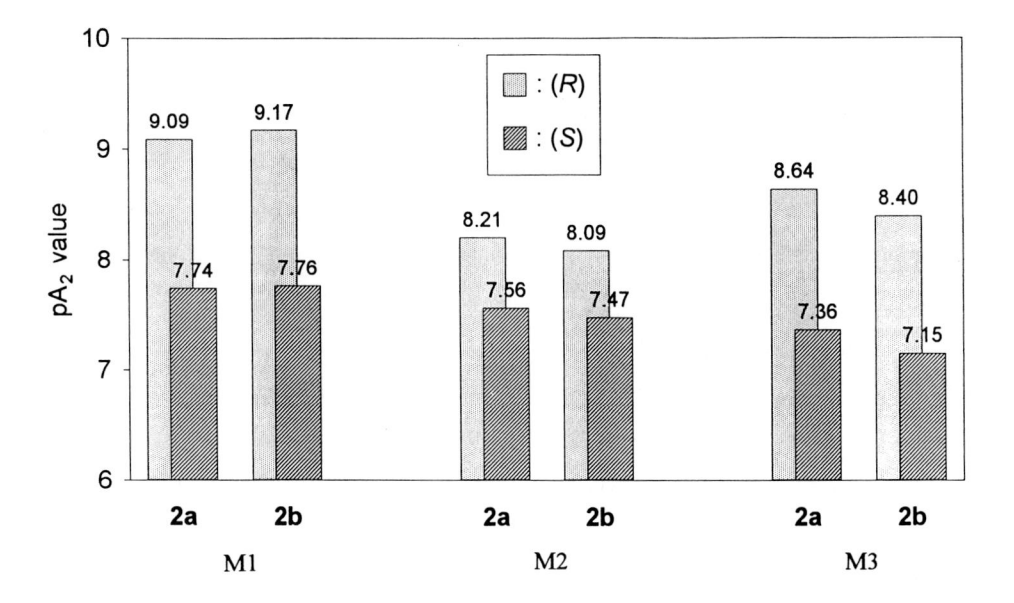

Fig 3. Affinities (pA_2 values) of the (*R*)- and (*S*)-enantiomers of **2a** and **2b** for muscarinic M1, M2, and M3 receptors

The (*R*)-enantiomers of **1a**, **1b**, **2a**, and **2b** were found to display higher affinities for all three muscarinic receptor subtypes than their corresponding (*S*)-antipodes, the highest stereoselectivities being observed for **2a** [(*R*)/(*S*) ratio: 22 (M1), 4.5 (M2), 19 (M3)] and **2b** [(*R*)/(*S*) ratio: 26 (M1), 4.2 (M2), 18 (M3)]. For the (*R*)-enantiomers, receptor selectivity was observed [(*R*)-**1a**, (*R*)-**1b**: M1 ≈ M3 > M2; (*R*)-**2a**, (*R*)-**2b**: M1 > M3 > M2]. As can be seen from Table 1 and Fig. 3, the pharmacological properties (pA_2 values; stereoselectivities and receptor selectivities) of the respective Si/Ge analogues are very similar, indicating a distinct Si/Ge bioisosterism. Because of their high antimuscarinic potency and their receptor selectivity, the silicon compound (*R*)-**2a** and its germanium analogue (*R*)-**2b** are considered to be interesting leads for the development of new receptor selective muscarinic antagonists.

Acknowledgement: Financial support of this work by the *Volkswagen-Stiftung* and the *Fonds der Chemischen Industrie* is gratefully acknowledged.

References:

[1] G. Lambrecht, R. Feifel, M. Wagner-Röder, C. Strohmann, H. Zilch, R. Tacke, M. Waelbroeck, J. Christophe, H. Boddeke, E. Mutschler, *Eur. J. Pharmacol.* **1989**, *168*, 71; and references cited therein (instead of 4-Cl–McN–A-343, 4-F–PyMcN$^+$ was used as M1-selective agonist).

[2] M. Waelbroeck, J. Camus, M. Tastenoy, G. Lambrecht, E. Mutschler, M. Kropfgans, J. Sperlich, F. Wiesenberger, R. Tacke, J. Christophe, *Br. J. Pharmacol.* **1993**, *109*, 360; and references cited therein.

[3] R. Tacke, J. Pikies, F. Wiesenberger, L. Ernst, D. Schomburg, M. Waelbroeck, J. Christophe, G. Lambrecht, J. Gross, E. Mutschler, *J. Organomet. Chem.* **1994**, *466*, 15; and references cited therein.

[4] R. Tacke, M. Kropfgans, A. Tafel, F. Wiesenberger, W. S. Sheldrick, E. Mutschler, H. Egerer, N. Rettenmayr, J. Gross, M. Waelbroeck, G. Lambrecht, *Z. Naturforsch.* **1994**, *49B*, 898; and references cited therein.

[5] R. Tacke, B. Becker, D. Berg, W. Brandes, S. Dutzmann, S. Schaller, *J. Organomet. Chem.* **1992**, *438*, 45; and references cited therein.

[6] R. Tacke, M. Strecker, G. Lambrecht, U. Moser, E. Mutschler, *Liebigs Ann. Chem.* **1983**, 922.

Biotransformation as a Preparative Method for the Synthesis of Optically Active Silanes, Germanes, and Digermanes: Studies on the (*R*)-Selective Microbial Reduction of MePh(Me$_3$C)ElC(O)Me (El = Si, Ge), MePh(Me$_3$Ge)GeC(O)Me, and MePh(Me$_3$Si)GeC(O)Me Using Resting Cells of *Saccharomyces cerevisiae* (DHW S-3) as Biocatalyst

Stephan A. Wagner, Susanne Brakmann, Reinhold Tacke*

Institut für Anorganische Chemie
Universität Fridericiana zu Karlsruhe (TH)
Engesserstr., Geb. 30.45, D-76128 Karlsruhe, Germany

Summary: The racemic compounds MePh(Me$_3$C)SiC(O)Me (*rac-1*), MePh(Me$_3$C)GeC(O)Me (*rac-2*), MePh(Me$_3$Ge)GeC(O)Me (*rac-3*), and MePh(Me$_3$Si)GeC(O)Me (*rac-4*) were synthesized as substrates for stereoselective microbial reductions. Resting cells of the yeast *Saccharomyces cerevisiae* (DHW S-3) were found to reduce *rac-1*, *rac-2*, and *rac-3* (*R*)-selectively [El–C(O)Me → El–CH(OH)Me; El = Si, Ge] to yield 1:1 mixtures of the respective diastereomeric reduction products with (*R*)-configuration at the carbon atoms (enantiomeric purities of the diastereomers ≥ 98 % *ee*). Under the same conditions, the acetyl(silyl)germane *rac-4* was converted into a 1:1 mixture of the respective diastereomeric (1-hydroxyethyl)hydridogermanes MePhHGeCH(OH)Me with (*R*)-configuration at the carbon atoms. In the course of this bioconversion, an additional (probably chemically induced) Si–Ge cleavage was observed.

Introduction

In previous publications we have demonstrated that biotransformations with whole microbial cells or free enzymes can be used for the preparation of optically active silicon and germanium compounds [1]. In continuation of these studies, we synthesized a series of optically active silanes, germanes, and digermanes (center of chirality: Si, Ge) using stereoselective microbial reductions of appropriate organosilicon or organogermanium substrates with El–C(O)Me units [El = Si, Ge; stereoselective reductions of the carbonyl groups: → El–CH(OH)Me]. We report here on studies on the (R)-selective microbial reduction of rac-1, rac-2, rac-3, and rac-4 (Fig. 1) using resting cells of the commercially available yeast Saccharomyces cerevisiae (DHW S-3) as biocatalyst.

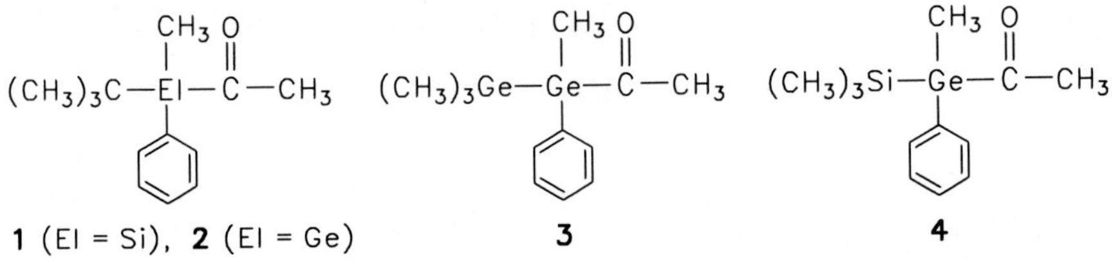

Fig. 1.

This microorganism has recently been shown to convert acetyldimethyl(phenyl)silane [Me$_2$PhSiC(O)Me] with high enantioselectivity (> 99 % ee) into (R)-(1-hydroxyethyl)dimethyl(phenyl)silane [(R)-Me$_2$PhSiCH(OH)Me] [2].

Results and Discussion

The acetylsilane rac-1 was synthesized according to the literature [3]. The germanium analogue rac-2 was prepared analogously as shown in Scheme 1.

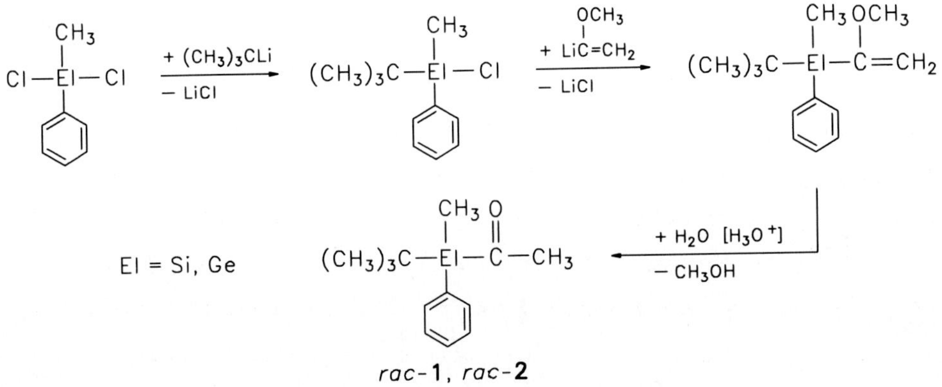

Scheme 1.

The acetyldigermane *rac*-3 and the related acetyl(silyl)germane *rac*-4 were obtained according to Scheme 2. Compounds *rac*-1, *rac*-2, *rac*-3, and *rac*-4 were isolated as liquids; their identity was established by elemental analyses, NMR-spectroscopic studies, and mass-spectrometric investigations.

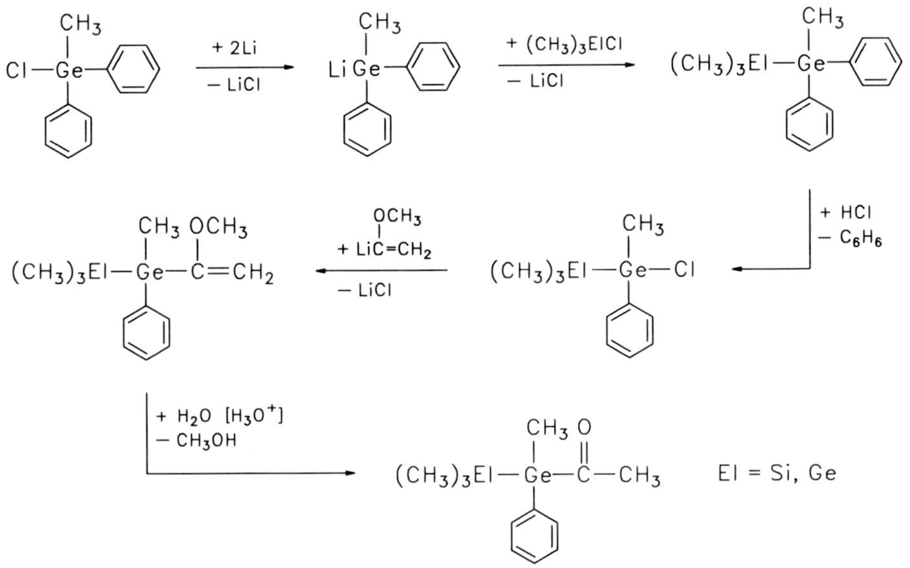

Scheme 2.

The acetylsilane *rac*-1 and the acetylgermane *rac*-2 were found to be reduced (*R*)-selectively by resting cells of *Saccharomyces cerevisiae* (DHW S-3) to yield 1:1 mixtures each of the diastereomeric (1-hydroxyethyl)silanes (SiR,CR)-5 and (SiS,CR)-5 and of the diastereomeric (1-hydroxyethyl)germanes (GeR,CR)-6 and (GeS,CR)-6 (Scheme 3). The bioconversions were carried out on a preparative scale [4] and the products isolated as 1:1 mixtures in 43 % yield (reduction of *rac*-1; substrate concentration 0.30 g/L, incubation time 30 h) or 60 % yield (reduction of *rac*-2; substrate concentration 0.25 g/L, incubation time 3 d). For both bioconversions, high (*R*)-selectivity was observed; the enantiomeric purities of (SiR,CR)-5, (SiS,CR)-5, (GeR,CR)-6, and (GeS,CR)-6 were ≥ 98 % *ee* [5]. The diastereomeric (1-hydroxyethyl)silanes (SiR,CR)-5 and (SiS,CR)-5 were separated by column chromatography on silica gel and then transformed by oxidation (DMSO, DCC, pyridinium trifluoroacetate) into the enantiomerically pure (≥ 98 % *ee*) acetylsilanes (*R*)-1 and (*S*)-1 (Scheme 3). The diastereomeric (1-hydroxyethyl)germanes (GeR,CR)-6 and (GeS,CR)-6 were also separated by column chromatography on silica gel (MPLC) and transformed analogously into the enantiomerically pure (≥ 98 % *ee*) acetylgermanes (*R*)-2 and (*S*)-2 (Scheme 3).

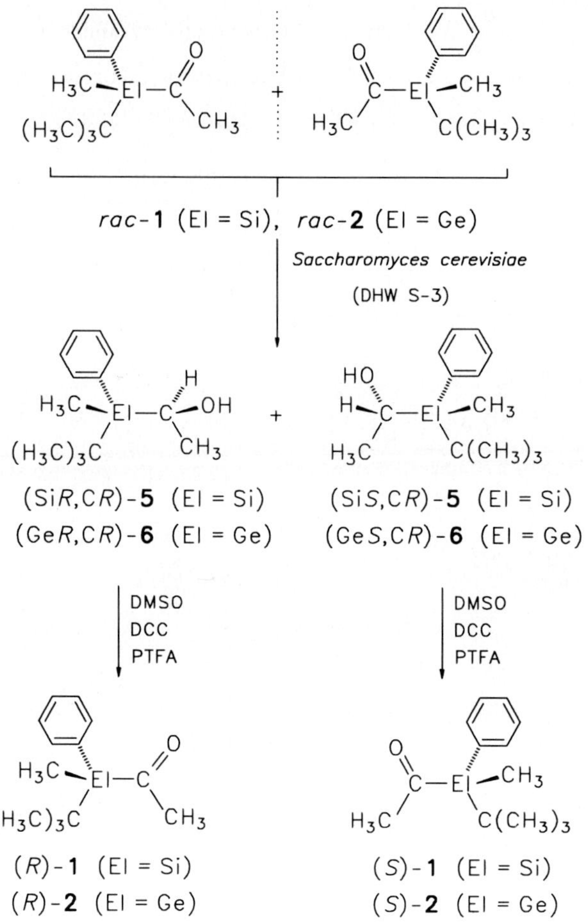

Scheme 3.

As shown in Scheme 4, the acetyldigermane *rac*-3 was also found to be reduced (*R*)-selectively using resting cells of *Saccharomyces cerevisiae* (DHW S-3). This bioconversion was performed on a preparative scale [4] and yielded a 1:1 mixture of the diastereomeric (1-hydroxyethyl)digermanes (Ge*R*,C*R*)-7 and (Ge*S*,C*R*)-7 (yield 62 %; substrate concentration 0.20 g/L, incubation time 4 d). The enantiomeric purities of these compounds were ≥ 98 % *ee* [5], indicating again a high (*R*)-selectivity of this type of bioconversion. Obviously, the Ge–Ge bonds of the substrate and product are rather stable against hydrolytic cleavage under the bioconversion conditions used.

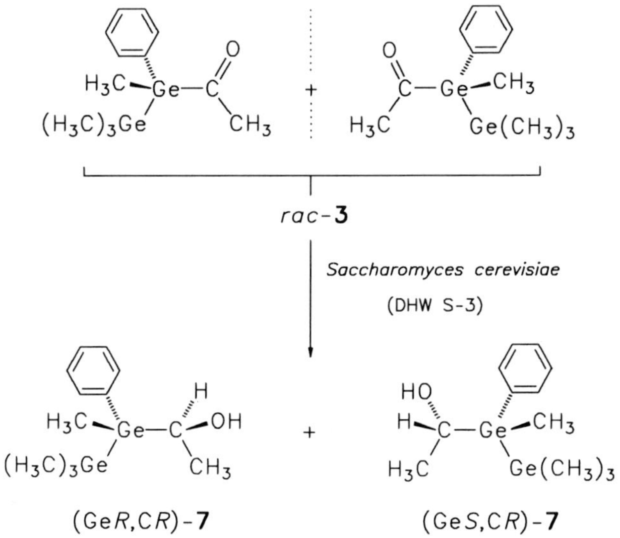

Scheme 4.

A quite different result was obtained when incubating the acetyl(silyl)germane *rac*-4 with resting cells of *Saccharomyces cerevisiae* (DHW S-3). When using the same conditions as applied for the microbial reduction of *rac*-1, *rac*-2, and *rac*-3 [4], cleavage of the Si–Ge bond was observed.

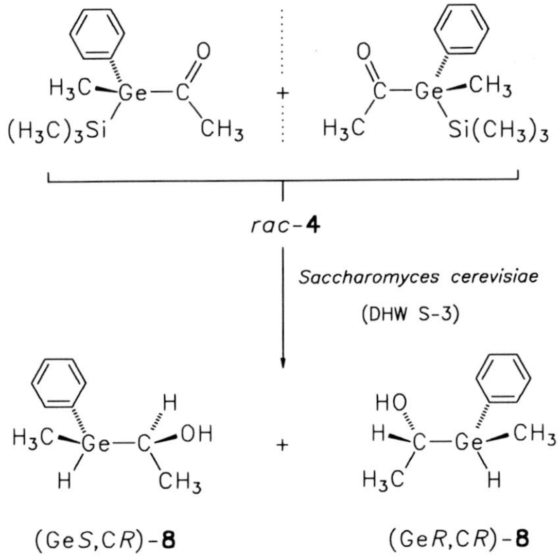

Scheme 5.

As shown in Scheme 5, rac-**4** was converted into a 1:1 mixture of the diastereomeric (1-hydroxyethyl)hydridogermanes (Ge*R*,C*R*)-**8** and (Ge*S*, C*R*)-**8** (yield 50 %; substrate concentration 0.20 g/L, incubation time 24 h). The enantiomeric purities of these bioconversion products were ≥ 98 % ee [5]. The Si–Ge bond cleavage is probably chemically induced.

In conclusion, enantioselective microbial reductions of silicon and germanium compounds containing an El–C(O)Me (El = Si, Ge) moiety [→ El–CH(OH)Me] proved to be an efficient preparative method for the synthesis of optically active silanes, germanes, and digermanes. Furthermore, the commercially available yeast *Saccharomyces cerevisiae* (DHW S-3) is considered to be an efficient biocatalyst for this particular type of bioconversion.

Acknowledgement: Financial support of this work by the *Deutsche Forschungsgemeinschaft* and the *Fonds der Chemischen Industrie* is gratefully acknowledged.

References:

[1] R. Tacke, S. Brakmann, F. Wuttke, J. Fooladi, C. Syldatk, D. Schomburg, *J. Organomet. Chem.* **1991**, *403*, 29; and references cited therein; R. Tacke, S. A. Wagner, J. Sperlich, *Chem. Ber.* **1994**, *127*, 639; and references cited therein.

[2] L. Fischer, S. A. Wagner, R. Tacke, *Appl. Microbiol. Biotechnol.* **1995**, *42*, 671.

[3] R. Tacke, K. Fritsche, A. Tafel, F. Wuttke, *J. Organomet. Chem.* **1990**, *388*, 47.

[4] The biotransformations were carried out in shaking flasks (110 rpm) at 30 °C (composition of the aqueous reaction medium: 12.0 g/L NaH_2PO_4, 30.0 g/L glucose, 50 g/L washed yeast cells). The bioconversion products were isolated by extraction of the culture broths with ethyl acetate, followed by chromatographic purification (silica gel) and Kugelrohr distillation. For monitoring of the bioconversions, extracts of the culture broths were analyzed by gas chromatography.

[5] The enantiomeric purities of the respective diastereomeric biotransformation products were determined by ^1H NMR spectroscopy, after derivatization with (*R*)-2-methoxy-2-(trifluoromethyl)-2-phenylacetyl chloride. These derivatives were also used to determine the absolute configurations at the carbon atoms of the respective El–CH(OH)Me (El = Si, Ge) moieties, following the strategy described in [3].

Aldehyde-functionalized Silanes: New Compounds to Improve the Immobilization of Biomolecules

Joseph Grobe*, Claudia Brüning, Manfred Wessels

Anorganisch-Chemisches Institut

Westfälische Wilhelms-Universität

Wilhelm-Klemm-Str. 8, D-48149 Münster, Germany

Summary: Aldehyde-functionalized ethoxysilanes were successfully used to attach glucose oxidase (GOD) on controlled pore glass (CPG). In contrast to other immobilization techniques, no bifunctional crosslinker is required.

The immobilization of biomolecules on solid surfaces offers considerable advantages to enzymes in solution. Immobilization via physical attraction is often not reliable because of problems concerning leaching and loss of the biomolecule [1]. Therefore, covalent attachment of the enzymes to solid surfaces is advantageous, especially in biotechnology, where the biomolecule should be separated from the reaction mixture, or for the construction of biosensors, where a biocatalyst would be used in a detector system.

In many cases, silicon organic compounds are used as connecting links between a silica surface and a biomolecule. The most popular technique utilizes the modification of a hydroxy-functionalized surface with 3-aminopropyltriethoxysilane (APTES) [2] followed by crosslinking with glutaraldehyde (GA) (Fig. 1A).

We have synthesized aldehyde-functionalized silanes as a new class of spacer molecules for silica surfaces. The aldehyde group can be used to attach biomolecules directly to the surface, with no bifunctional crosslinking species required (Fig. 1B) [3]. As many immobilization problems are closely related to the use of the crosslinking agent (e.g., glutaraldehyde tends to form undesirable polymers) [4], the development of a one-step immobilization method is a significant improvement.

In both, the traditional and the newly reported process, the first step consists of reaction of the ethoxy group of the silicon organic compound with hydroxy functions on the surface to form Si–O–Si bonds. The traditional APTES method, using the amino-functionalized silane, however, requires the surface be treated with glutaraldehyde after the silanization. Free amino groups of the silane react with aldehyde functions of the crosslinking species to form azomethines (Fig. 1A, 2.). Using the same reaction, free amino groups of the biomolecules may react with the second function of glutaraldehyde (Fig. 1A, 3.). In the new aldehyde-functionalized silane technique, free amino functions of the biomolecule react directly with the aldehyde group of the silane to form azomethines (Fig. 1B, 2.), thus avoiding a crosslinking agent such as GA.

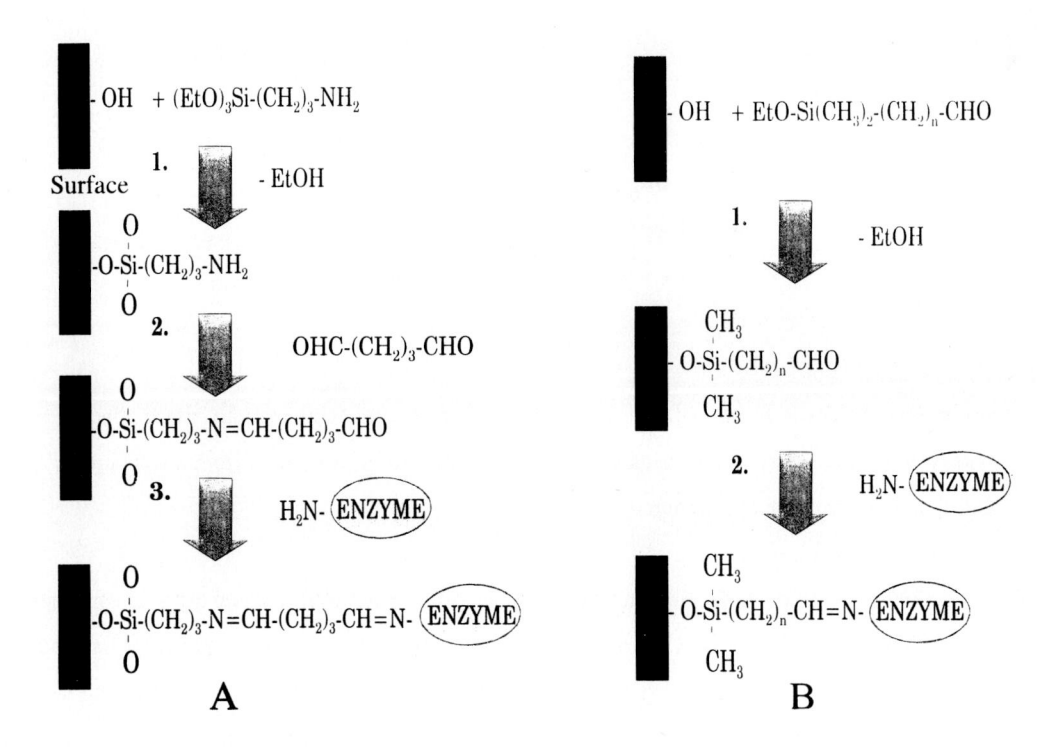

Fig. 1. Scheme of the different immobilization techniques:
A: Immobilization with 3-aminopropyltriethoxysilane (APTES) and glutaraldehyd (GA)
B: Immobilization using aldehyd-functionalized ethoxysilanes

A series of new immobilization reagents based on silanes with aldehyde functions, e.g., monoethoxy-dimethylsilylheptanal and triethoxy-silylheptanal, were synthesized. To test the effectiveness of the new compounds they have been applied to controlled pore glass as a model substrate, and their aldehyde functions were used after the silanization process to immobilize glucose oxidase as a test molecule. The efficiency of the immobilization of GOD on CPG using different techniques and reagents was compared using a photometric activity test of the enzyme, which is schematically shown in Fig. 2.

In this test, oxygen from air is used for the enzymatic oxidation of β-D-glucose in the presence of immobilized GOD to gluconic acid and hydrogen peroxide. Hydrogen peroxide can be determined by a second enzyme reaction. With horse radish peroxidase as catalyst o-phenylendiamine is oxidized by H_2O_2 to 2,3-diaminophenazine, which can be photometrically determined at 490 nm, thus establishing a quantitative relationship between active GOD sites and the intensity of the absorption band.

-O-Si(CH$_3$)$_2$-(CH$_2$)$_n$-CH=N-(GOD) ⟵ ß-D-glucose, oxygen

⟶ gluconic acid, hydrogen peroxide

(POD) peroxidase ⟵ hydrogen peroxide, o-phenylendiamine

⟶ 2,3-diaminophenazine (orange dye)

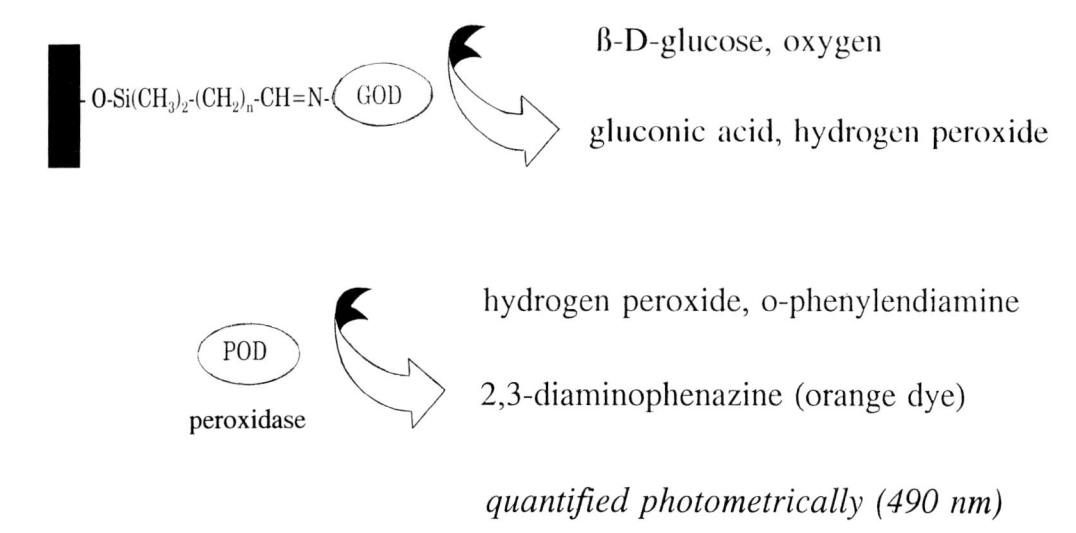

quantified photometrically (490 nm)

Fig. 2. Main reaction steps of the photometrical activity test

The results obtained for three different immobilization methods are summarized in Fig. 3.

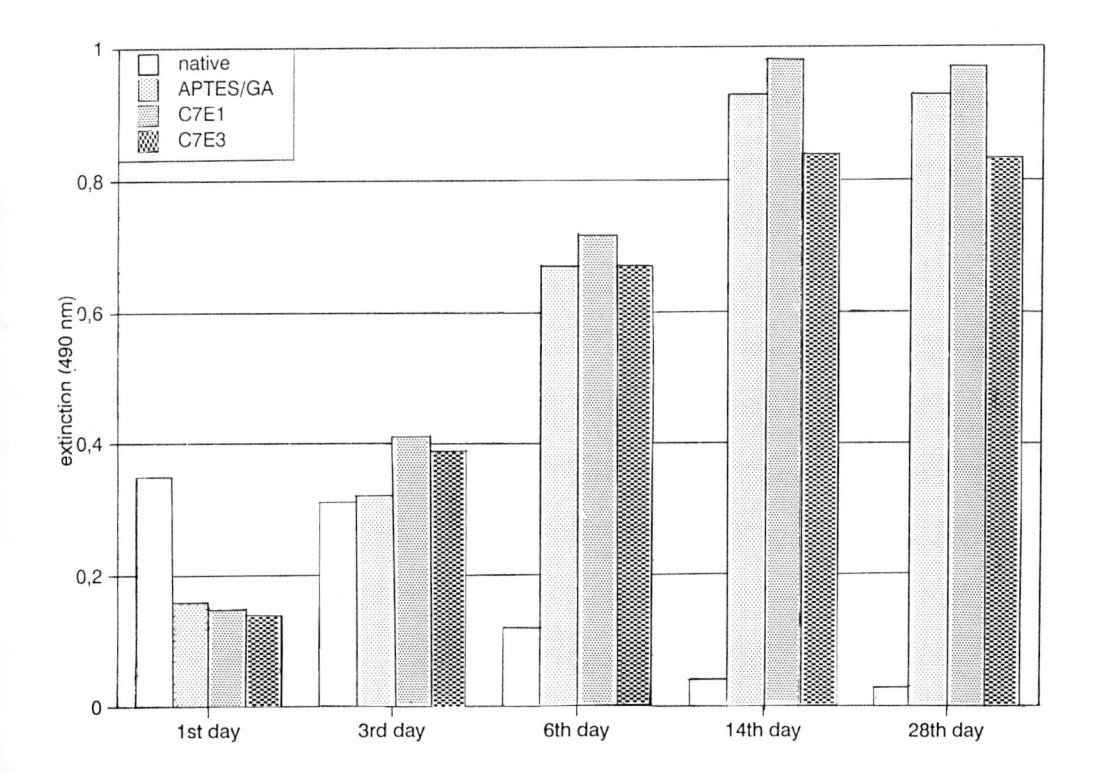

Fig. 3. Comparison of the activity of glucose oxidase (GOD) immobilized with various techniques

It presents a comparison of the observed GOD activities with activity of GOD physisorbed on native, underivatized CPG. An increase of enzyme activity over a time of four weeks was observed. The first bar shows the absorption of the sample immobilized without any reagent on native CPG. The second bar results from immobilization using APTES/GA, and the last two bars show the activity of GOD immobilized using aldehyde-functionalized silanes with the same chain length (C7) but differing in the number of substituent ethoxy groups.

On the first day, all samples show measurable catalytic activity, including the samples where the enzyme was coupled to the native, underivatized CPG. For the latter, however, the activity decreases almost to zero over the following weeks. This can be explained by desorption of the physisorbed GOD from unmodified CPG. In contrast, immobilizations using functionalized silanes provided increasing activities arriving at a constant value after two weeks.

The best results were obtained using aldehyde-functionalized ethoxysilanes for the immobilization of GOD indicated by a higher activity in the first days when compared to the APTES/GA-method. Even after 28 days, the activity of the immobilized GOD is very high in all cases.

The following advantages of the immobilization using aldehyde-functionalized silanes have been demonstrated in our preliminary study:

- The new immobilization technique requires only one step for surface activation, while at least two steps are neccessary in traditional literature methods.
- The use of glutaraldehyde is avoided. Therefore, the problems related to the use of this crosslinking reagent, e.g., undesireable polymerization reactions, do not occur.

Conclusion

One can state that the new anchor molecules provide an effective alternative to the APTES/GA-procedure for the immobilization of GOD and very probably of other biomolecules. Further investigations will focus on their application to low surface area substrates like silicon wafers or electrodes.

References:

[1] W. Hartmeier, *Immobilisierte Biokatalysatoren*, Springer, Berlin, **1986**.

[2] H. H. Weetall, R. D. Mason, *Biotechnol. Bioeng.* **1973**, *15*, 455.

[3] C. Brüning, J. Grobe, *Biosensor 92, The Second World Congress on Biosensors*, Proceedings, Elsevier Advanced Technology Oxford, **1992**, 509.

[4] S. K. Bhatia, L. C. Shriver-Lake, K. J. Prior, J. H. Georger, J. M. Calvert, R. Bredehorst, F. S. Ligler, *Anal. Biochem.* **1989**, *178*, 408.

Organosubstituted Silicon with Unusual Coordination Numbers

Norbert Auner, Johann Weis

The chemistry of silicon with unusual coordination numbers has developed in a remarkable progress. While short living silylenes (coordination number 2) were discussed as intermediates in gas phase reactions of suitable precursors already in the seventies, followed by investigations on derivatives with bulky substituents at silicon and thus stable in solution in the early eighties, n- and π-donor stabilized silylenes were synthesized in the solid state already in the late eighties.

Today X-ray structures of pretty stable cyclic silylenes are available: in these compounds silicon is bound to two nitrogen neighbours as part of a five membered ring skeleton resulting in the first stable silylene with a pure coordination number two!

Consequently the corresponding silylene transition metal complexes with coordination number three at silicon are accessible. In contrast the reaction of dichlorosilanes with carbonylate dianions in the presence of a donor solvent yields donor stabilized silylene complexes (coordination number at silicon 3+1).

A similar coordination sphere at silicon is found in unsaturated Si=E compounds (E = element of group 14–16) and their donor adducts (coordination numbers 3 and 3+1). Attached to coordination number three at silicon the existence of silicenium cations is still controversely discussed in the literature. While some experiments are interpreted as evidence for their existence in the condensed phase, theoretical calculations do not support the formulation of donor free silicenium cations. However, there seems to be no doubt about the verfication of inter- and intramolecular donor stabilized cationic species (coordination numbers 3+1 and 3+2, respectively).

In the field of high coordination numbers at silicon mechanisms of nucleophilic substitution reactions at these centers are of interest, whereas the crystralline, zwitterionic λ^5-silicates may be regarded as "soluble" SiO_2 derivatives. Their chemistry might answer the questions of silicon uptake and transport in living organisms some day. For further information we recommend to refer to the following literature.

References:

[1] J. L. Margrave, D. L. Perry, *Inorg. Chem.* **1977**, *16*, 1820.

[2] M. Weidenbruch, in: *Organosilicon Chemistry – From Molecules to Materials* (Eds.: N. Auner, J. Weis), VCH, Weinheim, **1994**, p. 125.

[3] R. West, M. J. Fink, J. Michl, *Science* **1981**, *214*, 1343.

[4] H. H. Karsch, in: *Organosilicon Chemistry – From Molecules to Materials* (Eds.: N. Auner, J. Weis), VCH, Weinheim, **1994**, p. 95.

[5] P. Jutzi, in: *Organosilicon Chemistry – From Molecules to Materials* (Eds.: N. Auner, J. Weis), VCH, Weinheim, **1994**, p. 87.

[6] M. Denk, R. West, R. Hayashi, Y. Apeloig, R. Pauncz, M. Karni, in: *Organosilicon Chemistry II* (Eds.: N. Auner, J. Weis), VCH, Weinheim, **1995**, p. 251.

Silylenes, Stable and Unstable

Michael Denk, Robert West*, Randy Hayashi

Department of Chemistry

University of Wisconsin

Madison, Wisconsin 53706 USA

Yitzhak Apeloig, Reuben Pauncz, Miriam Karni

Department of Chemistry

Technion - Israel Institute of Technology

Technion City, 32000 Haifa, Israel

Summary: Silylenes, R_2Si: have singlet ground states, unlike simple carbenes which are ground-state triplet species. The reason for this difference between silylenes and carbenes is considered in the light of SCF MO calculations. A stable silylene **5** has been synthesized and its properties investigated. Theoretical and experimental data suggest that the unprecedented stability of **5** results in part from aromatic stabilization in the 6π-electron ring. Novel reactions of **5** are described. Silylene **13**, the saturated analog of **5**, has also been isolated but is less stable than **5**.

Introduction

Silylenes: Divalent, dicoordinate silicon compounds, are the silicon counterparts to the carbenes well known in organic chemistry. Since silylenes are frequent intermediates in both thermal and photochemical reactions, their importance in organosilicon chemistry is great [1]. There is recent evidence that even the direct reaction of methyl chloride with silicon, the foundation stone of the worldwide silicone industry, may proceed through the formation of silylene intermediates [2].

It often happens that novel molecules are first proposed as transient intermediates, then later isolated in matrix at very low temperature and still later obtained (with appropriate substitution) as stable species. The chemistry of silylenes followed exactly this classic historical pattern. Silylenes were first proposed as intermediates in organosilicon chemistry in the 1960's. Organosilylenes were obtained in argon or hydrocarbon matrices beginning in 1979 from photochemical reactions like those shown in Eq. 1 and 2 [3, 4] (SiF_2 and a few other silylenes were matrix-isolated even earlier). Stable compounds of divalent silicon were isolated in the late 1980's; notable examples are decamethylsilicocene, prepared by Jutzi et al. [5], and the tetracoordinated phosphine derivative made by Karsch et al. [6]. Finally, within the last year, stable dicoordinated silylenes have become available [7].

These most recent developments will be the main focus of this paper. First, however, the electronic nature of silylenes will be briefly considered.

Silylenes as Ground-State Singlets

The prototypical carbene, CH_2, and other simple alkyl and aryl carbenes have triplet ground states, with two unpaired electrons. In striking contrast, all silylenes known to date have singlet ground states, with the two nonbonding electrons paired. This difference is of the greatest importance for the properties, spectra, and chemical behavior of silylenes.

The energies of the singlet and triplet states of the simple molecules CH_2 [8, 9] and SiH_2 [10, 11] are well established from both theoretical and experimental studies. The singlet-triplet promotion energy, E_{S-T}, is +8.7 for CH_2 and −21 kcal mol^{-1} for SiH_2, a difference of −30 kcal mol^{-1}.

What are the factors underlying this large difference between SiH_2 and CH_2? We have been considering this problem now for several years. Recently, at the Technion, ab initio calculations on both molecules in the singlet and triplet states have been carried out within the framework of a UHF single-determinant SCF method [12]. These calculations cannot give precise energies, because electron correlation effects are neglected, but they have the advantage that they are readily decomposed in such a manner that at least one of the basic factors contributing to the energy difference can be identified.

We will consider three possible factors:

1) Pairing energy, required to place two electrons in the same orbital in the singlet
2) Electrostatic stabilization of the singlet
3) Differences in *s-p* mixing in Si and C

1. Pairing energy: In the singlet species the two electrons must be paired, in the same in-plane σ orbital. The repulsion between the two electrons is much greater in the singlet than in the triplet, so the "pairing energy" going from the triplet to the singlet is an endothermic term. Since the frontier orbitals are more diffuse for silicon than for carbon, the endothermic pairing energy should be smaller for Si than for C. The pairing energy can be calculated in our SCF framework. As expected, the energy required is greater for CH_2 than for SiH_2 by 12.7 kcal mol^{-1} at the present level of theory. This is about 40 % of the total difference in the singlet-triplet stabilization energy.

2. *Electrostatic stabilization:* It has been argued by Harrison [13] that for :MX_2 species, electropositive substituents X favor triplet carbenes as silylenes, while electronegative substituents stabilize the singlet state. One way to view this is in terms of Bents rule; electronegative substituent tend to bind to orbitals having greater p character, leaving greater s character in the lone pair and lowering energy.

In CH_2 the C–H bonds are nearly nonpolar, so any electrostatic stabilization effects will be slight. However, in SiH_2, and likewise organosilylenes R_2Si, the bonds are significantly polarized, Si^+–X^- and stabilization of the singlet state should result.

The electrostatic stabilization of the singlet silylene cannot be calculated directly, but to estimate this effect the H atoms in SiH_2 can be replaced by H_3Si groups, to make the Si-substituent bonds nearly nonpolar. *MO* calculations lead to estimates for E_{S-T} of $(H_3Si)Si$ which are about 12 kcal mol^{-1} less than for SiH_2. The electrostatic effect thus may account for approximately another 40 % of the E_{S-T} difference between CH_2 and SiH_2. Added to the difference in pairing energy, about 80 % of the total difference between E of CH_2 and SiH_2 can be accounted for.

3. *s-p Mixing:* The radii of maximum overlap are nearly the same for $2s$ and $2p$ orbitals, so that mixing of s and p orbitals in forming chemical bonds is energetically favorable for first-row atoms like carbon. For second-row atoms like silicon, *s-p* mixing is less favorable; bonds tend to have lower p character, and lone pairs more s character, lowering their energy [14]. This effect probably also contributes to the E_{S-T} difference between CH_2 and SiH_2 but is not easily calculated.

Molecular orbital calculations agree with the conclusions from factors 2 and 3, concerning the nature of the lone pair orbital in the singlet states. The lone pair is estimated to have 89 % s character in SiH_2 compared with 52 % in singlet CH_2 [15]. The difference in orbital occupancy is reflected in the bond angles, calculated to be 92.5 ° in SiH_2 and 102.4 ° in singlet CH_2.

Stable Silylenes

Silylenes can be matrix-isolated not only in argon at very low temperature but also in hydrocarbon glasses at 77 K [16]. Since they are singlets, silylenes do not abstract hydrogen from the hydrocarbon matrix. However, in general silylenes disappear immediately upon melting of the matrix. The usual first step is dimerization to the disilene; if the substituents are sufficiently large, this product may be stable, as shown for an example in Eq. 3 [4].

$$(Me_2Si)_6 \xrightarrow{h\upsilon} (Me_2Si)_5 + [Me_2Si:] \quad (1)$$

$$ArSiR(SiMe_3)_2 \xrightarrow{h\upsilon} Me_3SiSiMe_3 + [Ar(R)Si:] \quad (2)$$

$$Mes_2Si(SiMe_3)_2 \xrightarrow[-Me_3SiSiMe_3]{h\upsilon} [Mes_2Si:] \longrightarrow 1/2\ Mes_2Si=SiMes_2 \quad (3)$$

More commonly the disilene is also unstable and undergoes further dimerization or polymerization below room temperature. Attempts to stabilize silylenes sterically have met with only limited success. The best example is the silylene **1**, which could be observed in solution but not isolated [17]. For the heavier elements of group 14, dicoordinate compounds are well-known, and in 1991, a stable dicoordinate carbene **2** was reported [18] by Arduengo. The latter compound provided the key to the synthesis of a stable silylene [7].

Fig. 1.

The preparation starts from glyoxal and *t*butylamine (Eq. 4). Lithiation of the resulting diimine **3** and coupling to silicon tetrachloride led to the five-membered ring **4**. The final step was dehalogenation of **4**. To our surprise this required vigorous reaction conditions: contact with liquid potassium metal in THF under reflux for 3 days (Eq. 5). Under these conditions, stable silylene **5** is obtained in >80 % conversion as colorless crystals.

Eq. 4.

Proof of the structure of **5** follows from several kinds of evidence.

Eq. 5.

Crystals of **5** so far have always been twinned, so accurate bond lengths could not be determined, but the X-ray analysis showed definitely that the compound is monomeric. The structure was confirmed by an electron-diffraction study, carried out at the University of Oslo. A representation is shown in Fig. 2; the molecule is planar, with bond distance Si–N 175.3, N–C 140.0, C=C 134.2 pm. The ^{29}Si NMR spectrum shows that the silicon atom is deshielded (δ = 78.3 ppm) consistent with low coordination at Si.

Silylene **5** is remarkably stable, in sharp contrast to previous silylenes. It was purified by vacuum distillation at 85 °C (1 Torr) and survives heating in toluene solution to 150 °C for many months. The pure compound decomposes only at its melting point, 220 °C. Compound **5** is also less reactive than usual silylenes. It is inert to triethylsilane, diphenylacetylene, or 2,3-dimethylbutadiene, all of which react rapidly with conventional silylenes. Moreover, it does not form acid-base complexes with Lewis bases such as THF and pyridine, although normal silylenes do [19]. However, **5** does react with methanol, water, and dioxygen.

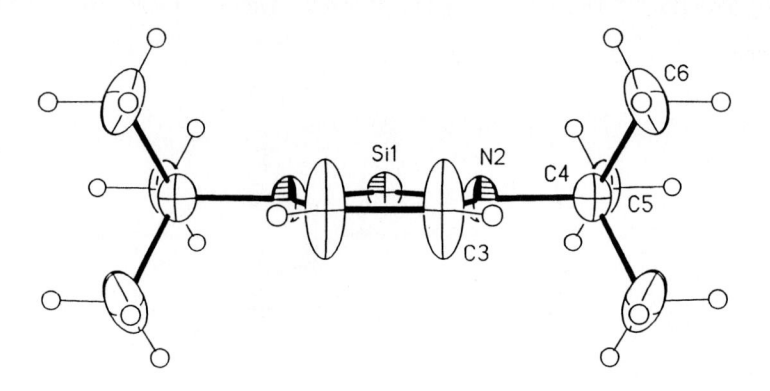

Fig. 2. Structure of silylene **5** from electron diffraction

In addition, **5** undergoes many reactions which are not possible to carry out using transient silylenes. Some of these lead to novel products not obtainable by other syntheses. For example, **5** reacts with many metal carbonyls to give base-free silylene-transition metal complexes. Two examples are shown in Eq. 6 and 7 (LSi: = **5**).

$$LSi: + Fe_2(CO)_9 \longrightarrow LSi=Fe(CO)_4$$
$$\mathbf{\underset{\sim}{6}}$$

$$LSi: + Ni(CO)_4 \longrightarrow LSi=Ni=SiL$$
with CO ligands above and below Ni
$$\mathbf{\underset{\sim}{7}}$$

Eq. 6/7 (Genuine drawing by the author)

An X-ray crystal structure has been determined for **7** and is shown in Fig. 3. This compound is the first silylene complex of nickel, as well as the first bis-silylene metal complex of any kind without base stabilization [20].

Surprise products sometimes appear in reactions with metal carbonyl derivatives. For instance in the reaction of **5** with bis(norbornadiene) chromium tetracarbonyl, the product isolated **8**, is one in which the silylene has become incorporated into the bicyclic ring structure.

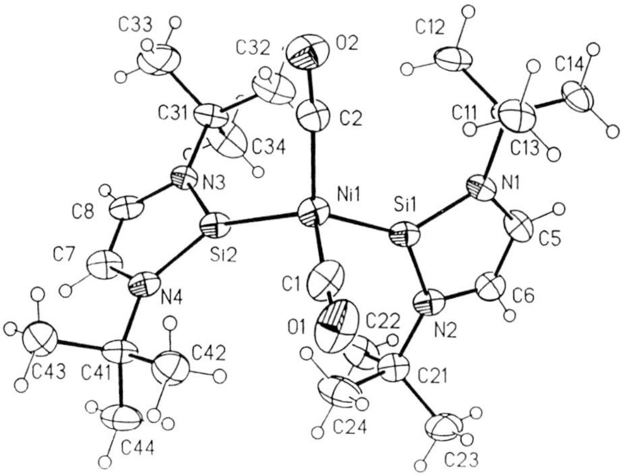

Fig. 3. X-ray crystal structure of (LSi)$_2$Ni(CO)$_2$ (**7**)

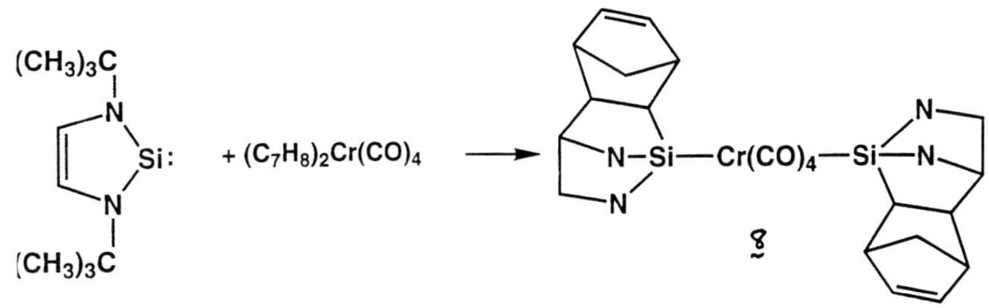

Eq. 8.

A true silylene-chromium complex was obtained however from the reaction shown in Eq. 9.

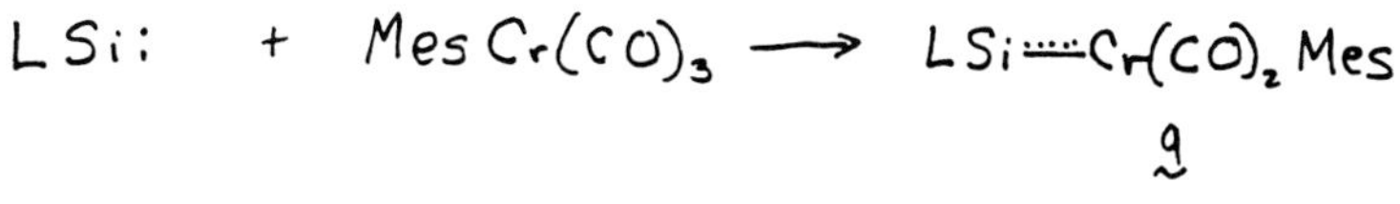

$$LSi: + Mes Cr(CO)_3 \longrightarrow LSi\cdots Cr(CO)_2 Mes$$
$$\underset{\sim}{9}$$

Eq. 9. (Genuine drawing by the author)

Unusual products are also obtained in the reaction of **5** with azides [21]. With trimethylsilyl azide, the isolated product **10**, is complex. The production of **10** can be rationalized, however, as due to the initial formation of a silanimine **10a**, followed by addition of a second equivalent of Me$_3$SiN$_3$ across the Si=N double bond (Eq. 10). An equally unexpected 2:1 product was obtained from the reaction of **5** with

1-adamantyl azide (Eq. 11). With triphenylmethyl azide, however, the silanimine is produced as the THF complex (Eq. 12). Its X-ray structure is shown in Fig. 4.

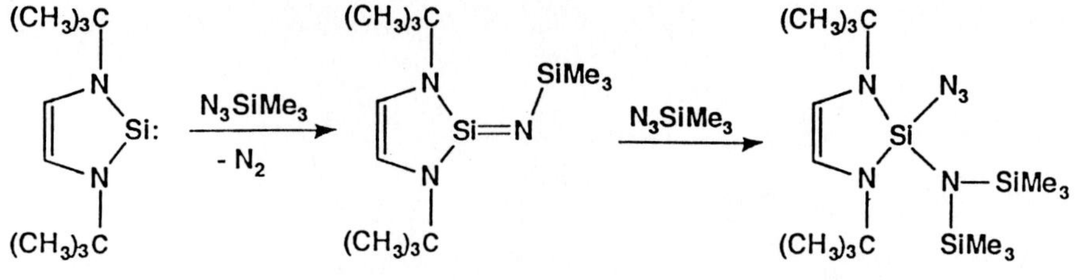

Eq. 10.

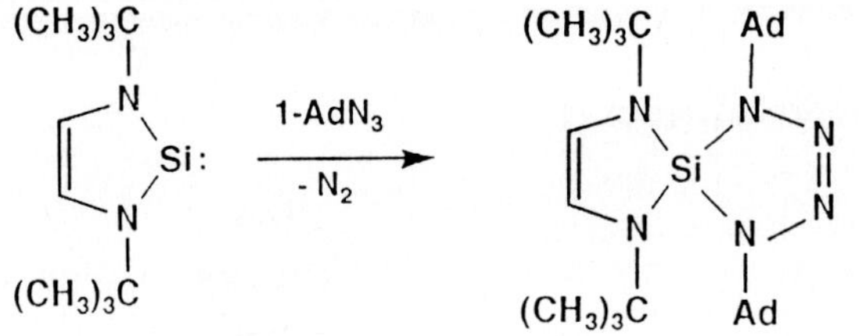

Eq. 11.

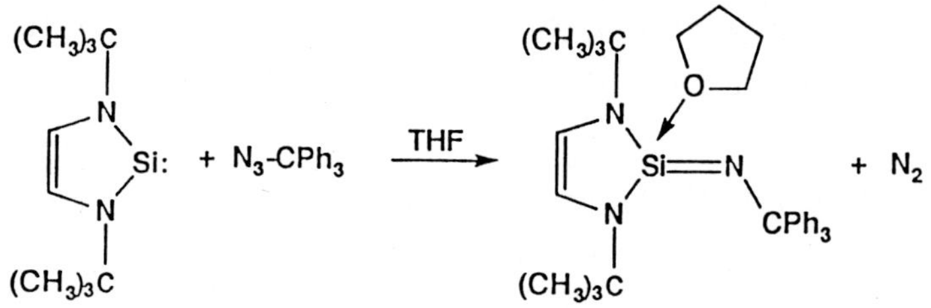

Eq. 12.

Why is **5** so much more stable than normal silylenes? In part the stability probably results from electron-donation by the nitrogen atoms into the vacant p orbital on silicon, making the silicon distinctly less electrophilic. But another source of stability may be aromatic delocalization in the five-membered ring. Silylene **5** has six π electrons: two from each nitrogen atom, one from each carbon and zero from silicon (since the two nobonding electrons are in an in-plane, σ-type orbital).

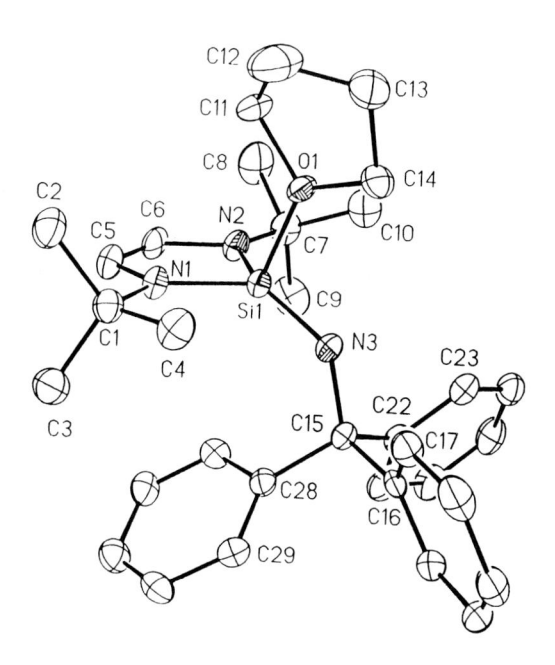

Fig. 4. X-ray crystal structure of LSi=NCPh$_3$ · THF, **12**

The ^{1}H NMR spectrum provides some evidence for aromatic stabilization in **5**. The important number is the chemical shift for the ring C–H protons, which falls at 6.75 ppm. This is significantly deshielded compared with the same protons in the precursor **4**, 5.73 ppm, or the corresponding dihydride (LSiH$_2$), 6.00 ppm. Theoretical calculations also support the idea that **5** is an aromatic molecule, with aromatic resonance energy of 12 ≅ kcal mol^{-1}. Particularly convincing, however, is a comparison of the properties of **5** with those of its saturated analog [22]. Silylene **13** was made by a reaction sequence similar to that for **5** in Eq. 4 and Eq. 5, except that for the final step it was necessary to use the dibromo rather than the dichloro compound (Eq. 13). For **13** an X-ray crystal structure could be determined; it is shown in Fig. 5.

Eq. 13.

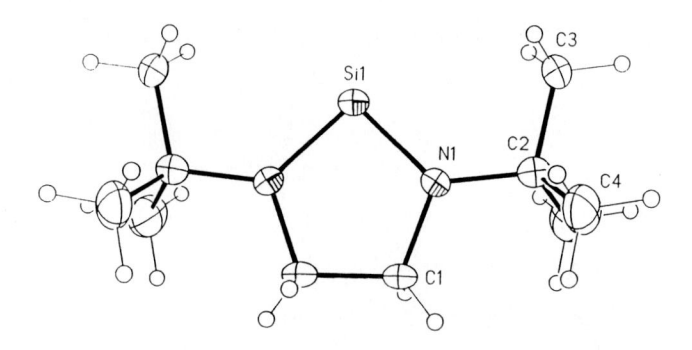

Fig. 5. X-ray crystal structure of silylene **13**

Chemical reactions of **13** have not yet been studied, but it is clear that **13** is far less stable than **5**. Although **13** can be isolated and purified, it slowly decomposes at room temperature, whereas **5** survives heating to 150 °C indefinitely.

The study of stable silylenes is in its infancy, but it is already apparent that a rich chemistry of these new species can be developed. We anticipate that additional stable silylenes will soon be made and that the synthesis of new kinds of organosilicon derivatives will be possible starting from the stable silylenes.

Acknowledgements: The authors acknowledge partial support from the *Alexander von Humboldt Foundation*, the *Lady Davis Foundation*, and the *National Science Foundation*, and valuable assistance by A. Haaland, A. V. Belyakov, and H. P. Verne, *University of Oslo*, M. Wagner, *Oxford University*, N. Metzler, *Universität München*, and R. Lennon, *University of Wisconsin*.

References:

[1] For a review on the subject see: P. P. Gaspar, K. L. Bobbit, M. E. Lee, D. Lei, M. Maloney, D. H. Pae, M. Xiao, *Front. Organosilicon Chem., Proc. Int. Symp. Organosilicon Chem. 9th*, **1990**, (Pub. **1991**), p. 110.

[2] Y. Ono, M. Okamoto, N. Watanabe, E. Suzuki, in: *Silicon for the Chemical Industry II* (Eds.: H. Øye, H. M. Rong, L. Nygard, G. Schüssler, J. K. Tuset), Tapir Forlag, Trondheim, Norway, **1994**, p. 185.

[3]   T. J. Drahnak, J. Michl, R. West, *J. Am. Chem. Soc.* **1979**, *101*, 5427.
[4]   R. West, M. J. Fink, J. Michl, *Science*, **1981**, *214*, 1343.
[5]   P. Jutzi, U. Holtmann, D. Kanne, C. Krüger, R. Blom, R. Gleiter, I. Hyla-Kryspin, *Chem. Ber.* **1989**, *122*, 1629.
[6]   H. H. Karsch, U. Keller, S. Gamper, G. Müller, *Angew. Chem., Int. Ed. Engl.* **1990**, *29*, 295.
[7]   M. Denk, R. Lennon, R. Hayashi, R. West, A. V. Belyakov, H. P. Verne, A. Haaland, M. Wagner, N. Metzler, *J. Am. Chem. Soc.* **1994**, *116*, 2691.
[8]   C. Wentrup, *Reactive Molecules*, Wiley, Chichester, **1984**.
[9]   T. J. Sears, P. R. Bunker, *J. Chem. Phys.* **1983**, *79*, 5265.
[10] J. L. Berkowitz, J. P. Greene, H. Cho, R. Ruscic, *J. Chem. Phys.* **1987**, *86*, 1235.
[11] K. Balasumbramian, A. D. McLean, *J. Chem. Phys.* **1986**, *85*, 5117.
[12] Y. Apeloig, R. Pauncz, M. Karni, N. Moisiyev, R. West, submitted for publication.
[13] J. F. Harrison, R. C. Liedtke, J. F. Liebman, *J. Am. Chem. Soc.* **1979**, *101*, 7162.
[14] W. Kutzelnigg, *Angew. Chem., Int. Ed. Engl.* **1984**, *21*, 272.
[15] B. T. Luke, J. A. Pople, M.-B. Krogh-Jespersen, Y. Apeloig, J. Chandrashekar, P. v. R. Schleyer, *J. Am. Chem. Soc.* **1986**, *108*, 266.
[16] M. J. Michalczyk, M. J. Fink, D. J. DeYoung, C. W. Carlson, *Silicon, Germanium, Tin, Lead Com.* **1986**, *9*, 75.
[17] D. B. Puranik, M. J. Fink, *J. Am. Chem. Soc.* **1989**, *111*, 5951.
[18] J. A. Arduengo III, R. L. Harlow, M. Kline, *J. Am. Chem. Soc.* **1991**, *113*, 361.
[19] G. R. Gillette, G. Noren, R. West, *Organometallics*, **1989**, *8*, 487.
[20] M. Denk, R. R. Hayashi, R. West, *J. Chem. Soc., Chem. Commun.* **1994**, 33.
[21] M. Denk, R. K. Hayashi, R. West, *J. Am. Chem. Soc.* **1994**, *116*.
[22] M. Denk, R. K. Hayashi, R. West, unpublished studies.

Silylenes and Multiple Bonds to Silicon: Synergism between Theory and Experiment

Yitzhak Apeloig*, Miriam Karni, Thomas Müller

Department of Chemistry

Technion - Israel Institute of Technology

Technion City, 32000 Haifa, Israel

Summary: Molecular orbital ab initio calculations were used to reproduce and to predict various properties of silylenes and disilenes:

1) MP4SDTQ/6–31G*//6–31G* calculations coupled with the spin-projection method have been used to calculate the energy of the first electronic transition of a variety of silylenes. Very good agreement is generally found between the calculated and the measured transitions. It is found that substituents which are *n*-donors (e.g., OH) induce large "blue" shifts, while π-donors such as vinyl and phenyl groups induce a "red" shift relative to either H or methyl substitution. Ethynyl substitution has a relatively small effect, it induces small "blue"- and "red"-shifts relative to methyl and H, respectively. The calculations can be used to predict the electronic transitions for silylenes which have not been yet characterized (e.g., for $(Me_3Si)_2Si$:) and to point-out erroneous interpretations (e.g., for $MesSiSiMe_3$).

2) GIAO(MP2) calculations reproduce excellently the experimental ^{29}Si chemical shifts of the stable silylenes **4a** and **5a** reported recently by West *et al*. The effects of the nitrogen substituents and of π-conjugation on the ^{29}Si chemical shifts and on the chemical shift anisotropy of silylenes are evaluated computationally.

3) Calculations show that substituents strongly effect the geometries and the dissociation energies of disilenes. Disilenes substituted by electronegative π-donors, such as NH_2, OH, or F are nonplanar with bending angles at silicon of up to 60 °. Furthermore, the Si=Si dissociation energy is strongly reduced by R = NH_2, OH, and F substituents and

disilenes of the type RHSi=SiHR are predicted to be very weakly bonded (by ca. 20 kcal mol^{-1} at 25 °C), relative to two separate silylenes making these silylenes difficult targets for synthesis.

4) Cyclic species of the type c-(R´Si(µ-R)$_2$SiR´) where R = F, OH, NH$_2$, and R´ = H, CH$_3$, are minima on the potential energy surface and they are close in energy to the corresponding isomeric disilenes, RR´Si=SiR´R, (ΔE = + 9.0 kcal mol^{-1} for R = F, + 3.0 kcal mol^{-1} for R = OH, −10 kcal mol^{-1} for NH$_2$, all with R´ = H). The cyclic (R´Si(µ-R)$_2$SiR´) have unusual structures and bonding, with the silicon atoms being 3-coordinated and the heteroatoms being hypercoordinated (e.g. 3-coordinated oxygen in c-(HSi[µ-(OH)$_2$SiH)). For R = OH and NH$_2$ the bridged isomers are predicted to be kinetically stable as they are separated by large energy barriers from the substantially more stable cyclic disilazanes which can be obtained by 1,2-hydrogen shifts from R to the silicon atoms. This suggests that the bridged c-(R´Si(µ-R)$_2$SiR´) for R = NH$_2$, OH, F are amenable to experimental observation. We suggest that a bridged R´SiF$_2$SiR´ isomer has actually been observed by Jutzi *et al.* [23], although originally the observed species was interpreted to be the isomeric disilene, R´FSi=SiFR´.

Introduction

The field of silicon chemistry has enjoyed a very fast development in the last two decades with many novel significant discoveries being made [1]. Of particular interest in the context of this paper is the synthesis and characterization of a variety of reactive intermediates such as silylenes [2] and compounds with multiple bonds to silicon [3]. These exciting developments were occurring at the time when theory, in particular ab initio molecular orbital theory, was reaching "maturity"; i.e. at the time when these methods could be used routinely to calculate reliably the properties of a variety of molecules, including silicon compounds [4].

This has provided computational chemistry with a unique opportunity to influence the development of silicon chemistry not only by providing interpretations to experimental findings but more importantly by making *predictions* and *directing future experiments*. Indeed in the last decade theoretical calculations played a major and sometimes even a crucial role in the development of silicon chemistry [5]. In this paper we hope to demonstrate the importance of theory for understanding and predicting the properties and chemistry of silylenes and of disilenes.

Method

In general the reported calculations used standard *ab initio* methods as implemented in the Gaussian series of programs. Details can be found in [4] and in our papers quoted in the text.

Results and Discussion

1 Silylenes

1.1 UV-visible Spectra

Following the first direct observation of an organosilylene in 1979 by Drahnak, Michl, and West [6] many other silylenes have been generated (usually by photolysis of the corresponding trisilanes) and their spectra were recorded in argon or hydrocarbon matrices [2, 7]. In order to be able to use the electronic spectra as a "spectroscopic fingerprint" to establish the presence of a silylene it is necessary to understand the effect of substituents on their spectra and to be able to predict the precise absorption wavelength (λ_{max}) of unknown silylenes. In this paper, we use molecular orbital *ab initio* theory [4] to calculate the absorption maxima (λ_{max}) for various silylenes. As will be demonstrated below this approach has proved to be extremely successful in reproducing known experimental results and in predicting the spectra of novel silylenes, as well as in understanding qualitatively and quantitatively the effect of substituents on λ_{max}.

In the $^1A'$ ground state of simple silylenes two electrons occupy an orbital of σ-symmetry [n(Si)], while the 3p-orbital on silicon [3p(Si)] is vacant (see the Scheme below). The lowest electronic transition involves the promotion of one of the n(Si) electrons to the empty 3p(Si) orbital (i.e., **1→2**), forming the $^1A''$ excited state. The energy of the first transition in the UV-visible spectra of silylenes, ΔE, is calculated using UHF ab initio methods, from the energy difference between the total energies of the $^1A'$ ground state singlet and the "vertical" (i.e. confined to have the geometry of the ground state) $^1A''$ excited state singlet, i.e., $\Delta E = E(^1A'') - E(^1A')$.

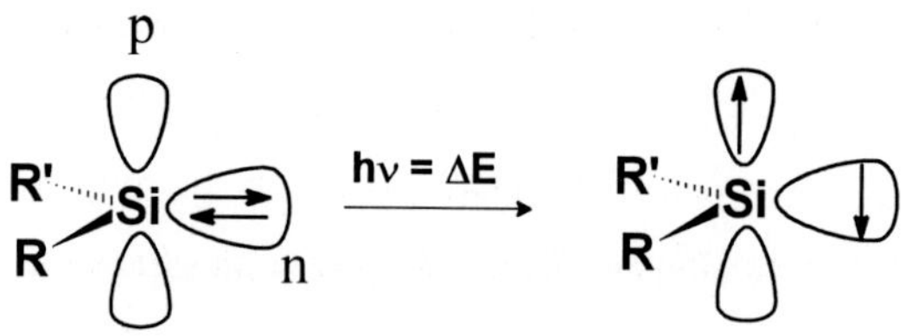

Fig. 1.

The theoretical calculations proved to be very helpful already at the very early stages of our study. A typical example is the successful prediction of the spectrum of Me_2Si [8], which at that time was controversial [9]. In 1983, West and Michl reported that in hydrocarbon matrices λ_{max} of Me_2Si is 450 nm [9a]. In 1984 this assignment was challenged by Griller and coworkers who reported for Me_2Si a λ_{max} of 350 nm in hydrocarbons and 300 nm in THF [9b]. Our calculations, although at a lower level than the current calculations, predicted λ_{max} 478 nm [8] strongly supporting the value of 450 nm reported by West et al. [9a]. A year later, in 1986, the discrepancy was resolved when Griller reported that the 350 nm absorption is not due to Me_2Si, but rather to the photolysis of impurities [9c]. Subsequent high-level calculations gave λ_{max} 456 nm [9d] in excellent agreement with West's measurements [9a].

In the current study [11] the energies of the two states of the silylenes of interest were calculated at the $MP4SDTQ/6-31G^*//6-31G^*$ level of theory (denoting a single-point $MP4SDTQ/6-31G^*$ calculation at the $6-31G^*$ optimized geometry [4]) coupled with the spin projection method of Schlegel [10] for annihilating spin contaminations from high-spin states. These calculations are superior to our previous calculations [8], which did not use spin-projection methods, in particular for silylenes substituted with π-donor groups [11].

Calculations for a variety of substituted silylenes reveal two major effects which determine the spectra of silylenes [8]:

1) Geometrical effects is shifted to longer wavelengths (red-shift) – as the RSiR′ bond angle increases λ_{max} and vice versa. The red-shift observed along the series: Ph_2Si (495 nm) < PhSiMes (530 nm) < Mes_2Si (577 nm) [7], where Mes = 2,4,6-trimethylphenyl, and the blue-shift observed for the series: Et_2Si (469 nm) > c-$C_5H_{10}Si$ (449 nm) > c-C_4H_8Si (436 nm) [7], agree nicely with this theoretical prediction.

2) Electronic effects *n*-donors (e.g., NH$_2$, OH, SH, CH$_3$) and σ-acceptors (e.g., F, CF$_3$) induce blue-shifts, while σ-donors (e.g., SiH$_3$) induce red-shifts. In cases where experimental data is available good agreement was found with the theoretical predictions, although usually the calculations were found to overestimate the substituent effects [8]. These substituent effects can be understood in terms of the effects of the substituents on the energies of the *n(Si)* and *3p(Si)* frontier orbitals of the silylene, and a reasonably good correlation is found between the calculated excitation energy, Δ*E*, and the calculated energy differences between the HOMO (*n*) and the LUMO (*3p*) orbitals of the silylenes.

n-Donors induce blue-shifts because the interaction of the substituents lone-pair and the formally empty *3p(Si)* orbital "pushes" the *3p(Si)* orbital to a higher energy – increasing the *n*-*3p* energy gap. These qualitative considerations are shown for an alkoxy substituted silylene in Fig. 2. The calculated blue-shift in HSiOH relative to H$_2$Si is 170 nm.

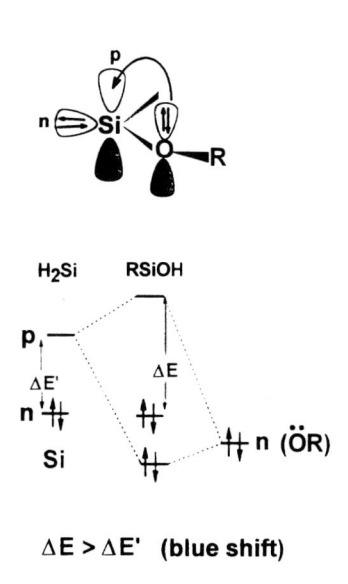

ΔE > ΔE' (blue shift)

Fig. 2.   Schematic orbital interaction diagram for an alkoxy silylene

What is expected for π-donors such as ethynyl and vinyl groups? Are they expected, in analogy to *n*-donors, also to induce blue-shifts? At the time when we became interested in this question the electronic spectra of vinyl – or of ethynyl-silylenes were not known and the study was carried out in collaboration with the group of R. West in Madison [11]. Measurements by West *et al.* showed that MeSiC≡CSiMe$_3$ absorbs at 473 nm, red-shifted from λ$_{max}$ of Me$_2$Si by 20 nm. Attempts to prepare MeSiC≡CH were unsuccessful and experimentally it was therefore not possible to establish whether the observed 20 nm red-shift in MeSiC≡CSiMe$_3$ relative to Me$_2$Si is due to the presence of the ethynyl group or to the presence of the Me$_3$Si-substituent.

Can theory separate the effects of the two substituents? It is first necessary to establish that the calculations can reproduce the experimental results. As shown in Fig. 3, the computational-experimental agreement for MeSiC≡CSiMe$_3$ is excellent. Both theory and experiment predict a red-shift of 20 nm relative to Me$_2$Si and a blue-shift of 7–9 nm relative to MeSiH. Calculations for the experimentally unknown MeSiC≡CH show that the red-shift of 20 nm observed for MeSiC≡CSiMe$_3$ relative to Me$_2$Si is composed of a red-shift of 11 nm due to substitution of a methyl by an ethynyl group and a 9 nm red-shift due to substitution of the acetylenic hydrogen by a SiMe$_3$ group (Fig. 3). Comparison of MeSiC≡CH with MeSiH shows that substitution of the silylene hydrogen by an ethynyl group induces a blue-shift of 18 nm in contrast to the red-shift relative to methyl. Most recently, actually in a poster presented in this meeting, G. Maier *et al.* reported on the generation of HSiC≡CH in a matrix and they have measured a λ_{max} of 500 nm [12], red-shifted from MeSiH by 20 nm, which is in good agreement with our theoretical prediction of 13 nm.

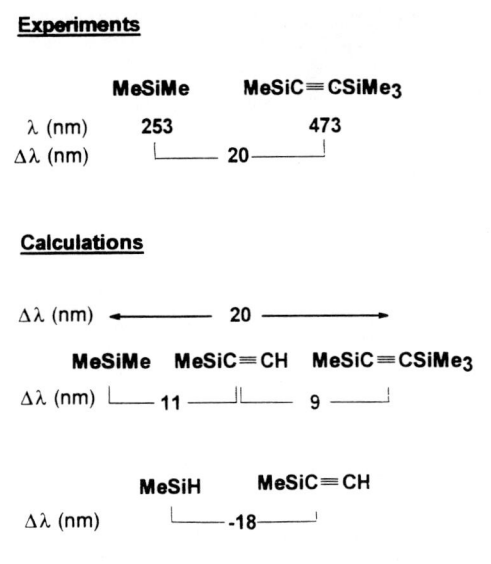

Fig. 3. $n \rightarrow 3p$ electronic transition energies for ethynylsilylenes

What is the effect of vinyl substitution on the first electronic transition of a silylene? The relevant calculations and the available experimental data for vinyl substituted silylenes are presented in Fig. 4. The calculations predict that substitution of a methyl group in Me$_2$Si by a vinyl group induces a red-shift of 47 nm, significantly larger than the red-shift of only 11 nm predicted for a similar ethynyl substitution. Substitution of the second methyl group by a vinyl group causes an additional red-shift of 38 nm.

Substitution of hydrogen in H_2Si by a vinyl group induces a red-shift of 31 nm, while a blue-shift of 12 nm was calculated for the ethynyl group. Recently, Kira et al. succeeded in generating and measuring the spectra of the cyclic vinylsilylenes shown in Fig. 4 [13].

Calculations

	MeSiMe	MeSiCH=CH$_2$	(CH$_2$=CH)$_2$Si:
λ (nm)	453	500	538
Δλ (nm)	⎣— 47 —⎦⎣— 38 —⎦		
	⎣——— 85 ———⎦		

Experiments [13]

	MeSiMe	αSi (ring)	Si (ring)	Si (ring)
λ (nm)	453	436	475	505
Δλ (nm)	⎣— -17 —⎦⎣— -39 —⎦⎣— 30 —⎦			
	⎣——— 22 ———⎦			

Calculations

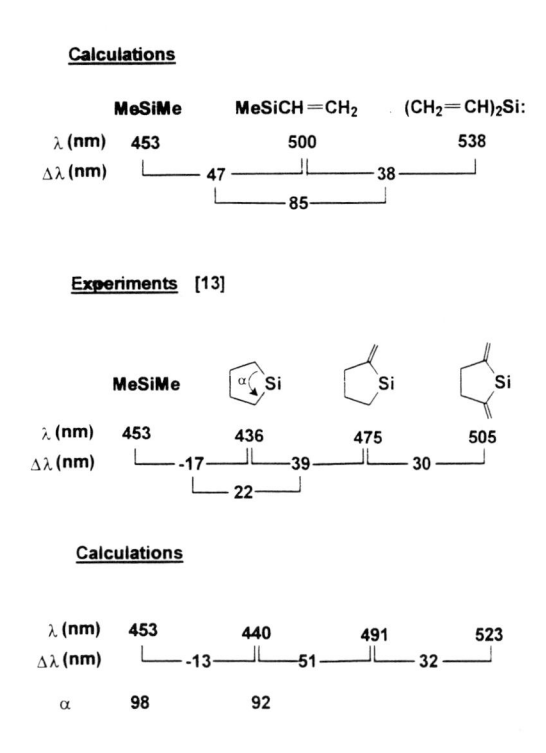

λ (nm)	453	440	491	523
Δλ (nm)	⎣— -13 —⎦⎣— -51 —⎦⎣— 32 —⎦			
α	98	92		

Fig. 4. $n \rightarrow 3p$ electronic transition energies for vinylsilylenes

Calculations for these silylenes, as well as the calculations for the simpler model systems discussed above, are in good agreement with Kira's experimental spectra (Fig. 4). The calculations also reproduce (Fig. 4) the small blue-shift of 17 nm observed on going from Me$_2$Si to cyclic-C$_4$H$_8$Si [13], showing that it can be attributed to the contraction of the bond angle at silicon from 98 ° in Me$_2$Si to 92 ° in cyclic-C$_4$H$_8$Si.

A phenyl group has a similar effect to that of a vinyl group. Substitution of H in H_2Si by a phenyl group is calculated to induce a red-shift of 36 nm, only 5 nm larger than the effect of a vinyl group. Available experimental data [7] support this conclusion.

Why do n-donors lead to blue-shifts while π-donors lead to red-shifts? This can be easily understood in terms of the qualitative orbital interaction diagram shown in Fig. 5. The qualitative considerations in Fig. 5 are supported by quantitative calculations, presented elsewhere [11], on the effect of such substituents on the energies of the ground and the excited states of these silylenes.

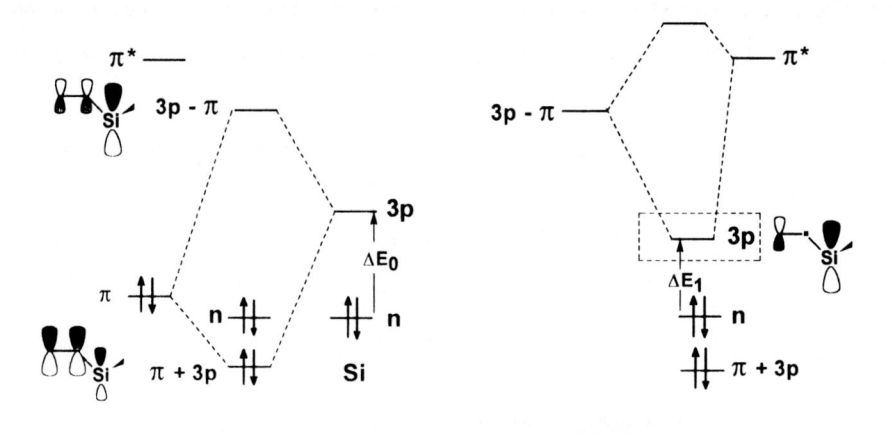

Fig. 5. Schematic 3-orbital interaction diagram for HSiCH=CH$_2$

The major difference between *n*-donors and π-donors is the presence of a low-lying π*-orbital in the latter family of substituents. Thus, as shown in Fig. 2, interaction of a silylene center with *n*-donors leads to *n*(donor)–3*p*(Si) orbital mixing which "pushes" the empty 3*p*(Si) orbital to a higher energy, resulting in a blue-shift of the *n*(Si)→3*p*(Si) transition. With π-donors a relatively low-lying π* orbital, which can interact with the 3*p*(Si) orbital, is present. The 3-orbital interaction, involving π, π*, and 3*p*(Si), leads to a significant lowering in the energy of the 3*p*(Si) orbital relative to its position in the unperturbed silylene, resulting in a red-shift of the *n*(Si)→3*p*(Si) transition (Fig. 5). For clarity, Fig. 5 dissects the 3-orbital interaction into two steps:

1) Interaction between the filled π and the empty 3*p*(Si) orbitals (with no involvement of π*), generating new bonding and antibonding π-3*p*(Si) combinations.
2) Interaction of π* with the antibonding [3*p*(Si)-π] combination, leading to a significant lowering of the energy of this orbital.

Once confidence was gained in the ability of the calculations to reproduce the spectra of silylenes we have proceeded to critically examine reported experimental λ_{max} values and to predict the λ_{max} values of as yet unknown silylenes. The following two examples will serve to demonstrate the predictive and analytical power of the calculations.

1) West, Michl and co-workers assigned the λ_{max} of 368 nm observed in the photolysis of MesSi(SiMe$_3$)$_3$ to MesSiSiMe$_3$ [7]. Calculations show that λ_{max} of MeSiSiMe$_3$ is red-shifted by 200 nm relative to Me$_2$Si. Known experimental data for Me$_2$Si and MeSiMes [7] can be used to place the computationally predicted λ_{max} of MesSiSiMe$_3$ at 720 nm, far remote from the λ_{max} of 368 nm assigned experimentally to the same species [7]. We have, therefore, concluded [11] that the species observed in the photolysis of MesSiSiMe$_3$ [7] is *not* MesSiSiMe$_3$. Indeed, most recently Kira *et al.* have also generated MesSiSiMe$_3$ and reported λ_{max} 760 nm [14], in good agreement with our theoretical prediction.

2) A second silyl substitution induces according to the calculations an additional large red-shift; (H$_3$Si)$_2$Si is calculated to have λ_{max} 810 nm. Substitution of the hydrogens in (H$_3$Si)$_2$Si by methyl groups induces a further red-shift of 177 nm, resulting from a widening of the bond angle around silicon from 95.4° in (H$_3$Si)$_2$Si to 106° in (Me$_3$Si)$_2$Si. The predicted λ_{max} of (Me$_3$Si)$_2$Si [15a] is 886 nm [15b] – to our best knowledge the longest wavelength absorption known todate for any silylene. The computational prediction about the very long wavelength absorption of disilylsilylenes was recently confirmed by experiments carried out at Haifa, and shown in Scheme 1 [15c].

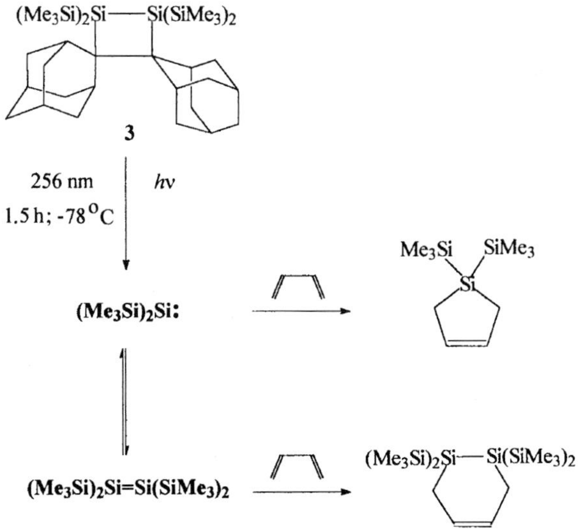

Scheme 1. The photolytic generation of (Me$_3$Si)$_2$Si and (Me$_3$Si)$_2$Si=Si(SiMe$_3$)$_2$ and their trapping products

Irriadiation at 256 nm of a hexane solution of the 1,2-disilacyclobutane **3**, produced the silylene $(Me_3Si)_2Si$, as confirmed by its expected trapping product with butadiene (Scheme 1). In agreement with the calculations the silylene solution has a strong violet color, with λ_{max} higher than 750 nm (which is the upper measurement limit of our spectrophotometer).

In conclusion, we have demonstrated that the calculations fully reproduce the experimentally observed spectral trends of silylenes. Lone-pair donors induce strong blue-shifts while vinyl and phenyl substituents induce significant red-shifts (relative to methyl). Ethynyl has a much smaller effect. The quantitative agreement between theory and experiment is not perfect and the calculations usually overestimate the shifts due to substitution [15a]. Yet, the calculations can be used reliably to predict the spectra of unknown silylenes and to evaluate the validity of existing interpretations, as demonstrated above for $Me_3SiSiMes$ and $(Me_3Si)_2Si$. Substantial discrepancy between calculated and measured transition energies probably points to an erroneous assignment of a certain absorption bond to the presence of a silylene (e.g., $Me_3SiSiMes$).

1.2 Stable Silylenes

A major recent achievement in the study of silylenes is the synthesis and isolation by West and Denk of the first "indefinitely" stable silylene **4a** [16]. The latter was reported to be colorless [16] in agreement with its calculated λ_{max} of 222 nm. The X-ray structure of **4a** [16] is also in excellent agreement with ab initio calculations for the model non-substituted silylene **4**, R = H [15c, 16].

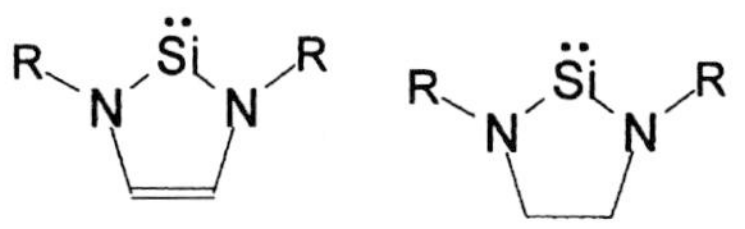

R = *t*Bu (**4a**), H (**4b**)   R = *t*Bu (**5a**), H (**5b**)

Fig. 6.

The isolation of **4a** allowed also to record for the first time the NMR spectra of a silylene and the measured chemical shifts are shown in Fig. 7.

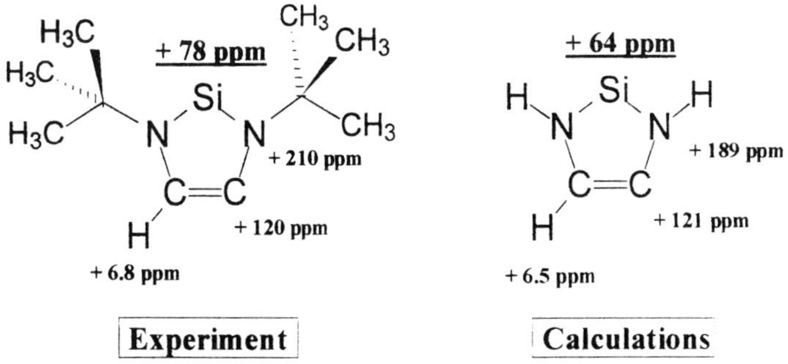

Fig. 7. Experimental and calculated (GIAO(MP2/tz2p(Si)tzp(C)dz(H))) chemical shifts in **4b** and **4a**, respectively; both the calculated and the experimental ^{15}N chemical shifts are relative to NH$_3$; the reported experimental ^{15}N chemical shift (relative to CH$_3$NO$_2$ [16]) was corrected using the experimental difference of 380 ppm between the ^{15}N chemical shifts of CH$_3$NO$_2$ and NH$_3$

Can calculations reproduce the NMR data? GIAO(MP2/tz2p(Si)tzp(C)dz(H)) calculations [7] for the model unsubstituted silylene **4b** predict chemical shifts (Fig. 7) which are in excellent agreement with experiment (Fig. 7). In particular, the ^{29}Si NMR chemical shift is calculated to be 64 ppm compared to the experimental value of 78 ppm. The agreement for the carbon and the nitrogen chemical shifts is also very good. Substitution of N–H by *t*Bu–N induces a deshielding of the ^{15}N chemical shift by ca. 30 ppm (e.g., Me$_2$NH vs *t*BuNMe$_2$) – bringing the calculated ^{15}N chemical shift for **4a** to 219 ppm very close to the experimental value.

During this meeting West reported the synthesis and the NMR spectra of **5a** [18] – the saturated analogue of **4a**. Calculations performed beforehand and independently in Haifa predicted for the non-substituted model, **5b**, a ^{29}Si NMR chemical shift of 117 ppm, almost in perfect agreement with the experimental value for **5a** of 119 ppm [18].

The success of the GIAO calculations in predicting correctly the NMR spectrum of **4a** and **5a** suggests that this method can be used to study also the NMR spectra of transient silylenes which cannot yet be studied experimentally. Such studies can provide important fundamental information on the electronic structure of silylenes.

GIAO calculations for the parent H$_2$Si reveal highly interesting results. First, the Si nucleus is calculated to be very strongly deshielded, its predicted ^{29}Si NMR chemical shift is 817 ppm relative to TMS. Second, the anisotropy of the ^{29}Si NMR chemical shift, $\Delta\sigma$, is very large, 1516 ppm. The strong deshielding and the large chemical shift anisotropy (CSA) of the Si nucleus result from a large paramagnetic σ-tensor lying perpendicular to the empty 3p-orbital of the singlet silylene. The following values have been calculated for the principal chemical shift tensors of H$_2$Si: $\sigma_{xx} = -51$ ppm, $\sigma_{yy} = -1832$ ppm, $\sigma_{zz} = 574$ ppm. The tensor directions are given in Fig. 8.

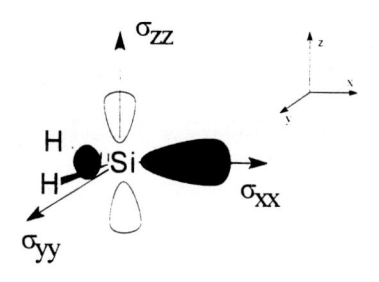

Fig. 8. Coordinate system for the chemical shift tensors of H_2Si

Why is the ^{29}Si NMR chemical shift in **4a** or **5a** so strongly shielded relative to H_2Si? Why is the calculated chemical shift anisotropy in **4b** only ca. 11 % of the CSA in H_2Si (experimental measurements of CSAs of silylenes are not yet available)? The series of calculations for aminosilylenes presented in Table 1 explains the origin of these observations.

Silylene	$\delta\,(^{29}Si)$[a]	$\Delta\sigma\,(^{29}Si)$
H_2Si	817	1516
$(H_2N)_2Si$, per.[b]	421	558
$(H_2N)_2Si$, pl.[c]	108	214
5b	117	73
4b	64	165

Table 1. Calculated ^{29}Si chemical shifts ($\delta\,(^{29}Si)$) and ^{29}Si chemical shifts anisotropies ($\Delta\sigma\,(^{29}Si)$), in ppm: [a] Relative to $(CH_3)_4Si$; [b] Perpendicular conformation, i.e., HNSiN = 90 °; [c] Planar conformation, i.e., HNSiN = 0 °

Substitution of the two hydrogens of H_2Si by two NH_2 groups held in a perpendicular conformation, i.e., a conformation in which the nitrogen lone-pairs are in the NSiN plane, leads to a strong shielding of the Si nucleus and $\delta\,(^{29}Si)$ is calculated to be 421 ppm. The CSA is also strongly reduced ($\Delta\sigma$ = 558 ppm).

As in this conformation the nitrogen lone-pairs cannot interact with the formally empty $3p(Si)$ orbital it is reasonable to conclude that the calculated change in the ^{29}Si NMR chemical shift and in the CSA are due to the inductive and polarizing effects of the amino-substituents. Rotation of the NH_2 groups to an all-planar conformation brings the nitrogen lone-pairs into conjugation with the empty $3p(Si)$ orbital. As a result, δ (^{29}Si) is strongly shifted upfield to 108 ppm and Δσ (^{29}Si) decreases further to 214 ppm. Thus, conjugation decreases strongly both the ^{29}Si NMR chemical shift and the CSA. The endocyclic π-bond in **4b** causes a further upfield shift of 44 ppm in δ ^{29}Si NMR chemical shift, relative to that of the saturated model, **5b**, and to a significant increase in the CSA, from 73 ppm in **5b** to 165 ppm in **4b**. These changes are consistent with the suggestion that the "aromatic" resonance structure **6** contributes significantly to the total wave-function describing **4**. Thermodynamic considerations also support a description of silylene **4** as being stabilized by "6π-aromaticity", where 4 electrons come from the two N lone-pairs and 2 electrons from the π(C=C) bond with no contribution from silicon [15, 16].

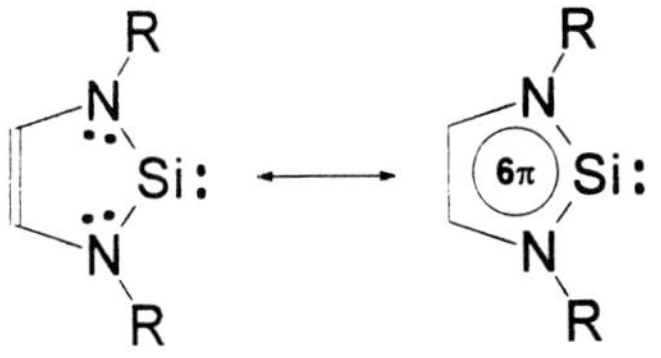

Fig. 9.

Calculations of the chemical shifts and the CSA of a variety of silylenes are in progress in our laboratory, and we are confident that they will provide valuable information as well as important insights into the electronic structure of these interesting species. We also hope that the calculations will prompt the experimental measurements of chemical shift anisotropies for stable silylenes.

2 Disilenes and their Novel Doubly Bridged Isomers

Since the spectacular isolation of the first stable disilene by West, Fink, and Michl in 1981 [19] this field has enjoyed a surge of activity [3]. Yet, despite the many important discoveries that have been made, the study of compounds containing double bonds to silicon is still in its infancy. In particular, relatively little is known on the effect of substituents, especially heteroatom substituents, on the properties of disilenes as only aryl-, alkyl-, (Me$_3$Si)$_2$N-, and Me$_3$Si-substituted disilenes have been isolated to date [3]. With experimental progress being relatively slow, *ab initio* calculations provide a reliable, fast and economical method to extend our knowledge on these novel compounds.

Calculations (mostly at the correlated MP3/6–31G* level) of mono-substituted disilenes, RSiH=SiH$_2$, and disubstituted disilenes, RHSi=SiHR, where R spans over the entire range of first-row substituents, i.e., Li to F (H$_3$Si was also included), reveal many interesting and unique properties of the Si=Si bond

[20]. We discuss below shortly the effect of these substituents on the geometry and the bond dissociation energy of disilenes.

Calculations for the parent $H_2C=CH_2$ and $H_2Si=SiH_2$ reveal a very different behaviour of these isoelectronic molecules with respect to pyramidalization, as demonstrated in Fig. 10. While for $H_2C=CH_2$ the pyramidalization potential in the range of $\theta = 0-40°$ is very steep for $H_2Si=SiH_2$ it is very flat.

Thus, bending $H_2Si=SiH_2$ away from planarity by 40° requires only 2–3 kcal mol^{-1}. This is in line with the experimental reports that some of the aryl- or alkyl-substituted disilenes characterized by X-ray crystallography are not planar [3]. Furthermore, due to the very soft pyramidalization potential it might be expected that the bending angle in disilenes will depend strongly on the electronic nature of the substituent. Indeed, the calculated structures of substituted disilenes are strongly dependent on the substituent (Fig. 11).

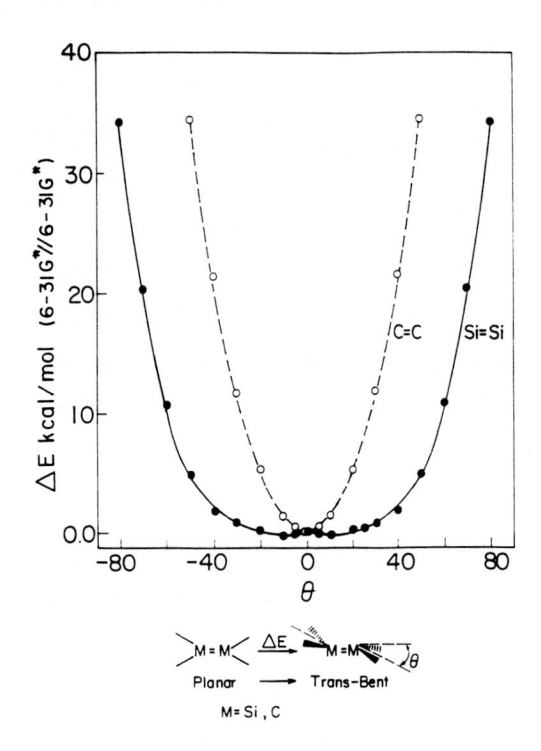

Fig. 10. Calculated pyramidalization potentials for the H_2M groups in $H_2C=CH_2$ and $H_2Si=SiH_2$ (6–31G*//3-21G$^{(*)}$)

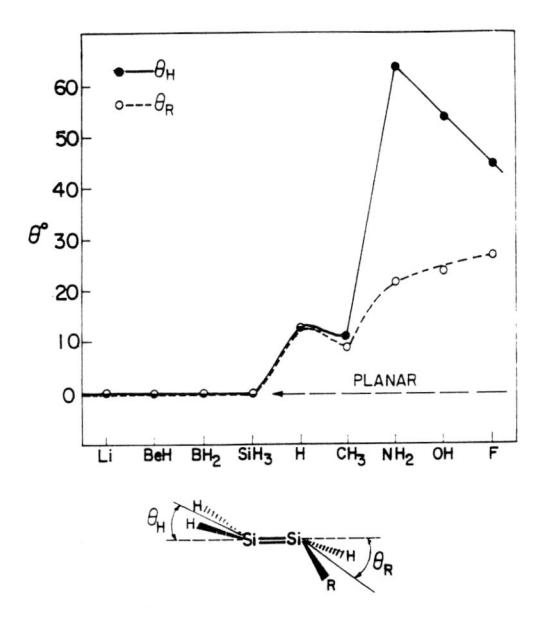

Fig. 11. Calculated pyramidalization angles at the substituted (θ_R) and unsubstituted (θ_H) silicon atoms in monosubstituted silenes, $H_2Si=SiHR$ (6-31G*//6-31G*)

With electropositive substituents, (e.g., R = Li, BeH, BH_2, and SiH_3) the disilenes are calculated to be planar and the Si=Si bond length is similar to that in $H_2Si=SiH_2$ [20]. In contrast, electronegative and π-donating substituents (e.g., NH_2, OH, F) induce very large deviations from planarity (Fig. 10); e.g., in $H_2Si=SiH(NH_2)$, the bending angles at the substituted and at the unsubstituted Si atom are: 22.1 ° and 64.1 °, respectively [20]! The Si=Si bond length in these disilenes is strongly elongated, e.g., to 2.25 Å in $H_2Si=SiHNH_2$ relative to 2.13 Å in $H_2Si=SiH_2$ [20].

Substituents effect dramatically also the bond dissociation energies of disilenes to the corresponding silylenes (Eq. 1), as is shown in Fig 12. which describes the correlation of the calculated dissociation energies for a series of substituted disilenes vs the sum of the singlet-triplet energy differences in the corresponding silylenes [20]. The theoretical background for the existence of such a correlation is discussed elsewhere [21].

$$RHSi=SiHR' \longrightarrow RHSi: + R'HSi:$$

Eq. 1.

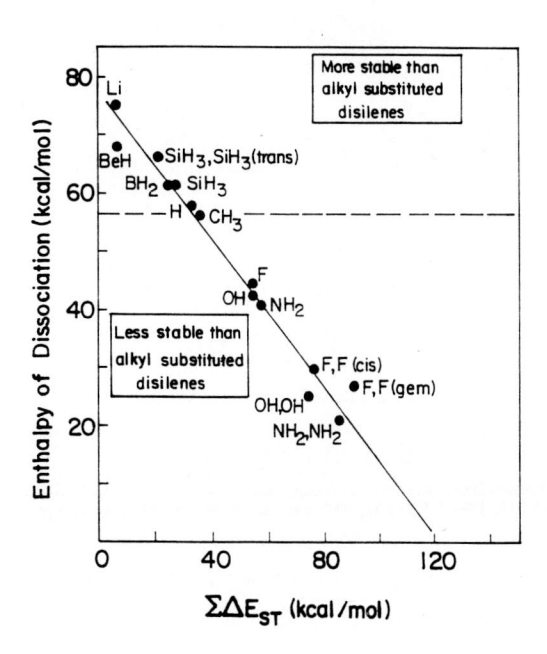

Fig. 12. Plot of calculated dissociation energies of substituted disilenes to the corresponding singlet silylenes (MP3/6–31G*//6–31G* + ZPE at 6–31G*) vs $\Sigma \Delta E_{ST}$ (S = singlet, T = triplet) of the corresponding silylenes (MP4SDTQ/6–31G*//3–21G$^{(*)}$)

Fig. 12 can be used to predict the thermodynamic stabilities toward dissociation of unknown disilenes and for deciding which disilenes are reasonable targets for synthesis. Examination of Fig. 12 shows that electropositive substituents, such as silyl groups, increase the thermodynamic stability of the Si=Si double bond, relative to the corresponding silylenes, making silyl-substituted disilenes an interesting synthetic target. The synthesis of (Me$_3$Si)$_2$Si=Si(SiMe$_3$)$_2$ was recently achieved independently by Sakurai et al. [22a] and by our group [22b]. In contrast, electronegative substituents weaken considerably the Si=Si bond. For example, the calculated bond dissociation enthalpy ($\Delta H°$) of (HO)HSi=SiH(OH) is only ca. 25 kcal mol^{-1}. As entropy favors dissociation, $\Delta G°$ for the dissociation process is lower than $\Delta H°$, e.g., $\Delta G°$ for the dissociation of (HO)HSi=SiH(OH) is calculated to be only ca. 15 kcal mol^{-1} at 298 K (ca. 20 kcal mol^{-1} at 150 K). Additional substitution of the disilene with π-donors weakens the Si=Si bond even further and (H$_2$N)$_2$Si=Si(NH$_2$)$_2$ is calculated not to be a minimum on the 6–31G* potential energy surface [15]. Thus, the calculations suggest that disilenes which are substituted with two or more electronegative heteroatoms such as NH$_2$, OH, or F will be thermodynamically relatively unstable and they are therefore expected to be difficult to observe even at relatively low temperatures [20], and in extreme cases not to exist at all (e.g., (H$_2$N)$_2$Si=Si(NH$_2$)$_2$ [15]).

In view of the predictions in Fig. 11 we were intrigued by the report of Jutzi et al. [23] on the detection of a 1,2-difluorodisilene, **7**, which was suggested to be formed by dimerization of the fluorosilylene, **8** generated in the reaction of decamethylsiliocene, $(C_5Me_5)_2Si$ with HBF_4, as shown in Scheme 2 [23].

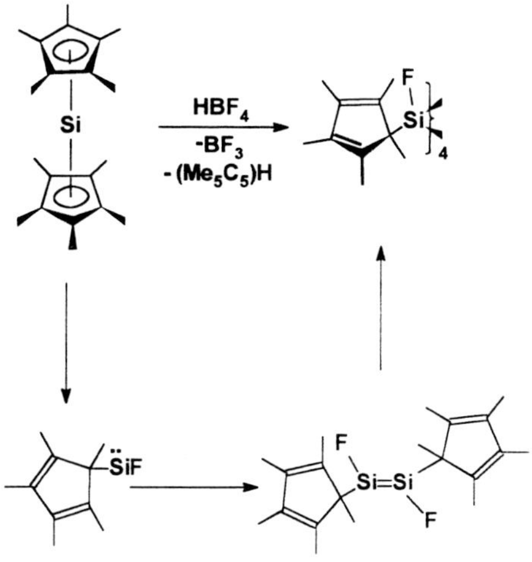

Scheme 2. Suggested reaction pathway of $(C_5Me_5)_2Si$ with HBF_4 [23]

In view of the predicted low bond dissociation energy of **7** (Fig. 12) we have decided to look for possible alternative structures to **7** [24]. The calculations revealed that the potential energy surfaces (PES) of the isoelectronic C_2H_4 and Si_2H_4 differ dramatically and that unusual bridged-structures unprecedented in carbon chemistry may be important in the chemistry of disilenes. Thus, at HF/6–31G*//6–31G* ethylene lies in a very deep energy minimum and it is more stable than two singlet carbenes and than the non-classical bridged structure **9** (which is not a minimum on the PES) by 179.6 and 172.5 kcal mol^{-1}, respectively. In contrast, the potential energy surface of disilene is much flatter and the analogous energy gaps are much smaller. At MP3/6–31G*//6–31G*, two silylenes, H_2Si, are only 50.3 kcal mol^{-1} higher in energy than $H_2Si=SiH_2$. Furthermore, the non classical bridged structure $HSi(\mu-H)_2SiH$, **10**, is a minimum on the PES and it is only 28.2 kcal mol^{-1} (at MP3/6–31G*// 6–31G*, 21.6 kcal mol^{-1} at the G2 level of theory) higher in energy than $H_2Si=SiH_2$.

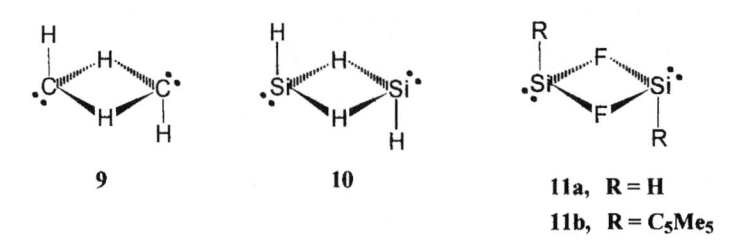

Fig. 13.

In the fluoro-substituted systems the energy difference between the classical FHSi=SiHF structure and the non classical isomer **11a** (the energy difference between the *syn* and *anti* isomers of **11a** is only 0.2 kcal mol^{-1}) is even smaller, **11a** being only 10.5 kcal mol^{-1} higher in energy than FHSi=SiHF at MP4SDTQ/6–31G*//6–31G*. Dissociation of HFSi=SiHF to two HFSi silylenes requires only 33.3 kcal/mol at the same level of theory.

The MP2(full)/6–31G* calculated structures of the two H$_2$Si$_2$F$_2$ isomers are given in Fig. 14. The *E*-1,2-difluorodisilene is non planar with a pyramidalization angle of 46 ° at each of the silicon atoms and with a relatively long Si=Si bond of 2.245 Å. The bridged isomer **11a**, has a planar four-membered Si(μ-F)$_2$Si ring with relatively long Si–F bonds of 1.861 Å (1.625 Å in FHSi=SiHF) and a Si···Si non bonding distance of 2.907 Å. The hydrogen atoms are nearly perpendicular to the Si(μ-F)$_2$Si ring and exhibit slightly elongated Si–H bonds. The silicon atoms in **11a** are each tricoordinated possessing also a non-coordinated lone-pair. According to NBO analysis there are no bonding interactions between the silicon atoms and the hybridization of the lone-pairs on silicon is sp$^{0.5}$ in HSi(μ-H)$_2$SiH and sp$^{0.33}$ in HSi(μ-F)$_2$SiH. Thus, **11a** can be classified as a novel isomer of difluorodisilene, exhibiting unique coordination at silicon and interesting bonding characteristics.

In view of these computational results, we believe that it is very likely that the species observed by low temperature NMR by Jutzi *et al.* [23] is *not* 1,2-difluorodisilene but rather the fluorine-bridged isomer **11b**. The molecular structure of **11b**, having magnetically identical silicon and fluorine atoms, is fully consistent with the published NMR-splitting data [23] which showed a single ^{29}Si resonance split into a triplet (J$_{Si-F}$ = 341 Hz). This new interpretation [24] of Jutzi's experiments [23] makes unnecessary the unlikely *ad hoc* assumption that in **7** ^{1}J$_{Si-F}$ = ^{2}J$_{Si-F}$ [23].

How is the bridged **11b** formed? Calculations show that two silylenes (e.g., HFSi:) can dimerize in two different modes, as shown in Scheme 3. Approach of two silylenes according to path a would lead to the disilene while coupling along path b would give the fluorine-bridged isomer. Approach of type b is driven by interactions between a lone-pair on fluorine (or a different heteroatom) of one silylene and the empty 3*p*-orbital on silicon of the second silylene. Note that the lone-pairs at the silicon atoms do not participate in the process b. In the case of HFSi: the calculations show that dimerization along either path a or path b proceeds without an energy barrier. The bulky pentamethylcyclopentadienyl substituents present in **8** might be responsible for the fact that it dimerizes to give the bridged **11b** in preference to the disilene **7**.

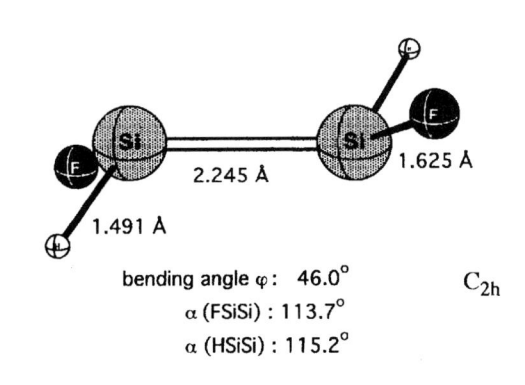

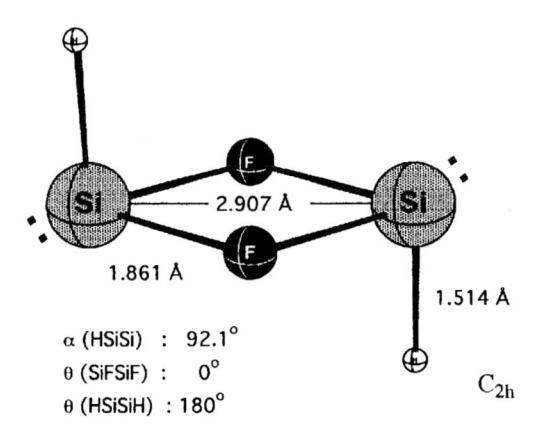

bending angle φ: 46.0°
α (FSiSi) : 113.7°
α (HSiSi) : 115.2°

C_{2h}

α (HSiSi) : 92.1°
θ (SiFSiF) : 0°
θ (HSiSiH) : 180°

C_{2h}

Fig. 14. Calculated geometries of FHSi=SiHF and HSi(μ-F)$_2$SiH (MP2(fu)/6–31G*)

Scheme 3. Possible dimerization pathways for silylenes

Can one find systems, in which the bridged non conventional isomer is more stable than the conventional disilene structure? Examination of the two dimerization paths in Scheme 3 suggests that a higher donor ability of the lone-pair on the substituent should favor the bridged structure.

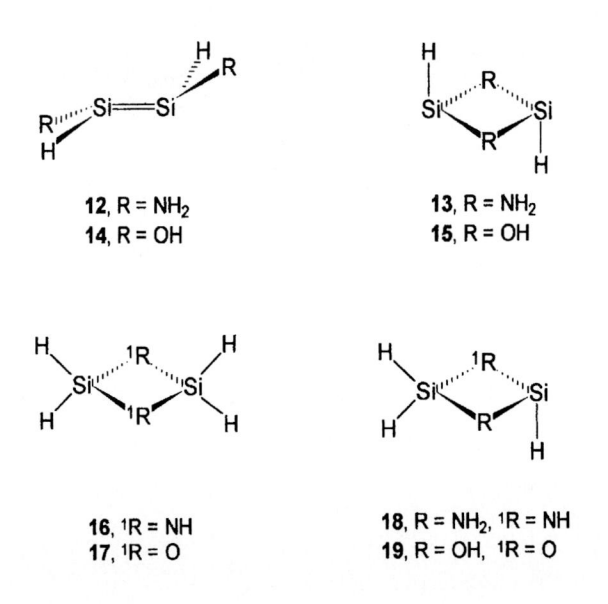

12, R = NH$_2$
14, R = OH

13, R = NH$_2$
15, R = OH

16, ^{1}R = NH
17, ^{1}R = O

18, R = NH$_2$, ^{1}R = NH
19, R = OH, ^{1}R = O

Fig. 15.

We have therefore studied computationally the diamino- and dihydroxy-substituted disilenes and their isomeric bridged structures **12–15**. The calculated relative energies of the isomeric pairs are given in Scheme 4.

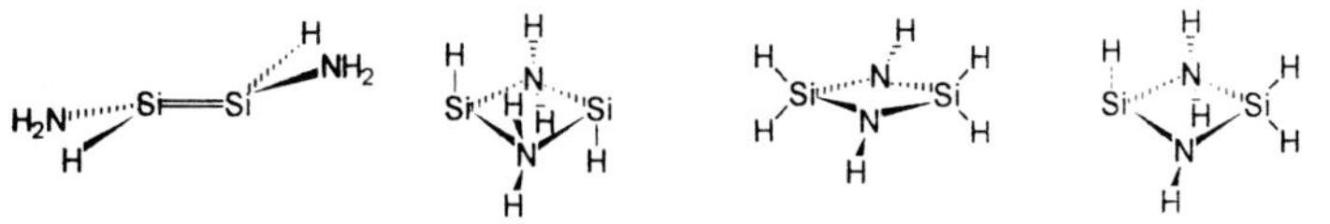

	12	**13**	**16**	**18**
rel. energy (kcal/mol):	0.0	-10.0	-56.0	-23.4

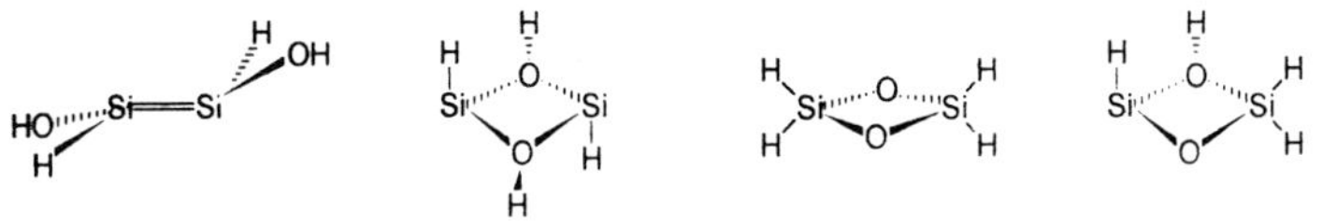

	14	**15**	**17**	**19**
rel. energy (kcal/mol):	0.0	3.2	-75.4	-29.4

Scheme 4. Calculated relative energies (kcal mol^{-1}) of isomeric $(H_2N)_2Si_2H_2$ and $(HO)_2Si_2H_2$ structures at MP4SDTQ/6–311G**//6–31G** + ZPE and MP4SDTQ/6–311G*// 6–31G* + ZPE, respectively

At the MP4SDTQ/6–311G**//6–31G** level of theory, the unconventional nitrogen-bridged isomer, **13**, is indeed calculated to be by 10.0 kcal mol^{-1} *more stable* than the isomeric 1,2-diaminodisilene **12**. The corresponding dihydroxy isomers **14** and **15** are very close in energy. Can one hope to isolate bridged compounds of type **13** or **15**? In answering this question one should note that in contrast to the bridged difluoro derivatives, the *N*- and *O*-bridged molecules **13** and **15** can rearrange via simple 1,2-hydrogen shifts to the "conventional" isomeric 1,3-disilazane **16** and 1,3-disiloxane **17**, respectively, which are substantially more stable (i.e., by 46 and 79 kcal mol^{-1}, respectively, at MP4SDTQ/6–311G**//6–31G*, Scheme 4). However, calculations at a fairly high level show that despite the high thermodynamic driving force and the perfect orientation of the migrating hydrogen atoms the calculated barriers for the 1,2-hydrogen rearrangements are very high, i.e., 53.8 kcal mol^{-1} for **13** and 39.5 kcal mol^{-1} for **15**. These extremely high barriers reflect the fact that the 1,2-hydrogen shift is symmetry-forbidden as it involves 4 electrons, two on Si and two of the N–H or O–H bonds.

Furthermore, the rearrangement of **13** to **16** or of **15** to **17** occur via two consecutive exothermic 1,2-hydrogen shifts, leading to the formation of an additional novel group of bridged isomers, i.e., **18** and **19**, which are thermodynamically more stable than **13** or **15**, respectively, (Scheme 4). In **18** and **19** one silicon atom is 4-coordinated, while the second silicon atom is still 3-coordinated. **18** and **19** are also separated by substantial barriers of 38.6 and 29.9 kcal mol^{-1} from the corresponding conventional 1,3-disilazane **16** and 1,3-disilaoxane **17**, respectively, and they are therefore predicted to be kinetically stable. The potential energy surface for the rearrangements of the nitrogen-substituted system is shown in Fig. 16. The calculated geometries of the diaminodisilene, **12**, of its nitrogen-bridged isomers **13**, **16**, and **18** and the transition states connecting them are shown in Fig.17.

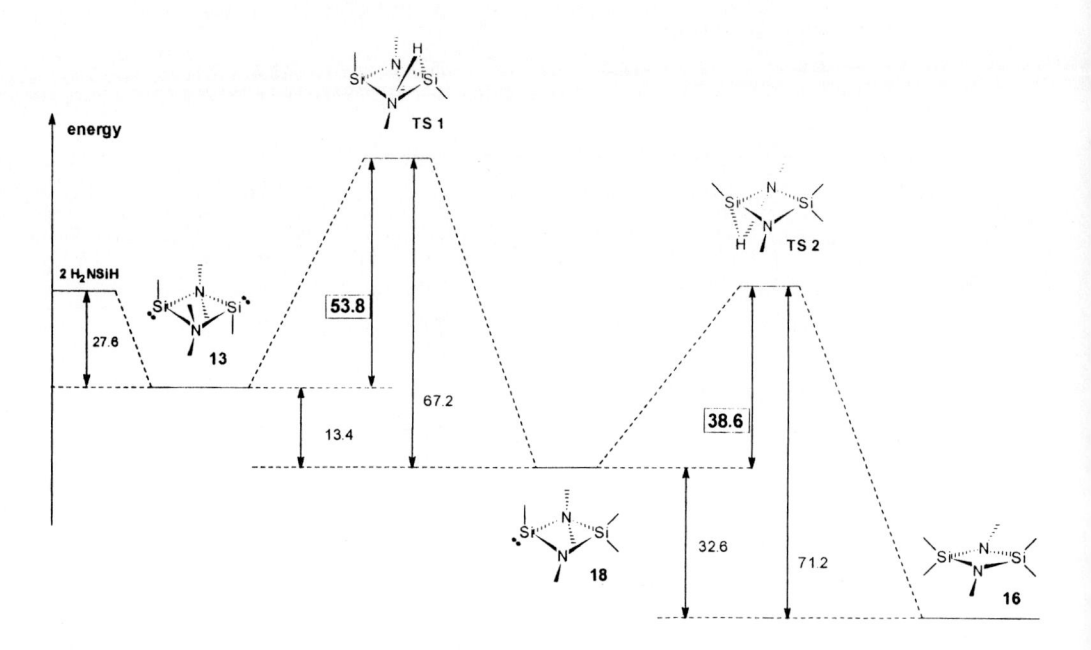

Fig. 16. The calculated potential energy surface for the interconversion of HSi[(μ-(NH$_2$)]$_2$SiH, **13**, and H$_2$Si[(μ-(NH)]$_2$SiH$_2$, **16** (MP4/6–311G**//6–31G** + ZPE); relative energies are given in kcal mol^{-1}

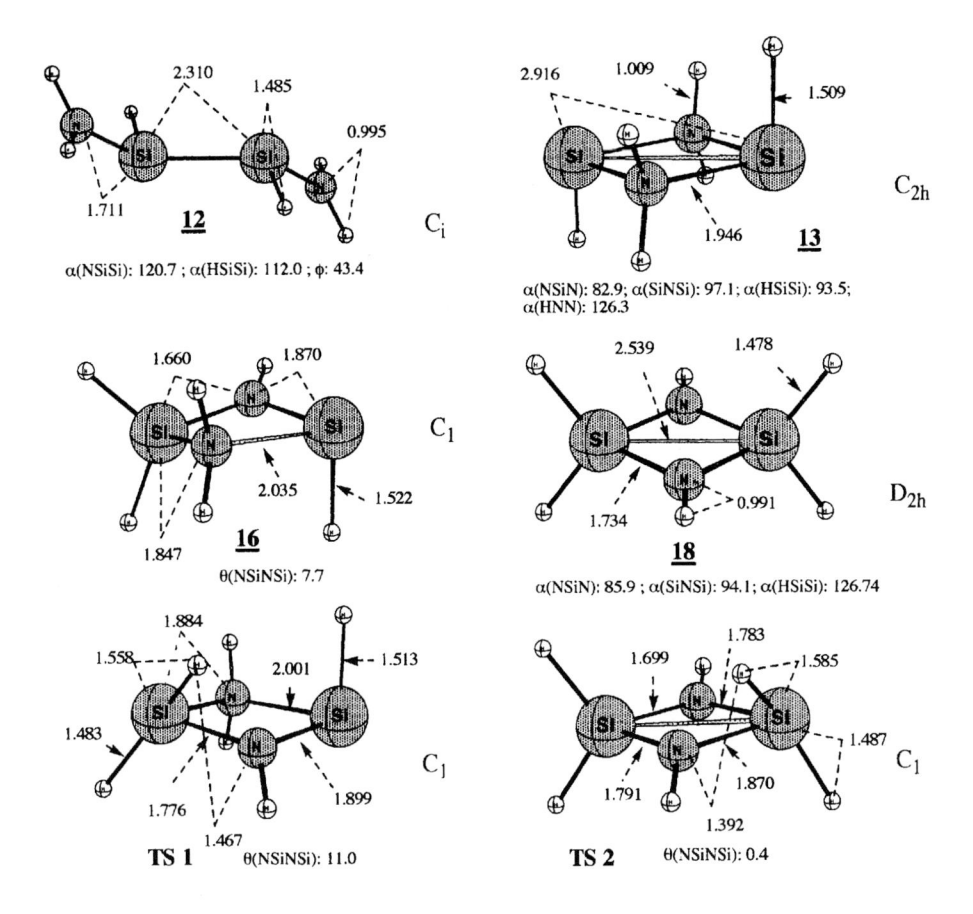

Fig. 17. Calculated geometries of diaminodisilene and its bridged isomers and the transition structures connecting them (HF/6–31G**); bond lengths in Å, bond angles in degrees

On the basis of these computational results we believe that bridged structures such as **13, 15, 18,** and **19** are kinetically stable (at least in the gas-phase to which the calculations correspond directly) and we therefore encourage experimentalists to design synthetic schemes for synthesizing and identifying these very interesting novel isomers of disilenes. Substitution of the hydrogens on the heteroatoms by alkyl groups is expected to increase even further the rearrangement barriers and therefore the kinetic stability of the bridged isomers. Due to their relatively low energies, bridged species can play an important role in various reactions of disilenes, even in cases where the bridging atoms have relatively poor bridging abilities, such as carbon. A bridged species similar to **13**, with aryl bridging groups, has been suggested as a possible transition state in the dyatropic rearrangement of unsymmetrically substituted disilenes [26].

A major question which we still cannot answer is, how to direct the dimerization reactions of silylenes along path b in Scheme 2 to yield the bridged isomers in preference to the corresponding disilenes. For example, why does MesSiN(SiMe$_3$)$_2$ dimerize to produce only the disilene, Mes[(Me$_3$Si)$_2$N]Si=Si[N(SiMe$_3$)$_2$]Mes [25]? Calculations aimed at resolving this question are in progress. An interesting system to study in this connection is the novel stable silylene **5**, because as pointed out above the corresponding disilene bearing four nitrogen substituents is predicted not to be a minimum on the potential energy surface and it may consequently dimerize to the corresponding bridged isomer of type **13**. The dimerization of (H$_2$N)$_2$Si to the bridged (H$_2$N)Si[μ-(NH$_2$)]$_2$Si(NH$_2$) is calculated to be exothermic by 15.9 kcal mol^{-1} (at MP2/6–31G**//6–31G** + ZPE at 6–31 G*). Finally we note that very recently Veith and Zimmer reported the synthesis of a germanium compound which possesses a cyclic-(R'NGe[μ-(NR$_2$)]$_2$GeNR') skeleton [27] analogous to that in **13**.

Conclusions

We have shown in this paper that molecular orbital calculations at the ab initio level can be used to predict reliably the spectral transitions in silylenes, to evaluate the effects of substituents on the Si=Si multiple bond, to shed new light on existing experimental data and to direct future work towards the synthesis of novel isomers of disilenes. Although carbon and silicon are isoelectronic, multiple bonds to silicon differ dramatically from multiple bonds to carbon and analogies from carbon chemistry might therefore be entirely misleading when applied to silicon compounds. We believe that our studies have demonstrated the enormous power of modern computational methods and hope that this paper will prompt future theoretical studies and more importantly, theoretical-experimental collaborations in the field of organosilicon chemistry.

Acknowledgments: Our interest in silicon chemistry was stimulated by the beautiful work of Professor R. West and his coworkers and by his scientific enthusiasm. We are indebted to Professor West for a long and stimulating collaboration between the Haifa and the Madison research groups and for the numerous stimulating discussions. This collaboration was supported over the last decade by grants from the *United States – Israel Binational Science Foundation (BSF)*. Y. Apeloig would like to acknowledge the *Alexander von Humboldt-Stiftung* for a Senior Scientist Award and to thank Professor H. Schwarz for his warm hospitality during his stay in Berlin when this work was completed. T. Müller is grateful to the *MINERVA-Foundation* for a post doctoral scholarship.

References:

[1] For a comprehensive review of recent developments in silicon chemistry, see: *The Chemistry of Organic Silicon Compounds*, (Eds.: S. Patai, Z. Rappoport), Wiley, New York, **1989**.

[2] For a recent review, see: R. West, *Chem. Rev.*, in press.

[3] For a recent review, see: G. Raabe, J. Michl, Chapt. 17 in [1].

[4] W. J. Hehre, L. Radom, P. v. R. Schleyer, J. A. Pople, *Ab Initio Molecular Orbital Theory*, Wiley, New York, **1986**.

[5] For comprehensive reviews of molecular orbital calculations of silicon compounds, see: Y. Apeloig, Chapt. 2 in [1]; Y. Apeloig, M. Karni, *Chem. Rev.* in press.

[6] T. J. Drahnak, J. Michl, R. West, *J. Am. Chem. Soc.* **1979**, *101*, 5427.

[7] M. J. Michalczyk, M. J. Fink, D. J. DeYoung, C. W. Carlson, K. M. Welsh, J. Michl, *Silicon, Germanium, Tin and Lead Compd.* **1986**, *9*, 75.

[8] Y. Apeloig, M. Karni, *J. Chem. Soc. Chem. Commun.* **1985**, 1018.

[9] a) C. A. Arrington, R. West, J. Michl, *J. Am. Chem. Soc.* **1983**, *105*, 6176.

b) A. S. Nazran, J. A. Hawari, D. Griller, I. S. Alnaimi, W. P. Weber, *J. Am. Chem. Soc.* **1984**, *106*, 7267; J. A. Hawan, D. Griller, *Organometallics* **1984**, *3*, 1123.

c) A. S. Nazran, J. A. Hawari, D. Griller, I. S. Alniami, W. P. Weber, *J. Am. Chem. Soc.* **1986**, *108*, 5041.

d) R. S. Grev, H. F. Schaefer III, *J. Am. Chem. Soc.* **1986**, *108*, 5804.

[10] H. B. Schlegel, *J. Phys. Chem.*, **1988**, *92*, 3075; *J. Phys. Chem.*, **1986**, *84*, 4530.

[11] Y. Apeloig, M. Karni, R. West, K. Welsh, *J. Am. Chem. Soc.*, in press.

[12] G. Maier, H. P. Reisenauer, H. Pack, *Angew. Chem.* **1994**, *106*, 1347; *Angew. Chem., Int. Ed. Engl.* **1994**, *33*, 1248.

[13] M. Kira, T. Maruyama, H. Sakurai, *Tetrahedron Lett.* **1992**, *33*, 243.

[14] M. Kira, personal communication.

[15] a) While the spectral shifts are very well reproduced by the calculations an empirical correction has to be introduced in order to predict correctly the precise λ_{max}; for details see footnote 15b in [11].

b) Y. Apeloig, D. Bravo-Zhivotovskii, M. Karni, T. Müller, unpublished results.

[16] M. Denk, K. Lennon, R. Hayashi, H. Haaland, R. West, A. V. Belyakov, A. V. Verne, M. Wagner, N. Metzler, *J. Am. Chem. Soc.* **1994**, *116*, 2961.

[17] a) R. Ditchfield, *Mol. Phys.* **1974**, *27*, 789.

b) J. Gauss, *J. Chem. Phys.* **1993**, *99*, 3629; and references therein.

[18] M. Denk, R. West, R. Hayashi, Y. Apeloig, R. Pauncz, M. Karni, in: *Organosilicon Chemistry II* (Eds.: N. Auner, J. Weis), VCH, Weinheim, **1995**, p. 251.

[19] R. West, M. Fink, J. Michl, *Science* **1981**, *214*, 1343.

[20] M. Karni, Y. Apeloig, *J. Am. Chem. Soc.* **1990**, *112*, 8589.

[21] E. A. Carter, W. A. Goddard III, *J. Phys. Chem.* **1986**, *90*, 998; E. A. Carter, W. A. Goddard III, **1988**, *88*, 1752.

[22] a) M. Kira, T. Maruyama, C. Kabuto, K. Ebata, H. Sakurai, *Angew. Chem.* **1994**, *106*, 1575; *Angew. Chem., Int. Ed. Engl.* **1994**, *33*, 1489.

b) Dimerization of $(Me_3Si)_2Si$ leads to $(Me_3Si)_2Si=Si(SiMe_3)_2$ which was characterized by trapping with butadiene (see Scheme 1) [15].

[23] P. Jutzi, U. Holtmann, H. Bögge, A. Müller, *J. Chem. Soc., Chem. Commun.* **1988**, 305.

[24] J. Maxka, Y. Apeloig, *J. Chem. Soc., Chem. Commun.* **1990**, 737.

[25] M. J. Michalczyk, R. West, J. Michl, *Organometallics* **1985**, *4*, 826.

[26] H. B. Yokelson, J. Maxka, D. A. Siegel, R. West, *J. Am. Chem. Soc.* **1986**, *108*, 4239.

[27] M. Veith, M. Zimmer, *Main Group Chemistry News* **1994**, *2*, 12; see also: M. Veith, *Chem. Rev.* **1990**, *90*, 3.

Heterosubstituted Silylenes:
Cycloadditions with Heterodienes and -trienes

Joachim Heinicke*, Barbara Gehrhus, Susanne Meinel

Institut für Organische Chemie

Ernst-Moritz-Arndt-Universität Greifswald

Soldtmannstr. 16, D-17487 Greifswald, Germany

Summary: Copyrolytic gas-phase reactions of 1,1,2,2-tetra(alkoxy)- and tetrakis-(dimethylamino)dimethyldisilanes with dienes, heterodienes, and heterotrienes in a flow-system furnish a number of unsaturated silicon heterocycles via silylene-intermediates with the exception of 1,4-diaza- and 1,4-oxazadienes, yielding 40–65 % of 1,3,2-diaza- and 1,3,2-oxazasilacyclopent-4-enes, respectively, mixtures of diastereoisomeric 1-oxa- or 1-aza-2-silacyclopent-4-enes and -3-enes are formed. Conjugated oxatrienes were found to undergo additionally [6+1]-cycloadditions giving 1-oxa-2-silacyclohepta-4,6-dienes.

Silylenes, like other low-coordinated and highly reactive species have found considerable interest during the last two decades. Main emphasis is focused on studies of their reactivity and synthetic use [1] as well as, in the last few years, on the search for stable silylenes [2]. The aim of the presented work was to investigate the reaction behaviour of silylenes which may be generated from technical available or slightly modified disilanes [3].

$Me_2Si_2Cl_4$, contaminated by (usually about 20 %) $Me_2ClSi–SiCl_2Me$, was obtained by repeated fractionation (at last by use of a spinning band column) of higher boiling products of the direct synthesis. Conversion of the chlorodisilane by reaction with alcohols (in presence of urea) or dimethylamine gave tetra(alkoxy)- or tetrakis(dimethylamino)disilanes with the corresponding impurities of tris(alkoxy)- and tris(dimethylamino)disilane respectively. All of these heterosubstituted disilanes undergo α-eliminations at 400–500 °C (residence time 40–70 s) to give heterosubstituted methylsilylenes (Scheme 1). The highly selective formation of MeSiX: from $Me_3Si_2X_3$, allowing the use of mixtures of $Me_2Si_2X_4/Me_3Si_2X_3$, may be attributed to a considerable π-stabilization (calculated values in Fig. 1 [3]). This seems to overcompensate the slightly stronger contribution of the intramolecular p(X)–d(*Si*X₂Me) compared to the p(X)–d(*Si*XMe₂) interaction in the transition state.

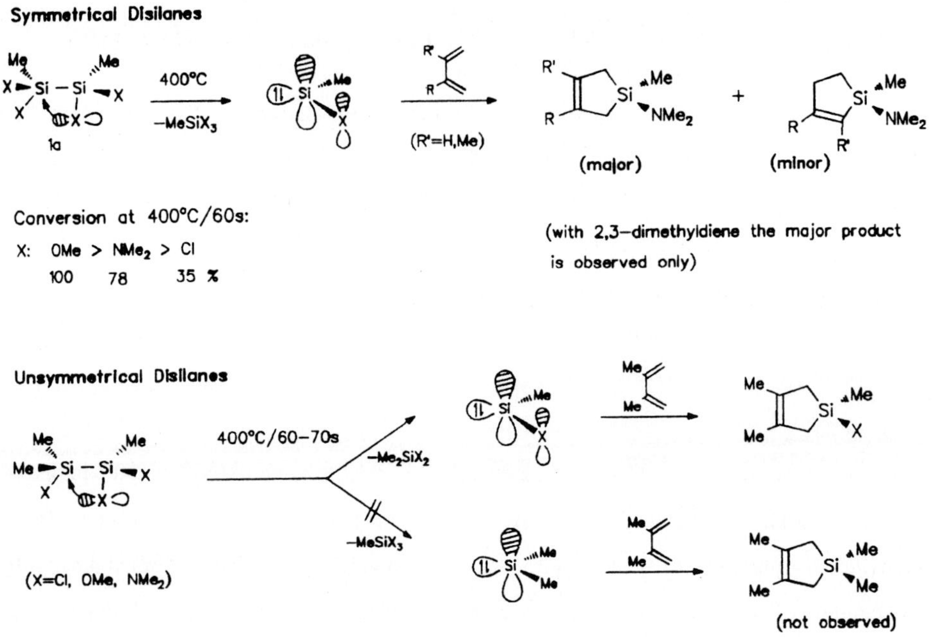

Symmetrical Disilanes
400°C
−MeSiX$_3$
1a
(R'=H,Me)
(major)
(minor)
Conversion at 400°C/60s:
X: OMe > NMe$_2$ > Cl
100 78 35 %
(with 2,3−dimethyldiene the major product is observed only)
Unsymmetrical Disilanes
400°C/60−70s
−Me$_2$SiX$_2$
−MeSiX$_3$
(X=Cl, OMe, NMe$_2$)
(not observed)

Scheme 1.

$$HSiX + SiH_4 = H_3SiX + SiH_2$$

MP4 6–31G*//HF 6–31G* [kcal/mol^{-1}] [4]:

X	H_2	OH	F	PH_2	SH	Cl	$(NH_2)_2$
E	22.5	15.0	9.3	7.0	15.1	7.0	37.8

Fig. 1. Gain of energy according to the isodesmic equation

In presence of dienes (Scheme 1), heterodienes, and heterotrienes (Scheme 2) rapid cycloadditions take place which prevent other reaction modes of the silylenes (insertions, polymerization, or HX-elimination-polymerization). Although a concerted [1+4]-cycloaddition is symmetry-allowed a stepwise mechanism (Scheme 3) via a three-membered intermediate is prefered or at least partly be competing to account for the formation of double-bond isomers. Ususally the formal [4+1]-cycloadducts (allylsilane-type) are the main products while the isomers with vinylsilane-units are side-products (< 30 %). Exceptions are

2,3-dimethyl-butadiene, 1,4-diaza-, -oxaza-, or dioxadienes forming the "normal" cycloadducts only and N-phenyl-1-azadienes with a preference of the vinylsilaneisomer (ca. 40:60). The former exceptions do not give evidence for a concerted mechanism since they are consistent also with the stepwise model – prefered α-cleavage in a primary [1+2]-cycloadduct of dimethylbutadiene by sterical strain and instable "side products" (N–N or N–O bonds) in the cycloaddition of 1,4-diheterodienes. The latter exception may be explained by an easier cleavage of the C–N compared to the C–Si bond of the azasiliridine-intermediate. Since N-isopropyl-1-azadienes, however, behave "normally" (analogously to oxadienes) discussions using tabulated bond energies of unstrained compounds may be misleading.

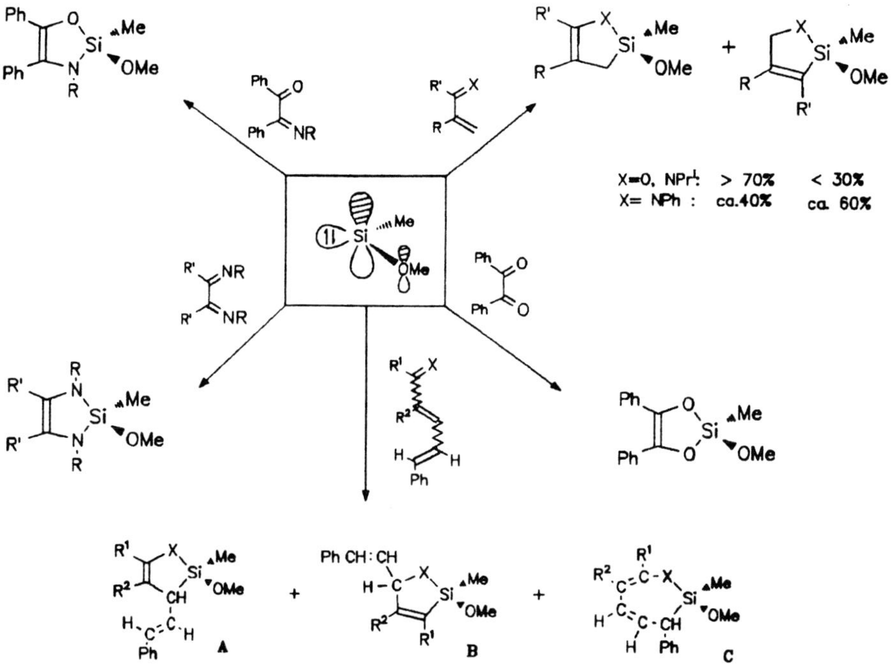

Ratio of (diastereo)isomers	X	(R¹/R²):	A/A'	B/B'	C/C"
	NPr^i	(H /Me)	30/40	5/5	not det.
	O	(H /Me)	–	29/26	27/18
	O	(Me/Me)	28/12	24/22	8/6
	O	(t–Bu/H)	31/17	2/1	25/24
	O	(Ph /H)	–	(low)	53/47

Scheme 2.

Dienes and heterodienes differ in the regioselectivity of their cycloadditions. In case of dienes they are controlled by substituent effects, in case of heterodienes by the strongest primary Lewis acid-base interaction [silylene to lone-pair of heteroatom with O > N (Scheme 3)].

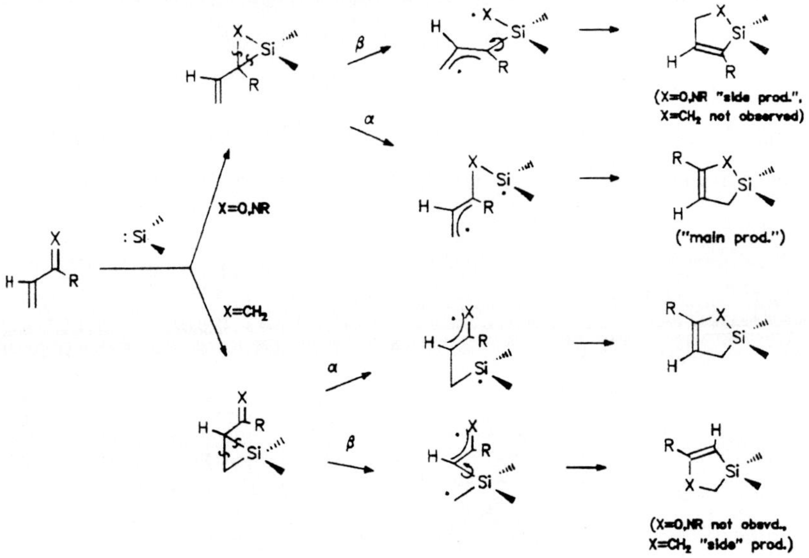

Scheme 3. Regioselectivity of diene- vs heterodiene-cycloaddition

Conjugated oxatrienes undergo frequently, sometimes preferred, [1+6]-cycloadditions besides [1+4]-cycloadditions. The observed 1-oxa-2-silahepta-4,6-dienes correspond a "normal" [1+6]-cycloaddition-mode and they repress the "normal" [1+4]-cycloaddition products. Seven-membered analogues of the [1+4]-"side-mode" products (5-styryl-1-oxa-2-silacyclopent-3-enes), which have been detected as by-products by characteristic NMR signals, could not yet be identified. However, it should be stated at this stage that there are further up-to-now unknown minor side products and that the separation is difficult. The formation of the above mentioned compounds is interpreted by a stepwise cycloaddition via a three-membered intermediate which undergoes a biradical ring-opening ring-closure rearrangement [5]. In this way a five-centered radical site may be formed from trienes, planarization provided, which can undergo intramolecular recombination with the Si-radical site in 1-, 3-, and 5-position to give either back reaction, 5- or 7-membered rings. In analogous recombinations with harder O-radicals after β-cleavage of the oxasilirane-intermediates the soft five-center radicals seem to be less favoured (Scheme 4).

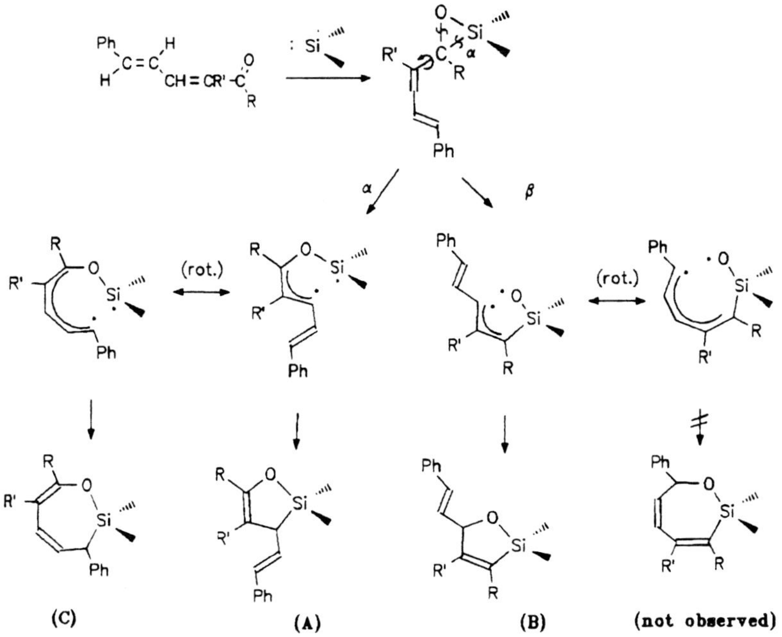

Scheme 4. Possible mechanism of heterotriene-cycloaddition

Acknowledgement: We appreciate the *Deutsche Forschungsgemeinschaft* for support of this work and a postdoc-stipendium for B. G.

References:

[1] P. P. Gaspar, in: *Reactive Intermediates* (Eds.: M. Jones Jr., R. A. Moss) **1985**, *3*, 333; E. A. Chernyshev, N. G. Komalenkova, *Sov. Sci. Rev. B, Chem.* **1988**, *12*, 107; M. Weidenbruch, *Coord. Chem. Rev.*, **1994**, *130*, 275.

[2] M. Denk, R. Lennon, R. Hayashi, R. West, A. V. Belyakov, H. P. Verne, A. Haaland, M. Wagner, N. Metzler, *J. Am. Chem. Soc.* **1994**, *116*, 2691; and references therein.

[3] J. Heinicke, B. Gehrhus, *J. Organomet. Chem.* **1992**, *423*, 13; J. Heinicke, B. Gehrhus, S. Meinel, *J. Organomet. Chem.* **1994**, *474*, 71; J. Heinicke, D. Vorwerk, G. Zimmermann, *J. Analyt. Appl. Pyrol.* **1994**, *28*, 93.

[4] B. T. Luke, J. A. Pople, M. B. Krogh-Jespersen, Y. Apeloig, J. Chandrasekar, P. v. R. Schleyer, *J. Am. Chem. Soc.* **1986**, *108*, 260; L. Nyulaszi, A. Belghazi, S. Kis-Szetsi, T. Veszpremi, J. Heinicke, *J. Mol. Struct. (THEOCHEM)*, in press.

[5] D. Lei, P. P. Gaspar, *J. Chem. Soc., Chem. Commun.* **1985**, 1149.

Decomposition Products of Methyltrichlorosilane

S. Leistner*, S. Baumann, G. Marx
Fachbereich Chemie, Lehrstuhl Physikalische Chemie
Technische Universität Chemnitz-Zwickau
Straße der Nationen 62, D-09111 Chemnitz, Germany

Introduction

Methyltrichlorosilane (CH_3SiCl_3, MTS) is an appropriate precursor to deposit SiC-layers on various carrier materials using chemical vapour deposition. MTS is decomposed according to the following overall equation of reaction:

$$CH_3SiCl_3 \longrightarrow SiC + 3\,HCl \quad (1)$$

However the thermal cracking of MTS can not be described correctly with this simple equation, therefore the decomposition reaction was investigated by an *in situ* IR spectroscopic method to identify additional decomposition products. The analysis of the formed, at 20 °C stable, products was achieved by gas chromatography as well as IR spectroscopy. The deposited solids were investigated both by X-ray diffractometry and glow discharge optical spectroscopy.

Argon or hydrogen at various amount of substance flow rate relations α ($n_{H_2}:n_{MTS}$) or β ($n_{Ar}:n_{MTS}$) was used as carrier gas for the MTS. The influence of temperature, amount of substance flow rate relation α or β and residence time τ was also investigated. The experimental identification of the reaction components is required to formulate a reaction mechanism and to compare with thermodynamic calculated gasphase composition. Knowledge of the individual running reactions should make the specific influence on the deposition of SiC layers possible.

Gas Chromatography

Investigations by gas chromatography have shown the decomposition of MTS to start at 600 °C when the residence time is 14.4 s. The complete decomposition occurs at 1000 °C. Start of cracking is proven by the first appearance of the SiCl$_4$-peaks. The absence of the MTS-peak proves the completeness of decomposition.

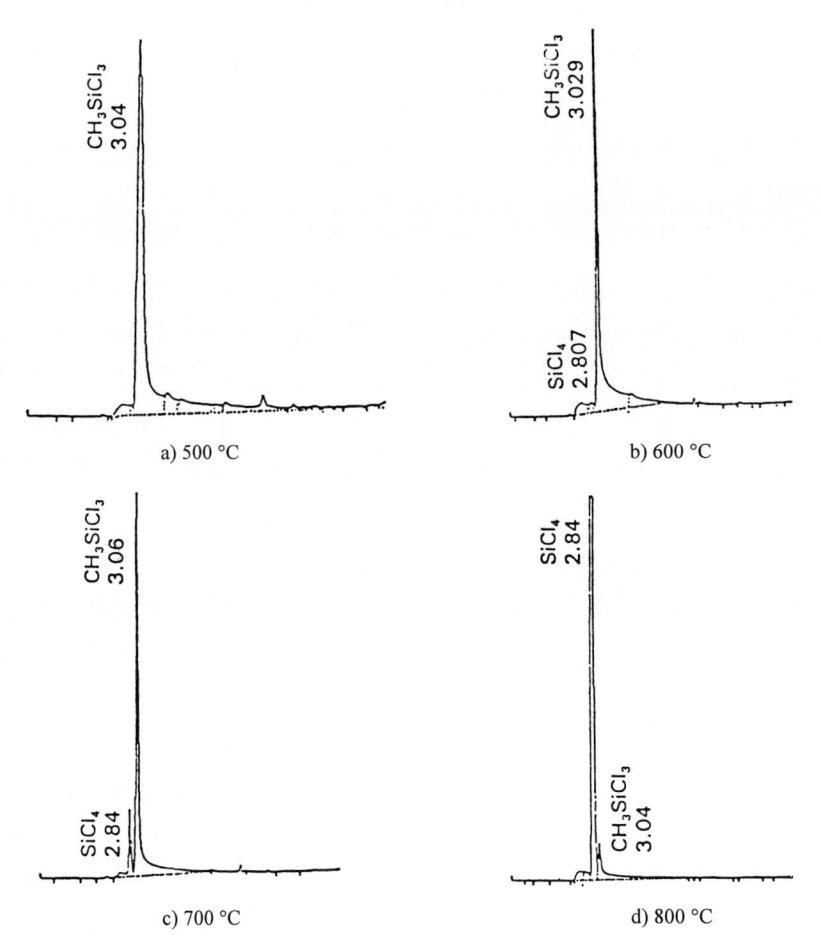

Fig. 1a-d. Change of relation between MTS- and SiCl$_4$-peaks at various temperatures

HCl and CH$_4$ also were identified as decomposition products of MTS by gas chromatography.

X-Ray Diffractometry and Glow Discharge Optical Spectroscopy

Carbon, silica, and siliconcarbid were detected as solid decomposition products by X-ray diffractometry of the layers deposited at various amount of substance flow rate relations and temperatures. The carrier material was graphite, therefore the detection of carbon can not be traced back to the decomposition reaction clearly.

The deep profile of a coated graphite probe, which was taken by glow discharge optical spectroscopy, shows that considerably more Si than C was coated. It means that no pure SiC was formed under the given conditions. It is shown at the deep profile (Fig. 2) the temperature has a significant influence on the composition of layer and the kinetics of deposition (temperature can be entered on the deep axis).

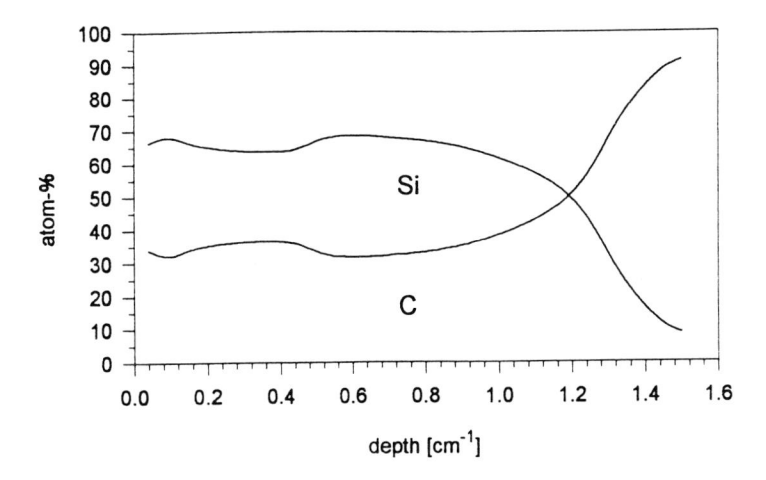

Fig. 2. Deep profile of a coated graphite probe

IR Spectroscopy

The preparation of a specific cuvet was required for the *in situ* IR spectroscopical observation of the gas-phase formed by decomposition of MTS (Fig. 3).

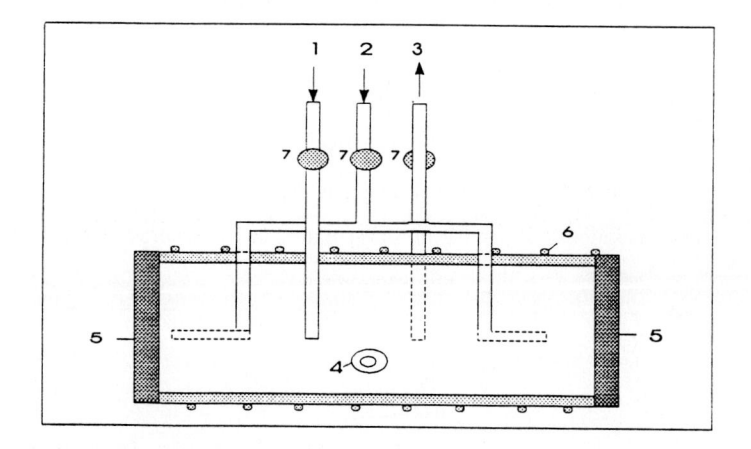

Fig. 3. Cuvet for *in situ* IR measurement:
1) hydrogen + MTS (or Argon + MTS) – feed
2) argon feed to window flushing
3) waste gas line
4) measuring point for thermocouple
5) optical window (material KBr)
6) resistance heating
7) teflon stopcocks

Identified decomposition products by IR spectroscopy are: CH_4, HCl, $SiCl_4$, $HSiCl_3$, H_2SiCl_2, and SiC. At 940 °C the bands at 470 cm^{-1} and 582 cm^{-1} were classified as the radical $SiCl_3$ and the bands at 513 cm^{-1}, 569 cm^{-1}, 675 cm^{-1}, and 763 cm^{-1} as the radical $HSiCl_2$.

At a residence time of 33.6 s and in pure Ar-atmosphere the decomposition of MTS was first detected at 820 °C, however in Ar/H_2-atmosphere at 720 °C (Fig. 4). This shows the influence of hydrogen on the mechanism of decomposition. At increasing residence time (above 30 min) the decomposition starts at 540 °C in both systems.

The bands of CH_4, HCl, and $SiCl_4$ increase significantly if the reaction gas is cooled down to room temperature (Fig. 5). The conclusion is the increasing of concentration of this compounds caused by the recombination of radicals.

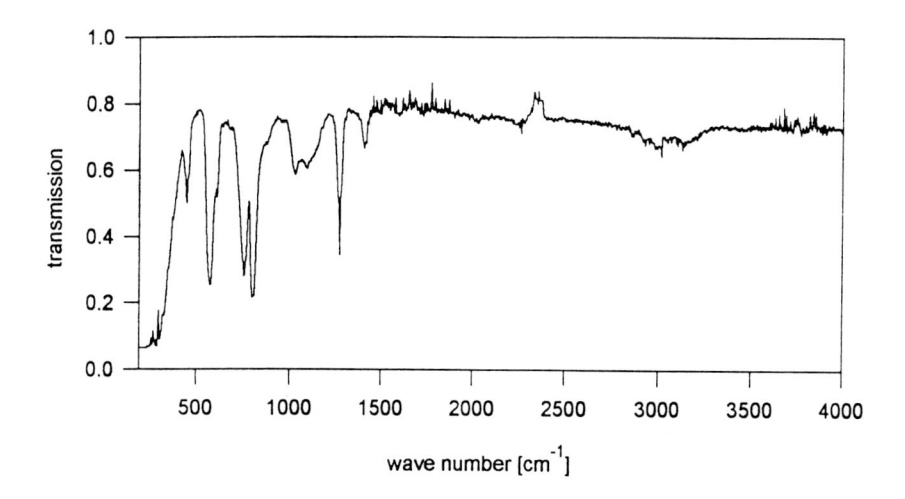

Fig. 4. IR Spectrum of the reaction gas at 720 °C

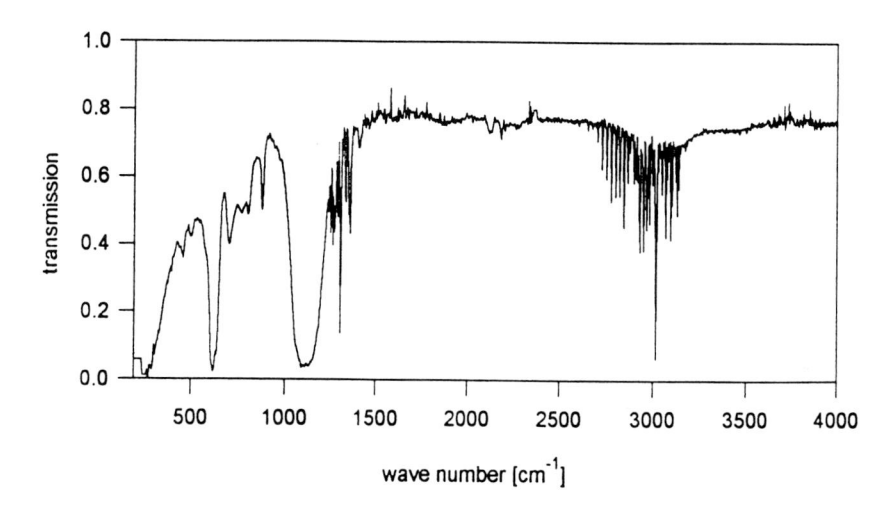

Fig. 5. IR Spectrum of the reaction gas at 20°C (cooled down)

Discussion of Results

Radicals $SiCl_3$ and $SiHCl_2$ could be formed corresponding to the following equations:

$$CH_3SiCl_3 \longrightarrow SiCl_3 + CH_3 \quad (2)$$

$$SiCl_3 + H_2 \longrightarrow HSiCl_2 + HCl \quad (3)$$

The formation of the CH_3 radical is only an assumption, because the used analytical method is not sensitive enough and there are no exact descriptions of the CH_3 radical in other publications. Following equations are devised to describe the increase of the substances methane, hydrogen chloride and silicon tetrachloride at the change from high to low temperature (cooling of the reaction gas).

$$2\ CH_3 + H_2 \longrightarrow 2\ CH_4 \quad (4)$$

$$CH_3 + SiCl_3 \longrightarrow SiC + 3\ HCl \quad (5)$$

$$2\ SiHCl_2 + C_2H_2 \longrightarrow 2\ SiC + 4\ HCl \quad (6)$$

$$SiH_2Cl_2 + 2\ HCl \longrightarrow 2\ H_2 + SiCl_4 \quad (7)$$

The hydrocarbons ethane, ethene, and ethyne were not identified. This could mean, that they are not present at detectable concentrations or that they react further immediately. It is conceivable that ethyne reacts with $SiHCl_2$ or H_2:

$$C_2H_2 \longrightarrow 2\ C + H_2 \quad (8)$$

$$C_2H_2 + 2\ SiHCl_2 \longrightarrow 2\ SiC + 4\ HCl \quad (9)$$

$$C_2H_2 + 3\ H_2 \longrightarrow 2\ CH_4 \quad (10)$$

Consequently the following reaction mechanism is proposed:

$$CH_3SiCl_3 \longrightarrow SiCl_3 + CH_3 \quad (11)$$

$$2\ SiCl_3 \longrightarrow {:}SiCl_2 + SiCl_4 \quad (12)$$

$$2\ CH_3 \longrightarrow C_2H_2 + 2\ H_2 \quad (13)$$

$$C_2H_2 \longrightarrow 2\ C_{(s)} + H_2 \quad (14)$$

$$2\ {:}SiCl_2 + H_2 \longrightarrow 2\ SiHCl_2 \quad (15)$$

$$2\ SiHCl_2 + C_2H_2 \longrightarrow 2\ SiC + 4\ HCl \quad (16)$$

$$2\,HSiCl_2 + HCl \longrightarrow H_2SiCl_2 + HSiCl_3 \qquad (17)$$

$$SiHCl_3 + HCl \longrightarrow H_2 + SiCl_4 \qquad (18)$$

$$SiH_2Cl_2 \longrightarrow Si_{(s)} + 2\,HCl \qquad (19)$$

$$CH_3SiCl_3 \longrightarrow SiC_{(s)} + 3\,HCl \qquad (20)$$

$$CH_3SiCl_3 + H_2 \longrightarrow CH_4 + HSiCl_3 \qquad (21)$$

$$CH_3SiCl_3 + 2\,H_2 \longrightarrow CH_4 + H_2SiCl_2 + HCl \qquad (22)$$

$$CH_3SiCl_3 + HCl \longrightarrow CH_4 + SiCl_4 \qquad (23)$$

It is assumed, that the principal mechanism of the MTS decomposition starts without hydrogen. That means H_2 is forming only just during the decomposition reaction. Hydrogen has an accelerating effect if it is a parent substance.

The identification of $SiCl_4$ by gas chromatography is a signal of the starting decomposition of MTS and SiC/Si-coating. This is a possibility to pursue a signal, which is proportional to SiC. The control of the SiC-coating process is conceivable by measurement of $SiCl_4$ concentration. However continuing extensive investigations are required to quantify the signal of $SiCl_4$. The amount of substance flow rate relation α or β has no detectable influence on the kind of decomposition products.

C_2H_2Si Isomers: Generation by Pulsed Flash Pyrolysis and Matrix-Spectroscopic Identification

Günther Maier*, Hans Peter Reisenauer, Harald Pacl

Institut für Organische Chemie

Justus-Liebig-Universität Gießen

Heinrich-Buff-Ring 58, D-35392 Gießen, Germany

Summary: The matrix-spectroscopic identification and photochemical interconversion of the isomeric silylenes **3–5**, and silacyclopropyne (**6**) are of interest in many ways. For one, their isolation serves to illustrate the potential of matrix isolation spectroscopy. In addition, the structural assignments for these species are based on the comparison of the experimentally observed and calculated IR spectra and therefore emphasize the importance of simultaneously applying quantum chemical calculations and spectroscopic measurements. Moreover, practically no examples exist for this class of silylene rearrangements. Lastly, the C_2H_2Si isomers eventually play a decisive role in the chemistry of interstellar clouds.

According to calculations by H. F. Schaefer *et al.* [1], 1-silacyclopropenylidene **3** is expected to be the most stable C_2H_2Si species. Thus, it has been discussed as the adduct of a Si atom with acetylene [2, 3]. H. Schwarz *et al.* have shown [4] that if one ionizes chlorotrimethylsilane in the gas phase, neutralization reionization mass spectroscopy allows detection of a particle whose connectivities are indicative of structure **3**.

Better experimental access to the C_2H_2Si potential energy surface is offered by pyrolysis of 2-ethynyl-1,1,1-trimethyldisilane (**1**). Compound **1** can be prepared by reaction of 1,1,1-trimethyl-2-phenyldisilane [5] with trifluoromethanesulfonic acid to give the corresponding triflate, which on treatment with sodium acetylide yields **1**.

Gaseous mixtures of disilane **1** and argon (1:1000–2000) were subjected to flash pyrolysis at various temperatures and pressures. After leaving the hot zone the reaction products were directly condensed onto a CsI or BaF_2 window at 10 K. The matrix-isolated products were studied by IR and UV/VIS spectroscopy. Under the conditions of *high-vacuum* flash pyrolysis only trimethylsilane (**2**) and small amounts of acetylene were detected. Any C_2H_2Si isomer that might have formed was too unstable to be detected under these pyrolysis conditions.

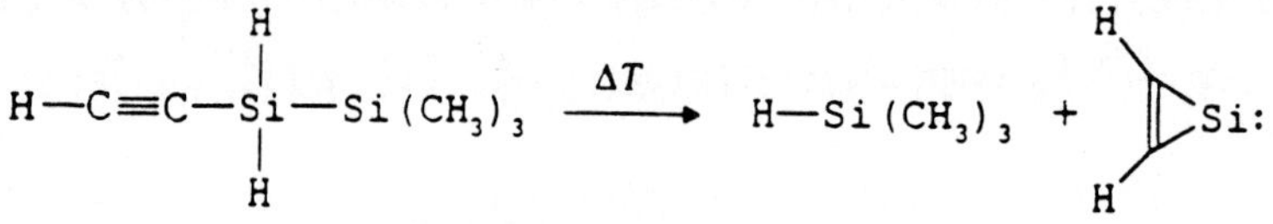

$$H-C\equiv C-\underset{\underset{H}{|}}{\overset{\overset{H}{|}}{Si}}-Si(CH_3)_3 \quad \xrightarrow{\Delta T} \quad H-Si(CH_3)_3 \quad + \quad \text{(silacyclopropenylidene)}$$

1   **2**   **3**

Scheme 1.

Pulsed flash pyrolysis proved more successful. The gaseous mixture (regulated by a pulsed magnetic valve) was expanded through a corundum tube directly into the high vacuum of the cryostat. The products were condensed onto the matrix window at 10 K.

Under these pyrolysis conditions the IR spectrum shows, apart from the bands for trimethylsilane (**2**), those of another compound. This compound is identified as 1-silacyclopropenylidene (**3**) – even though the structure of reactant **1** suggests formation of ethynylsilanediyl (ethynylsilylene) (**4**) – by comparison with the IR spectra for **3**, **4**, and **5** calculated by ab initio methods [1a]. Furthermore, **3** exhibits a weak, broad UV absorption between 320 and 260 nm (λ_{max} = 286 nm).

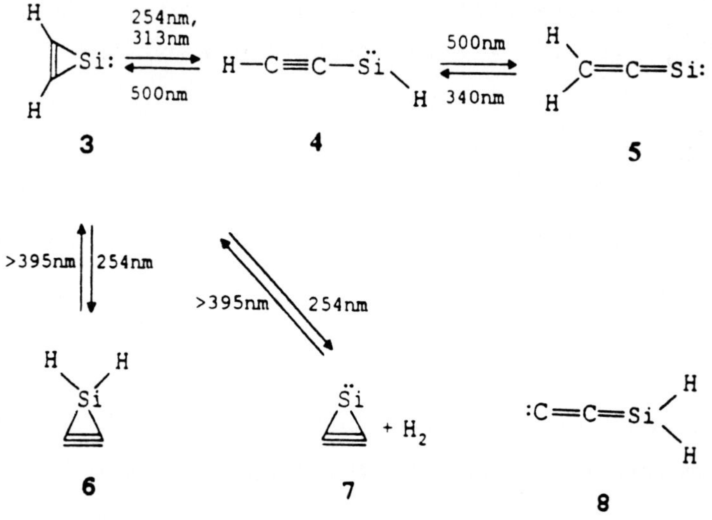

Scheme 2.

Irradiation into this band with monochromatic light of wavelength 313 nm results in a rearrangement to give ethynylsilanediyl (**4**). During this photoreaction only the absorbances of **3** and **4** change, and thus the IR and UV/VIS spectra of both compounds can be determined by subtraction. Fig. 1 shows the difference IR spectrum and the calculated spectra [1a] for **3** and **4**.

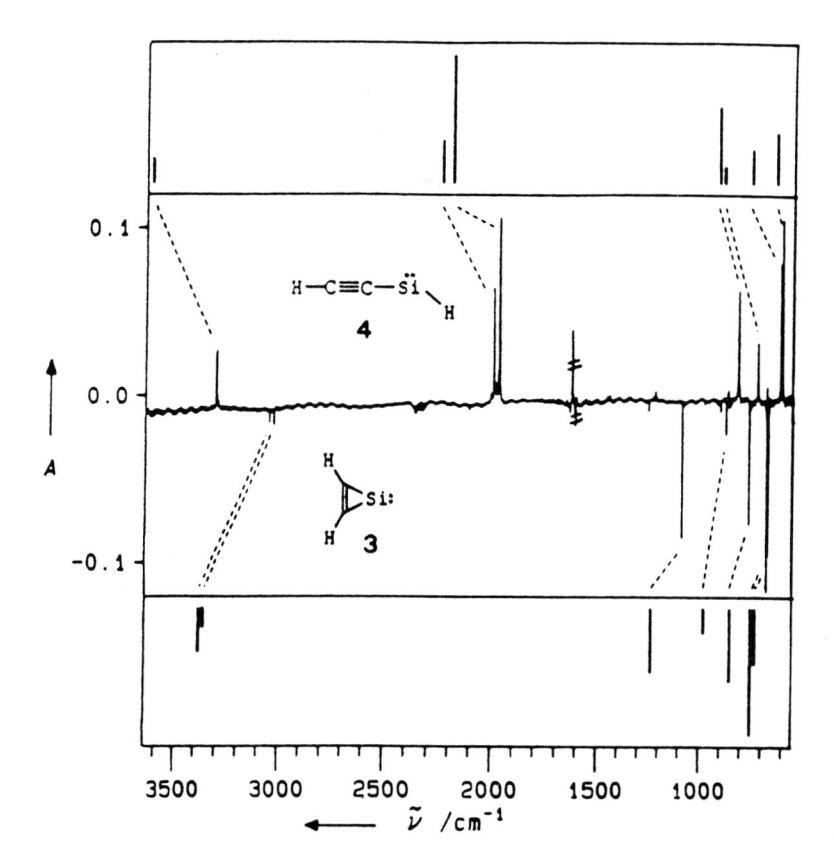

Fig. 1. IR spectra of **3** and **4**:
middle: difference IR spectrum for the photoreaction **3**→**4** (Ar matrix, 10 K)
top: calculated [1a] IR spectrum for **4**
bottom: calculated [1a] IR spectrum for **3**
IR bands due to water have been crossed; for silanediyl (**4**) a broad absorption band is recorded in the visible region of the spectrum (λ_{max} = 500 nm)

It is therefore not surprising that irradiation of **4** with visible light of wavelength 500 nm mainly leads to reisomerization to give 1-silacyclopropenylidene (**3**). A small amount of a new species is also formed, and two weak IR bands at 1667.5 and 957.7 cm^{-1} and a UV band with fine structure (λ = 340, 325, and 310 nm) are recorded. Irradiation into this absorption (λ = 340 nm) leads to reisomerization to **4**; hence, this species is another isomer of C$_2$H$_2$Si. By comparison with the calculated IR spectra this isomer is identified as vinylidenesilanediyl (**5**).

Irradiation of **3** with light of wavelength λ = 254 nm leads to the formation of **4** and two new compounds. One of these can easily be identified by its characteristic IR bands [6a–6c] and by its electronic transition band [6c] as the well-known, cyclic C_2Si (**7**). The second compound shows IR absorptions at 2228.9 and 2214.4 cm^{-1}, which are typical for SiH_2 groups. Furthermore, a weak band at 1769.9 cm^{-1} is observed. Therefore, the fourth isomer must either be of constitution **6** or **8**. Silacyclopropyne (**6**) was initially dismissed as an unlikely candidate, as calculations [1a] indicate that this structure represents a transition state. Thus, it was important to calculate IR spectra for **6** and **8**, something that had not previously been done. Calculations at the MP2/6–31G** level led to two important findings. Firstly, when the electron correlation energy is taken into account **6** is a minimum. Secondly, the experimentally measured and calculated spectra are in satisfactory agreement for **6**, but not for **8**. We therefore prefer the structure of silacyclopropyne (**6**) for the fourth isomer. To the best of our knowledge **6** would represent the first example of a cyclopropyne.

Upon irradiation (λ > 395 nm) compound **6** is transformed into **3** (therefore, both compounds are indeed isomers). Simultaneously, C_2Si (**7**), which is formed concurrently with **6**, reacts with eliminated hydrogen that is trapped in the same matrix cage to also give **3**. Trapping reactions of this kind have also been observed for various other unsaturated silicon compounds [7].

References:

[1] a) G. Frenking, R. B. Remington, H. F. Schaefer III, *J. Am.Chem. Soc.* **1986**, *108*, 2169.

b) G. Vacek, B. T. Colgrave, H. F. Schaefer III, *J. Am.Chem. Soc.* **1991**, *113*, 3192.

c) For the calculation of energies and rotational constants see: D. L. Cooper, *Astrophys. J.* **1990**, *354*, 229; for additional references see [4].

[2] a) D. Hussain, P. E. Norris, *J. Chem. Soc. Faraday Trans. 2* **1978**, *74*, 106.

b) S. C. Basu, D. Husian, *J. Photochem. Photobiol. A* **1988**, *42*, 1.

[3] M.-D. Su, R. D. Amos, N. C. Handy, *J. Am. Chem. Soc.* **1990**, *112*, 1499.

[4] R. Srimivas, D. Sülzle, T. Weiske, H. Schwarz, *Int. J. Mass Spectrom. Ion Processes* **1991**, *107*, 369.

[5] a) E. Hengge, G. Bauer, H. Marketz, *Z. Anorg. Allg. Chem.* **1972**, *394*, 93.

b) D. Littmann, *Ph. D. Thesis, Universität Gießen,* **1985**.

[6] a) R. A. Sheperd, W. R. M. Graham, *J. Chem. Phys.* **1985**, *82*, 4788.

b) R. A. Sheperd, W. R. M. Graham, *J. Chem. Phys.* **1988**, *88*, 3399.

c) W. Weltner Jr., D. McLeod Jr., *J. Chem. Phys.* **1964**, *41*, 235.

d) B. Kleman, *Astrophys. J.* **1956**, *123*, 162.

[7] G. Maier, J. Glatthaar, *Angew. Chem.* **1994**, *106*, 486; *Angew. Chem., Int. Ed. Engl.* **1994**, *33*, 473.

Silylene and Disilene Additions to the Double Bonds of Alkenes, 1,3-Dienes, and Hetero-1,3-dienes

Edwin Kroke, Peter Will, Manfred Weidenbruch*

Fachbereich Chemie
Carl-von-Ossietzky-Universität Oldenburg
Carl-von-Ossietzky-Str. 9, D-26111 Oldenburg, Germany

Summary: Di-*t*butylsilylene **2**, which is formed with tetra-*t*butyldisilene **3** on photolysis of hexa-*t*butylcyclotrisilane **1**, reacts with the C=C double bonds of acyclic and cyclic alkenes to furnish the corresponding siliranes. Reactions of **2** with 1,3-dienes lead preferentially to vinylsiliranes as the product of a [2+1]-cycloaddition at one of the double bonds. The disilene **3** only undergoes [2+2]- or [4+2]-cycloadditions with alkenes or 1,3-dienes, respectively, in exceptional cases. In those cases where addition of **3** to the C=C double bonds does not occur, the highly strained hepta-*t*butylcyclotetrasilane is obtained, presumably by way of cyclodimerization and concomitant cleavage of *i*butene. Compound **3** reacts smoothly with electron-poor 1,4-heterodienes such as 1,4-diazabutadienes or α-ketoimines to yield the corresponding six-membered ring products. On the other hand, **1** reacts with the electron-poor 3,6-bis(trifluormethyl)-1,2,4,5-tetrazine through complete degradation of one CF_3 group to provide the rearrangement product **19**.

Additions to Alkenes

In spite of their bulky substituents, the presence of which is necessary to prevent oligomerization reactions, disilenes undergo numerous addition and cycloaddition reactions; these have been summarized in several review articles [1]. For example, stable or marginally stable disilenes react with the C=O and C=S groups of ketones [2–4] and thioketones [5] as well as with the triple bonds of acetylenes [2, 3] and nitriles [6]. Surprisingly, however, no cycloaddition reactions of disilenes with simple alkenes have yet been reported [1].

Hexa-*t*butylcyclotrisilane **1** [7] which, upon photolysis, furnishes di-*t*butylsilylene **2** and tetra-*t*butyldisilene **3** by concomitant cleavage of two Si–Si bonds [8] reacts with *ortho*-methylstyrene to produce not only the silirane **4** but also the 1,2-disiletane **5** as the first example of a [2+2]-cycloadduct of a disilene to an alkene [9].

Eq. 1.

The unexpectedly smooth formation of the unequivocally characterized disiletane **5** is presumably attributable to the unsymmetrical substitution at the C=C double bond in combination with the aryl substituent. In support of this hypothesis, reactions of **1** with the symmetrically substituted olefins cyclopentene and cyclohexene give rise only to the siliranes **6** and **7** [10, 11], which were in part characterized previously, together with the cyclotetrasilane **8**.

Eq. 2.

The silylene **2**, formed in the photolysis of **1**, apparently undergoes ready [2+1]-cycloadditions with the C=C double bonds of the cyclic olefins to furnish the adducts in good yields while the simultaneously formed disilene does not show any tendency to participate in [2+2]-cycloadditions with these bond systems. The disilene **3** remaining in the reaction mixture after the trapping of **2** cannot undergo cyclodimerization for steric reasons. However, the formation of hepta-*t*butylcyclotetrasilane **8** from **3** is possible via cleavage of isobutene. The compound **8** probably exhibits the highest ring strain yet found in the series of isolated cyclooligosilanes. In spite of the conversion of one Si-tbutyl unit into an Si–H moiety, six tbutyl groups remain in direct proximity to each other on the cyclotetrasilane skeleton. Whereas **1**, having a similar substitution pattern, possesses the small endocyclic angle of 60 ° which allows an optimal arrangement of the bulky tbutyl groups, this angle (α) in **8** is increased to almost 90 ° and thus causes strong van der Waals interactions between the hydrogen atoms of directly adjacent tbutyl groups. This, in turn, leads to a dramatic lengthening of the Si–Si bonds in the $tBu_2Si–SitBu_2$ fragments up to 254 pm. This value is even larger than the corresponding bond length in **1** which was previously considered to possess the greatest known endocyclic Si–Si bond length [12].

Additions to 1,3-Dienes

The corresponding reactions in the presence of conjugated dienes are more complicated than those of the photolysis of **1** in the presence of alkenes. For example, disilene **3** reacts with 2,3-dimethylbutadiene to furnish a Diels-Alder product and an ene adduct [13]. On the other hand, silylene **2** undergoes [2+1]-cycloaddition with formation of the silirane **9** in preference to the competing [4+1]-cycloaddition process. This is somewhat surprising since the sterically encumbered product **9** experiences an appreciable hindrance to rotation about the central C–C bond as reflected in the line broadening in its ^1H and ^{13}C NMR spectra. The reaction with isoprene, in which one of the C-methyl units of 2,3-dimethylbutadiene is replaced by a hydrogen atom, proceeds regioselectively to furnish the silirane **10**; here, the [2+1]-cycloaddition has occurred exclusively at the less sterically hindered C=C double bond. This slight structural difference between **9** and **10** is sufficient to remove all restrictions to rotation in the product **10**.

Fig. 1.

The cyclic dienes cyclopenta-1,3-diene and cyclohexa-1,3-diene behave similarly to their acyclic counterparts. Photolyses in the presence of **1** lead mainly to the [2+1]-cycloaddition products **11** and **12**, respectively. It has not been possible to date to detect either [4+2]- or [2+2]-cycloadducts in the reaction mixtures.

Eq. 3.

The photolysis of **1** in the presence of 2,3-dimethoxy-1,3-butadiene proceeds differently and both the [2+1]- as well as the [2+2]-cycloadducts **13** and **14**, respectively, can be isolated from the reaction.

Eq. 4.

Although the currently available results still do not provide a uniform scheme, they do clearly indicate that silylenes bearing bulky substituents such as **2**, and also dimesitylsilylene [14], undergo [2+1]-cycloadditions to the double bonds of 1,3-dienes rather than [4+1]-cycloadditions. In contrast, the behavior of disilenes towards the C=C double bonds of alkenes and conjugated dienes is still not clear. While additions of the stable tetraaryldisilenes to such double bond systems are still unknown [1], the marginally stable disilene **3** is able, in individual cases, to take part in both [2+2]- and [4+2]-cycloaddition reactions.

Additions to 1,4-Dihetero-1,3-dienes

Although Diels-Alder reactions of disilenes to 1,3-dienes rather represent an exception, [2+4]-cycloaddition is the preferred reaction route for the photolysis of **1** in the presence of 1,4-heterodienes. Thus, for example, 1,4-diaza-1,3-butadienes react smoothly to yield the six-membered ring products **15** when the spatial demands of the substituents at nitrogen are not too great.

Fig. 2.

This reaction course is somewhat unexpected since the six-membered ring exhibits a considerable ring strain as manifested in the case of, for example, **15b** by a marked increase in the length of the Si–Si bond to 246.9(1) pm and a pronounced deviation of the two silicon atoms out of the N_2C_2 plane [15]. α-Ketoimines behave similarly, albeit only when the substituents on the nitrogen atoms are sterically undemanding, and also provide cyclic products of the type **16** by [4+2]-cycloaddition. The preference for the [4+2]-cycloaddition route over the competing [4+1] route is understandable in terms of the electronic properties of the 1,4-heterodienes. Both 1,4-diazabutadienes and α-ketoimines are typical electron-poor dienes that prefer to participate in Diels-Alder reactions with inverse electron demand upon treatment with strained, electron-rich alkenes [17]. The HOMO of the disilene is higher in energy in comparison to that of alkenes so that the Si=Si double bond is predestined for Diels-Alder reactions of this type [1a, 18].

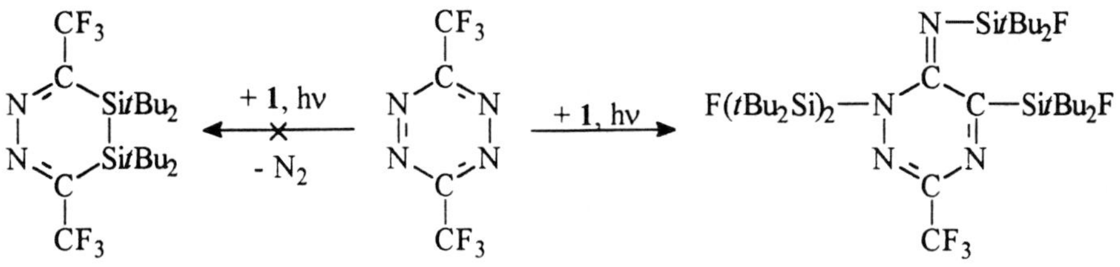

Eq. 5.

1,2,4,5-Tetrazines with electron-withdrawing substituents in the 3 and 6 positions constitute extremely electron-poor dienes [19]. In order to test their reactivity towards the disilene **3** we have selected the compound **18** because its CF_3 groups should not be so susceptible to the otherwise readily occurring halogen abstraction by **1** and its photolysis products. However, irradiation of **18** in the presence of an excess of **1** did not lead to the expected product **17**; instead the compound **19** was isolated and its structure confirmed by X-ray crystallography. The structure of **19** clearly reveals that not only has one

CF$_3$ group been completely degraded by the attack of **2** and **3** but that also a partial rearrangement of the six-membered ring skeleton has occurred.

Acknowledgements: Financial support of our work by the *Deutsche Forschungsgemeinschaft*, the *Volkswagen Stiftung*, and the *Fonds der Chemischen Industrie* is gratefully acknowledged.

References:

[1] Reviews: R. West, *Angew. Chem.* **1987**, *99*, 1231; *Angew. Chem., Int. Ed. Engl.* **1987**, *26*, 1202; G. Raabe, J. Michl, in: *The Chemistry of Organic Silicon Compounds*, Part 2 (Eds.: S. Patai, Z. Rappoport), Wiley, Chichester, **1989**, p. 1015; T. Tsumuraya, S. A. Batcheller, S. Masamune, *Angew. Chem.* **1991**, *103*, 916; *Angew. Chem., Int. Ed. Engl.* **1991**, *30*, 902; M. Weidenbruch, *Coord. Chem. Rev.* **1994**, *130*, 275.

[2] M. J. Fink, D. J. DeYoung, R. West, J. Michl, *J. Am. Chem. Soc.* **1983**, *105*, 1070.

[3] A. Schäfer, M. Weidenbruch, S. Pohl, *J. Organomet. Chem.* **1985**, *282*, 305.

[4] A. D. Fanta, D. J. DeYoung, J. Belzner, R. West, *Organometallics* **1991**, *10*, 3466.

[5] K. Kabeta, D. R. Powell, J. Hanson, R. West, *Organometallics* **1991**, *10*, 827.

[6] M. Weidenbruch, B. Flintjer, S. Pohl, W. Saak, *Angew. Chem.* **1989**, *101*, 89; *Angew. Chem., Int. Ed. Engl.* **1989**, *28*, 95.

[7] A. Schäfer, M. Weidenbruch, K. Peters, H. G. von Schnering, *Angew. Chem.* **1984**, *96*, 311; *Angew. Chem., Int. Ed. Engl.* **1984**, *23*, 302.

[8] M. Weidenbruch, in: *Frontiers of Organosilicon Chemistry* (Eds.: A. R. Bassindale, P. P. Gaspar), Royal Society of Chemistry, Cambridge, **1991**, p. 122.

[9] M. Weidenbruch, E. Kroke, H. Marsmann, S. Pohl, W. Saak, *J. Chem. Soc., Chem. Commun.* **1994**, 1233.

[10] R. Kumarathasan, P. Boudjouk, *Tetrahedron Lett.* **1990**, *31*, 3987.

[11] P. Boudjouk, E. Black, R. Kumarathasan, *Organometallics* **1991**, *10*, 2095.

[12] M. Weidenbruch, E. Kroke, S. Pohl, W. Saak, unpublished.

[13] S. Masamune, S. Murakami, H. Tobita, *Organometallics* **1983**, *2*, 1465.

[14] S. Zhang, R. T. Conlin, *J. Am. Chem. Soc.* **1991**, *113*, 4272.

[15] M. Weidenbruch, A. Lesch, K. Peters, *J. Organomet. Chem.* **1991**, *407*, 31.

[16] M. Weidenbruch, H. Piel, A. Lesch, K. Peters, H. G. von Schnering, *J. Organomet. Chem.* **1993**, *454*, 35.

[17] D. L. Boger, S. M. Weinreb, in: *Hetero Diels-Alder-Methodology in Organic Synthesis, Vol. 47* (Ed.: H. H. Wasserman), Academic Press, San Diego, **1987**.

[18] W. W. Schoeller, *J. Chem. Soc., Chem. Commun.* **1985**, 334.

[19] Review: J. Sauer, *Angew. Chem.* **1967**, *79*, 76; *Angew. Chem., Int. Ed. Engl.* **1967**, *6*, 16.

[20] M. Weidenbruch, P. Will, K. Peters, H. G. von Schnering, unpublished.

Generation of Silylenes and Silaethenes by Dehalogenation of Chlorosilanes

Joseph Grobe*, Thomas Schierholt

Anorganisch-Chemisches Institut

Westfälische Wilhelms-Universität

Wilhelm-Klemm-Str. 8, D-48149 Münster, Germany

Summary: The formation of reactive intermediates via dehalogenation of chlorosilanes was investigated by using lithium powder and sonication. Whereas in the absence of a diene substrate mainly polysilanes are obtained, reactions with, e.g., dimethylbutadiene, yield the corresponding cycloaddition products, indicating silylenes and silaethenes as intermediates.

Here we report on the synthesis of several silacyclopentenes and silacyclohexenes, which can be considered as proof for the intermediate formation of reactive silylenes and silaethenes during the reaction of suitable chlorosilane precursors with lithium. In particular, we have investigated the effects of sonic waves and of the Na-content in the lithium powder used on the product yield. Increasing Na-content in the lithium powder causes higher reactivity and a variety of by-products. The use of ultrasonic waves permits mild reaction conditions and leads to higher yields.

The formation of silylenes and silaethenes by dehalogenation of dichloro- and chloromethylchlorosilanes was previously described using the reaction of Na/K vapor in the gas-phase at 300 °C [1]. In the presence of butadiene the reaction with RR'SiCl$_2$ leads to silacyclopentenes.

A similar result was observed in the sonochemical reaction of di-*t*butyldihalosilane in the presence of lithium wire [2]. On the other hand, the ultrasound promoted reaction of dimethyldichlorosilane with lithium is known to generate dodecamethylcyclohexasilane **1** [3]. Ultraviolet irradiation of this compound in the presence of 2,3-dimethyl-1,3-butadiene (DMB) indicates the formation of 1,1,3,4-tetramethyl-1-silacyclopent-3-ene **2** according to Eq. 1.

In the course of our investigations we have tried to synthesize cyclization products of silylenes and DMB directly from chlorosilanes without using UV light. After that, we extended our studies to the formation of silaethenes and their trapping products with dienes by dehalogenation of chloromethylchlorosilanes.

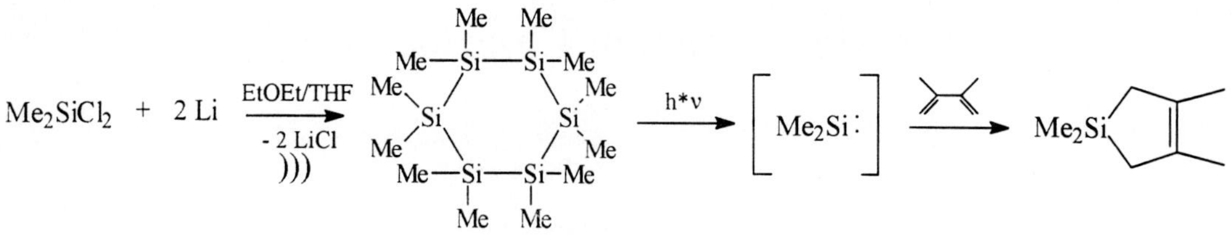

Eq. 1.

Sonication of a diethylether solution of dichlorosilanes and DMB in the presence of lithium powder leads to silaycyclopentenes (Eq. 2).

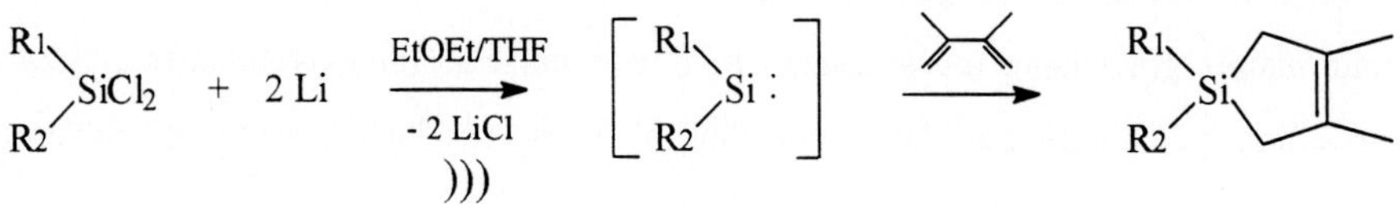

R1, R2 = H, Me, Ph

Eq. 2.

The increase in yield from R = H to R = Ph demonstrates the effect of steric stabilization of the silylene. Most of the polymeric by-products are due to the polymerisation of DMB. Changing the trapping agent to isoprene leads to the corresponding polymer as the major product and only small amounts of the silacyclopentene derivative.

In the same manner the dechlorination of chloromethylchlorosilanes in the presence of DMB produces silacyclohexenes (Eq. 3). In the absence of DMB the major product is the corresponding 1,3-disilacyclobutane.

Again, the application of isoprene as the quenching substrate is less effective and only low yields of the expected silacyclohexenes were obtained. Slight variations of reaction conditions lead to the formation of additional products. For example, a small amount of the 1,3,5-trisilacyclohexane **3** can be detected within 12 h in addition to tetramethyl-1,3-disilacyclobutane as the main product if the reaction of Me_2SiCl_2 is carried out in THF at −78 °C using lithium powder with a higher content of sodium (Eq. 4).

Eq. 3.

R = Me, H

25 - 30 %

55 %

Eq. 4.

The presence of very reactive intermediates is obvious from the formation of a variety of non identified by-products. However, the reactions with DMB as the diene substrate provide evidence for the formation of silylene and silaethene intermediates. Further investigations are necessary to elucidate the reaction pathways.

Acknowledgement: The authors thank the *Deutsche Forschungsgemeinschaft* for financial support.

References:

[1] P. Boudjouk, B. Hahn, *J. Org. Chem.* **1982**, *47*, 751.
[2] E. Gusel'nikov, Y. P. Polyakov, E. A. Volnina, N. S. Nametkin, *J. Organomet. Chem.* **1985**, *292*, 189.
[3] P. Boudjouk, U. Samaraweera, R. Sooriyakumaran, J. Chrisciel, K. R. Anderson, *Angew. Chem.* **1988**, *100*, 1406; *Angew. Chem., Int. Ed. Engl.* **1988**, *27*, 1355.

Triple Bonds to Silicon: Substituted Silanitriles Versus Silaisonitriles – A Theoretical Study

Karsten Albrecht, Yitzhak Apeloig*

Department of Chemistry

Technion - Israel Institute of Technology

Technion City, 32000 Haifa, Israel

Summary: Ab initio molecular orbital calculations for the RSiN ⇌ RNSi potential energy surface with R = H, Li, BeH, BH_2, CH_3, SiH_3, NH_2, PH_2, OH, SH, F, and Cl are reported. For R = H, Li, BeH, BH_2, CH_3, NH_2, PH_2, SH, and Cl the calculations predict that the silaisonitriles, RNSi, are thermodynamically more stable by 8–76 kcal mol^{-1} (at MP4/6–311+G^*//MP2(fu)/6–31G^*) than the corresponding silanitriles, RSiN. However, FSiN and HOSiN are more stable than FNSi and HONSi, and the isomers are separated by high barriers of 21.8 kcal mol^{-1} (FNSi→FSiN) and 31.1 kcal mol^{-1} (HONSi→HOSiN) at QCISD(T)/6–311G^{**}//QCISD/6–31G^*. Both isomers should therefore be observable.

Introduction

There has been considerable interest during the past decade in the study of compounds with multiple bonds to silicon. In particular Si=Si and Si=C double bonds have been studied extensively both experimentally and theoretically [1]. In contrast, relatively little is known about triple bonds to silicon and a stable compound of this family has not been prepared yet. However, there are three reports on the spectroscopic identification of transient compounds with Si–N triple bonds, the silanitrile, HSiN [2], the silaisonitrile, HNSi [3], and its phenyl substituted derivative, PhNSi [4] (the latter formally possess an Si=N double bond). Calculations for the parent silanitrile/silaisonitrile system [3d, 5] and for their phenyl- and methyl-substituted derivatives [6] predict that RSi≡N, which formally contains a Si≡N triple bond, is significantly less stable than the isomeric silaisonitrile, RN=Si, and that the rearrangement barriers of RSiN to RNSi are relatively small: e.g., for R = H the barrier is only 8.3 kcal mol^{-1} at G2 [3d]. This stability order is opposite to that in the isovalent carbon analogs HCN/HNC [7]. However, despite the unfavourable thermodynamics and the low rearrangement barrier, the parent silanitrile has most recently been characterized in an argon matrix [2].

In search for kinetically stable compounds with triple-bonds to silicon we examine in this paper, using up to date ab initio methods, the influence of various substituents spanning the first and second-row of the Periodic Table on the relative stability of silanitriles versus isosilanitriles and on the barriers that separate them. We were especially interested in finding substituted silanitriles, RSi≡N, which are more stable than the corresponding silaisonitriles, Si=NR, and which are separated from each other by significant energy barriers.

$$R-Si\equiv N \rightleftharpoons \left[\begin{array}{c} R \\ Si=N \end{array} \right]^{\neq} \rightleftharpoons Si=N-R$$

Scheme 1. R = H, Li, BeH, BeH$_2$, CH$_3$, SiH$_3$, NH$_2$, PH$_2$, OH, SH, F, Cl

Results and Discussion

The equilibrium structures of the substituted silanitriles, RSi≡N, and silaisonitriles, RN=Si, with R = H, Li, BeH, BH$_2$, CH$_3$, SiH$_3$, NH$_2$, PH$_2$, OH, SH, F, and Cl were optimized at the MP2(fu)/6–31G* [8] level of theory [9]. For R = H, CH$_3$, OH, F, and Cl we carried out geometry optimizations also with the CASSCF/6–31G* [10] and QCISD/6–31G* [11] methods, which are believed to produce more reliable structures. Comparison of the known experimental ν(Si–N) frequencies for HSiN and HNSi shows reasonable agreement with the values calculated at QCISD/6–31G* and CASSCF/6–31G* whereas the MP2(fu)/6–31G* frequencies are too low [12]. In all cases, the Si–N bond lengths are shorter in RNSi (1.571–1.607 Å) than in the isomeric RSiN (1.596–1.651 Å), indicating a higher bond order in the former, although this contradicts the calculated Wiberg bond indices [13] which are ca. 1.6 for the silaisonitriles and 2.5 for the silanitriles.

The calculated relative energies of the various RSiN and RNSi isomers at MP4/6–311+G**//MP2(fu)/6–31G* and for R = H, CH$_3$, OH, Cl, and F also at G2 [14], QCISD(T)/6–311G**//QCISD/6–31G* and CASSCF//6–31G* are presented in Table 1. The parent HN=Si and most of the substituted silaisonitriles, i.e., with R = Li, BeH, BH$_2$, CH$_3$, SiH$_3$, PH$_2$, and SH are substantially more stable than the corresponding silanitriles. RSiN, i.e., by 21.8–76.2 kcal mol^{-1}. However, for R = NH$_2$ the difference in energy decreases to only 10.5 kcal mol^{-1} and for R = Cl the two isomers are nearly equal in energy.

R	$\Delta E^{[a]}$	R	$\Delta E^{[a]}$
H	52.2 ($64.2^{[b]}$; $64.4^{[c]}$; $59.8^{[d]}$)	NH_2	10.5
Li	39.3	PH_2	43.7
BeH	68.9	OH	–13.2 ($-3.3^{[b]}$; $-1.8^{[c]}$; $-11.4^{[d]}$)
BH_2	76.2	SH	21.8
CH_3	42.0 ($49.6^{[b]}$; $51.5^{[c]}$; $48.6^{[d]}$)	F	–34.1 ($-22.6^{[b]}$; $-31.1^{[c]}$; $-35.5^{[d]}$)
SiH_3	61.2	Cl	–0.5 ($9.4^{[b]}$; $8.2^{[c]}$; $-0.4^{[d]}$)

Table 1. Relative energies (kcal mol^{-1}) of the RN=Si (taken as zero) and RSi≡N isomers
[a] At MP4/6–311G**//MP2(fu)/6–31G*; [b] At QCISD(T)/6–311G**//QCISD/6–31G*; [c] At G2; [d] At CASSCF//6–31G

The stability order is reversed for R = F and OH where the silanitriles FSi≡N and HOSiN are calculated to be the global minimum on the potential energy surface, being more stable than the isomeric RN=Si by 34.1 and 13.2 kcal mol^{-1}, respectively. The more reliable CASSCF and QCISD calculations for R = H, CH_3, F, Cl, and OH show (see Table 1) that the MP4/6–311+G** calculations somewhat overestimate the stability of the RNSi isomers relative to the RSiN isomers, but these calculations also support the conclusion that FSiN is substantially (i.e., 23–35 kcal mol^{-1}) more stable than FNSi and that HOSiN is more stable than HONSi, although in the latter case the energy difference is small.

The relative energies of the RSi≡N and RN=Si isomers are best interpreted in terms of the R–Si vs R–N bond energies. The thermodynamic stability of RSiN relative to RNSi increases on changing R from BH_2 to F parallel to the increase in the R–Si bond energy relative to the N–R bond energy (dissociation energies: Si–O: 129 kcal mol^{-1}, Si–F: 152 kcal mol^{-1}, N–O: 48 kcal mol^{-1}, N–F: 68 kcal mol^{-1}) [15]. For R = OH and F this effect is so large that it overrides the large intrinsic preference of HNSi over HSiN.

The transition structures for the interconversion of RSiN and RNSi were optimized at MP2(fu)/6–31G* and for R = H, CH_3, OH, F, and Cl they were also located at the CASSCF/6–31G* and QCISD/6–31G* levels (Fig. 1). In all cases except for R = F the transition state for the isomerization occurs "early" along the interconversion from RSi≡N to RN=Si as indicated by the short Si⋯R and relatively long R⋯N distances; i.e., for R = H, Si⋯H = 1.643 Å, N⋯H = 2.057 Å in the transition structure, compared with Si-H of 1.493 Å in H–Si≡N and H–N of 1.006 Å in H–N=Si; for R = SiH_3, Si⋯Si = 2.359 Å, Si⋯N = 2.957 Å in the transition structure vs Si–Si of 2.344 Å in H_3Si–Si≡N and Si–N of 1.735 Å in H_3Si–N=Si. This is in agreement with the Hammond postulate for exothermic processes.

In contrast, the transition state occurs "late" along the FNSi→NSiF reaction coordinate; $r(F \cdots Si)$ in the transition state is elongated to 2.399 Å from 1.603 Å in FSiN and $r(N \cdots F)$ is relatively short (N$\cdots$F = 1.480 Å, 1.346 Å in FNSi). This again is in accord with the Hammond postulate as the FNSi→NSiF rearrangement is endothermic.

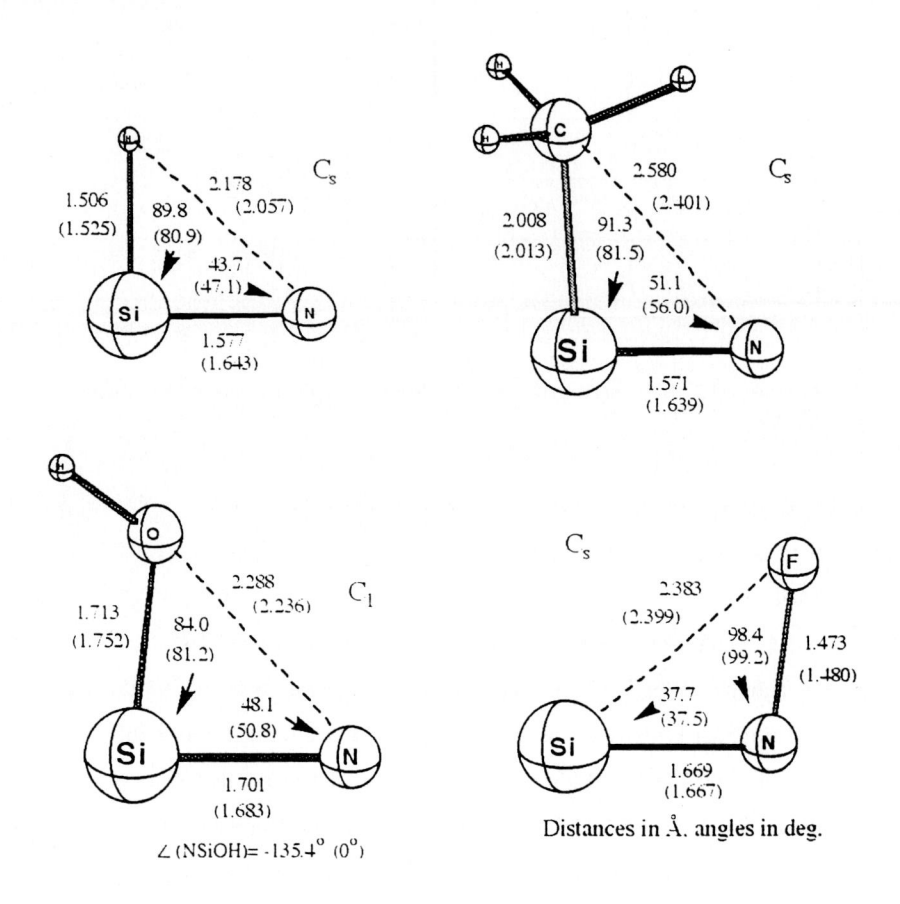

Fig. 1. Geometries of several transition structures for the RSiN→RNSi rearrangement at MP2(fu)/6–31 G*, values in parenthesis are at QCISD/6–31 G*

The calculated barriers separating RSiN and RNSi are presented in Table 2. At MP4/6–311+G**//MP2(fu)/6–31G* the barriers separating the two isomers are relatively small (1–17 kcal mol^{-1}) for R = Li, BeH, and BH$_2$ but significantly larger for R = H, CH$_3$, SiH$_3$, NH$_2$, PH$_2$, OH, SH, F, and Cl (21–56 kcal mol^{-1}). However, using G2 and the more reliable QCISD and CASSCF calculations, the barriers for rearrangement decrease significantly to ca. 10–20 kcal mol^{-1} for R = H, CH$_3$, in the range calculated by previous high-level studies [3d, 5]. In contrast, for R = OH, F the barrier heights are not very sensitive to the computational method, remaining high at all levels of theory. The isomerization of FNSi to the more stable FSiN requires an activation energy of 18.5–29.3 kcal mol^{-1} and the

HONSi→HOSiN rearrangement has a barrier of 31.1–42.1 kcal mol^{-1}. These high barriers suggest that *in the case of FSiN and HOSiN both the silanitrile and the silaisonitrile isomers should be kinetically stable and therefore isolable.*

	RSiN →RNSi				RNSi→ SiN			
R	MP4[b]	G2	QCISD[c]	CASSCF[d]	MP4[b]	G2	QCISD[c]	CASSCF[d]
H	32.9[d]	8.3	11.9	19.0	85.1	73.3	76.1	78.9
Li	1.8				41.1			
BeH	13.2				82.1			
BH$_2$	17.3				93.5			
CH$_3$	34.7	13.6	18.7	27.8	76.6	65.1	68.3	76.4
SiH$_3$	21.1				82.3			
NH$_2$	47.5				58.0			
PH$_2$	27.5				71.2			
OH	55.3	37.5	34.2	45.3	42.1	35.7	31.1	33.8
SH	42.1				63.9			
F	54.7	46.2	44.3	64.8	20.6	18.5	21.8	29.3
Cl	56.5	19.4	26.3	36.5	56.0	27.6	35.7	36.1

Table 2. Calculated energy barriers (kcal mol^{-1}) for the isomerization RSiN→SiNR[a]
[a] The smaller barriers are given in bold; [b] MP4/6–311+G**//MP2(fu)/6–31G*;
[c] QCISD(T)/6–311G**//QCISD/6–31G*; [d] CASSCF//6–31G*

Conclusion

The silanitriles FSi≡N and HOSi≡N are the global minima on their PES and they are separated from the isomeric silaisonitriles FN=Si and HON=Si by substantial barriers, suggesting that both FSiN and HOSiN should be accessible to detection, isolation, and characterization, at least in the gas phase and in matrices. We hope that this theoretical study will promote new experimental efforts towards the synthesis and characterization of FSiN and HOSiN.

References:

[1] G. Raabe, J. Michl, in: *The Chemistry of Organic Silicon Compounds* (Eds: S. Patai, Z. Rappoport), Wiley, Chichester, **1989**, p. 1015; Y. Apeloig, in: *The Chemistry of Organic Silicon Compounds* (Eds: S. Patai, Z. Rappoport), Wiley, Chichester, **1989**, p. 57; R. West, *Angew. Chem.* **1987**, *99*, 1231; *Angew. Chem., Int. Ed. Engl.* **1987**, *26*, 1201.

[2] G. Maier, J. Glatthaar, *Angew. Chem.* **1994**, *106*, 486; *Angew. Chem., Int. Ed. Engl.* **1994**, *33*, 473.

[3] a) J. F. Ogilvie, S. Cradock, *J. Chem. Soc., Chem. Commun.* **1966**, 364.

b) M. Bogey, C. Demuynck, J. L. Destombes, A. Walters, *Astron. Astroph.* **1991**, *244*, L47.

c) M. Elhanine, R. Farrena, G. Guelichvili, *J. Chem. Phys.* **1991**, *94*, 2529.

d) N. Goldberg, M. Iraqi, J. Hrusak, H. Schwarz, *Int. J. Mass Spectrom. Ion Processes* **1993**, *125*, 267.

[4] H. Bock, R. Dammel, *Angew. Chem.* **1985**, *97*, 128; *Angew. Chem., Int. Ed. Engl.* **1985**, *24*, 111; G. Gran, *J. Michl. Chem. Eng. News* **1985**, 12.

[5] R. Preuss, R. J. Buenker, S. D. Peyerimhoff, *J. Mol. Struct.* **1978**, *49*, 171.

[6] M. Samy El-Shall, *Chem. Phys. Lett.* **1989**, *159*, 21.

[7] C. Rüchardt, M. Meier, K. Haaf, J. Pakusch, E. K. A. Wolber, B. Müller, *Angew. Chem.* **1991**, *103*, 907; *Angew. Chem., Int. Ed. Engl.* **1991**, *30*, 893.

[8] M. S. Gordon, J. S. Binkley, J. A. Pople, W. J. Pietro, W. J. Hehre, *J. Am. Chem. Soc.* **1982**, *104*, 2797; J. S. Binkley, J. A. Pople, W. J. Hehre, *J. Am. Chem. Soc.* **1980**, *102*, 939; W. J. Pietro, M. M. Francl, W. J. Hehre, D. J. DeFrees, J. A. Pople, J. S. Binkley, *J. Am. Chem. Soc.* **1982**, *104*, 5039.

[9] Ab initio orbital calculations were performed using the GAUSSIAN 92 series of programs: M. J. Frisch, G. W. Trucks, M. Head-Gordon, P. M. W. Gill, M. W. Wong, J. B. Foresman, B. G. Johnson, H. B. Schlegel, M. A. Robb, E. S. Replogle, R. Goperts, J. L. Andres, K. Raghavachari, J. S. Binkley, C. Gonzalez, R. L. Martin, D. J. Fox, D. J. DeFrees, J. Baker, J. J. P. Stewart, J. A. Pople, GAUSSIAN 92, *Revision C, Gaussian Inc.*, Pittsburgh (**1992**).

[10] R. H. E. Eade, M. A. Robb, *Chem. Phys. Lett.* **1981**, *83*, 362.

[11] J. A. Pople, M. Head-Gordon, K. Raghavachari, G. W. Trucks, *J. Chem. Phys.* **1987**, *87*, 5968.

[12] Experimental and calculated n(Si–N): For HSiN: 1161 cm^{-1} (exp.), 936 cm^{-1} (MP2(fu)/6–31G* corrected by a factor of 0.9427), 1126 cm^{-1} (CASSCF/6–31G*), 1178 cm^{-1} (QCISD/6–31G*). For HNSi: 1198 cm^{-1} (exp.), 1115 cm^{-1} (MP2(fu)/6–31G* corrected by a factor of 0.9427), 1222 cm^{-1} (CASSCF/6–31G*), 1221 cm^{-1} (QCISD/6–31G*).

[13] K. B. Wiberg, *Tetrahedron* **1968**, *24*, 1083.

[14] J. A. Pople, M. Head-Gordon, D. J. Fox, K. Raghavachari, L. A. Curtiss, *J. Chem. Phys.* **1989**, *90* (10), 5622; L. A. Curtiss, K. Raghacachari, G. W. Trucks, J. A. Pople, *J. Chem. Phys.* **1991**, *94* (11), 7221.

[15] R. Walsh, in: *The Chemistry of Organic Silicon Compounds* (Eds: S. Patai, Z. Rappaport), Wiley, Chichester, **1989**, p. 371; W. E. Dasent, *Inorganic Energetics*, Penguin, Harmondsworth, England.

The Nature of Organosilicon Cations and Their Interactions

Christoph Maerker, Jürgen Kapp, Paul von Ragué Schleyer*
Institut für Organische Chemie
Friedrich-Alexander Universität Erlangen-Nürnberg
Henkestr. 42, D-91052 Erlangen, Germany

In sharp contrast to the many hypercoordinate silicon compounds, [1] trivalent organosilicon cationic species, called silyl, silicenium, or silylenium ions, have eluded isolation in condensed phases for more than 50 years [2]. Recently, both Lambert's group at Northwestern [3] and Reed's at USC [4] have made significant progress by solving X-ray structures of silyl cation-like moieties coordinated with rather weakly nucleophilic solvents or anions. These advances deserve the widespread publicity they received, [5] but have been beclouded by the controversies concerning the interpretations [6]. What is the nature of the silicon species in the X-ray structures? How "free" are these "silyl cations"? How should the degree of freedom be judged? The possibility that "free" silyl cations might be involved at all has been challenged by Schleyer et al. [7], Olsson and Cremer [8], Pauling (in his last scientific contribution) [9], and Olah et al. [6]. Houk provides an excellent review [10].

The elusiveness of "free" silyl cations in condensed phases stands in remarkable contrast to the behavior of the isoelectronic carbocations [11] which Olah's school has explored so extensively in super acid media [12]. Laube's splendid recent series of carbocation X-ray structures [13, 14] illustrate what to expect: truly "free" cations will be separated from the nearest atoms of the counteranions by 3 Å or so. Despite the larger size of silicon, the Si-ligand distances in the Lambert-Reed structures are only modestly longer than typical covalent bonds, and are in the range expected for dative bonds [15].

The physicochemical properties of silicon (Table 1) – the lower electronegativity in particular – should favor the generation of silicenium over carbonium ions. Indeed, this is exactly the situation in the gas phase where silyl cations have been well-characterized experimentally [16]. The ionization potentials of silyl radicals are much less than their alkyl radical conterparts. However, the association energies, not only with oxygen bases, but also with arenes are rather large (24–50 kcal mol^{-1}) [17]. Long before the interpretation of Lambert's crystal structure [3] was discussed, gas phase chemists had concluded that $R_3Si[arene]^+$ species were σ rather than π-complexes [18].

	C	Si
electronegativity according to		
Allred-Rochow[52]	2.5	1.7
Pauling[53]	2.5	1.8
covalent radius[19]	0.77 Å	1.17 Å
van-der-Waals radius[19]	1.85 Å	2.00 Å
Bond[19]	**Bond Length (Å)**	**Bond Energy (kcal mol^{-1})**
C–C	1.54	83
C–Si	1.87	72
Si–Si	2.32	54
C–H	1.09	81
Si–H	1.48	78
C–F[53]	1.37	105
Si–F	1.55	139
Ionization Potentials (eV) of R_3X° Radicals		
R = H	9.84 + 0.01[20]	8.01 + 0.02[21]
R = Me	6.34[a]	5.93[b]

Table 1. Physicochemical properties of carbon and silicon; [a] MP2–FU/6–31G*; [b] MP2–FC/6–31G*

Silyl cations, as well as many simple, but poorly known silicon species, can also be studied effectively computationally. Our current investigations afford new examples of how the chemical behavior of silicon and carbon can differ dramatically. For example, hypercoordinate silicon compounds like SiF_6^{2-} are well known, but have no carbon analogs. The lower electronegativity and larger size of silicon are responsible; "octet rule violations" are not involved [23]. The $SiX_3^+ - SiX_4 - SiX_5^- - SiX_6^{2-}$ -series (X = F, H) reveals no d-orbital participation in Si–X bonds, and NPA [22] orbital occupancy ratios of tetracoordinated SiX_4

compounds do not even follow the expected sp³ hybridization (Fig. 1). Even the hypercoordinate silicon compounds are approximately sp² hybridized. Silicon bonding is partially ionic. Note the large positive charges (NPA) on silicon in SiF_4, and even in the anions SiF_5^- and SiF_6^{2-}.

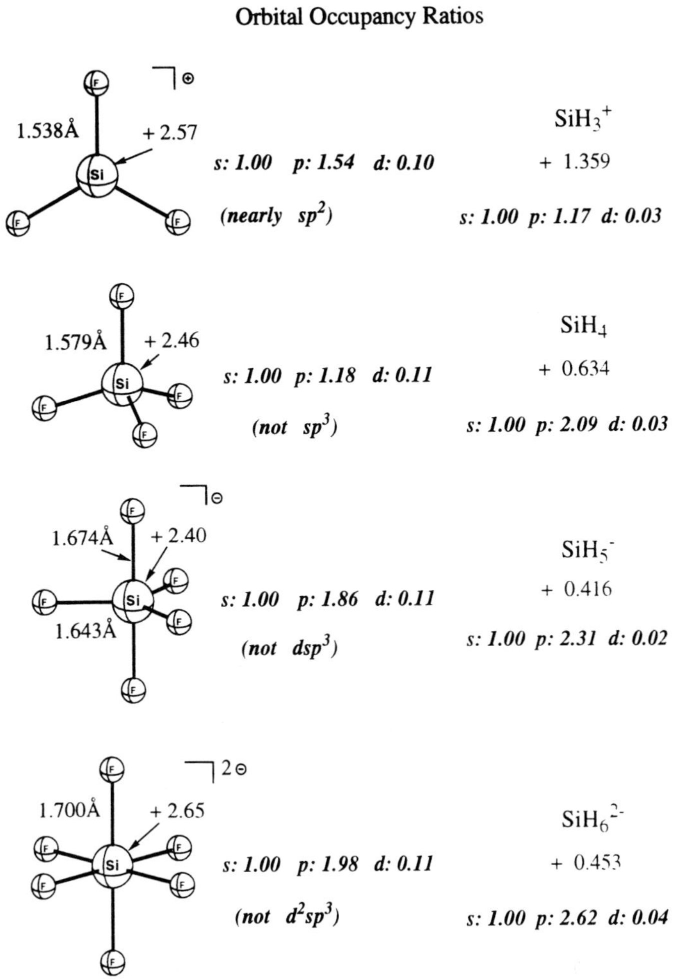

Orbital Occupancy Ratios

SiH_3^+ +2.57
s: 1.00 p: 1.54 d: 0.10
(nearly sp²)

SiH_3^+ +1.359
s: 1.00 p: 1.17 d: 0.03

SiH_4 +2.46
s: 1.00 p: 1.18 d: 0.11
(not sp³)

SiH_4 +0.634
s: 1.00 p: 2.09 d: 0.03

SiH_5^- +2.40
s: 1.00 p: 1.86 d: 0.11
(not dsp³)

SiH_5^- +0.416
s: 1.00 p: 2.31 d: 0.02

SiH_6^{2-} +2.65
s: 1.00 p: 1.98 d: 0.11
(not d²sp³)

SiH_6^{2-} +0.453
s: 1.00 p: 2.62 d: 0.04

Fig. 1. Silicon hybridizations and Si NPA charges

Differences even between the smallest silicon and carbon molecules can be quite startling. The CH_4 potential energy surface has only one minimum (methane, T_d), whereas SiH_4 has two: silane (T_d) and a silylene-hydrogen complex (C_s) [24]. The dissociation of SiH_4 into free hydrogen and silylene (with a *singlet* ground state) is much less endothermic than the methane dissociation into methylene (*triplet* ground state) and H_2 (Fig. 2). The trend towards the decreasing stability of the T_d AH_4 forms toward dissociation becomes more pronounced for the heavier group 14 metal hydrides. Likewise, the second minima, corresponding to singlet AH$_2$–H$_2$ complexes, become increasingly closer in energy to the tetrahedral form. Plumbane (T_d) is only 2.6 kcal mol^{-1} more stable than separated PbH$_2$ and H$_2$; their complex has nearly the same energy as the T_d minimum. While relativistic effects may contribute, [25], this trend can be attributed to the "inert pair effect", which favors the heavier A(II) species [26, 27].

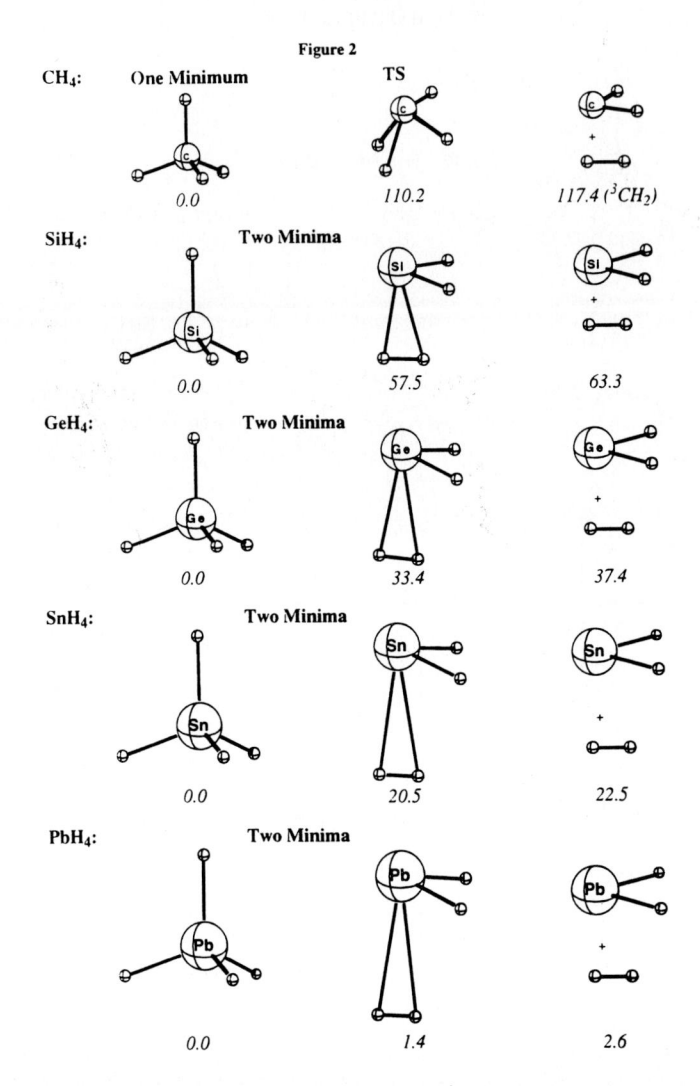

Figure 2

CH$_4$: One Minimum — 0.0 ; TS — 110.2 ; 117.4 (^{3}CH$_2$)

SiH$_4$: Two Minima — 0.0 ; 57.5 ; 63.3

GeH$_4$: Two Minima — 0.0 ; 33.4 ; 37.4

SnH$_4$: Two Minima — 0.0 ; 20.5 ; 22.5

PbH$_4$: Two Minima — 0.0 ; 1.4 ; 2.6

Fig. 2. CH$_4$⇒PbH$_4$: comparison of minima, rel. energies in kcal mol^{-1}; C, Si: MP2 all-electron calculations; Ge–Pb: Becke 3LYP pseudopotential calculations

The existence of this second AH$_4$ minimum also differentiates the stereomutation mechanisms of silane (as well as those of its heavier homologues) from methane. CH$_4$ can invert only via a single transition state (C_s) which is close in energy to complete dissociation into CH$_2$ and H$_2$ and higher in energy than CH$_3$ and H$^\bullet$ (Fig. 3) [28].

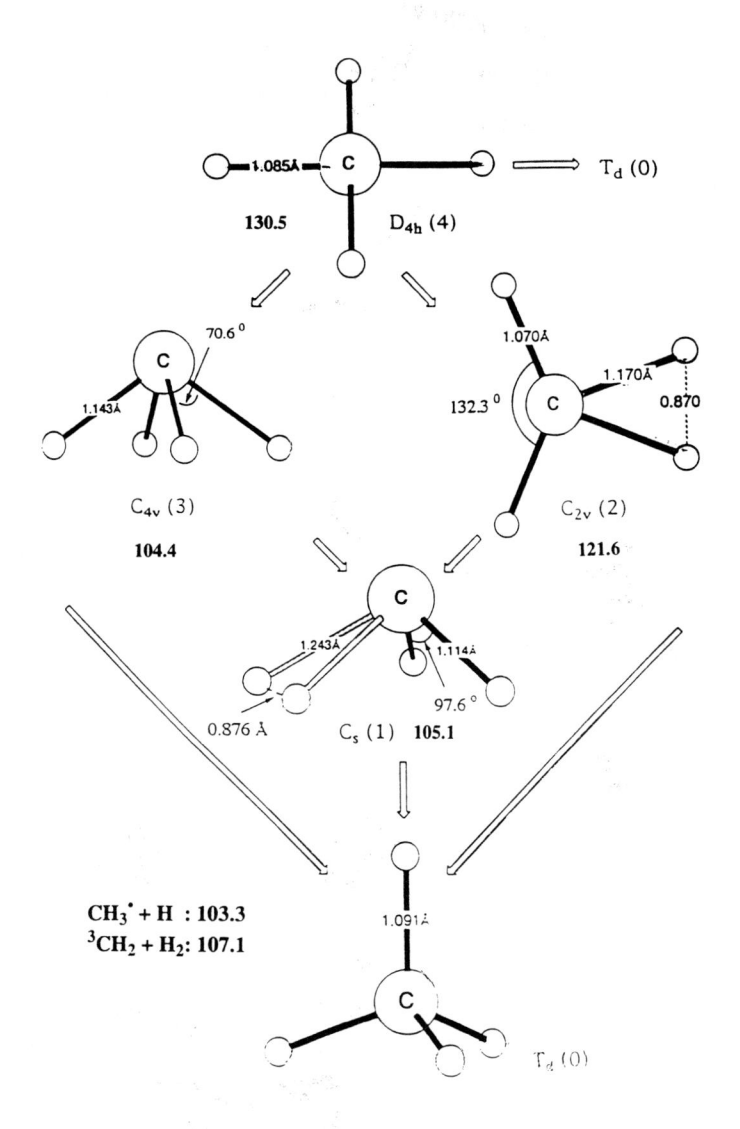

Fig. 3. Methane inversion at QCISD(T)/6–311+G(3df,2p)//CISD/6–311G**, rel. energies in kcal mol^{-1}; number of imaginary frequencies (NIMAG) in parentheses; NIMAG = 0 (minimum), 1 (transition state), 2 (second order saddle point)

In contrast, several 'routes' through different transition structures are possible for silane (Fig. 4):

1) Via square planar SiH$_4$ **SP** (but this process is unfavorable relative to the homolytic dissociation into SiH$_3^{\bullet}$ and H$^{\bullet}$),

2) Via the first C_s transition state **TS** to the SiH_2–H_2 complex minimum **COMP**. This may be followed by (a) a dissociation – recombination process **DISS**, by (b) rotation of bound H_2 **ROT**, or by (c) exchange of an H_2 and a SiH_2 hydrogen atom **EXC**.

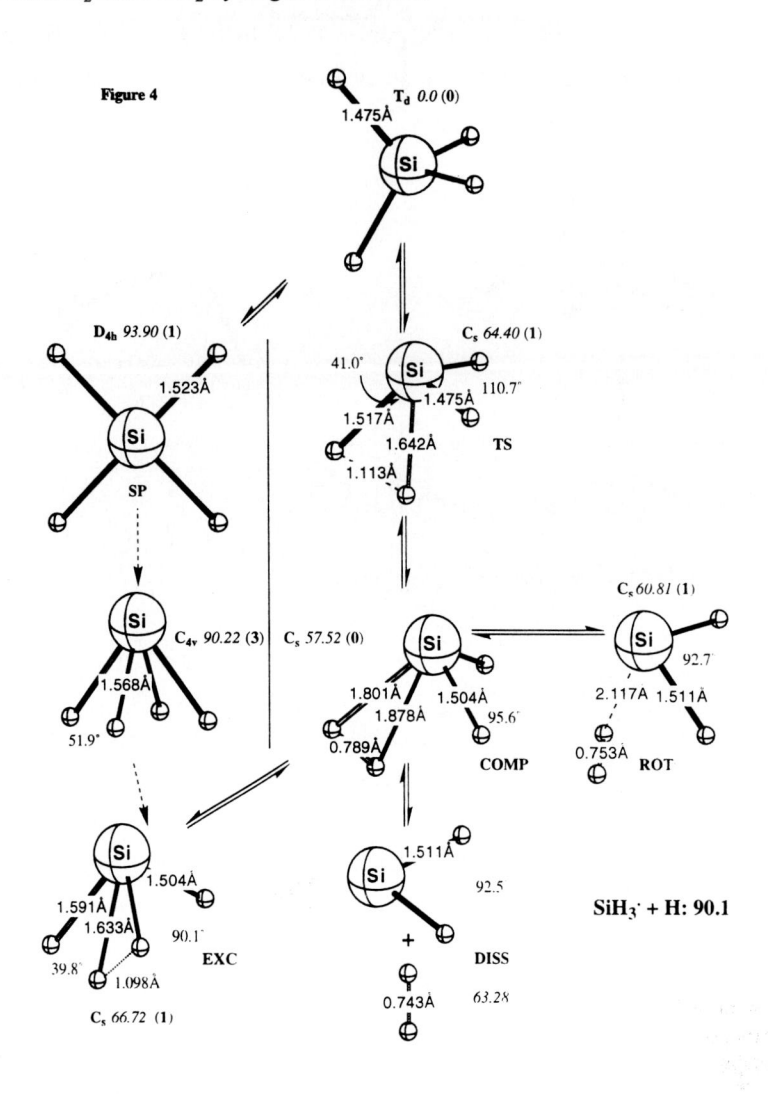

Figure 4

T_d 0.0 (0), 1.475Å

D_{4h} 93.90 (1), 1.523Å, **SP**

C_s 64.40 (1), 41.0°, 1.475Å, 1.517Å, 110.7°, 1.642Å, 1.113Å, **TS**

C_{4v} 90.22 (3), 1.568Å, 51.9°

C_s 57.52 (0), 1.801Å, 1.878Å, 0.789Å, 1.504Å, 95.6°, **COMP**

C_s 60.81 (1), 2.117Å, 1.511Å, 92.7°, 0.753Å, **ROT**

1.504Å, 1.591Å, 1.633Å, 90.1°, 39.8°, 1.098Å, **EXC**, C_s 66.72 (1)

1.511Å, 92.5°, 0.743Å, 63.28, **DISS**

$SiH_3^+ + H$: 90.1

Fig. 4. SiH_4 inversion at MP2(FC)/6–311G**, rel. energies in kcal mol^{-1}; number of imaginary frequencies (NIMAG) in parentheses; NIMAG = 0 (minimum), 1 (transition state), 2 (second order saddle point)

Not only the basic XH_4 neutral molecules, but also the XH_3^+ cations exhibit different structural preferences. Again the carbon case is the simplest: the methyl cation CH_3^+ exists only as a single D_{3h} symmetric isomer. Dissociation into H_2 and CH^+ is ca. 137 kcal mol^{-1} endothermic. However, the CH_3^+ PES does have a second minimum. This is a weakly bound (1.8 kcal mol^{-1}) side-on complex between the C–H$^+$ hydrogen and H_2.

SiH_3^+, on the other hand, possesses two minima; in both, all three hydrogens are bound to silicon. The D_{3h} silyl cation is ca. 33 kcal mol^{-1} lower in energy than a side-on HSi^+-H_2 complex. The binding energy of the latter is 7 kcal mol^{-1} relative to the dissociated $SiH^+ + H_2$ species. As with the neutral XH_4 compounds, the H_2-AH^+ complex becomes increasingly stable, relative to the D_{3h} AH_3^+ isomer, in going down group 14. The tin complex, H_2-SnH^+, is the absolute minimum; D_{3h} symmetric PbH_3^+ is even unstable toward dissociation (Fig. 5).

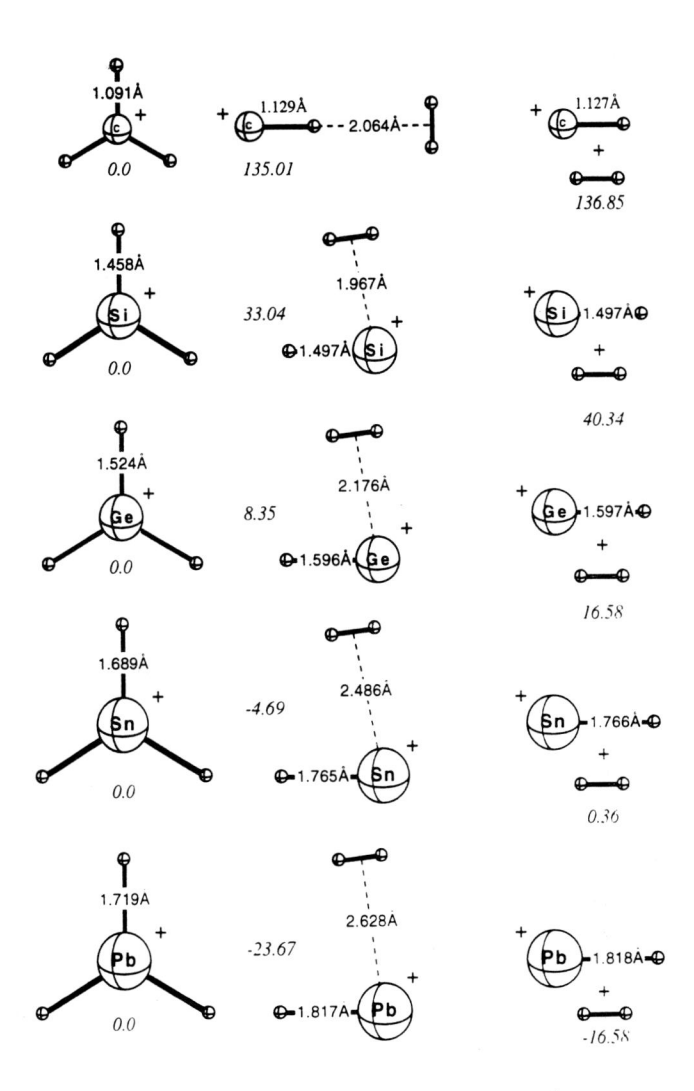

Fig. 5. $CH_3^+ \Rightarrow PbH_3^+$: comparison of both singlet minima, rel. energies in kcal mol^{-1}; C, Si: MP2 all–electron; Ge – Pb: Becke3LYP pseudopotential calculations

The preference for nonclassical structures is even more evident in the complex forms of the $X_3H_3^+$ cations (Fig. 6) [29]. There is only one cyclic minimum for $C_3H_3^+$, the well-known cyclopropenyl cation – but twelve minima have been located for $Si_3H_3^+$ [30]! When carbon compounds are forced to adopt unusual structures, the energies rise dramatically. But the carbon stability orders reverse for heavier group 14 elements and the hydrogen bridged structures become more stable than the "classical" forms. $Pb_3H_3^+$ can be described as a Pb_3^{2-} ring, in which the Pb–Pb bonds are protonated rather than the "inert" Pb(II) lone pairs. For better overlap, the 6p orbitals bend together and form a $2e^-$-3center (π) bond above the center of the ring, resulting in a C_{3v} symmetric minimum structure.

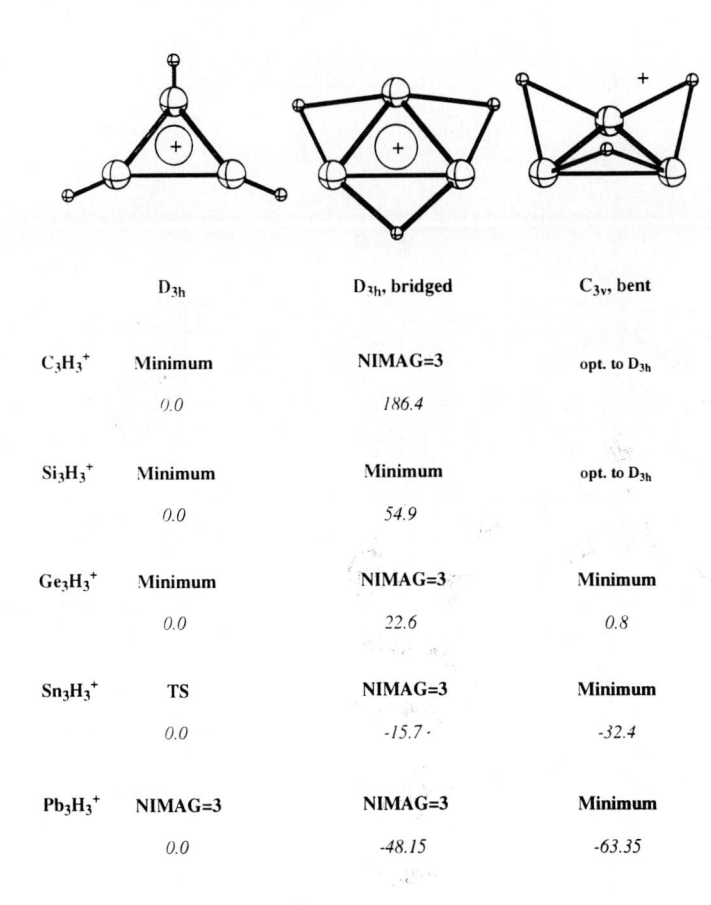

	D_{3h}	D_{3h}, bridged	C_{3v}, bent
$C_3H_3^+$	Minimum 0.0	NIMAG=3 186.4	opt. to D_{3h}
$Si_3H_3^+$	Minimum 0.0	Minimum 54.9	opt. to D_{3h}
$Ge_3H_3^+$	Minimum 0.0	NIMAG=3 22.6	Minimum 0.8
$Sn_3H_3^+$	TS 0.0	NIMAG=3 -15.7	Minimum -32.4
$Pb_3H_3^+$	NIMAG=3 0.0	NIMAG=3 -48.15	Minimum -63.35

Fig. 6. $A_3H_3^+$-Molecules, rel. energies in kcal mol^{-1}; C, Si: MP2 all-electron; Ge – Pb: Becke3LYP pseudopotential calculations

The differences in the chemical behavior of silicon, relative to carbon, is responsible for the elusiveness of SiR_3^+ compounds in condensed phases. Just is it that what hampers the observation of silyl cations, although they are intrinsically thermodynamically more stable than carbenium ions?

This question is hard to answer experimentally; the results of investigations of silyl cations (which cannot be observed directly in condensed phases) are ambiguous and difficult to interpret. In contrast, theoretical investigation even of highly reactive silyl species only is limited by the available programs and computer facilities. Apeloig has reviewed [31] the many applications of electronic structure methods [32] to organosilicon compounds, including silyl cations. A wealth of structural, energetic, and bonding information facilitates analysis, interpretation, and comparisons.

What are the relative thermodynamic stabilities of silyl cations vs carbenium ions? Eq. 1–3 provide estimates of the relative stabilities of carbenium and silicenium ions (silyl cations):

$$R-CH_2^+ + CH_4 \longrightarrow R-CH_3 + CH_3^+ \quad (1)$$
$$R-SiH_2^+ + SiH_4 \longrightarrow R-SiH_3 + SiH_3^+ \quad (2)$$
$$R-SiH_2^+ + R-CH_3 \longrightarrow R-SiH_3 + R-CH_2^+ \quad (3)$$

Previous theoretical and experimental gas phase investigations showed that SiH_3^+ is 53.7 kcal mol^{-1} more stable than CH_3^+ [31]. However, substitution, particularly by π-donors, reduces this difference: Table 2 updates prior results, [33] at the higher MP2(FC)/6–31G* level of theory, with a set of second and third period groups.

Table 3 summarizes additional stabilization energies, produced by multiple substitution, and also for individual systems discussed in the text. Multiple substitution is less effective in silyl cations. Note the different trend in stabilization energies of $R_n-SiH_{3-n}^+$ species vs SiH_3^+ and vs $R_n-CH_{3-n}^+$. Again, NH_2 groups stabilize carbenium ions to such a great extent that the inherent energetic advantages of silyl cations are overcome.

Eq. 1–3 can be rewritten in generalized form (1a, 2a, 3a) (n = 0–3; where n number of substituents).

$$R_n-CH_{3-n}^+ + CH_4 \longrightarrow R_n-CH_{4-n} + CH_3^+ \quad (1a)$$
$$R_n-SiH_{3-n}^+ + SiH_4 \longrightarrow R_n-SiH_{4-n} + SiH_3^+ \quad (2a)$$
$$R_n-SiH_{3-n}^+ + R_n-CH_{4-n} \longrightarrow R_n-SiH_{4-n} + R_n-CH_{3-n}^+ \quad (3a)$$

R	Stabilization Energy According to		
	Eq. 1	Eq. 2	Eq. 3
H	0.0	0.0	57.4
Li	75.3	57.8	39.8
BeH	15.1	14.3	56.5
BH_2, perp.	25.9	13.7	45.1
CH_3	40.6	15.1	31.8
NH_2, planar	100.5	37.5	−5.6
OH	66.3	19.1	10.1
F	25.3	−1.1	31.0
SiH_3	16.5	12.3	53.2
PH_2	63.4	17.6	11.6
SH	63.6	18.5	12.2
Cl	29.2	2.2	30.3
Ph	76.2[a]	31.5[b]	
Ph, perp.	27.2[a]	16.5[b]	
cyclopropyl	72.4	27.7	12.6

Table 2. Stabilizing effects of first and some second row substituents on carbenium ions and silyl cations (kcal mol^{-1}); MP2–FC/6–31G*); [a] MP2–FULL/6–31G*; [b] HF/6–31G

R_n	Stabilization Energy According to		
	Eq. (1a)	Eq. (2a)	Eq. (3a)
CH_3	40.6	15.1	31.8
$(CH_3)_2$	58.7	27.9	26.6
$(CH_3)_3$	74.8	38.4	21.0
SiH_3	16.5	12.3	53.2
$(SiH_3)_2$		20.0	
$(SiH_3)_3$		24.9	
NH_2	100.5	37.5	−5.6
$(NH_2)_3$	135.0[a]	63.5 (69.4)[a]	−22.8[a]
2–norbornyl	81.9	42.6	18.1
7–norbornadienyl (C_s)	96.1	53.8	15.1

Table 3. Stabilizing effects of multiple substitution and special environments on carbenium ions and silyl cations; (kcal mol^{-1}; MP2–FC/6–31G*); [a] HF/6–31G*

Besides being stabilized to a smaller extent by substituents, silyl cations are far less shielded sterically by bulky groups than carbocations (Fig. 7).

Fig. 7. CMe$_3^+$ vs SiMe$_3^+$: MP2(FC)/6–31G* geometries steric protection of the cationic center; note the greater exposure at Si$^+$

The much more highly charged silicon atom can interact far more readily with nucleophiles. Silyl cations may even be complexed simultaneously and symmetrically by two electron pair donors (hypercoordination), in contrast to carbocations. With ammonia, the methyl cation gives the very stable protonated methyl amine, $H_3C-NH_3^+$; a second ammonia molecule is only weakly bound to this complex. If both NH_3 groups are forced to be equidistant from carbon, a S_N2 transition state results, 20 kcal mol^{-1} higher in energy than the minimum.

In contrast, two NH_3 molecules add to SiH_3^+; the energy of the second addition to $H_3NSiH_3^+$ is smaller than that of the first, but is still very large (ca. 40 kcal mol^{-1}). The pentacoordinated silicon minimum possesses D_{3h} symmetry, i.e. both Si–N distances are equal (Fig. 8).

Energies in kcal/mol

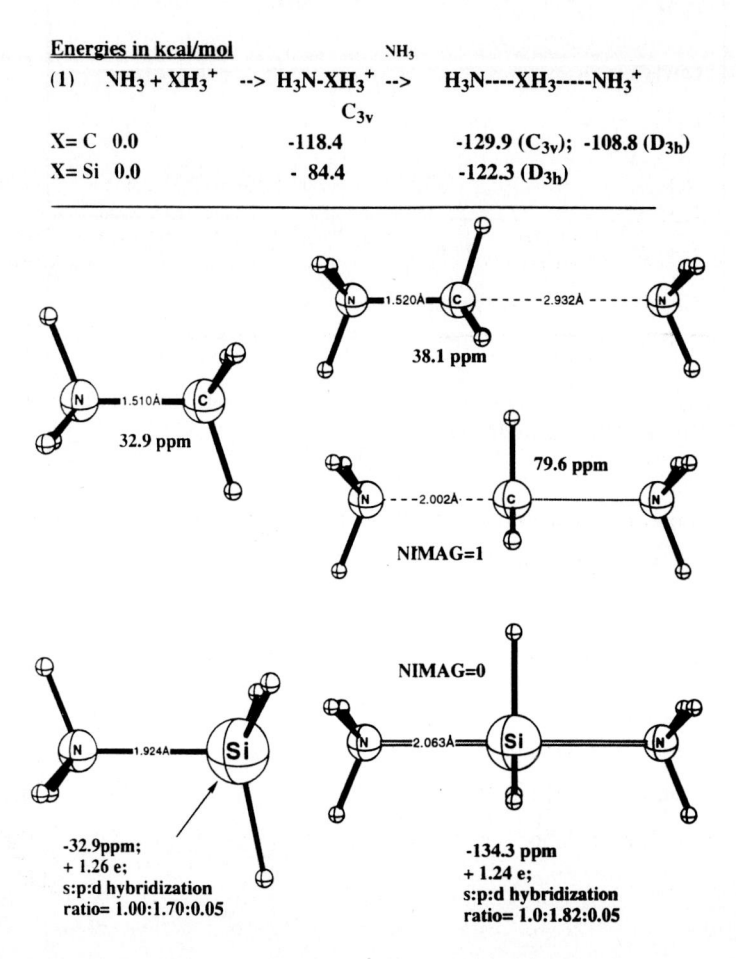

(1) $NH_3 + XH_3^+$ --> $H_3N\text{-}XH_3^+$ --> $H_3N\text{----}XH_3\text{-----}NH_3^+$
                              C_{3v}

X= C 0.0	-118.4	-129.9 (C_{3v}); -108.8 (D_{3h})
X= Si 0.0	- 84.4	-122.3 (D_{3h})

Fig. 8. NH_3 complexes of CH_3^+ and SiH_3^+ complexation energies in kcal mol^{-1}; (IGLO/II//MP2(FC)/6–31G* δ ^{29}Si NMR vs TMS, IGLO/DZ//MP2(FC)/6–31G* δ ^{13}C NMR vs CH_4)

Double coordination by Lewis bases also is known experimentally: SiH_3^+ binds two H_2 molecules symmetrically to form SiH_7^+ [34]. Complexation energies are 14 kcal mol^{-1} for the first hydrogen and 7 kcal mol^{-1} for the second (Fig. 9) [35]. The binding energy of H_2 with the methyl cation to give CH_5^+ is much larger, but a second H_2 forms only a weakly bound $CH_5^+ \cdots H_2$ complex [36].

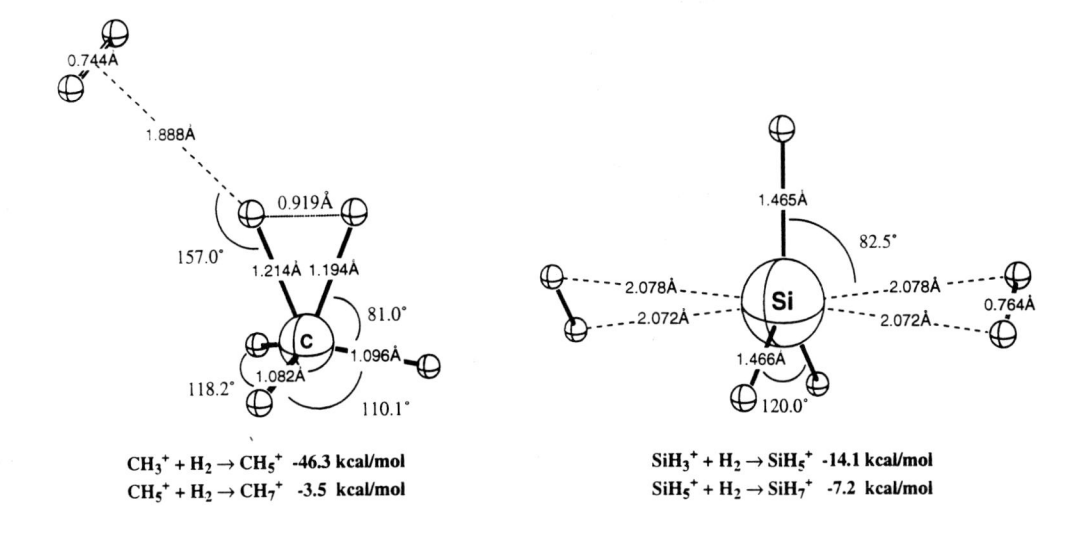

$CH_3^+ + H_2 \rightarrow CH_5^+$ -46.3 kcal/mol
$CH_5^+ + H_2 \rightarrow CH_7^+$ -3.5 kcal/mol

$SiH_3^+ + H_2 \rightarrow SiH_5^+$ -14.1 kcal/mol
$SiH_5^+ + H_2 \rightarrow SiH_7^+$ -7.2 kcal/mol

Fig. 9. CH_7^+ vs SiH_7^+: $CH_5^+\cdots H_2$ and $H_2-SiH_3^+-H_2$ structures; experimental complexation energies and computed geometries [34–36] saddle point

Silicon compounds are easily ionized, but as both these experimental and theoretical findings demonstrate, the resulting silyl cations are highly prone towards complexation. While silyl cations are thermodynamically stable, it is very difficult to obtain free SiR_3^+ ions, not coordinated to any counterions or solvents in the environment.

Experimental observation of silyl ions in condensed phases is difficult because:

1) Substituent stabilization is less effective
2) Steric shielding requires *very* large groups
3) The "appetite" for nucleophiles (leaving groups, counterions and solvent molecules) is voracious [10]
4) Addition-elimination involving higher coordinated silicon intermediates offers alternative mechanisms

There are the additional experimental problems of solubility as well as the difficulties in generating silyl cations from, e.g., hindered precursors, in condensed phases.

How can be assertained whether a silyl cation is "free" or not? The ^{29}Si NMR chemical shift may be the best criterion. While chemical shielding is influenced by other factors, there appears to be at least a rough relationship between the association energies of silyl cations and ^{29}Si chemical shifts. Indeed, the *ab initio* computation of chemical shifts has been proved a powerful tool for structure determination, [37, 38]. "The *ab initio*/IGLO/NMR approach rivals modern-day X-ray determinations of carborane structures in accuracy" [39]. IGLO calculated ^{29}Si NMR relative chemical shifts [38] nicely reproduce the experimental values over the whole range of organosilicon compounds (Table 4).

Species	IGLO[38]	Experiment[54]
SiH_4	−103.3	−93.1
$MeSiH_3$	−70.0	−65.2
$(Me)_2SiH_2$	−42.4	−41.5
$(Me)_3SiH$	−18.9	−18.5
$(Me)_4Si$	0.0	0.0
H_3SiF	−16.9	−17.4
H_2SiF_2	−41.6	−28.5
$HSiF_3$	−81.3	−77.8
SiF_4	−111.8	−109.9
Si_2H_6	−113.8	−104.8
Si_2F_6	−74.6	−77.9

Table 4. IGLO/II computed ^{29}Si NMR relative chemical shifts of several organosilicon compounds vs TMS

The calculated ^{29}Si chemical shifts of alkyl substituted Si cations are about 350 ppm, and only depend somewhat on the basis sets and level of geometry optimization [40]. The experimentally obtained shifts on "silyl cation" complexes can be compared with those calculated shifts for model structures.

Distances – both, calculated and those from crystal structures – are ambiguous, since significant bonding to silicon can occur at large separations (e.g., SiR_3^+ ···toluene complex, 2.18 Å) [3, 7]. Atomic charges depend on the method of calculation and are not easily interpreted. Silicon is much more 'cationic' than carbon in similar structures. Electropositive silicon is always quite charged in neutral and even in the hypervalent anions and dianions (Fig. 1).

To find out how free a SiR$_3^+$ cation would be in contact with inert solvents or even noble gas matrices, we optimized the geometries of complexes of SiH$_3^+$ with methane, an "inert" aliphatic solvent model, as well as the noble gases, He, Ne, and Ar, and computed the Si chemical shifts.

The noble gas complexes emphasize the problems of obtaining truly "free" silyl cations. Despite the large Ng···SiH$_3^+$ separations as well as the small complexation energies and degree of charge transfer, the ^{29}Si chemical shifts deviate strongly from the free SiH$_3^+$ value (Fig. 10).

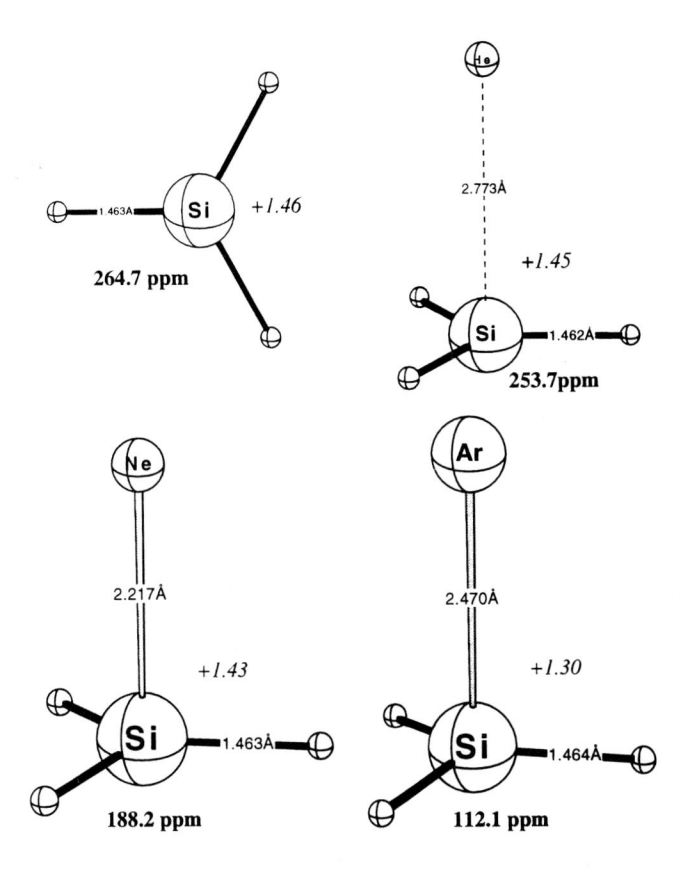

Fig. 10. SiH$_3^+$···noble gas complexes; IGLO/II//MP2(FC)/6–31G* δ ^{29}Si NMR vs TMS, *NPA, charges*

These deviations are still appreciable for the more realistic Me$_3$Si$^+$···noble gas complexes (Fig. 11). It is not likely that "free" silyl cations will be present in matrix isolation, in argon and the heavier noble gases.

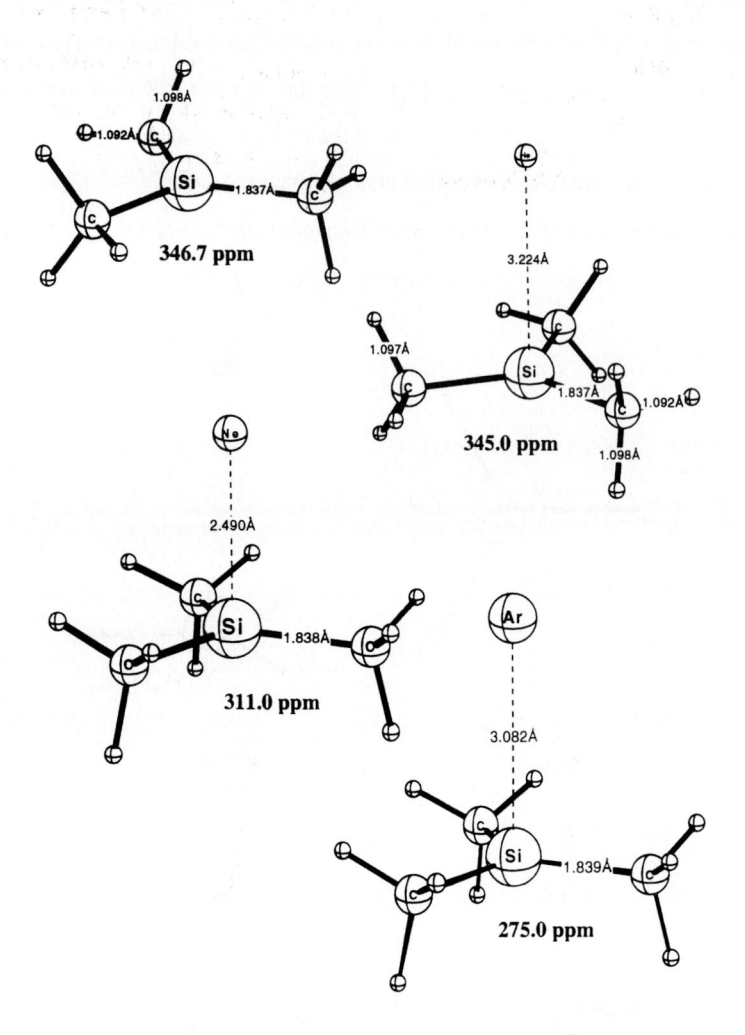

Fig. 11. SiMe₃⁺···noble gas complexes; IGLO/II/MP2(FC)/6–31G* δ ²⁹Si NMR vs TMS

Methane, a model for an aliphatic hydrocarbon solvent, is more strongly bound than argon. Accordingly, SiH₃⁺···CH₄ complexes have ^{29}Si chemical shifts far away from the free silyl cation value (Fig. 12).

When the Si···C distance is increased systematically (as might happen with bulky substituents), the computed ^{29}Si shifts approach the "free" value. But even at the large separation of 3 Å, the deviation from that of free SiH₃⁺ (Fig. 13) is still appreciable. Thus, it seems unlikely that free silyl cations can exist in solution, not even in the most non-nucleophilic solvents, unless, perhaps, very bulky substituents hinder coordination. But these bulky substituents, as we show below, are likely to coordinate with the Si⁺ centers intramolecularly.

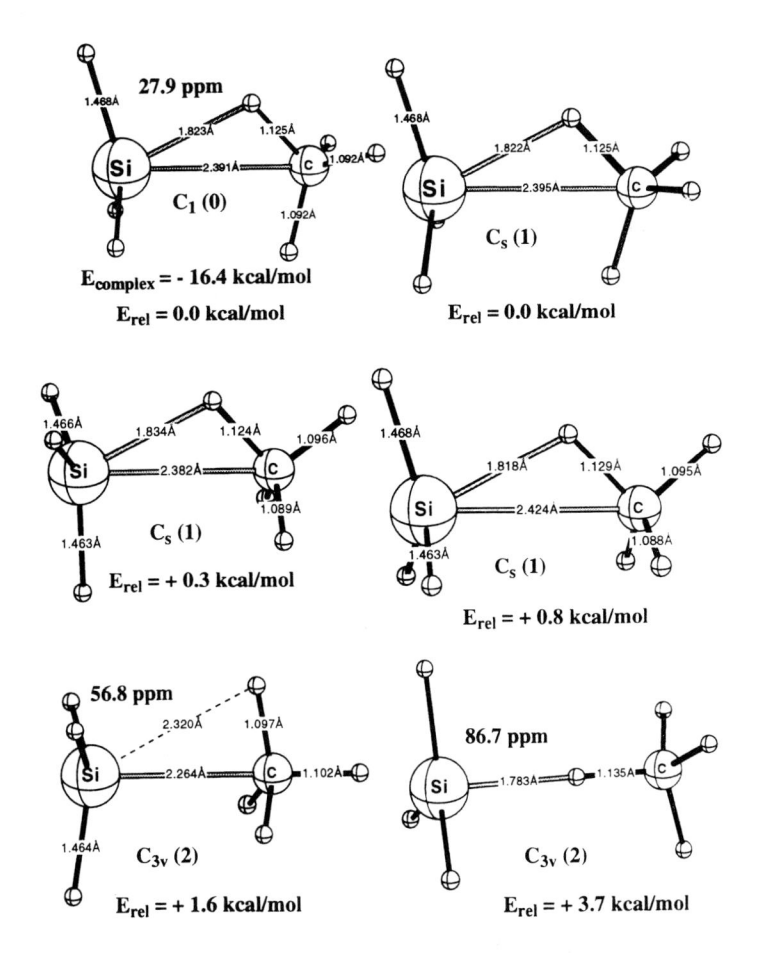

Fig. 12. $SiH_3^+ \cdots CH_4$ complexes; IGLO/II//MP2(FC)/6-31G* δ ^{29}Si NMR vs TMS; number of imaginary frequencies (NIMAG) in parentheses: NIMAG = 0 (minimum), 1 (transition state), 2 (second order)

Since silyl cations coordinate to "inert" species like methane or the noble gases, it is little wonder that the organosilicon species which have been observed in condensed phases are not really silyl cations. Strong interaction with solvent molecules or counterions always are present.

Last year we modeled Lambert's triisopropylsilyl system [3] theoretically and showed such systems to be toluene$\cdots SiR_3^+$ complexes [7]. The computed Si$\cdots$ring distance, 2.14 Å (Fig. 14), was close to the X-ray value of 2.18 Å. The Wiberg bond index between silicon and the *ipso* toluene carbon is appreciable (0.44), as is the charge transfer of +0.29 to the ring system. As already deduced from gas-phase determinations of the association energies, [18] the interaction between the fragments of the complex is strong. The earlier conclusions of the gas phase chemists were correct: Such species are best described as σ rather than as π complexes. The excellent agreement between the computed geometries and the ^{29}Si chemical shifts with experimental results guarantees that the real system is correctly represented by the *ab initio* calculation. Independent theoretical investigations by Olsson and Cremer [8], and by Olah's group [6], as well as an elaborate recent gas-phase study by Fornarini *et al.* [18] support our conclusions.

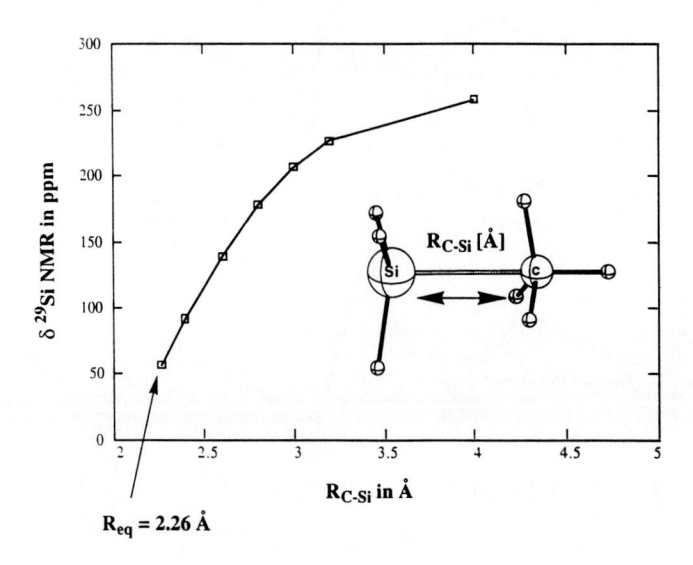

Fig. 13. SiH$_3^+$···CH$_4$ complex (face coordination), dependance of δ ^{29}Si NMR on the heavy atom distance; (IGLO/II//MP2(FC)/6-31G* vs TMS)

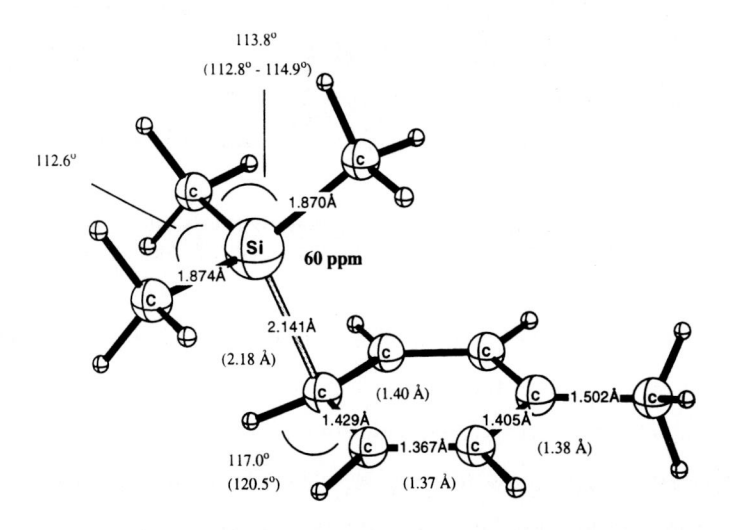

Fig. 14. SiMe$_3^+$···toluene complex; HF/6-31G* geometry; experimental values in parentheses

The latest X-ray structures pertaining to the "silyl cation" problem were reported by Reed [4]. He employed the "perhaps most non-nucleophilic anion" Br$_6$CB$_{11}$H$_6^-$, a polybrominated carborane, as the counterion. Nevertheless, the distance between the silicon and a bromine atom of the counterion was 2.48 Å (appropriate for a dative complex [15]), and the ^{29}Si chemical shift was 109.8 ppm, far from the

350 ppm value expected for a free trialkyl silyl cation. Reed's experimental chemical shift as well as his Si–Br distance are matched well by our calculated values for the $Me_3Si^+ \cdots BrCH_3$ model system, 102.1 ppm and 2.46 Å, respectively (Fig. 15).

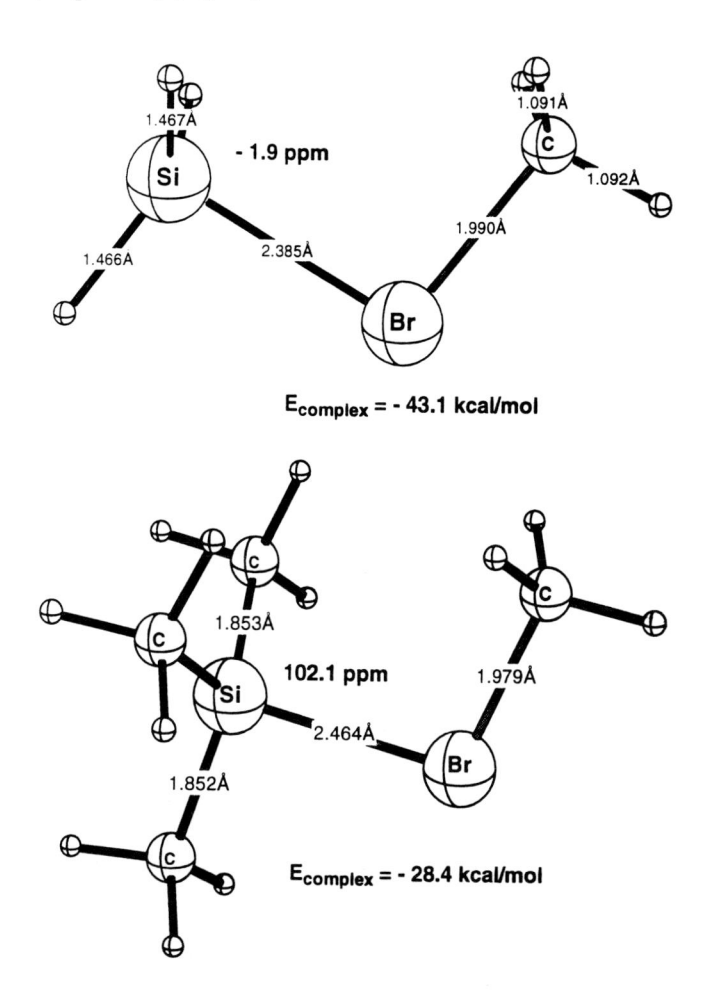

Fig. 15. $SiR_3^+ \cdots Br-CH_3$ complexes; IGLO/II//MP2/C, Si, Br (4,7–Ve–MWB–ECP, dz+p) H(dz)

The same silyl ion also was examined in condensed phases by Lambert with his perfluorinated tetraphenylborate counterion [41]. While the solid state structure of this complex is not known, its ^{29}Si shift is 107.6 ppm, close to the Reed's value. As the silicon atom might be coordinated to one or to two fluorine atoms [42], we calculated model complexes of $SiMe_3^+$ with methyl fluoride and with cis-difluorethene. The smaller fluorine atoms are able to approach closer to silicon than bromine (Fig. 16), but the shift in electron density towards the cationic moiety is negligible.

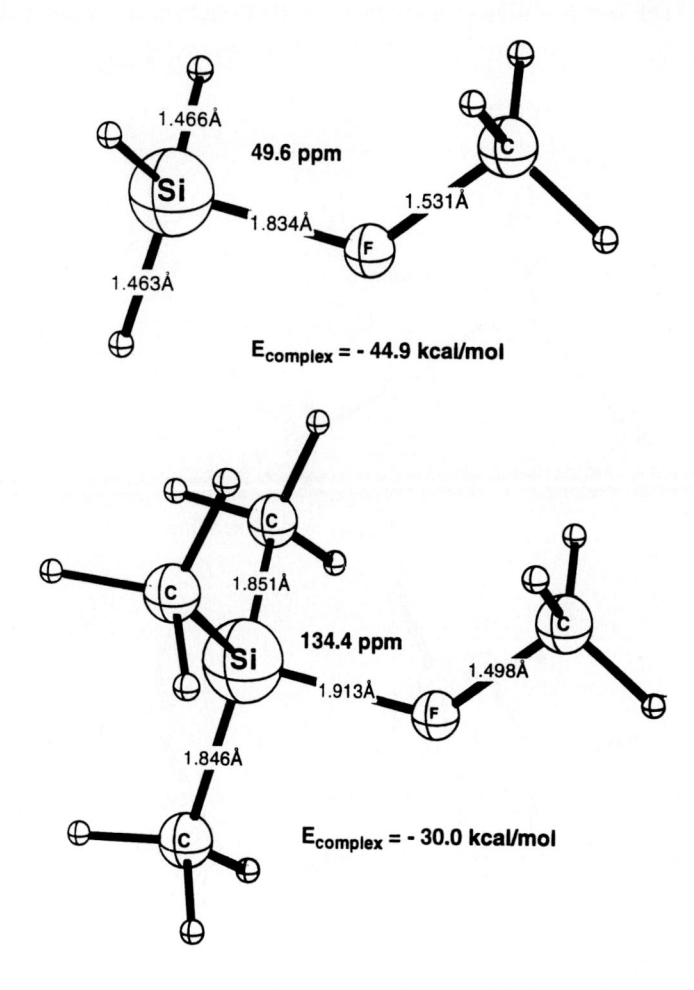

Fig. 16. SiR$_3^+$···F–CH$_3$ complexes; IGLO/II//MP2(FC)/6–31+G*

Thus the fluorine complexes are less covalent and more ionic than the bromine coordinated species due to the larger F/Si electronegativity difference. However, the complexation energies of CH$_3$F are similar to those of CH$_3$Br to Me$_3$Si$^+$ (ca. 30 kcal mol^{-1}, Table 5). The overview of complexation energies in Table 5 indicates that benzene and toluene compete well with these alkyl halides.

	CH_3^+	tBu^+	SiH_3^+	$SiMe_3^+$
He	−0.7		−0.3	−0.4
Ne	−5.9		−6.8	−4.4
Ar	−12.3		−7.1	−2.6
CH_4	−37.6[a]		−16.4	
CH_3F			−44.9	−30.0
CH_2F_2			−36.5	
CH_3Br			−43.1	−28.4
$CH_2=CH_2$				−47.9
cis-CHF=CHF			−36.5	−23.7
NH_3	−103.9[b]	−40.8[b]	−84.4	
	−105.8[c]	−40.0[d]		
2x NH_3	−129.9		−122.3	
C_6H_6	−86.4		−54.8[e]	−31.1[f]
				−23.9[g]
toluene				−34.2[f]
				−28.4[h]
H_2O	−71.2[b]	−13.5[b]		−62.8
	−68.9[c]	−11.2[d]		

Table 5. Complexation energies (kcal mol^{-1}) of CH_3^+, SiH_3^+, and $SiMe_3^+$ with several nucleophiles (MP2–FC/6–31G*); [a] MP2–FU/6–31G*, [55]; [b] MP4SDQ–FC/6–311+G*//MP2–FC/6–31G* + ZPVE, [57]; [c] experimental value, [20]; [d] experimental value, [58]; [e, f] MP2–FC/6–31G*//HF/6–31G*, [7]; [g] experimental value, [17]; [h] experimental value, [56]

The IGLO chemical shifts for these fluorine models were at least 25 ppm down field from the experimental ^{29}Si chemical shift (Fig. 16, 17). Nevertheless Lambert's system does not involve free silyl cation either. The experimental system is even more strongly coordinated. Furthermore, our calculations indicate that coordination by a single fluorine atom is preferred, since double coordinated structures have larger down field shifts and smaller complexation energies (Fig. 17). Difluorethene cannot approach the SiR_3^+ fragment as closely as methyl fluoride.

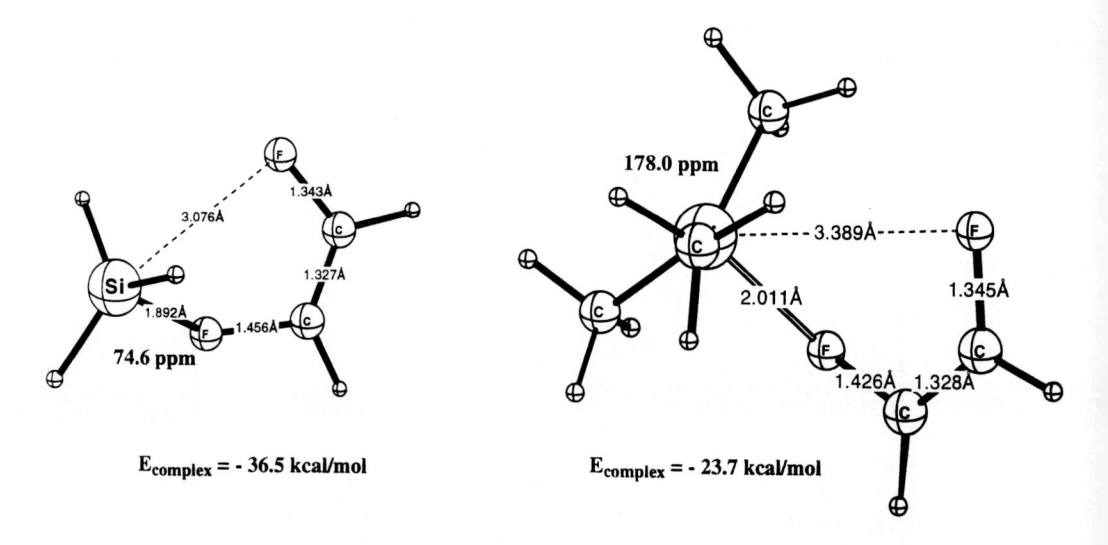

Fig. 17. $SiR_3^+ \cdots$ cis-difluorethene complexes; IGLO/II//MP2(FC)/6–31+G*

Since the *intermolecular* associations are so strong, *intramolecular* interactions also can be expected. Three famous carbocations having pronounced *intramolecular* interactions provide analogies: the 2-norbornyl cation, the 7-norbornadienyl cation, and the bisected cyclopropylcarbinyl cation [43]. To what extent do their sila congeners benefit from *intramolecular* stabilization of the cationic center? The 2-norbornyl carbocation was the focus of the nonclassical carbocation debate [44]. Both theoretical and experimental evidence has established its symmetrically bridged structure [45]. All three carbocations show remarkably shielded ^{13}C NMR signals for the carbocationic centers due to the participation of neighboring bonds. Are the structures of the silicon analogs nonclassical? Do they also show unusually highfield ^{29}Si NMR chemical shifts?

MP2–FC/6–31G* calculations reveal the symmetrically bridged 6-sila-2-norbornyl cation **1** not only to be a local minimum (Fig. 18), but also to be 17.2 kcal mol^{-1} more stable than the 2-norbornyl cation (Eq. 3) at MP2–FC/6–31G* + ΔZPE(SCF/6–31G*$\cdots$0.89). However, the inherently greater stability of silyl cations contributes to this difference. The ^{29}Si NMR chemical shift of the bridging silicon atom, ca 1 ppm vs TMS (IGLO/II'//MP2–FC/6–31G*), is very strongly shielded in comparison with ca 300 ppm expected for a free $RSiH_2^+$ species [38]. Thus, the sila congener of the 2-norbornyl carbocation also possesses a nonclassical structure which is reflected by its structure as well as its NMR properties.

The theoretical energies indicate that the 6-sila-2-norbornyl cation **1** might be prepared from cyclopentenyl precursors (π route, Fig. 18). The nonclassical structure is favored thermodynamically by ca 36 kcal mol^{-1} (MP2–FC/6–31G*) over **2**.

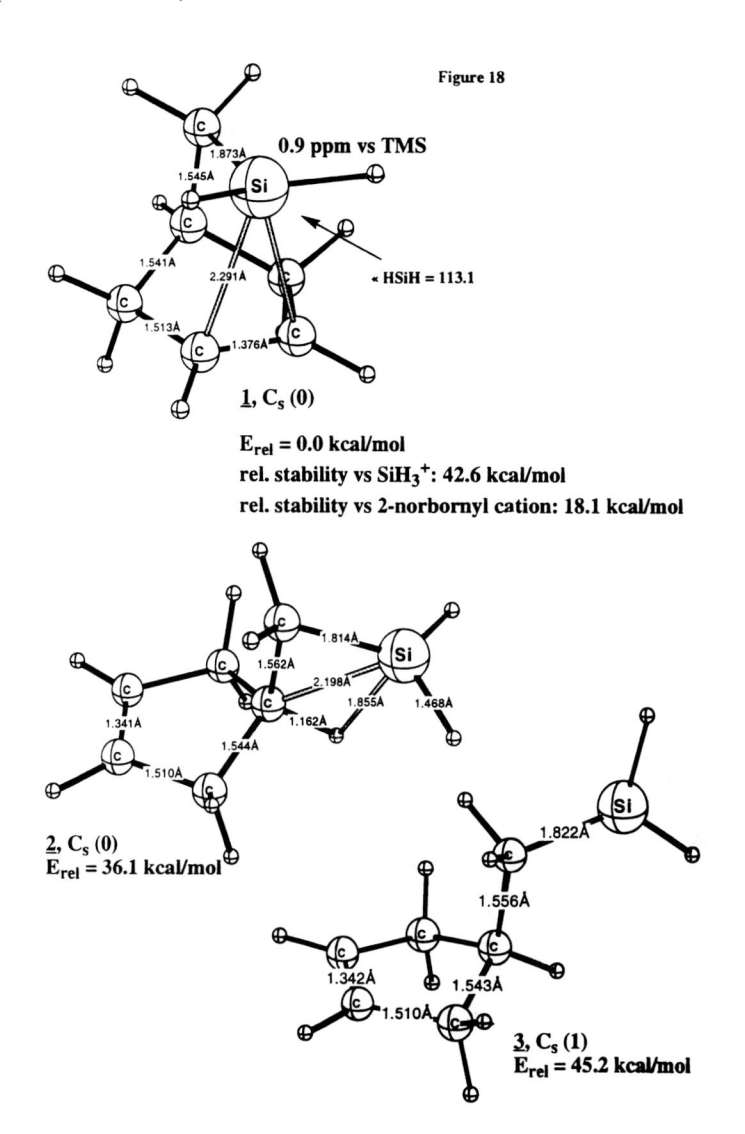

Figure 18

0.9 ppm vs TMS

∢HSiH = 113.1

1, C$_s$ (0)

E$_{rel}$ = 0.0 kcal/mol
rel. stability vs SiH$_3^+$: 42.6 kcal/mol
rel. stability vs 2-norbornyl cation: 18.1 kcal/mol

2, C$_s$ (0)
E$_{rel}$ = 36.1 kcal/mol

3, C$_s$ (1)
E$_{rel}$ = 45.2 kcal/mol

Fig. 18. Silicon congener of the 2-norbornyl cation; IGLO/II'//MP2(FC)/6–31G* δ ^{29}Si NMR vs TMS

Auner et al. described experiments which could provide another route to **1** [46]. Joint efforts are now in progress to provide evidence for the intermediacy of this cation.

In 1960, Winstein reported a record-breaking anchimerically assisted solvolysis rate acceleration: 7-norbornadienyl precursors reacted 10^{14} times faster than their saturated 7-norbornyl analogs [47]. The cation favors a C_s rather than a C_{2v} structure, [47, 48] since interaction between the cationic center and only one double bond is more effective than the simultaneous interation with both double bonds. The alternative C_{2v} structure serves as transition state for bridge flipping.

The 7-sila-7-norbornadienylcation congener, **4**, has a bent structure similar to the 7-norbornadienyl cation (Fig. 19). The silicon chemical shift is strongly shielded in the equilibrium C_s structure, but it is deshielded in the C_{2v} transition state **5** (while the chemical shifts of transition structures cannot be observed, the calculated values provide useful interpretive information [49]). The relative stability of **4** vs the 7-norbornadienyl cation is 15.1 kcal mol^{-1} at MP2–FC/6–31G* according to Eq. 3.

4, C_s (0)
E_{rel} = 0.0 kcal/mol
rel. stability vs SiH$_3^+$ = 53.8 kcal/mol
rel. stability vs 7-norbornadienyl cation (C_s)= 15.1kcal/mol

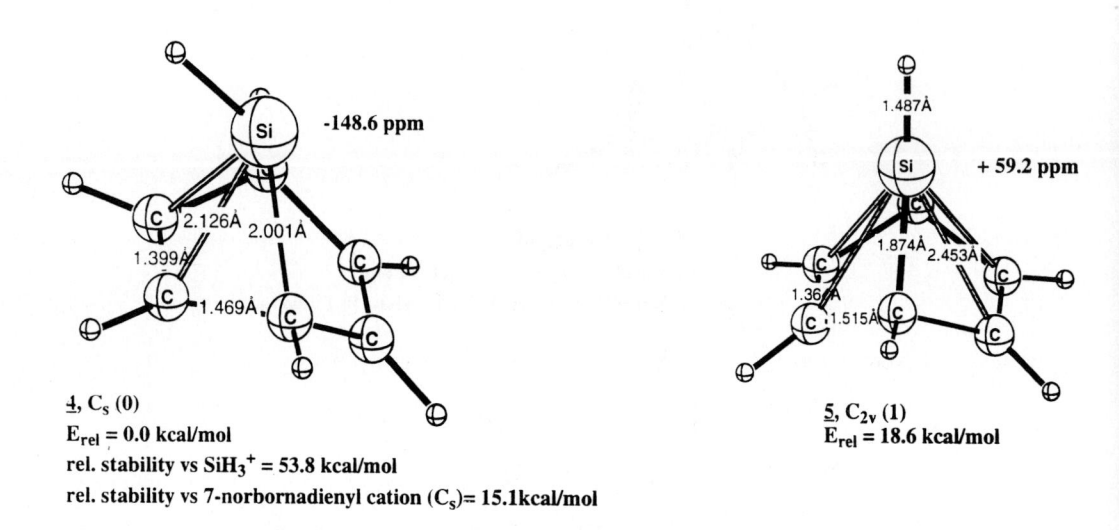

5, C_{2v} (1)
E_{rel} = 18.6 kcal/mol

Fig. 19. 7-Sila-7-norbornadienyl cation IGLO/II'//MP2(FC)/6–31G* δ ^{29}Si NMR vs TMS

Hyperconjugative stabilization of silyl cations is less effective (Fig. 20; Table 2). Geometrical distortions of the cyclopropyl group are smaller than in the related carbocation, and the ^{29}Si NMR chemical shift of 255.7 ppm not far from the ca 300 ppm for RSiH$_2^+$. The cyclopropyl hyperconjugative stabilization is 27.7 kcal mol^{-1} vs the parent silyl cation (MP2–FC/6–31G*), but only 12.6 kcal mol^{-1} more than for a methyl group (Table 2). However, hyperconjugation by a cyclopropyl group is competitive with phenyl conjugation (31.5 kcal mol^{-1} vs SiH$_3^+$ at SCF/6–31G*). Due to the silyl cation preference, the cyclopropylsilyl cation **7** is 12.6 kcal mol^{-1} more stable than the cyclopropylcarbinyl cation at MP2–FC/6–31G*. Hyperconjugation by Si–Si bonds in the tetrasilacyclopropylcarbinyl cation **8** appears to be quite ineffective; note the smaller Si–Si bond distances (Fig. 20).

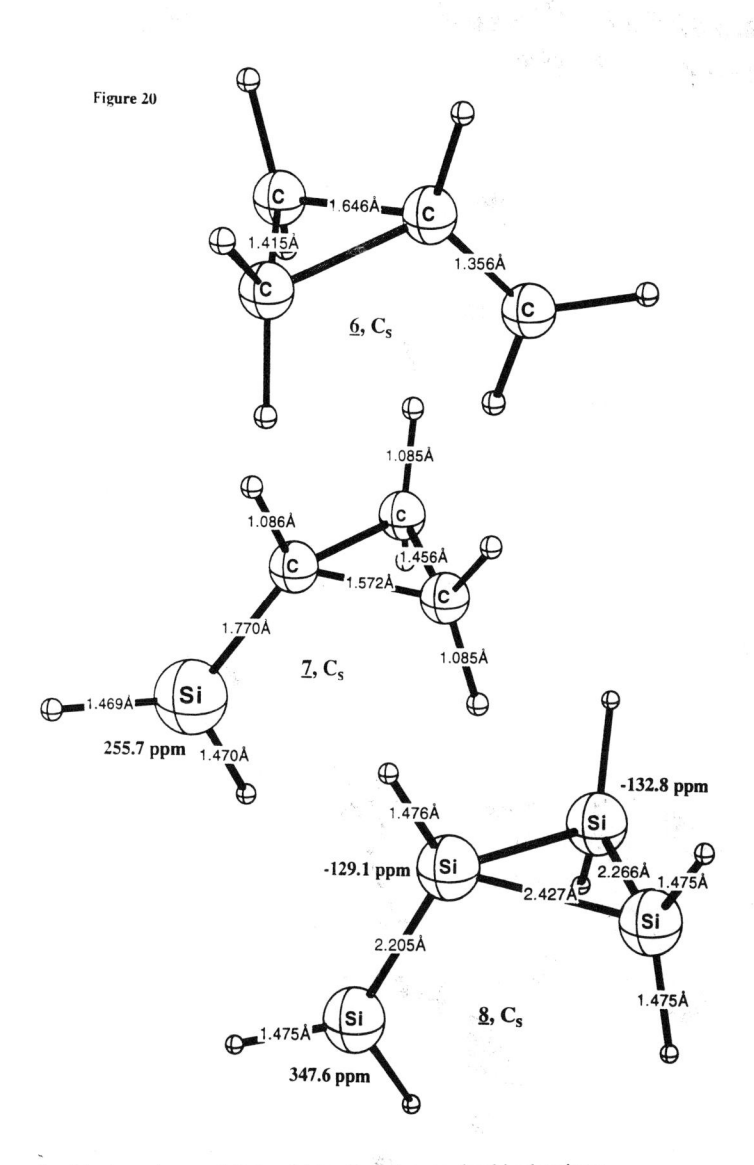

Fig. 20. Cyclopropylcarbinyl, cyclopropylsilyl and tetrasilacyclopropylcarbinyl cations; IGLO/II′//MP2(FC)/6–31G* δ ^{29}Si NMR vs TMS

There is compelling evidence for alkyl- and phenyl-bridged silyl cations from Eaborn's "S$_N$2-intermediate solvolysis" studies of organosilicon iodides [50]. The concept also is applicable to other systems were groups migrate during reaction [51]. We chose two simple model systems, with a methyl **9** and a phenyl **10** bridging between the two silicon atoms (Fig. 21). At MP2–FC/6–31G*, **9** (which also can be considered to be a C-protonated 1,3-disilacyclobutane) is a local minimum (lowest frequency 108 cm^{-1} A″). The phenyl-bridged **10** (C_s) also is a minimum. **10** in C_{2v} (not shown) is a transition structure for ring inversion with quite a low barrier (0.8 kcal mol^{-1} at MP2–FC/6–31G* + ΔZPE(SCF/6–

31G*···0.89). The theoretical ^{29}Si NMR chemical shifts of both **9** and **10** are slightly upfield of TMS, despite the large NPA positive charge of silicon (+1.44).

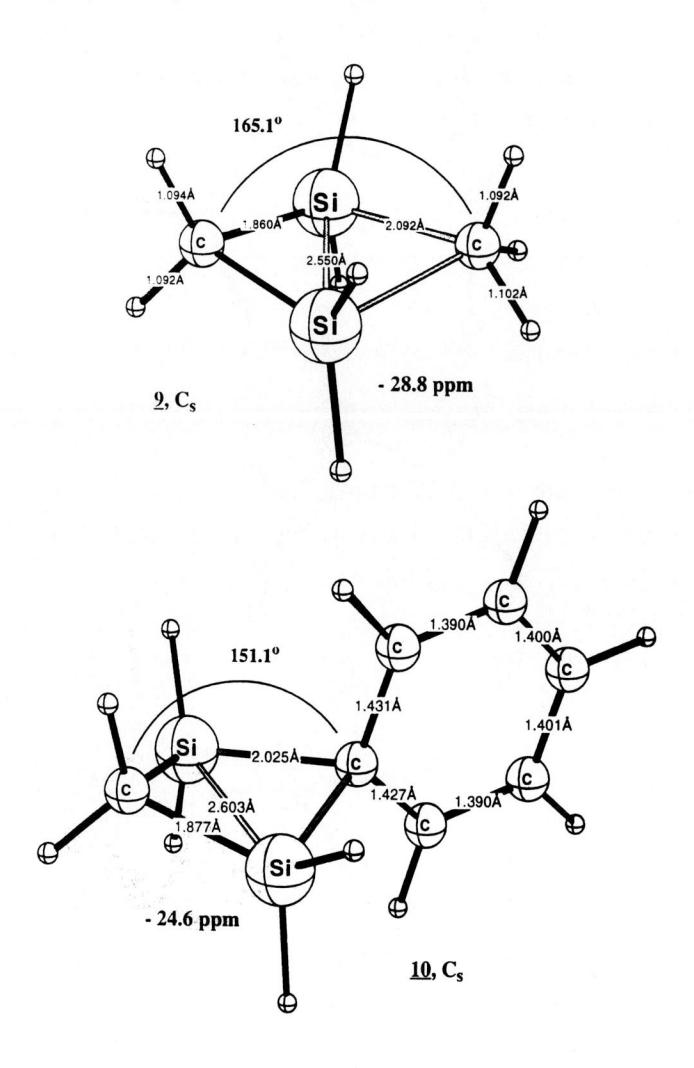

Fig. 21. Bridged silyl cations (S_N2 intermediates); IGLO/II'//MP2(FC)/6–31G* δ ^{29}Si NMR vs TMS

Conclusions

One major conclusion can be drawn: the prospects for obtaining and observing truly "free" silyl cations in condensed phases are very poor. Sterically unhindered systems will interact with all conceiveable solvents, media, and counterions, whereas strongly hindered substituents may interact *intramolecularly*.

Furthermore, electronically stabilizing substituents are not very effective, although the high positive charge on silicon may be retained.

The experimental search for the "naked" silyl cation in condensed phases will continue, but theory predicts what can be expected. *Ab initio* calculations indicate that alkanes interact strongly with silyl cations and influence the ^{29}Si NMR shifts dramatically, and that silyl cations will not be "free" even in argon matrices.

^{29}Si NMR chemical shifts of truly "free" trialkylsilyl cations will be greater than 300 ppm. While both Lambert's and Reed's silyl cation systems are quite far from meeting this criterion, they may be about as "free" as one can hope to achieve in condensed phase. This is the important development. The contributions of both groups to silicon chemistry deserve applause.

Acknowledgements: This work was supported by the *Deutsche Forschungsgemeinschaft*, the *Volkswagen Stiftung*, and the *Convex Computer Cooperation*. C. M. and J. K. gratefully acknowledge grants from the *Fond der Chemischen Industrie* (Kekulé-Stipendium) and the *Studienstiftung des Deutschen Volkes*, respectively.

References:

[1] C. Chuit, R. J. P. Corriu, C. Reye, J. C. Young, *Chem. Rev.* **1993**, *93*, 1371.

[2] R. J. P. Corriu, M. Henner, *J. Organomet. Chem.* **1974**, *74*, 1; J. B. Lambert, W. J. Schulz, in: *The Chemistry of Organic Silicon Compounds* (Eds.: S. Patai, Z. Rappoport), Wiley, Chichester, **1989**, p. 1007.

[3] J. B. Lambert, S. Zhang, C. L. Stern, J. C. Huffman, *Science* **1993**, *260*, 1917; J. B. Lambert, S. Zhang, S. M. Ciro, *Organomet.* **1994**, *13*, 2430.

[4] C. A. Reed, Z. Xie, R. Bau, A. Benesi, *Science* **1993**, *262*, 402; Z. Xie, R. Bau, C. A. Reed, *J. Chem. Soc., Chem. Commun.* **1994**, 2519; Z. Xie, R. Bau, A. Benesi, C. A. Reed, *Chemistry*, submitted for publication.

[5] J. Haggin, *Chem. Eng. News* **1993**, *June 28th*, 7.

[6] G. A. Olah, G. Rasul, X.-Y. Li, H. A. Buchholz, G. Sandford, G. K. S. Prakash, *Science* **1994**, *263*, 983.

[7] P. v. R. Schleyer, P. Buzek, T. Müller, Y. Apeloig, H.-U. Siehl, *Angew. Chem.* **1993**, *105*, 1558; *Angew. Chem., Int. Ed. Engl.* **1993**, *32*, 1471.

[8] L. Olsson, D. Cremer, *Chem. Phys. Lett.* **1993**, *215*, 433.

[9] L. Pauling, *Science* **1994**, *263*, 983.

[10] K. N. Houk, *Chemtracts Org. Chem.* **1993**, *6*, 360.

[11] P. Buzek, P. v. R. Schleyer, S. Sieber, *Chem. Unserer Zeit* **1992**, *26*, 116; M. Saunders, H. A. Jimenez-Vazquez, *Chem. Rev.*, **1991**, *91*, 375; and earlier references cited therein; G. A. Olah, G. K. S. Prakash, R. E. Williams, L. D. Field, K. Wade, *Hypercarbon Chemistry*, Wiley, New York, **1977**; M. Saunders, J. Chandrasekhar, P. v. R. Schleyer, in: *Rearrangements in Ground and Excited States*, Vol.1 (Ed.: P. de Mayo), Academic Press, New York, **1980**, p. 1; R. D. Bowen, D. H. Williams, in: *Rearrangements in Ground and Excited States*, Vol.1 (Ed.: P. de Mayo), Academic Press, New York, **1980**, p. 55; K. Lammertsma, P. v. R. Schleyer, H. Schwarz, *Angew. Chem.* **1989**, *101*, 1313; *Angew. Chem., Int. Ed. Engl.* **1989**, *28*, 1321; G. A. Olah, *Angew. Chem.* **1993**, *105*, 805; *Angew. Chem., Int. Ed. Engl.* **1993**, *32*, 767.

[12] G. A. Olah, Nobel Lecture, **1994**.

[13] T. Laube, C. Lohse, *J. Am. Chem. Soc.* **1994**, *116*, 9001; S. Hollenstein, T. Laube, *J. Am. Chem. Soc.* **1993**, *115*, 7240; T. Laube, *J. Am. Chem. Soc.* **1989**, *111*, 9224; T. Laube, *Angew. Chem.* **1987**, *99*, 580; *Angew. Chem., Int. Ed. Engl.* **1987**, *26*, 560; T. Laube, *Angew. Chem.* **1986**, *98*, 368; T. Laube, *Angew. Chem., Int. Ed. Engl.* **1986**, *25*, 349; T. Laube, S. Hollenstein, *Helv. Chim. Acta* **1994**, *77*, 1773; T. Laube, *Helv. Chim. Acta* **1994**, *77*, 943.

[14] P. v. R. Schleyer, C. Maerker, *Pure Appl. Chem.*, in press.

[15] A. Haaland, *Angew. Chem.* **1989**, *101*, 1017; *Angew. Chem., Int. Ed. Engl.* **1989**, *28*, 992.

[16] H. Schwarz, in: *The Chemistry of Organic Silicon Compounds* (Eds.: S. Patai, Z. Rappoport), Wiley, Chichester, **1989**, p. 445.

[17] A. C. M. Wojtyniak, J. A. Stone, *J. Int. Mass. Spectrum Ion Processes* **1986**, *74*, 59.

[18] M. E. Crestoni, S. Fornarini, *Angew. Chem.* **1994**, *106*, 1157; *Angew. Chem., Int. Ed. Engl.* **1994**, *33*, 1094; F. Cacace, M. E. Crestoni, S. Fornarini, R. Gabriell, *J. Int. Mass. Spectrum Ion Processes* **1988**, *84*, 17; S. Fornarini, *J. Org. Chem.* **1988**, *53*, 1314; F. Cacace, M. E. Crestoni, S. Fornarini, *J. Am. Chem. Soc.* **1992**, *114*, 6776.

[19] J. Emsley, *The Elements*, Clarendon Press, Oxford, **1989**.

[20] S. G. Lias, J. E. Bartmess, J. F. Liebman, J. L. Holmes, R. D. Levin, N. G. Mallard, *J. Phys. Chem. Ref. Data* **1988**, *17*, Suppl. 1.

[21] J. Berkowitz, J. P. Greene, H. Cho, R. Ruscic, *J. Chem. Phys.* **1987**, *86*, 1235.

[22] A. E. Reed, P. v. R. Schleyer, *J. Am. Chem. Soc.* **1990**, *112*, 1434; D. L. Cooper, T. P. Cunningham, J. Gerratt, P. B. Karadakov, M. Raimond, *J. Am. Chem. Soc.* **1994**, *116*, 4414.

[23] A. E. Reed, L. A. Curtiss, F. Weinhold *Chem. Rev.* **1988**, *88*, 899.

[24] M. S. Gordon, D. R. Gano, J. S. Binkley, M. J. Frisch, *J. Am. Chem. Soc.* **1986**, *108*, 2191; M.-D. Su, H. B. Schlegel, *J. Phy. Chem.* **1993**, *97*, 9981.

[25] P. Pyykkö, *Chem. Rev.* **1988**, *88*, 563.

[26] M. Kaupp, P. v. R. Schleyer, *J. Am. Chem. Soc.* **1993**, *115*, 1061.

[27] K. G. Dyall, *J. Chem. Phys.* **1992**, *96*, 1210.

[28] M. S. Gordon, M. W. Schmidt, *J. Am. Chem. Soc.* **1993**, *115*, 7486; M. J. M. Pepper, I. Shavitt, P. v. R. Schleyer, M. N. Glukhovtsev, R. Janoschek, M. Quack, *J. Comp. Chem.*, in press.

[29] E. D. Jemmis, G. N. Srinivas, J. Leszczynski, J. Kapp, A. A. Korkin, P. v. R. Schleyer, *J. Am. Chem. Soc.*, submitted for publication.

[30] A. A. Korkin, P. v. R. Schleyer, to be published.

[31] Y. Apeloig, in: *The Chemistry of Organic Silicon Compounds*, Vol. 1 (Eds.: S. Patai, Z. Rappoport), Wiley, New York, **1989**, p. 57.

[32] W. J. Hehre, L. Radom, P. v. R. Schleyer, J. A. Pople, *Ab Initio Molecular Orbital Theory*, Wiley, New York, **1986**; J. B. Foresman, Æ. Frisch, *Exploring Chemistry with Electronic Structure Methods: A Guide to Using Gaussian*, Gaussian, Inc., Pittsburgh, **1993**.

[33] Y. Apeloig, P. v. R. Schleyer, *Tetrahedron Lett.* **1977**, *52*, 4647; Y. Apeloig, S. A. Godleski, D. J. Heacock, J. M. McKelvey, *Tetrahedron Lett.* **1981**, *22*, 3297.

[34] Y. Cao, J.-H. Choi, B.-M. Haas, M. S. Johnson, M. Okumura, *J. Phys. Chem.* **1993**, *97*, 5215.

[35] C.-H. Hu, P. R. Schreiner, P. v. R. Schleyer, H. F. Schaefer III, *J. Phys. Chem.* **1994**, *98*, 5040.

[36] D. W. Boo, Y. T. Lee, *Chem. Phys. Lett* **1993**, *211*, 358; S.-J. Kim, P. R. Schreiner, P. v. R. Schleyer, H. F. Schaefer III, *J. Phys. Chem.* **1993**, *97*, 12232.

[37] J. Gauss, *Chem. Phys. Lett.* **1994**, in press; J. Gauss, *Chem. Phys. Lett.* **1992**, *191*, 614; K. Wolinski, J. F. Hinton, P. Pulay, *J. Am. Chem. Soc.* **1990**, *112*, 8251; H. Fukui, K. Miura, H. Yamakaki, T. Nosaka, *J. Chem. Phys.* **1985**, *82*, 1410; R. Ditchfield, *Mol. Phys.* **1974**, *27*, 789.

[38] W. Kutzelnigg, *Isr. J. Chem.* **1980**, *19*, 193; W. Kutzelnigg, U. Fleischer, M. Schindler, *NMR Basic Principles and Progress.*, *23*, Springer, Berlin, **1990**, p. 165.

[39] T. Onak, *Inorg. Chem.* **1993**, *32*, 487.

[40] G. A. Olah, G. Rasul, L Heiliger, J. Bausch, G. K. S. Prakash, *J. Am. Chem. Soc.* **1992**, *114*, 7737.

[41] J. B. Lambert, S. Zhang, *J. Chem. Soc., Chem. Commun.* **1993**, 393.

[42] X. Yang, C. L. Stern, T. J. Marks, *Organometallics* **1991**, *10*, 840.

[43] *Carbonium Ions, Vols. I-V* (Eds., G. A. Olah, P. v. R. Schleyer), Wiley, New York, **1968-1976**.

[44] H. C. Brown, *The Nonclassical Ion Problem, with comments by P. v. R. Schleyer*, Plenum Press, New York, **1977**.

[45] P. v. R. Schleyer, S. Sieber, *Angew. Chem.* **1993**, *105*, 1676; *Angew. Chem., Int. Ed. Engl.* **1993**, *32*, 1606; W. Koch, L. Bowen, D. J. DeFrees, D. E. Sunko, H. Vancik, *Angew. Chem.* **1990**, *102*, 198; *Angew. Chem., Int. Ed. Engl.* **1990**, *29*, 183; P. C. Myhre, G. G. Webb, C. S. Yannoni, *J. Am. Chem. Soc* **1990**, *112*, 8991; W. Koch, B. Liu, D. J. DeFrees, *J. Am. Chem. Soc.* **1989**, *111*, 1527; K. Raghavachari, R. C. Haddon, P. v. R. Schleyer, H. F. Schaefer III, *J. Am. Chem. Soc.* **1983**, *105*, 5915; M. Yoshimine, A. D. McLean, B. Liu, D. J. DeFrees, J. S. Binkley, *J. Am. Chem. Soc.* **1983**, *105*, 6185; C. S. Yannoni, V. Macho, P. C. Myhre, *J. Am. Chem. Soc.* **1982**, *104*, 907.

[46] H.-U. Steinberger, N. Auner, in: *Organosilicon Chemistry II* (Eds.: N. Auner, J. Weis), VCH, Weinheim, **1995**, p. 49; N. Auner, H.-U. Steinberger, *Z. Naturforsch.* **1994**, *49B*, 1743.

[47] S. Winstein, C. Ordronneau, *J. Am. Chem. Soc.* **1960**, *82*, 2084.

[48] M. Bremer, K. Schötz, P. v. R. Schleyer, U. Fleischer, M. Schindler, W. Kutzelnigg, W. Koch, P. Pulay, *Angew. Chem.* **1989**, *101*, 1063; *Angew. Chem., Int. Ed. Engl.* **1989**, *28*, 1042.

[49] H. Jiao, P. v. R. Schleyer, *Angew. Chem.* **1993**, *105*, 1833; *Angew. Chem., Int. Ed. Engl.* **1993**, *32*, 1763; H. Jiao, P. v. R. Schleyer, *J. Chem. Soc. Faraday Trans.* **1994**, *90*, 1559.

[50] S. A. I. Al-Shali, C. Eaborn, F. A. Fattah, S. T. Najim, *J. Chem. Soc., Chem. Commun.* **1984**, *5*, 318.

[51] C. Eaborn, P. D. Lickiss, S. T. Najim, *J. Chem. Soc., Perkin Trans.* **1993**, *2*, 391; C. Eaborn, P. D. Lickiss, S. T. Najim, W. A. Stanczyk, *J. Chem. Soc., Perkin Trans.* **1993**, *2*, 59; A. I. Al-Mansour, M. A. M. R. Al-Gurashi, C. Eaborn, F. A. Fattah, P. D. Lickiss, *J. Organomet. Chem.* **1990**, *393*, 27; G. A. Ayoko, C. Eaborn, *J. Chem. Soc., Perkin Trans.* **1987**, *2*, 1047; N. H. Buttrus, C. Eaborn, P. B. Hitchcock, P. D. Lickiss, S. T. Najim, *J. Chem. Soc., Perkin Trans.* **1987**, *2*, 1753; N. H. Buttrus, C. Eaborn, P. B. Hitchcock, P. D. Lickiss, S. T. Najim, *J. Chem. Soc., Perkin Trans.* **1987**, *2*, 891; C. Eaborn, P. D. Lickiss, S. T. Najim, W. A. Stanczyk, *J. Chem. Soc., Chem. Commun.* **1987**, *19*, 1461; C. Eaborn, A. K. Saxena, *J. Chem. Soc., Perkin Trans.* **1987**, *2*, 779; C. Eaborn, P. D. Lickiss, S. T. Najim, M. N. Romanelli, *J. Organomet. Chem.* **1986**, *315*, C5; C. Eaborn, A. K. Saxena, *J. Chem. Soc., Chem. Commun.* **1984**, *22*, 1482; R. Damrauer, C. Eaborn, D. A. R. Happer, A. I. Mansour, *J. Chem. Soc., Chem. Commun.* **1983**, *7*, 348.

[52] A.L. Allred, E. J. Rochow, *J. Inorg. Nuc. Chem.* **1958**, *5*, 264.

[53] L. Pauling, in: *The Nature of the Chemical Bond and the Structure of Molecules and Crystals*, Cornell University Press, Ithaca NY, **1986**.

[54] H. C. Marsmann, *NMR Basic Principles and Progress*, *17*, Springer, Berlin, **1981**.

[55] J. W. de M. Carneiro, P. v. R. Schleyer, M. Saunders, R. Remington, H. F. Schaefer III, A. Rauk, T. S. Sorensen, *J. Am. Chem. Soc.* **1994**, *116*, 3483.

[56] J. M. Stone, J. A. Stone, *J. Int. Mass. Spectrum Ion Processes* **1991**, *109*, 247.

[57] P. R. Schreiner, *Ph. D. Thesis*, Universität Erlangen-Nürnberg, **1994**.

[58] K. Hiraoka, P. Kebarle, *J. Am. Chem. Soc.* **1977**, *99*, 360.

NMR Spectroscopic Investigation of the β-Silyl Effect in Carbocations

Hans-Ullrich Siehl*, Bernhard Müller, Martin Fuß

Institut für Organische Chemie

Eberhard-Karls-Universität Tübingen

Auf der Morgenstelle 18, D-72076 Tübingen, Germany

Yutaka Tsuji

Institute for Fundamental Research of Organic Chemistry

Kyushu University

Fukuoka, 813 Japan

Summary: 1-Aryl-2-trialkylsilyl-substituted vinyl cations are characterized in solution by NMR spectroscopy. The NMR chemical shift data reveal the stabilization of the positive charge by a β-silyl substituent. The order of hyperconjugative stabilization of a positive charge by β-substituents is H < alkyl < silyl. The β-silyl effect is dependent on the electron demand of the carbocation and decreases with better electron donating α-substituents. NMR spectroscopy is a suitable tool to investigate the competition between π-resonance and σ-hyperconjugation in these type of carbocations.

Introduction

Organosilicon compounds are extensively used in chemical synthesis. The investigation of the mechanistic implications of silicon chemistry is therefore of utmost importance. Computational and experimental investigations have shown that carbocations are stabilized by β-silyl groups. All attempts to prepare β-silyl-substituted carbocations in superacid solution however have been unsuccessful so far. Nucleophilic attack of halogen or oxygen nucleophiles at silicon and concomitant C–Si bond cleavage occur easily because of the higher bond energy of Si–O and Si–Hal bonds compared to Si–C bonds. Even the low nucleophilic anions of superacids can attack the silicon in silyl-substituted carbocations and this leads to rapid β-C–Si bond fission to yield silicon-free carbocations.

We have successfully prepared the first β-silyl-substituted carbocations as persistent species in solution using progenitors with bulky alkyl groups at silicon [1–3]. The C–Si bond fragmentation can be completely suppressed by three isopropyl substituents at silicon.

The effect of β-silyl groups was investigated in α-aryl-substituted vinyl cations. In these disubstituted carbocations the silyl substituent is attached at the doubly bonded β-carbon. The C_β–Si-bond is thus fixed in plane with the vacant orbital at the C^+-carbon allowing maximum overlap of the β-σ-C–Si bond and the vacant $2p(C^+)\pi$-orbital.

Results and Discussion

The α-aryl-β-triisopropylsilyl-substituted vinyl cations **1–3** are generated by protonation of the corresponding 1-aryl-2-silyl-alkynes with superacids at low temperatures (Eq. 1). For the protonation of α-ferrocenyl-β-(triisopropyl)silylethyne to yield the 1-ferrocenyl-2-(triisopropyl)silylvinyl cation **4** the weaker acid CF_3COOH is sufficient (Eq. 2).

1	$R_1, R_2, R_3 = H$
2	$R_1, R_2 = H, R_3 = CH_3$
3	$R_1, R_2, R_3 = CH_3$

Eq. 1.

The vinyl cations were characterized by ^1H, ^{13}C, and ^{29}Si NMR spectroscopy at low temperatures. Fig. 1 shows as a typical example the ^{13}C NMR spectrum of the 1-mesityl-2-(triisopropyl)silyl vinyl cation **3**.

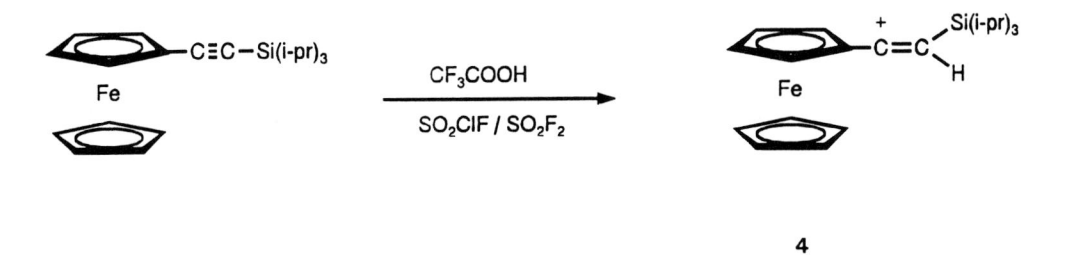

Eq. 2.

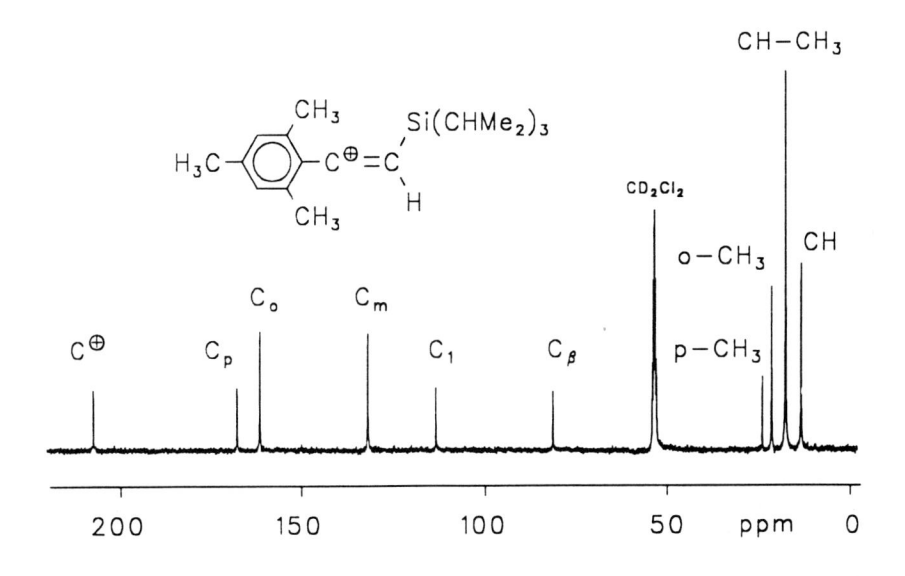

Fig. 1. ^{13}C NMR spectrum of 1-mesityl-2-(triisopropyl)silylvinyl cation **3**

In these type of carbocations the positive charge can be delocalized by α-aryl π-conjugation as in **5a ↔ 5b** and by β-silyl σ-hyperconjugation depicted in valence bond formalism as **6a ↔ 6b**.

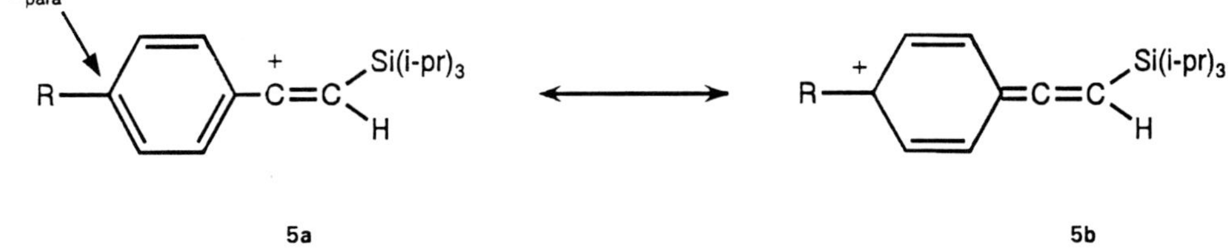

Fig. 2.

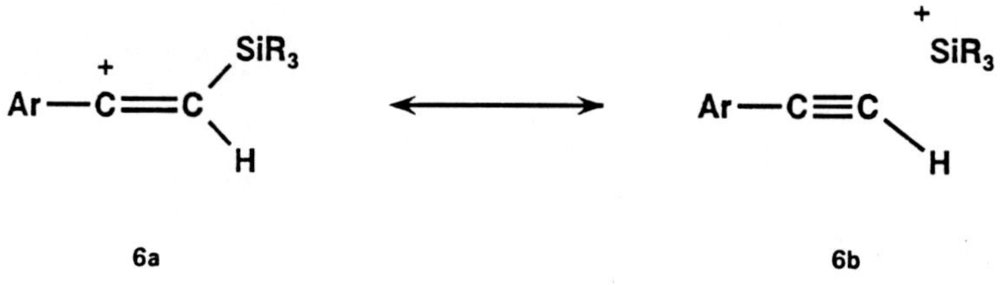

$$\text{Ar}-\overset{+}{\text{C}}=\text{C}\begin{smallmatrix}\diagup\text{SiR}_3\\\diagdown\text{H}\end{smallmatrix} \quad \longleftrightarrow \quad \text{Ar}-\text{C}\equiv\text{C}\begin{smallmatrix}\diagdown\overset{+}{\text{SiR}_3}\\\diagup\text{H}\end{smallmatrix}$$

            6a   6b

Fig. 3.

The demand for α-aryl π-resonance stabilization (**5a** ↔ **5b**) is dependent on the ability of the β-substituent to stabilize the positive charge by β-σ-hyperconjugation (**6a** ↔ **6b**) and vice versa, i.e., the more stabilizing the β-substituent the less is the demand for π-delocalization of the positive charge into the aromatic ring.

The ^{13}C NMR chemical shift of the *para*-carbon can be used as a probe to monitor the effect of different β-substituents on the π-conjugative charge delocalization. Comparing the *para*-carbon shift of α-mesityl vinyl cations Mes-C$^+$=CH(R) with different β-substituents (R = H, 180.0 ppm; *t*butyl, 178.5 ppm; Si(*i*Pr)$_3$, 168.5 ppm) shows that β-Si–C hyperconjugation is more efficient than β-C–H or β-C–C hyperconjugation.

Comparing β-silyl-substituted vinyl cations however with *different* α-aryl substituents shows that the stabilizing effect of a β-silyl group is not constant but is dependent on the electron demand of the carbocation. Fig. 4 shows the *para*-carbon ^{13}C NMR chemical shift difference Δδ between the β-silyl and β-unsubstituted vinyl cations (For the α-ferrocenyl substituted cation Δδ C3,4 is used to probe the silyl effect).

The chemical shift difference Δδ ^{13}C of the para position for carbocations with better electron donating α-aryl substituents is smaller. The π-resonance stabilization is increasing from tolyl to mesityl, anisyl, and ferrocenyl substituted cations and the contribution of β-silyl stabilization becomes concomitantly less important. ^{29}Si NMR chemical shifts are used as a additional tool to probe the magnitude of the β-silyl effect [4]. Table 1 shows ^{29}Si NMR chemical shift data for a series of α-aryl β-silyl-substituted vinyl cations and the corresponding precursor alkynes. For vinyl cations with better electron donating α-aryl substituents less positive charge is hyperconjugatively delocalized to the silyl group and consequently the silicon is less shielded.

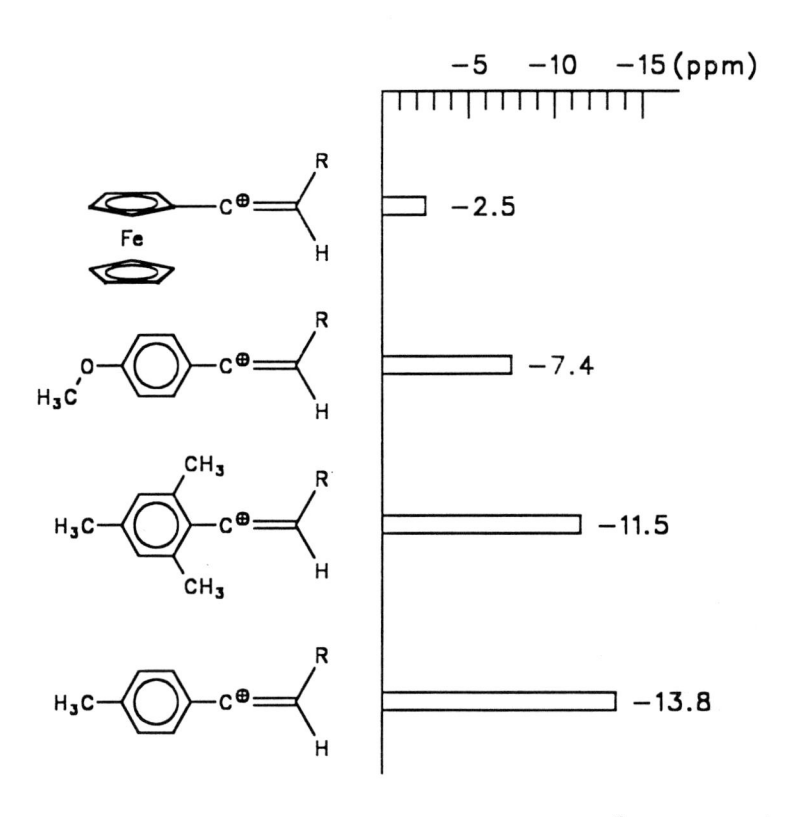

Fig. 4. Variable demand for β-silyl stabilization in α-aryl vinyl cations; Δδ of C *para* ^{13}C shift for Ar–C$^+$=C(H)R for R = H and R = Si(*i*Pr)$_3$

	α-Substituent	Ar–C$^+$=CHSiR$_3$	Ar–C≡CSiR$_3$
1	Phenyl	46.8	−1.82
2	Tolyl	0.7	−1.97
3	Mesityl	37.6	−2.10
4	Ferrocenyl	24.0	−2.83

Table 1: ^{29}Si NMR chemical shift [ppm] of α-aryl-β-(triisopropyl)silyl-substituted vinyl cations Ar–C$^+$=CHSiR$_3$ (1–4) and the corresponding alkynes Ar–C≡CSiR$_3$ [R = Si(CHMe$_2$)$_3$]

Acknowledgement: This work was supported by the *Deutsche Forschungsgemeinschaft*. H.-U. S. thanks the *Japanese Society* for the Promotion of Science for support of a visiting professorship at *Kyushu University*. H.-U. S. and Y. T. thank Prof. Y. Tsuno, *Institute for Fundamental Research of Organic Chemistry*, *Kyushu University*, Japan, for support.

References:

[1] H.-U. Siehl, F. P. Kaufmann, Y. Apeloig, D. Danovich, A. Berndt, N. Stamatis, *Angew. Chem.* **1991**, *103*, 1546; *Angew. Chem., Int. Ed. Engl.* **1991**, *30*, 1479.

[2] H.-U. Siehl, F. P. Kaufmann, *J. Am. Chem. Soc.* **1992**, *114*, 4937.

[3] H.-U. Siehl, F. P. Kaufmann, K. Hori, *J. Am. Chem. Soc.* **1992**, *114*, 9343.

[4] B. Müller, *Diploma Thesis*, Universität Tübingen, **1993**.

Underloaded and Overloaded Unsaturated Silicon Compounds

Nils Wiberg

Institut für Anorganische Chemie

Ludwig-Maximilians-Universität München

Meiserstr. 1, D-80333 München, Germany

Summary: After a short comment about the preparation and chemical properties of unsaturated compounds >E=Y with E = Si, Ge, Sn and Y = C<, N-, P-, O, S, it is stated that the production of dimers of $Me_2Si=C(SiMe_3)_2$ by LiBr elimination of $Me_2SiBr–CLi(SiMe_3)_2$ goes by way of $Me_2Si=C(SiMe_2)_2$, which does not dimerize but inserts in the CLi bond of the silaethene source. The same holds for bulkier silaethenes, whereas salt elimination sources of very bulky silaethenes like $Me_2Si=C(SiMe_3)(SiMeBu_2)$ stops at the silaethene step. Salt elimination reactions may also lead by condensation reactions to dimers of >E=Y without intermediate formation of >E=Y. On the whole, lithium organyls add to $Me_2Si=C(SiMe_2)_2$, whereby the reactivity of RLi decreases with increasing bulkiness of RLi. Donors D add to silenes >Si=Y under formation of adducts D→Si=Y with long DSi coordination bonds, elongated SiY double bonds, and pyramidalized Si atoms, whereby the adduct formation plays a central role for the reactivity of the silenes. The donor electron pair might be a nonbonding electron pair of halogenides, ethers, amines (formation of n-complexes; reaction of $tBu_2Si=N–SitBu_3$ with CO described in more detail), but possibly also a bonding σ electron pair (formation of σ-complexes; reaction of $tBu_2Si=N–SiCltBu_2$ with $EtNMe_2$ described in more detail), or a bonding π electron pair (formation of π-complexes; reaction of $Me_2Si=C(SiMe_3)_2$ with organic dienes described in more detail). The π-bond strengths of $Ph_2Si=C(SiMe_3)_2$ and $tBu_2Si=C(SiMe_3)_2$ are less than those of the isomers $Me_2Si=C(SiMe_2Ph)_2$ and $Me_2Si=C(SiMe_3)(SiMetBu_2)$ for the first example as a result of electronic effect and because of steric effects for the second example. The attempt to synthesize overcrowded $tBu_3Si–Si≡Si–SitBu_3$ by debromination of $tBu_3Si–SiBr_2–SiBr_2–SitBu_3$ with tBu_3SiNa leads to the tetrasilatetrahedrane $(tBu_3Si)_4Si_4$. Certainly, the disilyne is not a reaction intermediate; the same does not hold for the disilene $(t\text{Bu}_3\text{Si})\text{BrSi}=\text{SiBr}(\text{Si}t\text{Bu}_3)$, which could be trapped by PhC≡CPh.

Introduction

In our research group we study under- and overloaded silicon compounds, that is to say compounds in which a silicon atom is undercrowded with regard to the number of substituents, or is overcrowded in respect of the spatial extent of its substituents. Although the former compounds – like the silaethene shown in Scheme 1 – are manifested by their chemical reactivity even at low temperature and were not at all isolable under normal conditions [1], the latter compounds may have an unexpected thermal stability.

In fact, an isolation of compounds – such as the substituted tetrasilatetrahedrane shown in Scheme 1 – is in many cases not possible without introducing overloaded groups like the $t\text{Bu}_3\text{Si}$ group, that is the supersilyl group [2]. In addition, compounds which possess underloaded as well as overloaded silicon atoms may become isolable but retain – like the silaneimine shown in Scheme 1 – their chemical potency [3].

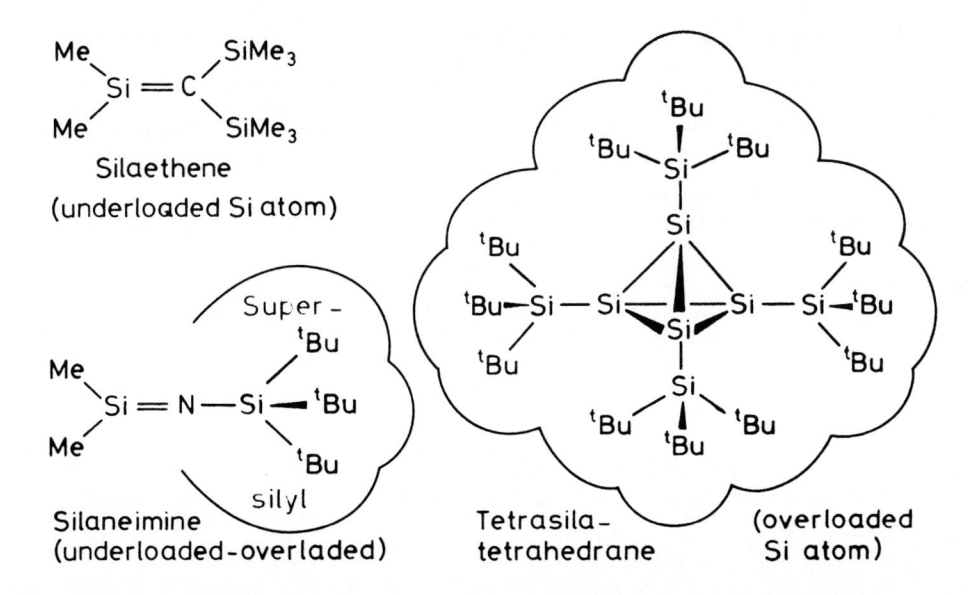

Scheme 1. Underloaded and overloaded silicon compounds

In this connection I intend to discuss three themes concerning unsaturated silicon compounds. But before I enter into the essential points, I will offer some comments about the unsaturated silicon compounds in question, as well as about the homologous unsaturated germanium and tin compounds. These unsaturated systems have the general composition >E=Y, shown in the middle of Scheme 2.

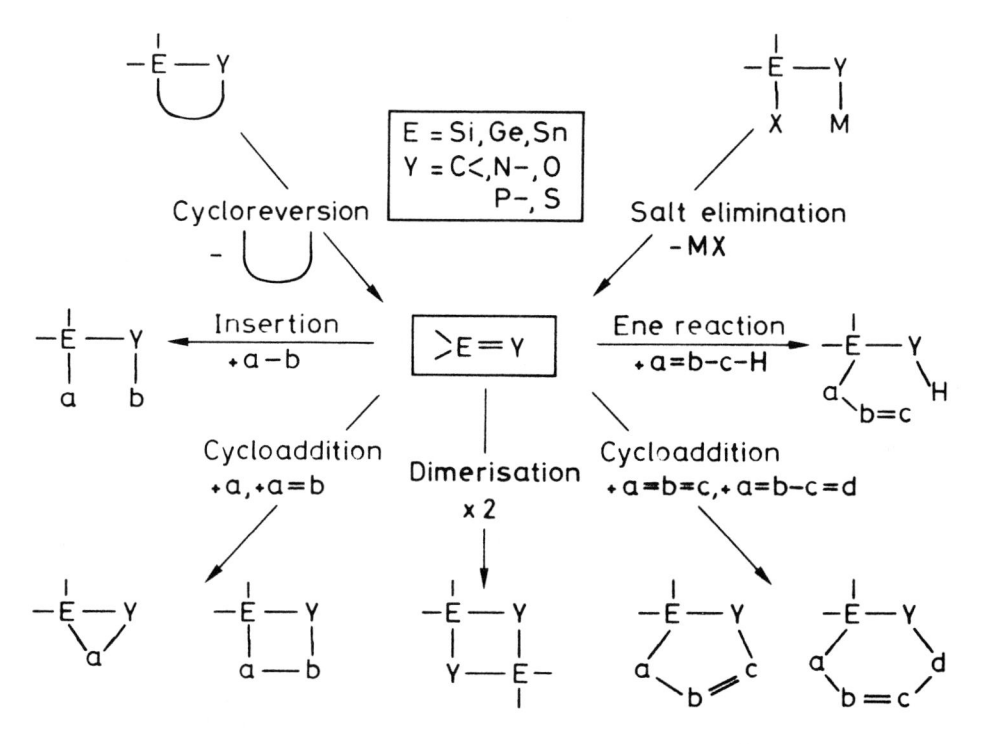

Scheme 2. Preparation, stabilization, and reactions of silenes, germenes, and stannenes

Because their structural characteristic is an unsaturated silicon, germanium, or tin atom >E=, they may, in the widest sense, be named *silenes*, *germenes*, or *stannenes*. As a rule, they are unstable under normal conditions and isolable only as dimers, but they are formed as intermediates from suitable initial stages by *cycloreversion* or *salt elimination* (Scheme 2) [1, 3–11].

The silenes, germenes, and stannenes so produced may be trapped by active reactants: They undergo *insertions* into a–b simple bonds, react with enes a=b–c–H under *ene reactions* and form *cycloadducts* with unsaturated systems a, a=b, a=b=c and a=b–c=d, respectively (Scheme 2) [1, 3–11].

The questions now to be answered in connection with silenes are:

1) By which route are the dimers of silenes formed?
2) What causes the reactivity of silenes?
3) How stable are silenes?

1 By which Routes are the Dimers of Silenes Formed?

The Woodward-Hoffman rules of *conservation of orbital symmetry* forbid a concerted thermal [2+2] cycloaddition of two ethene molecules. As is shown on Scheme 3 (left-hand side), no effective overlapping of HOMO and LUMO is possible in this case and, as a result, the activation energies of dimerization are very high. On the other hand, by lowering the symmetry of the double bond, that is by going from ethenes to silenes, germenes, or stannenes, the prohibition for dimerization of the unsaturated system is more or less reduced. In this case, as demonstrated in Scheme 3 (right-hand side), a certain overlapping of HOMO and LUMO is possible with the result that the activation energies of dimerization are also more or less reduced. And this holds especially when the difference in electronegativity of E and Y is large and when the substituents of the double bond are small. On the other hand, the activation energies remain greater than zero. Therefore, silene, germene, and stannene intermediates may be trapped by active reactants (Scheme 2) [1, 3–11].

In the past, many chemists have concluded not only an intermediate existence of silenes, germenes, and stannenes from the formation of products which formally were dimers of these unsaturated systems but also that such dimers may be formed by [2+2] cycloaddition of the silenes, germenes, and stannenes. In fact, these conclusions might be wrong as will be demonstrated in the light of dimer formation as a result of thermal salt elimination.

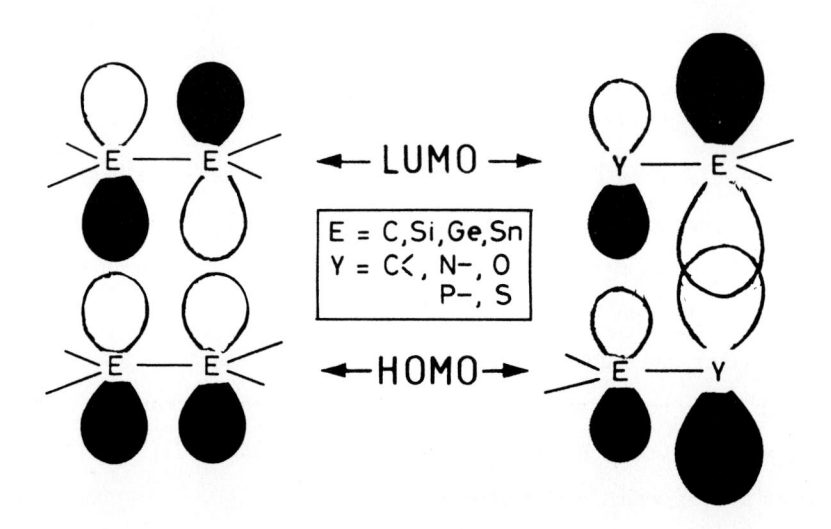

Scheme 3. HOMO and LUMO which characterize dimerization of ethenes, silenes, germenes, and stannenes

To produce sila-, germa-, or stannaethenes one starts advantageously with dibromides as shown at the middle left-hand side of Scheme 4. They transform by reaction with lithium organyls at low temperature under bromine/lithium exchange into lithium containing sources of the unsaturated systems. The latter eliminate LiBr under formation of products, which appear as dimers of silenes, germenes, and stannenes, that is for example as dimers of the systems shown at the bottom of Scheme 4 [1, 4, 5, 11]. To simplify matters, I now restrict the discussion on the whole to the silene $Me_2Si=C(SiMe_3)_2$ (Scheme 4), which will be named "standard silaethene" [1].

In principle, two routes are available which lead from the mentioned sources to the dimer of the standard silaethene (Scheme 5; methyl and trimethylsilyl groups not explicitly shown): one route (a) involves condensation steps without any intermediacy of the standard silaethene, another route (b) has elimination steps followed by dimerization of two intermediately formed silaethene molecules. Because thermolyses of the standard silaethene sources lead in the presence of *one* or *two* active silaethene traps, as we found, exclusively to products of trapped standard silaethenes, whereby in the latter case the yields of the two trapping products are independent of the nature of the source employed, route (a) must be ruled out and the intermediate formation of the standard silaethene is proved. As we discovered, MX elimination in other cases leads, of course, to dimers of silenes by condensation route (a) without any intermediacy of silenes. An example is the formation of $(tBu_2SiO)_2$ by thermolysis of $tBu_2SiX-OLi$ (X = triflate) [12].

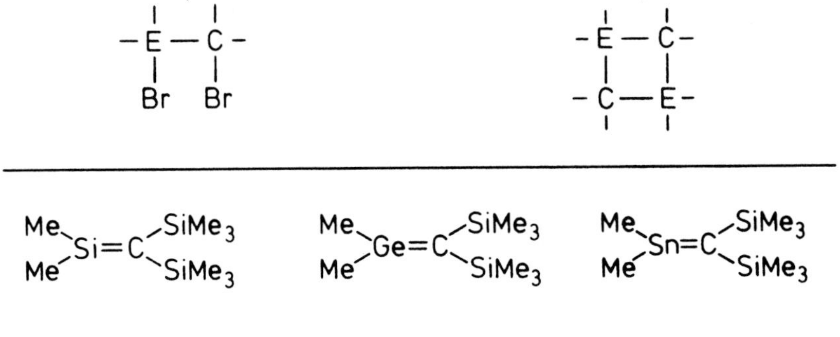

Scheme 4: Preparation and stabilization of sila-, germa-, and stannaethenes.

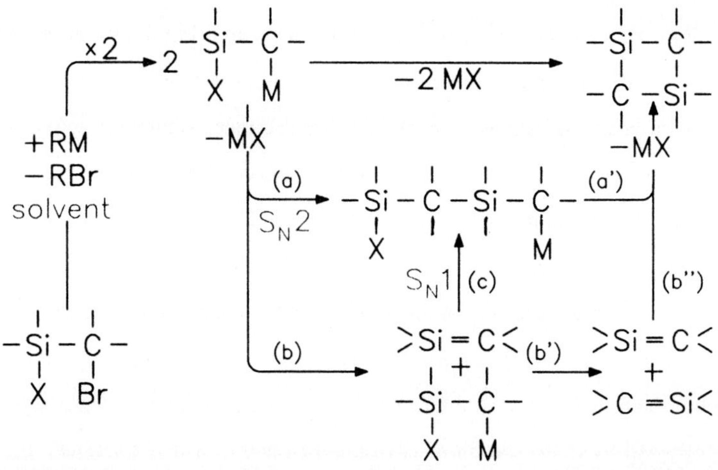

Scheme 5. Mechanism of the formation of dimers of silaethenes from silene sources >SiX–CM< (X = electronegative group as Br; M = electropositive metal as Li; an analogous scheme applies for other silene sources with N-, P-, O, S instead of C<)

To prove or disprove route (b) we carried out many chemical studies [13] but I will not enter into details and only give an account of the results. Obviously, the dimer of the standard silaethene is formed neither by condensation route (a) alone nor by silaethene dimerization route (b) alone, but half by elimination way (b) than by insertion way (c) and finally half by condensation way (a') (Scheme 5). In other words, the product in the middle of Scheme 5 is not formed by an S_N2-reaction – usual in silicon chemistry – via route (a), but by a S_N1-reaction – uncommon in silicon chemistry – via routes (b, c). In fact, two reactants compete for the silaethene intermediates so formed: the silaethene source, that is an organometal compound, and the silaethene itself. The latter is defeated. In this connection, it should be remembered that metal organyls are used for the preparation of the sources of standard silaethene. Naturally, these compete, too, for the silaethene and may suppress reaction (c).

To gain insight in the trapping activities of lithium organyls and also of ketones, azides, or organic dimers, we produced the standard silaethene in the presence of two different traps in diethyl ether at −78 °C (Eq. 1).

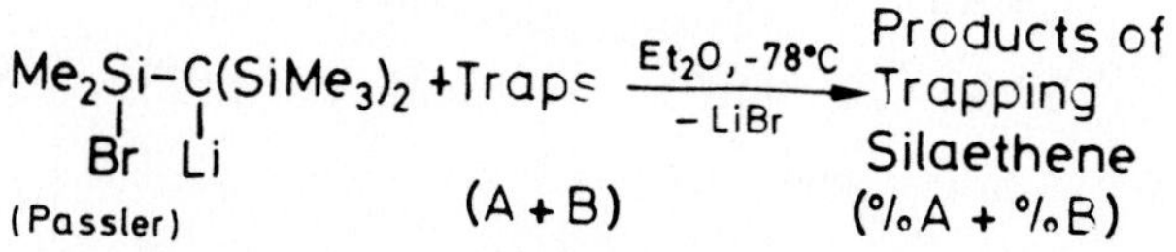

Eq. 1.

Then we determined the relative yields of trapping products. From that we calculated the relative reactivities of traps against the standard silaethene taking statistical factors into consideration [13]. Some data connected with the trap butadiene (relative reactivity ≡1) are placed together in Fig. 1, which shows that ketones are more reactive than azides and lithium organyls, which themselves are more reactive than organic dienes. The difference in rate between the reaction of benzophenone and butadiene, respectively, with the standard silaethene amounts to over a million. Indeed, butadiene stands at the end of the scale, but nevertheless reacts with the standard silaethene considerably better than an analogous substituted ethene. In fact, silenes – being underloaded systems – are extremely reactive.

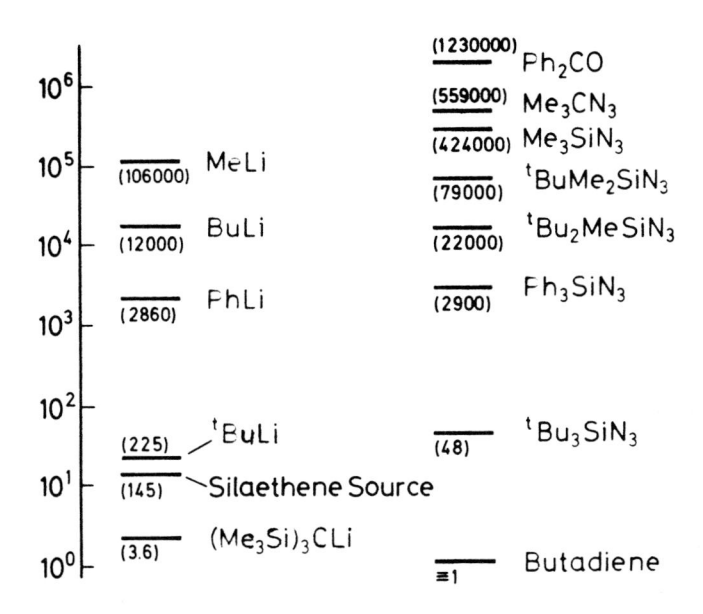

10^6		(1230000) Ph_2CO
		(559000) Me_3CN_3
		(424000) Me_3SiN_3
10^5	(106000) MeLi	(79000) $^tBuMe_2SiN_3$
10^4	(12000) BuLi	(22000) $^tBu_2MeSiN_3$
10^3	(2860) PhLi	(2900) Fh_3SiN_3
10^2		
	(225) ^tBuLi	(48) $^tBu_3SiN_3$
10^1	(145) Silaethene Source	
	(3.6) $(Me_3Si)_3CLi$	
10^0		≡1 Butadiene

Fig. 1. Relative reactivity of traps against the silaethene $Me_2Si=C(SiMe_3)_2$

With regard to the rate of insertion of the standard silaethene into the CLi bond of lithium organyls, the following can be learnt from Fig. 1:

1) For *steric* reasons the reactivity of lithium organyl decreases with increasing bulkiness of the latter, that is in the direction methyl, butyl, phenyl, *t*butyl, trisyl lithium.
2) The source $Me_2SiBr-CLi(SiMe_3)_2$ of standard silaethene reacts ca. 100 times faster with the standard silaethene than analogously crowded trisyl lithium $Me_3Si-CLi(SiMe_3)_2$. The reason for this *electronic* acceleration appears to be different structures of the lithium organyls.
3) Butyl and phenyl lithium, which typically were used for the synthesis of the silaethenes, react faster with the standard silaethene than the silaethene source itself.

Therefore, the production of the standard silaethene can only work well if the formation of the source by bromine/lithium exchange takes place faster than the thermal LiBr elimination. This is the case in diethyl ether, but not in pentane as *solvent*.

Besides the solvent, the *substituents* of the silaethene double bond influence the reaction scene. In this connection, we asked ourselves if it would be possible to force silaethenes, which were prepared by thermal salt elimination, to dimerize by [2+2] cycloaddition (Scheme 5, (b")). We therefore have prepared the silaethene source, shown on the left-hand side in the first row of Scheme 6.

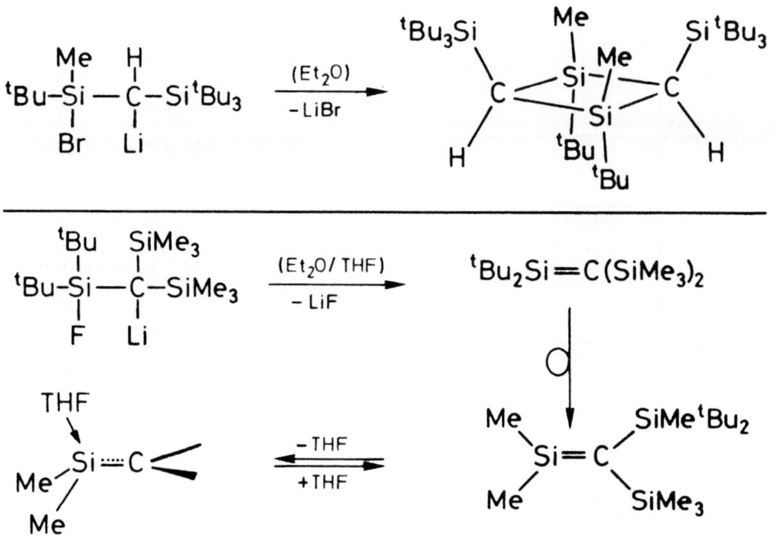

Scheme 6. Formation of an overcrowded silaethene dimer from an isolable silaethene as well as from its tetrahydrofuran adduct

Thermolysis of this in diethyl ether leads to a silaethene dimer, shown on the right-hand side in the first row of Scheme 6 as the only diastereomer. From the X-ray structure determination [14], both supersilyl groups are located on the same side of the four-membered ring. This fact speaks against the formation of the dimer by [2+2] cycloaddition.

If the bulkiness of the silaethene to be produced is further increased, the formation of dimers finally stops. Accordingly, LiF elimination of the silaethene source shown on the left-hand side in the second row of Scheme 6 leads to the silaethene $Me_2Si=C(SiMe_3)(SiMetBu_2)$. With the latter compound we were able for the first time to isolate a system with a normal Si=C bond and to solve its structure [15]. It stabilizes itself above room temperature by elimination of isobutene and not by dimerization. Obviously, the formation of dimers by [2+2] cycloaddition of two molecules of silenes is not the rule.

By crystallizing the stable silaethene from diethyl ether/THF we isolated – according to the X-ray structure determination – a tetrahydrofuran adduct (Scheme 6) [16]. This adduct, which could be transformed into the "naked" silaethene, represents the first example of *base adducts of silenes* which themselves play a central rôle in my following statements.

2 What Causes the Reactivity of Silenes?

Following studies of our research group [16–19] and later of other work groups [20], silenes with *polar double bonds* form adducts with donors D, which like halogenides, ethers, or amines possess atoms with a free, that is a nonbonding electron pair (Eq. 2):

$$\underset{X}{\overset{X}{\diagdown}}\underset{Y}{\overset{\beta}{\underset{\alpha}{Si}}}=Y \quad \xrightarrow[\text{Formation}]{+D \atop n\text{-Complex}} \quad \underset{X}{\overset{X}{\diagdown}}\underset{Y}{\overset{D\downarrow}{\underset{}{Si}}}\overset{d}{\cdots\cdots}Y \qquad \begin{array}{l} D = Hal^-, OR_2, NR_3 \\ Y = CR_2, NR, O, S, ML_n \\ X = R, NR_2, OR, SR, Hal \end{array}$$

Eq. 2.

In these *n-complexes* of silenes the double bonded group Y may be, among others, a methylene or imino group, an oxygen or sulfur atom, and a transition metal complex fragment, respectively. The SiD coordination bond is relatively long. Furthermore, *n*-complex formation is accompanied by an elongation of the double bond and a pyramidalization of the formerly planar unsaturated silicon atom. The SiY distance in the adducts however never reaches the length of a single bond. Over and above that, the pyramidalization does not lead to a tetrahedral silicon atom. With other words, the donor addition rests halfway. Probably this results from the tendency of Si and Y to preserve the Si=Y bond. So the unsaturated carbon atom does not change its planar conformation in the course of *n*-complex formation.

As Table 1 demonstrates, the SiY bond length and the Si pyramidalization increases with increasing basicity of dimers, that is in direction THF, ethyldimethyl amine, fluoride.

	$R_2Si=C(SiR_3)_2$		$R_2Si=N(SiR_3)$	
	d [Å]	$\alpha+\beta+\gamma$ [°]	d [Å]	$\alpha+\beta+\gamma$ [°]
Double Bond	1.702	360 [a]	1.568	360 [c]
THF-Adduct	1.747	348.7 [a]	1.588	349.5 [d]
NR_3-Adduct	1.761	341.8 [b]	1.604	344.4 [d]
F^--Adduct	1.777	341.7 [a]	1.619	342.6 [e]
Single Bond	≈1.87	≈328.5	≈1.73	≈328.5

a) $Me_2Si=C(SiMe_3)(SiMe^tBu_2)$ (Wagner, Müller). - b) $Me_2Si=C(SiMe_2Ph)_2$ (Joo, Polborn).
c) $^tBu_2Si=NSi^tBu_3$ (Schurz, Müller). - d) $Me_2Si=NSi^tBu_3$ (Schurz, Lerner, Müller, Polborn).
e) $^iPr_2Si=NMes^*$ (Klingebiel, Boese, JOM **315** (986) C17).

Table 1. Geometries of n-complexes of silaethenes and silaneimines (for d, α, β, γ; Eq. 2)

Furthermore, the geometry of adducts is influenced by the silicon bonded X groups. It should be pointed out that the rotation about the SiY bond is obviously not much hindered; therefore, the substituents at silicon are always oriented in such a way in the crystalline phase as to minimize repulsive intramolecular interactions with substituents at Y. Over and above that, it should be noticed that the formation of n-complexes of silenes stabilize the unsaturated systems.

The formation of n-complexes of silenes, as well as of germenes or stannenes is of central importance to the reactivity of the unsaturated systems. I will only offer one example out of many to illustrate the facts. As we found out, our stable silaneimine $tBu_2Si=NSitBu_3$ takes up carbon monoxide under formation of supersilyl isocyanide as well as cyclosiloxanes (Scheme 7).

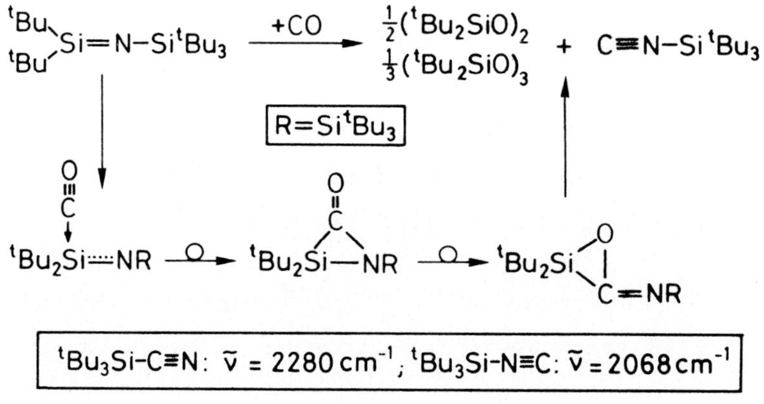

Scheme 7. Reaction of stable silaneimine $tBu_2Si=NSitBu_3$ with carbon monoxide

Obviously, the silaneimine like many transition metal complexes behaves like a Lewis acid even with regard to CO. Certainly, the CO adduct is unstable and transforms into the reaction products. Possibly the products with three-membered rings, shown on Scheme 7, are reaction intermediates. Supersilyl isocyanide could be prepared on this way for the first time because the reaction of cyanide with supersilyl halides exclusively gives supersilyl cyanide [21].

Altogether, the reaction with CO demonstrates impressively the overloaded-underloaded principle, mentioned at the beginning. The unsaturated silicon atom is, on the one hand, coordinatively underloaded. This, at first, makes the reaction with CO possible (catalytic processes at the centre of transition metal complexes). On the other hand, the overloaded supersilyl group shields the unsaturated silicon atom and therefore lays open the reactivity of the latter atom against CO. In normal cases, other reactions which will be operative never give the adduct formation with CO a chance.

The formation of n-complexes of silenes with donors like ethers or amines is understandable within the scope of effective HOMO/LUMO overlapping (Scheme 8).

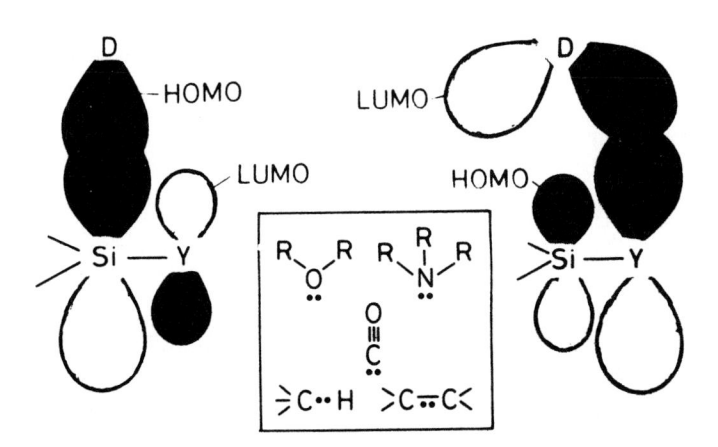

Scheme 8. HOMO and LUMO which characterize adduct formation of silenes and donors

Of course, this simple bonding model, which implies a certain similarity of silenes with compounds of aluminium, explains inadequately the intermediate formation of a complex with carbon monoxide. As is indeed known, π backbonding is essential for the stability of CO complexes. In contrast to aluminium compounds, silenes possess an occupied π molecular orbital which may interfere with a π^* state of CO (Scheme 8). Should then silenes also be qualified for intermediate formation of σ- and π-complexes, respectively? I will now present two reaction examples on this theme:

1) Silene Insertion into CH Bonds: By thermolyzing complexes of silaneimines with ethyldimethyl amine in solvents, one surprisingly observes the formation of insertion products of the silenes into α-C–H bonds of the donor [3]. For example, the silaneimine $tBu_2Si=NSiCltBu_2$ slowly transforms in benzene even at room temperature into the mentioned insertion product (Scheme 9) [12].

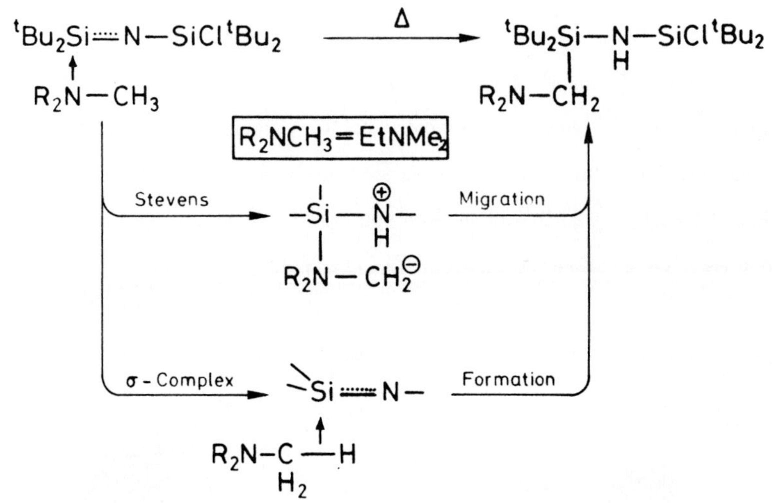

Scheme 9. Insertion of silenes in C–H bonds; σ-Complex formation

Its formation may be explained within the scope of a Stevens migration: in the first stage a methyl proton shifts to the unsaturated nitrogen atom, then the disilazyl group so formed shifts from the amine nitrogen to the deprotonated methyl group (Scheme 9). Indeed, in silaneimine adducts the silicon donor bond is not very strong and consequently the increase of acidity of the methyl group and basicity of the unsaturated nitrogen, respectively, not very large. Therefore, the basic requirements for a Stevens migration are unfavorable. On the other hand, a transformation of the silaneimine adduct into the product by way of an intermediate formation of a σ-*complex* of the silaneimine with the amine is more reasonable. Over and above that, as is known, the formation of weak σ-complexes is connected with a strong CH activation, that means a strong increase of CH acidity [22].

2) Silene Reactions with Organic Dienes. The second example of reactions will concern Diels-Alder reactions of silenes. Certainly, as was already mentioned, the diene additions to silaethenes occur much faster than those to ethenes. Over and above that, in most cases other reactions operate in addition. As is shown in Scheme 10, the [4+4] *cycloaddition* of silaethenes with 2,3-dimethyl-1,3-butadiene (DMB) leads – independently of the direction of silaethene addition to DMB – to one and the same [4+2] cycloadduct. On the other hand, the silaethenes may react with DMB also by an *ene reaction* under formation of the ene product shown in the middle of Scheme 10. This silylated DMB itself then undergoes a [4+2] cycloaddition with the silaethene in one or the other direction under formation of cycloadducts with different constitutions. Again, the silylated DMB may also react with the silaethene under formation of an ene product, shown in the middle of the bottom of Scheme 10.

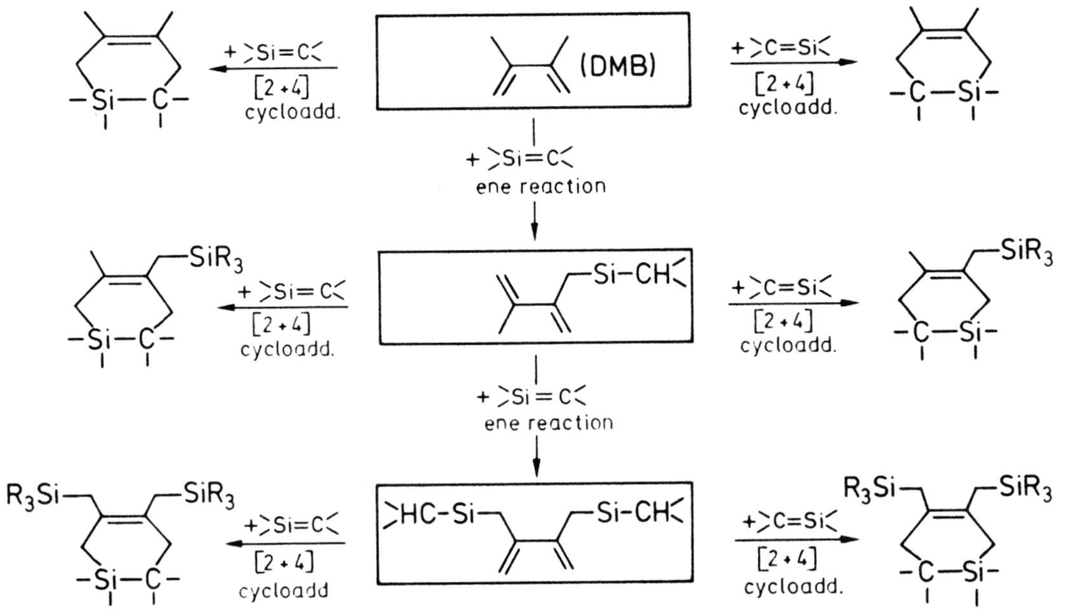

Scheme 10. Reactions of silaethene with 2,3-dimethyl-1,3-butadiene (DMB)

This disilylated DMB as well may undergo a [4+2] cycloaddition. With the standard silaethene all mentioned products are identified [1, 23]. Furthermore, a *[2+2] cycloaddition* of dienes and silaethenes – not shown on Scheme 10 – may be observed under special circumstances.

To simplify matters, I only will look at primary Diels-Alder reactions of the already mentioned standard silaethene with organic dienes. As is shown in Scheme 11, the [4+2] cycloaddition of this silaethene occurs with isoprene to a high degree *regioselective* and with *trans*-piperylene even more regiospecific [24]. The preferred direction of silaethene addition, which is meta to silicon with regard to the methyl group of the dienes, is explained by making use of the frontier orbital model of Houk [25]. All this is in agreement with organic Diels-Alder reactions.

The rate of cycloaddition increases – also in agreement with organic Diels-Alder reactions – when electron donating substituents like methyl groups are introduced into the dienes. For example, isoprene is four times, trans-piperylene three times as reactive as butadiene (Scheme 11) [23]. Of course, these substituents can be rate retarding on steric grounds. For this reason, trans-piperylene reacts somewhat slower than isoprene.

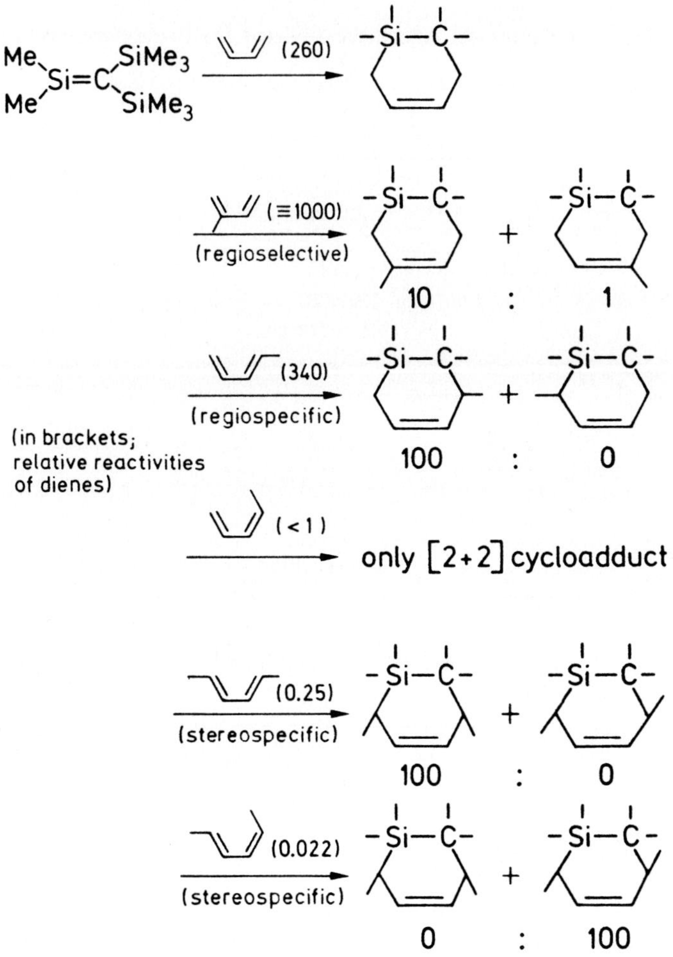

Scheme 11. Regioselectivity, stereospecifity, conformative effects and reaction rates of [4+2] cycloadditions (Diels-Alder reaction) of the silaethene Me$_2$Si=C(SiMe$_3$)$_2$ with organic dienes

Whereas in the case of *trans*-piperylene a [4+2] cycloadduct is formed with the standard silaethene, no such product is formed in the case of *cis*-piperylene but only a [2+2] cycloadduct (Scheme 11) [26]. Obviously, the [4+2] cycloaddition proceeds here more slowly than the unfavored [2+2] cycloaddition. In this sense, silico Diels-Alder reactions are influenced by *conformative effects* of the organic dienes like Diels-Alder reactions [27]: a displacement of the *transoid/cisoid* equilibrium in the direction of the *transoid* isomer on steric grounds decreases the rate of cycloadditions.

The [4+2] cycloaddition of the standard silaethene with *trans,trans-* and *cis,trans-*hexadiene occurs exclusively with *stereospecifity*, whereby the latter diene reacts 10 times more slowly than the former on grounds of unfavorable *cisoid* conformation [28].

In general, the influence of diene conformation on the rate of silico Diels-Alder reaction speaks for a concerted mechanism of the [4+2] cycloaddition. This mechanism is confirmed by the small influence of solvent polarity and radical traps on the *reaction rate*, which speaks against a pronounced ionic or radical multiple step mechanism. On the other hand, trans-piperylene reacts over 1000 times faster with the standard silaethene than does *trans,trans-*hexadiene, whereas the opposite is true for ethenes [27]. This fact contradicts dramatically the carbon-analogy of silaethenes, from which either *trans-*piperylene reacts too fast, or *trans,trans-*hexadiene too slow with the silaethene.

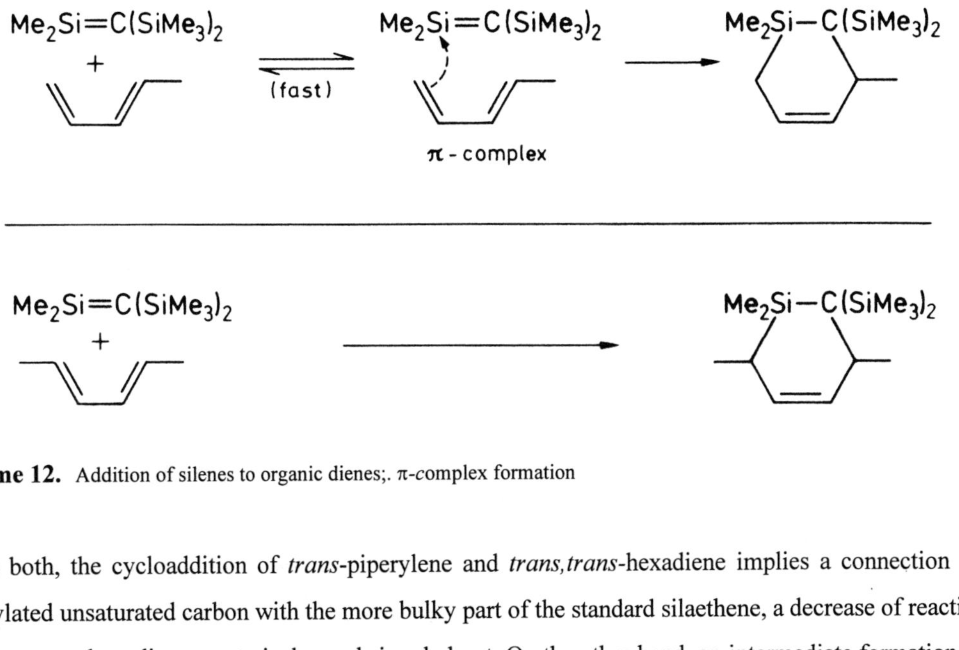

Scheme 12. Addition of silenes to organic dienes;. π-complex formation

As both, the cycloaddition of *trans-*piperylene and *trans,trans-*hexadiene implies a connection of a methylated unsaturated carbon with the more bulky part of the standard silaethene, a decrease of reactivity of *trans,trans-*hexadiene on sterical grouds is ruled out. On the other hand, an intermediate formation of a weak π-complex of *trans-*piperylene but not of bulkier *trans,trans-*hexadiene with the unsaturated silicon atom of silaethene would explain an increased reactivity of *trans-*piperylene (Scheme 12). In a way, the silaethene would then catalyze its own [4+2] cycloaddition with *trans-*piperylene but not with *trans,trans-*hexadiene. This mechanism can also explain the transformation of the [4+2] into a [2+2] cycloaddition by going from *trans-* to *cis-*piperylene but not from *trans,trans-* to *cis,trans-*hexadiene.

Now I will go over to the third question to be answered:

3 How Stable are Silenes?

Certainly, the stability of chemical compounds has many facets. In the case of silenes, germenes, and stannenes it involves among others, the kinetic or thermodynamic stability against intermolecular dimerization or intramolecular migrations.

The dimerization of the unsaturated systems is connected with an increase in coordination of the unsaturated atoms by one number in each case. As a result, increasing the bulkiness of substituents at the double bond leads to an increase in their *kinetic* and *thermodynamic* stabilities with regard to dimerization. Ultimately, the unsaturated systems like the silaethene $Me_2Si=C(SiMe_3)(SiMetBu_2)$ [15] or the silaneimine $tBu_2Si=NSitBu_3$ [3] became isolable. Now, regarding the *kinetic stability* of unstable silenes, germenes and stannenes in more detail, it can be said that the dimerization on the one hand is a process which is more or less hindered because of violation of orbital symmetry, and on the other hand, it is probably accelerated by small amounts of donors and certainly retarded by large amounts of donors.

Another aspect of stability of the unsaturated systems concerns the *π-bond strength* of the double bonded fourth group elements, the experimental determination of which certainly represents problems. Nevertheless, from π-bond migration processes some insights may be derived about the relative stability of these π-bonds as a function of *the double bond substituents*. In this connection, I turn first to the question: Are SiC-π-bonds stabilized by mesomeric effects of *silicon bonded phenyl groups*?

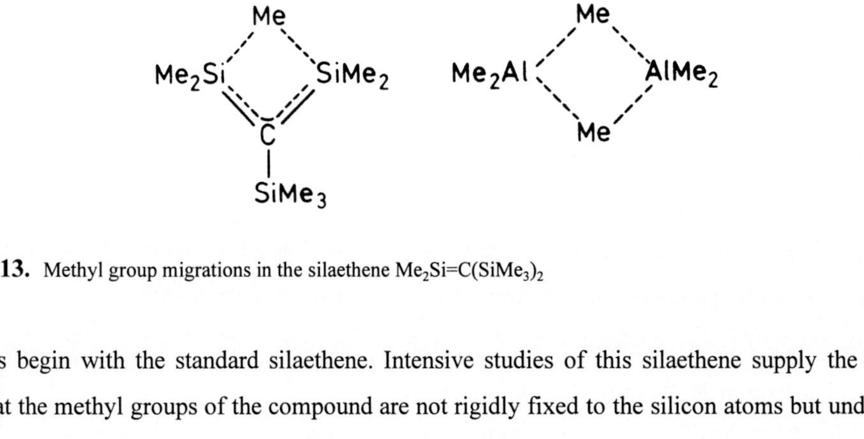

Scheme 13. Methyl group migrations in the silaethene $Me_2Si=C(SiMe_3)_2$

Let us begin with the standard silaethene. Intensive studies of this silaethene supply the surprising result that the methyl groups of the compound are not rigidly fixed to the silicon atoms but undergo rapid migration from one silicon atom to another (Scheme 13) [29]. In the sense: "a sorrow shared is a sorrow halved" every unsaturated silicon atom in the compound now and then yearns for saturation. Obviously, in the transition state of methyl migration a methyl group – as in dimeric trimethyl aluminium – is simultaneously bound to two metal centers. This state is best described in the sense of a σ-complex.

Each fluctomer of the standard silaethene has an equal energy content. The same is not valid for its diphenyl derivative, shown in Scheme 14 at the top left-hand side [9], which in contrast to the standard silaethene is metastable in diethyl ethers at −78 °C. At about −50 °C the mentioned silaethene transforms first under methyl then under phenyl and finally again under methyl migration into the silaethenes, shown in the second, third and fourth columns of Scheme 14. Now each of the four isomers, which in every case were independently prepared by salt elimination methods and are chemically characterized [30], has a different energy content. Silaethenes with phenyl groups at the unsaturated silicon atom are obviously richer in energy than those with methyl groups at this atom. Therefore, the expectation that unsaturated silicon will be stabilized by resonance with phenyl groups clearly is not the case. Phenyl groups act only as electron-withdrawing substituents which make the unsaturated silicon more positive, that is more Lewis acidic. Certainly, the tendency for group migrations of this compound originates in the tendency of the unsaturated silicon atom to lower its acidity. The transformation of $Ph_2Si=C(SiMe_3)_2$ into $Me_2Si=C(SiMe_2Ph)_2$ is in a way an intramolecular neutralization reaction.

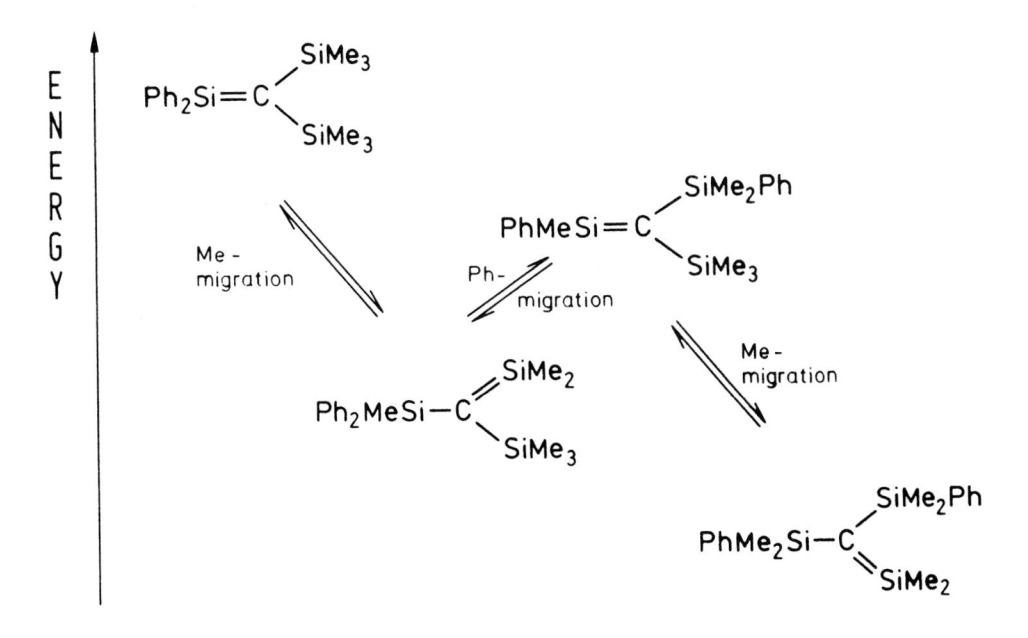

Scheme 14. Methyl and phenyl group migrations in the silaethene $Ph_2Si=C(SiMe_3)_2$

The reason of the methyl migration in the silaethene tBu$_2$Si=C(SiMe$_3$)$_2$ with bulky *silicon bonded tertiary butyl groups* is less due to *electronic* effects like in the silene Ph$_2$Si=C(SiMe$_3$)$_2$ than to *steric* effects: The bulkiness around the Si=C bond decreases somewhat, as is demonstrated by space-filling models, by going from this silaethene with tertiary butyl groups at unsaturated silicon to the isomer with methyl groups at this atom (Scheme 15). In fact, by generating tBu$_2$Si=C(SiMe$_3$)$_2$ by salt elimination, only the isomer Me$_2$Si=C(SiMe$_3$)(SiMetBu$_2$) is isolated [15]. With benzophenone as reagent the latter silaethene forms at −78 °C a yellow [4+2] cycloadduct. At −50 °C this adduct transforms itself into a colorless [2+2] cycloadduct of Me$_2$Si=C(SiMe$_3$)(SiMetBu$_2$) by way of cycloreversion and cycloaddition (cf. Scheme 15), as was proved by trapping the silaethene, formed as an intermediate. At −20 °C the [2+2] cycloadduct finally changes into a yellow [4+2] cycloadduct of the silene tBu$_2$Si=C(SiMe$_3$)$_2$ (Scheme 15) [31]. It follows from this that the migration equilibrium, shown on the first line of Scheme 15, really exists. Obviously, benzophenone reacts better with the silene tBu$_2$Si=C(SiMe$_3$)$_2$ on sterical grounds than with its isomer; the former, therefore, is trapped. The [4+2] cycloadduct itself works – when heated – as a source of tBu$_2$Si=C(SiMe$_3$)$_2$, which may be trapped by very active reactants like acetone.

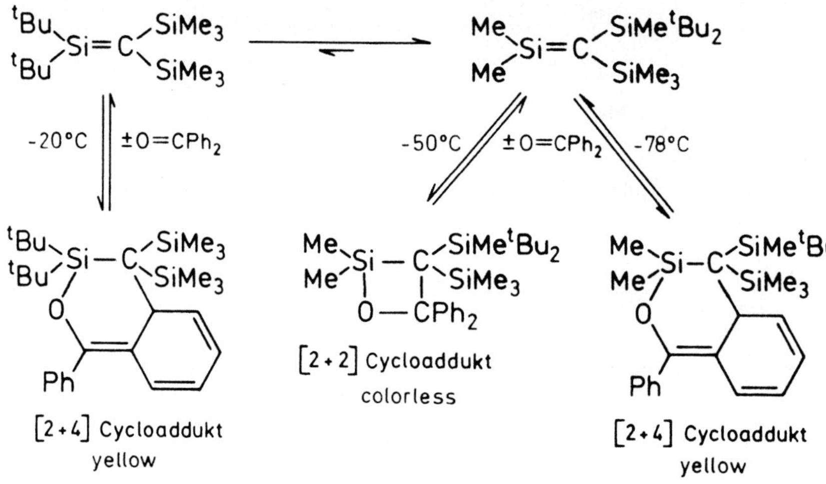

Scheme 15. Methyl group migration in the silaethene tBu$_2$Si=C(SiMe$_3$) and trapping of this silene and its isomer with benzophenone

The π-bond strength of unsaturated silicon compounds does not only depend on the double bond substituents but also on the nature of the *multiple bonded groups Y*. According to ab initio calculations, it increases by going from silaethenes >Si=C< through silaneimines >Si=N– to silanones >Si=O, and it decreases by going from ethenes >C=C< through silaethenes >Si=C< to disilenes >Si=Si<, or from ethynes –C≡C– through silaethynes –Si≡C– to disilynes –Si≡Si–, respectively. In fact, in the ground state, the underloaded disilyne Si_2H_2 has not a linear structure as acetylene C_2H_2 but exists as a folded four-membered ring which contains by turns silicon and hydrogen atoms in vertex positions and over and above that an SiSi single bond [32].

In this connection, we asked ourselves if disilyne with overloaded substituents has a linear structure analogous to acetylenes and if such an unsaturated system could be chemically demonstrated as a reaction intermediate or could even be isolated in substance. As overcrowded substituents we chose supersilyl groups, and as the method of preparation thermal salt elimination.

Accordingly, we first synthesized bis(supersilyl)tetrabromodisilane $(tBu_3Si)SiBr_2–SiBr_2(SitBu_3)$ (Scheme 16). It transforms by reaction with supersilyl sodium into tetrakis(supersilyl)tetrasilatetrahedrane in high yields, shown at the right-hand bottom of Scheme 16 [2].

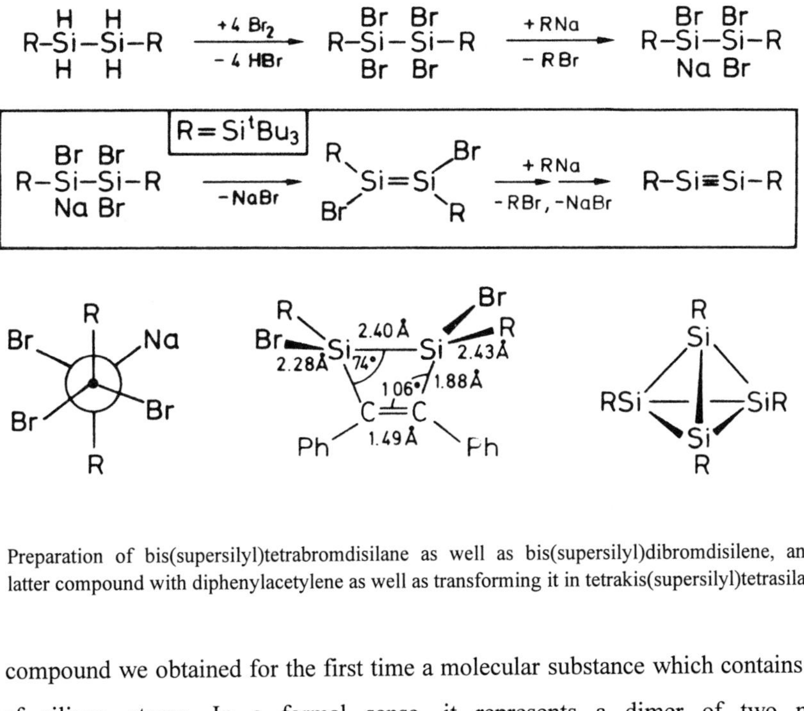

Scheme 16. Preparation of bis(supersilyl)tetrabromdisilane as well as bis(supersilyl)dibromdisilene, and trapping the latter compound with diphenylacetylene as well as transforming it in tetrakis(supersilyl)tetrasilatetrahedrane

With this compound we obtained for the first time a molecular substance which contains a tetrahedral framework of silicon atoms. In a formal sense, it represents a dimer of two molecules of bis(supersilyl)disilyne $tBu_3Si–Si≡Si–SitBu_3$. Certainly, this fact allows no conclusion about the real existence of a disilyne as a reaction intermediate (the opening remarks of this review).

It seems very probable that supersilyl sodium reacts with the mentioned bis(supersilyl)tetrabromdisilane by bromine/sodium exchange. The sodium containing disilane so formed should have a conformation with the overcrowded supersilyl groups in *trans*-positions (Scheme 16).

Therefore, NaBr elimination must produce a disilene with *trans*-configured supersilyl groups, which cannot have any tendency for dimerization because of its overloaded supersilyl substituents. Its intermediate existence could be proved by trapping it with diphenyl acetylene. Hereby, the compound shown at the middle of the bottom of Scheme 16 is found. As expected, according to an X-ray structure determination, it contains supersilyl groups in *trans*-positions [14].

The disilene (tBu$_3$Si)BrSi=SiBr(SitBu$_3$), found as an intermediate, probably reacts with excess supersilyl sodium under bromine/sodium exchange. Whether the so formed intermediate (tBu$_3$Si)BrSi=SiNa(SitBu$_3$) transforms by NaBr elimination into the disilyne tBu$_3$Si–Si≡Si–SitBu$_3$, which dimerizes, or – more probable – this intermediate adds to (tBu$_3$Si)BrSi=SiBr(SitBu$_3$) under formation of an insertion product, which ultimately transforms by intramolecular condensation reactions into the tetrasilatetrahedrane, must still be proved. We have not yet been successful in trapping the postulated disilyne.

4 Concluding Remarks

Obviously, supersilyl substituents are not sufficient for stabilizing a disilyne. Here, only substituents which are much more overcrowded than supersilyl groups ("megasilyl groups") may help on. Therefore, we learn that with regard to its stabilizing effect one and the same bulky group may act as an overloaded substituent for one compound but an underloaded substituent for another substance. With other words, like many other things, the bulkiness of a group is a very relative quantity.

Acknowledgement: We thank the *Deutsche Forschungsgemeinschaft* and the *Fonds der Chemischen Industrie* for generously supporting these investigations.

References:

[1] N. Wiberg, G. Preiner, O. Schieda, G. Fischer, *Chem. Ber.* **1981**, *114*, 2087, 3505, 3518.
[2] N. Wiberg, Ch. M. M. Finger, K. Polborn, *Angew. Chem.* **1993**, *105*, 1140; *Angew. Chem., Int. Ed. Engl.* **1993**, *32*, 923.
[3] N. Wiberg, K. Schurz, *Chem. Ber.* **1988**, *121*, 581.
[4] N. Wiberg, *J. Organomet. Chem.* **1984**, *273*, 141.
[5] N. Wiberg, Ch.-K. Kim, *Chem. Ber.* **1986**, *119*, 2966, 2980.

[6] N. Wiberg, P. Karampatses, Ch.-K. Kim, G. Preiner, K. Schurz, *Chem. Ber.* **1987**, *120*, 1203, 1213, 1357.
[7] N. Wiberg, G. Preiner, G. Wagner, H. Köpf, G. Fischer, *Z. Naturforsch.* **1987**, *42B*, 1055, 1062.
[7a] N. Wiberg, G. Preiner, K. Schurz, *Chem. Ber.* **1988**, *121*, 1407.
[8] N. Wiberg, G. Preiner, K. Schurz, G. Fischer, *Z. Naturforsch.* **1988**, *43B*, 1468.
[9] N. Wiberg, M. Link, G. Fischer, *Chem. Ber.* **1989**, *122*, 409.
[10] N. Wiberg, H. Schuster, *Chem. Ber.* **1991**, *124*, 93.
[11] N. Wiberg, S.-K. Vasisht, *Angew. Chem.* **1991**, *103*, 105; *Angew. Chem., Int. Ed. Engl.* **1991**, *30*, 93.
[12] N. Wiberg, H. W. Lerner, unpublished results.
[13] N. Wiberg, T. Passler, unpublished results.
[14] N. Wiberg, Ch. M. M. Finger, K. Polborn, unpublished results.
[15] N. Wiberg, G. Wagner, *Chem. Ber.* **1986**, *119*, 1455, 1467; N. Wiberg, G. Wagner, J. Riede, G. Müller, *Organometallics* **1987**, *6*, 32.
[16] N. Wiberg, G. Wagner, G. Müller, J. Riede, *J. Organomet. Chem.* **1984**, *271*, 381.
[17] N. Wiberg, H. Köpf, *J. Organomet. Chem.* **1986**, *315*, 9.
[18] N. Wiberg, K. Schurz, *J. Organomet. Chem.* **1988**, *341*, 145; G. Reber, J. Riede, N. Wiberg, K. Schurz, G. Müller, *Z. Naturforsch.* **1989**, *44B*, 786.
[19] N. Wiberg, K.-S. Joo, K. Polborn, *Chem. Ber.* **1993**, *126*, 67.
[20] Ch. Zybill, H. Handwerker, H. Friedrich, *Adv. Organomet. Chem.* **1994**, *36*, 229.
[21] M. Weidenbruch, H. Pesel, *Z. Naturforsch.* **1987**, *33B*, 1465.
[22] R. H. Crabtree, *Angew. Chem.* **1993**, *105*, 828; *Angew. Chem., Int. Ed. Engl.* **1993**, *32*, 789.
[23] N. Wiberg, S. Wagner, unpublished results.
[24] N. Wiberg, K. Schurz, G. Fischer, *Chem. Ber.* **1986**, *119*, 3498.
[25] K. N. Houk, *J. Am. Chem. Soc.* **1973**, *95*, 4092.
[26] N. Wiberg, G. Fischer, K. Schurz, *Chem. Ber.* **1987**, *120*, 1605.
[27] S. Sauer, R. Sustmann, *Angew. Chem.* **1980**, *92*, 773; *Angew. Chem., Int. Ed. Engl.* **1980**, *19*, 779.
[28] N. Wiberg, G. Fischer, S. Wagner, *Chem. Ber.* **1991**, *124*, 769.
[29] N. Wiberg, H. Köpf, *Chem. Ber.* **1987**, *120*, 653.
[30] N. Wiberg, K.-S. Joo, unpublished results.
[31] N. Wiberg, H.-S. Hwang-Park, unpublished results.
[32] R. G. Grev, *Adv. Organomet. Chem.* **1991**, *33*, 125.

The Synthesis of Transient Silenes using the Principle of the Peterson Reaction

Clemens Krempner, Helmut Reinke, Hartmut Oehme*

Fachbereich Chemie, Abteilung Anorganische Chemie

Universität Rostock

Buchbinderstr. 9, D-18051 Rostock, Germany

Summary: Deprotonation of 1-hydroxyalkyltris(trimethylsilyl)silanes, $(Me_3Si)_3Si-C(OH)R^1R^2$ (**1**) with methyl lithium in ether at low temperature leads to transient silenes, $(Me_3Si)_2Si=CR^1R^2$ (**2a**: $R^1 = R^2 = Me$; **2b**: $R^1 = H$, $R^2 = tBu$; **2c**: $R^1 = H$; $R^2 = Mes$), which dimerize in absence of trapping agents to give 1-isopropenyl-2-isopropyl-1,1,2,2-tetrakis-(trimethylsilyl)disilane **3**, (E)-3,4-di-tbutyl-1,1,2,2-tetrakis(trimethylsilyl)-1,2-disilacyclo-butane **4** and (E)-1,2,3,8α-tetrahydro-1-mesityl-5,7,8α-trimethyl-2,3,3-tetrakis(trimethyl-silyl)-2,3-disilanaphthalene **5**, resp. Deprotonation of **1a**–**1c** with excess organolithium reagents leads to trisilanes **6**, the addition products of the nucleophilic organolithium compounds across the Si=C bond of the intermediate silenes **2a**–**2c**. Compounds **2b** and **2c** can also be trapped with 2,3-dimethyl-1,3-butadiene to give the silacyclohex-3-enes **9a** and **9b**. The interaction of **1a**–**1c** with methyllithium or sodium hydride in THF causes a 1,3-Si–O trimethylsilyl migration resulting in formation of trimethylsiloxybis(trimethylsilyl)silyl-alkanes $(Me_3Si)_2SiH-C(OSiMe_3)R^1R^2$ **10a**–**10c**. Deprotonation of **1c** with methylmagnesium-bromide in THF gives (E)-2,4-dimesityl-1,1,3,3-tetrakis(trimethylsilyl)-1,3-disilacyclobutane **12**.

Introduction

The Peterson reaction, modified in a way demonstrated in Eq. 1, has proven to be a suitable method for the preparation of transient silaethenes [1–4].

$$\underset{\text{Me}_3\text{Si R}^2}{\overset{\text{Me}_3\text{Si OH}}{\text{Me}_3\text{Si-Si-C-R}^1}} \xrightarrow[\text{-Me}_3\text{SiOH}]{\text{base}} \underset{\text{Me}_3\text{Si}'}{\overset{\text{Me}_3\text{Si}}{\text{Si}}}{=}\underset{R^2}{\overset{R^1}{}}$$

Eq. 1.

But till now only methods involving an *in situ* formation of deprotonated **1** have been described. In this paper we present a procedure leading to pure, isolated 1-hydroxyalkyl-polysilanes **1**. In presence of base they are easily converted into transient silenes, which were characterized by various dimerization and addition reactions. The availability of isolated **1** in the synthesis of silenes according to the Peterson concept offers the possibility of a free choice of the solvent and the base used to initiate the silanolate elimination. With respect to the significance of the reaction medium for the silene generation and its subsequent reactions, this is of particular importance.

The Synthesis of the 1-Hydroxyalkyl-tris(trimethylsilyl)silanes 1a–1c

$$\underset{\text{Me}_3\text{Si}}{\overset{\text{Me}_3\text{Si}}{\text{Me}_3\text{Si-Si-MgBr}}} + R^1\overset{O}{\underset{\|}{C}}R^2 \longrightarrow \left[\underset{\text{Me}_3\text{Si R}^2}{\overset{\text{Me}_3\text{Si OMgBr}}{\text{Me}_3\text{Si-Si-C-R}^1}}\right]$$

$$\downarrow (\text{H}^+)$$

$$\underset{\text{Me}_3\text{Si R}^2}{\overset{\text{Me}_3\text{Si OH}}{\text{Me}_3\text{Si-Si-C-R}^1}}$$

1a–c

1a: R^1 = Me R^2 = Me
1b: R^1 = H R^2 = tert.Bu
1c: R^1 = H R^2 = Mes

Eq. 2.

Tris(trimethylsilyl)silyl-magnesiumbromide, made *in situ* by interaction of tris(trimethylsilyl)silyl-lithium [5] with anhydrous magnesium bromide in ether, reacts with aldehydes and ketones under addition of the magnesium silanide to the C=O double bond.

During hydrolytic work up the intermediate magnesium alkoxides, which at low temperature do not eliminate silanolate, are converted into the alkohols **1a–1c** [Eq. 2]. **1a–1c** are stable, crystalline, waxy solids [6].

The Generation and Conversion of the Transient Silenes 2a–2c

Deprotonation of **1a–1c** with methyllithium in ether at −78 °C leads to spontaneous elimination of lithiumtrimethylsiloxide according to a modified Peterson mechanism and formation of the unstable silenes **2a–2c**. In absence of trapping agents these undergo dimerization reactions, **2a** leading to 1-isopropenyl-2-isopropyl-1,1,2,2-tetrakis(trimethylsilyl)disilane (**3**) (Scheme 1).

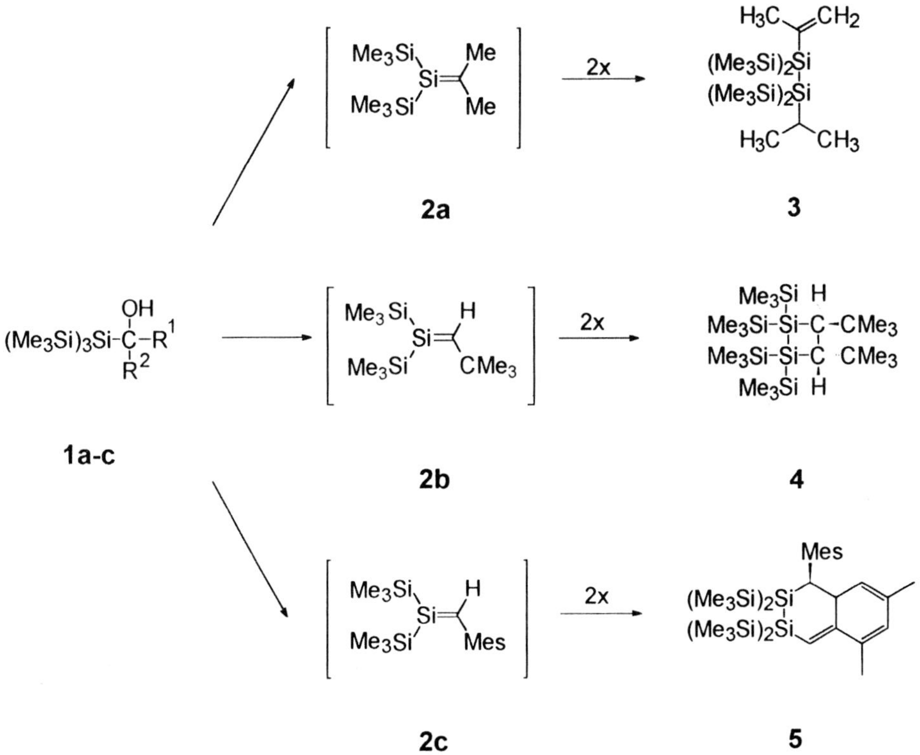

Scheme 1.

Compound **3** has also been obtained by Ishikawa *et al.* by interaction of acetyltris(trimethylsilyl)silane with methyllithium [3]. Deprotonation of **1b** under the same conditions gives (*E*)-3,4-di-*t*butyl-1,1,2,2-tetrakis(trimethylsilyl)-1,2-disilacyclobutane **4**, the head-to-head [2+2] cycloaddition product of silene **2b** (Scheme 1). The constitution of **4** is proved also by an X-Ray crystal structure analysis (Fig. 1).

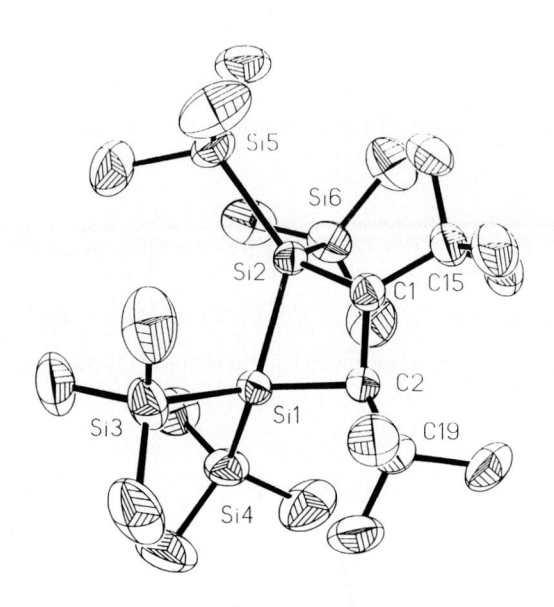

Fig. 1.

The four-membered ring is slightly folded. The planes through the atoms Si2–Si1–C2 and Si2–C1–C2 meet at an angle of 35.9 °. The ring C–C bond distance is 157.7 pm [6].

An unexpected und unusual dimerization behaviour is observed for 2-mesityl-1,1-bis(trimethylsilyl)-silene (**2c**), formed by deprotonation of **1c**. The structure of the product, obtained in more than 70 % yield, was revealed as (*E*)-1,2,3,8α-tetrahydro-1-mesityl-5,7,8α -trimethyl-2,2,3,3-tetrakis(trimethylsilyl)-2,3-disilanaphthalene (**5**) (Scheme 1). The formation of this unexpected structure is interpreted as the result of a head-to-head-dimerization of the transient silene **2c**, which in an unusual [2+4]-cycloaddition reaction is formally acting as the monoene and – involving the aromatic substituent – also as the diene (Eq. 3).

Eq. 3.

The result of the X-Ray analysis is shown in Fig. 2.

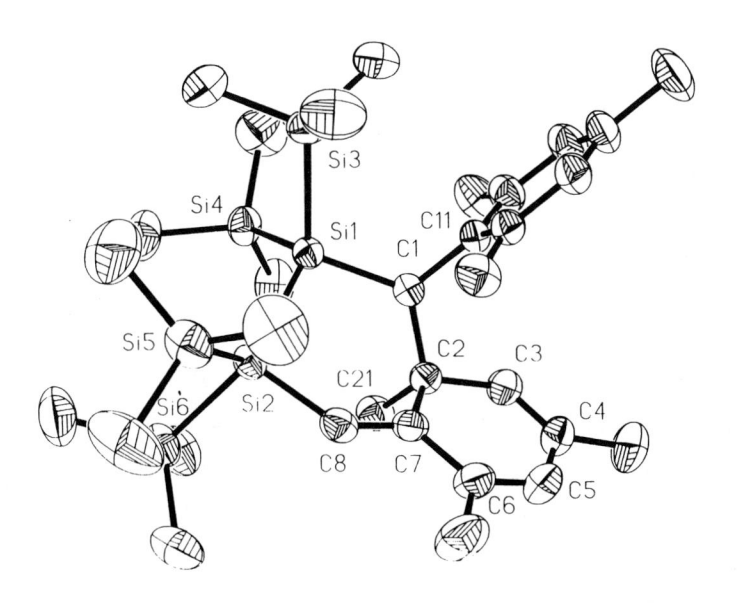

Fig. 2.

For the disilacyclohexene substructure a boat form is found, the intersection angle of the plane Si1–Si2–C8 being 16.8 ° and that of the plane C1–C2–C7 43.6 ° with respect to the plane Si1–C1–C7–C8. The bond distance C1–C2 (159.0 pm) is comparatively long [6].

The dimerization of the transient silenes **2a–2c** is discussed in terms of a radical process, which was proved by Brook *et al.* for some similarly structured 2-alkyl- and 2-phenyl-2-trimethylsiloxy-1,1-

bis(trimethylsilyl)silenes [7, 8]. After Si–Si bond formation the resulting 1,4-biradicals stabilize usually to give 1,2-disilacyclobutanes, such as in the case of silene **2b**, which is the typical behavior of sterically congested unstable silenes. Sterically crowded silenes bearing „allylic" protons prefer – after Si–Si-bond formation and hydrogen transfer – a linear dimerization, as observed for **2b** and related silenes [2, 3, 7, 9]. In case of **2c** the mesityl substituent allows the extension of the conjugation system of the biradical intermediate and the observed [2+4]-reaction can be understood as the cyclization of a transient 1,6-biradical. This deviation from the standard behaviour of sterically crowded silenes may be due to the bulkyness of the mesityl substituents and the expected weakness of the ring C–C bond in a hypothetic 1,2-disilacyclobutane, the [2+2]-dimer of **2c**.

Deprotonation of the hydroxyalkylpolysilanes **1a–1c** with two or more equivalents methyllithium, *t*butyllithium, or phenyllithium, respectively, leads to the trisilanes **6a–6e**, which are formed by addition of the excess organolithium reagent to the polar Si=C-bond (Eq. 4).

	R^1	R^2	R^3
6a:	Me	Me	Me
6b:	H	t-Bu	t-Bu
6c:	H	t-Bu	Ph
6d:	H	Mes	Me
6e:	H	Mes	Ph

Eq. 4.

Besides its synthetic significance, this result is at the same time a valuable proof for the intermediate existence of transient silenes also in the formation of **3–5**. Whereas the addition of the organolithium compounds to the silicon carbon double bond with the observed regiospecifity is obvious, the formation of **6a–6e** by interaction of the nucleophilic organolithium derivatives with any precursor of the silenes **2a–2c** is hardly conceivable.

Similarly, when silene **2a** is generated in presence of excess tris(trimethylsilyl)silyllithium, the lithium silanide is added across the silicon-carbon double bond to give an organolithium intermediate, which undergoes a rearrangement, a 1,3-Si,C-trimethylsilyl migration, resulting in formation of a lithium silanide, which is trapped with chlorotrimethylsilane to yield the polysilane **7**. The H-silane **8** is obtained as the protonation product after usual hydrolytic work up (Eq. 4–5).

$$\begin{bmatrix} Me_3Si & Me \\ Si{=} \\ Me_3Si & Me \end{bmatrix} + (Me_3Si)_3SiLi \longrightarrow \begin{bmatrix} Me_3Si\diagdown\diagup SiMe_3 \\ Si \\ (Me_3Si)_2Si\diagup\diagdown CMe_2 \\ Me_3Si \quad\ Li \end{bmatrix}$$

2a

$$\downarrow$$

$$\begin{bmatrix} Me_3Si\diagdown\diagup SiMe_3 \\ Si \\ (Me_3Si)_2Si\diagup\diagdown CMe_2 \\ Li \quad\ SiMe_3 \end{bmatrix}$$

(Me$_3$SiCl) ↙   ↓ (H$^+$)

Me$_3$Si SiMe$_3$ Me$_3$Si SiMe$_3$
Me$_3$Si–Si–Si–CMe$_2$SiMe$_3$ Me$_3$Si–Si–Si–CMe$_2$SiMe$_3$
Me$_3$Si SiMe$_3$ H   SiMe$_3$

7   **8**

Eq. 5.

Deprotonation of **1b** or **1c** with methyllithium in presence of 2,3-dimethyl-1,3-butadiene gives the silacyclohexenes **9a** and **9b**, the expected [2+4] cycloaddition products of the respective silene and the diene (Eq. 6). Attempts to perform the same reaction with **1a** failed.

$$(Me_3Si)_3Si-\underset{H}{\underset{|}{C}}-R \quad + \quad \text{[methylenecyclohexane]} \quad \xrightarrow{MeLi} \quad \text{9a,b}$$

1b,c

9a,b structure: Me₃Si, Me₃Si-Si, R, H on cyclohexene ring with two Me groups

9a: R = tertBu
9b: R = Mes

Eq. 6.

Deprotonation of the Hydroxyalkylpolysilanes 1a–1c in THF

The outcome of the reactions of **1a–1c** initiated by deprotonation, is clearly dependent on the reaction medium and the applied base, i.e., the metal cation acting as the counter ion in the intermediate alkoxides. Changing the solvent, from ether to THF, the reaction of **1a–1c** with lithium methyl at low temperature leads to trimethylsiloxybis(trimethylsilyl)silylalkanes **10a–10c** (Eq. 7).

$$\underset{\textbf{1a-c}}{\underset{Me_3Si\ R^2}{\underset{Me_3Si-Si-C-R^1}{Me_3Si\ OH}}} \quad \xrightarrow[(H_2O)]{\text{MeLi in THF} \atop \text{NaH in ether or THF}} \quad \underset{\textbf{10a-c}}{\underset{Me_3Si\ R^2}{\underset{Me_3Si-Si-C-R^1}{H\ OSiMe_3}}}$$

10a: R = Me   R = Me
10b: R = H   R = tertBu
10c: R = H   R = Mes

Eq. 7.

The same products are obtained, when sodium hydride is used as the base in ether as well as in THF. This is the result of just a 1,3-Si,O-trimethylsilyl shift. Under these conditions no trimethylsiloxide is eliminated; **3**, **4**, or **5**, respectively, cannot be identified.

As the result of the interaction of **1c** with methylmagnesiumbromide (made from MeLi and MgBr$_2$) in THF at 0 °C besides large quantities of **10c** a solid product is obtained in a 15 % yield, which was very surprisingly revealed as (*E*)-2,4-dimesityl-1,1,3,3-tetrakis(trimethylsilyl)-1,3-disilacyclobutane **12** (Eq. 8).

Eq. 8.

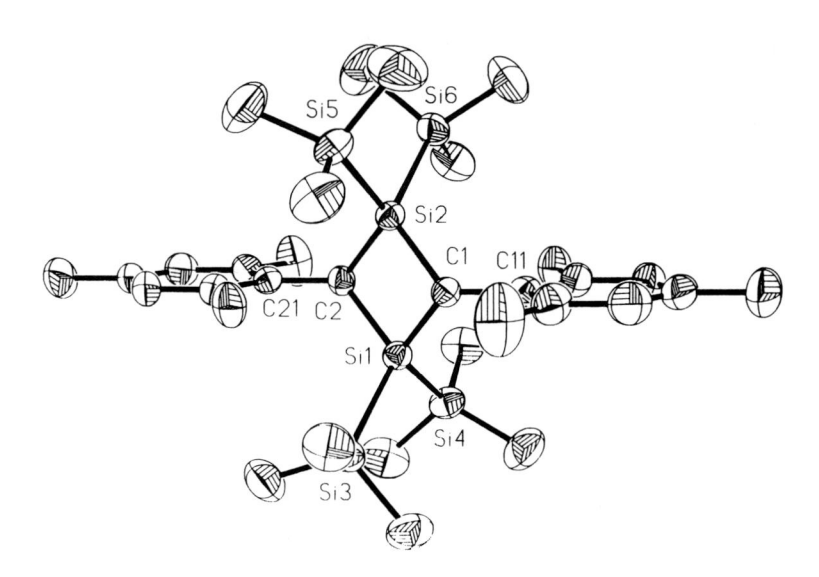

Fig. 3.

Fig. 3 shows the result of the X-ray structure analysis. The 1,3-disilacyclobutane ring is almost planar. The distance between the two ring silicon atoms is 274.4 pm [10].

12 is formally the head to tail [2+2] cyclodimere of the transient silene **2c**. But it is supposed that the change of the reaction medium and particularly the presence of magnesium as the counter ion of the intermediate alkoxide or silanide, respectively, may effect a complete change of the reaction mechanism. Whereas the formation of all products obtained by conversion of **1a–1c** with methyllithium in ether is best interpreted by the intermediate existence of respective transient silenes, it is supposed **12** to be the result of an intermolecular trimethylsilanolate elimination from the intermediate magnesium silanide **11**.

Acknowledgement: We gratefully acknowledge the support of our research by the *Deutsche Forschungsgemeinschaft* and the *Fonds der Chemischen Industrie*.

References:

[1] R. Wustrack, H. Oehme, *J. Organomet. Chem.* **1988**, *352*, 95.
[2] D. Bravo-Zhivotovskii, V. Braude, A. Stanger, M. Kapon, Y. Apeloig, *Organometallics*, **1992**, *11*, 2326.
[3] J. Ohshita, Y. Masaoka, M. Ishikawa, *Organometallics* **1991**, *10*, 3775.
[4] H. Oehme, R. Wustrack, A. Heine, G. M. Sheldrick, D. Stalke, *J. Organomet. Chem.* **1993**, *452*, 33.
[5] G. Gutekunst, A. G. Brook, *J. Organomet. Chem.* **1982**, *225*, 1.
[6] C. Krempner, H. Reinke, H. Oehme, *Chem. Ber.*, submitted for publication.
[7] A. G. Brook, J. W. Harris, J. Lennon, M. El Sheikh, *J. Am. Chem. Soc.* **1979**, *101*, 83.
[8] A. G. Brook, K. M. Baines, *Adv. Organomet. Chem.* **1986**, *25*, 26.
[9] K. Baines, A. G. Brook, *Organometallics* **1987**, *6*, 692.
[10] C. Krempner, H. Reinke, H. Oehme, *Angew. Chem.*, in press.

Cycloaddition Reactions of 1,1-Dichloro-2-neopentylsilene with Pentafulvenes

Claus-Rüdiger Heikenwälder, Norbert Auner*

Fachinstitut für Anorganische und Allgemeine Chemie

Humboldt-Universität zu Berlin

Hessische Str. 1-2, D-10115 Berlin, Germany

Summary: The reaction between in situ formed 1,1-dichloro-2-neopentylsilene, $Cl_2Si=CHCH_2tBu$ (**3**), and 1,1-dimethylpentafulvene (**8**) leads to the formation of *exo/endo*-isomeric [4+2] cycloadducts **9** and [2+2] stereoisomers **10** in good yields. 2D-NMR spectroscopic investigations on the product mixture prove the different mode of the silene cycloaddition reactions ([4+2] vs [2+2] addition).

The product formation from reactions of **9** and **10** with MeLi and LiAlH$_4$ yields the stereo- and regioisomeric substituted derivatives (MeLi: **11**, **12**; LiAlH$_4$: **14**, **15**), whereas PhMgBr reacts selectively with **10** to give the silacyclobutane **13**. The reactions of silene **3** with other pentafulvenes (e.g., **16**, **17** and **18**) lead to similar results.

1,1-Dichloro-2-neopentylsilene (**3**) is obtained by the reaction of trichlorovinylsilane (**1**) and Li*t*Bu via the α-lithiated species **2** [1]. The formation and reactivity of various neopentylsilenes, $R^1R^2Si=CR^3CH_2tBu$ ($R^1 = R^2$ = Cl (**3**), OSiMe$_3$, Me, *t*Bu, Ph; R^3 = H, SiMe$_3$, Ph), has been studied extensively [1, 2] and showed that the cycloaddition potential of **3** is quite different from that of diorganosubstituted derivatives; its tendency to preferentially form [2+2] as well as [4+2] cycloadducts in the presence of dienes has aroused interest concerning its electronic structure.

Cycloaddition reactions with anthracene [3], cyclopentadiene [3] and dienes of lower activity, such as naphthalene [4] and furans [6], are well known. In this contribution, we describe the competitive formation of [4+2] and [2+2] cycloaddition compounds resulting from $Cl_2Si=CHCH_2tBu$ (**3**) and 6,6-dimethylpentafulvene.

Scheme 1. Facile synthesis of [4+2] cycloaddition compounds from silene **3**

Fulvenes are cyclic, cross-conjugated molecules with an odd number of carbon atoms in the ring. According to the size of the ring skeleton they are named tria- (**A**), penta- (**B**), hepta- (**C**), and nonafulvenes (**D**) (Fig. 1) [6].

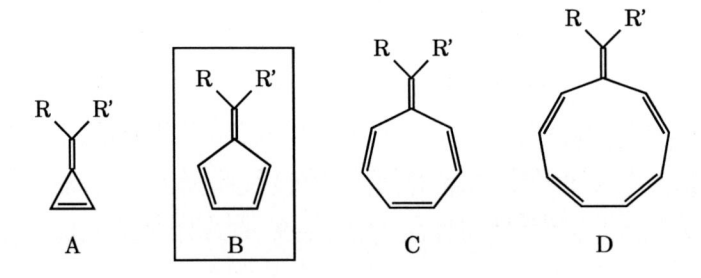

Fig. 1.

From this series, pentafulvenes have been most intensively investigated; they were first prepared and named (latin: *fulvus*, meaning yellow) by Thiele in 1900 [7]. Since then a large number of pentafulvene syntheses were developed [8], and their reactivity pattern is well known today:

They react easily with electrophiles and add nucleophiles at C-6. In cycloaddition reactions they may react as 2π, 4π, or 6π compounds. According to frontier orbital considerations they readily react with electron-deficient dienophiles (e.g., silenes) in Diels-Alder reactions; this is due to the strong interaction between the fulvene HOMO and dienophile LUMO [9]. Although the π and π^* orbitals of silenes are generally 1–2.5 eV higher in energy than is the case for the alkene congeners [10] a normal [4+2] cycloaddition behaviour for **3** is observed in earlier works [3–5].

This frontier orbital consideration fits well with our investigations [11]: Reaction of **3** with pentafulvene **8** [12] leads to the [4+2] *exo/endo*-isomers **9**. Compared to reactions of **3** with cyclopentadiene [3] it is surprising that the *E/Z* [2+2] cycloadducts **10** are formed (*exo/endo:E/Z* = 45/25:29/11) additionally.

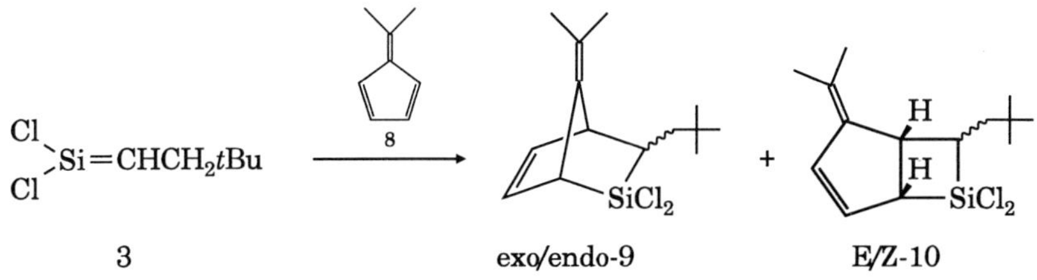

Scheme 2. Cycloaddition reactions of silene **3** with fulvene **8**

The oily mixture **9/10** cannot be separated by distillation. NMR investigations prove the presence of both isomeric products [13].

A useful method to separate regioisomeric cycloadducts is based on their derivatization by organo groups [1, 14]. Thus, the reaction of **9/10** with two equivalents of methyllithium or LiAlH$_4$ yields compounds **11** and **12** (average ratio of *exo/endo:E/Z* = 45/25:11/29) or **14** and **15**, whereas the reaction with phenyl Grignard reagent leads to a product mixture containing the *exo/endo*-isomeric [4+2] cycloadducts **9** and the diphenylated products *E/Z*-**13**, which can be separated.

Fig. 2.

Furthermore cycloaddition reactions of silene **3** with other pentafulvenes (e.g., **16**, **17**, and **18**) [15] lead to similar results, which will be published elsewhere [11].

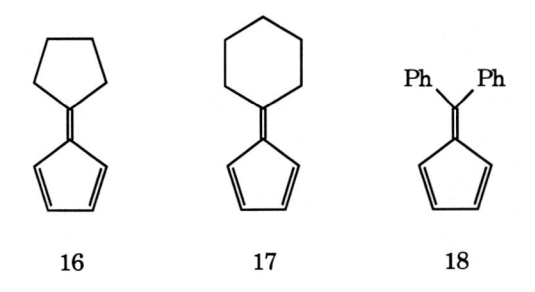

Fig. 3.

Acknowledgment: This work has been supported by *Fonds der Chemischen Industrie, Stiftung Volkswagenwerk, Wacker-Chemie GmbH, Chemetall GmbH,* and *Dow Corning Ltd.* (Barry).

References:

[1] For recent reviews on 1,1-dichloro-2-neopentylsilene see: N. Auner, *"Neopentylsilenes – Laboratory Curiosities or Useful Building Blocks for the Synthesis of Silaheterocycles"*, in: *Organosilicon Chemistry – From Molecules to Materials* (Eds.: N. Auner, J. Weis), VCH, Weinheim, **1994**, p. 103; N. Auner, *J. Prakt. Chem.* **1995**, *337*, 79.

[2] For formation and reactivity see: $(Me_3SiO)_2Si=CHCH_2tBu$: N. Auner, C.-R. Heikenwälder, W. Ziche, *Chem. Ber.* **1993**, *126*, 2177; N. Auner, C.-R. Heikenwälder, C. Wagner, *Organometallics* **1993**, *12*, 4135; $Me_2Si=CHCH_2tBu$: P. R. Jones, T. F. O. Lim, *J. Am. Chem. Soc* **1977**, *99*, 2013; P. R. Jones, T. F. O. Lim, *J. Am. Chem. Soc.* **1977**, *99*, 8447; P. R. Jones, T. F. O. Lim, R. A. Pierce, *J. Am. Chem. Soc.* **1980**, *102*, 4970; $tBu_2Si=CHCH_2tBu$: N. Auner, *Z. Anorg. Allg. Chem.* **1988**, *558*, 87; $Ph_2Si=CHCH_2tBu$: N. Auner, W. Ziche, E. Herdtweck, *J. Organomet. Chem.* **1992**, *142*, 1; $Cl_2Si=C(SiMe_3)CH_2tBu$: N. Auner, W. Ziche, J. Behm, *Organometallics* **1992**, *11*, 2494; $Cl_2Si=C(Ph)CH_2tBu$: N. Auner, C. Wagner, W. Ziche, *Z. Naturforsch.* **1994**, *49b*, 831.

[3] N. Auner, *J. Organomet. Chem.* **1988**, *353*, 275.

[4] N. Auner, C. Seidenschwarz, N. Sewald, E. Herdtweck, *Angew. Chem.* **1991**, *103*, 425; *Angew. Chem., Int. Ed. Engl.* **1991**, *30*, 444.

[5] N. Auner, A. Wolff, *Chem. Ber.* **1993**, *126*, 575.

[6] For reviews on fulvenes see: J. H. Day, *Chem. Rev.* **1953**, *53*, 167; E. D. Bergmann, *Prog. Org. Chem.* **1955**, *3*, 81; P. Yates, *Fulvenes*, in: *Advances in Alicyclic Chemistry, Vol 2*, (Ed.: H. Hart), Academic Press, New York, **1968**, p. 59; E. D. Bergmann, *Chem. Rev.* **1968**, *68*, 41; M. Neuenschwander, *Fulvenes*, in: *Supplement A: The Chemistry of Double-Bonded Functional Groups, Vol 2, Part 2* (Eds.: S. Patai, Z. Rappaport), Wiley, Chichester, **1989**, p. 1131.

[7] J. Thiele, *Ber.* **1900**, *33*, 666.

[8] Fulvene syntheses have been thoroughly reviewed in: K. P. Zeller, *Pentafulvenes*, in: *Houben Weyl, Methoden der organischen Chemie, Vol 5/2c* (Eds.: H. Kropf, K. H. Bückel), Thieme, Stuttgart, New York, **1985**, p. 476.

[9] K. N. Houk, J. K. George, R. E. Duke, Jr., *Tetrahedron* **1974**, *30*, 523.

[10] Y. Apeloig, M. Karni, *J. Am. Chem. Soc* **1984**, *106*, 6676.

[11] N. Auner, C.-R. Heikenwälder, submitted.

[12] For the synthesis of 6,6-dimethylfulvene (**8**) see: K. J. Stone, R. D. Little, *J. Org. Chem.* **1984**, *49*, 1849.

[13] A full NMR spectroscopic assigment ($^1H^1H$-COSY and $^1H^{13}C$ correlation spectra) is given in [11].

[14] N. Auner, C. Seidenschwarz, N. Sewald, *Organometallics* **1992**, *11*, 1137.

[15] For the synthesis of 6,6-tetramethylenefulvene (**16**) and 6,6-pentamethylenefulvene (**17**) see: K. J. Stone, R. D. Little, *J. Org. Chem.* **1984**, *49*, 1849; for the synthesis of 6,6-diphenylfulvene (**18**) see: J. L. Kice, F. M. Parham, *J. Am. Chem. Soc.* **1958**, *80*, 3792.

Silaneimines

Nils Wiberg*, Hans-Wolfram Lerner

Institut für Anorganische Chemie

Ludwig-Maximilians-Universität München

Meiserstrasse 1, D-80333 München, Germany

Preparation of Silaneimines

The reaction of one equivalent tBu_2XSiNa (X = tBu, Ph) with one equivalent tBu_2SiClN_3 leads to silaneimine $tBu_2Si=N-SiXtBu_2$ possibly by the way of triazene $tBu_2ClSi-N=N-NNa-SiXtBu_2$ (Scheme 1 and Table 1). Over and above that silatetrazolines of type $tBu_2Si=N-SiR_3$ X tBu_2SiXN_3 (R= tBu, Ph; X= H, Me, F, Cl, Br) thermolyze quantitatively by [3+2]-cycloreversion in silaneimines and silyl azides (Scheme 1 and Table 1). Kinetic studies show, that this decomposition follows first order kinetics [1].

Modified salt elimination:

$$tBu_2SiClN_3 + tBu_2XSiNa \xrightarrow[\text{(Bu}_2\text{O},-78°C)]{-NaCl, -N_2} tBu_2Si=N-SiXtBu_2$$

Thermolysis of silatetrazolines:

$$tBu_2Si=N-SitBu_3 \text{ X } tBu_2SiXN_3 \xrightarrow[\text{(benzene, r.t)}]{-tBu_3SiN_3} tBu_2Si=N-SiXtBu_2$$

Scheme 1.   Preparation of silaneimines

Table 1. ^{29}Si NMR data of silaneimines

$\delta\,^{29}$Si of tBu$_2$Si=N–Si'XtBu$_2$		
X =	Si	Si'
tBu	78.4	−7.7
Ph	81.8	−13.1
Cl	3.0	3.0

Preparation of Silaneimine Donor Adducts

Salt elimination in presence of a donor

R$_2$SiCl–NM–SiR$_3$

route A
−MCl, + D

Modified salt elimination in presence of a donor

R$_2$SiClN$_3$ + R$_3$SiNa

route B
−NaCl, + D

D→Si=N-Si-

−R$_3$SiN$_3$, +D
route C

R$_2$Si=N–SiR$_3$ X R$_3$SiN$_3$

Thermolysis of silatetrazolines in presence of a donor

+ D
route D

R$_2$Si=N–SiR$_3$

Reaction of silaneimines with donors

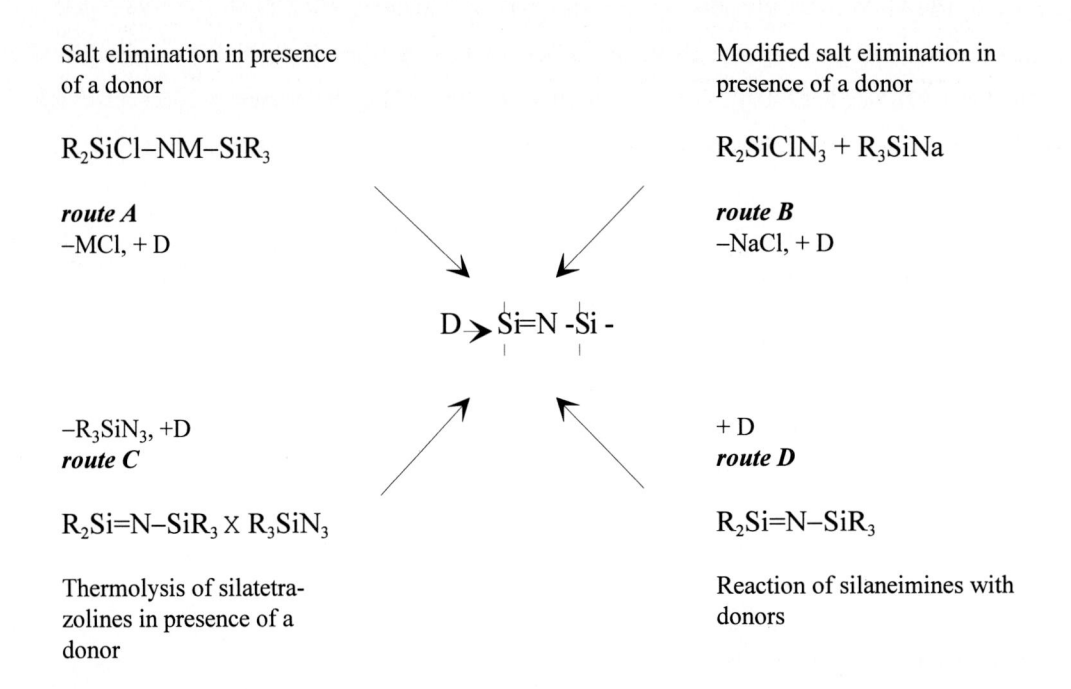

Scheme 2. Preparation of silaneimine donor adducts

	Preparation Route	δ ^{29}Si	δ ^{29}Si'
Me$_2$Si=N–SitBu$_3$·Me$_2$NEt	A	−8.9	−10.3
Et$_2$Si=N–SitBu$_3$·Me$_2$NEt	A	−8.8	−12.3
tBu$_2$Si=N–SiCltBu$_2$·Me$_2$NEt	C	−3.4	−6.4
tBu$_2$Si=N–SitBu$_3$·Me$_2$NEt	D	18.1	−13.6
Me$_2$Si=N–SitBu$_3$·NEt$_3$	A	−11.2	−11.1
tBu$_2$Si=N–SitBu$_3$·NMe$_3$	D	−12.5	−12.6
Me$_2$Si=N–SitBu$_3$·Et$_2$O	A	−1.5	−11.1
tBu$_2$Si=N–SiCltBu$_2$·Et$_2$O	C	−0.2	−6.5
Me$_2$Si=N–SitBu$_3$·THF	A	−4.4	−11.1
tBu$_2$Si=N–SiCltBu$_2$·THF	A, C	2.9	−5.2
tBu$_2$Si=N–SitBu$_3$·THF	A, B, D	1.1	−14.7

Table 2. Preparation and ^{29}Si NMR data of silaneimine donor adducts, R$_2$Si=N–Si'R$_3$·D

Reactivity of Silaneimines

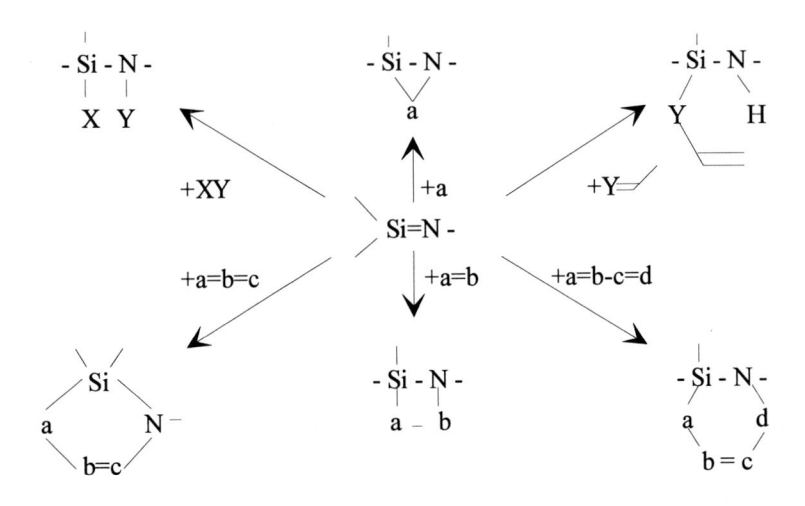

Scheme 3. Reactivity of silaneimines

Silaneimines undergo insertions into X–Y single bonds, ene reactions with enes $CH_3–CR=Y$ and [2+1], [2+2], [3+2], and [4+2] cycloadditions with unsaturated systems a, a=b, a=b=c, and a=b–c=d [2] (Scheme 3 and Table 3).

Insertion reactions		Ene reactions		[2+2] Cycloaddition:	
Silaneimine:	XY:	Silaneimine:	Enophile:	Silaneimine:	a=b
$Me_2Si=N–SitBu_3$	$Me_2SiOHNHSitBu_3$	$Me_2Si=N–SitBu_3$	CH_3COCH_3	$tBu_2Si=N–SiCltBu_2$	$CH_2=CH–OMe$
$Et_2Si=N–SitBu_3$	$Me_2SiOHNHSitBu_3$	$Et_2Si=N–SitBu_3$	CH_3COCH_3	$tBu_2Si=N–SitBu_3$	$CH_2=CH–OMe$
$tBu_2Si=N–SiCltBu_2$	$tBu_2SiOHNHSiCltBu_2$	$tBu_2Si=N–SiFtBu_2$	CH_3COCH_3	$tBu_2Si=N–SitBu_3$	$PhCHO$
$tBu_2Si=N–SitBu_3$	$tBu_2SiOHNHSitBu_3$	$tBu_2Si=N–SiCltBu_2$	CH_3COCH_3	$tBu_2Si=N–SitBu_3$	CO_2
$tBu_2Si=N–SiPhtBu_2$	$tBu_2SiOHNHSiPhtBu_2$	$tBu_2Si=N–SiBrtBu_2$	CH_3COCH_3	[3+2] Cycloaddition:	
$Me_2Si=N–SitBu_3$	H_2O	$tBu_2Si=N–SitBu_3$	CH_3COCH_3	Silaneimine:	a=b=c
$tBu_2Si=N–SiCltBu_2$	H_2O	$tBu_2Si=N–SiPhtBu_2$	CH_3COCH_3	$tBu_2Si=N–SitBu_3$	$N_2O^{[a]}$
$tBu_2Si=N–SitBu_3$	H_2O	$tBu_2Si=N–SiPh_3$	CH_3COCH_3	$tBu_2Si=N–SiPh_3$	$N_2O^{[a]}$
$tBu_2Si=N–SiPh_3$	H_2O	$Me_2Si=N–SitBu_3$	$CH_2=CH–CH_3$	$tBu_2Si=N–SitBu_3$	tBu_2SiHN_3
$Me_2Si=N–SitBu_3$	$MeOH$	$tBu_2Si=N–SitBu_3$	$CH_2=CH–CH_3$	$tBu_2Si=N–SitBu_3$	tBu_2SiMeN_3
$tBu_2Si=N–SiCltBu_2$	$MeOH$	$tBu_2Si=N–SiCltBu_2$	$CH_2=CH–CH_2–CH_3$	$tBu_2Si=N–SitBu_3$	tBu_2SiFN_3
$tBu_2Si=N–SitBu_3$	$MeOH$	$tBu_2Si=N–SiPhtBu_2$	$CH_2=CH–CH_2–CH_3$	$tBu_2Si=N–SitBu_3$	tBu_2SiClN_3
$tBu_2Si=N–SiPh_3$	$MeOH$	$Me_2Si=N–SitBu_3$	$CH_2=CMe_2$	$tBu_2Si=N–SitBu_3$	tBu_2SiBrN_3
$Me_2Si=N–SitBu_3$	$nBuLi$	$tBu_2Si=N–SiCltBu_2$	$CH_2=CMe_2$	$tBu_2Si=N–SiCltBu_2$	tBu_2SiMeN_3
$tBu_2Si=N–SiCltBu_2$	$nBuLi$	$tBu_2Si=N–SitBu_3$	$CH_2=CMe_2^{[a]}$	$tBu_2Si=N–SiPhtBu_2$	tBu_2SiClN_3
$tBu_2Si=N–SiPh_3$	$nBuLi$	$tBu_2Si=N–SiPhtBu_2$	$CH_2=CMe_2$	$tBu_2Si=N–SitBu_3$	Ph_3SiN_3
$tBu_2Si=N–SiPh_3$	tBu_3SiNa	$Me_2Si=N–SitBu_3$	$CH_2=CMeMeC=CH_2$	[4+2] Cycloaddition:	
$tBu_2Si=N–SiPh_3$	$(Me_3Si)_2CHLi$	$tBu_2Si=N–SiFtBu_2$	$CH_2=CMeMeC=CH_2$	Silaneimine:	a=b–c=d
$tBu_2Si=N–SiCltBu_2$	Et_3NHF	$tBu_2Si=N–SiCltBu_2$	$CH_2=CMeMeC=CH_2$	$tBu_2Si=N–SiPh_3$	$CH_2=CH–HC=CH_2$
$tBu_2Si=N–SiCltBu_2$	Me_3NHCl	$tBu_2Si=N–SiBrtBu_2$	$CH_2=CMeMeC=CH_2$	$tBu_2Si=N–SiPh_3$	$CH_2=CMeMeC=CH_2^{[a]}$
[2+1] Cycloaddition:		$tBu_2Si=N–SitBu_3$	$CH_2=CMeMeC=CH_2$	$tBu_2Si=N–SitBu_3$	$CH_2=CH–HC=CH_2^{[a]}$
Silaneimine	a:	$tBu_2Si=N–SiPhtBu_2$	$CH_2=CMeMeC=CH_2$		
$tBu_2Si=N–SitBu_3$	$CO^{[a]}$	$tBu_2Si=N–SiPh_3$	$CH_2=CMeMeC=CH_2$		

Table 3. Reactivity of silaneimines;
[a] this reaction leads to more than one product or the product is not stable

Structure of Silaneimine Amine Adduct Me$_2$Si=N–SitBu$_3$·Me$_2$NEt

The geometry of the silaneimine amine adduct Me$_2$Si=N–SitBu$_3$·Me$_2$NEt (Fig. 1) differs like that of the silaneimine tetrahydrofuran adduct Me$_2$Si=N–SitBu$_3$·THF from the geometry of the donor free silaneimine tBu$_2$Si=N–SitBu$_3$ by elongation of the double bond and a pyramidalization of the unsaturated silicon atom (Table 4) [3].

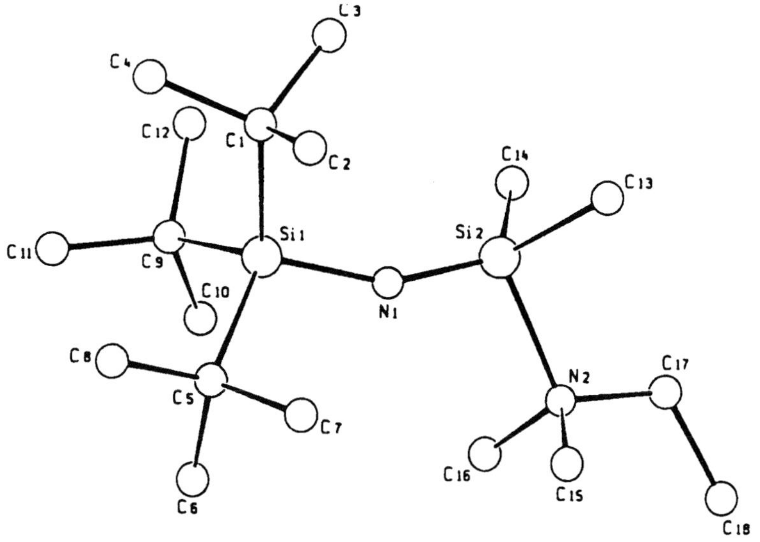

Fig. 1. Structure of Me$_2$Si=N–SitBu$_3$ · Me$_2$NEt

	Me$_2$Si=N–SitBu$_3$· Me$_2$NEt	Me$_2$Si=N–SitBu$_3$ THF[4]	tBu$_2$Si=N– SitBu$_3$[4]	H$_2$Si=N–SiH$_3$ (calc.)[5]
Si=N–distance	1.6036	1.588[a]	1.568	1.549
Si–N–distance	1.6600	1.654[a]	1.695	1.688
SiNSi–angle	156.44	161.5[a]	177.8	175.6

Table 4. Comparison of principal distances and angles
[a] two crystallographically independent molecules

References:

[1] N. Wiberg, H.-W. Lerner, unpublished results.
[2] N. Wiberg, K. Schurz, H.-W. Lerner, unpublished results.
[3] N. Wiberg, H.-W. Lerner, D. Fenske, publication in preparation.
[4] N. Wiberg, K. Schurz, G. Reber, G. Müller, *J. Chem. Soc., Chem. Commun.* **1986**, 591.
[5] P. von Ragué Schleyer, P. D. Stunt, *J. Chem. Soc., Chem. Commun.* **1986**, 1373.

Recent Explorations of the Chemistry of Pentacoordinate Silicon

Alan R. Bassindale*, Simon G. Glynn, Jianxiong Jiang, David J. Parker,
Robert Turtle, Peter G. Taylor

Department of Chemistry

The Open University

Walton Hall, Milton Keynes MK7 6AA, United Kingdom

Scott S. D. Brown

Central Resarch and Development

Dow Corning Ltd

Cardiff Road, Barry CF63 2YL, South Glamorgan, Wales, United Kingdom

Summary: The rich variety of the coordination chemistry of silicon is discussed and some theoretical issues are raised. In an attempt to understand further the underlying chemistry, some thermodynamic and kinetic parameters for the formation and substitution of pentacoordinate silicon compounds have been measured by NMR methods. Values of -31 ± 3 kJ mol^{-1} for ΔH and -100 ± 10 J K^{-1}mol^{-1} for ΔS were measured for the intramolecular coordination of a pyridine ligand to a chlorosilane moiety. A detailed kinetic analysis of a nucleophilic substitution at pentacoordinate silicon in a chelated complex revealed that substitution both with inversion and retention of configuration at silicon are taking place on the NMR time-scale. The substitution with inversion of configuration is zero order in nucleophile but a retentive route is zero order in nucleophile at low temperature but shows an increasing dependence on nucleophile at higher temperatures. These results are analysed and mechanistic hypotheses are proposed. Some tentative conclusions are drawn about the nature of reactivity in pentacoordinate silicon compounds.

Introduction

In recent years the chemistry of so-called hypervalent silicon compounds has attracted a great deal of attention both for its intrinsic interest and the possibility of discovering new materials and chemical reagents [1–3]. Although silicon compounds with coordination number four are still the most numerous, it is now no longer considered that four coordination is the "normal" state for silicon. It is well recognised that coordination numbers at silicon of five and six, and possibly higher, are quite usual and do not require special or forcing conditions for their preparation. Despite this, much of the research effort on penta- and hexavalent silicon has been directed towards their synthesis and structural investigations [4].

More recently the chemistry of such species has started to be developed and, as examples, penta- and hexacoordinate silicon have found use in organic synthesis [5], and diaryltrifluorosilicates have been studied extensively for mechanistic pathways and synthetic utility [6]. Corriu has also made many fundamental and stimulating studies concerning the mechanisms of reactions of hypercoordinate silicon and their role in substitution reactions [1].

The work reported here is a continuation of our attempts to understand the basis of the reactivity of hypervalent compounds of silicon [7–8]. A particular emphasis of the current work is investigation of reactivity by quantitative methods such as thermodynamic and kinetic investigations.

First, some indications and illustrations from our work of the complexity and variety of the interaction of silanes with nucleophiles: Scheme 1 shows the possible outcomes of the reaction of a tetracoordinate neutral silicon compound with a neutral nucleophile, Nu, such as N-methylimidazole. The compound on the lower left can be any tetracoordinate compound of silicon that has at least one labile group, X, like a halide or trifluoromethylsulfonate (triflate, OTf). Reaction of nucleophiles with that compound can follow a variety of routes depending on the number and nature of X and the nature and amount of the nucleophile.

Reactions that take place along a row are those in which a leaving group is substituted by a nucleophile. Assuming the leaving group departs as X^- and the nucleophile is neutral then the reactions along a row are substitutions, S, in which the charge on the complex silicon species increases by one unit for each substitution. Vertical arrows represent additions, A, increasing the coordination number of silicon by one for each addition. There is no charge difference between the reagent and product in these addition reactions. Reactions taking place down a row, in the opposite sense to the arrows, or following the diagonal arrows are dissociations, D, in which the coordination number decreases by one. For reaction of SiX_4 with up to six equivalents of nucleophile any, or all, of the fifteen species are possible. As the number of X groups decreases so the number of options decreases, but even for $RRSiX_2$ the number of possible species is nine. We have studied many of the reactions on this manifold and observed almost all of the complex types by examining the solution NMR spectra and electrical conductivity of mixtures of silanes and nucleophiles [9–11].

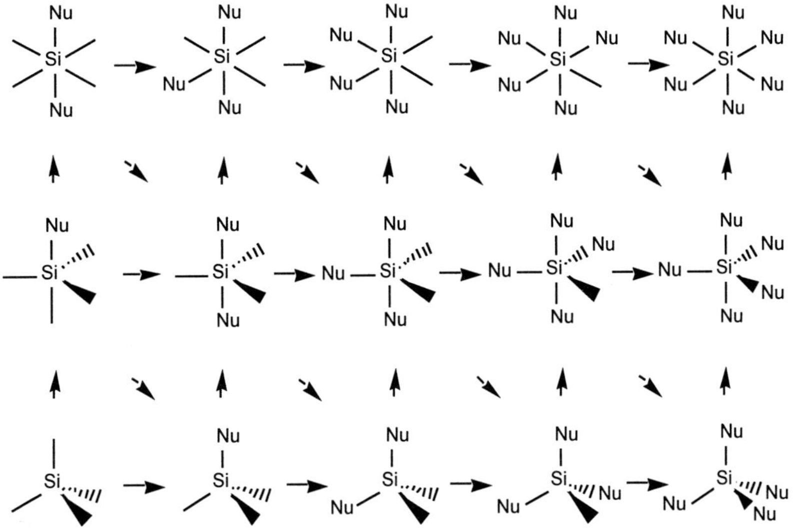

Scheme 1. A matrix of possible products from silane – nucleophile interactions

One example is given by the addition of increasing amounts of hexamethylphoshoramide (HMPA) to a solution of methylditriflatosilane, MeHSi(OTf)$_2$ **1** in dichloromethane. The species present may be identified with a good degree of certainty by measuring the electrical conductivity and ^{29}Si NMR spectra as the nucleophile concentration increased. The sequence is shown in Scheme 2.

The ^{29}Si NMR spectrum of **1** is a singlet appearing at δ = –14 ppm and that resonance remained apparent, but with decreasing intensity up to a silane nucleophile ratio of 1:1 while a new doublet resonance at δ = –62 ppm simultaneously grew. This is interpreted as an addition reaction of HMPA to the silane to give **2** in which the coupling between silicon and phosphorus was the cause of the doublet pattern. During this process the electrical conductivity barely increased. As the ratio of nucleophile to silane increased towards 2:1 another new resonance grew while the electrical conductivity increased sharply. The new resonance which was a triplet at δ = –24 ppm clearly has two HMPA molecules coordinated to give the triplet coupling pattern, and both the chemical shift and conductivity measurements lead to the conclusion that the species is **4**, which was formed by the addition of a further molecule of HMPA to **2** to form **3** which dissociated rapidly to form **4**. As soon as the ratio of nucleophile to silane increased beyond 2:1 the triplet pattern of **4** collapsed to a singlet as the lifetime of individual Si–HMPA bonds became short on the NMR time scale, through exchange processes of one kind or another.

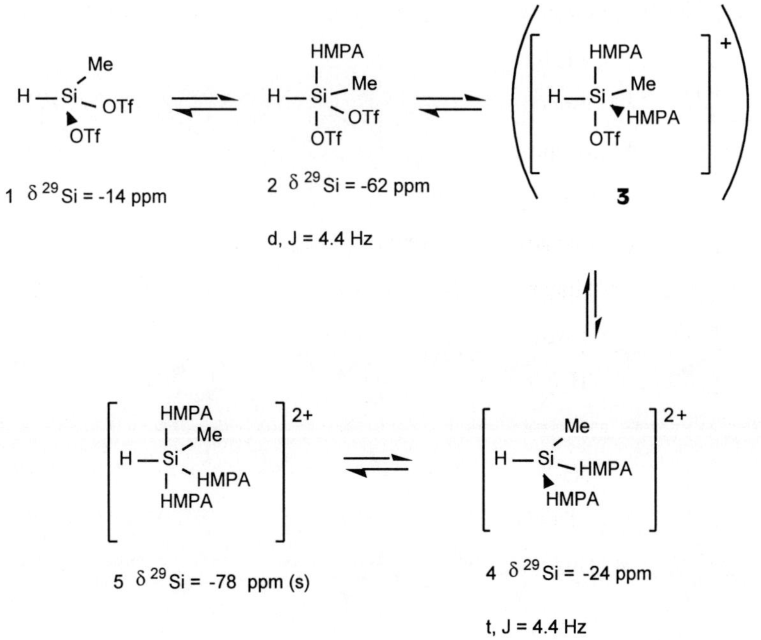

Scheme 2. The course of the reaction of MeHSi(OTf)$_2$ with HMPA

The single ^{29}Si NMR resonance moved to low frequency with each addition of HMPA, and finally remained as a singlet at δ = −78 ppm at all ratios of HMPA to silane of 3:1 or greater. The only reasonable structure for this new species is **5**. These complex changes for a relatively simple system illustrate the subtle relationship between coordination and reactivity for silicon. It was observations such as these and others that stimulated us to try to make quantitative measures on hypervalent silicon compounds.

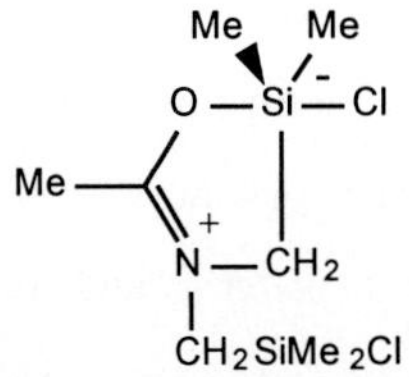

Fig. 1.

The type of compounds that we use in our investigations are almost all chelated amides or related compounds based on the pentacoordinated acetamide **6**, first synthesised by Lasocki and finally characterised by Yoder and McPhail [12, 13].

The formation of such compounds from the *N*-trimethylsilylamide and $ClCH_2SiMe_2Cl$ has been studied by Pestunovich, Baukov and co-workers using NMR spectroscopy [14]. They postulated that the first step in the formation of the pentacoordinate silicon compound is silyl exchange to give the *N*-chloromethyldimethylsilyl derivative which rearranges to an OCH_2SiMe_2Cl species in which the imido nitrogen is coordinated to the silicon in a pentacoordinate silicon chelate. That species was suggested to be thermodynamically unstable with respect to rearrangement to the final product, analogous to **6**, by a Chapman-type OCH_2R to NCH_2R rearrangement. None of the intermediate species were isolable but NMR resonances were entirely compatible with the suggested structures. We have recently discovered that for urea derivatives the NCH_2 derivative is the thermodynamically less-favored form and under some circumstances we can isolate the NCH_2 isomer and show that with gentle warming it irreversibly rearranges to the OCH_2 form. This is shown in Scheme 3.

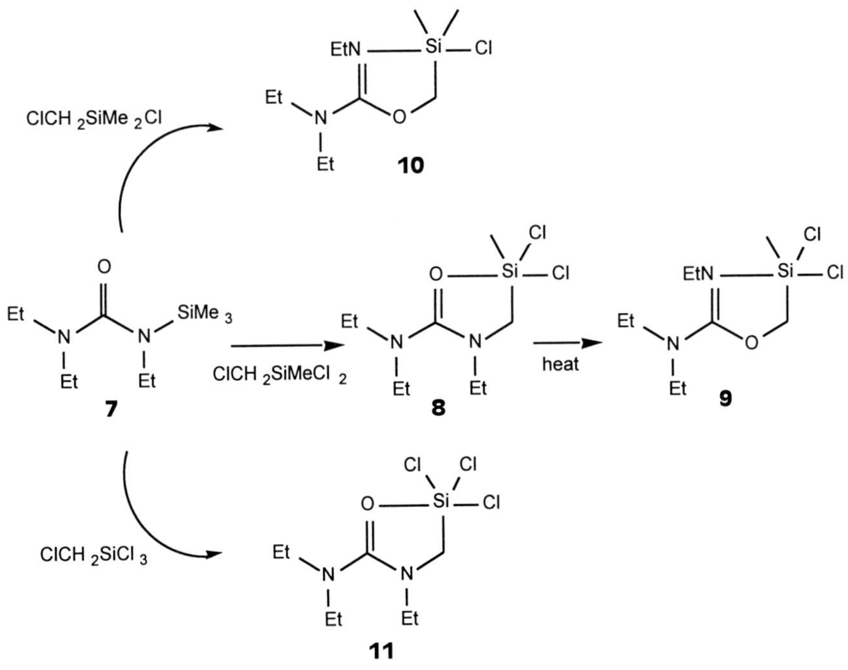

Scheme 3. The synthesis of some pentacoordinate silicon compounds from a silyl urea

When **7** was treated with one molar equivalent of ClCH$_2$SiMeCl$_2$ in the cold needle crystals of **8** were obtained in high yield. On warming in solution **8** rearranged to **9** quantitatively and irreversibly. The final N···Si product **10** was the only one that could be observed when **7** was treated with ClCH$_2$SiMe$_2$Cl, even when the reaction was carried out at a temperature lower than 0 °C. By contrast, when **7** was treated with ClCH$_2$SiCl$_3$ the product was the kinetically favored **11**. This compound underwent partial rearrangement to the thermodynamic N···Si product on heating at temperatures greater than 100 °C but the conditions were so harsh that decomposition accompanied rearrangement. An explanation for the difference in behaviour between these similar reactions may be found in the relative strengths of the initially-formed O···Si bonds. The rearrangement is assumed to take place through a series of steps like those shown in Scheme 4.

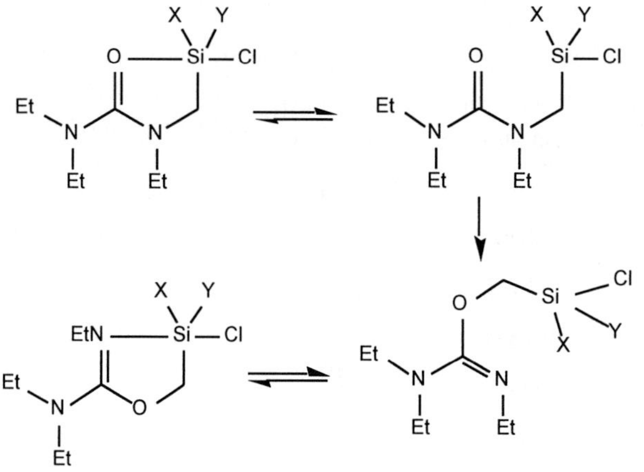

Scheme 4. Steps in the rearrangement of initially-formed silyl urea chelates

If, as seems reasonable, the first dissociative step is rate limiting then the weakest O···Si coordination bond would allow rearrangement to take place at the greatest rate. The O···Si coordination bond is weakest when the substituents on Si are two methyl groups and one chlorine, as the coordination bond to silicon is generally recognised to be strongest when the silicon bears the greatest number of electron-withdrawing substituents. This simple rationalisation satisfactorily accounts for the increasing rate of rearrangement as the number of chlorine atoms attached to silicon decreases. Once again the experimental observations stimulated our interest in finding simple methods for gaining quantitative information.

The next section shows how we have used NMR spectroscopy to measure the energetics of one particular chelation in which a pentacoordinate silicon compound is produced.

A thermodynamic Investigation of Pentacoordination at Silicon

We have reported a new technique for mapping the progress of nucleophilic substitution at silicon, through a pentacoordinate silicon [7]. The method requires the synthesis of a series of model compounds, each of which corresponds to a point on the reaction profile. The first example was the pyridone series represented in Scheme 5.

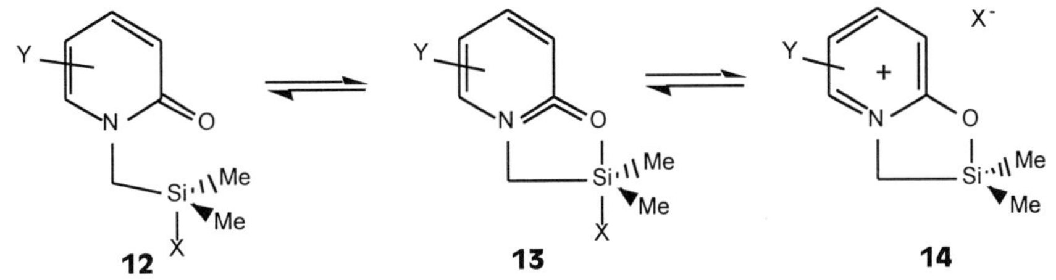

Scheme 5. A model system for mapping nucleophilic substitution at silicon in solution

For that series it was shown that, by variation of X and Y, compounds could be prepared that were disposed at points along the **12–13–14**, manifold. Each compound was analysed by NMR spectroscopy with the ^{29}Si NMR chemical shift giving a good indication of the extent of pentacoordination at silicon and the ^{13}C NMR spectrum of the aromatic ring being responsive to the degree of O···Si bond making.

The procedure has been modified to enable the thermodynamic parameters for coordination to be estimated. A new mapping system based on thiopyridones has been developed and the compound **15** was studied by variable temperature NMR spectroscopy.

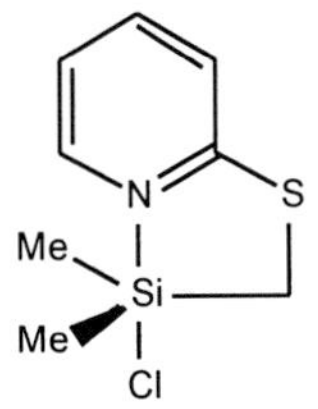

Fig. 2.

As the temperature was varied between +50 °C and −50 °C the ^{29}Si and ^{13}C NMR spectra of **15** varied in a way entirely consistent with a greater extent of pentacoordination at the lower temperature. The

^{29}Si NMR chemical shift varied from $\delta = -8$ ppm at +50 °C to $\delta = 31$ ppm at −50 °C, which is appropriate for increasing pentacoordination as the nucleophile attacks the silicon center [15].

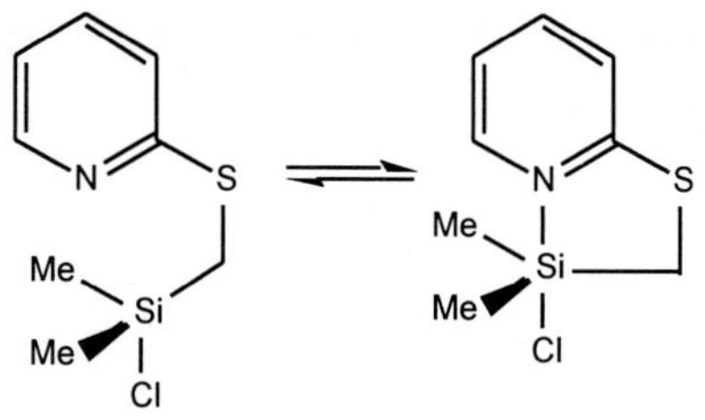

Scheme 6. The equilibrium reaction used for thermodynamic calculations

The changes that are observed are equivalent to the reaction shown in Scheme 6 although we suggest that what is actually being observed is not an equilibrium but a steady increase in N$\cdots$Si bond-making and Si$\cdots$Cl bond lengthening. Nevertheless, for the purposes of calculation we assume that an equilibrium is observed. For the calculation of ΔH and ΔS a plot of K vs $1/T$ was used, where $K = [\mathbf{17}]/[\mathbf{16}]$. The relative amounts of **17** and **16** at a particular temperature were estimated from the ^{29}Si NMR chemical shift. It was assumed, for the purposes of calculation, that the two species were present and undergoing rapid interconversion on the NMR time scale. A limiting value of the ^{29}Si NMR chemical shift was assigned to each of **16** and **17** based on extrapolation of the changes in chemical shift with temperature and our expectation of the limiting values. As there is a degree of uncertainty in these procedures the calculations were carried out using a variety of assumed limiting values. The position of the observed chemical shift relative to the limiting values by simple arithmetic gives a value for [**17**]/[**16**] and therefore K. Some applications of these calculations to the equilibrium in Scheme 6 are shown in Fig. 3.

Although the thermodynamic parameters are dependent on the limiting values the variation, for sensible limits, does not invalidate the method, which is not claimed to give highly accurate data. Values of -31 ± 3 kJ mol^{-1} for ΔH and -100 ± 10 JK^{-1}mol^{-1} for ΔS were measured assuming limiting values of $\delta = -31.5$ to -35 for the pentacoordinate limit and $\delta = +5$ to $+20$ for the tetracoordinate limit. These values show that the coordination in **15** is exothermic to a relatively small extent but as expected there is a significant decrease in entropy consistent with the loss of degrees of freedom on chelation. These are preliminary data that we have not previously reported and work is in progress to carry out refined calculations on analogues of **15** in which the groups on silicon are varied between SiMe$_2$Cl and SiCl$_3$.

Those results should be useful in evaluating our explanation for the relative rates of rearrangement of the silylureas described earlier in this chapter.

different chemical shift assumptions

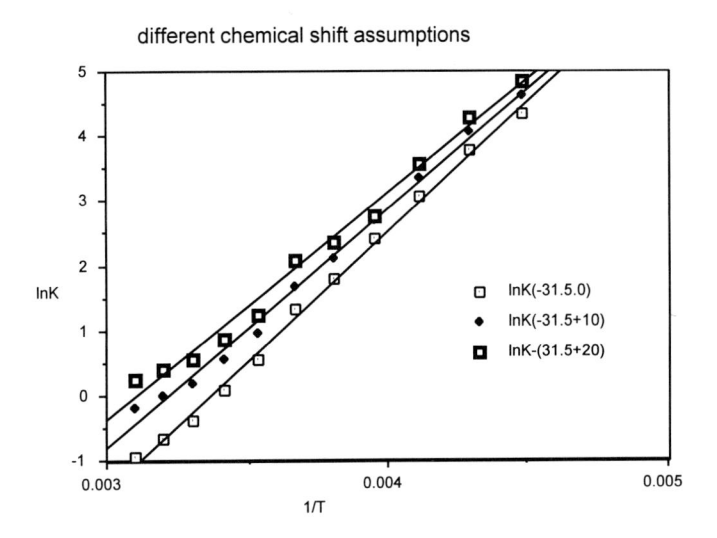

Fig. 3. lnK vs 1/T for coordination of **15**, with diffeent chemical shift assumptions

A Kinetic and Stereochemical Study on Nucleophilic Substitution at Pentacoordinate Silicon

It is well recognised that, in some circumstances, the prior coordination of a nucleophile to tetracoordinate silicon can produce a pentacoordinate silicon centre that is activated to nucleophilic attack [1, 8]. We have previously published preliminary data on the use of chelated pentacoordinate silylamides as reactivity probes for nucleophilic activation [8]. The work presented here is an extension of those studies and incorporates a stereochemical and kinetic investigation in a single series of experiments. The compound chosen for the experiments was **18** which has several advantages for a kinetic and stereochemical study. The coordinated N-methylimidazole (NMI) group is readily displaced by another NMI molecule in a degenerate substitution reaction.

This provides a particularly convenient way of measuring reaction rates, simply by running the NMR spectra of a series of mixtures and analysing the results. The total rate of nucleophilic substitution at the pentacoordinate silicon can be measured by using dynamic NMR programs such as DNMR5 (QCPE) to model the resonances from the bound and non-coordinated NMI and their interconversions. At the slow exchange limit sharp, separate resonances are observed for the coordinated and non-coordinated NMI.

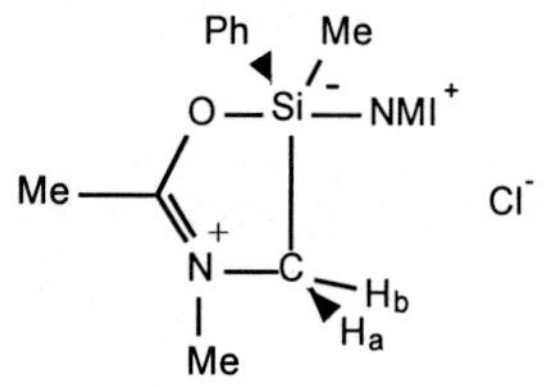

Fig. 4.

[Structure 18]

When the two are rapidly exchanging environments a single averaged spectrum appears for the NMI. The kinetics of total NMI exchange (i.e., all nucleophilic substitution processes of all possible stereochemical outcomes and kinetic orders in NMI) may be studied in a straightforward way. Kinetic orders in nucleophile (NMI) may be obtained by keeping the concentration of **18** constant and making a series of mixtures of **18** and NMI in which the concentration of NMI varies. The methylene protons H_a and H_b provide a stereochemical probe, as they are diastereotopic by virtue of the chirality of the silicon centre. In the ^{1}H NMR spectrum the resonances for H_a and H_b appear as a doublet of doublets. If exchange of NMI takes place with retention of configuration then the AB pattern for H_a and H_b persists.

On the other hand if exchange of NMI takes place with inversion of configuration then the AB pattern collapses to a singlet as the protons become homotopic. Using this information the rate of substitution at silicon with inversion of configuration may be measured. For a series of nucleophile concentrations it is therefore possible to measure the total rate of NMI exchange and the rate of inversion of configuration at silicon accompanying the exchange. From these data the individual rates and orders for substitution at pentacoordinate silicon with both inversion and retention of configuration should be calculable. Accordingly, kinetic and thermodynamic studies were carried out on the reaction shown in Eq. 1.

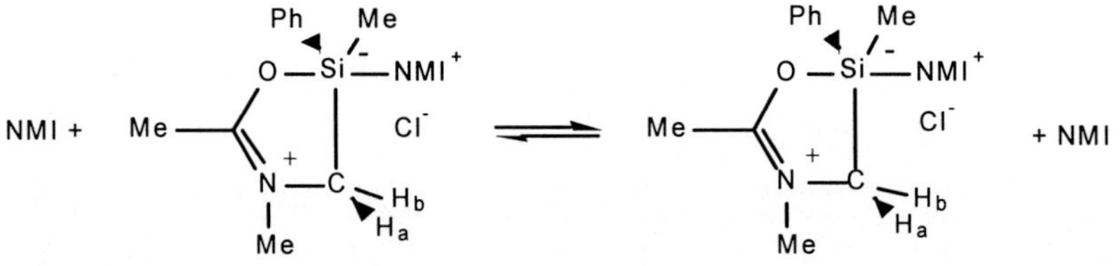

Eq. 1.

The initial results have produced some interesting data. For the reaction with inversion of configuration there does not appear to be any rate dependence on the concentration of NMI. The inversion reaction, being zero order in NMI cannot take place by an elementary reaction with direct attack of NMI at the pentacoordinate silicon. The rate determining step must involve only **18** and we postulate that the inversion mechanism follows the path shown in Scheme 7, in which the breaking of the O···Si bond is the rate limiting step, followed by fast attack of NMI on the open chain four coordinate silicon compound.

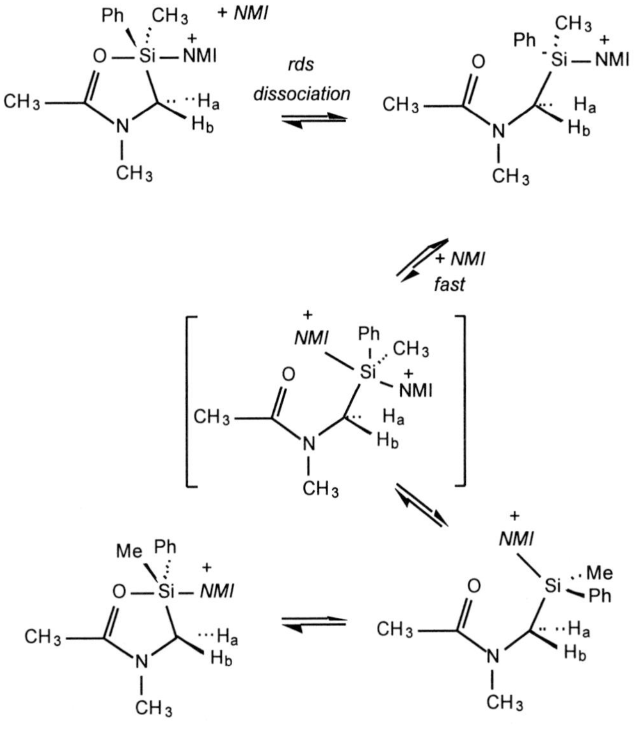

Scheme 7. A possible mechanism for inversion of configuration at a pentacoordinate silicon center

Variable temperature studies on inversion of configuration give an activation energy, ΔH of +67 kJ mol^{-1} and the entropy of activation ΔS was calculated to be + 64 J K^{-1} mol^{-1}. The rate determining step of this proposed process is the reverse of the process for the equilibrium, **16 ↔ 17** which is discussed above. Although these are two different systems the thermodynamic data are highly compatible and in particular the relatively large positive entropy of activation for the inversion process is of the same order of magnitude as the dissociation reaction **17 ↔ 16**.

The total exchange of NMI for the reaction shown in Eq. 1 is also interesting. At low temperatures the rate of exchange is almost independent of the concentration of NMI, but as the temperature increases so the dependence of the rate on the NMI concentration increases. This is shown in Fig. 5.

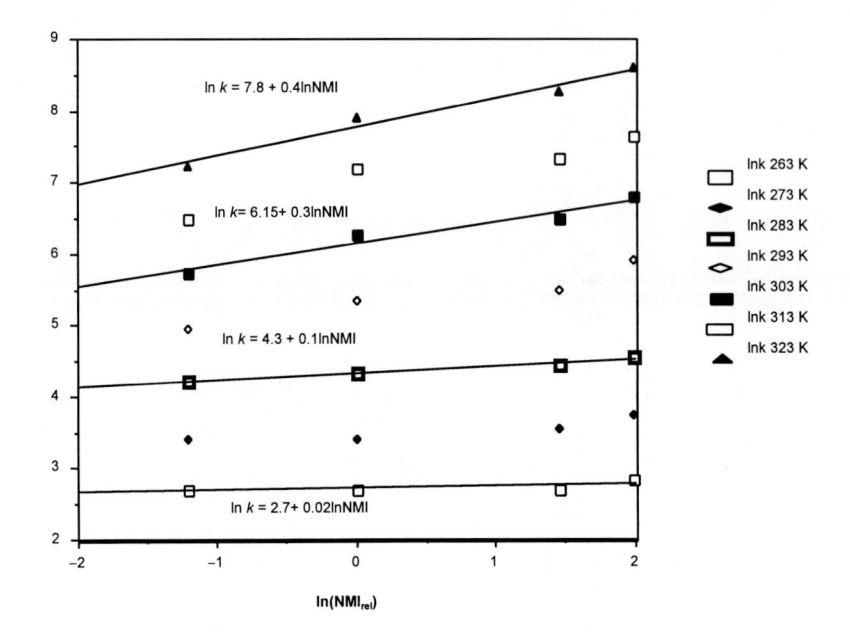

Fig. 5. Variation of ln k with ln NMI at various temperatures for the reaction of NMI with **18**

This observation is compatible with either a change in rate limiting step with temperature or with two competing mechanisms for retention of configuration (the inversion mechanism is always zero order in NMI). Both of these possibilities are expressed in Scheme 8.

The zero order mechanism for retention of configuration has a rate limiting step of dissociation of the other coordinate bond to pentacoordinate silicon, the Si⋯N bond. Recoordination of another NMI to the tetracoordinate silicon completes that process. The Scheme shows one plausible route for attack with retention of configuration in which NMI is involved in the rate determining step. That route is the anticlockwise series of four steps starting at the top left of the Scheme. We have been unable to envisage a single step displacement of NMI that involves a stereochemically reasonable route. As there is clearly more than one retentive process taking place it is not particularly revealing to calculate thermodynamic parameters for the combination of reaction mechanisms. It is, however, possible to obtain an estimate for the rate constant for the Si⋯NMI dissociative mechanism for retention.

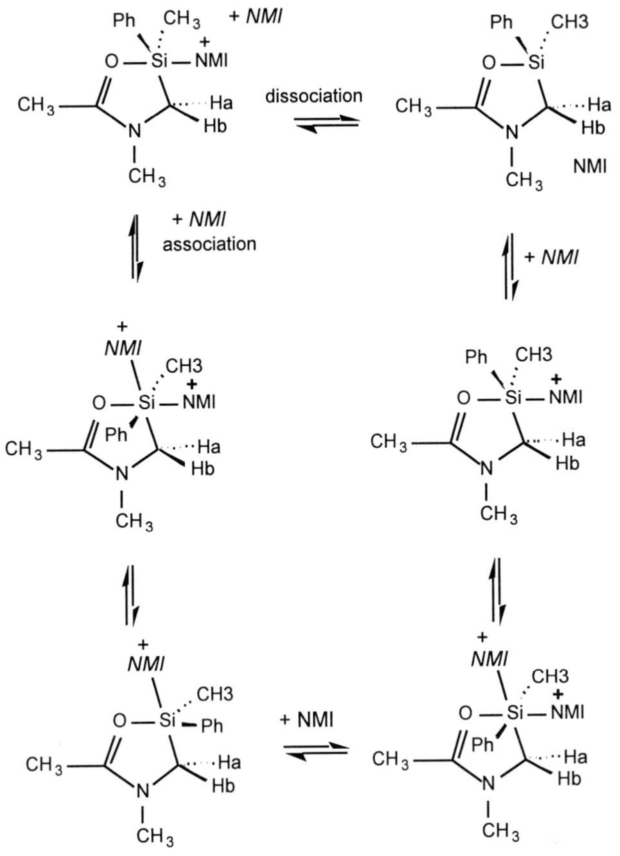

Scheme 8. Two possible mechanisms for nucleophilic attack at silicon with retention of configuration; the first two steps in a clockwise direction, starting at the top left represent the reaction that is zero order in NMI

If we assume that for every inversion taking place through rate determining O⋯Si dissociation there is also likely to be a retention (this is very likely as attack by NMI on the four coordinate salt is very fast on the NMR time scale, [9]) then the rate constant for retention by SiN cleavage, taking place through a zero order process in NMI (the first two clockwise steps in Scheme 8) is given by Eq. 2.

$$k_{ret(NSi)} = k_{totalNMI} - 2\,k_{(invOSi)}$$

Eq. 2.

When this data manipulation was carried out for the reaction under consideration a good straight line was obtained for ln$k$ vs $1/T$ and the activation parameters ΔH = 42 kJ mol^{-1} and ΔS = −20 J K^{-1}mol^{-1} were calculated in the usual way.

In conclusion, our data show that at low temperature there are two competitive mechanisms for nucleophilic substitution at pentacoordinte silicon in our chosen system (**1**). Neither of these mechanisms involves direct attack by nucleophile at the pentacoordinate silicon. However, at higher temperature a reaction involving nucleophile in the rate limiting step and taking place with retention of configuration at pentacoordinate silicon becomes important. We suggest that the reaction may be a direct attack by nucleophile on pentacoordinate silicon. The mechanistic pathways may be different or have different weight for compounds with differently substituted silicon.

We have shown in this article that quantitative kinetic data can be obtained for reaction at pentacoordinate silicon by NMR methods. These data throw some light on the reactivity of hypercoordinate silicon and with further elaboration could make a useful contribution to the debate on the relationship between coordination and reactivity.

Acknowledgements: We would like to acknowledge generous funding from *Dow Corning Ltd., SERC*, and the *Open University*.

References:

[1] C. Chuit, R. J. P. Corriu, C. Reye, J. C. Young, *Chem. Rev.* **1993**, *93*, 1371.

[2] R. J. P. Corriu, C. J. Young, in: *The Chemistry of Organic Silicon Compounds* (Eds.: S. Patai, Z. Rappoport), Wiley, Chichester, **1989**, p. 1241.

[3] R. R. Holmes, *Chem. Revs.* **1990**, *90*, 17.

[4] S. N. Tandura, N. V. Alekseev, M. G. Voronkov, *Top. Curr. Chem.* **1986**, *131*, 99.

[5] K. Tamao, T. Hayashi, Y. Ito, in: *Frontiers of Organosilicon Chemistry* (Eds.: A.R. Bassindale, P. P. Gaspar), Royal Society of Chemistry, Cambridge, **1991**, p. 197.

[6] A. Boudin, G. Cerveau, C. Chuit, R. Corriu, C. Reye, *Angew. Chem., Int. Ed. Engl.* **1989**, *25*, 473.

[7] A. R. Bassindale, M. Borbaruah, *J. Chem. Soc., Chem. Commun.* **1991**, 1499; **1991**, 1501.

[8] A. R. Bassindale, M. Borbaruah, *J. Chem. Soc., Chem. Commun.* **1993**, 352.

[9] A. R. Bassindale, T. Stout, *J. Chem. Soc., Perkin Trans.* **1986**, *2*, 221.

[10] A. R. Bassindale, T. Stout, *Tetrahedron Lett.* **1986**, *26*, 3403.

[11] A. R. Bassindale, J. Jiang, *J. Organomet. Chem.* **1993/4**.
[12] J. Kowalski, Z. Lasocki, *J. Organomet. Chem.* **1976**, *116*, 75.
[13] K. D. Onan, A. T. McPhail, C. H. Yoder, R. W. Hillyard, *J. Chem. Soc., Chem. Commun.* **1978**, 209.
[14] I. D. Kalikhman, Yu. I. Baukov, V. A. Pestunovich, *J. Organomet. Chem.* **1989**, *361*, 147.
[15] A. R. Bassindale, T. Stout, *J. Chem. Soc., Chem. Commun.* **1984**, 1387.

Syntheses, Structures, and Properties of Molecular $\lambda^5 Si$-Silicates Containing Bidentate 1,2-Diolato(2−) Ligands Derived from α-Hydroxycarboxylic Acids, Acetohydroximic Acid, and Oxalic Acid: New Results in the Chemistry of Pentacoordinate Silicon

Reinhold Tacke*, Olaf Dannappel, Mathias Mühleisen

Institut für Anorganische Chemie
Universität Fridericiana zu Karlsruhe (TH)
Engesserstr., Geb. 30.45, D–76128 Karlsruhe, Germany

Summary: A series of zwitterionic (molecular) spirocyclic $\lambda^5 Si$-silicates (SiO_4C or SiO_5 frameworks) containing two unsymmetric bidentate diolato(2−) ligands derived from α-hydroxycarboxylic acids (such as glycolic acid, 2-methyllactic acid, benzilic acid, and citric acid) were synthesized and structurally characterized by single-crystal X-ray diffraction. In addition, some analogous zwitterionic $\lambda^5 Ge$-germanates (GeO_4C framework) were studied. Furthermore, a series of related zwitterionic $\lambda^5 Si,\lambda^5 Si'$-disilicates ($SiO_4C/Si'O_4C$ or $SiO_5/Si'O_5$ frameworks) containing four bidentate 2-methyllactato(2−) ligands or two tetradentate (R,R)- or (S,S)-tartrato(4−) ligands were prepared and their crystal structures studied by X-ray diffraction. Finally, a zwitterionic spirocyclic $\lambda^5 Si$-silicate (SiO_4C framework) containing two bidentate acetohydroximato(2−) ligands and a zwitterionic monocyclic $\lambda^5 Si$-silicate (SiO_2C_2F framework) containing one bidentate oxalato(2−) ligand were synthesized and structurally characterized by X-ray diffraction. All these compounds contain pentacoordinate (formally negatively charged) silicon atoms and tetracoordinate (formally positively charged) nitrogen atoms and therefore represent neutral molecular species. The nitrogen atoms belong to ammonio-substituted organo groups (compounds with an SiO_4C, GeO_4C, or SiO_2C_2F framework) or alkoxy groups (compounds with an SiO_5 framework) which are bound to the silicon(IV) coordination center.

1 Introduction

In the past few years, we have reported on the synthesis and structure of a series of novel zwitterionic (molecular) $\lambda^5 Si$-silicates [1]. The silicates **1** [1f], **2** [1d], **3**·CH_3CN [1c], **4** [1f], **4**·H_2O [1f], **5**·¼CH_3CN [1b, 1f], **6**·CH_3CN [1a], **7**·½CH_3CN [1e], **8**·CH_3CN [1g], and **9**·½CH_3CN [1g] are typical examples of this particular type of compound (Fig. 1). These molecular spirocyclic $\lambda^5 Si$-silicates are characterized by the presence of a pentacoordinate (formally negatively charged) silicon atom and a tetracoordinate (formally positively charged) nitrogen atom. In these molecules, two symmetric bidentate 1,2-diolato(2–) ligands (derived from diols of the pyrocatechol type) and one ammonio-substituted organic group are bound to the silicon(IV) coordination center.

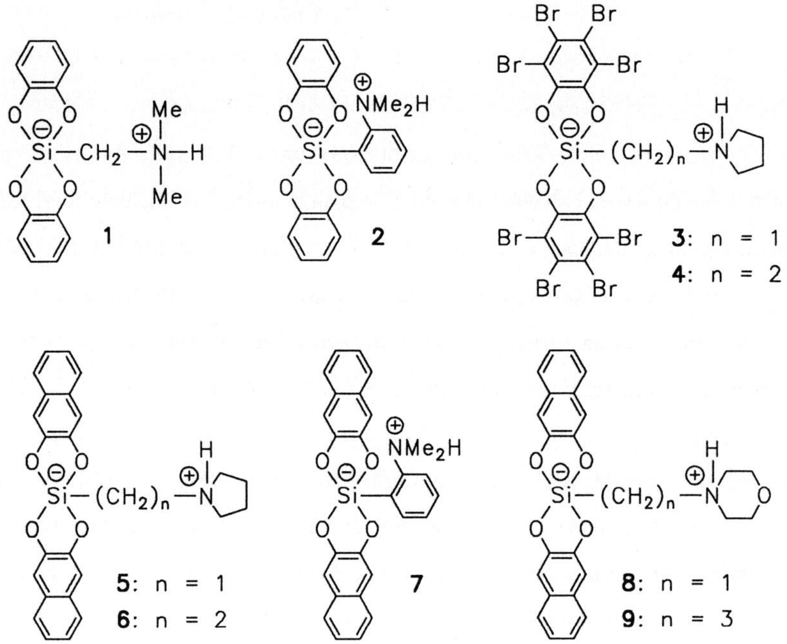

Fig. 1.

In this paper, we report on the synthesis and structural characterization of a series of related zwitterionic (molecular) spirocyclic $\lambda^5 Si$-silicates (mononuclear $\lambda^5 Si$-silicon(IV) complexes) containing two bidentate 1,2-diolato(2–) ligands derived from α-hydroxycarboxylic acids, such as glycolic acid, 2-methyllactic acid, benzilic acid, and citric acid. In addition, some analogous zwitterionic $\lambda^5 Ge$-germanates (mononuclear $\lambda^5 Ge$-germanium(IV) complexes) are described. Furthermore, we report on the synthesis and structural characterization of related zwitterionic $\lambda^5 Si,\lambda^5 Si'$-disilicates (binuclear $\lambda^5 Si$-

silicon(IV) complexes) which contain four bidentate 2-methyllactato(2–)-O^1,O^2 ligands or two tetradentate (R,R)- or (S,S)-tartrato(4–)-O^1,O^2:O^3,O^4 ligands. Finally, mononuclear λ^5Si-silicon(IV) complexes containing two bidentate acetohydroximato(2–)-O^1,O^2 ligands or one bidentate oxalato(2–)-O^1,O^2 ligand are described. The investigations presented here were carried out as a part of our systematic studies on the chemistry of molecular λ^5Si-silicon(IV) complexes (for short reviews, [2]; for reviews on pentacoordinate silicon, see [3]).

2 Results and Discussion

2.1 Zwitterionic λ^5Si-Silicates Containing Two Bidentate Diolato(2–) Ligands Derived from α-Hydroxycarboxylic Acids

Following the strategy described for the synthesis of the zwitterions **1–9** [1], the spirocyclic zwitterionic λ^5Si-silicates **10–12** [4], **13** [5], **14** [6], and **15·CH$_3$CN** [7] were synthesized according to Scheme 1.

These compounds contain two bidentate glycolato(2–)-O^1,O^2 (**10**), 2-methyllactato(2–)-O^1,O^2 (**11, 13, 14**), or benzilato(2–)-O^1,O^2 (**12, 15·CH$_3$CN**) ligands. The syntheses were carried out in acetonitrile at room temperature, and **10–14** and **15·CH$_3$CN** were isolated in high yields (≥ 80%) as crystalline solids.

As shown in Scheme 2, compounds **10** and **14** were additionally obtained by alternative methods involving Si–C bond cleavage reactions. These syntheses were performed in acetonitrile (**10, 14**) or water (**14**) at room temperature.

As demonstrated by single-crystal X-ray diffraction, the coordination polyhedron surrounding the silicon atom of **10** can be described either as a strongly distorted trigonal bipyramid (Berry distortion [8] 51.7 %), with the carboxylate oxygen atoms in the axial sites, or as a strongly distorted square pyramid, in which the carbon atom occupies the apical position [4]. In contrast, the silicon coordination polyhedra of **11** [4], **12** [4], **13** [5], **14** [6], and **15·CH$_3$CN** [7] can be best described as distorted trigonal bipyramids, with the carboxylate oxygen atoms in the axial sites. The respective Berry distortions [8] for these compounds amount to 38.4 % (**11**), 20.6 % (**12**), 5.4 % (**13**), 2.2 % (**14**), and 16.0 % (**15·CH$_3$CN**).

Because of the chiral nature of the zwitterions **10–15**, two enantiomeric species each were found in the crystal of these compounds. As examples of the structural chemistry of this particular type of compound, the molecular structures of **10**, **13**, and **15** in the crystal are depicted in Fig. 2–4.

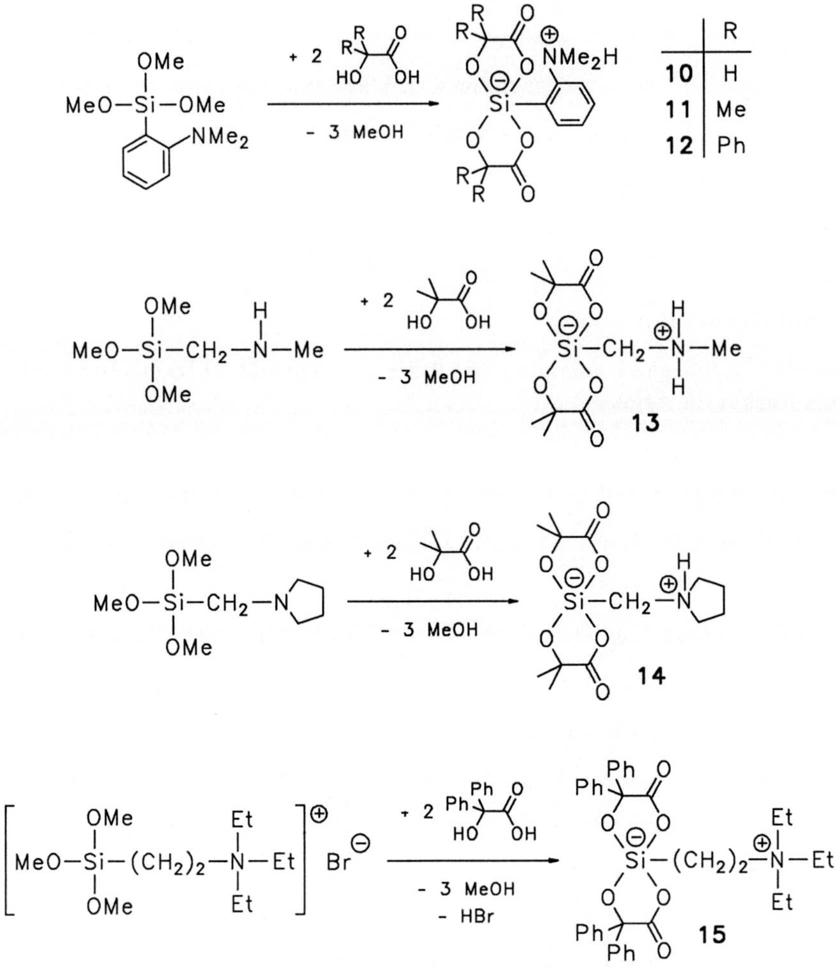

Scheme 1.

Scheme 2.

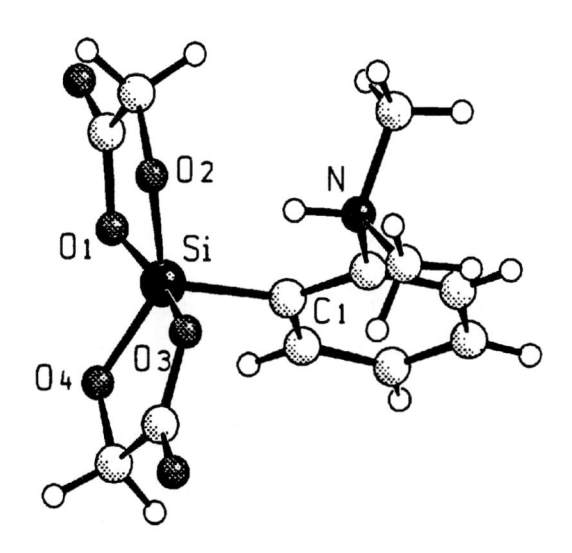

Fig. 2. Molecular structure of **10** in the crystal (SCHAKAL plot); selected geometric parameters: Si–O(1) 178.18(12), Si–O(2) 168.27(11), Si–O(3) 180.85(12), Si–O(4) 167.31(11), Si–C(1) 187.41(14) pm; O(1)–Si–O(3) 165.14(5)°

Citric acid was found to react with (aminoorgano)trimethoxysilanes in the same manner as observed for glycolic acid, 2-methyllactic acid, and benzilic acid. Thus, treatment of $(MeO)_3SiCH_2NMe_2$ with two mole equivalents of citric acid in acetonitrile at room temperature gave the zwitterionic spirocyclic bis[citrato(2–)-O^3,O^4]silicate **16** (Scheme 3) [9].

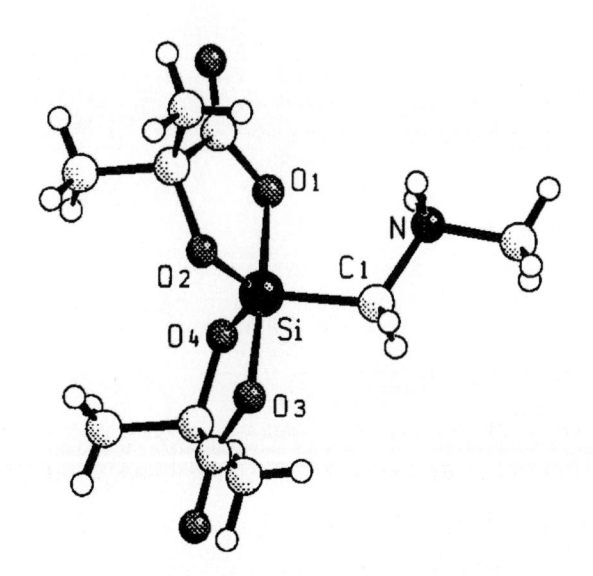

Fig. 3. Molecular structure of **13** in the crystal (SCHAKAL plot); selected geometric parameters: Si–O(1) 179.45(14), Si–O(2) 165.6(2), Si–O(3) 181.33(14), Si–O(4) 167.32(14), Si–C(1) 188.9(2) pm; O(1)–Si–O(3) 176.98(5)°

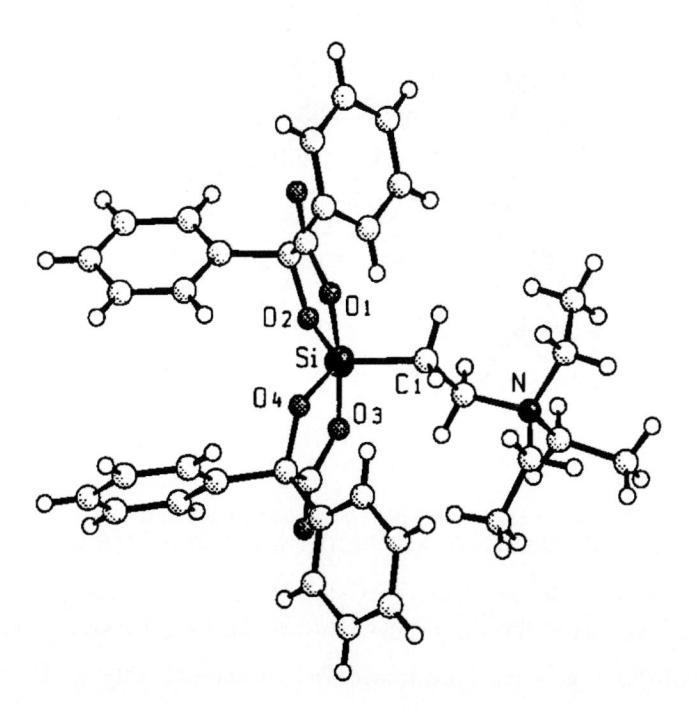

Fig. 4. Molecular structure of **15** in the crystal of **15·CH$_3$CN** (SCHAKAL plot); selected geometric parameters: Si–O(1) 179.96(14), Si–O(2) 166.57(14), Si–O(3) 179.89(14), Si–O(4) 167.49(14), Si–C(1) 188.2(2) pm; O(1)–Si–O(3) 172.40(7)°

[Scheme 3 structures]

Scheme 3.

Recrystallization of this compound from water yielded the monohydrate **16·H₂O** (yield 81 %) which was studied by single-crystal X-ray diffraction [9]. The coordination polyhedron around the silicon atom of **16** was found to be a distorted trigonal bipyramid (Berry distortion [8] 18.7 %), with the carboxylate oxygen atoms in the axial positions (Fig. 5).

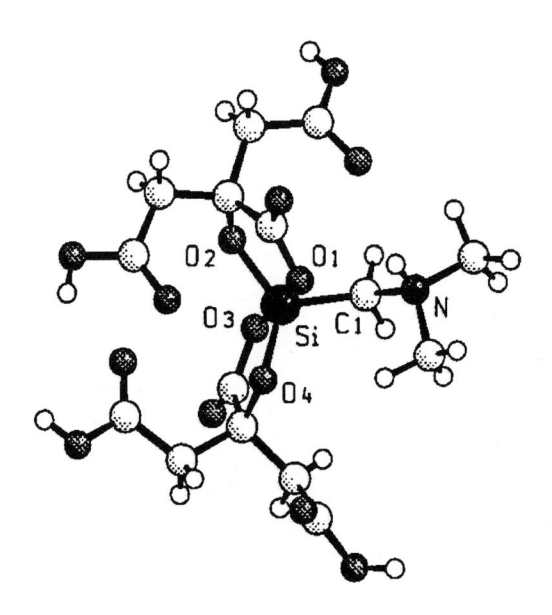

Fig. 5. Molecular structure of **16** in the crystal of **16·H₂O** (SCHAKAL plot); selected geometric parameters: Si–O(1) 179.4(2), Si–O(2) 166.6(2), Si–O(3) 180.2(2), Si–O(4) 166.2(2), Si–C(1) 188.8(4) pm; O(1)–Si–O(3) 175.07(11)°

In Scheme 4, the syntheses of the zwitterionic spirocyclic bis[glycolato(2−)-O^1,O^2]silicate **17** [10] and the related bis[benzilato(2−)-O^1,O^2]silicate **18** [11] are shown.

Scheme 4.

In contrast to compounds **10–16** (SiO_4C framework), the λ^5Si-silicates **17** and **18** contain a SiO_5 framework. The latter compounds were obtained by reaction of Si(OMe)$_4$ with two mole equivalents of the respective α-hydroxycarboxylic acid and one mole equivalent of 2-(dimethylamino)ethanol in boiling acetonitrile (reaction time 50 h; yield 61 % (**17**) or 77 % (**18**)). Recrystallization of **18** from dimethylformamide yielded the **18·DMF** solvate which was structurally characterized by single-crystal X-ray diffraction [11]. As can be seen from Fig. 6, the coordination polyhedron surrounding the silicon atom of **18** is a distorted trigonal bipyramid (Berry distortion [8] 13.9 %), in which the carboxylate oxygen atoms occupy the axial sites.

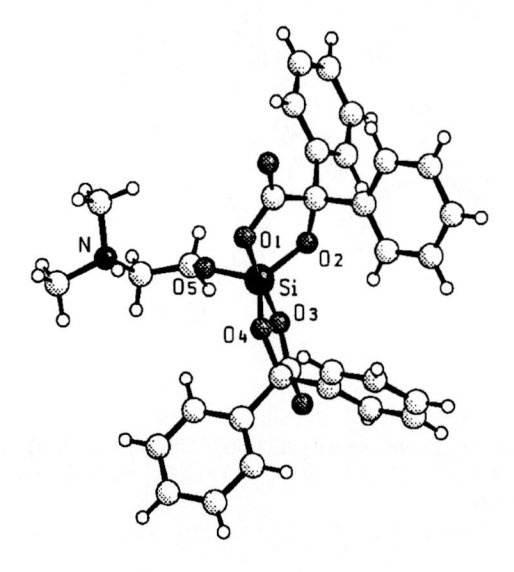

Fig. 6. Molecular structure of **18** in the crystal of **18·DMF** (SCHAKAL plot); selected geometric parameters: Si–O(1) 179.3(3), Si–O(2) 167.3(2), Si–O(3) 179.5(3), Si–O(4) 166.3(2), Si–O(5) 165.2(3) pm; O(1)–Si–O(3) 174.96(8)°

NMR-spectroscopic studies demonstrated that **10–18** also exist in solution ([D$_6$]-DMSO); however, no information about the solution-state geometry of these λ^5Si-silicates could be obtained so far. In the crystal, similar structural characteristics for **10–18** were found. With the exception of **10** (Berry distortion [8] 51.7 %), the silicon-coordination polyhedra of all compounds are best described as more or less distorted trigonal bipyramids, with the carboxylate oxygen atoms in the axial positions. This finding is in excellent agreement with the results obtained in ab initio studies of the bis[glycolato(2–)-O^1,O^2]-hydridosilicate(1–) ion (**19**) which served as a simple model for **10–18** [5]. SCF geometry optimizations (SVP basis sets) for the five isomers **I–V** of the anion **19** demonstrated that isomer **I** is the most stable one and corresponds best to the geometries observed for the coordination polyhedra of **11–18** in the crystal (Fig. 7). In contrast, the geometry of **10** can be alternatively described with that of isomer **I** or isomer **IV**.

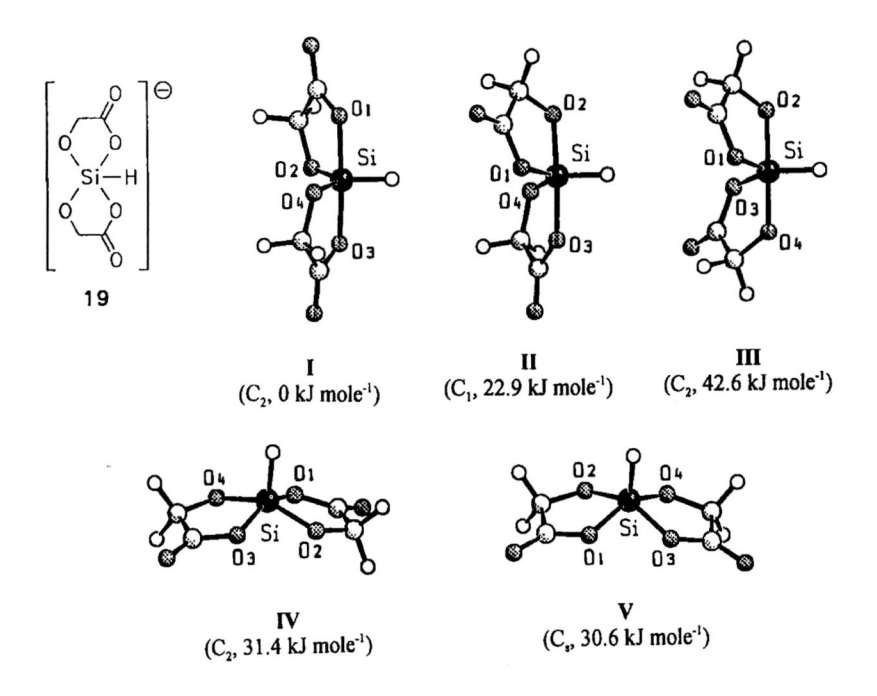

Fig. 7. Calculated structures and relative energies for the isomers **I–V** of the anion **19** obtained by SCF geometry optimizations (SVP basis sets); in contrast to the achiral species **V**, the isomers **I–IV** are chiral

The solution-state NMR spectra of **10–18** (solvent [D$_6$]-DMSO, room temperature) are compatible only with the presence of one particular species or with a rapid ligand exchange leading to an interconversion of some of the different isomers. In general, the Berry distortions observed for **10–18** in the crystal can all be regarded as transitions from isomer **I** to isomer **IV**. It is likely to assume that the respective degrees of distortion are influenced, inter alia, by the intra- and/or intermolecular N–H···O hydrogen bonds present in the crystal of these compounds (except for **15** which does not contain an NH donor function). In addition, further lattice effects may also be of importance.

2.2 Zwitterionic λ^5Ge-Germanates Containing Two Bidentate Diolato(2–) Ligands Derived from α-Hydroxycarboxylic Acids

By analogy to (aminoorgano)trimethoxysilanes, (aminoorgano)trimethoxygermanes were also found to react with α-hydroxycarboxylic acids to yield related zwitterionic spirocyclic λ^5Ge-germanates. Thus, reaction of $(MeO)_3GeCH_2NMe_2$ with two mole equivalents of glycolic acid, 2-methyllactic acid, or benzilic acid in acetonitrile at room temperature gave the corresponding λ^5Ge-germanium(IV) complexes **20–22** which were isolated in high yields (≥ 80 %) as crystalline solids (Scheme 5) [12].

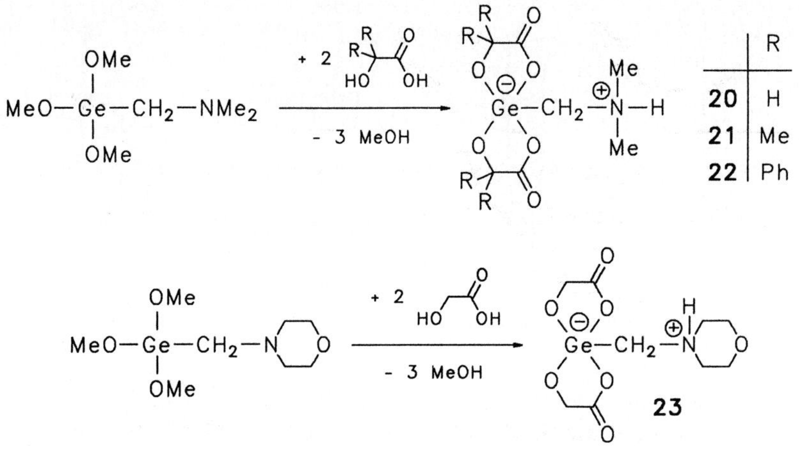

Scheme 5.

Compound **23** was obtained in an analogous manner (Scheme 5). Recrystallization of **23** from water gave the monohydrate **23·H$_2$O** which was structurally characterized by single-crystal X-ray diffraction (Fig. 8) [12].

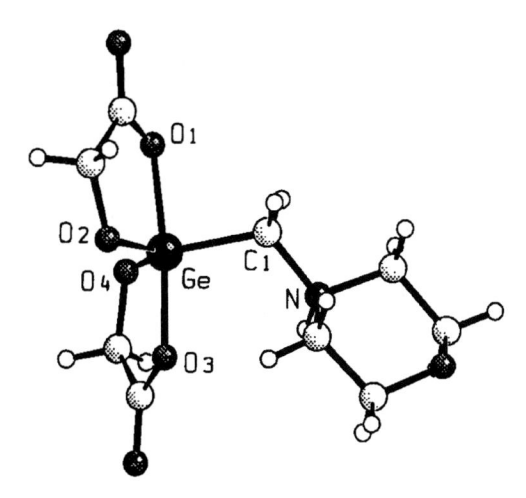

Fig. 8. Molecular structure of **23** in the crystal of **23**·H_2O (SCHAKAL plot); selected geometric parameters: Ge–O(1) 192.14(13), Ge–O(2) 178.33(14), Ge–O(3) 195.15(13), Ge–O(4) 178.35(13), Ge–C(1) 195.4(2) pm; O(1)–Ge–O(3) 173.09(6)°

By analogy to the structures of the related zwitterionic λ^5Si-silicates described in Chapter 2.1, the coordination polyhedron surrounding the germanium atom of the λ^5Ge-germanate **23**·H_2O is a distorted trigonal bipyramid (Berry distortion [8] 11.4 %), in which the carboxylate oxygen atoms occupy the axial sites.

2.3 Zwitterionic λ^5Si,λ^5Si'-Disilicates Containing Four Bidentate 2-Methyllactato(2–)-O^1,O^2 Ligands or Two Tetradentate (R,R)- or (S,S)-Tartrato(4–)-$O^1,O^2:O^3,O^4$ Ligands

The zwitterionic λ^5Si-silicates **10–18** described in Chapter 2.1 can be regarded as molecular mononuclear λ^5Si-silicon(IV) complexes, whereas the zwitterionic λ^5Si,λ^5Si'-disilicates *meso*-**24** [13], (+)-**25** [14], **26** [15], and (+)-**27** [10] (Scheme 6, 7) represent molecular binuclear λ^5Si-silicon(IV) complexes.

These compounds are characterized by the presence of two pentacoordinate (formally negatively charged) silicon atoms and two tetracoordinate (formally positively charged) nitrogen atoms. Because of the chiral nature of the trigonal-bipyramidal silicate moieties (see Chapter 2.1), special stereochemical features for these binuclear complexes have to be considered.

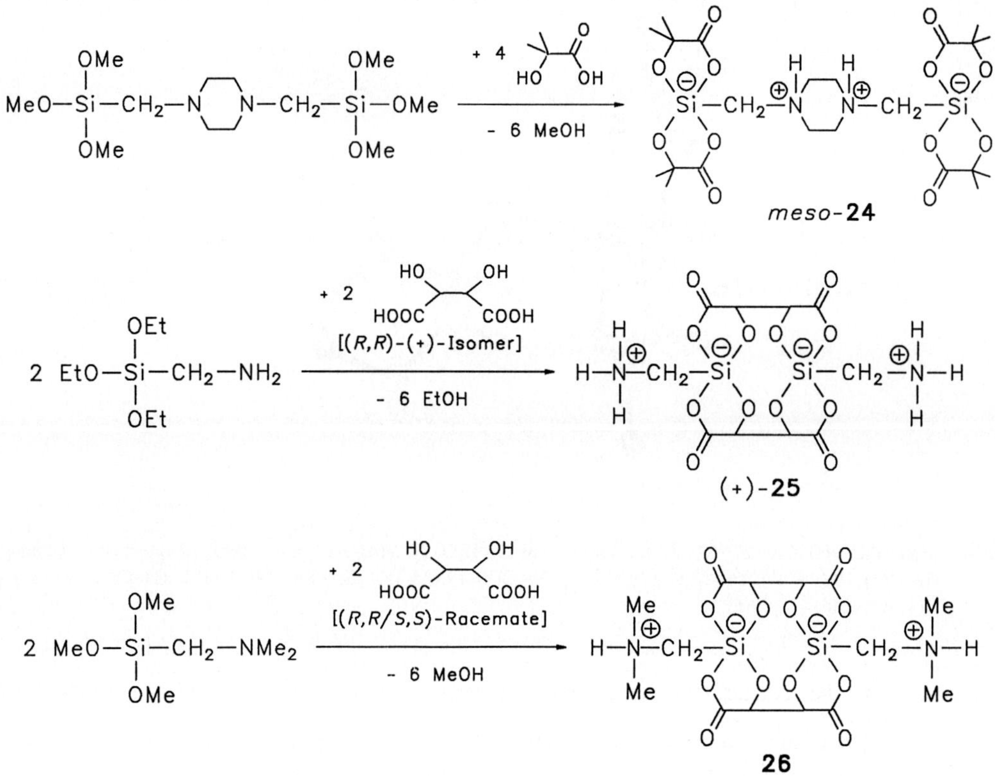

Scheme 6.

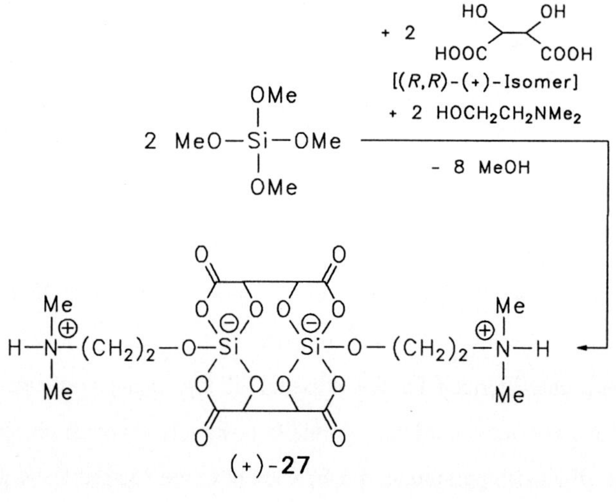

Scheme 7.

Compound *meso*-24 was synthesized according to Scheme 6 by reaction of bis-1,4-[(trimethoxysilyl)methyl]piperazine with four mole equivalents of 2-methyllactic acid in water/acetone at room temperature and isolated in 82 % yield as the crystalline octahydrate *meso*-24·8H$_2$O. Because of the presence of the two chiral trigonal-bipyramidal silicate moieties in **24** (carboxylate oxygen atoms in the axial positions), isomeric structures should be possible. In principle, the synthesis outlined in Scheme 6 could yield either a racemic mixture (identical absolute configurations of both silicate moieties in each of the two enantiomers) or the corresponding *meso*-isomer (opposite absolute configurations of the silicate moieties). As shown by single-crystal X-ray diffraction of **24**·8H$_2$O, the product isolated was the *meso*-isomer (Fig. 9) [13], the molecular structure of which is characterized by a crystallographic center of inversion. The coordination polyhedra around the two crystallographically equivalent silicon atoms are distorted trigonal bipyramids (Berry distortions [8] 16.5 %), with the carboxylate oxygen atoms in the axial positions.

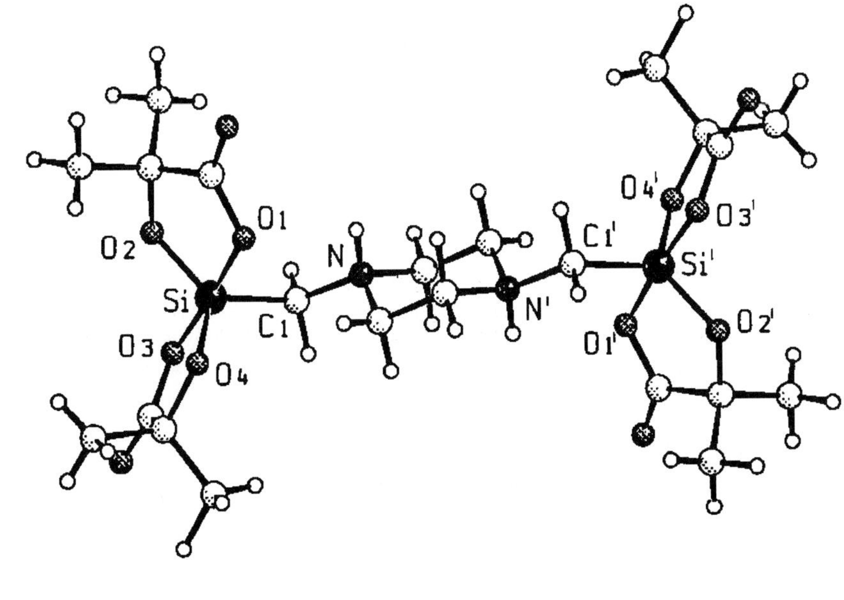

Fig. 9. Molecular structure of *meso*-24 in the crystal of *meso*-24·8H$_2$O (SCHAKAL plot); selected geometric parameters: Si–O(1) 179.4(4), Si–O(2) 167.1(3), Si–O(3) 177.7(4), Si–O(4) 165.8(4), Si–C(1) 188.4(6) pm; O(1)–Si–O(3) 176.1(2)° (the molecular structure is characterized by a crystallographic center of inversion)

Compound (+)-**25** was prepared according to Scheme 6 by reaction of (EtO)$_3$SiCH$_2$NH$_2$ with (*R,R*)-tartaric acid (molar ratio 1:1) in aqueous solution at room temperature and isolated as a 2:1 mixture of (+)-**25** and (+)-**25**·3H$_2$O (yield ca. 84 %) [14]. Both compounds were obtained in pure form by mechanical sorting and then characterized. The trihydrate was studied by single-crystal X-ray diffraction [14]. The binuclear λ^5Si-silicon(IV) complex (+)-25 is optically active; its formation occurred stereospecifically (100 % ee, 100 % de). The optical activity results from the presence of the two chiral silicate moieties and the two chiral (R,R)-tartrato(4–) ligands. Treatment of $(EtO)_3SiCH_2NH_2$ with (S,S)-tartaric acid yielded the corresponding antipode (–)-25 [14].

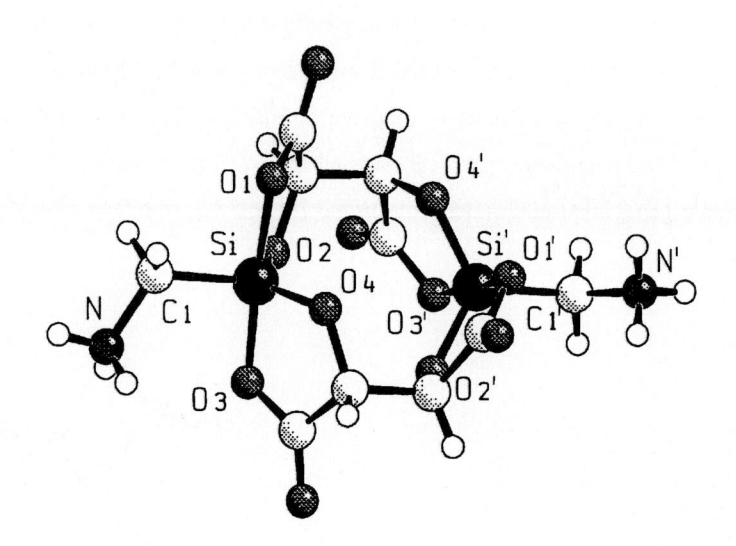

Fig. 10. Molecular structure of (+)-25 in the crystal of (+)-25·3H$_2$O (SCHAKAL plot); selected geometric parameters: Si–O(1) 182.0(2), Si–O(2) 166.5(2), Si–O(3) 179.1(2), Si–O(4) 167.5(2), Si–C(1) 188.5(3), Si'–O(1') 179.7(2), Si'–O(2') 166.6(2), Si'–O(3') 182.5(2), Si'–O(4') 166.9(2), Si'–C(1') 188.4(2) pm; O(1)–Si–O(3) 175.56(9), O(1')–Si'–O(3') 175.13(9)°

The molecular structure of (+)-25 in the crystal of the trihydrate is depicted in Fig. 10. The coordination polyhedra surrounding the silicon atoms Si and Si' are distorted trigonal bipyramids (Berry distortions [8] 7.7 % and 10.6 %, respectively), in which each of the axial positions are occupied by carboxylate oxygen atoms. The molecular symmetry of the pentacyclic molecular framework of (+)-25 may be described as approximately D_2 if the two ammoniomethyl groups are ignored. Thus, the absolute configurations of the two chiral silicate moieties are identical.

Racemic tartaric acid could also be successfully used for the synthesis of binuclear complexes. Thus, reaction of $(MeO)_3SiCH_2NMe_2$ with one mole equivalent of racemic tartaric acid in aqueous solution at room temperature gave the λ^5Si,λ^5Si'-disilicate 26 which was isolated in 84 % yield as the crystalline monohydrate 26·H$_2$O (Scheme 6) [10]. As shown by single-crystal X-ray diffraction [15], the two enantiomers of **26** in the crystal of the monohydrate have a similar structure (Fig. 11) as observed for (+)-**25** in the crystal of the trihydrate (Fig. 10).

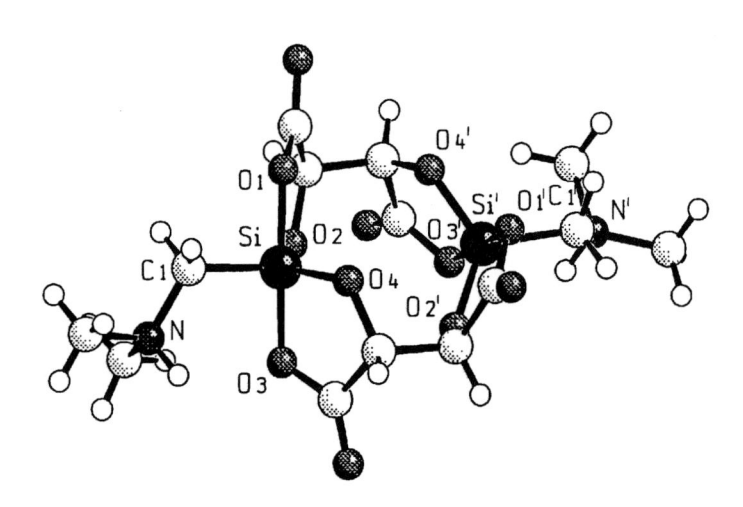

Fig. 11. Molecular structure of **26** in the cyrstal of **26·H₂O** (SCHAKAL plot); selected geometric parameters: Si–O(1) 182.90(13), Si–O(2) 166.01(13), Si–O(3) 178.64(14), Si–O(4) 165.98(14), Si–C(1) 188.7(2), Si'–O(1') 180.43(14), Si'–O(2') 165.77(13), Si'–O(3') 179.43(14), Si'–O(4') 167.66(13), Si'–C(1') 189.4(2) pm; O(1)–Si–O(3) 177.66(6), O(1')–Si'–O(3') 175.97(6)°.

The coordination polyhedra surrounding the silicon atoms Si and Si' of **26** are distorted trigonal bipyramids (Berry distortions [8] 6.1 % and 10.9 %, respectively), in which each of the axial sites are occupied by carboxylate oxygen atoms.

Attempts to synthesize related binuclear λ^5Si-silicon(IV) complexes containing *meso*-tartrato-(4–)-$O^1,O^2:O^3,O^4$ ligands failed [10]. Obviously, for geometrical reasons this particular type of ligand does not allow the formation of such compounds.

Compound (+)-**27** was synthesized according to Scheme 7 by reaction of Si(OMe)₄ with one mole equivalent each of (*R,R*)-tartaric acid and 2-(dimethylamino)ethanol in boiling dimethylformamide and isolated in 79 % yield as a crystalline solid [10]. Its identity was established by elemental analyses, solution-state (^{1}H, ^{13}C, ^{29}Si) and solid-state (^{29}Si CP/MAS) NMR studies, and FAB-MS investigations. By analogy to (+)-**25**, compound (+)-**27** is optically active.

2.4 Zwitterionic λ^5Si-Silicates Containing Two Bidentate Acetohydroximato(2–)-O^1,O^2 Ligands or One Bidentate Oxalato(2–)-O^1,O^2 Ligand

The zwitterionic λ^5Si-silicate **28** is, to the best of our knowledge, the first pentacoordinate silicon species containing diolato(2–) ligands derived from a hydroximic acid (Scheme 8).

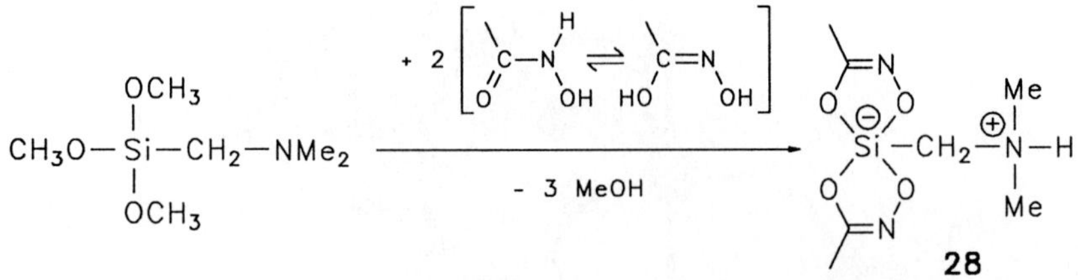

Scheme 8.

The two five-membered rings of this spirocyclic λ^5Si-silicon(IV) complex each contain two oxygen atoms, one nitrogen atom, one carbon atom, and one (pentacordinate) silicon atom. As shown in Scheme 8, this compound was prepared by analogy to the synthesis of the related λ^5Si-silicates **10–16**. Thus, treatment of (MeO)$_3$SiCH$_2$NMe$_2$ with two mole equivalents of acetohydroxamic acid in acetonitrile at room temperature yielded **28** in 80 % yield as a crystalline solid [5]. The formation of this compound can be understood in terms of a reaction of (MeO)$_3$SiCH$_2$NMe$_2$ with the 1,2-diol acetohydroximic acid, which is a tautomer of acetohydroxamic acid.

Compound **28** was structurally characterized by single-crystal X-ray diffraction [5]. As can be seen from Fig. 12, the coordination polyhedron surrounding the silicon atom can be described as a distorted trigonal bipyramid (Berry distortion [8] 17.6 %), in which the oxygen atoms bound to nitrogen occupy the axial sites.

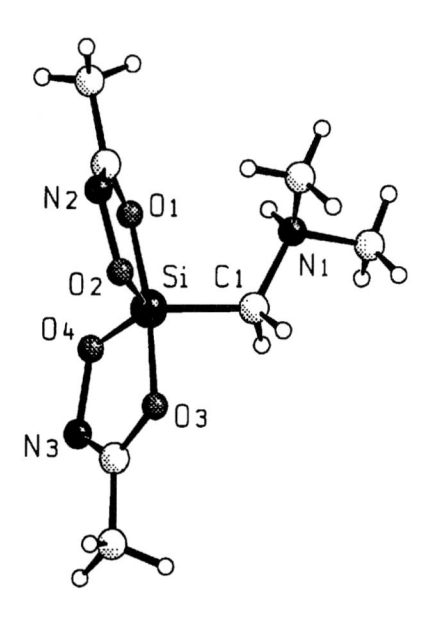

Fig. 12. Molecular structure of **28** in the crystal (SCHAKAL plot); selected geometric parameters: Si–O(1) 176.6(2), Si–O(2) 169.1(2), Si–O(3) 177.0(2), Si–O(4) 168.3(2), Si–C(1) 188.3(2) pm; O(1)–Si–O(3) 172.17(7)°

The zwitterionic monocyclic λ^5Si-silicate **29** is a molecular λ^5Si-silicon(IV) complex containing a SiO_2C_2F framework (Scheme 9).

Scheme 9.

In this molecule, a bidentate oxalato(2–)-O^1,O^2 ligand is bound to the silicon(IV) coordination center. Compound **29** was prepared by a two-step synthesis according to Scheme 9 and isolated as a crystalline solid. Thus, treatment of Me(MeO)$_2$SiCH$_2$NMe$_2$ with HF in ethanol/water yielded the zwitterionic λ^5Si-trifluorosilicate MeF$_3$SiCH$_2$NMe$_2$H, which upon reaction with bis(trimethylsilyl) oxalate (molar ratio 1:1) in boiling acetonitrile gave **29** (total yield 70 %) [5].

Compound **29** was structurally characterized by single-crystal X-ray diffraction [5]. As can be seen from Fig. 13, the coordination polyhedron surrounding the silicon atom is a distorted trigonal bipyramid (Berry distortion [8] 3.3 %), in which the fluorine atom and one of the two carboxylate oxygen atoms occupy the axial sites.

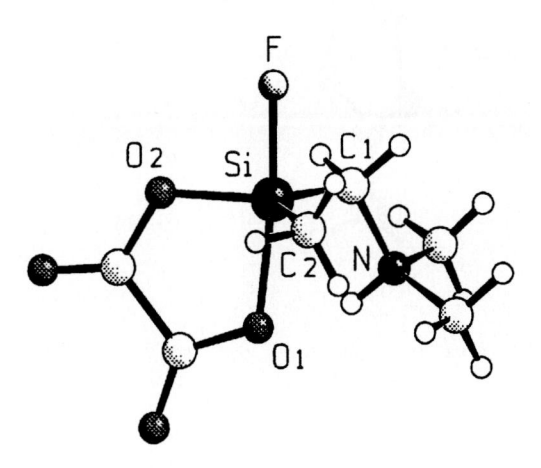

Fig. 13. Molecular structure of **29** in the crystal (SCHAKAL plot); selected geometric parameters: Si–F 166.63(14), Si–O(1) 191.7(2), Si–O(2) 173.9(2), Si–C(1) 190.5(2), Si–C(2) 185.6(2) pm; F–Si–O(1) 172.42(5)°

3 Conclusions

The results presented here suggest that a rich complex chemistry of pentacoordinate silicon with ligands derived from α-hydroxycarboxylic acids (including tartaric acid), hydroximic acids, and oxalic acid may be developed. As most of these ligands derive from natural products and as some of these λ^5Si-silicon(IV) complexes were shown to exist in aqueous solution, compounds of this formula type are of particular interest: it has been speculated in the literature [16] that silicon transport in biological systems might be based on higher coordinate Si species, and complexes such as the title compounds could be of interest as model systems in this respect.

Acknowledgement: R. T. wishes to express his sincere thanks to his coworkers without whose contributions this article could not have been written; their names are cited in the references. In addition, the authors thank Mr. G. Mattern (*Institut für Kristallographie, Universität Karlsruhe*) for collecting some of the X-ray diffraction data sets, and Professor P. G. Jones (*Technische Universität Braunschweig*) for performing some of the crystal structure analyses. Furthermore, assistance of our theoretical studies by the group of Professor Ahlrichs (*Lehrstuhl für Theoretische Chemie, Universität Karlsruhe*) is gratefully acknowledged. Finally, financial support of this work by the *Deutsche Forschungsgemeinschaft* and the *Fonds der Chemischen Industrie* is gratefully acknowledged.

References:

[1] a) C. Strohmann, R. Tacke, G. Mattern, W. F. Kuhs, *J. Organomet. Chem.* **1991**, *403*, 63.

b) R. Tacke, J. Sperlich, C. Strohmann, G. Mattern, *Chem. Ber.* **1991**, *124*, 1491.

c) R. Tacke, J. Sperlich, C. Strohmann, B. Frank, G. Mattern, *Z Kristallogr.* **1992**, *199*, 91.

d) R. Tacke, F. Wiesenberger, A. Lopez-Mras, J. Sperlich, G. Mattern, *Z. Naturforsch.* **1992**, *47B*, 1370.

e) R. Tacke, A. Lopez-Mras, W. S. Sheldrick, A. Sebald, *Z. Anorg. Allg. Chem.* **1993**, *619*, 347.

f) R. Tacke, A. Lopez-Mras, J. Sperlich, C. Strohmann, W. F. Kuhs, G. Mattern, A. Sebald, *Chem. Ber.* **1993**, *126*, 851.

g) J. Sperlich, J. Becht, M. Mühleisen, S. A. Wagner, G. Mattern, R. Tacke, *Z. Naturforsch.* **1993**, *48B*, 1693.

[2] R. Tacke, J. Becht, A. Lopez-Mras, J. Sperlich, *J. Organomet. Chem.* **1993**, *446*, 1; R. Tacke, J. Becht, O. Dannappel, M. Kropfgans, A. Lopez-Mras, M. Mühleisen, J. Sperlich, in: *Progress in Organosilicon Chemistry* (Eds.: B. Marciniec, J. Chojnowski), Science Publishers Gordon and Breach, in press.

[3] S. N. Tandura, M. G. Voronkov, N. V. Alekseev, *Top. Curr. Chem.* **1986**, *131*, 99; R. R. Holmes, *Chem. Rev.* **1990**, *90*, 17; C. Chuit, R. J. P. Corriu, C. Reye, J. C. Joung, *Chem. Rev.* **1993**, *93*, 1371.

[4] R. Tacke, A. Lopez-Mras, P. G. Jones, *Organometallics*, **1994**, *13*, 1617.

[5] O. Dannappel, R. Tacke, unpublished results.

[6] J. Becht, R. Tacke, unpublished results.

[7] B. Pfrommer, P. G. Jones, R. Tacke, unpublished results.

[8] R. R. Holmes, *Progress Inorg. Chem.* **1984**, *32*, 119; and references cited therein.

[9] M. Mühleisen, R. Tacke, *Chem. Ber.* **1994**, *127*, 1615.

[10] M. Mühleisen, R. Tacke, unpublished results.

[11] R. Tacke, M. Mühleisen, *Inorg. Chem.* **1994**, *33*, 4191.

[12] J. Heermann, P. G. Jones, R. Tacke, unpublished results.

[13] M. Mühleisen, R. Tacke, *Organometallics* **1994**, *13*, 3740.

[14] R. Tacke, M. Mühleisen, P. G. Jones, *Angew. Chem.* **1994**, *106*, 1250; *Angew. Chem., Int. Ed. Engl.* **1994**, *33*, 1186.

[15] M. Mühleisen, P. G. Jones, R. Tacke, unpublished results.

[16] C. W. Sullivan, in: *Silicon Biochemistry* (Eds.: D. Evered, M. O'Connor), Wiley, Chichester, **1986**, p. 59; and references therein.

Syntheses and Solution-State NMR Studies of Zwitterionic Spirocyclic λ^5Si-Organosilicates Containing Two Identical Unsymmetrically Substituted 1,2-Benzenediolato(2−) Ligands

Mathias Mühleisen, Reinhold Tacke*

Institut für Anorganische Chemie
Universität Fridericiana zu Karlsruhe (TH)
Engesserstr., Geb. 30.45, D-76128 Karlsruhe, Germany

Summary: The zwitterionic spirocyclic λ^5Si-organosilicates bis[4-nitro-1,2-benzenediolato(2−)](morpholiniomethyl)silicate (**4**), bis[3,5-di-*t*butyl-1,2-benzenediolato(2−)](morpholiniomethyl)silicate (**5**), and bis[4-nitro-1,2-benzenediolato(2−)][(1-methylmorpholinio)methyl]silicate (**6**) were synthesized. At room temperature, two NMR spectroscopically (^{1}H, ^{13}C, ^{29}Si) distinguishable pentacoordinate Si species in solution were observed for **4−6** (concentration ratio ca. 1:1.1; [D$_6$]-DMSO). At higher temperatures, coalescence phonenema were observed (^{1}H NMR), indicating an interconversion of these species. The energy barrier of this isomerization was determined to be $\Delta G^* =$ ca. 70 kJ mol^{-1} for all three compounds (^{1}H NMR). The isomerization is discussed in terms of an intramolecular ligand-exchange process involving a trigonal-bipyramidal transition state with the respective carbon atom in an axial position.

Introduction

Over the past few years, a series of zwitterionic spirocyclic bis[1,2-benzenediolato(2−)]organosilicates and related zwitterions with 2,3-naphthalenediolato(2−) and symmetrically substituted 1,2-benzenediolato(2−) ligands have been synthesized and structurally characterized [1]. The zwitterionic λ^5Si-silicates **1−3** are typical examples of this particular type of compound (Fig. 1). In contrast to the well established solid-state structures of these zwitterions, only little is known about their structure in solution. NMR spectroscopic studies have clearly demonstrated that these species exist in solution ([D$_6$]-DMSO)

[1]; however, almost nothing is known about their geometry in solution. In order to obtain more information about the structural chemistry of such compounds, we prepared the derivatives **4–6** and studied them by temperature-dependent solution-state NMR experiments (Fig.1). In contrast to the zwitterionic spirocyclic λ^5Si-silicates described in the literature (such as **1–3**) [1], compounds **4–6** contain two identical unsymmetrically substituted bidentate diolato(2–) ligands. This particular substitution pattern was expected to give rise to the formation of NMR-spectroscopically detectable isomers in solution.

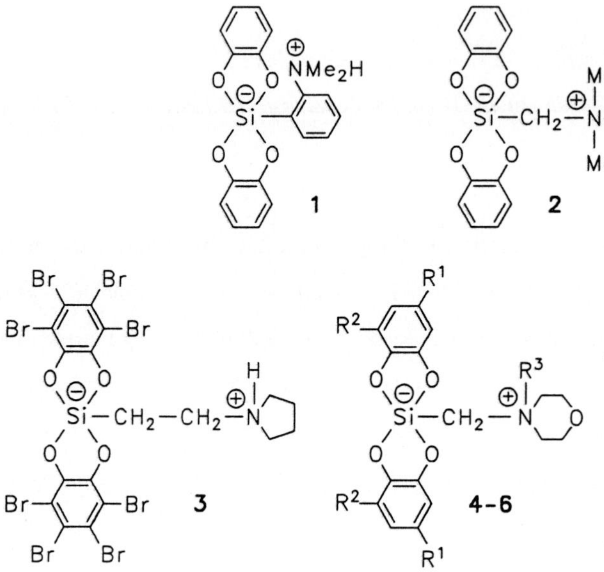

	R^1	R^2	R^3
4	NO_2	H	H
5	*t*-Bu	*t*-Bu	H
6	NO_2	H	Me

Fig. 1.

Results and Discussion

Compounds **4** and **5** were synthesized by treatment of trimethoxy(morpholiniomethyl)silane with two mole equivalents of 1,2-dihydroxy-4-nitrobenzene and 3,5-di-*t*butyl-1,2-dihydroxybenzene, respectively, in acetonitrile at room temperature (Scheme 1).

The zwitterion **6** was obtained similarly by reaction of 1-[(tri-ethoxysilyl)methyl]-1-methyl-morpholinium iodide with 1,2-dihydroxy-4-nitrobenzene in boiling acetonitrile in the presence of KOH (Scheme 1). Compounds **4–6** were isolated in 75–85 % yield as crystalline solids; their identity was established by elemental analyses, NMR experiments, and mass-spectrometric investigations.

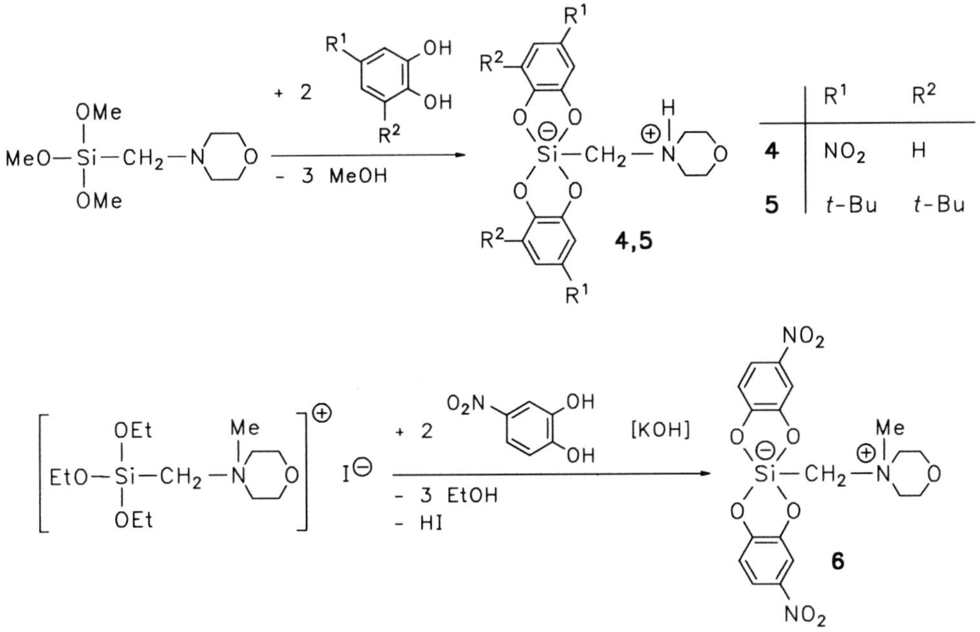

Scheme 1.

The zwitterionic λ⁵Si-silicates **4–6** were studied by ^{1}H, ^{13}C, and ^{29}Si NMR experiments using [D$_6$]-DMSO as solvent. For all three compounds, two sets of resonance signals (intensity ratio ca. 1:1.1) were observed in the ^{1}H, ^{13}C, and ^{29}Si NMR spectra at room temperature, indicating the presence of two NMR-spectroscopically distinguishable species in solution (Fig. 3) (in this context, see also [3]). In contrast, in the NMR spectra of related compounds containing symmetrical 1,2-benzenediolato(2–) ligands (such as **1–3**) only one set of resonance signals was observed [1].

As shown by temperature-dependent ^{1}H NMR experiments, the two distinguishable species present in solutions of **4–6** were found to interconvert into each other as demonstrated by the detection of coalescence phenonema (coalescence temperature (^{1}H NMR) = 39 °C (**4**), 49 °C (**5**), 34 °C (**6**)) (Fig. 3). Provided that this isomerization is based on an intramolecular ligand exchange, the free energy of activation for this process amounts to $\Delta G^* = 70$ (**4**), 72 (**5**), or 70 (**6**) kJ mol^{-1}. These data are in accordance with the results obtained in ab initio studies with the related bis[1,2-benzenediolato(2–)]-hydrido(1–) ion (**7**) [2] (Fig. 2).

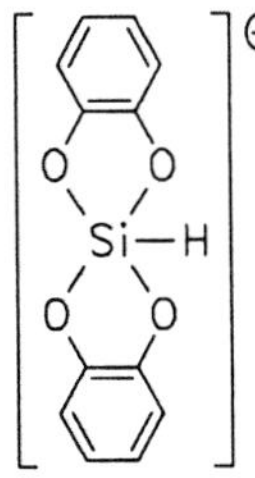

Fig. 2.

This anion was shown to undergo an intramolecularligand via a trigonal-bipyramidal transition state with the hydrogen atom in an axial position. The energy barrier for this process was calculated to be $\Delta G^* = 66.5$ kJ mol^{-1} [2]. The coalescence phenonema observed in the NMR studies of **4–6** might be described by an analogous ligand-exchange process via a trigonal-bipyramidal transition state with the respective carbon atom in an axial position.

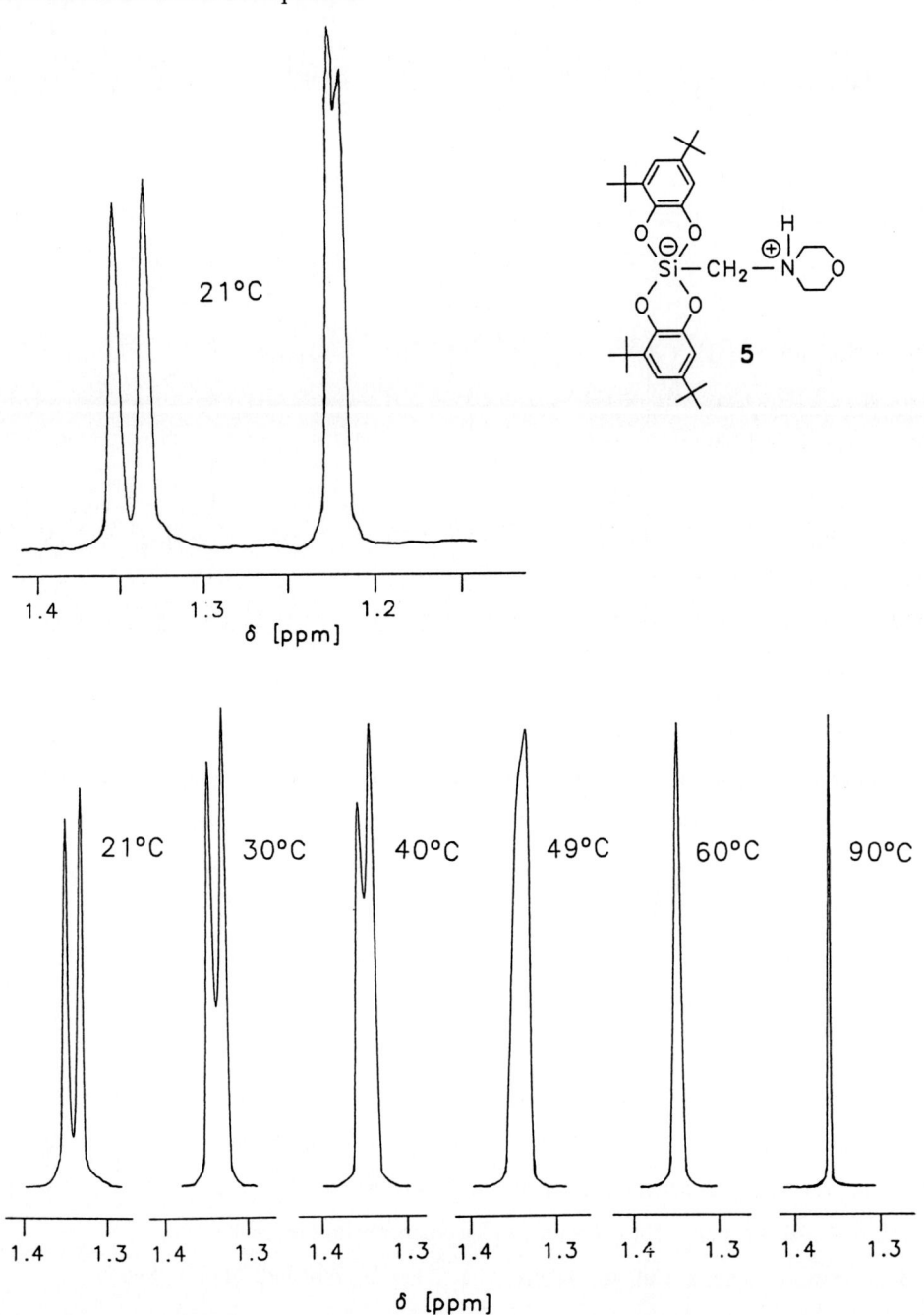

Fig. 3. *above:* Partial ^{1}H NMR spectrum (δ = 1.2–1.4) of **5** showing two sets of signals (two singlets each for *t*Bu) for the 3,5-di-*t*butyl-1,2-benzenediolato(2–) ligand ([D$_6$]-DMSO, 21 °C, 250.1 MHz);
below: Partial ^{1}H NMR spectra (δ = 1.3–1.4) of **5** showing the temperature dependence of one of the two sets of signals ([D$_6$]-DMSO, 21–90 °C, 250.1 MHz; coalescence temperature 49 °C)

Acknowledgement: Financial support of this work by the *Deutsche Forschungsgemeinschaft* and the *Fonds der Chemischen Industrie* is gratefully acknowledged.

References:

[1] R. Tacke, F. Wiesenberger, A. Lopez-Mras, J. Sperlich, G. Mattern, *Z. Naturforsch.* **1992**, *47B*, 1370; R. Tacke, J. Sperlich, C. Strohmann, B. Frank, G. Mattern, *Z. Kristallogr.* **1992**, *199*, 91; R. Tacke, A. Lopez-Mras, J. Sperlich, C. Strohmann, W. F. Kuhs, G. Mattern, A. Sebald, *Chem. Ber.* **1993**, *126*, 851; R. Tacke, J. Becht, A. Lopez-Mras, J. Sperlich, *J. Organomet. Chem.* **1993**, *446*, 1; R. Tacke, A. Lopez-Mras, W. S. Sheldrick, A. Sebald, *Z. Anorg. Allg. Chem.* **1993**, *619*, 347; J. Sperlich, J. Becht, M. Mühleisen, S. A. Wagner, G. Mattern, R. Tacke, *Z. Naturforsch.* **1993**, *48B*, 1693.

[2] O. Dannappel, R. Tacke, in: *Organosilicon Chemistry II* (Eds.: N. Auner, J. Weis), VCH, Weinheim, **1995**, p. 453.

[3] D. F. Evans, A. M. Z. Slawin, D. J. Williams, C. Y. Wong, J. D. Woollins, *J. Chem. Soc., Dalton Trans.* **1992**, 2383.

Intramolecular Ligand Exchange of Zwitterionic Spirocyclic Bis[1,2-benzenediolato(2–)]organosilicates: Ab Initio Studies of the Bis[1,2-benzenediolato(2–)]-hydridosilicate(1–) Ion

Olaf Dannappel, Reinhold Tacke*

Institut für Anorganische Chemie
Universität Fridericiana zu Karlsruhe (TH)
Engesserstr., Geb. 30.45, D-76128 Karlsruhe, Germany

Summary: In order to get more information about the structural chemistry of zwitterionic spirocyclic bis[1,2-benzenediolato(2–)]organosilicates, ab initio studies (RHF/SCF level, optimized SVP basis sets) were performed using the bis[1,2-benzenediolato(2–)]-hydridosilicate(1–) ion as a model. The geometry of the energetically preferred Si-coordination polyhedron of this anion was found to be a slightly distorted trigonal bipyramid (C_2 symmetry). Different ligand-exchange processes for this species are postulated, an undistorted square pyramid (C_{2v} symmetry) and a distorted trigonal bipyramid (C_s symmetry) being the transition states for these processes (with energy barriers of 5.82 or 66.5 kJ mol^{-1}). These results are compared with experimental data obtained in temperature-dependent solution-state NMR studies of some of the zwitterionic title compounds containing two identical unsymmetrically substituted 1,2-benzenediolato(2–) ligands.

Introduction

Over the past few years, a series of zwitterionic spirocyclic bis[1,2-benzenediolato(2–)]-organosilicates, such as compounds **1** [1c] and **2** [1a] (Fig. 1), and related zwitterions with 2,3-naphthalene-diolato(2–) and symmetrically substituted 1,2-benzenediolato(2–) ligands have been structurally charac-terized by X-ray diffraction [1]. The coordination polyhedra around the pentacoordinate silicon atoms of such compounds can be described using the idealized geometries of the trigonal bipyramid (TBP) and the square pyramid (SP). In the solid state, almost all transitions between a nearly ideal TBP and a nearly ideal SP were observed [1]. For example, for compounds **1** and **2** a nearly ideal TBP and a distorted SP, respectively, were found. In most cases, these geometries are located on the Berry-pseudorotation coordinate of these species. In contrast to the well established solid-state structures of the title compounds, only little is known about their structure in solution. In order to get more information about the structural chemistry of compounds of this particular formula type, ab initio studies were performed using the related bis[1,2-benzenediolato(2–)]hydridosilicate(1–) ion (**3**) as a model.

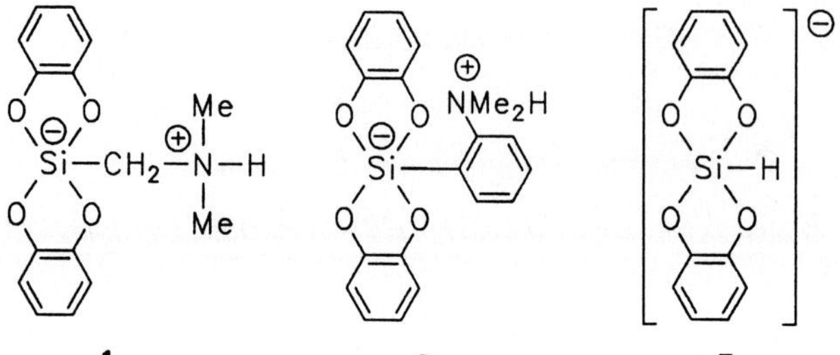

Fig. 1.

Results and Discussion

The SCF geometry optimization of **3** demonstrated a slightly distorted TBP (3_{TBP}, C_2 symmetry) to be the energetically preferred geometry (Fig. 2, Table 1) [2]. In addition, two further geometries were found ($3\ddagger^1_{SP}$, $3\ddagger^2_{TBP}$; Fig. 2, Table 1) which are transition states of intramolecular ligand-exchange processes [2]. Two different pathways (*path 1*, *path 2*) can be assumed (Fig. 2).

Path 1 ($3_{TBP} \rightleftharpoons 3\ddagger^1_{SP} \rightleftharpoons 3_{TBP}'$) is characterized by a Berry-type pseudorotation with the hydrogen atom H(1) as the pivot ligand. In this process, both pairs of axial and equatorial oxygen atoms change their positions in the TBP, the energy barrier between 3_{TBP} ($3_{TBP}'$) and $3_{SP}\ddagger^1$ amounting to 5.82 kJ mol^{-1}. The geometry of the transition state $3\ddagger^1_{SP}$ is an undistorted SP (C_{2v} symmetry).

Path 2 (example: $3_{TBP} \rightleftharpoons 3\ddagger^2_{TBP} \rightleftharpoons 3_{TBP}''$) can be described by two subsequent pseudorotation processes with an equatorial oxygen atom each as the pivot ligand [3]. As an example for *path 2*, the first pseudorotation (O(2) as pivot ligand) leads from 3_{TBP} to the transition state $3\ddagger^2_{TBP}$ (a distorted TBP, C_s symmetry) which upon a further pseudorotion (O(1) as pivot ligand) yields $3_{TBP}''$. Thus, as a result of the ligand exchange according to *path 2*, only the oxygen atoms of one of the two bidentate ligands (O(1), O(2)) have changed their positions in the TBP. The energy barrier between 3_{TBP} ($3_{TPB}''$) and the transition state $3_{TBP}\ddagger^2$ amounts to 66.5 kJ mol^{-1} and is therefore significantly higher than that calculated for *path 1*.

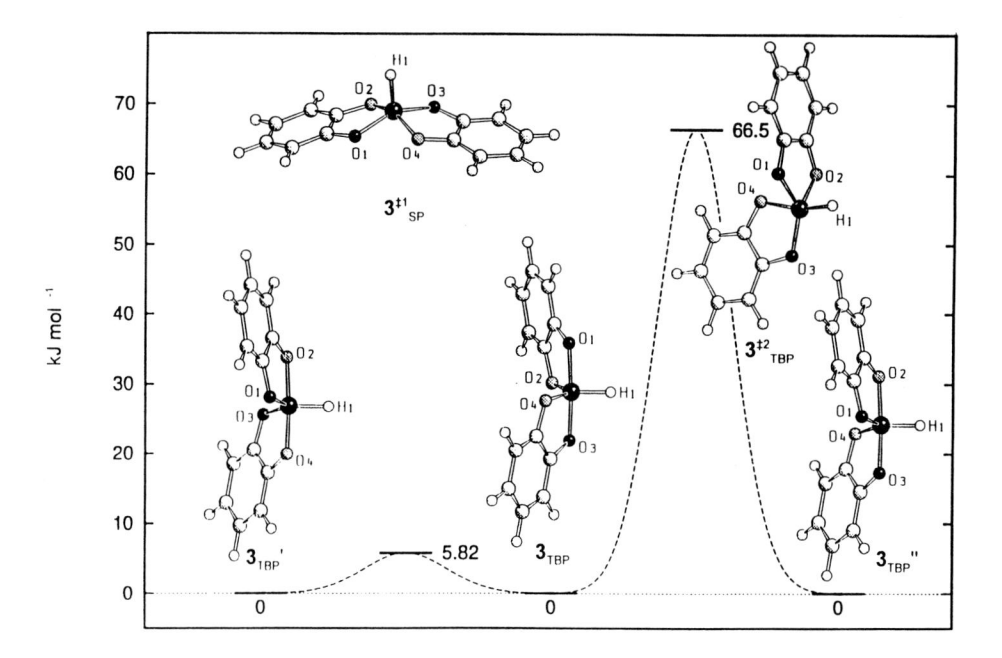

Fig. 2. Calculated structures of 3 (3_{TBP}, $3^{\ddagger 1}_{SP}$, $3^{\ddagger 2}_{TBP}$) and illustration of the different ligand-exchange processes of this anion (*path 1, path 2*; see text); for reasons of symmetry, 3_{TBP}, $3_{TBP}'$, and $3_{TBP}''$ are energetically equivalent species.

	3_{TBP}	$3^{\ddagger 1}_{SP}$	$3^{\ddagger 2}_{TBP}$		3_{TBP}	$3^{\ddagger 1}_{SP}$	$3^{\ddagger 2}_{TBP}$
Si–O(1)	178.1	174.1	172.0	O(1)–Si–H(1)	92.4	104.2	95.8
Si–O(2)	178.1	174.1	172.0	O(2)–Si–O(3)	89.9	86.8	132.9
Si–O(3)	169.9	174.1	172.6	O(2)–Si–O(4)	121.1	151.6	90.7
Si–O(4)	169.9	14.1	178.7	O(2)–Si–H(1)	119.5	104.2	95.8
Si–H(1)	146.7	146.7	148.8	O(3)–Si–O(4)	87.7	86.3	84.8
O(1)–Si–O(2)	87.7	86.3	93.9	O(3)–Si–H(1)	92.4	104.2	85.8
O(1)–Si–O(3)	175.2	151.6	132.9	O(4)–Si–H(1)	119.5	104.2	170.6
O(1)–Si–O(4)	89.9	86.8	90.7				

Table 1. Selected calculated interatomic distances [pm] and angles [°] for 3_{TBP}, $3^{\ddagger 1}_{SP}$, and $3^{\ddagger 2}_{TBP}$

From the results obtained for the model species **3**, it is concluded that the TBP, with oxygen atoms in the axial positions, is the energetically preferred geometry of the Si coordination polyhedra of the title compounds. As the pseudorotation according to *path 1* needs only a small amount of energy, package effects in the crystal (which are individual for each compound) can easily cause distortions of the TBP toward the SP. This assumption is in agreement with the results obtained in experimental studies [1].

If the silicon atom of the title compounds is linked with two identical unsymmetrically substituted 1,2-benzenediolato(2–) ligands (see examples **4–6**; Fig. 4), the four TBPs **A–D** and their corresponding enantiomers **A'–D'** (Fig. 3) have to be considered to describe the aforementioned ligand-exchange processes (in the case of two unsymmetrically substituted diolato(2–) ligands, **C** and **D** (**C'** and **D'**) are identical species). The pairs **A/B'**, **B/A'**, **C/D'**, and **D/C'** each are connected by a pseudorotation according to *path 1*, whereas all other conversions of **A–D** (**A'–D'**) need at least one step according to *path 2*.

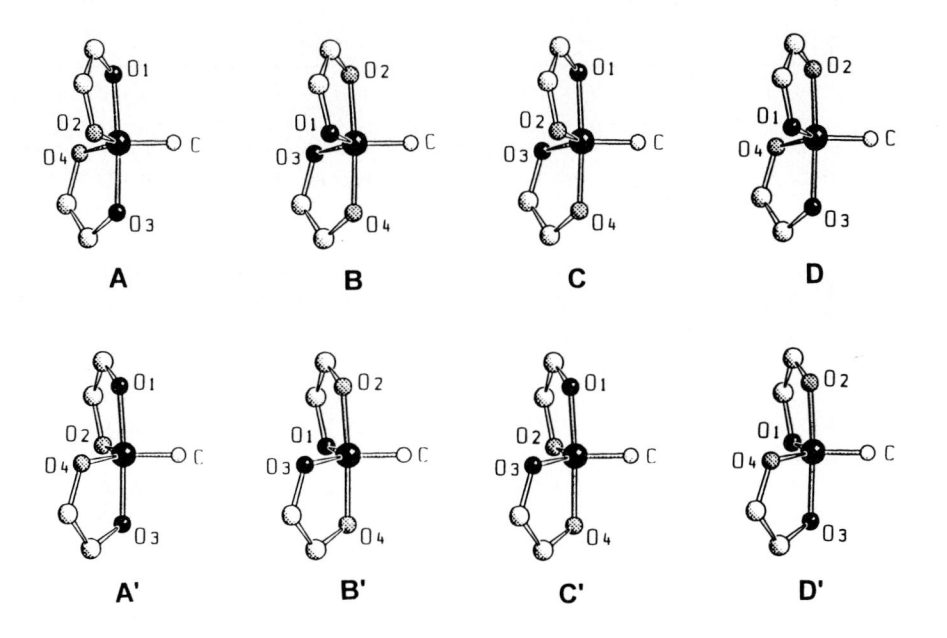

Fig. 3. Structure of the TBPs **A–D** and their corresponding enantiomers **A'–D'** containing two identical unsymmetrically substituted bidentate diolato(2-) ligands (models for **4–6**; see text); the constitutionally equivalent oxygen atoms of these ligands are depicted as black (O(1), O(3)) or spotted (O(2), O(4)) balls

For geometrical reasons, an NMR-spectroscopic differentiation between **A**, **B**, and **C/D** (**A'**, **B'**, and **C'/D'**) should be possible. However, because of the rapid isomerizations **A** ⇌ **B'** and **A'** ⇌ **B** (*path 1*) at room temperature, only two sets of resonance signals in the NMR spectra of **4–6** should be expected.

Provided that the energies of the species **A**, **B**, and **C/D** (**A'**, **B'**, and **C'/D'**) are almost the same (as may be expected for **4–6**), the intensity ratio of the respective resonance signals should be about 1:1 [4] (in this context, see also [5]). Coalescence of these signals at a temperature corresponding to an energy barrier of ca. 70 kJ mol^{-1} is expected. This result is in agreement with experimental data obtained for **4–6** in temperature-dependent ^{1}H NMR studies using [D$_6$]-DMSO as solvent [6]. Provided that the experimentally observed coalescence is due to such an intramolecular ligand-exchange process, the energy barrier for **4–6** amounts to 70–72 kJ mol^{-1}.

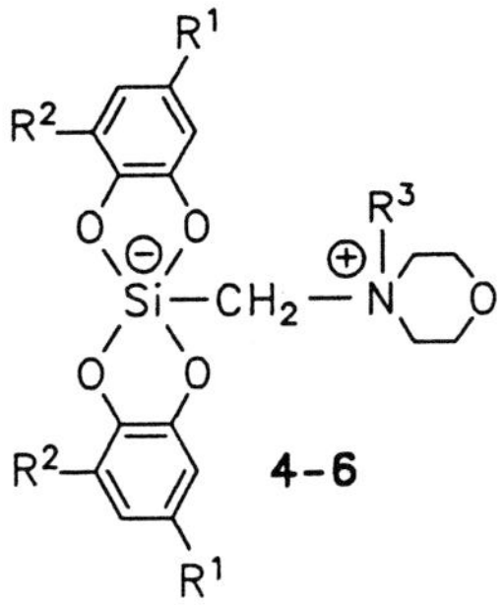

	R^1	R^2	R^3
4	NO$_2$	H	H
5	NO$_2$	H	Me
6	t-Bu	t-Bu	H

Fig. 4.

Acknowledgement: Financial support of this work by the *Deutsche Forschungsgemeinschaft* and the *Fonds der Chemischen Industrie* is gratefully acknowledged. In addition, we would like to thank Dr. U. Schneider and Prof. Dr. R. Ahlrichs, *Universität Karlsruhe*, for helpful discussions.

References:

[1] a) R. Tacke, F. Wiesenberger, A. Lopez-Mras, J. Sperlich, G. Mattern, *Z. Naturforsch.* **1992**, *47B*, 1370.

b) R. Tacke, J. Sperlich, C. Strohmann, B. Frank, G. Mattern, *Z. Kristallogr.* **1992**, *199*, 91.

c) R. Tacke, A. Lopez-Mras, J. Sperlich, C. Strohmann, W. F. Kuhs, G. Mattern, A. Sebald, *Chem. Ber.* **1993**, *126*, 851.

d) R. Tacke, J. Becht, A. Lopez-Mras, J. Sperlich, *J. Organomet. Chem.* **1993**, *446*, 1.

e) R. Tacke, A. Lopez-Mras, W. S. Sheldrick, A. Sebald, *Z. Anorg. Allg. Chem.* **1993**, *619*, 347.

f) J. Sperlich, J. Becht, M. Mühleisen, S. A. Wagner, G. Mattern, R. Tacke, *Z. Naturforsch.* **1993**, *48B*, 1693.

[2] All calculations were performed on IBM RISC/6000 workstations using the program package TURBOMOLE (R. Ahlrichs, M. Bär, M. Häser, H. Horn, C. Kölmel, *Chem. Phys. Lett.* **1989**, *162*, 165). For geometry optimizations (RHF/SCF level), completely optimized SVP basis sets were used (A. Schäfer, H. Horn, R. Ahlrichs, *J. Chem. Phys.* **1992**, *97*, 2571) [Si, (10s7p1d)/[4s3p1d]; C and O, (7s4p1d)/[3s2p1d]; H, (4s1p)/[2s1p] (p function only for the hydrogen atom bound to silicon)]. Calculated energies (E_{SCF} + "single point" E_{MP2} + $E_{vib(0)}$): −1049.69205 a.u. (3_{TBP}), −1049.68983 a.u. ($3\ddagger^1_{SP}$), −1049.66674 a.u. ($3\ddagger^2_{TBP}$); the existence of potential minima and transition states was verified by vibrational analysis.

[3] Alternatively, *path 2* might also be described as a one-step process with the transition coordinate [O(2)–Si–O(3) − O(1)–Si–O(3) + ½O(4)–Si–H(1)]. While the difference O(2)–Si–O(3) − O(1)–Si–O(3) changes from ca. −90 ° to ca. 90 °, the angle O(4)–Si–H(1) increases from ca. 120 ° up to ca. 240 °.

[4] If these species differ by an energy of 1.7 kJ mol^{-1}, an intensity ratio of the respective resonance signals of ca. 1:2 would be expected (instead of 1:1 for energetically equivalent species).

[5] D. F. Evans, A. M. Z. Slawin, D. J. Williams, C. Y. Wong, J. D. Woollins, *J. Chem. Soc., Dalton Trans.* **1992**, 2383.

[6] Two NMR-spectroscopically distinguishable species were detected at room temperature, and at higher temperatures coalescence of the respective resonance signals was observed: M. Mühleisen, R. Tacke, in: *Organosilicon Chemistry II* (Eds.: N. Auner, J. Weis), VCH, Weinheim, **1995**, p. 447.

Highly Coordinated Silicon Compounds –
Hydrazino Groups as Intramolecular Donors

Johannes Belzner*, Dirk Schär

Institut für Organische Chemie

Georg-August-Universität Göttingen

Tammannstr. 2, D-37077 Göttingen, Germany

Summary: 2-(Trimethylhydrazino)phenyl substituted organosilicon compounds **3**, **4**, and **5** were prepared. Their NMR spectra reflect the pentacoordination of the silicon center in solution at room temperature. Compound **5** is hexacoordinated in the solid state, whereas **4** shows pentacoordination. The phenyltrimethylhydrazino as well as the dimethylbenzylamino substituent proved to be highly suitable for the intramolecular stabilization of silyl cations.

Amino groups are well established as intramolecular donors for silicon centers [1]. Due to the enhanced nucleophilicity of hydrazines the intramolecular coordination ability of the terminal nitrogen center in the 2-(trimethylhydrazino)phenyl substituent (PTMH) is expected to be stronger than in the 2-(dimethyl-aminomethyl)phenyl substituent (DMBA).

When trimethylphenylhydrazine (**1**) is treated with *n*butyllithium in the presence of tetramethylethylenediamine, the *ortho*-lithiated compound **2** is formed, which reacts with tetrachlorosilane or [2-(dimethylaminomethyl)phenyl]trichlorosilane to yield dichlorosilanes **3** and **4**, respectively (Scheme 1). Both compounds can be transformed into the corresponding silanes and diethoxysilanes.

The ^{29}Si NMR shifts (Table 1) of **3** as well as **4** are significantly high field shifted in comparison to tetracoordinated dichlorodiphenylsilane (**3**: $\Delta\delta = -34.9$ ppm; **4**: $\Delta\delta = -37.1$ ppm), thus indicating a coordinative interaction between the silicon center and the hydrazino groups [2]. For the diethoxysilanes (**5**: $\Delta\delta = -13.1$ ppm; **6**: $\Delta\delta = -7.6$ ppm) and silanes (**7**: $\Delta\delta = -10.0$ ppm; **8**: $\Delta\delta = -12.3$ ppm) an analogous, but less marked high field shift in comparison to the corresponding phenyl substituted analogues is observed.

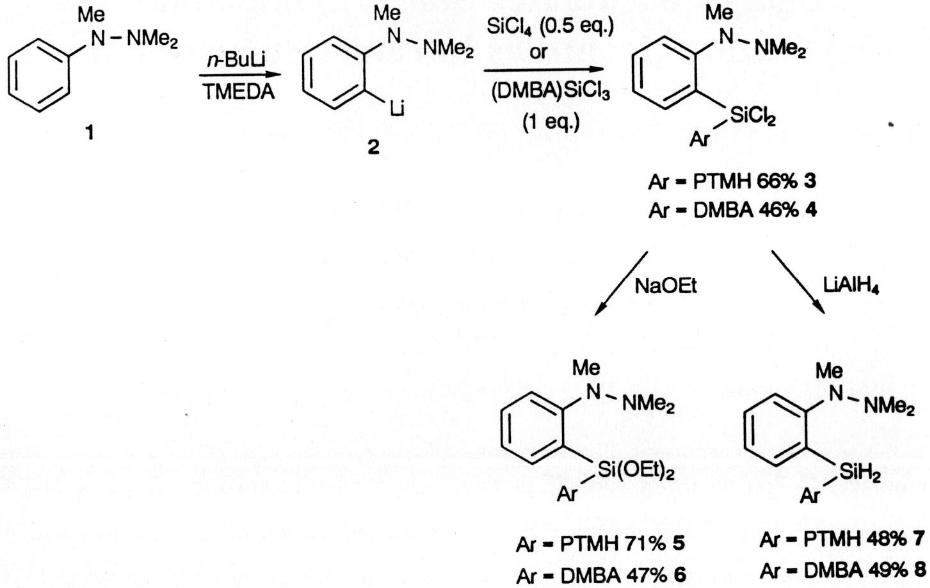

Scheme 1. Synthesis of PTMH substituted silanes

ArAr'SiX$_2$	δ [ppm]	Δδ[a]
(PTMH)$_2$SiCl$_2$ (**3**)	−28.7	−34.9
(PTMH)(DMBA)SiCl$_2$ (**4**)	−31.0	−37.1
(DMBA)$_2$SiCl$_2$ (**9**)	−30.1	−36.3
(PTMH)$_2$Si(OEt)$_2$ (**5**)	−47.6	−13.1
(PTMH)(DMBA)Si(OEt)$_2$ (**6**)	−42.1	−7.6
(DMBA)$_2$Si(OEt)$_2$	−35.6	−1.1
(PTMH)$_2$SiH$_2$ (**7**)	−43.8	−10.0
(PTMH)(DMBA)SiH$_2$ (**8**)	−46.1	−12.3
(DMBA)$_2$SiH$_2$	−45.0	−11.2

Table 1. ^{29}Si NMR data of PTMH- and DMBA-substituted organosilicon compounds; [a] Δδ = δ(ArAr'SiX$_2$) − δ(PhSiX$_2$)

The ^{1}H NMR spectra of **3** and **5** are temperature dependent: The analysis of these spectra argues for a mutual dynamic coordination of the terminal nitrogen atoms to the silicon center at room temperature [3] at lower temperature a rigid hexacoordinated complex is formed (Scheme 2). The values of $\Delta G^{\#}$ for this transition are of the same order of magnitude as that for similar processes [4].

X = Cl $\Delta G^{\#}$ = 48.2±0.2 kJ/mol
X = OEt $\Delta G^{\#}$ = 41.0±0.4 kJ/mol

Scheme 2. Dynamic process in PTMH substituted silanes

Hexacoordinate silicon is also found in the solid state structure of **5** (Fig. 1). The coordination geometry around the silicon center is that of a bicapped tetrahedron, showing Si–N distances of 268.9 and 277.2 pm, respectively.

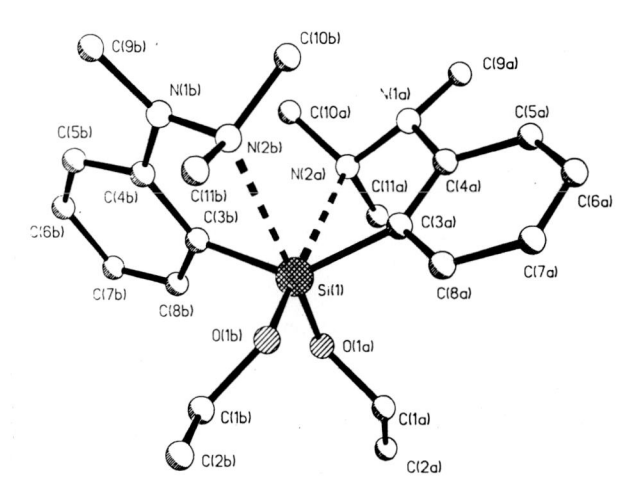

Fig. 1. X-ray structure of **5**; selected bond lengths [pm] and bond angles [°]: Si(1)···N(2a) 277.2(4), Si(1)···N(2b) 268.9(4), Si(1)–O(1a) 166.3(2), Si(1)–O(1b) 166.2(2), C(3a)–Si(1)–C(3b) 137.8(1), C(3b)–Si(1)–O(1a) 100.4(1), O(1a)–Si(1)–C(3a) 104.8(1), C(3b)–Si(1)–O(1b) 100.4(1), O(1a)–Si(1)–C(3b) 98.7(1)

In contrast, dichlorosilane **4** (Fig. 2), which differs from **3** formally by exchange of a PTMH substituent by a DMBA substituent, adopts a pentacoordinate silicon center in the solid state: Only the terminal nitrogen center of the hydrazino group is coordinating to silicon (Si–N: 256.4 pm).

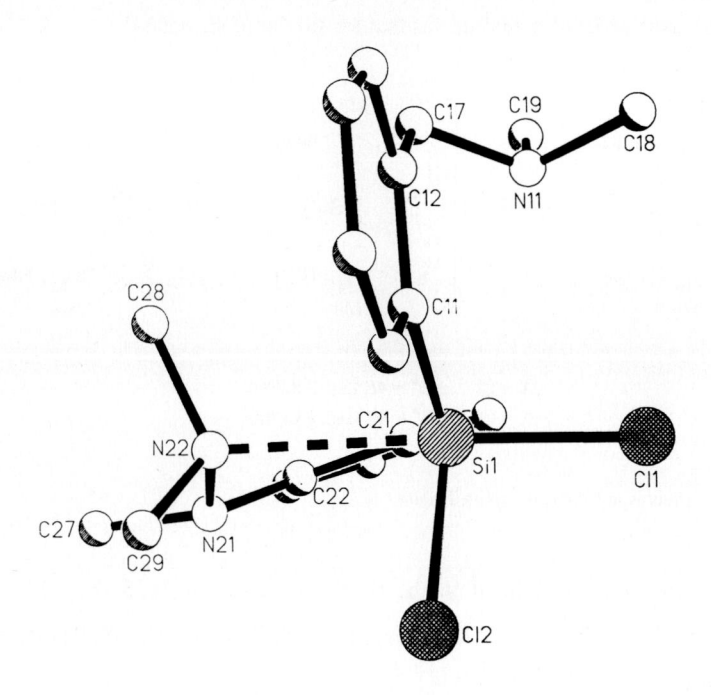

Fig. 2. X-ray structure of **4**; selected bond lengths [pm] and bond angles [°]: Si(1)···N(11) 315.9(2), Si(1)···N(22) 256.4(2), Si(1)–Cl(1) 211.8(1), Si(1)–Cl(2) 207.8(1), N(11)···Si(1)–Cl(1) 167.7(1), N(22)···Si(1)–Cl(2) 177.1(1), C(21)–Si(1)–C(11) 129.8(1), C(11)–Si(1)–Cl(1) 101.4(1), Cl(1)–Si(1)–C(21) 104.2(1), C(11)–Si(1)–Cl(2) 110.1(1), Cl(2)–Si(1)–C(21) 109.5(1)

These results show that, as anticipated, the hydrazino group of the PTMH substituent has a stronger coordination ability to silicon than the dimethylamino group of the DMBA substituent: Whereas dichlorosilane **9** [3] is pentacoordinated in solution at low temperature, the analogous dichlorosilane **3** is hexacoordinated under similar conditions. In addition, only the hydrazino group is coordinating to silicon in **4**, whereas the dimethylamino group of the DMBA substituent does not interact with the silicon center.

Intramolecularly coordinated silyl cations are known since the pioniering studies of Corriu [5] and Willcott [6] both using a tridentate substituent at silicon. Siliconium ions bearing the bidentate PTMH or DMBA substituent may be obtained via different pathways (Scheme 3).

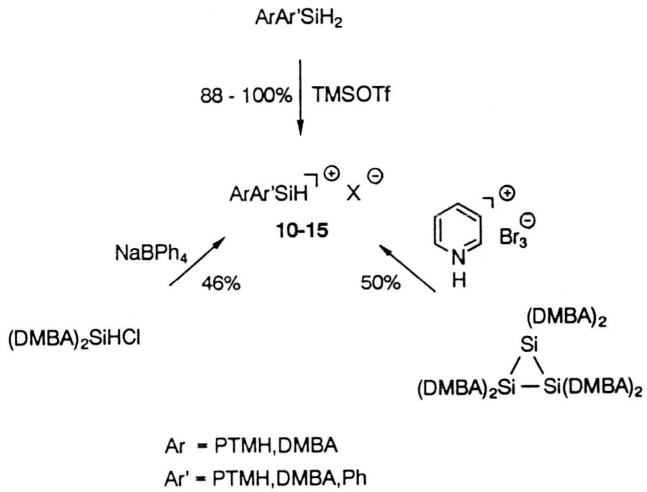

Scheme 3. Synthesis of siliconium ions

When silane **8** is treated with trimethylsilyl triflate, the siliconium ion **11** is formed quantitatively. The ^{1}H NMR spectrum, which is not effected by temperature changes up to 60 °C, provides evidence for strong coordination of both NMe$_2$ groups to the silicon center: All five methyl groups located at nitrogen as well as the benzylic protons are chemically inequivalent to each other, as shown by five singlets (δ = 2.44, 2.55, 2.86, 3.06, 3.09 ppm) and an AB pattern (δ = 3.81, 4.43 ppm). The ^{1}H^{29}Si coupling constant of **11** is 291 Hz, which is significantly increased (+79 Hz) in comparison to the coupling constant found for starting silane **8**. This seems to be characteristic for pentacoordinated siliconium ions (Table 2) [5].

ArAr'SiHX	δ [ppm]	$^1J_{SiH}$[Hz]	Δ^1J_{SiH}[a] [Hz]
(PTMH)$_2$SiH$^+$OTf$^-$ (**10**)	−54.5	283	70
(PTMH)(DMBA)SiH$^+$OTf$^-$ (**11**)	−53.6	291	79
(DMBA)$_2$SiH$^+$OTf$^-$ (**12**)	−51.6	272	63
(DMBA)$_2$SiH$^+$BPh$_4^-$ (**13**)	−51.7	271	62
(DMBA)$_2$SiH$^+$Br$^-$ (**14**)	−51.5	271	62
(DMBA)PhSiH$^+$OTf$^-$ (**15**)	−56.2	294	86

Table 2. ^{29}Si NMR data of intramolecularly coordinated silyl cations;
[a] $\Delta^1J_{SiH} = J(\text{ArAr'SiHX}) - J(\text{ArAr'SiH}_2)$

For the first time the proposed cationic pentacoordinated structure in solution could be confirmed for the solid state by a X-ray structure determination (Fig. 3), which shows clearly the ionic structure of bis[2-(dimethylaminomethyl)phenyl]silyl triflate **12**. The coordination geometry around silicon is that of a nearly ideal trigonal bipyramidal structure with both dimethylamino groups occupying the axial positions (Si–N: 205.2 and 207.2 pm).

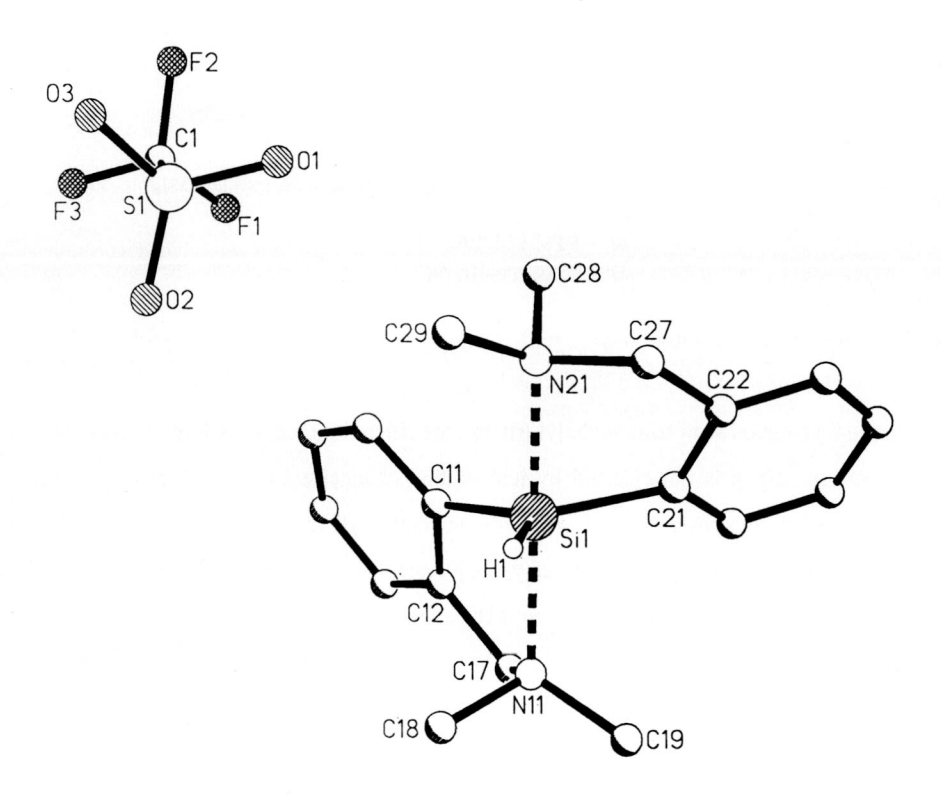

Fig. 3. X-ray structure of **12**; selected bond lengths [pm] and bond angles [°]: Si(1)···N(11) 205.2(2), Si(1)···N(21) 207.2(2), N(11)···Si(1)···N(21) 171.2(1), C(21)–Si(1)–C(11) 119.8(1), C(11)–Si(1)–H(1) 123.8(1), H(1)–Si(1)–C(21) 116.3(1)

Acknowledgement: This work was supported by the *Deutsche Forschungsgemeinschaft* (financial support, fellowship to J. B.), *Fonds der Chemischen Industrie* (financial support) and *Graduiertenförderung des Landes Niedersachsen* (fellowship to D. S.). We thank Dr. R. Herbst-Irmer, Dr. M. Noltemeier and cand.-chem. B. O. Kneisel for the determination of X-ray structures.

References:

[1] R. J. P. Corriu, J. C. Young, in: *The Chemistry of Organic Silicon Compounds* (Eds.: S. Patai, Z. Rappoport; Wiley, Chichester **1989**, 1241.

[2] J. Bradley, R. West, R. J. P. Corriu, M. Poirier, G. Royo, A. De Saxce, *J. Organomet. Chem.* **1983**, *251*, 295.

[3] compare: R. Probst, C. Leis, S. Gamper, E. Herdtweck, C. Zybill, N. Auner, *Angew. Chem.* **1991**, *103*, 1155; *Angew. Chem., Int. Ed. Eng.* **1991**, *31*, 1132.

[4] F. Carre, G. Cerreau, C. Chuit, R. J. P. Corriu, C. Reye, *Angew. Chem.* **1989**, *101*, 474; *Angew. Chem., Int. Ed. Eng.* **1989**, *28*, 489.

[5] C. Chuit, R. J. P. Corriu, A. Mehdi, C. Reye, *Angew. Chem.* **1993**, *105*, 1372; *Angew. Chem., Int. Ed. Eng.* **1993**, *33*, 1311.

[6] V. A. Benin, J. C. Martin, M. R. Willcott, *Tetrahedron Lett.* **1994**, *35*, 2133.

Organosilicon Metal Compounds: Their Use in Organosilicon Synthesis, Coordination Chemistry, and Catalysis

Norbert Auner, Johann Weis

The use of metals in organosilicon synthesis is based on the possibility of the elimination of metal halides MX (M = mostly metal of group 1, X = halogen) from the corresponding Si(X)–E(M) precursors (E = element of group 14–16, partly with substituents), in which the stability or reactivity of the E–M bond is mainly influenced by the silicon substituents, e.g., using the silicon α- and β-effect for the stabilization of carbanionic and -cationic (transient) centers. Also depending on steric effects these compounds are stable in the mono- or dimeric form or they eliminate MX intra- or intermolecularly yielding unsaturated, cyclic or polymeric species.

The use of transition metals in organosilicon synthesis may be discussed under various topics:

- *stabilization of silicon centers in low coordination numbers (e.g. silylene complexes)*
- *metal silanolates as model compounds for silicone precursors*
- *catalytic activities of organosilicon complexes*

Especially the latter subject is a strongly developing field of research, trying to find answers to such important questions as, e.g., the stereo- and enantioselective synthesis of organosilicon polymer precursors and the use of Si–M compounds for the production of new polymers. In addition, even in widely used industrial processes as, e.g. the hydrosilylation reaction, the respective mechanism is still under discussion and research is focused on the development of more active and less expensive catalysts (compared to the today used nobel metal complexes).

A totally different but nevertheless exciting field is the solid state chemistry of silicides and of Zintl anions aimed on the optimization of the physical properties (e.g. conductivity) and their derivatization to give molecular species.

References:

[1] T. D. Tilley, in: *The Chemistry of Organic Silicon Compounds* (Eds.: S. Patai, Z. Rappoport), Wiley, Chichester, **1989**, p. 1415.

[2] I. Ojima, in: *The Chemistry of Organic Silicon Compounds* (Eds.: S. Patai, Z. Rappoport), Wiley, Chichester, **1989**, p. 1479.

[3] T. D. Tilley, *Acc. Chem. Res.* **1993**, *26*, 22.

Silicon Frameworks and Electronic Structures of Novel Solid Silicides

Reinhard Nesper*, Antonio Currao, Steffen Wengert
Laboratorium für Anorganische Chemie
Eidgenössische Technische Hochschule Zürich
ETH-Zentrum, CH-8092 Zürich, Switzerland

1 Introduction

Around 1928, Zintl had begun to investigate binary intermetallic compounds, in which one component is a rather electropositive element, e.g., an alkali- or an alkaline earth metal [1, 2]. Zintl discovered that in cases for which the Hume-Rothery rules for metals do not hold, significant volume contractions are observed on compound formation, which can be traced back to contractions of the electropositive atoms [2]. He explained this by an electron transfer from the electropositive to the electronegative atoms. For example, the structure of NaTl [3] can easily be understood using the ionic formulation Na^+Tl^- where the poly- or Zintl anion $^3_\infty[Tl^-]$ forms a diamond-like partial structure – one of the preferred structures, for a four electron species [1, 2]. Zintl has defined a class of compounds, which, in the beginning, was a somewhat curious link between well-known valence compounds and somehow odd intermetallic phases.

Components of Zintl phases are metals A (Li–Cs, Mg–Ba) and semimetals X (B–Tl, Si–Pb, P–Bi, Te). The definition of the components of Zintl phases is not quite sharp, neither for the metals nor for the semimetals, until today. While the alkali- and alkaline earth metals are undoubtedly metals in the classical sense, there is still the vast field of transition metals T, which in Zintl type compounds, might also occur. Electropositive T atoms may replace the main group metal A and electronegative ones like a semimetal X. One of the wellknown examples here is CsAu (a diamagnetic semiconductor). This phase shows the typical properties of a valence compound according to the electron transfer description: Cs^+Au^- [4].

To separate Zintl phases from other well-known groups like, e.g., insulators and typical intermetallic phases, a more decisive definition has to be given. Zintl phases show characteristics of valence compounds. This means:

- There is a well defined relationship between the chemical structure of a Zintl phase and its electronic structure. For the majority of these compounds, AX_x homopolar X–X contacts are present and can be explained as two-electron, two-center bonds. The octet rule [5] is fulfilled for the A and for the X atoms. This is provided by a *formal* charge transfer of the valence electrons from A to X leading to $A^{\alpha+}(X^{\frac{\alpha}{x}-})_x$ (e_A, e_X = number of valence electrons of atoms A and X, respectively). The cation $A^{\alpha+}$ simply yields its full shell state by oxidation to the electronic configuration of the preceeding noble gas neighbor ($|\alpha| = e_A$). The anionic component $X^{\frac{\alpha}{x}-}$ yields a filled octet by the charge transfer from A and by X–X bonds.

- Zintl phases are semiconductors, look dark, mostly black, sometimes with an opaque lustre, and are not or hardly transparent in larger pieces.

- Zintl phases are mostly diamagnetic like the majority of the insulator compounds. This means they tend to form full shell systems by obeing the (8–N) rule.

Klemm proposed designating the charged $X^{\frac{\alpha}{x}-}$ unit as a pseudoatom (or a set of pseudoatoms) behaving structure-chemically like isoelectronic elements [6–8]. This was an extension of the old Grimm-Sommerfeld-rule for four-electron species [9].

The combination of Zintl's and Klemm's description is called Zintl-Klemm concept. It was given in a very similar way by Mooser and Pearson [10] around the time that Klemm and coworkers published their concept [7, 11].

A typical Zintl phase is a compound like NaSi [12]. It is described by the formulation Na^+Si^-. The anion Si^- has to be triply bonded, which is expressed by the symbol (3b) [13]. Si^- is isoelectronic with phosphorus and arsenic (diagonal relationship) and it is, therefore, according to Klemm, a pseudo element of group 15. This means it should accomodate a structure which is typical for group 15 elements. In NaSi (3b)Si^- anions form $[Si_4]^{4-}$ tetrahedra, like phosphorus and arsenic do in white phosphorus and yellow arsenic, respectively. In $CaSi_2$, a comparable situation occurs. According to the formulation $Ca^{2+}(Si^-)_2$, anions (3b)Si^- are expected as well.

A general relationship between Zintl-Klemm concept and defect formation has been formulated [15, 16] recently. For a compound $A_aB_b\square_d$ (A = cation, B = anion, $\square$ = defect; a, b, d = stoichiometric numbers) the total number of valence electrons E per formula unit relates to the average number of bonds per atom N and to the number of defects d.

Considering only the (poly)anionic partial structure one obtains

$$E = (8-N) \cdot b + N^* \cdot d$$

where N^* is the mean number of broken bonds per defect.

This formula allows to tackle a number of different problems concerning the relation of crystal structure, valence electron number, and basic structure type as given in [15]. The basic defect-free structure type to the compound $A_aB_b\square_d$ has the composition $A_aB_bC_d$ or A_aB_{b+d}, B and C being semimetallic component atoms.

In the Zintl-Klemm concept, the description of compounds like Li^+Al^-, Li^+Ga^-, and Li^+In^- is in principle the same. There is also a similar relation between MgGa and LiGe.

It is important to note that one is not really interested in effective charges which surely do change in such series, but only in the character and the number of occupied electronic states. This means that we are not counting transferred electrons but occupied electronic states showing the character of anion states. These are in general bonding and nonbonding with a contribution of one or two electrons per participating atom, respectively. In Zintl phases bonding and nonbonding orbital combinations derive to a large percentage from the atomic orbitals of the semimetals (B, C, X) while the antibonding states have their main contributions from the metal components A. If the antibonding states are not occupied and there is an energy gap separating bonding and nonbonding from antibonding states, the basic Zintl-Klemm concept is valid. This means that there are no bonding or nonbonding but only antibonding orbitals centered at the metal atoms A.

Numerous investigations over the recent 10–15 years demonstrate a real flourishing in the field of Zintl phases and reveal that these semiconductor compounds can contain beautiful homo- or heteroatomic polyanions [13, 17–22]. Zintl phases can mimic arrangements well known from the more classical insulator compounds but, in addition, they may occur with completely novel clusters and frameworks of the X components.

For a long time, binary Zintl phases of alkali- and alkaline earth metals like MSi (M = Na–Cs, Ca–Ba) [23–25] and MSi_2 (M = Mg–Ba) [14, 26–28] have been known. In general the structures of the alkali-mono-silicides MSi and earth-alkaline-disilicides MSi_2 contain isoelectronic species $(3b)Si^-$, as mentioned before. This means, the charge on the silicon atoms is one minus and each silicon has three silicon neighbors on the average. In all alkali metal compounds MSi and in ternary compounds

$M_xM'_{1-x}Si$ (M = Na–Cs, M' = Li) which have been synthesized and characterized in numerous different kinds quite recently, anions Si_4^{4-} are found [29, 30]. This is, true as well for $BaSi_2$ (cf. Fig. 1c). Peculiarly enough, K_7LiSi_8, which is a member of this family, is a red transparent metal [29, 30].

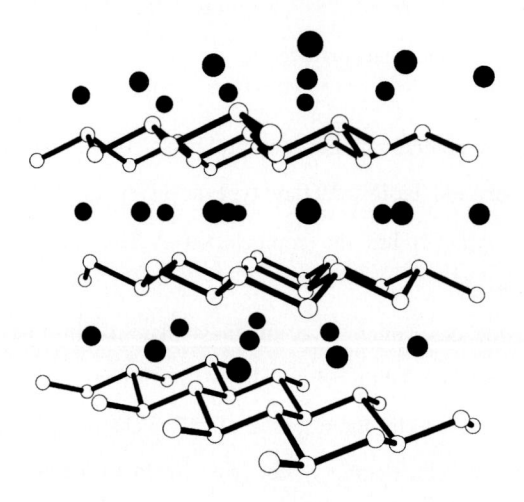

Fig. 1a. Structures of binary disilicides of alkaline earth and rare earth metals:
part of the structure of $CaSi_2$ showing layers of buckled six membered rings related to those in α-arsenic

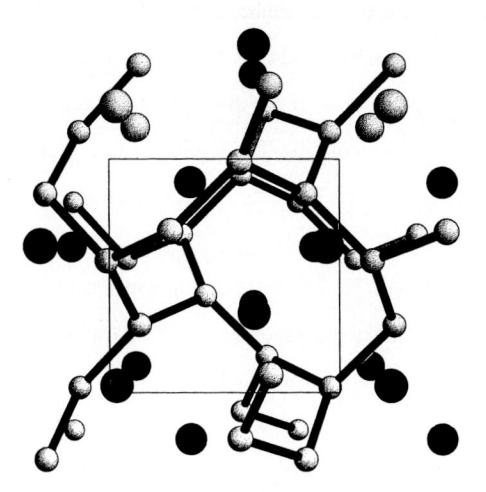

Fig. 1b. Structures of binary disilicides of alkaline earth and rare earth metals:
the cubic $SrSi_2$ arrangement with backbones of helical chains resembling those in the α-tellurium modification

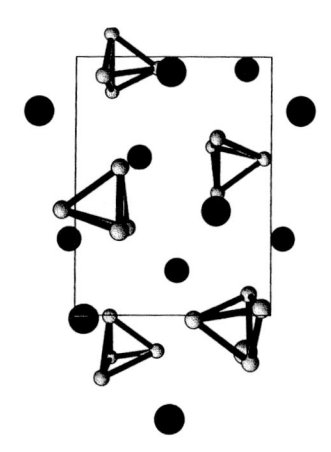

Fig. 1c. Structures of binary disilicides of alkaline earth and rare earth metals: structure of $BaSi_2$ with Si_4 tetrahedra which are closely related to those in white phosphorus

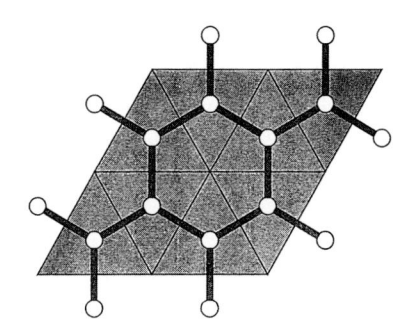

Fig. 1d. Structures of binary disilicides of alkaline earth and rare earth metals: $EuSi_{2-x}$ crystallizing with graphite-like silicon layers

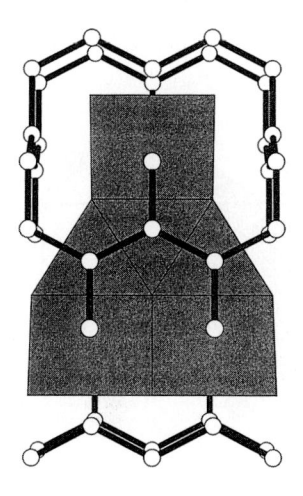

Fig. 1e. Structures of binary disilicides of alkaline earth and rare earth metals: the structure of $EuSi_2$ containing a threedimensional tetragonal silicon network which is related to the structure of B_2O_3

In CaSi$_2$, a layered polyanion $^2_\infty$[Si$^-$] is present which consists of puckered Si$_6$ rings and which is related to a layer in the α-arsenic structure (Fig. 1a, [14]). SrSi$_2$ forms a cubic arrangement where the triply silicon atoms [(3b)Si] connect to a threedimensional framework which is a type of its own (Fig. 1b, [26])

In spite of the fact that in many cases the larger alkaline earth metals can be exchanged for divalent rare earth metals, the latter ones form different varieties of three-dimensional or layered triply bonded disilicides with either the α-ThSi$_2$ or the AlB$_2$ structure. These two structure types are clearly dominated by a trigonal prismatic arrangement of the cations and contain silicon in exclusively trigonal planar coordination (Fig. 1e and 1f).

On the basis of the known variety of compounds and structures it was reasonable to expect new polyanionic Si arrangements in ternary silicides. The different interactions metal to silicon and the specific spatial requirements of different metal atoms was anticipated to have structure directing influences. We report here about compounds M$_x$M'$_y$Si with M = Ca–Ba and M' = Mg.

Magnesium seems to play a key role in the formation of such new silicides which show a series of novel polyanionic silicon arrangements with fascinating bonding patterns and electronic structures compared to those known hitherto.

2 Composition Close to MSi

The compounds CaSi, SrSi, BaSi, SrMgSi$_2$, and Ca$_7$Mg$_{7.25}$Si$_{14}$ are very close to being isoelectronic and thus, on the average, should contain twofold bonded doubly charged silicon atoms (2b)Si^{2-}, which are isoelectronic with sulfur. Some structural data are given in Table 1 and representative plots of the polyanions in Fig. 2a–2e.

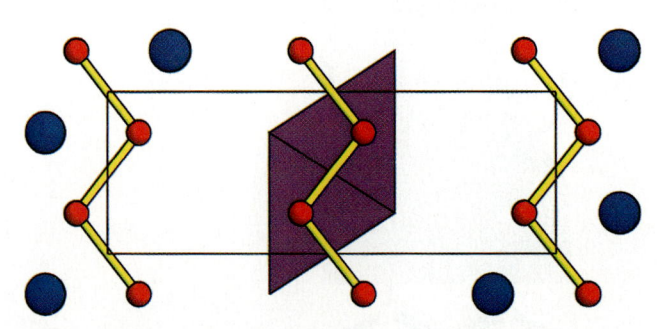

Fig. 2a. Structures of binary and ternary silicides close to the composition MSi:
CaSi, SrSi, and BaSi crystallize with the CrB structure type which has a trigonal prismatic arrangement of the metal atoms

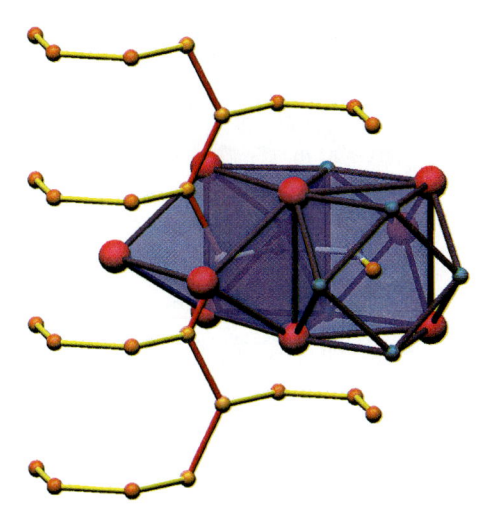

Fig. 2b. Structures of binary and ternary silicides close to the composition MSi: part of the structure of SrMgSi$_2$ (Sr: red spheres, Mg: small blue spheres, Si: small orange spheres); the bonds of the side chains are displayed in yellow those of the central backbone in red

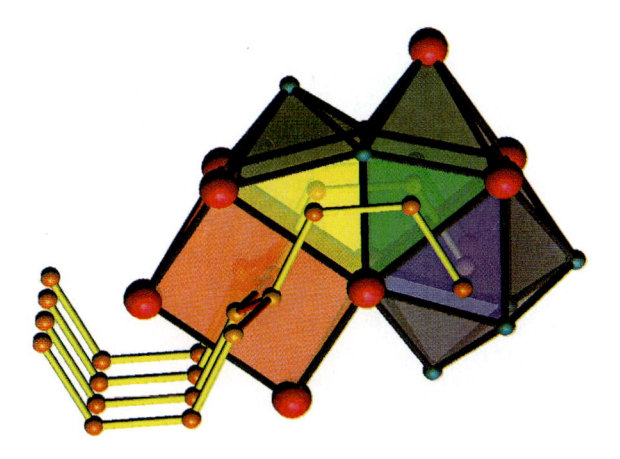

Fig. 2c. Structures of binary and ternary silicides close to the composition MSi: view along one chain of the polyanion $^1_\infty[\mathrm{Si(Si_3)}]^{8-}$ in SrMgSi$_2$ emphasizing the trigonal prismatic coordination of backbone and side chain silicon atoms

The three binary monosilicides crystallize with the same structure type. All silicon atoms in this structure are functionally equal. This means, there is no obvious disproportionation, the formal charges and the local coordinations are close to being equal for each compound. Still, it is remarkable that alkaline earth metals form the CrB structure with planar Si$_n$ chains and bond angles of 120 °(!) while in LiAs an arrangement with helical chains and bond angles close to classically expected ones (105 °, 108 °) are found.

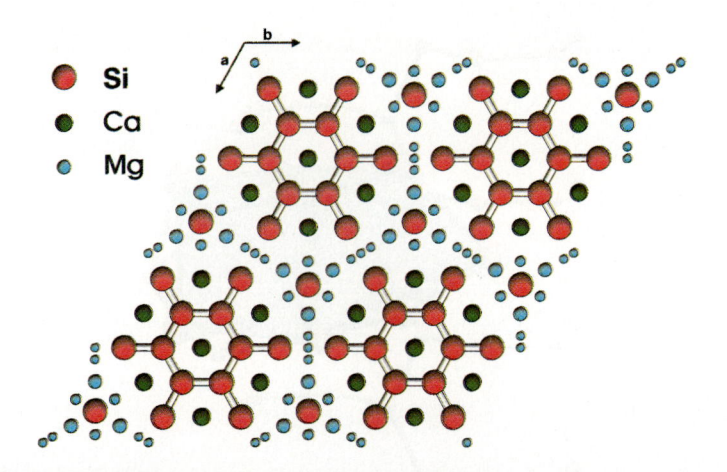

Fig. 2d. Structures of binary and ternary silicides close to the composition MSi:
Ca$_7$Mg$_{7.25}$Si$_{14}$ – view of the crystal structure along the [001] direction (Si: red spheres, Ca: green spheres, Mg: small blue spheres), disordered model; there are isolated Si atoms and planar six membered ring derivatives Si$_{12}$

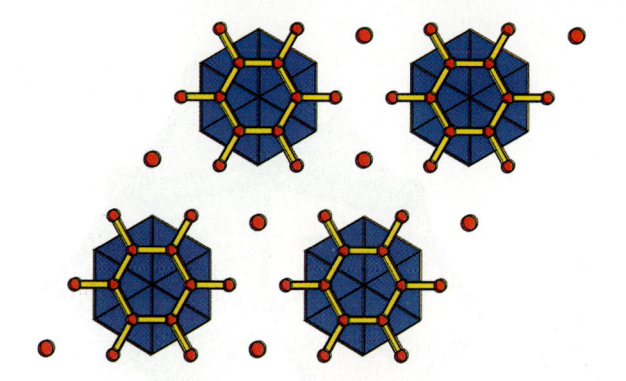

Fig. 2e. Structures of binary and ternary silicides close to the composition MSi:
the same as d) emphasizing the trigonal prismatic Ca coordination of the planar isometric Zintl anion Si$_{12}$ with D$_{6h}$ symmetry

The Si–Si bond lengths are quite large increasing slightly from the Ca over the Sr to the Ba compound. This may indicate a matrix effect due to the size of the earth alkaline metals. Quite frequently transition metals form trigonal prismatic arrangements, like in this case. However, structures with trigonal prismatic coordinations do not seem to be typical for compounds of main group metals.

Compound	Lattice Constants	Space Group	Distances	Ref.
CaSi	4.560(1), 10.731, 3.890(1)	Cmcm	245	[23, 31]
SrSi	4.826(8), 11.29(1), 4.042(3)	Cmcm	251	[23, 31]
BaSi	5.042(5), 11.97(1), 4.142(2)	Cmcm	251	[23, 25]
$SrMgSi_2$	14.374(1), 4.4512(5), 11.398(2)	Pnma	235 – 249	[31]
$Ca_7Mg_{7.25}Si_{14}$	12.696(2), 4.403(1)	P6/mmm	233 – 234	[31–33]
$Ca_{3.3}Li_{3.7}Si_8$	13.523(7), 10.262(5), 4.442(2)	Pnnm	232 – 236	[46]
$Sr_{11}Mg_2Si_{10}$	19.744(6), 4.754(1), 14.841(4)	C2/m	244 – 246	[31]
$Ca_7Li_xMg_{4-x}Si_8$	14.722(3), 4.438(4), 26.359(3)	Pnma	235 – 242	[31]
$Ca_{14}Si_{19}$	8.679(1), 68.53(1)	$R\bar{3}c$	235 – 256	[31]
$Ca_{3-x}Mg_xSi_4$	8.531(4), 14.894(2)	$P6_3/m$	238 – 256	[31]

Table 1. Comparison of binary and ternary silicides of alkaline earth metals. For the silicon-silicon distances [pm] only minimal and maximal values are given

The new ternary compound $SrMgSi_2$ shows a considerable disproportionation of the silicon atoms into terminal (1b)Si^{3-}, bridging (2b)Si^{2-}, and branching (3b)Si^- species (Fig. 2b, [31]). As shown in figure 2b the whole polyanion contains a one dimensional planar chain of silicon atoms with syndiotactically arranged side chains $^1_\infty[Si(Si_3)]^{8-}$. The Si–Si distances vary somewhat between 235 and 249 pm, with the shortest distances between the side chain bridging atoms and those of the main backbone.

The longer bonds occur at the end of the side chains (245 pm) and between the atoms of the central one dimensional backbone (249 pm). The "functionalisation" of the silicon atoms is not only reflected but seemingly initiated by the distribution of the Mg and Sr atoms. Clearly the latter coordinate exclusively the Si atoms of the central backbone of the polyanion. Those silicon atoms which are in the bridging side chain have a 4+3 coordination of Sr and Mg atoms while the terminal (1b)Si^{3-} groups have an outer coordination of four Mg atoms. The whole one dimensional polyanion looks like a „blossomed CaSi chain" with its syndiotactic side chains.

The Sr coordination of the central backbone Si atoms is trigonal prismatic contrary to close packing considerations of nonbonded cation interactions. Even for the side chain coordination by Mg and Sr atoms prismatic arrangements are found (Fig. 2c).

It seems to be a sort of a rule in these Zintl phases that whenever there is a trigonal prismatic cation coordination then the silicon atoms form planar fragments and vice versa. It is not yet understood why this is so. To none of the Si–Si-bonds in SrMgSi$_2$ a π-bonding contribution can be assigned neither in terms of bond distances nor by valence electron numbers and counting rules.

At the first glance it may be surprising that there is only the phase Mg$_2$Si in the binary system while the higher alkaline earth metals form a variety of different binary phases. A comparison with ternary A$_x$Mg$_y$Si phases, however, indicates that there is a tendency that Mg atoms prefer a large formal charge transfer to silicon which may be smeared out by the polarizing ability of Mg^{2+} cations.

The structure of SrMgSi$_2$ can be understood by assigning quite different qualities to Mg–Si and Sr–Si interactions, respectively. The discussion of the following compounds will show that this reasoning has some general meaning.

Ca$_7$Mg$_{7.25}$Si$_{14}$ is a new compound in the ternary system with an overall composition close to M:Si = 1:1 [31–33, 45]. This means that the average formal charge on silicon is again close to q = –2 and the mean bond order is close to two like in the previous compounds. The structure, however, is very different and a type of its own. There is an extreme disproportionation of the silicon atoms into isolated formal (0b)Si^{4-} ions and a planar Si$_{12}$ group of D$_{6h}$ symmetry which may be described as hexasilylhexasila-cyclo-"hexadiene" (Fig. 2d). Again the isolated and terminal silicon atoms have a strong or a predominant Mg coordination while atoms in the six membered ring are surrounded exclusively by Ca atoms in a trigonal prismatic way. In this case, there is no question that Si–Si π-bonding must occur, because the composition does not allow for a singly bonded ring system. Despite a certain disorder of the Mg atoms on different sites the composition can be fixed to Ca$_7$Mg$_{7.25}$Si$_{14}$ due to X-ray diffraction, chemical synthesis, and analysis. Subtraction of two Si^{4-} anions per formula unit leaves a net charge on the oligomeric moiety of q = –20.5 and thus yields a 68.5 electron system, which is neither compatible with a benzene-like (66 e) nor with a exclusively singly bonded unit (72 e). According to the filling scheme of the π-states for the planar system there are 1.75 antibonding states unoccupied. The bond distances in the ring (233 pm) and to the terminal Si atoms (234 pm) are only a little bit shorter than the ideal Si–Si single bond distance of α-silicon (236 pm) but still in accordance with 1.75 double bonds being well distributed over the whole anion.

Using Paulings estimation for bond orders an overall bond order of 1.15 per Si–Si vector would give rise to a shortening of 4 pm to 232 pm. As there are charge repulsion and cation matrix effects to be considered, too, the observed distances seem to fit quite well with the π filling scheme.

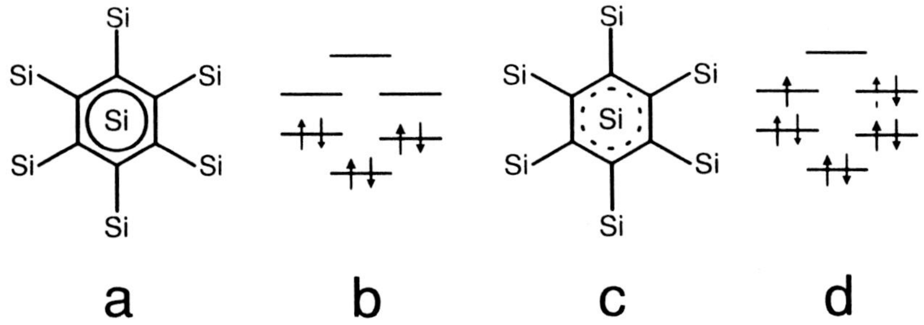

Scheme 1.

Quantum mechanical calculations by the Extended Hückel (EH) type [34] and the linear muffin tin orbital approach (LMTO) [35] show a fairly comparable result. The latter is a high quality self consistent field method (SCF) which, however, must be applied to the full unit cell, while EH calculations are of lower quality but allow to model charged entities and to project information much better into normal chemical language. For the LMTO description an ordered model was constructed in a larger elementary cell (2a · 2b · c) which avoids unreasonable distances and yields the correct cummulated occupancies and thus the correct stoichiometry (Fig. 3a). For the present discussion the quantum mechanical information was projected into the so called electron localization function (ELF) which emphasizes spaces of localized electrons in the total electron density [36–39]. The localization properties are coded by a color scale which ranges from dark blue over green to white. The latter color designates high localization, namely bonds and lone electron pairs. Medium localization is indicated by a green color and is comparable to an electron gas-like behaviour [36]. In Fig. 3b interatomic bonds and lone electron pairs are clearly discriminated from interstitial space, from mainly electrostatic M–Si interactions, and from the spherical atomic cores.

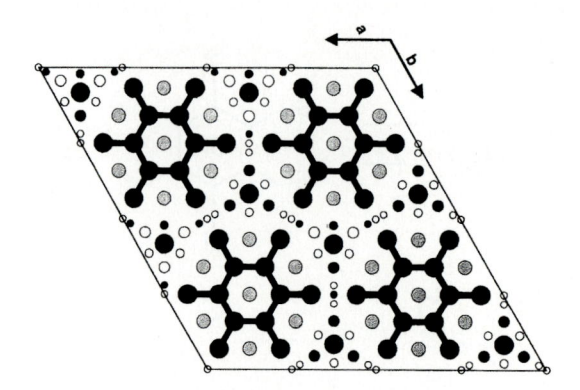

Fig. 3a. Ordered Mg distribution in $Ca_7Mg_{7.25}Si_{14}$ – view of a quarter of the ordered model structure along the [001] direction (Si: large grey circles, Ca: medium sized grey circles, Mg: open and black small circles; the filled circles show a consistent local ordering model for the Mg atoms avoiding exclusively short distances and preserving the overall composition)

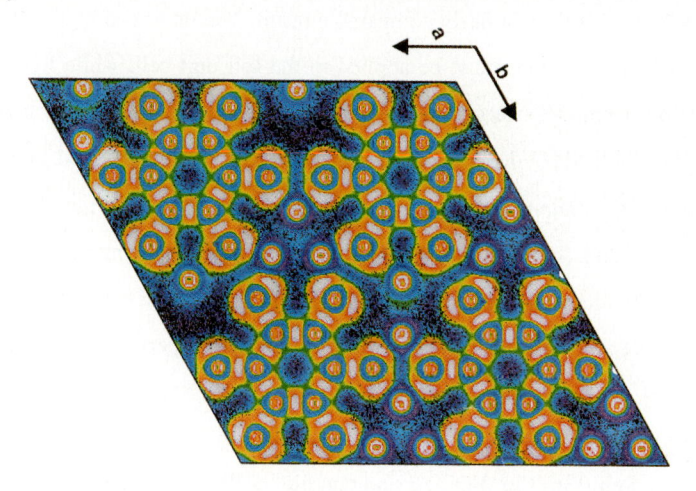

Fig. 3b. Electron localization function (ELF) in the plane of the Si_{12} entities; the white areas indicate large electron localization and denote regions of bonds and lone electron pairs; the small isolated circular regions between the ring systems are cores of magnesium atoms

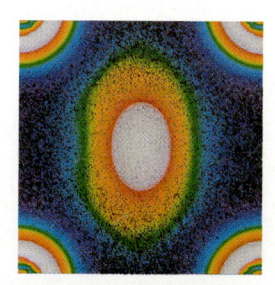

Fig. 3c. ELF section perpendicular to an intra ring bond; clearly, the shape of the white area (high ELF) indicates π-contributions

Fig. 3d. The same for a terminal bond with high σ-bond character

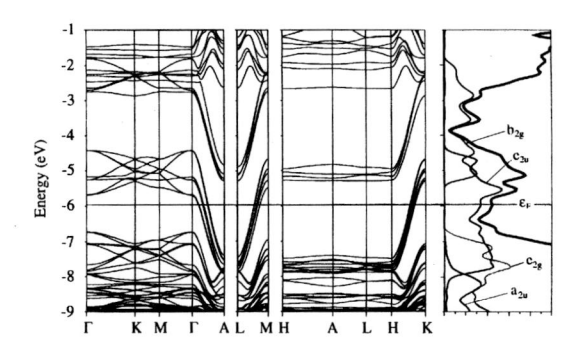

Fig. 3e. Selected section of the band structure shows pronounced dispersions for directions with strong z component, e.g., along the stacking direction of the planar ring systems; exclusively these dispersions are responsible for the metallic conductivity cutting through the Fermi level(horizonal line, ε_F)

Fig. 3f. Density of states plot and projected densities of states for the Si_{12} ring states; the major contribution at ε_F arises from the e_{2u} state which is a π-antibonding state within one ring but bonding between adjacent rings

Although some polarization effects of the lone electron pairs at the terminal silicon atoms are visible, there is no covalent contribution between Mg and Si to be seen as indicated by the blue color in the intermediate areas. The distribution of the Mg atoms breaks the observed symmetry due to the application of an ordered model in a fourfold unit cell. The bond sections show nicely that there is a considerable π-contribution to the intra ring bonds while the sections of the terminal bonds show a fairly circular ELF typical for σ-bonds (Fig. 3c and 3d).

At the first glance, it is surprising that the π-bonds are not localized to form double bonds. A similar system is K_4P_6, containing planar P_6 ring with D_{6h} symmetry which yield 34 electrons (P_6^{4-}) and thus are left with one π-bond which is essentially delocalized over the ring [40].

Both, EH and LMTO calculations yield band structures with band overlap at the Fermi level meaning that a metallic conduction is expected. This is displayed in Fig. 3e and 3f. Indeed, $Ca_7Mg_{7.25}Si_{14}$ shows metallic conductivity. Thus, there are no localized spins but the HOMO states form a conduction band which, according to LMTO-band structure calculations, is exclusively due to π-orbital overlapping between adjacent ecliptically arranged Si_{12} moieties along the stacking direction.

A slight locking-in of π-bonding is found in $Ca_{3.3}Li_{3.7}Si_8$ [46] where planar but not isometric Si_6 rings are connected via Si_2 links to a one dimensional planar polyanion $\frac{1}{\infty}[Si_6Si_2]^{10.3-}$.

The composition determines the broken formal charge of this partial structure. Out of the three reasonable bond distributions in Scheme 2 only (c) comes close to the required electron count.

a   b   c

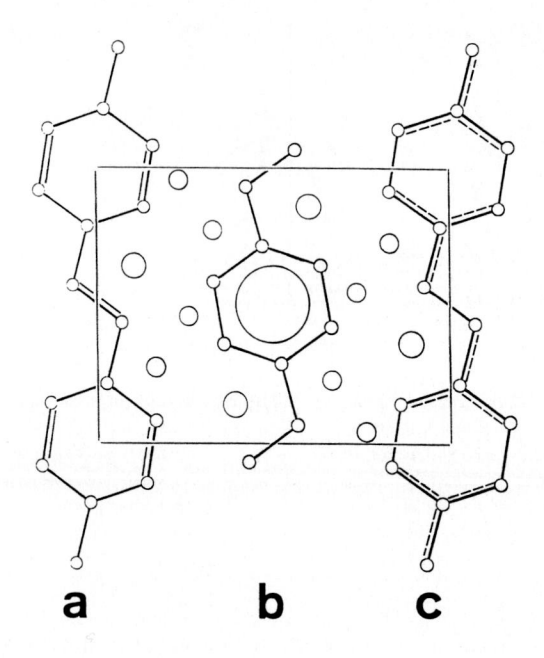

Scheme 2.

An extended Hückel (EH) tight binding band structure calculation reveals via the electron localization function (ELF) that π-bonding occurs in the CO_3-like fragments while the residual linking bonds in and between the six membered rings show essentially single bond behaviour. In Fig. 4a, a top view of the repetition unit of the polymer are shown together with five bond sections in form of ELF maps [36] shown. One can clearly discriminate between the ellipsoidal bond sections 1, 3, 4, and the nearly circular ones 2 and 5 which are typical of σ-bonds (Fig. 4c 1–5). There is no real discrimination on the basis of the bond distances (bonds 1–3: 232.2 to 233.8 pm; 4,5: 234.9 and 235.6 pm) although the distances in the carbonate-like fragment with the bonds 1, 3, and 4 are slightly shorter. According to the EH density of states there is no band gap and thus metallic conductivity is expected.

In both compounds, $Ca_7Mg_{7.25}Si_{14}$ and $Ca_{3.3}Li_{3.7}Si_8$, the fractional valence electron numbers and thus fractional charges on the Zintl anions seem to be due to the existence of metallic bonds arising mainly from π-states. These states are obviously filled until an optimal energy gain is reached independent of clearly defined π-bond structures in the ring systems. This must be accompanied by an uptake of "excess" metal atoms which may be controlled by the spatial requirements like optimal cation-cation distances. One of the main driving forces for the "excess" uptake of metal into the structure could be the coulombic interaction between cations and Zintl anions. Surprisingly enough, no aromatic π-systems have been observed in the silicides of the heavier alkaline earth metals contrary to lithium silicides like $Li_{12}Si_7$ and

Li_8MgSi_6 where cyclopentadienyl-like planar Si_5^{6-} rings have been found [41–43]. A strongly preferred tendency to form aromatic systems seems to be lacking in the Zintl anions of silicon.

The Ca cations again form trigonal prismatic arrangements which are centered by silicon and terminated by magnesium.

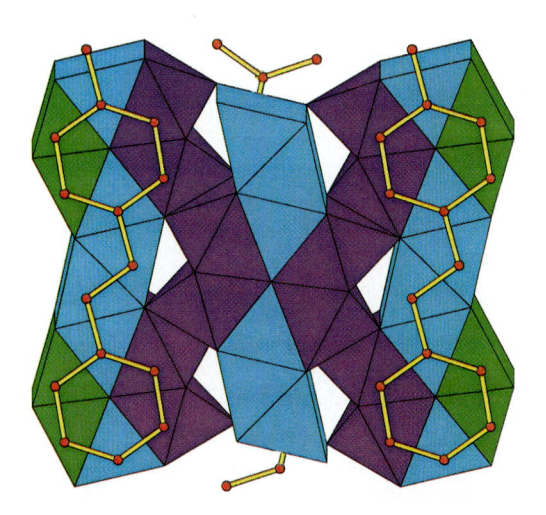

Fig. 4a. Structure of $Ca_{3.3}Li_{3.7}Si_8$ (skew [001] view) with planar silicon polyanions of para-linked flat six-membered rings; the trigonal prismatic cation array is indicated by the polyhedral representation; trigonal prisms and extended trigonal prismatic coordination of the Zintl anion (CN6 (Ca): light blue, CN6 (Li, Ca): green, CN7 (Li, Ca): purple); trigonal prisms coordinating the central part of the anion are exclusively composed of Ca atoms

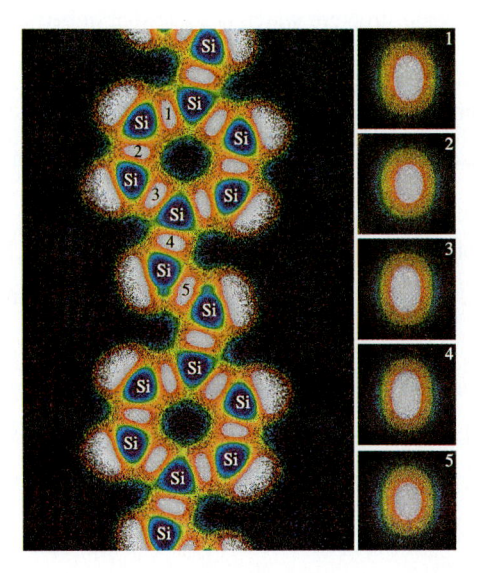

Fig. 4b. ELF in the plane of the silicon polyanion clearly showing bonding and lone electron pair regions (white areas)
Fig. 4c. ELF cross sections through the Si–Si bonds as numbered in 4b; the ELF is more ellipsoidal for the bonds 1, 3, and 4 indicating a π-type contribution while 2 and 5 tend to be more σ-like

3 Silicides Richer in Metal

$Ca_7Li_xMg_{4-x}Li_8$ with x = 1.44 was found by trying to synthesize a slightly oxidized $Ca_7Mg_{7.25}Si_{14}$ by replacing some of the Mg atoms by lithium [31]. The aim was to get a valence electron count which would allow for an aromatic system in the Si_{12} unit. The new compound has a completely different structure which, however, shows strong relations to $Ca_{3.3}Li_{3.7}Si_8$ and to $Sr_{11}Mg_2Si_{10}$, as well. This is an indicator for the enormously rich variational possibilities and the sensitive balances in the structure chemistry of these Zintl phases.

There is again a strong disproportionation of the silicon atoms into isolated, (1b), (2b), and (3b)Si. The overall formula can be rewritten as $(Ca_7Li_xMg_{4-x})^{20.6+} Si_6^{12.6-} (Si^{4-})_2$, according to the Zintl-Klemm concept. Again, there is no possibility of defining a clear electron distribution in terms of the Lewis notation.

Consequently there is an average formal charge on silicon of q = 2.82–, which means the average number of bonds per Si atoms is only slightly larger than one.

The unit cell contains two isolated (0b)Si^{4-} anions mainly coordinated by Mg^{2+} and a planar $Si_6^{12.6-}$ unit which is a derivative of a carbonate-like moiety (Fig. 5a). A comparison of the bond distances give a similar impression as for the other Zintl anions discussed before. In the carbonate-related fragment only slightly but significantly shorter distances are found, compared with the other Si–Si bonds.

The Zintl anion has an electronic structure which is somewhat in between a completely singly bonded (q = –14) and a carbonate related system (q = –12). On the basis of the available topological bond vectors and the actual electron count one would expect 2/3 of a π-bond in the trigonal part. A metallic conductivity is anticipated, e.g., overlap of unfilled π-type orbitals along the ecliptically stacked Si_6 moieties.

It may, however, not be surprising at this point that the carbonate like Si_4 fragment of the Si_6 unit is coordinated by Ca atoms in a trigonal prismatic way while the terminal Si atoms have some Mg coordination.

Slightly more reduced but smaller in Mg content is a new ternary strontium silicide with the composition $Sr_{11}Mg_2Si_{10}$ [31] and a mean formal charge of q = –2.6 on silicon. There are two isolated Si^{4-} anions per formula unit while the other eight silicon atoms form a unit of a linear chain Si_8^{18-}, which is again completely planar. The structure is displayed in Fig. 5b.

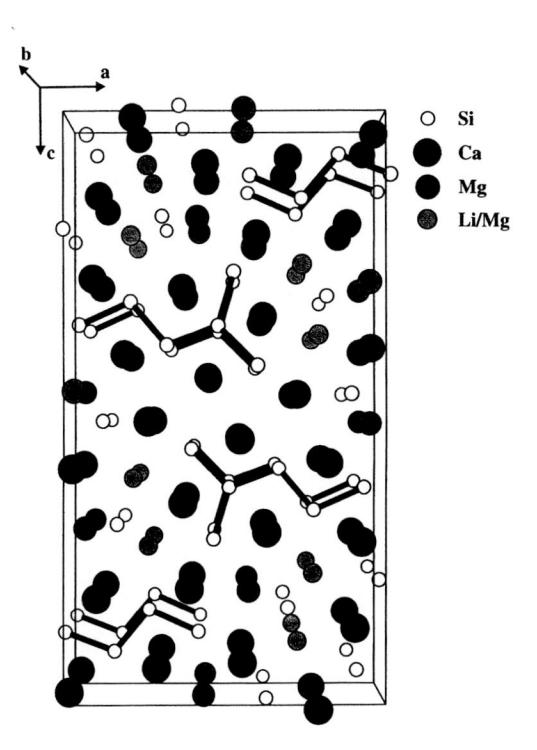

Fig. 5a. Perspective view of the $Ca_7Li_xMg_{4-x}Si_8$ structure along the [010] direction; the trigonal prismatic coordination of the central part of the planar Si_6 anion by Ca atoms is clearly visible

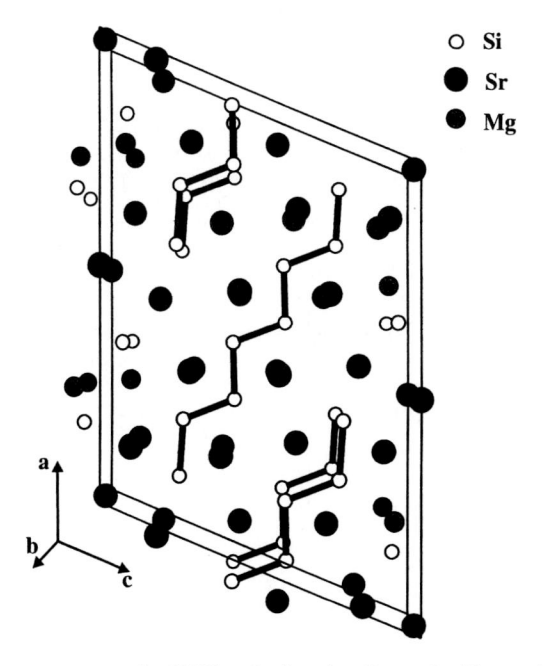

Fig. 5b. Structure of $Sr_{11}Mg_2Si_{10}$ in perspective [010] projection; the oligomeric silicon anion crystallizes like in **5a** with some isolated Si atoms; the planar Si_8 chain moiety is again covered by trigonal Sr prisms

The formal charges imply a "normal" singly bonded silicon strand. The bond distances hardly vary (244.0 – 245.8 pm), but are somewhat long compared to the standard single bond length though not exceptionally longer than "single" bonds that are found in some other Zintl phases mentioned in Table 1 and have been reported elsewhere for molecules containing Si–Si bonds [47].

This chain has a beautyful trigonal prismatic Sr coordination which is only varied at the two terminal silicon sites by additional Mg atom, as we would expect now.

4 Silicon-rich Silicides

On the silicon-rich side of the Ca/Mg/Si systems new phases have additionally been found. According to the Zintl-Klemm concept they must have a higher mean bond order than those discussed previously.

$Ca_{14}Si_{19}$ is an overlooked phase in the binary system with nearly an even number of (3b)- and (2b)-Si atoms per formula unit [31, 48]. It contains a novel two dimensional polyanion of silicon which is displayed in Fig. 6a. The central unit is a 3,3,3-barrelane Si_{11} which is linked via six Si_3 bridges to six neighboring barrelanes. The whole polyanion contains seven layers of silicon atoms and has a width of ca. 900 pm. The barrelane part is shown in Fig. 6b with its trigonal planar (3b)-silicon atoms. The calcium atoms are distributed not only between but also inside the layers.

A tiny addition of magnesium generates the closely related but significantly different ternary silicide $Ca_{3-x}Mg_xSi_4$ (x = 0.08) [31], which contains fairly similar 3,3,3-barrelanes (Fig. 6c) but a different type of linkage. The result is another two-dimensional silicon framework constituted of only five layers of silicon atoms. Comparison of the barrelane units of the two related structures shows some differences as displayed in Fig. 6b and 6d. The two top atoms are in a trigonal pyramidal coordination in the ternary compound and have bond angles of 110.8 ° instead of 120 °.

Obviously, this is a result of the incorporation of one Ca atom into the center of the cage which, has fairly short Ca–Si distances of 288 pm although two Si atoms are pushed outwards. The central barrelane unit in the binary compound is again trigonal pyramidally coordinated by calcium, but without a central Ca atom.

It remains an interesting question why the central Ca site is not coordinated directly by any silicon lone pair because geometrically there is none available in the direct vicinity. This might be taken as one reason for the extremely short distance Ca–Si distance of 288 pm, but the shortest ones in the binary $Ca_{14}Si_{19}$ are only a little longer. Comparably short distances have been observed in $Ca_{3.3}Li_{3.7}Si_4$ with the mixed Ca/Li site. Closer inspection of the structures reveals that in all cases there are no or too few neighboring lone pairs around the corresponding Ca sites.

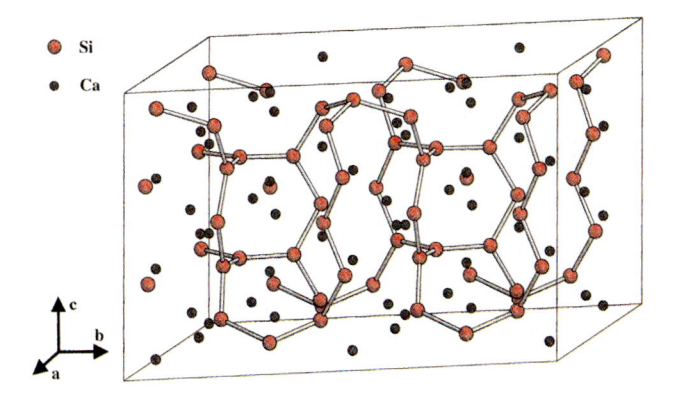

Fig. 6a. Structural motifs of novel silicon rich Zintl phases:
part of the structure of $Ca_{14}Si_{19}$ (Ca: green, Si: red)

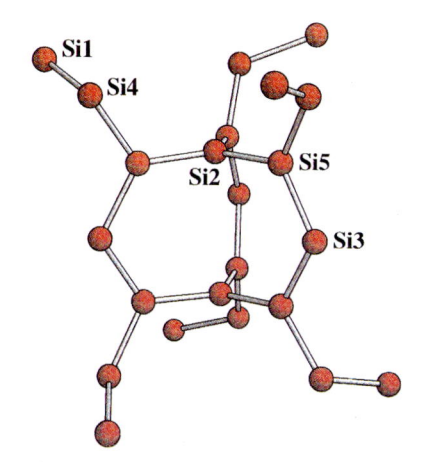

Fig. 6b. Structural motifs of novel silicon rich Zintl phases:
part of the polyanion $^{2}_{\infty}[Si_{19}^{28-}]$ with all of the (3b)Si atoms in a trigonal planar environment

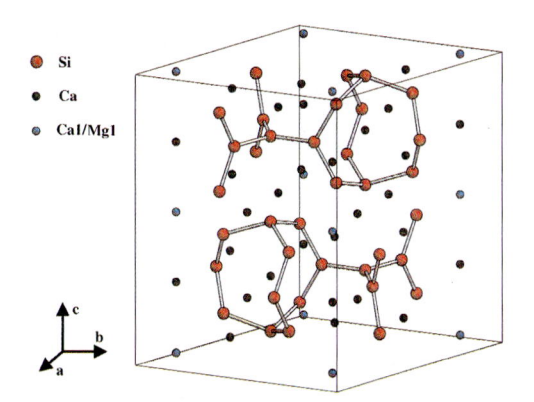

Fig. 6c. Structural motifs of novel silicon rich Zintl phases:
part of the structure of $Ca_{3-x}Mg_xSi_4$ (Ca: green, Ca/Mg (mixed occupation): blue, Si: red)

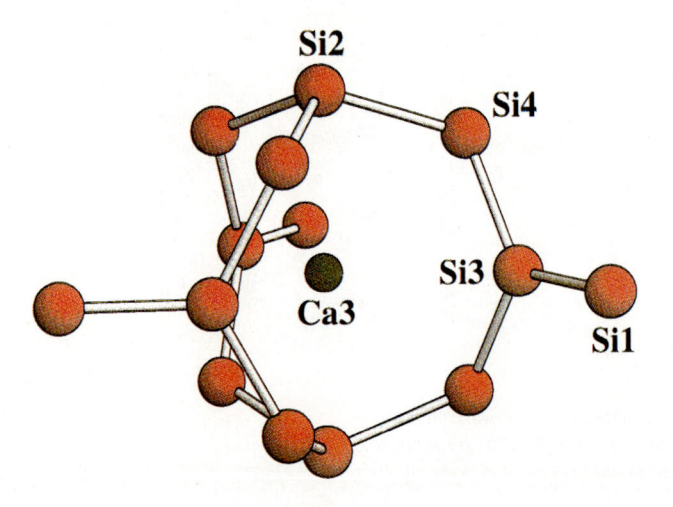

Fig. 6d. Structural motifs of novel silicon rich Zintl phases:
part of the polyanion $^{2}_{\infty}[Si_4^{8-}]$ with the encapsulated Ca atom in the center which pushes the top and the bottom Si atoms out

4.1 Reactivity

The large set of older and novel silicides provides a fascinating supply for silicon linkages from molecular to polymeric character in one, two, and three dimensions. Silicides have been used to perform topochemical reactions preserving at least partially the silicon framework. Numerous investigations on $CaSi_2$ date back more than 50 years until the late 70's [49, 50].

It is still an open question whether new pathways to generate silanes, siloxanes, or even silicones can be designed using specific solid silicides. This is especially interesting because it has been found that there is a direct dependence of the reactivity towards diluted mineral acids and the degree of polymerization of the silicon partial structure. The larger the linkage and the higher the number of bonds per silicon, the lower the reactivity [51–53]. Furthermore, a clear tendency has been observed that highly linked silicon polyanions are easily oxidized to a variety of forms of elementary silicon, from amorphous to more or less crystalline new modifications. The formation of allo-germanium and allo-silicon from phases Li_7Ge_{12} [52, 53] and Na_3LiSi_6 [29, 51], respectively, may serve as examples. Silicides with small Si_n clusters preferably react to form silanes by reaction with protic species. For intermediate cases mixtures of products can be expected but these still have to be investigated much more intensively.

5 Conclusions

Ternary silicides of main group metals form a fascinating family of new compounds. There are classical and surprisingly new bonding situations, sometimes even closely neighbored in one phase. A direct influence on the size, the structure, and the functionality of the Zintl anions can be obtained by careful inspection of the different cation-silicon interactions. Especially, magnesium provides a means for generating end groups in silicon clusters. In some cases the small size of Mg atoms seems to enhance the formation of high pressure phases. While magnesium acts purely as a main group metal there is a suspicion that the heavier alkaline earth metals may utilize other interaction types because of the dominance of trigonal prismatic metal packing in most of these phases. The quite frequent planar Si fragments may be stabilized due to the latter.

Silicides and other Zintl phases are a widely unused source of precursor compounds for a variety of reactions, although, to date, they are in many cases fairly difficult to control and to guide.

References:

[1] E. Zintl, *Angew. Chem.* **1939**, *52*, 1.

[2] E. Zintl, G. Brauer, *Z. Phys. Chem.* **1933**, *20b*, 245.

[3] E. Zintl, W. Dullenkopf, *Z. Phys. Chem.* **1931**, *A154*, 1; *Z. Phys. Chem.* **1932**, *B16*, 183.

[4] G. Kienast, J. Verma, *Z. Anorg. Allg. Chem.* **1961**, *310*, 143; G. A. Tinelli, D. F. Holcomb, *J. Solid State Chem.* **1987**, *25*, 157.

[5] W. Hume-Rothery, *J. Inst. Met.* **1926**, *35*, 307.

[6] W. Klemm, *Trab. Reun. Int. React. Solidos, 3rd* **1956**, *1*, 447.

[7] W. Klemm, *Proc. Chem. Soc. London* **1959**, 329.

[8] W. Klemm, *FIAT – Review of German Science, Naturforschung und Medizin in Deutschland, Anorganische Chemie, Teil IV* **1939–1946**, *26*, 103.

[9] H. O. Grimm, A. Sommerfeld, *Z. Phys.* **1926**, *36*, 36; H. O. Grimm, *Z. Elektrochem.* **1925**, *31*, 474; *Naturwissenschaften* **1929**, *17*, 535.

[10] E. Mooser, W. B. Pearson, *Phys. Rev.* **1956**, *101*, 1608; E. Mooser, W. B. Pearson, *Progress in Semiconductors*, Vol. 5, 103 (Eds.: Gibson, Kröger and Burgess), Wiley, Chichester, **1960**.

[11] W. Klemm, *Festkörperprobleme*, Vieweg, Braunschweig, **1963**; E. Busmann, *Z. Anorg. Allg. Chem.* **1961**, *313*, 90.

[12] W. Klemm, E. Hohmann, *Alkalisilizide, Alkaligermanide, ihre Darstellung und einige wichtige Eigenschaften*, report, Universität Münster, January **1947**; H. Hohmann, *Z. Anorg. Allg. Chem.* **1948**, *257*, 113; J. Witte, H. G. v. Schnering, *Z. Anorg. Allg. Chem.* **1964**, *327*, 260.

[13] H. G. v. Schnering, *Nova Acta Leopold.* **1985**, *59*, 165.

[14] J. Böhm, O. Hassel, *Z. Anorg. Allg. Chem.* **1927**, *160*, 152; J. Evers, *J. Solid State Chem.* **1979**, *28*, 369.

[15] R. Nesper, H. G. v. Schnering, *Tschermaks mineralog. petrogr. Mitt.* **1983**, *32*, 195.

[16] E. Parthé, *Elements of structural inorganic chemistry*, Pöge Druck, Leipzig, **1990**.

[17] H. G. v. Schnering, *Angew. Chem.* **1981**, *93*, 44; *Angew. Chem., Int. Ed. Engl.* **1981**, *20*, 33.

[18] H. G. v. Schnering, Rheinisch-Westfälische Akademie der Wissenschaften, Westdeutscher Verlag, Opladen, **1984**, p. 7.

[19] H. G. v. Schnering, W. Hönle, *Chem. Rev.* **1988**, *88*, 243.

[20] H. Schäfer, B. Eisenmann, *Rev. Inorg. Chem.* **1981**, *3*, 29.

[21] H. Schäfer, *Ann. Rev. Mater. Sci.* **1985**, *15*, 1.

[22] R. Nesper, *Prog. Solid State Chem.* **1990**, *20*, 1.

[23] W. Rieger, E. Parthé, *Acta Crystallogr.* **1967**, *22*, 919.

[24] B. Eisenmann, H. Schäfer, K. Turban, *Z. Naturforsch.* **1974**, *29b*, 464.

[25] F. Merlo, M. L. Fornasini, *J. Less-Common Met.* **1967**, *13*, 603.

[26] K. H. Janzon, H. Schäfer, A. Weiss, *J. Solid State Chem.* **1965**, *24*, 199.

[27] H. Schäfer, K. H. Janzon, A. Weiss, *Angew. Chem.* **1963**, *75*, 451; *Angew. Chem., Int. Ed. Engl.* **1963**, *2*, 393.

[28] J. Evers, G. Oehlinger, A. Weiss, *J. Solid State Chem.* **1977**, *20*, 173.

[29] M. Schwarz, *Ph. D. Thesis*, Universität Stuttgart, **1987**

[30] H. G. v. Schnering, M. Schwarz, R. Nesper, *Angew. Chem.* **1986**, *98*, 558; *Angew. Chem., Int. Ed. Engl.* **1986**, *25*, 556.

[31] A. Currao, *Ph. D. Thesis*, ETH Zürich, in preparation.

[32] A. Currao, *Diploma Thesis*, ETH Zürich, **1992**.

[33] R. Nesper, A. Currao, J. Curda; IVth *Europ. Conf. Solid State Chem.*, Dresden, **1992**, Book of Abstracts, B64, p. 215.

[34] Program EHMACC based on R. Hoffmann, W. N. Lipscomb: *J. Chem. Phys.* **1962**, *36*, 2179; M.-H. Whangbo, R. Hoffmann, R. B. Woodward, *Proc. R. Soc. London* **1979**, *A366*, 23.

[35] M. van Schilfgarde, T. A. Paxton, O. Jepsen, O. K. Andersen, *Programm TBLMTO, Max-Planck-Institut für Festkörperforschung*, Stuttgart, **1992**, unpublished.

[36] a) A. D. Becke, N. E. Edgecombe, *J. Chem. Phys.* **1990**, *92*, 5397.

b) A. Savin, H. J. Flad, J. Flad, H. Preuss, H. G. v. Schnering, *Angew. Chem.* **1992**, *104*, 185; *Angew. Chem., Int. Ed. Engl.* **1992**, *31*, 185.

c) A. Savin, O. Jepsen, J. Flad, O. Anderson, H. Preuss, H. G. v. Schnering, *Angew. Chem.* **1992**, *104*, 186; *Angew. Chem., Int. Ed. Engl.* **1992**, *31*, 187.

d) A. Burkhardt, U. Wedig, H. G. v. Schnering, *Z. Anorg. Allg. Chem.* **1993**, *619*, 437.

e) D. Seebach, H. M. Bürger, D. A. Plattner, R. Nesper, T. Fässler, *Helv. Chim. Acta.* **1993**, *76*, 2581.

[37] B. Silvi, A. Savin, *Nature* **1994**, *371*, 683.

[38] U. Häußermann, S. Wengert, P. Hofmann, A. Savin, O. Jepsen, R. Nesper, *Angew. Chem.* **1994**, *106*, 2147; *Angew. Chem., Int. Ed. Engl.* **1994**, *33*, 2069.

[39] U. Häußermann, S. Wengert, R. Nesper, *Angew. Chem.* **1994**, *106*, 2151; *Angew. Chem., Int. Ed. Engl.* **1994**, *33*, 2073.

[40] W. Schmettow, A. Lipka, H. G. v. Schnering, *Angew. Chem.* **1974**, *86*, 10; *Angew. Chem., Int. Ed. Engl.* **1974**, *13*, 5; H.-P. Abicht, W. Hönle, H. G. v. Schnering, *Z. Anorg. Allg. Chem.* **1984**, *519*, 7.

[41] H. G. v. Schnering, R. Nesper, J. Curda, K.-F. Tebbe, *Angew. Chem.* **1980**, *92*, 1070; *Angew. Chem., Int. Ed. Engl.* **1980**, *19*, 1033; R. Nesper, H. G. v. Schnering, J. Curda, *Chem. Ber.* **1986**, *119*, 3576.

[42] R. Nesper, J. Curda, H. G. v. Schnering, *J. Solid State Chem.* **1986**, *170*, 199.

[43] M. C. Böhm, R. Ramirez, R. Nesper, H. G. v. Schnering, *Ber. Bunsenges. Phys. Chem.* **1985**, *89*, 465.

[44] R. Ramirez, R. Nesper, H. G. v. Schnering, M. C. Böhm, *Chem. Phys.* **1985**, *95*, 17.

[45] R. Nesper, A. Currao, S. Wengert, *Chemistry*, in preparation.

[46] W. Müller, H. Schäfer, A. Weiss, *Z. Naturforsch.* **1970**, *25B*, 1371.

[47] H. Bock, *Angew. Chem.* **1993**, *105*, 413; *Angew. Chem., Int. Ed. Engl.* **1993**, *32*, 414.

[48] A. Currao, R. Nesper, J. Curda, B. Hillebrecht, unpublished results.

[49] H. Kautsky, G. Herzberg, *Z. Anorg. Allg. Chem.* **1925**, *147*, 81.

[50] A. Weiss, G. Beil, H. Meyer, *Z. Naturforsch.* **1979**, *34B*, 25.

[51] H. G. v. Schnering, M. Schwarz, R. Nesper, *J. Less-Common Met.* **1988**, *137*, 297.

[52] A. Grüttner, H. G. v. Schnering, M. Schwarz, R. Nesper, *J. Less-Common Met.* **1988**, *137*, 297.

[53] A. Grüttner, *Ph. D. Thesis*, Universität Stuttgart, **1982**.

Alkali Metal Derivatives of Tris(trimethylsilyl)silane – Syntheses and Molecular Structures

Karl Wilhelm Klinkhammer*, Gerd Becker, Wolfgang Schwarz

Institut für Anorganische Chemie

Universität Stuttgart

Pfaffenwaldring 55, D-70550 Stuttgart, Germany

Summary: The solvent-free alkali metal derivatives of tris(trimethylsilyl)silane (hypersilane [1]) were synthesized. The molecular structures of unsolvated hypersilyl lithium **1**, -sodium **2**, and -potassium **3** as well as of the solvates [(Me$_3$Si)$_3$SiRb]$_2$(toluene) (**4a**), [2] [(Me$_3$Si)$_3$SiCs]$_2$(toluene)$_3$ (**5a**) [2], [(Me$_3$Si)$_3$SiCs]$_2$(THF) (**5b**), and [(Me$_3$Si)$_3$SiCs]$_2$(biphenyl)(pentane) (**5c**) were determined by X-ray crystal structure analyses.

Recently we reported the syntheses of solvent-free potassium, rubidium, and cesium derivatives of tris(trimethylsilyl)silane (hypersilane): (Me$_3$Si)$_3$SiK (**3**), (Me$_3$Si)$_3$SiRb (**4**), and (Me$_3$Si)$_3$SiCs (**5**). The crystal structures of two toluene adducts, [(Me$_3$Si)$_3$SiRb]$_2$(toluene) (**4a**) and [(Me$_3$Si)$_3$SiCs]$_2$(toluene)$_3$ (**5a**) could be determined (Fig. 1 and Fig. 2) [2].

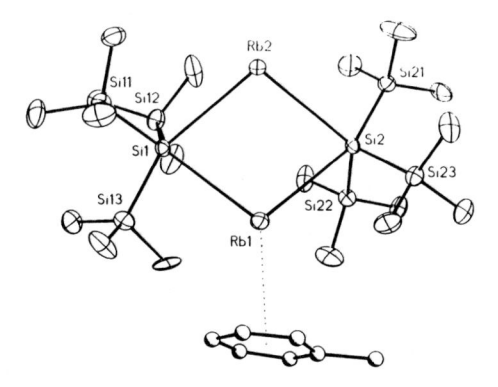

Fig. 1. Molecular structure of [(Me$_3$Si)$_3$SiRb]$_2$(toluene) (**4a**) (thermal ellipsoids at 30 % probability level); selected bond lengths [pm], angles and torsion angles [°]: Rb–Si 352.2(4)–361.6(4), Rb1···Rb2 430.5(2), Rb···C$_{arom}$ > 342, Rb···CH$_3$ > 337, Si–Si 232.7(6)–234.6(6), Rb–Si–Rb 74.3 (av.), Si–Rb–Si 96.9 (av.), Si–Si–Si 100.6(2)–104.2(2), Rb1–Si1–Rb2–Si2 31.8

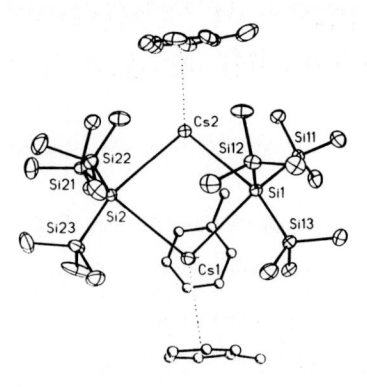

Fig. 2. Molecular structure of [(Me$_3$Si)$_3$SiCs]$_2$(toluene)$_3$ (**5a**) (thermal ellipsoids at 30 % probability level); selected bond lengths [pm], angles and torsion angles [°]: Cs–Si 377.4(2)–385.0(2), Cs1···Cs2 492.0(1), Cs···C$_{arom}$ > 358, Cs···CH$_3$ > 370, Si–Si 233.8(3)–234.8(3), Cs–Si–Cs 87.3 (av.), Si–Cs–Si 92.8 (av.), Si–Si–Si 99.7(1)–102.4(1), Cs1–Si1–Cs2–Si2 27.1

These adducts consist of cyclic dimers comprising folded M$_2$Si$_2$ backbones. The spectroscopic data reveal mainly ionic interactions between alkali metal cations and hypersilanide anions. Additionally, oligo-hapto coordination of the metal ions by the toluene molecules is observed. Intra- and intermolecular agostic interactions to C–H bonds finally govern the peculiarities of both, the molecular and the crystal structures of these compounds.

Solvent-free hypersilanides have been shown to be useful reagents for the preparation of other especially low coordinated hypersilyl derivatives. The thallium(II) compound **6** [1] or derivatives of divalent tin **7** [3] and lead **8** [3] were synthesized via the reactions of (Me$_3$Si)$_3$SiRb (**4**) with the bis(trimethylsilyl)amides of thallium(I), tin(II), and lead(II), respectively.

[(Me$_3$Si)$_3$Si]$_2$Tl–Tl[Si(SiMe$_3$)$_3$]$_2$ [(Me$_3$Si)$_3$Si]$_2$Sn···Sn[Si(SiMe$_3$)$_3$]$_2$ [(Me$_3$Si)$_3$Si]$_2$Pb

6   **7**   **8**

Here we report the syntheses of solvent-free hypersilyl derivatives of sodium and lithium – (Me$_3$Si)$_3$SiLi (**1**) and (Me$_3$Si)$_3$SiNa (**2**) – for the first time. Moreover, we are presenting the crystal structures of unsolvated lithium **1**, sodium **2**, and potassium **3** derivatives and of two additional solvent adducts of the cesium compound **5**.

Hypersilyl lithium **1** and sodium **2** can be easily synthesized by the reaction of an excess of molten sodium or lithium powder with bis(hypersilyl)zinc in boiling *n*heptane (Eq. 1). After removal of zinc and sodium by filtration and cooling down the reaction mixtures to room temperature, the less soluble sodium derivative **2** precipitates nearly quantitatively in colorless crystals. A similar procedure has been utilized for preparing crystalline samples of solvent-free hypersilyl potassium **3**. Hypersilyl lithium **1**, which is much more soluble, is obtained by further cooling to –60 °C.

$$Zn[Si(SiMe_3)_3]_2 + 2\,M \longrightarrow [MSi(SiMe_3)_3]_2 + Zn$$

Eq. 1. M = Li, Na

X-ray crystal structure analyses had been performed on all three solvent-free compounds [4]. A common feature of hypersilyl lithium, sodium and potassium molecules (Fig. 3–5) is a cyclic M_2Si_2 backbone, already found in the toluene solvates **4a** and **5a**. All observed M–Si distances are in a good agreement with the sum of the covalent radii of silicon and the appropriate alkali metal (M) as derived from X-ray data by Schade and Schleyer [5]. Remaining differences can easily be rationalized by different coordination numbers and types. While in the potassium and lithium derivatives the M_2Si_2 ring is found to be almost planar, it is somewhat folded in the case of the sodium derivative; a pronounced folding, however, is observed in the solvates of the heavier alkali metal derivatives **4a, 5a–c**. The reasons for these differences are not obvious, but we suggest that the energy surfaces for ring-folding are flat and thus even weak interactions of van der Waals or agostic type might lead to distortions of the ideal geometry.

In hypersilyl sodium and potassium both, intra- and intermolecular M···H–C agostic interactions occur, leading to short M···C distances. Interactions of this type were found in many other alkali metal compounds – a survey is given in reference [5]. In the hypersilyl lithium dimers only short *intra*molecular M···H–C contacts are observed (M···C 239, 245 pm), combined with a significant lengthening of the involved Si–C bonds (191 pm) and a pronounced tilting of the hypersilyl ligands (Fig. 3).

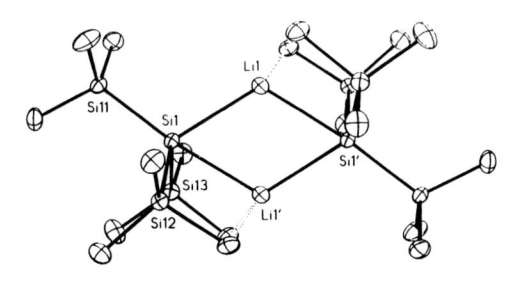

Fig. 3. Molecular structure of hypersilyl lithium **1** (thermal ellipsoids at 30 % probability level); selected bond lengths [pm], angles and torsion angles [°]: Li1–Si1 258.9(4), Li1–Si1' 260.1(4), Li1···Li1' 276.2(8), Si–Si 232.9(1)–234.4(1), Li1–Si1–Li1' 64.3(2), Si1–Li1–Si1' 115.7(2), Si–Si–Si 105.69(4)–109.21(3), Li1–Si1–Li1'–Si1' 0

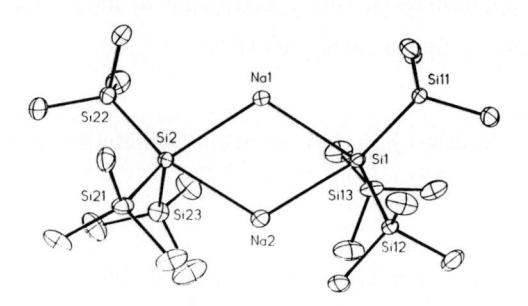

Fig. 4. Molecular structure of hypersilyl sodium **2** (thermal ellipsoids at 30 % probability level); selected bond lengths [pm], angles and torsion angles [°]: Na–Si 298.5(3)–302,7(3), Na1···Na2 316.9(3), Na···CH$_3$ > 280, Si–Si 233.1(3)–234.7(2), Na–Si–Na 64.1 (av.), Si–Na–Si 115.2 (av.), Si–Si–Si 100.92(8)–104.46(9), Na1–Si1–Na2–Si2 9.1

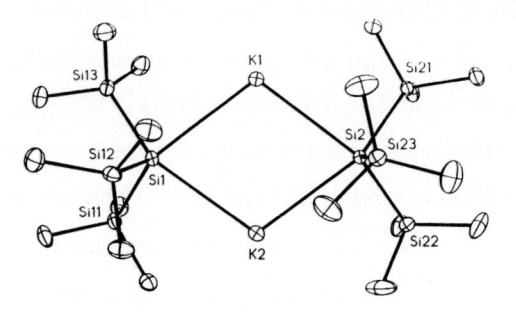

Fig. 5. Molecular structure of hypersilyl potassium **3** (thermal ellipsoids at 30 % probability level); selected bond lengths [pm], angles and torsion angles [°]: K–Si 336.8(1)–341,7(1), K1···K2 405.0(1), K···CH$_3$ > 325, Si–Si 232.6(1)–234.5(1), K–Si–K 73.2 (av.), Si–K–Si 106.7 (av.), Si–Si–Si 100.54(4)–104.06(4), K1–Si1–K2–Si2 2.8

At the moment we are investigating the influence of different donor ligands on the structural features especially of the rubidium and cesium derivatives. Such adducts are prepared by adding the corresponding ligand to a suspension of hypersilyl rubidium **4** or cesium **5** in *n*pentane until a clear solution is obtained. Although, in some cases side-reactions occur, such as deprotonation (diphenylmethane) or reduction (pyrene, bipyridine). In addition to the toluene solvates **4a** and **5a** already mentioned above, the only two crystalline derivatives isolated so far are the adducts with THF as a typical σ-donor and with biphenyl as a further π-donor (Fig. 6 and Fig. 7).

In both cases the resulting molecules are still dimers, comprising four-membered M$_2$Si$_2$ rings. Unexpectedly, the complex isolated from THF/pentane solutions turned out to be the *mono*-adduct **5b** (Fig. 7), in which THF occupies an unusual bridging position between the two alkali metal atoms of *one* molecule. The ligand serves as a classical σ-donor to one metal atom via its oxygen atom

(Cs–O 310 pm), whereas the second cesium ion is coordinated in a more side-on fashion, involving not only the oxygen atom (Cs–O 328 pm) but also two adjacent CH_2 groups (Cs–C 389 pm).

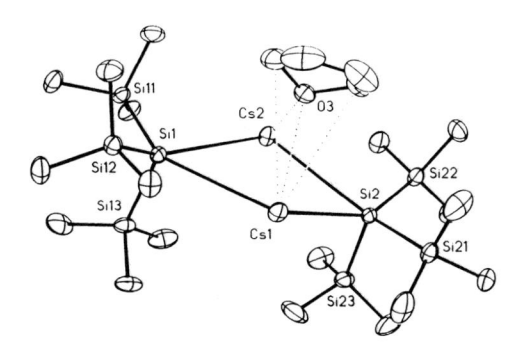

Fig. 6. Molecular structure of $[(Me_3Si)_3SiCs]_2(THF)$ **5b** (thermal ellipsoids at 30 % probability level); selected bond lengths [pm], angles and torsion angles [°]: Cs–Si 367.2(1)–373.2(1), Cs1···Cs2 419.7(1), Cs1–O3 328.0(3), Cs2–O3 309.8(3), Cs···CH_3 > 366, Si–Si 232.4(2)–234.1(2), Cs–Si–Cs 69.0 (av.), Si–Cs–Si 109.3 (av.), Si–Si–Si 100.91(6)–104.60(7), Cs1–Si1–Cs2–Si2 14.1

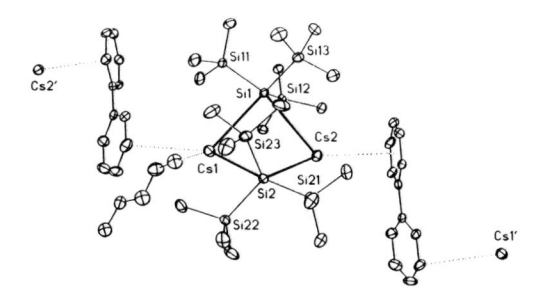

Fig. 7. Molecular structure of $[(Me_3Si)_3SiCs]_2$(biphenyl)(pentane) **5c** (thermal ellipsoids at 30 % probability level); selected bond lengths [pm], angles and torsion angles [°]: Cs–Si 367.7(2)–381.1(2), Cs1···Cs2 463.7(1), Cs···C_{arom} > 368, Cs···CH_3 > 363, Si–Si 232.9(2)–234.0(2), Cs–Si–Cs 76.4 (av.), Si–Cs–Si 91.8 (av.), Si–Si–Si 99.92(8)–103.79(8), Cs1–Si1–Cs2–Si2 35.5

The biphenyl solvate **5c** (Fig. 7) which contains one additional molecule of pentane per $[(Me_3Si)_3SiCs]_2$ dimer, crystallizes as a two-dimensional coordination polymer. The twisted biphenyl molecule acts as a bridging ligand. One phenyl ring is bound in an η^6-fashion to the Cs2 ion of one dimer, whereas the second ring exhibits just one short C–Cs contact to the Cs1 ion of another dimer. The coordination sphere of Cs1 is completed by agostic C–H···Cs interactions to the pentane molecule and to trimethylsilyl groups of the same and of adjacent $[(Me_3Si)_3SiCs]_2$ units. Four further contacts of this type are also found around Cs2.

The observed Cs-Si bond lengths in **5b** and **5c** are significantly shorter than those found in the tris(toluene) adduct **5a** (367–374 pm vs. 378–381 pm), indicating a more efficient coordination of the cesium ion by the electron-rich toluene molecules.

References:

[1] K. W. Klinkhammer, W. Schwarz, *Z. Anorg. Allg. Chem.* **1993**, *619*, 1777.
[2] S. Henkel, K. W. Klinkhammer, W. Schwarz, *Angew. Chem.* **1994**, *106*, 721; *Angew. Chem., Int. Ed. Engl.* **1994**, *33*, 681.
[3] K. W. Klinkhammer, W. Schwarz, *J. Chem. Soc., Chem. Commun.*, in press.
[4] Crystal data of **1, 2, 3, 5b,** and **5c**: Mo$_{K\alpha}$, –100 °C; **1**: monoclinic, space group $P2_1/n$, $a = 9.3851(9)$, $b = 15.7646(11)$, $c = 12.0793(8)$ A, $\beta = 104.315(5)$ °. 3904 independent reflections ($4.3 < 2\Theta < 55$ °), $R1 = 0.048$; **2**: triclinic, space group $P1$, $a = 12.490(1)$, $b = 13.105(1)$, $c = 23.232(1)$ A, $\alpha = 92.73(1)$, $\beta = 103.26$, $\gamma = 102.00$ °. 9210 independant reflections ($6.5 < 2\Theta < 45$ °), $R1 = 0.056$; **3**: monoclinic, space group $P2_1/c$, $a = 14.552(2)$, $b = 13.945(2)$, $c = 18.115(3)$ A, $\beta = 91.02(1)$ °. 4314 independent reflections ($4.3 < 2\Theta < 52$ °) $R1 = 0.036$; **5b**: monoclinic, space group $P2_1/c$, $a = 16.647(2)$, $b = 18.758(2)$, $c = 14.227(2)$ A, $\beta = 107.56(1)$ °. 8864 independent reflections ($6.5 < 2\Theta < 52$ °), $R1 = 0.040$; **5c**: monoclinic, space group $P2_1/n$, $a = 15.327(4)$, $b = 15.395(4)$, $c = 20.649(6)$ A, $\beta = 94.41(1)$ °. 7517 independent reflections ($6.5 < 2\Theta < 48$ °), $R1 = 0.041$.
[5] C. Schade, P. v. R. Schleyer, *Adv. Organomet. Chem.* **1987**, *27*, 169.

Mono-, Bis-, Tris-, and Tetrakis(lithiomethyl)silanes: New Building Blocks for Organosilicon Compounds

C. Strohmann*, S. Lüdtke

Institut für Anorganische Chemie

Universität des Saarlandes

Postfach 1150, D-66041 Saarbrücken, Germany

Summary: Mono-, bis-, tris-, and tetrakis(lithiomethyl)silanes were prepared by reductive cleavage of C–S bonds with lithium naphthalenide ($LiC_{10}H_8$) or lithium p,p'-di-*t*butylbiphenylide (LiDBB) and characterized by the reaction with Bu_3SnCl. The poly(lithiomethyl)silanes can be used as new building blocks for in α-position substituted silanes or, for example, silacyclobutanes.

Introduction

Alkyllithium compounds can be prepared by the following methods [1]:

1) Hydrogen-lithium exchange with lithium or lithium-bases
2) Halogen-lithium exchange with lithium or lithium-bases
3) Metal-lithium exchange with lithium or lithium-bases
4) Reductive addition of lithium or lithium-bases
5) Reductive cleavage of C–S bonds with lithium

For the synthesis of the known two poly(lithiomethyl)silanes the halogen-lithium exchange (for **1** [2]) and the reductive cleavage of C–S bonds with lithium (for **3** [3]) were used [4] (Scheme 1).

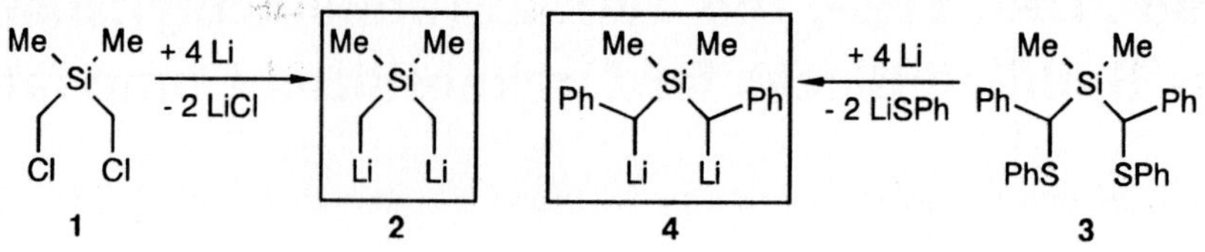

Scheme 1.

As a part of our systematic studies on the structure element "–CR_2–El–CR_2–" (El = element of group 14–16, partly with substituents; R = H, alkyl, aryl) we are investigating the synthetic potential of the reductive cleavage of C–S bonds with electron transfer reagents for the synthesis of poly(lithiomethyl)silanes with the structure element "LiCH$_2$–SiR$_2$–CH$_2$Li" (El = SiR$_2$).

Synthesis

The (phenylthiomethyl)silanes used for reductive cleavage were produced by the reaction of (phenylthiomethyl)lithium with the corresponding chlorosilanes (Scheme 2).

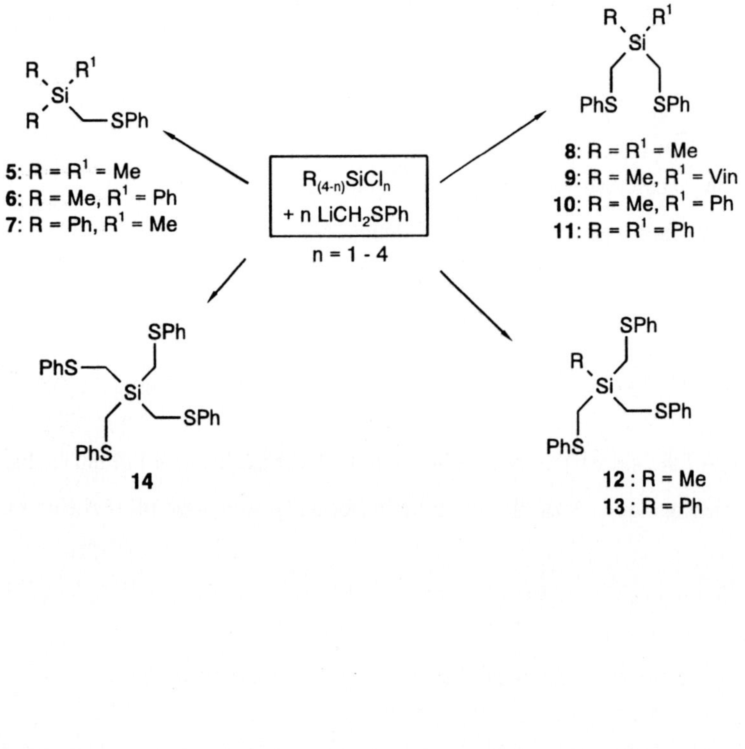

5: R = R' = Me
6: R = Me, R' = Ph
7: R = Ph, R' = Me

8: R = R' = Me
9: R = Me, R' = Vin
10: R = Me, R' = Ph
11: R = R' = Ph

12 : R = Me
13 : R = Ph

Scheme 2.

The reductive cleavage of just one C–S bond yields the mono(lithiomethyl)silanes. This transformation is well known and used in the Peterson olefination for the preparation of (lithiomethyl)trimethylsilane (**15**) [5]. We employed it as a method to exchange the thiophenyl-group in the (phenylthiomethyl)silanes **5–14** with lithium, thus creating the corresponding poly(lithiomethyl)silanes (see, e.g., Scheme 3 for methylsilanes) [4].

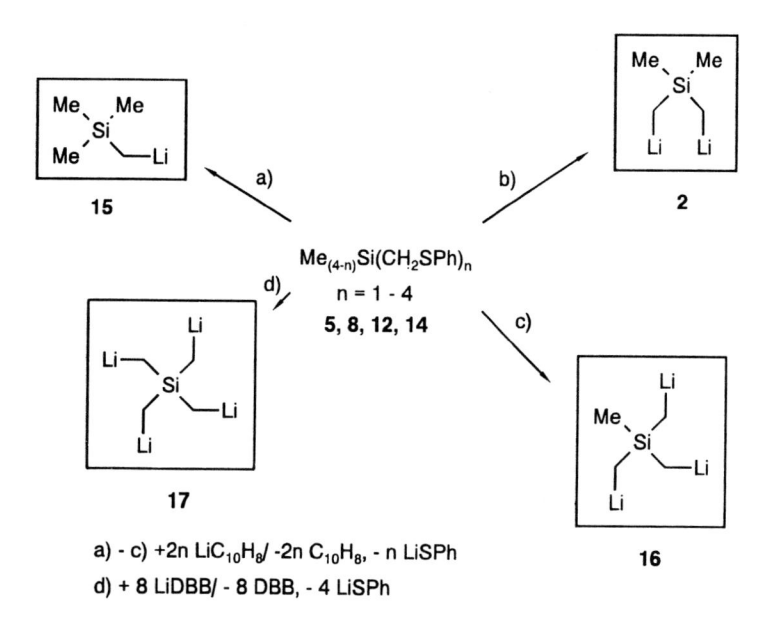

a) - c) +2n LiC$_{10}$H$_8$/ -2n C$_{10}$H$_8$, - n LiSPh
d) + 8 LiDBB/ - 8 DBB, - 4 LiSPh

Scheme 3.

Bis- and tris(lithiomethyl)silanes were prepared with lithium naphthalenide (LiC$_{10}$H$_8$) and tetrakis(lithiomethyl)silane (**17**) with lithium p,p'-di-*t*butylbiphenylide (LiDBB) as electron transfer reagent and characterized by reaction with Bu$_3$SnCl [>95 % overall yield by NMR; 42 % to 81 % yield of isolated pure (stannylmethyl)silanes] (Scheme 3 and 4).

Scheme 4.

We were not able to characterize the poly(lithiomethyl)silanes in the reaction mixture. Therefore the question remains: Were the poly(lithiomethyl)silanes formed before or during the reaction with the trapping reagent Bu$_3$SnCl. Two observations indicate that the poly(lithiomethyl)silanes were formed before the addition of the trapping reagent:

a) The green or green-blue color of the electron transfer reagent had disappeared, indicating complete reaction.

b) Only partly metallated species were formed at lower temperature or shorter reaction time and trapped with Bu$_3$SnCl.

Some Reactions of Bis(lithiomethyl)diphenylsilane

To demonstrate the synthetic potential of the new building blocks the bifunctional reagent bis(lithiomethyl)diphenylsilane (**18**) was used for some reactions with mono- and bifunctional substrates (Scheme 5).

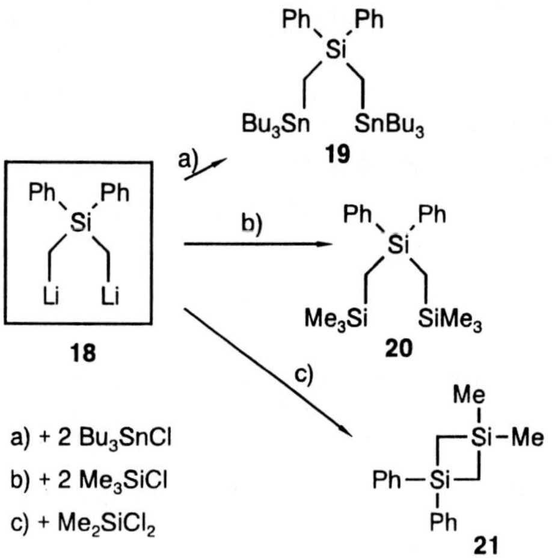

Scheme 5.

The 1,3-dilithio-compounds exhibit "normal" reactivity with monofunctional reagents (Scheme 5). 1,1- and 1,2-dilithio-compounds do not always react in the expected way of a alkyllithium-compound [6]. 1,3-Disilacyclobutanes can be synthesized by the reaction of **18** with dichlorosilanes (e.g., **21**, Scheme 5), illustrating the synthetic potential of these new building blocks for the preparation of silacyclobutanes.

Attempted Synthesis of Some Carbon Analogs

We tried to synthesize the carbon analogs of **14** and **12** by reductive cleavage of C–S bonds. The isolated products indicate, that 50 % of the C–S bonds were cleaved by electron transfer reactions and 50 % by LiSPh-elimination, probably after a H,Li-exchange (Scheme 6).

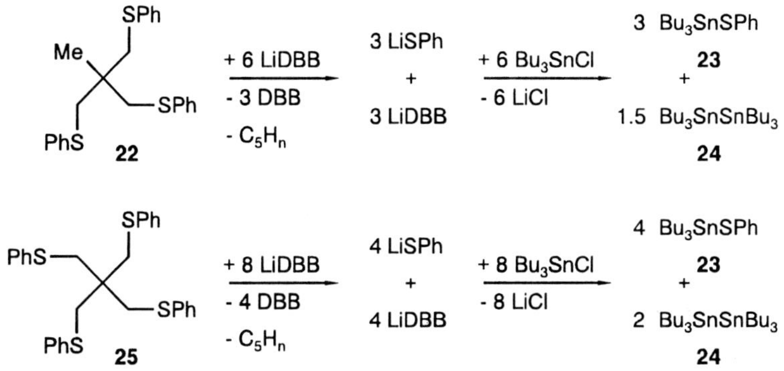

Scheme 6.

An application of the synthetic concept to carbon analogs is not possible. LiSPh-elimination and/or LiSPh-elimination after previous H,Li exchange occurred for alkyllithium-intermediates formed by reductive cleavage of C–S bonds. Reaction of this type have been observed for similar compounds [7]. The silicon atom obviously plays a central role in the reactivity of the presented structure types.

Conclusion

Poly(lithiomethyl)silanes can be prepared by the reaction of poly(phenylthiomethyl)silanes with electron transfer reagents. The carbon analogs of some (phenylthiomethyl)silanes show a different reactivity. The new poly(lithiomethyl)silanes can be used as building blocks for organosilicon compounds.

Further related work including investigations of the reductive cleavage of C–S bonds for the synthesis of other compounds with the structure element "Li–CR$_2$–El–CR$_2$–Li" (El = element of group 14–16, partly with substituents; R = H, alkyl, aryl) is in progress.

Acknowledgement: Financial support of this study by the *Deutsche Forschungsgemeinschaft* is gratefully acknowledged. The authors would like to thank Prof. M. Veith for supporting this work.

References:

[1] L. R. Subramanian, in: *Carbanionen, Houben-Weyl, Vol. E19d*, Thieme, Stuttgart, **1993**, p. 713.

[2] D. Seyferth, C. J. Attridge, *J. Organomet. Chem.* **1970**, *21*, 103; D. Seyferth, E. G. Rochow, *J. Am. Chem. Soc.* **1955**, *77*, 907.

[3] O. S. Akkerman, F. Bickelhaupt, *J. Organomet. Chem.* **1988**, *338*, 159.

[4] The illustrations of (lithiomethyl)silanes are not indicative of the real structure of these compounds in solution or the solid state.

[5] J. Backes, in: *Carbanionen, Houben-Weyl, Vol. E19d*, Thieme, Stuttgart, **1993**, p.825.

[6] A. Maercker, in: *Carbanionen, Houben-Weyl, Vol. E19d*, Thieme, Stuttgart, **1993**, p. 448.

[7] C. Rücker, *J. Organomet. Chem.* **1986**, *310*, 135; T. Cohen, R. H. Ritter, D. Qelette, *J. Am. Chem. Soc.* **1982**, *104*, 7142.

Neutral and Anionic Lithiumsilylamides as Precursors of Inorganic Ring Systems

Ina Hemme, Uwe Klingebiel*

Institut für Anorganische Chemie

Georg-August-Universität Göttingen

Tammannstr. 4, D-37077 Göttingen, Germany

The trilithium derivatives of tris(amino)silanes, synthesized so far, all have a dimeric structure [1–5]. By treating the tris(amino)silane with three equivalents of *n*BuLi in THF, however, we obtained colorless cystals of the monomeric trilithium compound (Fig. 1, **1**).

The crystal structure shows three four-membered (SiNLiN)-ring systems which are planar. Although the nitrogen silicon angles are normally near 120 °, the (SiNSi)-angles in this molecule are much larger.

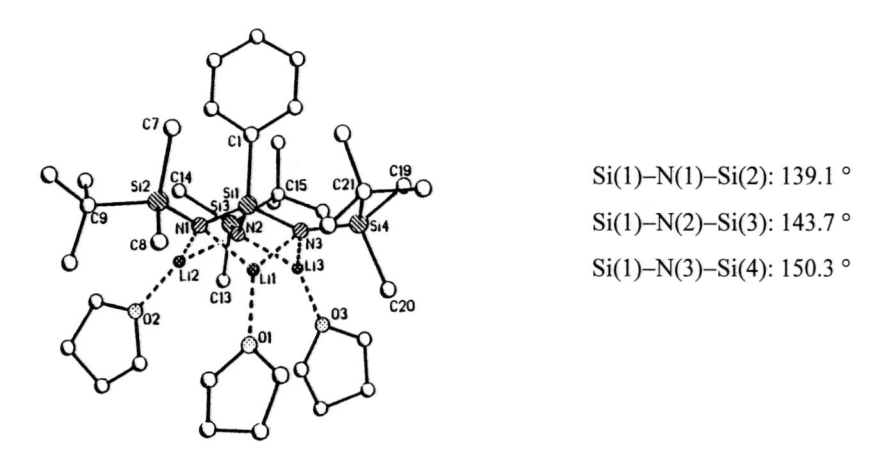

Si(1)–N(1)–Si(2): 139.1 °

Si(1)–N(2)–Si(3): 143.7 °

Si(1)–N(3)–Si(4): 150.3 °

Fig. 1. Crystal structure of **1**

The crystal structures of some dilithiated bis(alkylamino)- and bis(arylamino)silanes have been determined [2, 3, 6–8].

The reaction of a dilithiated fluorosilylhydrazine with an excess of BuLi gave the dilithium derivative of a bis(silylamino)silane [9]. This was the reason for the systematic preparation of other such compounds. We synthesized the dilithiated $Me_2Si(NLiSiCMe_3Me_2)_2$ and found an open cage structure with short Li···C-distances (Li(3)–C(51): 233.6 pm, Li(3)–C(32): 265.3 pm) (Fig. 2a, **2**). At the same time the dilithium derivative of $Me_2Si(NHSiMe_3)_2$ was prepared by Zimmer and Veith. It has the same new type of structure as ours [10]. Because a *trans*-conformation was found in the reaction of a hydrazine with BuLi we synthesized $(Me_2HC)_2Si(NHSiCMe_3Me_2)_2$, dilithiated the compound and determined the expected structure (Fig. 2b, **3**). The Li···C-distances in this molecule are Li(2)–C(31) = Li(2a)–C(31a): 278.7 pm.

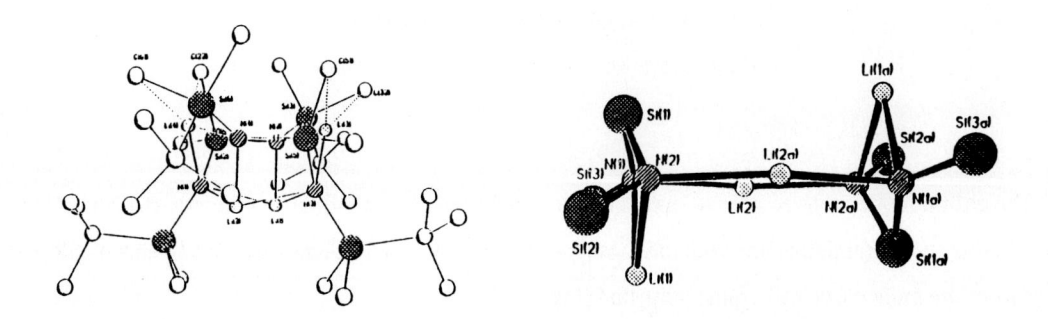

Fig. 2a. Crystal structure of **2**

Fig. 2b. Crystal structure of **3**

In the reaction of the dilithiated bis(amino)silanes with PF_3 or SiF_4, the inorganic four-membered ring systems are isolated [11] (Scheme 1).

Scheme 1.

The lithiation of sterically differently substituted bis(amino)fluorophenylsilanes leads to different results. With the bulky $Si(CMe_3)_2Me$, the dilithiated bis(amino)fluorophenylsilane is not formed; instead the lithium(lithiumsilyldiamide) **8** is obtained by a fluorine-CMe_3-exchange depending on the reaction conditions [12] (Scheme 2). By adding 12-crown-4, colorless crystals of **9** are isolated (Fig. 3, **9**).

Scheme 2.

$$R-N-\underset{\underset{H}{\overset{Ph}{|}}}{\overset{\overset{Ph}{|}}{Si}}-N-R$$
$$\quad\quad H \quad F \quad H$$

7

+ 2 t-BuLi
+ n thf
− 2 t-BuH

$\longrightarrow$ $\left[R-N-\underset{\underset{Li}{|}}{\overset{\overset{Ph}{|}}{Si}}-N-R\right]$ (thf)$_n$
$\quad\quad\quad\quad Li \quad F \quad Li$

+ 3 t-BuLi
+ n thf
− 2 t-BuH
− LiF

$\longrightarrow$ $\left[\begin{array}{c}Me_3CPh\\ \diagdown\diagup\\ Si\\ R-N\diagdown\diagup N-R\\ Li\\ (thf)\end{array}\right]^-$ $\left[Li(thf)_n\right]^+$

8

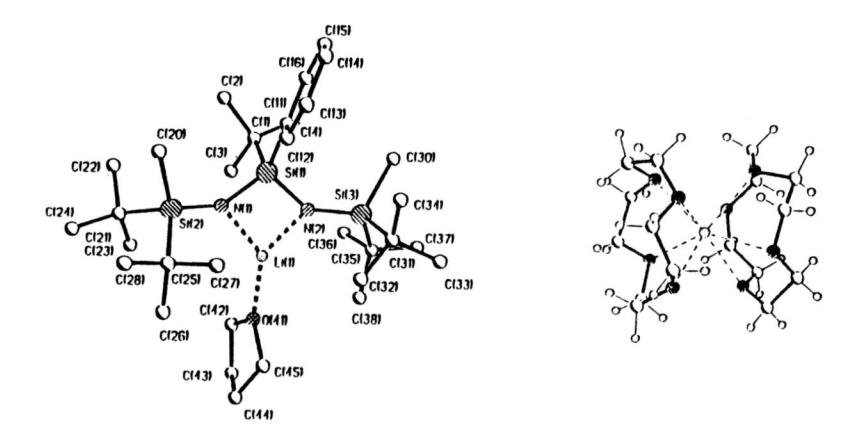

Fig. 3. Crystal structure of 9

When less bulky substituents like SiCMe$_3$Me$_2$ are present, the intermediate silaamidide dimerizes and the dimeric silaamidide **11** is formed by the following mechanism (Eq. 1).

After the addition of THF, LiF is eliminated. The intermediate silaamidide [13, 14] dimerizes in a [2+2]-cycloaddition to give the lithium(cyclodisilazane)diamide anion. The lithium which is part of the anion is only twofold coordinated. The Li$^+$ cation is complexed by THF.

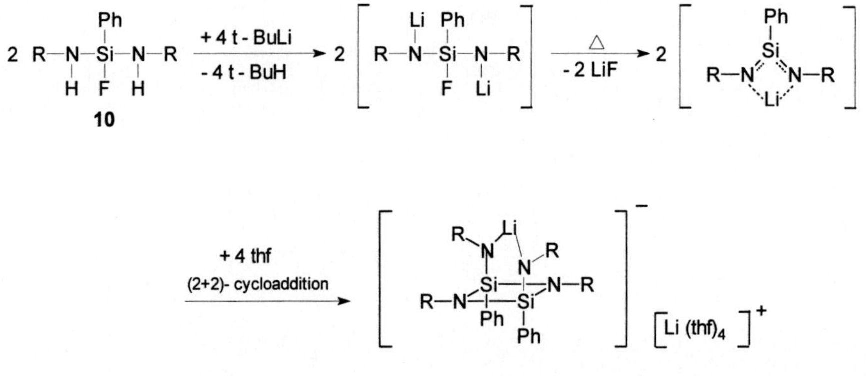

Eq. 1.

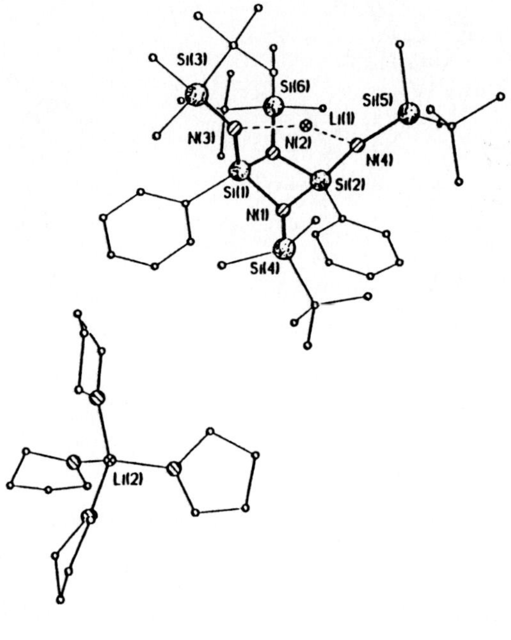

Fig. 4. Crystal structure of **11**

Acknowledgement: We thank the *Deutsche Forschungsgemeinschaft* and the *Fonds der Chemischen Industrie* for financial support.

References:

[1] D. J. Brauer, H. Bürger, G. R. Liewald, J. Wilke, *J. Organomet. Chem.* **1985**, *287*, 305.

[2] D. J. Brauer, H. Bürger, G. R. Liewald, *J. Organomet. Chem.* **1986**, *308*, 119.

[3] M. Veith, *Chem. Rev.* **1990**, *90*, 3.

[4] L. H. Gade, N. Mahr, *J. Chem. Soc., Dalton Trans.* **1993**, 489.

[5] L. H. Gade, Ch. Becker, J. W. Lauher, *Inorg. Chem.* **1993**, *32*, 2308.

[6] M. Veith, *Angew. Chem.* **1987**, *99*, 1; *Angew. Chem., Int. Ed. Engl.* **1987**, *26*, 1.

[7] H. Chen, R. A. Bartlett, H. V. Rasika Dias, M. M. Olmstead, P. P. Power, *Inorg. Chem.* **1991**, *30*, 2487.

[8] B. Tecklenburg, *Ph. D. Thesis*, Universität Göttingen, **1991**.

[9] Ch. Drost, *Ph. D. Thesis*, Universität Göttingen, **1992**.

[10] M. Veith, personal communication.

[11] I. Hemme, *Ph. D. Thesis*, Universität Göttingen, **1994**.

[12] B. Tecklenburg, U. Klingebiel, D. Schmidt-Bäse, *J. Organomet. Chem.* **1992**, *429*, 287.

[13] G. E. Underiner, R. West, *Angew. Chem.* **1990**, *102*, 579; *Angew. Chem., Int. Ed. Engl.* **1990**, *29*, 529.

[14] G. E. Underiner, R. P. Tan, D. R. Powell, R. West, *J. Am. Chem. Soc.* **1990**, *113*, 8437.

Lithium *t*Butylamidosilanes – Syntheses and Structures

G. Becker*, S. Abele, J. Dautel, G. Motz, W. Schwarz

Institut für Anorganische Chemie

Universität Stuttgart

Pfaffenwaldring 55, D-70550 Stuttgart, Germany

If sterically demanding substituents such as *t*butyl groups are bonded to the nitrogen atoms of aminosilanes, in the pertinent lithium amides lithium is not allowed to expand its coordination sphere by complexation of solvent molecules. As a consequence of its accordingly low coordination number agostic Li⋯H(C) and extremely strong Li⋯H(Si) interactions may occur. Starting with a very simple example this statement will be further verified by some more complicated structures.

1: At 70 °C equimolar amounts of hexaphenyl-*cyclo*-trisilazane [1] and *n*butyl lithium react in toluene to give initially a colorless powder, which may be redissolved at 40 °C in THF (Eq. 1).

Eq. 1.

If such a solution is cooled down to room temperature colorless squares precipitate. An X-ray structure determination {triclinic, $P\overline{1}$; $a = 997.1(2)$, $b = 1090.5(2)$, $c = 2037.2(4)$ pm; $\alpha = 75.98(1)°$, $\beta = 84.56(2)°$, $\gamma = 71.66(1)°$; $\vartheta = -100$ °C; $Z = 2$ formula units; $R = 0.078$} shows monomeric 1-bis(tetrahydrofuran)-lithium 2,2,4,4,6,6-hexaphenyl-*cyclo*-trisilazane **1** to be present in the solid state. Contrary to this result the analogous 2,2,4,4,6,6-hexamethyl derivative studied by Sheldrick and Haase [2] is dimeric with only one THF-substituent bound to the three coordinated lithium. In compound **1** all Si-C$_{phenyl}$ bonds adjacent to the N-Li(THF)$_2$ fragment are slightly elongated, whereas the corresponding distances in the opposite

Si(C_6H_5)$_2$ groups do not deviate from the standard value – in our opinion an experimental evidence for electronic $p(N)$–σ^*(Si–C_{phenyl}) interactions.

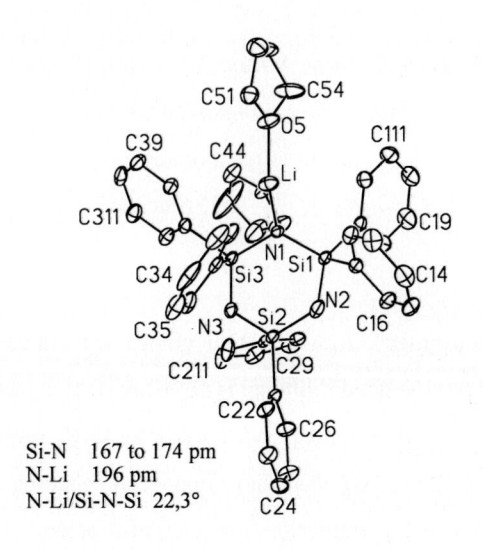

Fig. 1. Crystal structure of **1**

Si–N   167 to 174 pm
N–Li   196 pm
N–Li/Si–N–Si 22,3°

2: Bis(*t*butylamino)-methyl-vinylsilane, easily prepared from dichloro-methyl-vinylsilane and excess *t*butylamine, may be lithiated with *n*butyl lithium at –70 °C in *n*pentane.

Fig. 2. Structure of **2**: Si–N 170 pm, N–Li(THF) 199–205 pm, 194–196 pm

After recrystallization of the colorless product from a 1:1 mixture of toluene and THF (20/–60 °C) squares of [*N*-lithium*t*butylamido][*N*-(tetrahydrofuran)lithium*t*butylamido]methylvinylsilane **2** are obtained in 98 % yield. An X-ray structure determination [triclinic, P^-; a = 930.8(2), b = 933.2(2), c = 1165.5(2) pm; α = 81.30(1)°, β = 79.56(1)°, γ = 71.20(1)°; ϑ = −100 °C; Z = 2 formula units; R = 0.048] shows the compound to be a centrosymmetric dimer in the solid state. Whereas coordinated lithium is bonded to both nitrogen atoms of either monomeric unit acting as a chelate ligand, the remaining lithium atoms give rise to dimerization forming two N–Li–N bridges. Despite their low coordination number of 2, no electronic interactions between lithium and the vinyl groups could be ascertained.

3: Several years ago Bürger et al. [3] determined the structure of bis(lithium-*t*butylamido)-dimethylsilane **3a**. This compound, first prepared by Fink [4] in order to use it as starting material in the syntheses of *cyclo*-disilazanes, crystallizes as a dimer from petrolether; the molecular skeleton built up of lithium and nitrogen in an alternating sequence is best described as a slightly distorted dodecahedron. It, however, bis(*t*butylamino)methylsilane which differs from the previous educt by an exchange of either Si–CH$_3$ for an Si–H group only reacts with *n*butyl lithium in a mixture of toluene and tetrahydrofuran, a compound (**3b**) with a very complicated structure is obtained. As shown by an X-ray structure analysis {monoclinic, $P2_1/n$; a = 1176.9(3), b = 2392.3(7), c = 1923.9(5) pm; β = 100.91(1)°; ϑ = −100 °C; Z = 4 tetramers; R = 0.067} a tetramer has been formed.

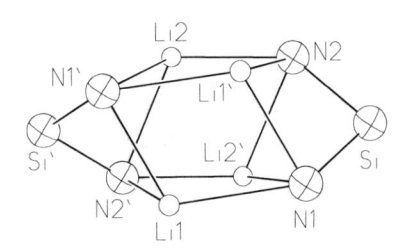

Fig. 3. Crystal structures of **3a**

Contours of the dodecahedron characteristic for the dimethyl derivative may still be discerned, but agostic Li⋯H(Si) interactions are responsible for a strong distortion of the molecular skeleton and especially for tetramerization. A detailed discussion of the structure leads to the conclusion that the newly formed Li⋯H interactions are at least as strong as or even stronger than the broken Li–N bonds.

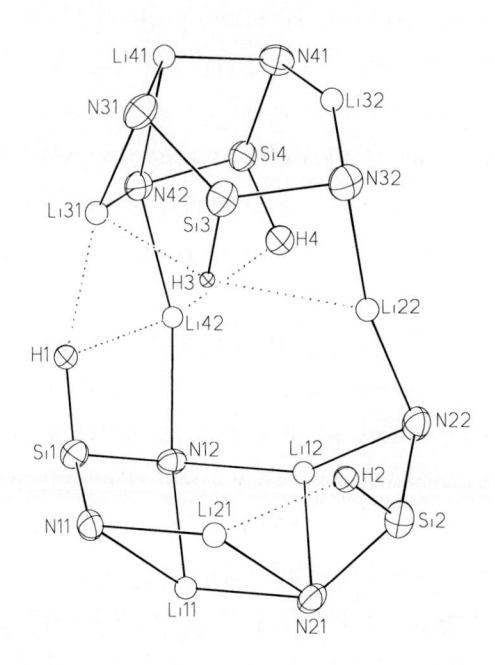

Fig. 4. Crystal structures of **3b**; characteristic structural parameters: Si–N 168–172 pm, N–Li 190–235 pm, Si–H 140–149 pm, Li···H 197–232 pm

4: If di(*t*butylamino)silane, a colorless liquid (bp: 50 °C/3 mbar) easily prepared in tetrahydrofuran from commercially available dichlorosilane and *t*butylamine in a molar ratio of 1:4, is treated at 0 °C with an equimolar amount of *n*butyl lithium in *n*pentane, lithiation of one N–H group as well as cyclization combined with elimination of hydrogen occurs (Eq. 2). Reactions of this type are known from the literature [4, 5] to proceed at higher temperature or in the presence of small amounts of an alkali metal or its hydride. We therefore assume that traces of lithium hydride formed in a side reaction might have catalyzed the formation of compound **4a**.

Eq. 2. Synthesis of **4a**

As an X-ray structure determination of 2,4-bis(*N*-lithium*t*butylamido)-1,3-di-*t*butyl-*cyclo*-disilazane (**4a**), obtained as large colorless crystals from an *n*pentane solution failed for still unknown reasons, the compound was treated with excess chlorotrimethylsilane. Depending on the reaction temperature and the solvent used – either refluxing THF with a subsequent exchange for *n*pentane or only *n*pentane at room temperature – *cis*- (**4b**) or *trans*-2,4-bis(*t*butyl-trimethylsilylamino)-1,3-di-*t*butyl-*cyclo*-disilazane (**4c**) is formed in a nearly quantitative yield.

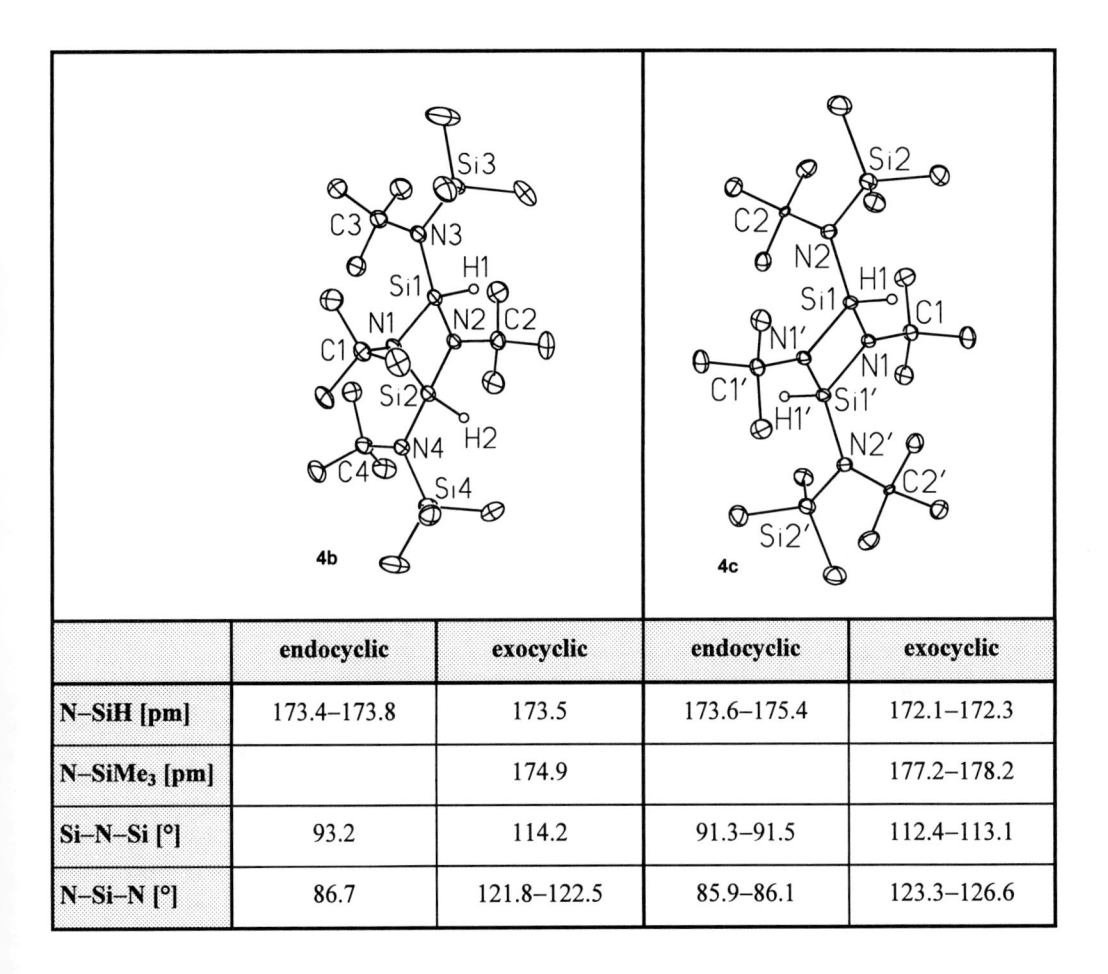

	endocyclic	exocyclic	endocyclic	exocyclic
N–SiH [pm]	173.4–173.8	173.5	173.6–175.4	172.1–172.3
N–SiMe₃ [pm]		174.9		177.2–178.2
Si–N–Si [°]	93.2	114.2	91.3–91.5	112.4–113.1
N–Si–N [°]	86.7	121.8–122.5	85.9–86.1	123.3–126.6

Fig. 5. Crystal structures of **4b** and **4c**

Both compounds isolated in the usual way from the corresponding reaction mixtures differ significantly in their reproducible melting points (149 vs. 204 °C) and their δ^{29}Si NMR chemical shifts (−55.8 vs −46.8 ppm). Conversion does not occur even by refluxing them in solvents such as THF, *n*pentane or toluene.

X-Ray structure determinations [**4b**: monoclinic, $P2_1/c$; $a = 1017.6(3)$, $b = 1115.0(3)$, $c = 2771.9(5)$ pm; $\beta = 100.46(4)°$; $\vartheta = -100$ °C; $Z = 4$; $R = 0.051$; **4c**: monoclinic, $P2_1/c$; $a = 1096.1(2)$, $b = 1351.5(3)$, $c = 1020.0(3)$ pm; $\beta = 100.97(3)°$; $\vartheta = -100$ °C; $Z = 2$; $R = 0.050$] show the four-membered rings either to be folded with an angle of 23.6° between the two N–Si–N planes or to be strictly planar due to a centre of inversion. Some bond lengths and angles are given above.

5: If, however, di(*t*butylamino)silane is treated with twice the amount of *n*butyl lithium at –60 °C in *n*pentane, both hydrogen atoms are removed from the SiH$_2$ groups of the starting material – one being evolved as gaseous hydrogen, the other being precipitated as lithiumhydride together with the lithiated di(*t*butylamido)silane (Eq. 3). As the two lithium compounds could not be separated from each other, a tetrahydrofuran solution of the initially formed white powder was treated at 0 °C with excess chlorotrimethylsilane to give trimethylsilane and lithium chloride. Finally large colorless crystals of [μ-(2,4-di-*n*butyl-1,3-*N*,*N'*-tetra*t*butyl-*cyclo*-disilazan-2*r*,4*c*-diaminato(2-)-N^1,N^2,N^4:N^2,N^3,N^4)]-bis(tetrahydro-furan-*O*)-dilithium **5** were isolated in 62 % yield from a filtered *n*pentane solution. An X-ray structure determination {monoclinic, $C2/c$; $a = 1722.3(2)$, $b = 1278.4(2)$, $c = 1772.1(2)$ pm; $\beta = 97.02(1)°$; $\vartheta = -100$ °C; $Z = 4$; $R = 0.055$} shows the skeleton of the neutral complex **5** to be a strongly distorted cube built up of a Si–N–Si–N and an N–Li–N–Li ring. Silicon, nitrogen, and lithium are bound additionally either to an *n*butyl or a *t*butyl group or to a tetrahydrofuran molecule, respectively. The Si–N bonds vary between 168.6 and 176.2 pm; the N–Li distances are in the range 213.0–233.0 pm.

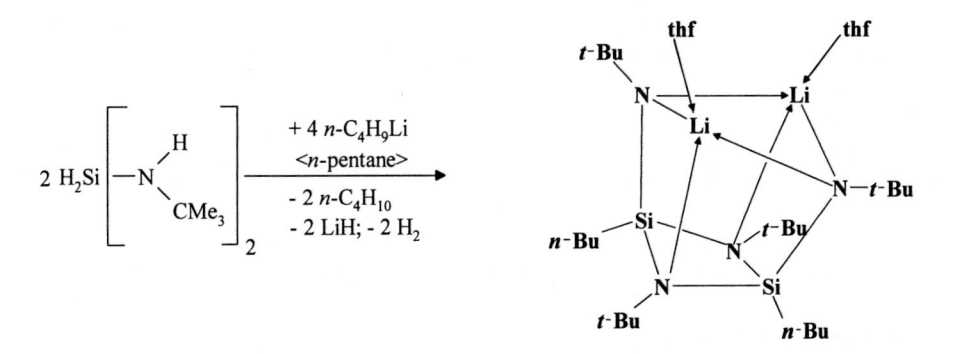

Eq. 3. Synthesis of **5**

References:

[1] E. Larsson, L. Bjellerup, *J. Am. Chem. Soc.* **1953**, *75*, 995.
[2] M. Haase, G. M. Sheldrick, *Acta Cryst.* **1986**, *C42*, 1009.
[3] D. J. Brauer, H. Bürger, G. R. Liewald, *J. Organomet. Chem.* **1986**, *308*, 119.
[4] W. Fink, *Helv. Chim. Acta* **1968**, *51*, 954.
[5] D. Seyferth, G. H. Wiseman, *Comm. Am. Ceram. Soc.* **1984**, C132.

Dilithiated Oligosilanes:
Synthesis, Structure, and Reactivity

J. Belzner*, U. Dehnert, D. Stalke

Institut für Organische Chemie

Georg-August-Universtät Göttingen

Tammannstr. 2, D-37077 Göttingen, Germany

Summary: The cleavage of cyclotrisilane **1** with lithium metal in etheral solvents yields, depending on the reaction conditions, either 1,3-dilithiotrisilane **2-Li** or vicinal 1,2-dilithiodisilane **3-Li**. Both of them are the first structurally characterized representatives of this class of compounds. In both **2-Li** and **3-Li** the lithium atoms are located at the termini of the silicon backbone. The ^{29}Si^{7}Li coupling constant of **3-Li** (36 Hz) evidences the partly covalent character of the Si–Li bond.

A well established method for the preparation of metalated silanes is the cleavage of Si–Si σ-bonds by alkali metals [1]. The reaction of arylsubstituted cyclosilanes yields the corresponding α,ω-dilithiated oligosilanes. Whereas the reaction of cyclopenta- and cyclotetrasilanes with lithium is well known [2, 3], a metal mediated cleavage of a cyclotrisilane was not described up to now. Here we report the reaction of cyclotrisilane **1** with lithium, which affords, depending on the conditions, either 1,3-dilithiotrisilane **2-Li** or 1,2-dilithiodisilane **3-Li**.

Stirring cyclotrisilane **1** with two equivalents lithium metal in 1,4-dioxane gave rise to the formation of 1,3-dilithiotrisilane **2-Li** (Scheme 1), which was isolated in 49 % yield. The linear trisilane **2-H** was obtained by subsequent protonation with cyclopentadiene.

The ^{1}H and ^{13}C NMR spectra, which are very sensitive towards changes in temperature or concentration, showed broad and partly overlapping lines and did not contribute much to the structural elucidation of **2-Li**. The same applied to the ^{29}Si NMR spectrum, in which two broadened signals at δ = –26.5 and –31.8 ppm were observed. Thus, the identification of the reaction product as the 1,3-dilithiotrisilane **2-Li** is based mainly on the result of the X-ray structure analysis (Fig. 1).

Scheme 1.

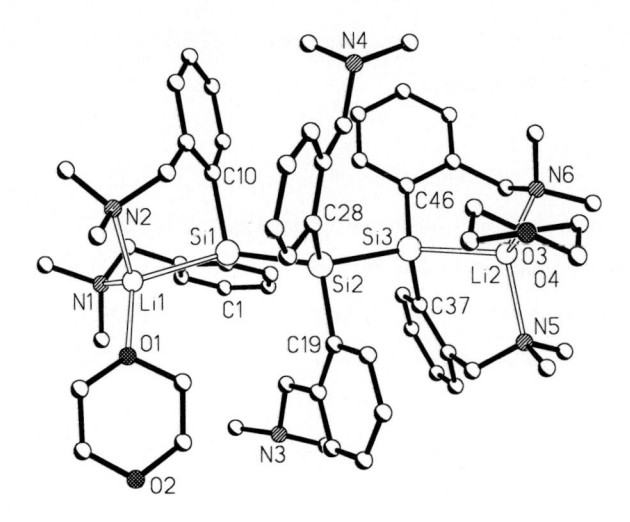

Fig. 1. Structure of **2-Li** in the solid; selected bond lengths [pm] and angles [°]: Si1–Si2 240.7(3), Si2–Si3 240.1(3), Si1–Li1 255(1), Si3–Li2 254(1), Li1–N1 214(1), Li1–N2 215(1), Li1–O1 197(1), Li2–N5 216(1), Li2–N6 211(1), Li2–O3 199(1); C1–Si1–C10 99.5(3), C19–Si2–C28 101.9(3), C37–Si3–C46 99.8(3), Si1–Si2–Si3 136.9(1), Li1–Si1–Si2 131.6(3), Si2–Si3–Li2 129.7(3)

The Li-atoms of **2-Li** are coordinated distorted tetrahedral via complexation by one molecule of dioxane and both amino nitrogen atoms of the terminal aryl substituents. The Si–Li distances are short in comparison with other mono- and dilithiated silanes, which range between 263 and 271 pm [3–5]. Because of the steric demand of the six aryl substituents, the LiSi$_3$Li backbone is, similar to the conformation of 1,4-dilithiooctaphenyltetrasilane [3], appreciably stretched and adopts an approximately antiperiplanar conformation.

Stirring **1** with excess Li in THF gave rise to the formation of 1,2-dilithiodisilane **3-Li**. Alternatively, **3-Li** was obtained by reaction of **2-Li** with Li (Scheme 1). The X-ray structure determination (Fig. 2) shows the lithium centers in **3-Li** to have a different environment: whereas Li1 is coordinated intramoleculary by *two* Me$_2$N-groups, Li2 interacts with only *one* amino function. The tetrahedral coordination sphere of the lithium centers is completed by one and two of THF molecules, respectively.

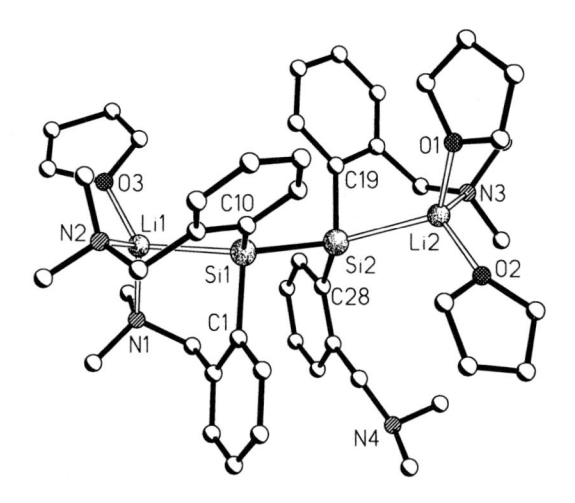

Fig. 2. Structure of **3-Li** in the solid; selected bond lengths [pm] and angles [°]; the structural parameters of the second, crystallographically independent molecule are given in brackets: Si1–Si2 238.0(2) (237.7(2)), Si1–Li1 254.2(8) (258.9(8)), Li1–O3 195(1) (196(1)), Li1–N1 217(1) (215(1)), Li1–N2 213(1) (215(1)), Li2–O1 198(1) (199(1)), Li2–O2 196(1) (197(1)), Li2–N3 213(1) (214(1)); C1–Si1–C10 100.7(2) (99.6(2)), C19–Si2–C28 102.9(2) (101.3(2)), Li1–Si1–Si2 143.7(2) (143.7(2)), Si1–Si2–Li2 134.9(2) (134.7(2))

Just as the ligand spheres of both lithium centers are different, the Si–Li distances are not identical. It appears that the coordination of the two amino "side-arms" induces a shorter Si–Li distance; this parallels the short Si–Li distance in **2-Li**, in which both Li atoms are coordinated by 2 amino groups. The Si–Si bond length of **2-Li** is 238.0 (237.7) pm – a typical value for a Si–Si single bond. The Li-atoms of substituted dilithioethanes show a side on coordination to the central C–C bond [6]; this is not observed for **2-Li**, which might be a consequence of the intramolecular complexation.

In solution, the NMR spectra of **3-Li** do not reflect the structural asymmetry which is characteristic for the solid state. The ^{29}Si NMR spectrum in [D$_8$]-THF shows just one 1:1:1:1 quadruplett due to the ^{29}Si^{7}Li coupling [7]. The coupling constant of 36 Hz [4, 5, 8] indicates the partially covalent character of the Si–Li bond.

As expected, **3-Li** reacts readily with electrophiles: protonation with cyclopentadiene or treatment with trimethylchlorosilane gave **3-H** and **3-TMS**, respectively (Schemes 1 and 2). Surprisingly, the reactions with 1,2-dibromoethane or diaryldichlorosilane gave rise to the formation of the cyclotrisilane **1** in moderate yield. Other silicon containing ring systems should be accessible via reaction with bis-electrophiles.

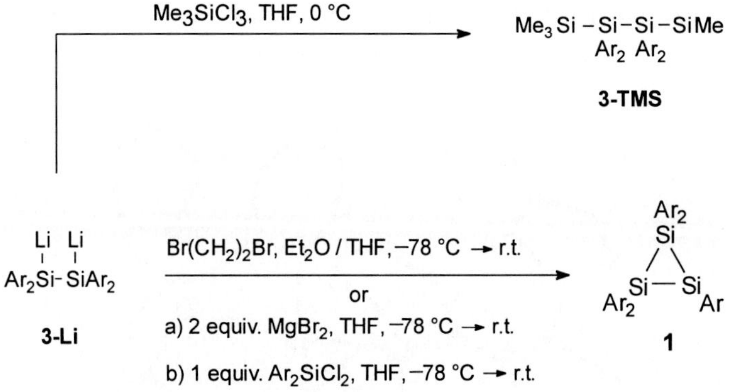

Scheme 2.

Acknowledgement: This work was supported by the *Deutsche Forschungsgemeinschaft* and the *Fonds der Chemischen Industrie*.

References:

[1] A. G. Brook, H. Gilman, *J. Am. Chem. Soc.* **1954**, *76*, 278; H. Gilman, G. D. Lichtenwalter, *J. Am. Chem. Soc.* **1958**, *80*, 608.

[2] H. Gilman, D. J. Peterson, A. W. Jarvie, H. J. S. Winkler, *J. Am. Chem. Soc.* **1960**, *82*, 2076; H. Gilman, G. L. Schwebke, *J. Am. Chem. Soc.* **1963**, *85*, 1016; E. Hengge, D. Wolfer, *J. Organomet. Chem.* **1974**, *66*, 413.

[3] G. Becker, H. M. Hartmann, E. Hengge, F. Schrank, *Z. Anorg. Allg. Chem.* **1989**, *572*, 63.

[4] A. Heine, R. Herbst-Irmer, G. M. Sheldrick, D. Stalke, *Inorg. Chem.* **1993**, *32*, 2694; and references cited therein.

[5] H. V. R. Dias, M. M. Olmstead, K. Ruhlandt-Senge, P. P. Power, *J. Organomet. Chem.* **1993**, *462*, 1.

[6] M. Walczak, G. D. Stucky, *J. Organomet. Chem.* **1975**, *97*, 313; M. Walczak, G. Stucky, *J. Am. Chem. Soc.* **1976**, *98*, 5531; A. Sekiguchi, T. Nakanishi, C. Kabuto, H. Sakurai, *J. Am. Chem. Soc.* **1989**, 111, 3748; A. Sekiguchi, T. Nakanishi, C. Kabuto, H. Sakurai, *Chem. Lett.* **1992**, 867.

[7] Attempts to verify by means of NMR spectroscopy the different environment of the lithium atoms of **3-Li** in less coordinating solvents, such as [D_8]-toluene or C_6D_6, were unsuccessful.

[8] U. Edlund, T. Lejon, T. K. Venkatachalam, E. Buncel, *J. Am. Chem. Soc.* **1985**, *107*, 6408.

Synthesis of Gallium Hypersilyl Derivatives

R. Frey, G. Linti*, K. Polborn, H. Schwenk

Institut für Anorganische Chemie
Ludwig-Maximilians-Universität München
Meiserstr. 1, D-80333 München, Germany

Summary: The gallium hypersilyl derivatives $(TMP)_2GaSi(SiMe_3)_3$ and $Me_2GaSi(SiMe_3)_3 \cdot THF$ have been prepared and structurally investigated. $(TMP)_2GaSi(SiMe_3)_3$ reacts with protic reagents under cleavage of the GaN bonds and not of the GaSi bond.

Introduction

Organyl compounds of gallium are well established and their synthetic use has been studied extensively. In contrast, the chemistry of gallium compounds with bonds to heavier elements of group 14 has gathered less attention. Rösch et al. [1] prepared tris(trimethylsilyl)gallium **1** (Eq. 1) and the gallate $Li[Ga(SiMe_3)_4]$. The only structurally investigated gallium silyl derivative so far has been Cowley's bis(hypersilyl)gallate **2** [2] (Eq. 2).

$$GaCl_3 + 3\ Me_3SiCl + 6\ Li \xrightarrow{THF} (Me_3Si)_3Ga \cdot THF \xrightarrow{50°C} (Me_3Si)_3Ga$$
$$\mathbf{1}$$

Eq. 1.

$$GaCl_3 + 2\ (Me_3Si)_3SiLi \cdot 3\ THF \xrightarrow{-\ LiCl} \begin{array}{c}(Me_3Si)_3Si \quad Cl \quad THF \\ \diagdown \quad \diagup \quad \diagdown \quad \diagup \\ Ga \qquad Li \\ \diagup \quad \diagdown \quad \diagup \quad \diagdown \\ (Me_3Si)_3Si \quad Cl \quad THF\end{array}$$
$$\mathbf{2}$$

Eq. 2.

Amberger [3] has described the synthesis of gallium germanium compounds Me_2GaGeR_3 (R = H, Ph). With the heavier element homologue tin only "ionic" species like $Li[GaMe_3(SnMe_3)]$ [4] are known. Here we describe the synthesis, structure and some reactions of $(TMP)_2GaSi(SiMe_3)_3$ (TMP = 2,2,6,6-tetramethylpiperidino) [5].

Results

Reaction of $(TMP)_2GaCl$ with $LiSi(SiMe_3)_3 \cdot 3THF$ [6] yielded the gallium silyl **3** as pale yellow crystals (Eq. 3). Compound **3** proved to be stable towards oxidation of the gallium silyl bond by molecular oxygen and sulfur. In addition CS_2 does not insert neither in the GaSi nor in the GaN bonds. But **3** is sensitive to moisture, hydrolysis with 2 equivalents of water produced the dimeric bishydroxide **4** and TMPH (Eq. 4). Obviously, the GaSi bond is not sensitive to water and oxygen due the steric shielding in **3**. The behavior of **3** towards more acidic reagents like phenol and hydrogen chloride is similiar but different in detail. As in the case of water, only the GaN bonds are cleaved, however, the amine is protonated, forming ammonium salts. These add to the primary Lewis acid protolysis products, yielding gallate structures **5** and **6**.

By reaction of Me_2GaCl with $LiSi(SiMe_3)_3 \cdot 3THF$, the diorganylgalliumsilyl **7** is synthesized as a THF adduct (Eq. 5).

$$\text{Me}_2\text{GaCl} + \text{LiSi(SiMe}_3)_3 \cdot 3\text{THF} \xrightarrow[\substack{-\text{LiCl} \\ -2\text{ THF}}]{\text{Et}_2\text{O}} \underset{\text{Me}}{\overset{\text{Me}}{\underset{|}{\text{Ga}}}}\overset{\text{THF}}{\underset{|}{-}}\text{Si(SiMe}_3)_3$$

$$\mathbf{8}$$

Eq. 5. synthesis of **7**

X-Ray Crystal Structures

The results of the crystal structure analyses of **3**, **6**, and **7** are given in Fig. 1–3. The GaSi bond length in **3** is rather long, compared with the calculated sum of covalent radii (240 pm). Compound **7** and the gallates **2** and **6** show shorter bond lengths near the calculated value, albeit the gallium atoms are tetracoordinated. That might be due to the steric strain in **3**. This is supported by molecular modelling studies. Force field calculations (MM2) reproduce the experimental geometry of **3** very exactly. Compound **6** is particularly interesting because of hydrogen bridging, which stabilizes an aggregate of gallate anion, tetramethylpiperidinium cation and an additional phenol molecule. In this context it is interesting to note, that the hydrogen linked ion pair gallate/ammonium accounts for the molecular ion peak in the mass spectrum of **6**.

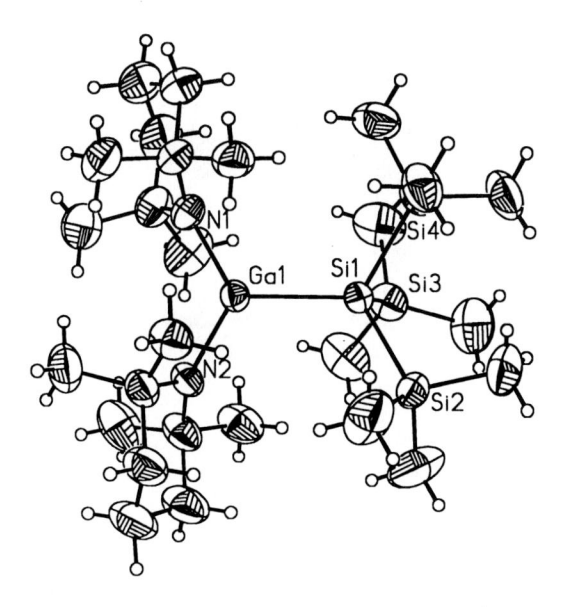

Fig. 1. Molecular structure of **3**; thermal ellipsoids represent 30 % probability; selected bond lengths [pm]: Ga1–N1 190.8 (3), Ga1–N2 191.3 (2), Ga1–Si1 246.8 (1), Si1–Si2 236.9 (1), Si1–Si3 236.8 (1), Si1–Si4 237.0 (1); selected bond angles: N1–Ga1–Si1 118.9 (1) °, N2–Ga1–Si1 119.2 (1) °, N1–Ga1–N2 121.8 (1) °, Ga–Si1–Sin 114 ° (av.), Sin–Si1–Sim 105 ° (av.)

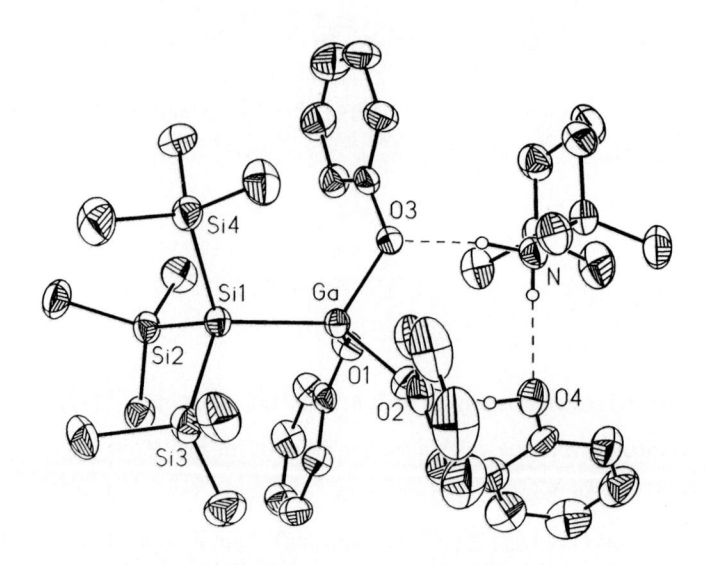

Fig. 2. Molecular structure of **6**; thermal ellipsoids represent 30 % probability; selected bond lengths [pm]: Ga–Si1 241.4 (3), Ga–O1 185.6 (6), Ga–O2 188.4 (7), Ga–O3 189.8 (6), Si1–Si2 235.9 (4), Si1–Si3 236.0 (4), Si1–Si4 234.9 (4); selected bond angles: O1–Ga–Si1 116.1 (2) °, O2–Ga–Si1 121.4 (2) °, O3–Ga–Si1 119.3 (2) °, On–Ga–Om 99 ° (av.), Ga–Si1–Sin 112 ° (av.), Sin–Si1–Sim 107 ° (av.)

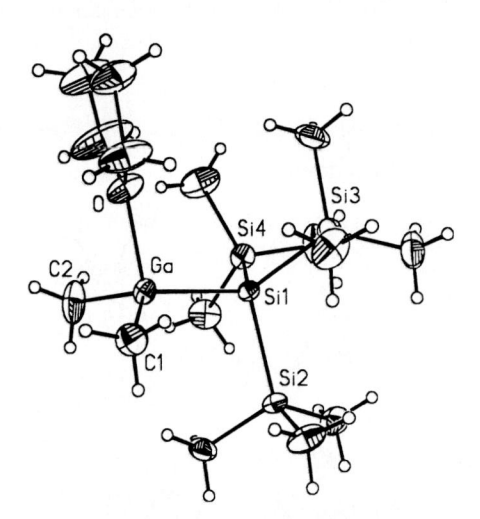

Fig. 3. Molecular structure of **7**; thermal ellipsoids represent 30 % probability; selected bond lengths [pm]: Ga–Si1 241.1 (3), Ga–O 210.3 (7), Ga–C1 199.9 (9), Ga–C2 195.4 (13), Si1–Si2 234.4 (4), Si1–Si3 232.4 (4), Si1–Si4 234.7 (4); selected bond angles: Sim–Si1–Sin 109 ° (av.), Si1–Ga–C1 117.4 (5) °, Si1–Ga–C2 117.7 (4) °, C1–Ga–C2 114.3 (7) °, Si1–Ga–O 106.0 (2) °, C1–Ga–O 98.1 (4) °, C2–Ga–O 98.6 (5) °, Ga–Sin–Sim 102.6 – 114.2 °

Conclusions

The hypersilyl group allows isolation of stable gallium silyl compounds with tri- and tetracoordinated gallium centres. The GaSi bond seems to be inert to protic reagents.

Acknowledgement: We thank Prof. Dr. H. Nöth and the *Deutsche Forschungsgemeinschaft* for supporting this work.

References:

[1] L. Rösch, H. Neumann, *Angew. Chem.* **1980**, *92*, 62; *Angew. Chem., Int. Ed. Engl.* **1980**, *19*, 55.

[2] A. M. Arif, A. H. Cowley, T. M. Elkins, R. A. Jones, *J. Chem. Soc., Chem. Commun.* **1986**, 1776.

[3] E. Amberger, W. Stoeger, J. Hönigschmid, *J. Organomet. Chem.* **1968**, *18*, 77.

[4] A. T. Weibel, J. P. Oliver, *J. Am. Chem. Soc.* **1972**, *94*, 8590; A. T. Weibel, J. P. Oliver, *J. Organomet. Chem.* **1974**, *74*, 155.

[5] R. Frey, G. Linti, *Chem. Ber.* **1994**, *101*, 101.

[6] A. Heine, R. Herbst-Irmer, G. M. Sheldrick, D. Stalke, *Inorg. Chem.* **1993**, *32*, 2694.

New Organometallic Silicon-Chalcogen Compounds

K. Merzweiler*, U. Linder

Institut für Anorganische Chemie
Martin-Luther-Universität Halle-Wittenberg
Weinbergweg 16, D-06120 Halle (Saale), Germany

The reaction of organometallic silicon halides $R_{4-x}SiCl_x$ (R = {Cp(CO)$_2$Fe}, x = 1–3) with sodium chalcogenides Na_2E (E = S, Se, Te) is a useful method for the synthesis of linear, cyclic, and polycyclic silicon-chalcogen compounds. Model experiments showed that [{Cp(CO)Fe}$_2$HSiCl] reacts with Na_2Se (THF as solvent) to form [{Cp(CO)$_2$Fe}$_4$H$_2$Si$_2$Se], **1** (Eq. 1).

$$2\,[\{Cp(CO)_2Fe\}_2HSiCl] + Na_2Se \longrightarrow [\{Cp(CO)_2Fe\}_4H_2Si_2Se] + 2\,NaCl$$

Eq. 1.

From the X-ray structure determination, **1** is found to contain a bent Si–Se–Si unit (Si–Se: 233.2(3)–234.2(3) pm) with an angle of 110.4(1) ° at the selenium atom (Fig. 1).

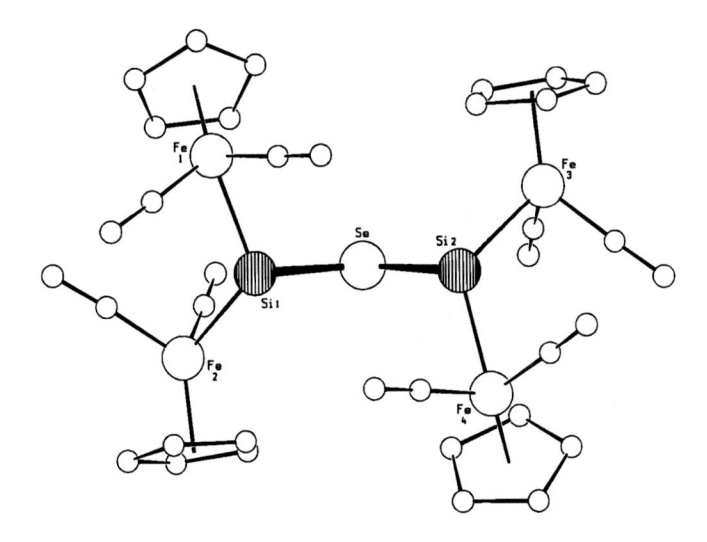

Fig. 1. Molecular structure of **1** in the crystal (hydrogen atoms not shown)

The silicon atoms are coordinated nearly tetrahedrally by a selenium atom, a hydrogen atom and two iron atoms.

Due to the reactivity of the Si–H bonds, **1** can be chlorinated by CCl_4 to form [{Cp(CO)$_2$Fe}$_4$Cl$_2$Si$_2$Se], **2**. The structure of **2** is very similar to that of **1**, but the Si–Se–Si angle is increased to 126.0(1) ° as a consequence of the larger steric demand of the chlorine atoms. In the presence of an excess of CCl_4, **2** decomposed to [{Cp(CO)$_2$Fe}$_2$SiCl$_2$], **3**, and elementary selenium.

$$[\{Cp(CO)_2Fe\}_4H_2Si_2Se] \xrightarrow{CCl_4} \underset{\textbf{2}}{[\{Cp(CO)_2Fe\}_4Cl_2Si_2Se]} \xrightarrow{CCl_4} \underset{\textbf{3}}{[\{Cp(CO)_2Fe\}_2SiCl_2]}$$

Eq. 2.

By analogy to the formation of **1**, heterocyclic silicon chalcogen derivatives can be synthesised by the reaction of bifunctional silicon halogen compounds RR'SiCl$_2$ with Na$_2$E (Eq. 3).

$$2\,[\{Cp'(CO)_2Fe\}MeSiCl_2] + 2\,Na_2E \longrightarrow [\{Cp'(CO)_2Fe\}_2Me_2Si_2E_2] + 4\,NaCl$$

Eq. 3.   E = Se (**4**), S (**5**)

Fig. 2 shows the result of the X-ray structure determination of **4**.

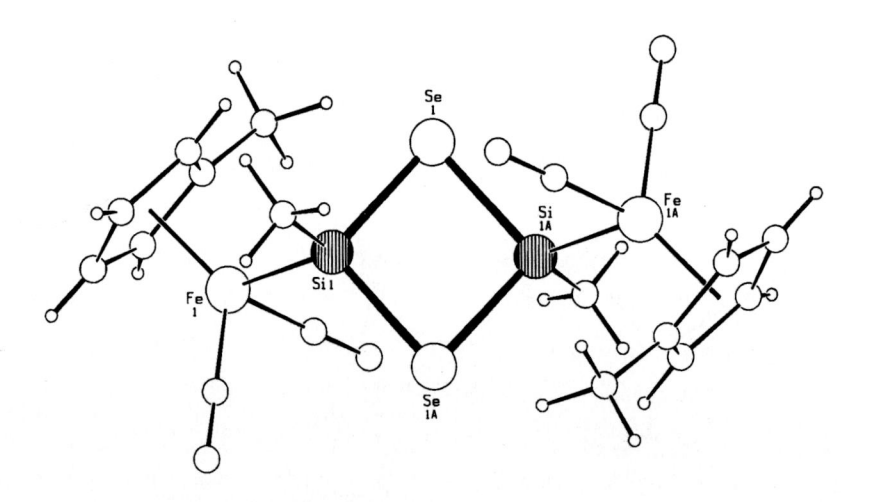

Fig. 2. The molecular structure of **4** in the crystal

It contains an exactly planar Si_2Se_2 ring lying on a crystallographic inversion center. The silicon atoms are coordinated roughly tetrahedrally by two selenium atoms, a methyl group and a {Cp'(CO)$_2$Fe} unit. The {Cp'(CO)$_2$Fe} groups are arranged trans with respect to the Si_2Se_2 plane. Compound 5 has an analogous structure but the Si–S bond lengths (217.0(1)–217.6(1) pm) are approx. 14 pm shorter than the Si–Se bond lengths in 4. This is a result of the different covalent radii of S (102 pm) and Se (117 pm) [1]. As in tin chalcogen rings, e. g., [{Cp(CO)$_2$Fe}$_4$Sn$_2$Se$_2$], [{(CO)$_4$Co}$_4$Sn$_2$E$_2$] (E = S, Se) [2], the Si–E–Si angles (4: 82.9(1) °, 5: 84.2(1) °) are smaller than the E–Si(Sn)–E angles (4: 95.8(1) °, 5: 97.1(1) °).

The reaction of the trifunctional derivatives RSiCl$_3$ with Na$_2$E in different molar ratios leads to a variety of silicon-chalcogen compounds (Scheme 1).

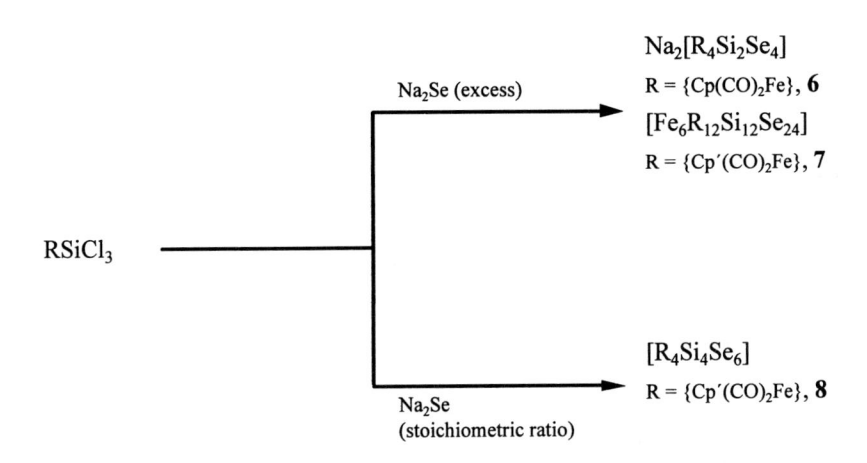

RSiCl$_3$

Na$_2$Se (excess) → Na$_2$[R$_4$Si$_2$Se$_4$]
R = {Cp(CO)$_2$Fe}, **6**
[Fe$_6$R$_{12}$Si$_{12}$Se$_{24}$]
R = {Cp'(CO)$_2$Fe}, **7**

Na$_2$Se (stoichiometric ratio) → [R$_4$Si$_4$Se$_6$]
R = {Cp'(CO)$_2$Fe}, **8**

Scheme 1.

Treatment of [{Cp(CO)$_2$Fe}SiCl$_3$] with excess Na$_2$Se yields Na$_2$[{Cp(CO)$_2$Fe}$_2$Si$_2$Se$_4$] **6**. The latter forms pale yellow crystals which are extremely air-sensitive. Compound **6** is ionic in character consisting of sodium cations and [{Cp(CO)$_2$Fe}$_2$Si$_2$Se$_4$]$^{2-}$ anions. The [{Cp(CO)$_2$Fe}$_2$Si$_2$Se$_4$]$^{2-}$ anions contain a four-membered Si_2Se_2 ring.

The silicon atoms are coordinated nearly tetrahedrally by a {Cp(CO)$_2$Fe} group, two bridging selenium atoms and a terminal selenium atom. Whereas the Si–Se bond lengths to the bridging Se atoms, Se1 and Se2 (234.6(2)–239.1(2) pm), are comparable to those of **1**, **2**, and **5**, the Si–Se distances to the terminal Se atoms, Se3 and Se4 (226.7(2)–228.0(2) pm), are significantly shorter. In contrast to **5**, the Si_2Se_2 ring is markedly puckered (32.3 ° along the Se–Se vector) and the {Cp(CO)$_2$Fe} groups are arranged *cis* with respect to the ring. In the solid state, two Na$^+$ cations complete the Si_2Se_4 framework to give a distorted heterocubane. The Na$_2$Si$_2$Se$_4$ heterocubanes are linked to dimers by additional Na–Se interactions (Fig. 3).

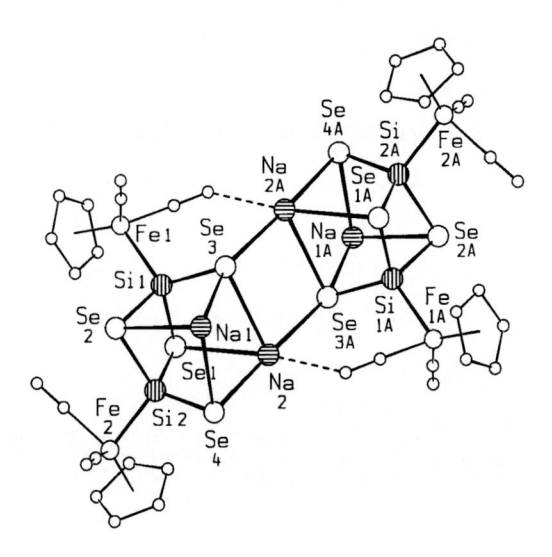

Fig. 3. The structure of the linked $Na_2[\{Cp(CO)_2Fe\}_2Si_2Se_4]$ units (the THF molecules coordinated to the Na^+ cations are not shown)

Both sodium atoms have a coordination number of six. Na2 is bonded to four selenium atoms, an oxygen atom of a carbonyl group and a THF molecule. Unlike Na2, Na1 is coordinated by three selenium atoms and three THF molecules. The THF molecules bonded to Na1 have a bridging function. This leads to a ribbon like arrangement of the heterocubane dimers along the crystallographic c axis (Fig. 4).

Besides the formation of the $\{Cp'(CO)_2Fe\}$ substituted analogue of **6**, the reaction of $[\{Cp'(CO)_2Fe\}SiCl_3]$ with an excess of Na_2Se leads to $[Fe_6\{Cp'(CO)_2Fe\}_{12}Si_{12}Se_{24}]$ **7** in low yield. Compound **7** forms dark brown crystals which are moderately air-stable. The X-ray structure analysis showed that **7** consists of six $[\{Cp'(CO)_2Fe\}_2Si_2Se_4Fe_2]$ heterocubane units which are linked by corner-sharing to a macrocycle (Fig. 5).

In analogy to **6** the heterocubane units consist of a Si_2Se_4 fragment and two iron atoms. Formally, **7** can be derived from **6** by the substitution of the sodium atoms by half the number of Fe^{2+} cations. As a consequence of the linking of the heterocubane units, the iron atoms in the corners of the $Si_2Se_4Fe_2$ units are coordinated by six selenium atoms. If only the central Fe_6Se_{24} framework is viewed, the structure of **7** can be derived from six edge shared $FeSe_6$ octahedra.

If the reaction of $[\{Cp'(CO)_2Fe\}SiCl_3]$ and Na_2Se is carried out with a slight excess of the chlorosilane, $[\{Cp'(CO)_2Fe\}_4Si_4Se_6]$ **8** is obtained in high yield. It is known from the corresponding tin sesquichalcogenides that Sn_4E_6 frameworks usually have an adamantane like structure.

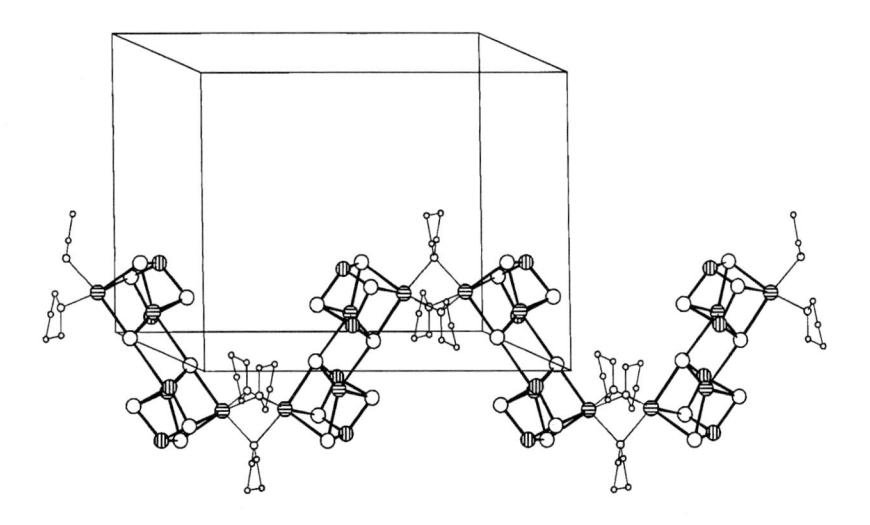

Fig. 4. Part of the crystal structure of **6**

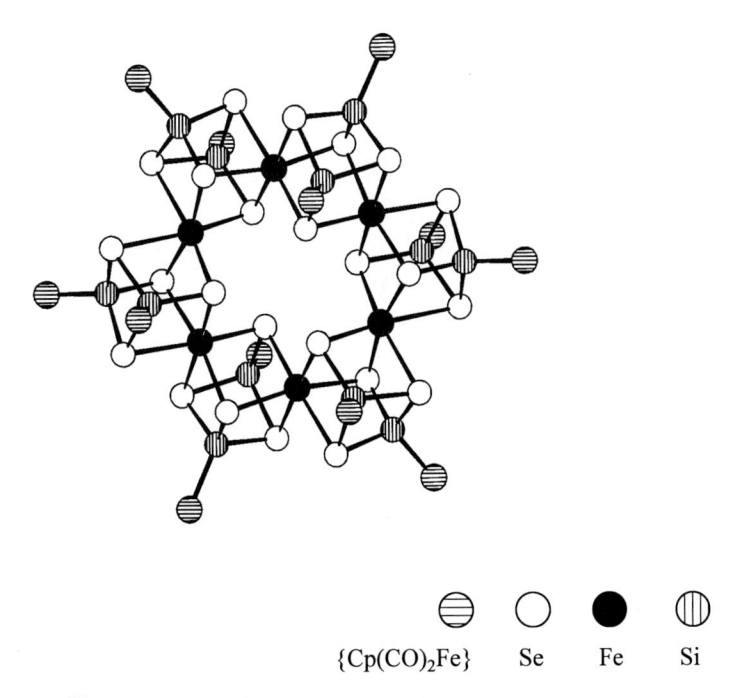

{Cp(CO)₂Fe} Se Fe Si

Fig. 5. The molecular structure of **7**

In contrast to this, the Si_4Se_6 core of **8** (Fig. 6) consists of two planar Si_2Se_2 rings which are linked by two bridging selenium atoms (Se3, Se3a).

This type of structure has already been observed in $tBu_4Ge_4S_6$ [3] and in $(iPrN)_6P_4$ [4] and is frequently called a "double decker" or "Candiani"-structure [5]. In **8**, the Si–Se distances vary in the expected range of 229.1(3)–230.6(3) pm. The Si–Se–Si angles can be divided into two groups. In the four-membered Si_2Se_2 rings the Si–Se–Si angles are 82.8(1) °. Similar Si–Se–Si angles have been observed in the monocyclic compounds **4** and **5**. In contrast to this the Si–Se–Si angle at Se3 is much larger (117.2(1) °).

This difference can be understood by taking into account the fact that Se3 is part of an eight-membered ring whereas Se1 and Se2 belong to a strained four-membered ring. The ^{77}Se NMR spectrum reflects the different bonding situation of the selenium atoms showing two singlet signals at 199.5 ppm (Se1, Se2) and 274.2 ppm (Se3) [6].

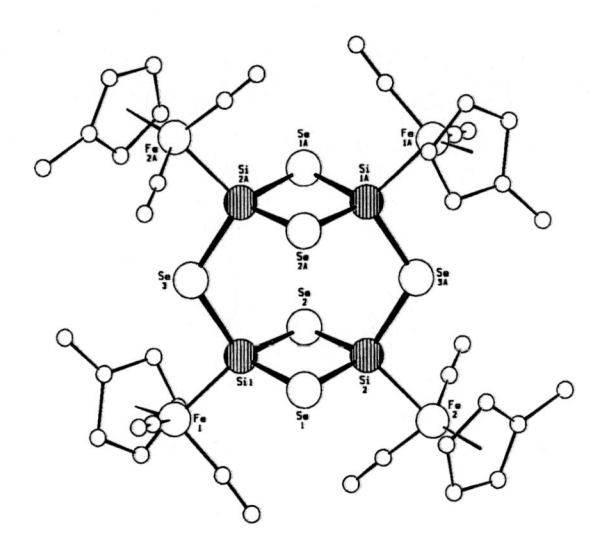

Fig. 6. The molecular structure of **8**

As in the case of the silicon-selenium compounds, different products are formed if $RSiCl_3$ is treated with Na_2S in stoichiometric amounts or in excess (Scheme 2). The reaction of $RSiCl_3$ (R = Cp'(CO)$_2$Fe}) with Na_2S in the molar ratio 2:3 (or a slight excess of $RSiCl_3$) leads to the formation of [{Cp'(CO)$_2$Fe}$_4Si_4S_6$] **9** in a high yield. Compound **9** forms yellow, moderately air-sensitive crystals. The X-ray structure determination revealed a Si_4S_6 core which corresponds to **8**. As a result of the smaller covalent radius of the sulfur atoms **9** exhibits silicon-chalcogen bond lengths that are approximately 14 pm shorter than in **8**.

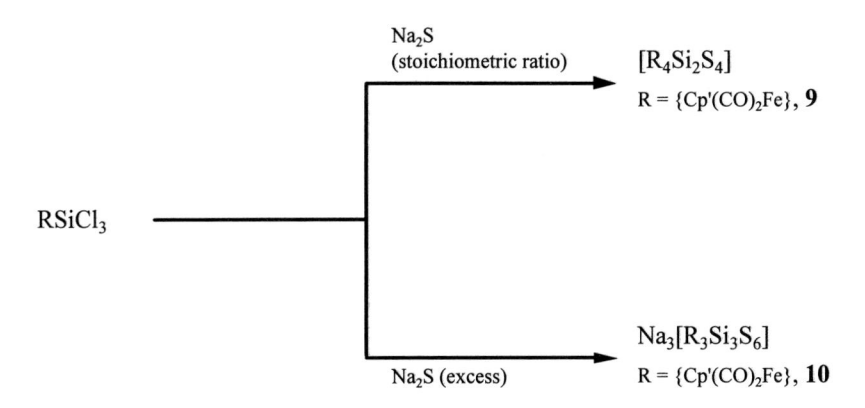

```
                    Na₂S
                    (stoichiometric ratio)    [R₄Si₂S₄]
                    ┌──────────────────────►  R = {Cp'(CO)₂Fe}, 9
                    │
RSiCl₃  ────────────┤
                    │
                    │                         Na₃[R₃Si₃S₆]
                    └──────────────────────►  R = {Cp'(CO)₂Fe}, 10
                    Na₂S (excess)
```

Scheme 2.

If the reaction between [{Cp'(CO)$_2$Fe}SiCl$_3$] and Na$_2$S is performed with an excess of Na$_2$S, **10** is obtained. Compound **10** consists of Na$^+$ cations and [{Cp'(CO)$_2$Fe}$_3$Si$_3$S$_6$]$^{3-}$ anions. The X-ray structure determination showed that the [{Cp'(CO)$_2$Fe}$_3$Si$_3$S$_6$]$^{3-}$ anions contain a six-membered Si$_3$S$_3$ heterocycle with an alternating arrangement of silicon and sulfur atoms (Fig. 7).

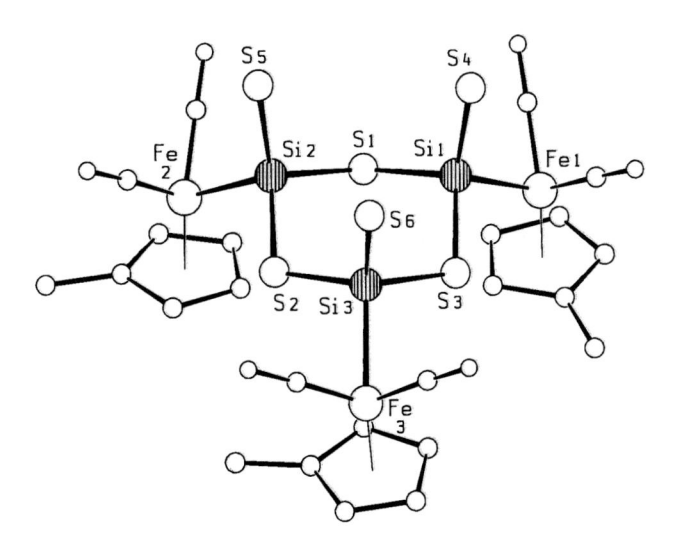

Fig. 7. The molecular structure of the [{Cp'CO)$_2$Fe}$_3$Si$_6$]$^{3-}$ anion

Each of the silicon atoms is coordinated almost tetrahedrally by three sulfur atoms and a {Cp'(CO)$_2$Fe} group. The Si$_3$S$_3$ ring adopts a chair conformation. Whereas the sterically demanding {Cp'(CO)$_2$Fe} groups occupy equatorial positions, the terminal sulfur atoms are arranged axially. As expected, the exocyclic Si–S bonds (207.3(3)–207.6(3) pm) are significantly shorter than the endocyclic Si–S distances (217.1(3)–219.6(3) pm). In the solid state two [{Cp'(CO)$_2$Fe}$_3$Si$_6$]$^{3-}$ anions form a sandwich-like structure (Fig. 8)

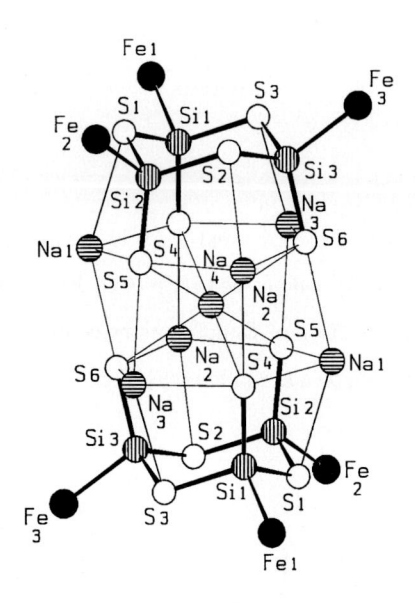

Fig. 8. The arrangement of Na$^+$ cations and [{Cp'(CO)$_2$Fe}$_3$Si$_3$S$_6$]$^{3-}$ anions of **10** in the solid state (the THF molecules coordinated to the Na$^+$ cations are not shown)

In this arrangement the terminal sulfur atoms of the two {Cp'(CO)$_2$Fe}$_3$Si$_3$S$_6$}$^{3-}$ anions define a distorted octahedron which is centered on a Na$^+$ cation (Na4). Each of the remaining Na$^+$ cations (Na1–Na3) is coordinated by three exocyclic sulfur atoms, an endocyclic sulfur atom and a THF molecule. As a consequence of crystallographic disorder, the positions of Na2 show half occupancy.

Acknowledgement: The authors thank the *Deutsche Forschungsgemeinschaft* and the *Fonds der Chemischen Industrie* for the generous financial support of this work.

References:

[1] J. E. Huheey, *Anorganische Chemie*, De Gruyter, Berlin, **1988**, p. 155.

[2] K. Merzweiler, L. Weisse, *Z. Naturforsch.* **1990**, *45B*, 971; K. Merzweiler, L. Weisse, H. Kraus, *Z. Naturforsch.* **1993**, *48B*, 287.

[3] W. Ando, T. Kadowaki, Y. Kabe, M. Ishii, *Angew. Chem.* **1992**, *104*, 84; *Angew. Chem., Int. Ed. Engl.* **1992**, *31*, 84.

[4] O. J. Scherer, K. Anders, C. Krüger, Y.-H. Tsay, G. Wolmershäuser, *Angew. Chem.* **1980**, *92*, 563; *Angew. Chem., Int. Ed. Engl.* **1980**, *19*, 571.

[5] P. Candiani, *Gazz. Chim. Ital.* **1985**, *25*, 81.

[6] The ^{77}Se NMR shifts are given relative to Me$_2$Se.

Alternative Ligands XXXIII[1]:
Heterobimetallic Donor-Acceptor Interactions in Si/Ni-Cages: Metallosilatranes

Joseph Grobe*, Hans-Hermann Niemeyer, Rudolf Wehmschulte

Anorganisch-Chemisches Institut

Westfälische Wilhelms-Universität

Wilhelm-Klemm-Str. 8, D-48149 Münster, Germany

Summary: Isolobal replacement of the NR_3-unit in silatranes **1** by a d^{10} ML_4-unit leads to metallosilatranes **2**. The cage compounds **2** show weak, attractive donor-acceptor interactions. As observed by X-ray diffraction there exist sub-van-der-Waals distances between silicon and nickel. The proposed 3-center 4-electron interaction is in good agreement with some unusual NMR data of **2**.

Silatranes **1** are meanwhile classical cage compounds with donor-acceptor interactions and represent examples of hypercoordinated silicon [2]. The donor-acceptor contact in **1** is formed by an interaction of the Lewis-basic amino group with the Lewis-acidic silicon center favored by the chelate effect. Numerous examples show that electron-rich transition metal complexes also possess Lewis-basic properties [3, 4]. Isolobal replacement of the NR_3-unit in **1** by a d^{10} ML_4-unit [5] leads to compounds of type **2** [1, 6, 7]. These Si/Ni-cages **2** can be regarded as metallosilatranes. Here we report on the synthesis, structure and bonding of **2**.

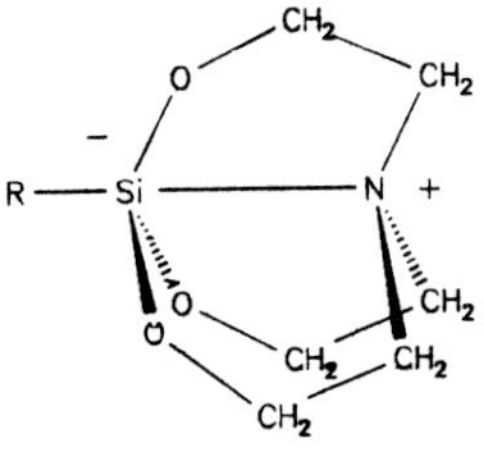

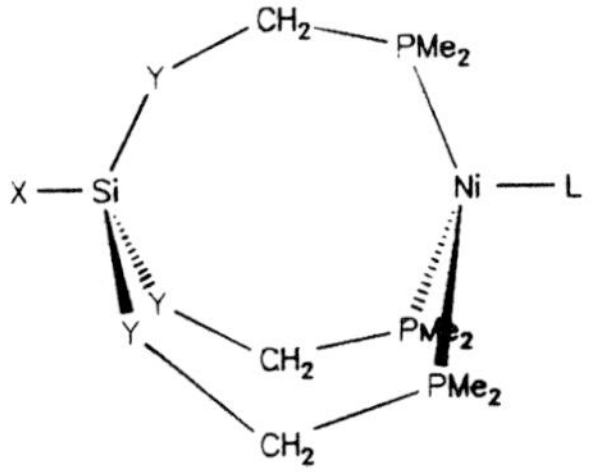

2a: X = CH_3, Y = O, L = PPh_3

2b: X = F, Y = CH_2, L = PPh_3

2c: X = F, Y = CH_2, L = CO

Fig. 1.

The triphenylphosphane derivatives **2a** and **2b** have been prepared by reaction of the corresponding tripod ligand **3** with $Ni(c-C_8H_{12})_2$ and triphenylphosphane (Eq. 1–2).

$$CH_3Si(OCH_2PMe_2)_3 + Ni(c-C_8H_{12})_2 + PPh_3 \longrightarrow CH_3Si(OCH_2PMe_2)_3Ni(PPh_3)$$
                3a   **2a**

Eq. 1.

$$FSi(CH_2CH_2PMe_2)_3 + Ni(c-C_8H_{12})_2 + PPh_3 \longrightarrow FSi(CH_2CH_2PMe_2)_3Ni(PPh_3)$$
                3b   **2b**

Eq. 2.

The related carbonyl derivative **2c** was obtained using the dilution principle for the reaction of $Ni(CO)_4$ with the tripod ligand **3b** under reflux in toluene (Eq. 3).

$$FSi(CH_2CH_2PMe_2)_3 + Ni(CO)_4 \longrightarrow FSi(CH_2CH_2PMe_2)_3Ni(CO)$$
                3b   **2c**

Eq. 3.

Metallosilatranes **2** show some unusual chemical shifts and coupling constants in the NMR spectra which find plausible explanations by assuming 3-center 4-electron interactions (see below). Compound **2b** will be discussed here as an example.

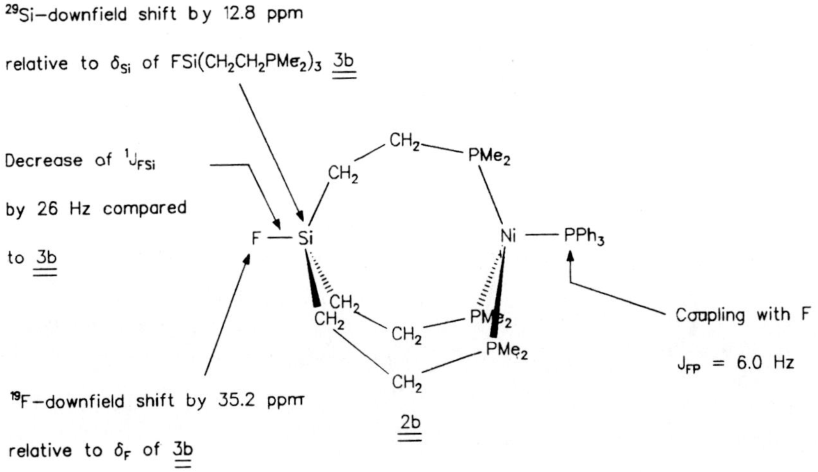

Fig. 2.

A strong coupling (J_{FP} = 6 Hz) is observed between fluorine at silicon and the phosphorus of the triphenylphosphane ligand of **2b**. Furthermore the ^{19}F signal (δ_F = –135.7 ppm) is shifted downfield by 35.2 ppm when compared with the free ligand **3b** (δ_F = –170.9 ppm). The ^{29}Si NMR signal of **2b** (δ_{Si} = –41.9 ppm) also is shifted to lower field by 12.8 ppm with respect to the resonance of **3b** (δ_{Si} = –29.1 ppm), and a considerable coordination effect is observed for the coupling constant $^1J_{FSi}$ resulting in a decrease from 292 Hz in **3b** to 266 Hz in **2b**.

In order to elucidate these interesting results X-ray diffraction studies were carried out for the derivatives **2a**, **2b**, and **2c**. Fig. 3 shows the molecular structure of **2c**.

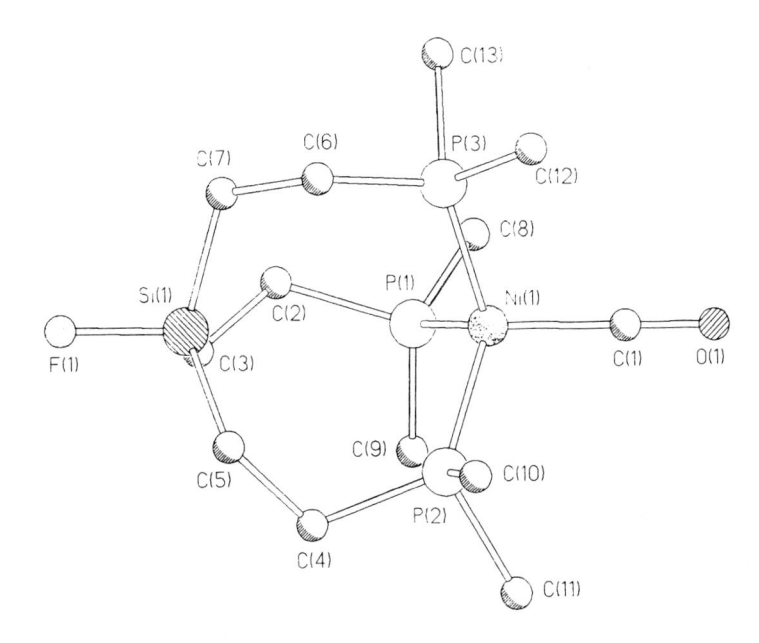

Fig. 3. Molecular structure of **2c**; a view perpendicular to C_3-axis is shown; the hydrogens are not presented; selected bonding distances [pm] and angles [°] (the numbers in parentheses are the standard deviations in units of the last cited decimal): F(1)–Si(1) 162.5(2), Si(1)–C(5) 187.4(4), C(5)–C(4) 152.9(6), C(4)–P(2) 183.2(4), P(2)–Ni(1) 211.17(10), Ni(1)–C(1) 175.3(4), C(1)–O(1) 114.8(4), P(2)–C(11) 183.1(5), F(1)Si(1)C(5) 102.6(2), Si(1)C(5)C(4) 120.2(3), C(5)C(4)P(2) 114.9(3), C(4)P(2)Ni(1) 121.96(13), P(2)Ni(1)C(1) 107.85(12), Ni(1)C(1)O(1) 178.5(3), C(11)P(2)Ni(1) 114.2(2), C(4)P(2)C(11) 99.2(2)

Strongest evidence for weak, but significant, Ni→Si interactions comes from the observed Si–Ni distances of 395 (**2a**), 391 (**2b**), and 389 pm (**2c**), which are smaller than the sum of the van der Waals radii of Si and Ni (about 410 pm) [8]. Further support is provided by a flattening of the cage and unusual bond angles within it.

The donor-acceptor contact in **2** can be discussed in terms of a 3-center 4-electron interaction. The HOMO of the $C_{3v}d^{10}$-ML$_4$ unit with symmetry a_1 acts as the donor orbital [5].

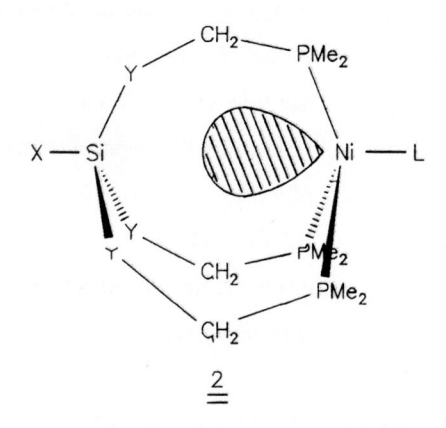

Fig. 4.

The combination of this donor orbital with the a_1 orbitals of the $XSiY_3$ units of **2** is symmetry-allowed and generates a 3-center 4-electron interaction analogously to that in classical silatranes **1** [9]. The formation of such a 3-center 4-electron-bond interaction is expected to cause a decrease of electron density at X, Si, and Ni and a weakening of the X-Si bond. Both expectations are in accord with the NMR results mentioned above.

Further synthetic and theoretical investigations to gain a better understanding of the donor-acceptor interactions in metallosilatranes are in progress.

Acknowledgement: The authors thank the *Deutsche Forschungsgemeinschaft* and the *Fonds der Chemischen Industrie* for financial support, M. Läge and B. Krebs for carrying out the X-ray diffraction analyses. R. W. is grateful for a Promotionsstipendium des *Landes Nordrhein-Westfalen*. H. H. N. thanks the *Deutsche Forschungsgemeinschaft* for a Graduiertenstipendium.

References:

[1] Communication XXXII: J. Grobe, R. Wehmschulte, B. Krebs, M. Läge, *Z. Anorg. Allg. Chem.*, in press.

[2] C. Chuit, R. J. P. Corriu, C. Reye, J. C. Young, *Chem. Rev.* **1993**, *14*, 1371.

[3] D. F. Shriver, *Acc. Chem. Res.* **1970**, *3*, 231.

[4] H. Werner, *Angew. Chem.* **1983**, *95*, 932; *Angew. Chem., Int Ed. Engl.* **1983**, *22*, 927.

[5] M. Elian, R. Hoffmann, *Inorg. Chem.* **1975**, *14*, 1058.

[6] J. Grobe, R. Wehmschulte, *Z. Anorg. Allg. Chem.* **1993**, *619*, 563.

[7] J. Grobe, N. Krummen, R. Wehmschulte, B. Krebs, M. Läge, *Z. Anorg. Allg. Chem.* **1994**, *620*, 1645.

[8] L. Pauling, *Die Natur der chemischen Bindung*, Verlag Chemie, Weinheim, **1976**.

[9] M. S. Gordon, M. T. Caroll, J. H. Jensen, L. P. Davis, L. W. Burggraf, R. M. Guidry, *Organometallics* **1993**, *10*, 2657.

Reactivity of π-Coordinated Chlorocyclopentadienylsilanes

S. Schubert, J. Hofmann, K. Sünkel*

Institut für Anorganische Chemie

Ludwig-Maximilians-Universität München

Meiserstr. 1, D-80333 München, Germany

Summary: In the ring-silylated cymantrenes $[C_5Y_{4-n}(SiMe_2H)_{1+n}]Mn(CO)_3$ with $n = 0-4$, $Y = Br$; $n = 0$, $Y = Cl$ the Si–H functionalities can be transformed into Si–Cl or Si–F groups by means of $PdCl_2$ and Ph_3CSbF_6, respectively. The reaction of the Si–Cl groups with alcohols ROH ($R = C_mH_{2m+1}$, $m = 1-10$) yields the corresponding alkoxysilanes ($n = 0, 1$) or bi- or tricyclic disiloxanes ($n > 1$).

Recently, we reported the synthesis of a series of cymantrene derivatives with one to five $SiMe_2H$ substituents at the cyclopentadienyl ring [1]. We could also show that $[C_5(SiMe_2H)_mH_{5-m}]Mn(CO)_3$ ($m = 4, 5$) can be chlorinated by means of $PdCl_2$ under careful exclusion of moisture and then be transformed into $SiMe_3$-derivatives by treatment with MeMgCl [2].

When $[C_5Br_{5-m}(SiMe_2H)_m]Mn(CO)_3$ ($m = 1$: **1a**; $m = 2$: **2**) or $[C_5Cl_4(SiMe_2H)]Mn(CO)_3$ (**1b**) are treated with $PdCl_2$, the corresponding chlorosilanes **3a**, **3b**, and **4** can be obtained (Fig. 1):

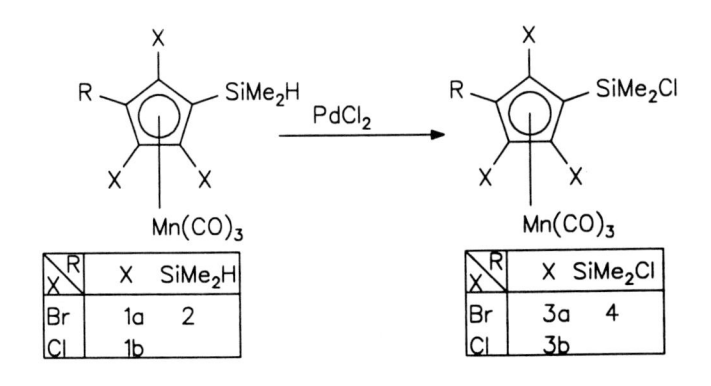

Fig. 1.

These chlorosilanes can be hydrolyzed by addition of water to give the corresponding hydroxosilanes **5a, 6a,** and **7a,** respectively. Under these conditions, no intermolecular condensation reactions are observed for **3a** and **4**. When alcohols $C_nH_{2n+1}OH$ ($n = 1–10$) are used instead of water, alkoxy-silanes **5b–5m, 6b–6m,** and **7b–7m** can be prepared.

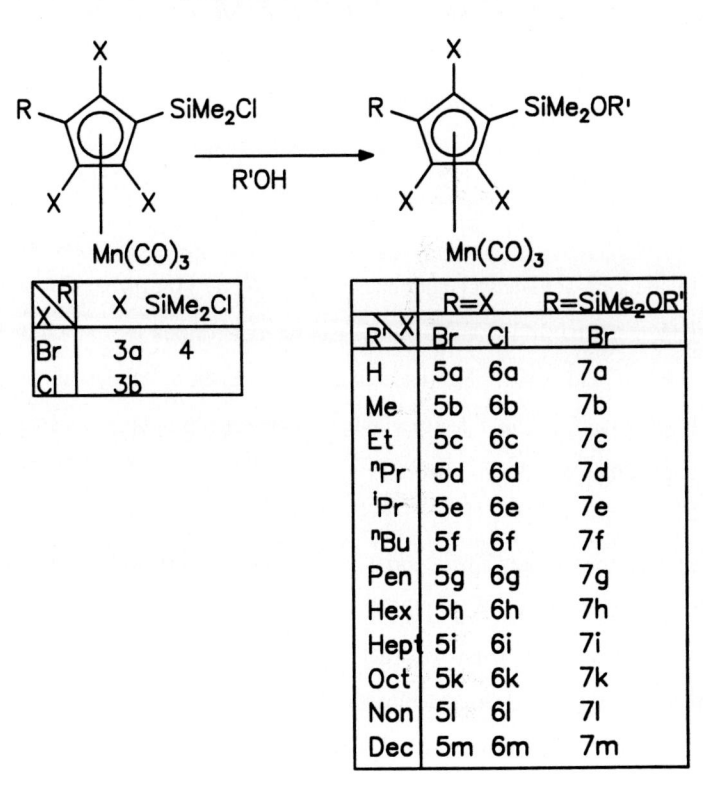

Fig. 2.

R\X	X SiMe$_2$Cl
Br	3a 4
Cl	3b

R\X	R=X		R=SiMe$_2$OR'
	Br	Cl	Br
H	5a	6a	7a
Me	5b	6b	7b
Et	5c	6c	7c
^{n}Pr	5d	6d	7d
^{i}Pr	5e	6e	7e
^{n}Bu	5f	6f	7f
Pen	5g	6g	7g
Hex	5h	6h	7h
Hept	5i		7i
Oct	5k	6k	7k
Non	5l	6l	7l
Dec	5m	6m	7m

All compounds **3–7** have been characterized by IR, ^{1}H, and ^{13}C NMR spectroscopy. They are obtained as yellow solids (**3, 4, 5a–5g, 6b–6f, 7a–7c, 7e**) or oils.

The trisilane [C$_5$Br$_2$(SiMe$_2$H)$_3$]Mn(CO)$_3$ (**8a**) also reacts with PdCl$_2$ to give [C$_5$Br$_2$(SiMe$_2$Cl)$_3$]Mn(CO)$_3$ (**9a**). This compound, however, is extremely moisture-sensitive, due to the fact that two SiMe$_2$Cl groups are next to each other. Consequently, the first product of hydrolysis is the bicyclic disiloxane **9b**, which is slowly converted to the hydroxysilyl derivative **9c** by further contact with moisture. **9b** and **9c** crystallize together, as could be shown by X-ray crystallography (Fig. 4).

Fig. 3.

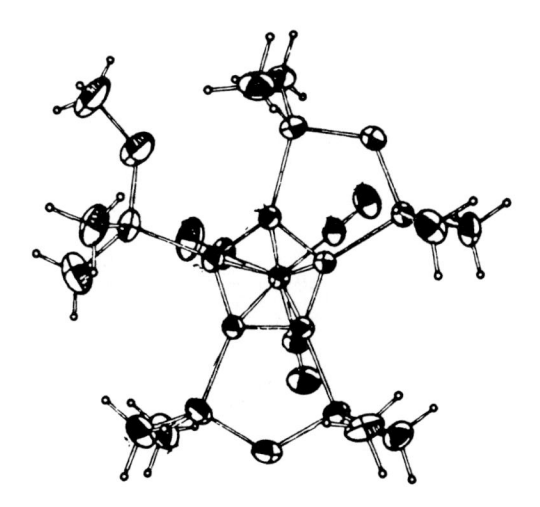

Fig. 4. Molecular structure of **9b**

When moisture is excluded thoroughly, alcoholysis of **9a** yields the tris-alkoxysilanes **10**, while with traces of moisture in the alcohol, alkoxysilyl-disiloxanes of type **9d** are formed.

After treatment of [C$_5$(SiMe$_2$H)$_5$]Mn(CO)$_3$ with excess PdCl$_2$, followed by addition of MeOH, crystals of a compound **11** can be obtained. An X-ray crystal structure determination showed this compound to be a tricyclic bis-disiloxane with one additional methoxysilyl substituent (Fig. 5).

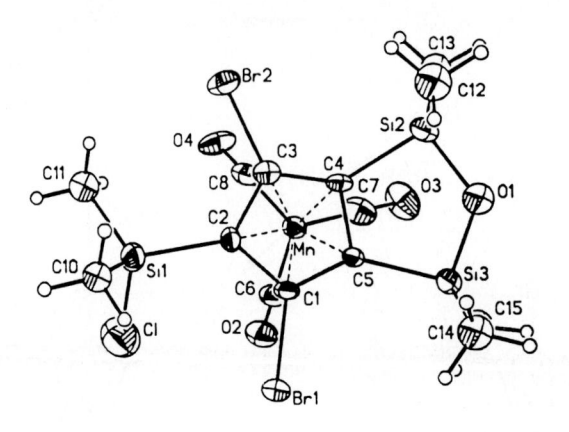

Fig. 5. Molecular structure of **11**

The hydride in silanes ArSiR$_2$H can be abstracted by the trityl cation, leaving principally a silicenium ion. This highly electrophilic species interacts with the anion associated with the trityl cation either by coordination (e.g., ClO$_4^-$) or by halide abstraction (e.g., BF$_4^-$) [3].

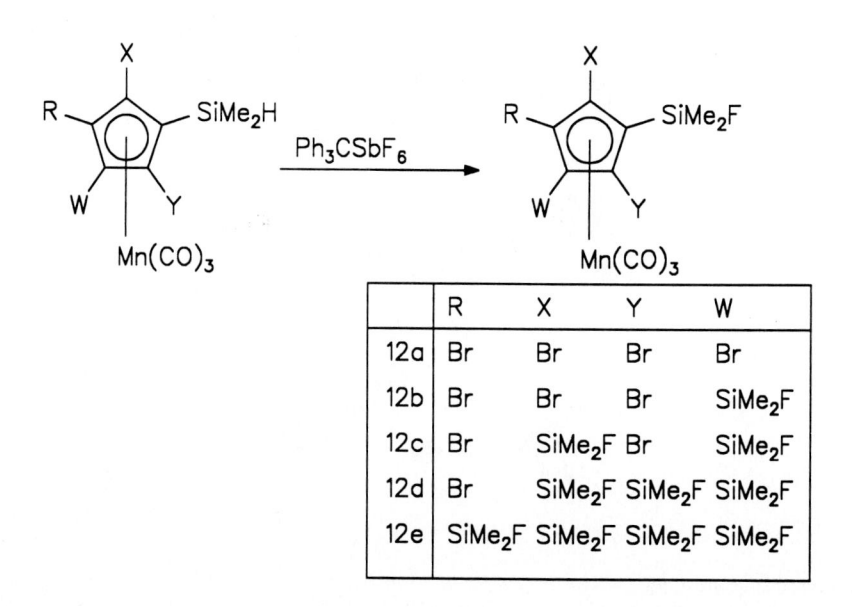

	R	X	Y	W
12a	Br	Br	Br	Br
12b	Br	Br	Br	SiMe$_2$F
12c	Br	SiMe$_2$F	Br	SiMe$_2$F
12d	Br	SiMe$_2$F	SiMe$_2$F	SiMe$_2$F
12e	SiMe$_2$F	SiMe$_2$F	SiMe$_2$F	SiMe$_2$F

Fig. 6.

The use of the Ph$_3$CSbF$_6$ reagent with our silanes [C$_5$Br$_{5-m}$(SiMe$_2$H)$_m$]Mn(CO)$_3$ (m = 1–5), leads to immediate formation of the corresponding fluorosilanes **12a–e**, which were characterized by ^{1}H, ^{13}C, and ^{19}F NMR and IR spectroscopy. There was no indication of an intermediate coordination of the SbF$_6^-$ anion.

Conclusion

The cymantrenyl-dimethylsilanes comprise an easily to handle, but still reactive class of organometallic compounds, which undergo many of the typical reactions "normal" hydrosilanes ArSiR$_2$H are known for. For simple geometric reasons, in a cyclopentadienyl derivative the silicon substituents come less close to each other than in the corresponding benzene derivatives, thus avoiding some difficulties by steric hindrances known in the latter class of compounds.

References:

[1] K. Sünkel, J. Hofmann, *Organometallics* **1992**, *11*, 3923.

[2] K. Sünkel, J. Hofmann, *J. Coord. Chem.* **1993**, *30*, 261; K. Sünkel, J. Hofmann, *Chem. Ber.* **1993**, *126*, 1791.

[3] P. D. Lickiss, *J. Chem. Soc., Dalton Trans.*, **1992**, 1333; and references cited therein.

CO-, Isonitrile-, and Phosphine-induced Silyl Migration Reactions in Heterobimetallic Fe–Pd and Fe–Pt Complexes

Pierre Braunstein*

Laboratoire de Chimie de Coordination, Associé au CNRS (URA 0416)
Université Louis Pasteur
4 Rue Blaise Pascal, F-67070 Strasbourg Cedex, France

Michael Knorr*

Institut für Anorganische Chemie
Universität des Saarlandes
Postfach 151150, D-66041 Saarbrücken, Germany

Summary: By using the assembling ligands bis(diphenylphosphino)methane (dppm) or diphenylphosphido, stable heterobimetallic Fe–Pd and Fe–Pt complexes containing a trialkoxysilyl ligand could be prepared and their reactivity was investigated. In the dppm complexes, a new bridging situation was encountered for the $-Si(OMe)_3$ ligand which forms a strong bond with Fe and a more labile dative interaction with Pd or Pt through an oxygen lone pair. This O···Metal bond may be displaced by donor ligands such as CO or isonitriles and this property was exploited in the case of heterobimetallic alkyls for the study of migratory insertion reactions, leading to, for example, acyl derivatives. In the presence of phosphine ligands, migration of the silyl ligand to the acyl oxygen was observed, which led to bridging siloxycarbenes. When the $-Si(OMe)_3$ ligand was replaced with the siloxane $-Si(OSiMe_3)_3$, the corresponding siloxycarbenes were formed spontaneously upon reaction of the bimetallic alkyls with CO. In the case of complexes containing a $\mu-PPh_2$ ligand, the silyl ligand was only found in a terminal position but can migrate from Fe to Pt under the influence of added ligands, therefore providing the first examples of intramolecular silicon migration from one metal to another.

The migratory insertion of carbon monoxide into the metal-carbon bond of mononuclear complexes plays a fundamental role in organometallic chemistry and has been studied in detail owing to its importance for C–C bond formation in homogeneous catalysis [1–3]. However studies dealing with CO insertion reactions into the metal carbon-bond of homo- and heterobimetallics are rather scarce, perhaps owing to the limited number of alkyl/aryl bimetallic complexes available, and/or their tendency to fragment on attempts to insert small molecules [4–9]. In the course of our studies on heterometallics containing a Fe–Si(OMe)$_3$ moiety we prepared a series of acyl complexes of the type [(OC)$_3$Fe{μ-Si(OMe)$_2$(OMe)}(μ-dppm)MC(O)R] (M = Pd, Pt; R = Me, Et, norbornyl) by carbonylation of the corresponding alkyl precursors [(OC)$_3$Fe{μ-Si(OMe)$_2$(OMe)}(μ-dppm)MR] and investigated their reactivity [10–12].

The stable alkyl complexes [(OC)$_3$Fe{μ-Si(OR')$_2$(OR')}(μ-dppm)MMe] (**1** M = Pd, **2** M = Pt) are easily prepared by reaction of the metalate K[Fe{Si(OR')$_3$}(CO)$_3$(η^1-dppm)] with [M(Cl)(Me)(1,5-COD)] (Eq.1).

1a M = Pd, R' = OMe
1b M = Pd, R' = OSiMe$_3$
2 M = Pt, R' = OMe

Eq. 1.

Facile insertion reaction of CO into the M–methyl bond of **1a** and **2** leads to the stable acyl complexes [(OC)$_3$Fe{μ-Si(OMe)$_2$(OMe)}(μ-dppm)MC(O)Me)] **3** (M = Pd) and **4** (M = Pt), in which the acyl ligand occupies a position *trans* to the metal-metal bond. This can be rationalized by assuming that opening of the μ-Si–O bridge liberates a vacant coordination site which is occupied by CO to give the CO adduct [(OC)$_3${(MeO)$_3$Si}Fe(μ-dppm)Pt(CO)(CH$_3$)] (Eq. 2), which was spectroscopically characterized in the case of Pt. Subsequent *cis*-migration of the methyl group should result in a complex with the acyl ligand *trans* to phosphorus and we believe that the very rapid isomerization leading to **3** and **4** must be driven, at least in part, by the tendency to restore the entropically favored μ-Si–O bridge.

Eq. 2.

Eq. 3.

In order to confirm the thermodynamic stability of **3** and **4**, we attempted their synthesis by another route: the reaction of the metalate K[Fe{Si(OMe)$_3$}(CO)$_3$(η^1-dppm)] with trans-[MCl{C(O)R}(PPh$_3$)$_2$] (Eq. 3). In the case of platinum, the reaction was performed at 293 K in THF for 24 h. However, the stable μ-carbene complexes [(OC)$_3$Fe{μ-C(R)OSi(OMe)$_3$}(μ-dppm)Pt(PPh$_3$)] **5a** (R = Me) and **5b** (R = Et) were isolated in ca. 75 % yield instead of the expected heterobimetallic acyls.

The structure suggested for complexes **5** was deduced in solution by IR and multinuclear NMR spectroscopic methods and further confirmed for the solid by a X-ray diffraction study of **5b** [13]. The existence of a metal-metal bond [d(Fe–Pt) = 2.5062(9) Å] in **5b** confers to the Pt and Fe centers their usual 16 and 18 e$^-$ count. The most salient feature is the bridging carbene ligand, which is symmetrically bonded between the metals, with Fe–C(1) and Pt–C(1) distances of 2.066(7) and 2.074(7) Å, respectively. This was not expected on the basis of our NMR spectroscopic data which suggested a stronger Pt–C interaction [^2J(P(1)–C(1)) = 14 Hz and ^2J(P(2)–C(1)) = 84 Hz] and differs from the situation found in the related carbene complex [(OC)Pt(μ-dppm){μ-C(OMe)(C$_6$H$_4$Me–4)}W(CO)$_4$] where a significant asymmetry was noted [14]. The overall geometry of complexes **5** is therefore similar to that of the μ-aminocarbyne complexes [(CO)$_3$Fe(μ-CNRR′)(μ-dppm)Pt(PPh$_3$)][BF$_4$] [15].

When a reaction similar to that of Eq. 3 was performed with the palladium acyl precursor (–10 to 25 °C, 3 h), no pure complex could be isolated but the palladium analogue of **5a** was observed spectroscopically and identified by comparison with an authentic sample prepared quantitatively according to Eq. 4. We felt that a reason for the formation of complexes **5** instead of **4** in reaction 3 could be the presence of PPh$_3$. Indeed, the direct reaction of **4** with one equivalent of PR$_3$ in CH$_2$Cl$_2$ at ambient temperature also yielded **5** (Eq. 4).

$$\mathbf{4} \xrightarrow[\text{CH}_2\text{Cl}_2]{L} (OC)_3Fe \begin{array}{c} P\frown P \\ \downarrow \quad \downarrow \\ \rule{1em}{0.4pt} Pt \leftarrow L \\ \diagdown \; C \diagup \\ (MeO)_3Si{-}O \quad Me \end{array}$$

5c L = PEt$_3$
5d L = P(p-tolyl)$_3$
5e L = P(OMe)$_3$

Eq. 4.

The derivatives **5c–5e** were obtained in nearly quantitative yields. The basicity of the PR$_3$ ligand seems to play a more important role than its cone angle on the rate of this transformation. IR monitoring of the reaction revealed completion with PEt$_3$ within ca. 20 min (disappearance of the acyl stretch at 1635 cm^{-1}), whereas in the case of P(OMe)$_3$, even in slight excess, more than 1 h was needed. With AsPh$_3$ (in slight excess) no reaction was observed even after 3 h. In an analogous manner, the more labile palladium derivatives **6a–6c** were isolated in high yields from the reactions of **3** with L.

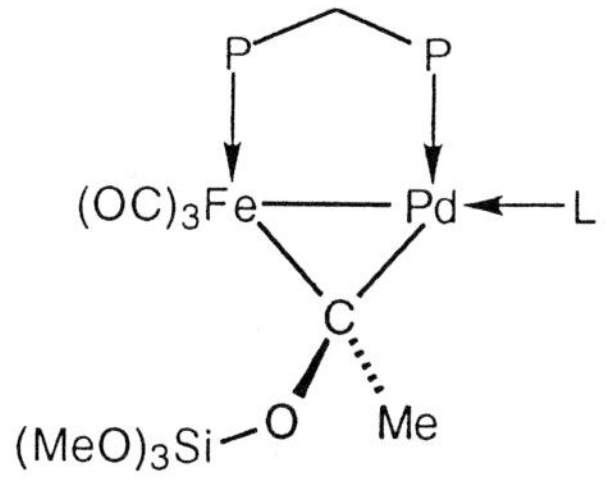

6a L = PEt$_3$
6b L = PPh$_3$
6c L = P(OMe)$_3$

Fig. 1.

These findings suggest that initial formation of a bimetallic acyl complex also occurs during the synthesis of **5** and **6**, which is followed by silyl migration from Fe to O (acyl). We have, however, no direct spectroscopic evidence for this acyl intermediate. Evidence that the C–O unit found in the carbene ligand stems from the acyl group was provided by a ^{13}C labelling experiment. Complex **4** was labelled on the acyl group (by purging a solution of [(OC)$_3$Fe{μ-Si(OMe)$_2$(OMe)}(μ-dppm)PtMe] **2** for 4 min. with 99 % enriched ^{13}CO) and reacted with tri(p-tolyl)phosphine. The labelled carbon appears in the ^{13}C NMR spectrum at δ 184.2 as a doublet of doublets with ^2J(P–C) = 13 and 85 Hz and shows a strong coupling of 860 Hz with the ^{195}Pt nucleus.

There are two possible roles for the phosphine ligand:

1) It renders the acyl oxygen more electron-rich, thus favouring Si-migration.
2) It may stabilize the acyl ligand in a *cis* position with respect to the metal-metal bond long enough to allow Si-migration to occur.

Conversely, the lack of Si-migration during the synthesis of **3** and **4** according to Eq. 2 would be due to the preferred μ-Si–O bridge formation. A related silicon-oxygen coupling has been reported in the mononuclear acyl, silyl complex *cis*-[(OC)$_4$Fe{C(O)Me}(SiMe$_3$)]. This labile compound rearranged rapidly via a 1,3-silatropic shift to afford the siloxycarbene complex [(OC)$_4$Fe=C(Me)(OSiMe$_3$)] [16].

Obviously the formation of a strong silicon-oxygen bond is also assumed to be the driving force in these rearrangements. In the presence of isonitriles, no silyl shift could be induced in complexes **3** and **4**.

Instead ring opening of the μ-Si–O bridge occurred and the stable isonitrile complexes **7** were formed in quantitative yield.

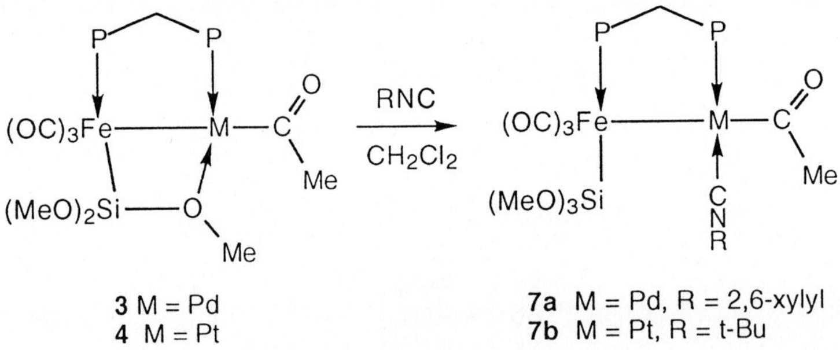

Eq. 5.

3 M = Pd
4 M = Pt

7a M = Pd, R = 2,6-xylyl
7b M = Pt, R = t-Bu

The strongly bound isonitrile ligand occupies a position *cis* to the metal-metal bond, thus preventing the isomerization of the acyl ligand. Since we have verified that PPh$_3$ does not replace the O-donor ligand in **1a**, we suggest that the phosphine-induced silyl migration is not initiated by the ring opening of the μ-Si–O bridge of **4** to give an intermediate with a square planar geometry around Pt. Rather, formation of a five-coordinated Pd or Pt intermediate would not only make the acyl oxygen more electron-rich and therefore more nucleophilic towards the silicon centre but also facilitate the isomerization of the acyl group required to bring it in closer proximity to the silicon atom.

Upon addition of MeO(O)C≡C(O)OMe (DMAD) to a solution of **4** a third type of reactivity was observed. In contrast to the situation with the isonitriles described above, the acetylene readily inserted into the Pt–acyl bond to afford the alkenyl complex **8** [17].

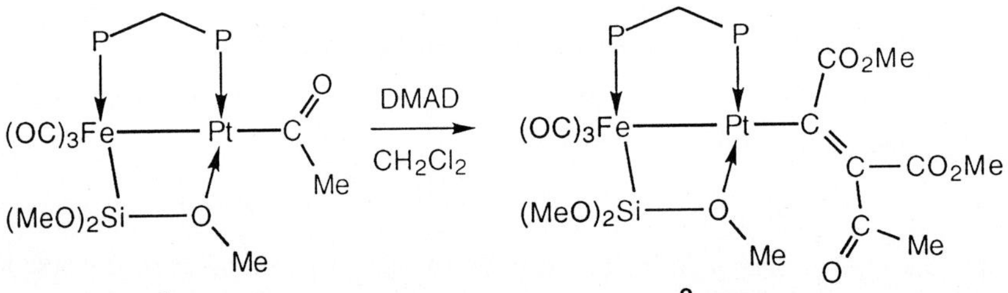

Eq. 6.

With the Fe–Pd complex **1b** which contains a less electron donating –Si(OSiMe$_3$)$_3$ siloxane ligand, no acyl formation was observed but instead the µ-carbene complex **9** was formed quantitatively upon bubbling of CO for 5 min. through a solution of **1b** (Eq. 7).

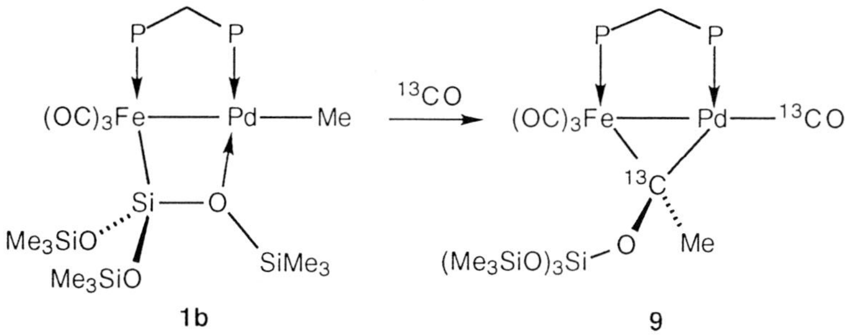

Eq. 7.

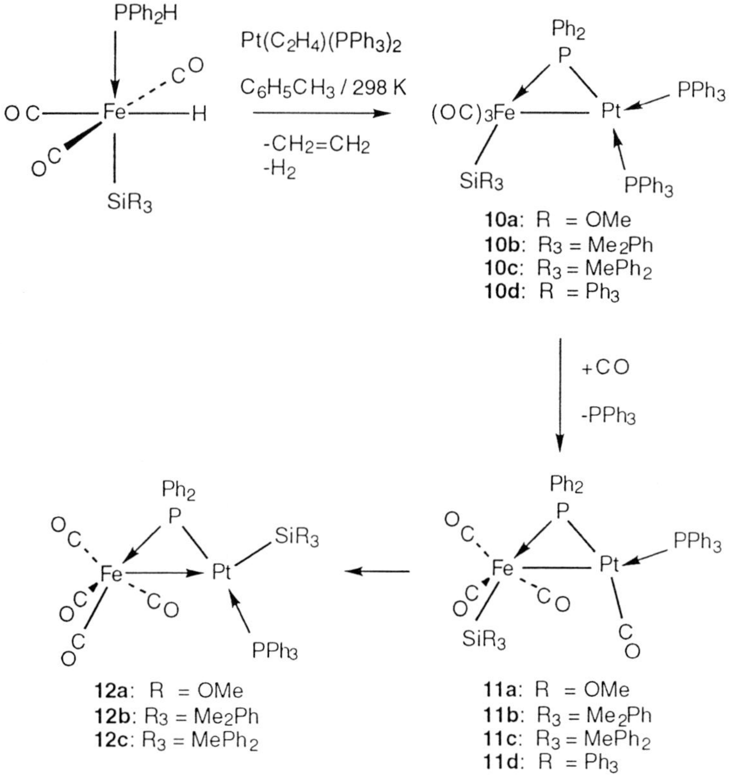

Scheme 1.

The NMR data obtained when using ^{13}C enriched (99 %) CO are consistent with an incorporation of the labelled ^{13}C nuclei as outlined in (Eq. 7). This reaction demonstrates interesting differences in the reactivity of closely related alkoxysilyl- and siloxane derivatives since in the former case only the acyl derivatives **3** and **4** were observed (Eq. 2). One of the reasons for this difference may be found in the fact that the central silicon atom of **1b** is more electropositive than in **1a**, thereby facilitating its migration.

Related Fe–Pt complexes were prepared (Scheme 1) which contain a three electron μ-diaryl(or dialkyl)phosphido bridging ligand, instead of the four-electron donor μ-dppm ligand. This modification led to a completely different behaviour for the silyl ligand: when CO was bubbled through a solution of complexes **10**, selective substitution of the PPh$_3$ ligand *trans* to the phosphido-bridge occurred, which led to the rapid and quantitative formation of **11** [18].

Unexpectedly, complexes **11a–11c** quantitatively rearranged in solution to their isomers **12**, in which the Si atom is now bonded to Pt, whereas the Pt-bound CO ligand has migrated to the Fe center, as established by the X-ray structure analysis of **12a** [18a]. The rate of this unprecedented silyl shift appears to depend mainly on steric requirements of the SiR$_3$ group. In the case of the Si(OMe)$_3$ ligand, the migration is completed within approximately 1 h, whereas for the SiMe$_2$Ph group ca. 5 h and for the even bulkier SiMePh$_2$ group more than 1 day was needed, respectively. Derivative **11d**, which was also structurally characterized, did not rearrange at all.

Oxidative-addition of the Fe–H and P–H bonds of *mer*-[HFe(SiR$_3$)(CO)$_3$(PPh$_2$H)] on [Pt(1,5-COD)$_2$] afforded the phosphido-bridged complexes [(OC)$_3$(R$_3$Si)Fe(μ-PPh$_2$)Pt(1,5-COD)] **13**, in which the Pt-bound PPh$_3$ ligands of **10** are replaced by a COD ligand. This modification of the ligand sphere of the platinum atom enhances the reactivity of the complex to the point that upon carbonylation, the COD ligand is substituted yielding the rearranged complexes [(OC)$_4$Fe(μ-PPh$_2$)Pt(CO)(SiR$_3$)] **14** within minutes, irrespective of the nature of the SiR$_3$ ligand: even SiPh$_3$ was now found to migrate from iron to platinum. NMR spectroscopy proved very valuable to monitor these reactions; in particular ^{29}Si NMR allowed the determination of ^1J(SiPt) couplings, e.g., 2117 Hz in complex **14a** (δ = −26.8 vs TMS) whereas a ^2J(SiPt) coupling of 49 Hz was found in **13a**.

13a R = OMe
13b R = Ph

14a R = OMe
14b R = Ph

Eq. 8.

We also examined the possibility of inducing this silyl shift in the phosphido-bridged complexes **10** and **13** by isonitriles instead of CO. In the case of **10a** we succeeded only in substituting the PPh$_3$ ligand *trans* to the phosphido bridge and isolated quantitatively **15**. We found no evidence for a silyl migration, irrespective of the of SiR$_3$ group and of the stoichiometry, steric, and electronic nature of the various isonitriles used. However, the enhanced reactivity of the COD complexes **13** allowed a silyl transfer of a Fe-bound SiR$_3$ group to the adjacent platinum center to be promoted under mild conditions: addition of two equivalents of *t*BuNC afforded [(OC)$_3$(*t*BuNC)Fe(μ–PPh$_2$)Pt(*t*BuNC)(SiR$_3$] **16** [18d].

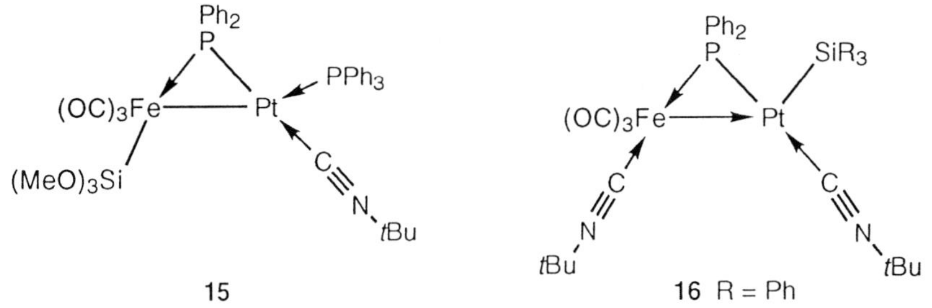

Fig. 2.

The study of the CO-induced silyl migration reaction in a 1:1 mixture of complexes **13a** and [(OC)$_3$(Ph$_3$Si)Fe(μ-PCy$_2$)Pt(1,5-COD)] showed that only **14a** and [(OC)$_4$Fe(μ-PCy$_2$)Pt(SiPh$_3$)(CO)] are formed. This indicates that the silyl transfer reaction occurs in an intramolecular manner [18d]. One may speculate about an intermediate (or transition state) possessing a bridging μ-SiR$_3$ group to account for this mutual ligand exchange. Such an unusual bridging bonding mode for a –SiR$_3$ group has already been structurally established in a borane and more recently in a polynuclear copper complex [19, 20]. Recent studies by Girolami *et al.* and Akita *et al.* in dinuclear SiMe$_3$-substituted Ru–Ru complexes have concluded that the silyl ligand could reversibly flip from one metal center to another via a μ-SiR$_3$ intermediate [21, 22].

In conclusion, we have observed in dppm-stabilized complexes that CO or PR$_3$ induced migration of the silyl group from iron to an acyl oxygen of a neighboring Pd or Pt center results in complexes with a bridging siloxycarbene ligand. Subtle electronic factors determine whether carbonylation of complexes **1** and **2** affords stable acyl or μ-siloxycarbene complexes, the latter resulting from CO-induced silyl migration. It is interesting to note that with phosphido-bridged complexes of the same metal couple, an unprecedented silyl migration from iron to platinum was observed under similiar conditions. Whereas in the rearrangement leading to the siloxy carbene complexes the formation of a strong Si–O bond may be the main driving force, the energy gained by the combined formation of a stronger Pt–Si bond and of a

Fe–CO (or Fe–CNR) bond must be larger than that the Fe–Si and Pt–CO (or Pt–CNR) bonds of the precursors which are broken and this drives the silyl migration reaction from iron to platinum to completion.

Acknowledgements: We would like to thank the *Deutsche Forschungsgemeinschaft* for a grant to M. K., the *Centre National de la Recherche Scientifique* for financial support of this research and *Johnson Matthey PLC* for a generous loan of $PdCl_2$ and $PtCl_2$.

References:

[1] E. Garrou, R. F. Heck, *J. Am. Chem. Soc.* **1976**, *98*, 4115.

[2] K. Anderson, R. J. Cross, *Acc. Chem. Res.* **1984**, *17*, 67.

[3] A. Yamamoto, in: *Organotransition Metal Chemistry*, Wiley, Chichester, 1986.

[4] B. Longato, J. R. Norton, J. C. Huffman, J. A. Marsella, K. G. Caulton, *J. Am. Chem. Soc.* **1981**, *103*, 209.

[5] S. J. Young, B. Kellenberger, J. H. Reibenspies, S. E. Himmel, M. Manning, O. P. Anderson, J. K Stille, *J. Am . Chem. Soc.* **1988**, *110*, 5745.

[6] M. Ferrer, O. Rosell, M. Seco, P. Braunstein, *J. Chem. Soc., Dalton Trans.* **1989**, 379.

[7] D. M. Antonelli, M. Cowie, *Organometallics* **1991**, *10*, 2550.

[8] F. Antwi-Nsiah, M. Cowie, *Organometallics* **1992**, *11*, 3157.

[9] A. Fukuoka, T. Sadashima, T. Sugiura, X. Wu, Y. Mizuho. S. Komiya, *J. Organomet. Chem.* **1994**, *473*, 139.

[10] P. Braunstein, T. Faure, M. Knorr, F. Balegroune, D. Grandjean, *J. Organomet. Chem.* **1993**, *462*, 71.

[11] P. Braunstein, T. Faure, M. Knorr, T. Stährfeldt, A. DeCian, J. Fischer, *Gazz. Chim. Ital.*, in press.

[12] P. Braunstein, M. Knorr, T. Stährfeldt, *J. Chem. Soc., Chem. Commun.*, in press.

[13] M. Knorr, P. Braunstein, manuscript in preparation.

[14] K. A. Mead, I. Moore, F. G. A. Stone, P. Woodward, *J. Chem. Soc. Dalton Trans.* **1983**, *9*, 2083.

[15] M. Knorr, T. Faure, P. Braunstein, *J. Organomet. Chem.* **1993**, *447*, C4.

[16] K. C. Brinkmann, A. J. Blakeny, W. Krone-Schmidt, J. A. Gladysz, *Organometallics* **1984**, *3*, 1325.

[17] Unpublished results.

[18] P. Braunstein, M. Knorr, B. Hirle, G. Reinhard, U. Schubert, *Angew. Chem., Int. Ed. Engl.* **1992**, *1*, 1583; G. Reinhard, M. Knorr, P. Braunstein, U. Schubert, S. Khan, C. E. Strouse, H. D. Kaesz, A. Zinn, *Chem. Ber.* **1993**, *126*, 17; M. Knorr, T. Stährfeldt, P. Braunstein, G. Reinhard, P. Hauenstein, B. Mayer, U. Schubert, S. Khan, H. D. Kaesz, *Chem. Ber.* **1994**, *127*, 295; T. Stährfeldt, M. Knorr, P. Braunstein, *Xth FECHEM Conference on Organometallic Chemistry*, Agia Pelagia, Crete (Greece), **1993**, Abstract P 46.

[19] J. C. Calabrese, L. F. Dahl, *J. Am. Chem. Soc.* **1971**, *93*, 6042.

[20] A. Heine, D. Stalke, *Angew. Chem., Int. Ed. Engl.* **1993**, *32*, 121.

[21] W. Lin, S. R. Wilson, G. S. Girolami, *Organometallics* **1994**, *13*, 2309.

[22] M. Akita, T. Oku, R. Hua, Y. Moro-oka, *J. Chem. Soc., Chem. Commun.* **1993**, 1670.

Novel Diazomethylsilyl Substituted Fischer Carbene Complexes

Dieter Mayer, Gerhard Maas*

Fachbereich Chemie

Universität Kaiserslautern

Erwin-Schrödinger-Str., D-67663 Kaiserslautern, Germany

Summary: The synthesis of various Fischer carbene complexes of manganese bearing a diazomethylsilyl group in α-position to the carbene carbon atom is described.

Introduction

After their discovery by E. O. Fischer and A. Maasböl in 1964 [1], a large number of carbene complexes with various transition metals such as Cr, Mo, W, Mn, and Fe were prepared [2]. Their synthetic applications in organometallic and organic chemistry increased rapidly, especially with respect to annelation reactions (Dötz reaction). Among all known Fischer carbene complexes there is no example featuring a diazo functionality. In this contribution we describe the synthesis of a new class of Fischer-type carbene complexes with a diazomethylsilyl substituent in α-position to the carbene carbon-atom.

Synthesis of Diazomethylsilyloxycarbene and Diazomethylsilylmethylcarbene Complexes

Diazomethylsilyloxycarbene complexes are prepared by following the traditional Fischer route starting from methylcyclopentadienyl- or cyclopentadienyl-tricarbonyl-manganese. After addition of an organolithium reagent to the carbonyl compound, the resulting anion is quenched with the (diisopropyltrifloxysilyl)diazoacetic ester **1** to give the O-silylated carbene complexes **2a–2k**, and the crude product is purified by chromatography on silica gel at –30 °C. The highly reactive silylating reagent **1** is obtained from diisopropylsilyl bistriflate and one equivalent of methyl diazoacetate in the presence of ethyldiisopropylamine.

Eq. 1.

2	L$_n$M	R	yield [%]	2	L$_n$M	R	yield [%]
a	Cp(CO)$_2$Mn	Me	15 [a]	h	Cp'(CO)$_2$Mn	(N-methylpyrrolyl)	72
b	Cp'(CO)$_2$Mn	Me	35				
c	Cp(CO)$_2$Mn	Bu	10 [a]				
d	Cp'(CO)$_2$Mn	Bu	47				
e	Cp(CO)$_2$Mn	Ph	28 [a]	i	Cp(CO)$_2$Mn	$-C\equiv C-Ph$	45
f	Cp'(CO)$_2$Mn	Ph	80				
				j	Cp(CO)$_2$Mn	$-C_6H_4-OMe$	80
g	Cp'(CO)$_2$Mn	(furyl)	23				
				k	Cp(CO)$_2$Mn	(propenyl)	15 [a]

[a] decomposition during chromatographic work up
Cp' = methylcyclopentadienyl

Table 1.

An analogous reaction sequence starting from Mo(CO)$_6$ was carried out. Thus, complex **3** containing the electron-rich and therefore stabilizing *p*-methoxyphenyl group was prepared and isolated (Eq. 2). Although **3** decomposes at –20 °C within several hours after work-up, it could be characterized by ^{13}C NMR and IR spectroscopy. Attempts to obtain analogous complexes of Mo and W with less electron-donating carbene substituents failed.

It is known that a methyl group attached to the metal carbene C-atom can be deprotonated easily [3]. If the Fischer carbene complex **4** is treated with butyllithium and the resulting anion is silylated with **1**, the diazomethylsilylmethyl substituted carbene complex **5** is obtained in 43 % yield (Eq. 3).

Eq. 2.

$Mo(CO)_6$ → 1. Li—C₆H₄—OMe 2. TfOSiiPr₂CN₂COOMe (1) → $(CO)_5Mo=C$(O-Si(iPr)₂-C(=N₂)-COOMe)(C₆H₄-OMe)

3

Eq. 3.

$L_nM=C$(OEt)(CH₃) → 1. BuLi 2. TfOSiiPr₂CN₂COOMe (1) → $L_nM=C$(OEt)(CH₂-Si(iPr)₂-C(=N₂)-COOMe)

4 **5**

$L_nM = Cp(CO)_2Mn$

Stability of the Novel Complexes

All Fischer carbene complexes presented in this work are air- and moisture-sensitive. The derivatives of **2** bearing electron-donating π-substituents and compound **5** do not decompose in substance and in dry solvents after a few weeks. The fact that corresponding Fischer carbenes of Mo and W prepared by the same method decompose even at low temperatures underlines once more the particular stability of manganese carbene complexes.

For example, in attempts to realize benzannelation reactions, alkyloxy aryl carbene complexes of manganese failed to react with alkynes even in refluxing toluene, and the starting compounds could be recovered [4]. The documented low reactivity of the Mn as opposed to Cr and Mo carbene complexes may in part explain why the electrophilic carbene C-atom and the nucleophilic diazo C-atom tolerate each other in the same molecule. Besides, the bulky substituents at the silicon atom protect it from being attacked by nucleophiles leading to desilylation as reported for trimethylsilyl substituted Cr carbene complexes [5].

Spectroscopic Characterization

All carbene complexes were characterized by their ^{1}H, ^{13}C NMR, and IR spectra. In the ^{13}C NMR spectra the resonances of the carbene C-atoms are found at very low field between 334.0 ppm (**2f**) and 348.3 ppm (**2g**). Most of the signals are broadened or occur twice. This phenomenon is due to the quadrupole moment of manganese and hindered rotation around the O–C(carbene) bond and O–Si bond as often described [6]. The IR spectra show the characteristic absorption bands of the diazo group ($\nu = 2100$ cm^{-1}) and the carbonyl ligands ($\nu = 1850–2000$ cm^{-1}).

As for any other new class of organometallic compounds, it should be interesting to examine the structure of these unusual substituted complexes by X-ray structure analysis, but we did not yet get suitable crystals. CI mass spectra of complexes **2f** and **2g** show the expected molecular ion peak.

Reactivity

Nitrogen extrusion from diazo compounds can be induced photochemically, thermally, or by transition-metal catalysis. Complex **2f** is decomposed neither by copper(I) triflate (CuOTf) nor by Rh$_2$(OAc)$_4$ at room temperature. The photochemical decomposition of **2f** leads to simultanous nitrogen and carbonyl extrusion, but no product could be characterized.

The alkinyl substituted carbene complex **2i** was refluxed several days in toluene without any reaction. Besides, diethylamine does not add to the C≡C triple bond in **2i**, a reaction that is reported for a number of Cr-carbene complexes [7].

References:

[1] E. O. Fischer, A. Maasböl, *Angew. Chem.* **1964**, *76*, 645; *Angew. Chem., Int. Ed. Engl.* **1964**, *3*, 580.

[2] K. H. Dötz, *Angew. Chem.* **1984**, *96*, 573; *Angew. Chem., Int Ed. Engl.* **1984**, *23*, 587.

[3] R. Aumann, M. Runge, *Chem. Ber.* **1992**, *125*, 259.

[4] B. L. Balzer, M. Cazanoue, M. G. Finn, *J. Am. Chem. Soc.* **1992**, *114*, 8735.

[5] U. Schubert, *J. Organomet. Chem.* **1988**, *358*, 215.

[6] K. H. Dötz, H. Larbig, *J. Organomet. Chem.* **1992**, *433*, 115.

[7] F. Stein, M. Duetsch, R. Lackmann, M. Noltemeyer, A. De Meijere, *Angew. Chem.* **1991**, *103*, 1669; *Angew. Chem., Int. Ed. Engl.* **1991**, *30*, 1658.

1-Metalla-2-sila-1,3-diene Compounds

Markus Weinmann, Heinrich Lang*, Olaf Walter, Michael Büchner

Anorganisch-Chemisches Institut

Ruprecht-Karls-Universität Heidelberg

Im Neuenheimer Feld 270, D-69120 Heidelberg, Germany

Summary: The synthesis and structural features of inter- and intramolecularly stabilized 1-metalla-2-sila-1,3-diene compounds will be discussed.

Introduction

σ^3,λ^4-Phosphanediyl complexes (neutral phosphenium ion complexes) of the type [(R)(R´)P=ML$_n$] (ML$_n$ = MnCp(CO)$_2$, Co(CO)$_3$; R = 2,4,6-tBu$_3$C$_6$H$_2$O; R´ = C≡CPh, CH=CHPh) have been shown to exhibit a versatile reaction chemistry at the phosphorus-metal and the carbon-carbon multiple bonds [1–4]. Formally, these compounds consist of a phosphenium ion (R)(R´)P$^+$ and an anionic organometallic ML$_n$ group [1–7]. In terms of the concept of isolobal analogy [8], electron sextet species R$_2$P$^+$ are equivalents of H$_2$C and R$_2$Si, as well as of organometallic 16-VE moieties. In this respect, 1-metalla-2-phospha-1,3-dienes (type **A** molecules) and 1-metalla-2-sila-1,3-dienes (type **B** molecules) can be considered as heterobutadienes.

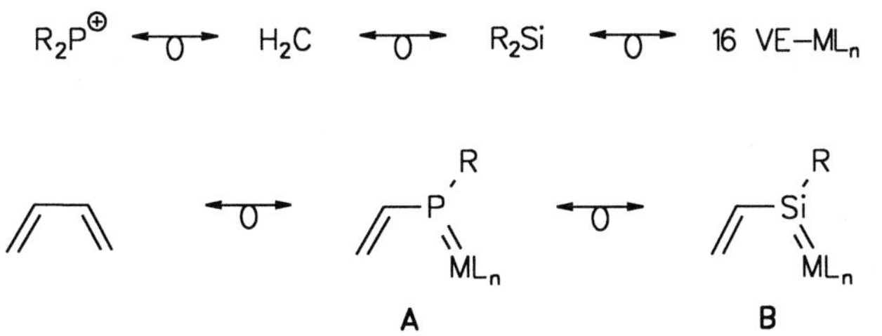

Scheme 1.

In this context we report on synthesis and structural features of intermolecular (type **C** molecule) and intramolecular (type **D** molecule) donor-stabilized acyclic 1-metalla-2-sila-1,3-diene complexes.

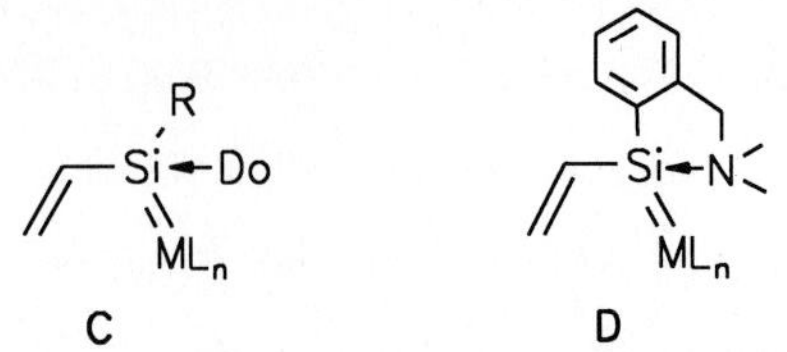

Fig. 1.

Syntheses

1 Intermolecular Stabilized 1-Chromium-2-sila-1,3-diene Complexes

According to the method used by Zybill et al. [9], structural type **C** compounds can be synthesized by the reaction of $(H_2C=CH)SiCl_3$ (**1**) with $K_2[Cr(CO)_5]$ (**2**) in THF at -30 °C in the presence of HMPA (Do) (HMPA = hexamethyl phosphorus-amide). The 1-chromium-2-sila-1,3-diene complex **3** can be isolated in 60 % yield after recrystallization from THF/$n$pentane solution.

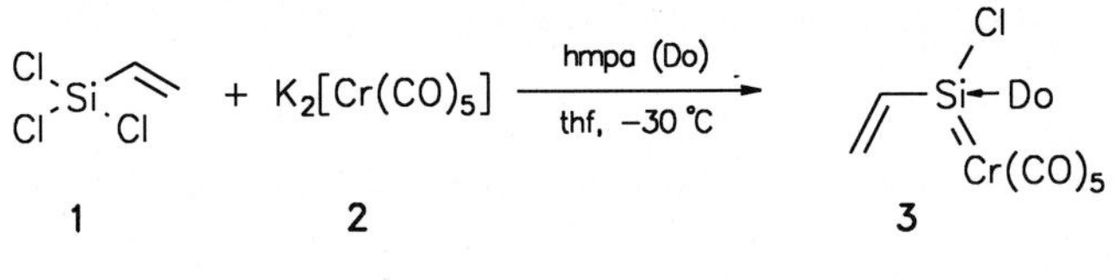

Eq. 1.

2 Intramolecular Stabilized 1-Chromium-2-sila-1,3-diene Complexes

A further possibility for the stabilization of low-valent silicon complexes is the intramolecular coordination exhibited by the DMBA ligand (DMBA = 2-dimethyl-benzyl-amine) [10–12].

Structural type **D** molecules can be synthesized in a two step reaction sequence: $K_2[Cr(CO)_5]$ (**2**) reacts with the hypervalent trichlorosilane $(2-Me_2NCH_2C_6H_4)SiCl_3$ (**4**) in THF at -25 °C to afford the chloro-functionalized silanediyl complex **5** in 55 % yield.

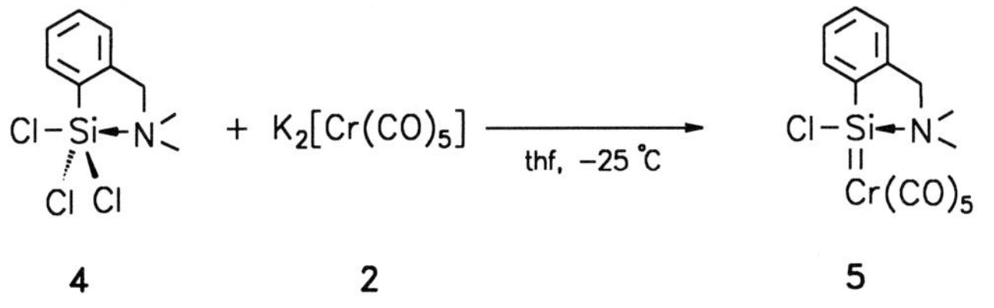

Eq. 2.

Nucleophilic substitution reaction of **5** with BrMgCH=CH$_2$ (**6**) leads to the formation of the acyclic heterobutadiene **7**.

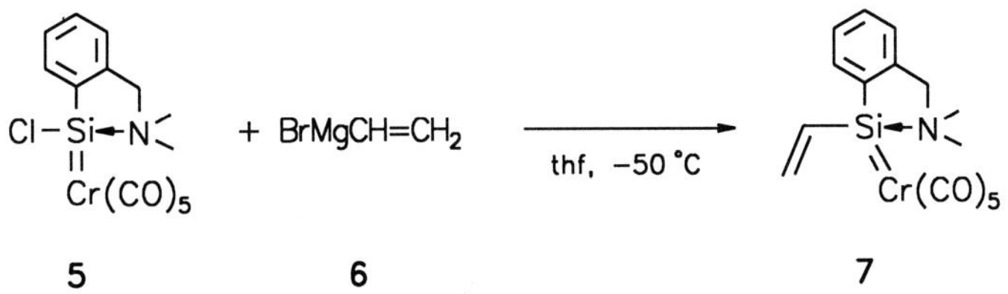

Eq. 3.

This type of reaction is stoichiometrically straightforward.

Spectroscopy and Bonding

On the basis of spectroscopic data (IR, ^{1}H, ^{13}C, ^{29}Si NMR, MS) the heterobutadiene structure could be unequivocally assigned to **3** and **7**. The most significant properties of the silanediyl complexes **3**, **5**, and **7** are their ^{29}Si NMR resonances, which are shifted to lower field (**3**: δ = 83.3 (d, $^2J_{SiP}$ = 35.4 Hz); **5**: δ = 120.4; **7**: δ = 114.9 ppm) compared to the starting compounds **1** and **4**. For compounds **5** and **7** diastereotopic CH$_3$ and the CH$_2$ groups of the amino ligand are characterize (Fig. 2).

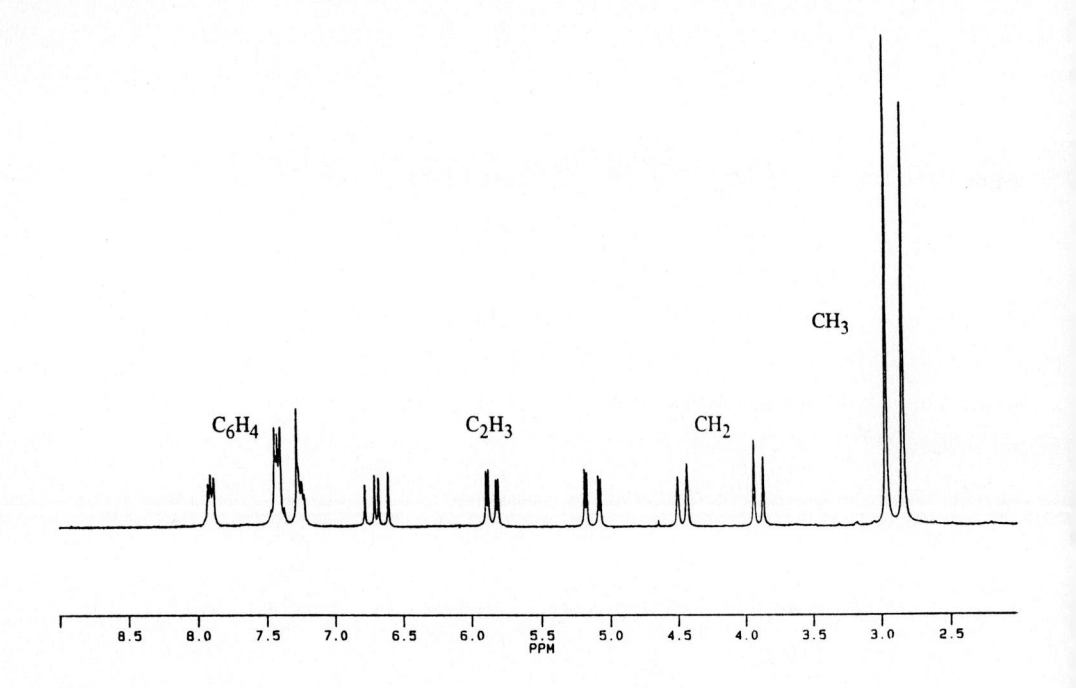

Fig. 2. ¹H NMR spectrum of **7**

This conclusion is confirmed by single crystal X-ray structure analysis of **7** (Fig. 3).

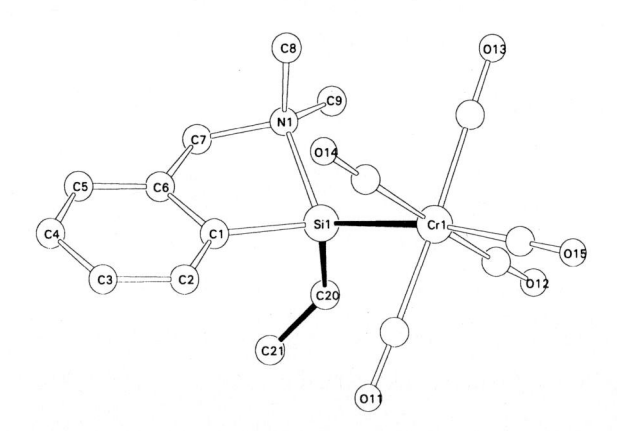

Fig. 3. Molecular geometry and atom labeling scheme for **7**; important interatomic distances (Å) and angles (°) are as follows: Cr1–Si1 2.385(1), Si1–N1 1.954(2), Si1–C1 1.878(3), Si1–C20 1.881(3), C20–C21 1.309(4); Cr1–Si1–N1 122.64(7), Cr1–Si1–C1 121.60(9), Cr1–Si1–C20 117.37(9), N1–Si1–C1 86.8(1), N1–Si1–C20 98.1(1), Si1–C20–C21 125.6(3)

Compound **7** crystallizes in the monoclinic space group $P2_1/n$, and shows a pseudo-tetrahedral geometry around the silicon atom. One notable feature of **7** is the short Cr1–Si1 interatomic distance of 2.385(1) Å, which corresponds to those found in other chromium-silicon multiple bond systems [9, 12]. In addition, the Si1–N1 interatomic distance of 1.954(2) Å is shorter than that found in hypervalent DMBA- substituted aryl- [12], alkyl-, and vinyl-dichlorosilanes [13]. This is in agreement with the high electrophilicity of the silicon atom in **7**.

A similarly distorted tetrahedral arrangement around the silicon atom is found in the intermolecular HMPA-stabilized acyclic heterobutadiene **3** (Fig. 4). Compound **3** crystallizes in the orthorhombic space group $P2_12_1$. The Cr1–Si1, Si1–Cl1, and Si1–O6 interatomic bond lengths of **3** are comparable with the distances found in other silanediyl compounds of type $[R_2Si=Cr(CO)_5]\cdot HMPA$ [9].

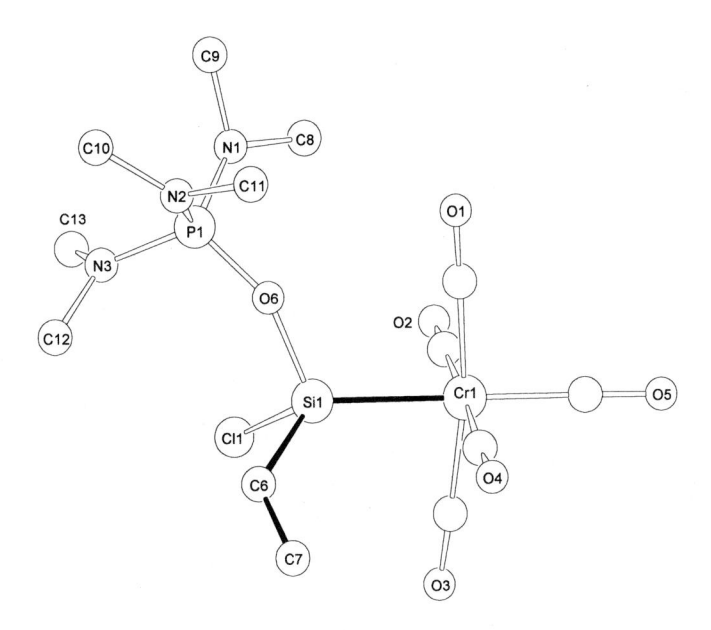

Fig. 4. Molecular geometry and atom labeling scheme for **3**; important interatomic distances (Å) and angles (°) are as follows: Cr1–Si1 2.373(2), Si1–Cl1 2.102(7), Si1–C6 1.825(25), Si1–O6 1.703(4), P1–O6 1.526(4), C6–C7 1.34(3); Cr1–Si1–Cl1 116.7(2), Cr1–Si1 C6 123.3(7), Cr1–Si1–O6 114.2(2), Si–O6–P1 152.5(3)

Acknowledgement: We thank the *Deutsche Forschungsgemeinschaft* and the *Fonds der Chemischen Industrie* for financial support. We are grateful to Prof. Dr. G. Huttner and Dr. E. Kaifer for fruitful discussions and to A. Gehrig for assistance in preparative work.

References:

[1] H. Lang, O. Orama, *J. Organomet. Chem.* **1989**, *371*, C48; H. Lang, M. Leise, L. Zsolnai, *J. Organomet. Chem.* **1990**, *386*, 349; H. Lang, M. Leise, L. Zsolnai, *J. Organomet. Chem.* **1991**, *410*, 379.

[2] H. Lang, M. Leise, L. Zsolnai, *J. Organomet. Chem.* **1990**, *389*, 325; H. Lang, M. Leise, L. Zsolnai, M. Fritz, *J. Organomet. Chem.* **1990**, *395*, C30; H. Lang, M. Leise, L. Zsolnai, *J. Organomet. Chem.* **1993**, *447*, C1.

[3] H. Lang, *Phosphorus, Sulfur, Silicon* **1993**, *77*, 9.

[4] H. Lang, M. Leise, L. Zsolnai, *Polyhedron* **1993**, *12*, 1257.

[5] W. Malisch, U. A. Hirth, T. A. Bright, H. Käb, T. S. Ertel, S. Hückmann, H. Bertagnolli, *Angew. Chem., Int. Ed. Engl.* **1992**, *31*, 1525; W. Malisch, K. Hindahl, R. Schemm, *Chem. Ber.* **1992**, *125*, 2027.

[6] L. D. Hutchins, H. U. Reisacher, G. L. Wood, E. N. Duesler, R. T. Paine, *J. Organomet. Chem.* **1987**, *335*, 229.

[7] A. H. Cowley, D. M. Giolando, C. M. Nunn, M. Pakulski, D. Westmoreland, N. C. Norman, *J. Chem. Soc., Dalton Trans.* **1988**, 2127; A. M. Arif, A. H. Cowley, C. M. Nunn, S. Quashie, *Organometallics* **1989**, *8*, 1878.

[8] R. Hoffmann, *Angew. Chem., Int. Ed. Engl.* **1982**, *21*, 711.

[9] C. Zybill, G. Müller, *Angew. Chem., Int. Ed. Engl.* **1987**, *26*, 669; C. Zybill, G. Müller, *Organometallics* **1988**, *7*, 1368; C. Zybill, D. L. Wilkinson, G. Müller, *Angew. Chem., Int. Ed. Engl.* **1988**, *27*, 583; C. Zybill, D. L. Wilkinson, C. Leis, G. Müller, *Angew. Chem., Int. Ed. Engl.* **1989**, *28*, 203; C. Leis, C. Zybill, J. Lachmann, G. Müller, *Polyhedron* **1991**, *10*, 1163; C. Leis, D. L. Wilkinson, H. Handwerker, C. Zybill, *Organometallics* **1992**, *11*, 514.

[10] G. van Koten, C. A. Schaap, J. G. Noltes, *J. Organomet. Chem.* **1975**, *99*, 157; G. van Koten, J. G. Noltes, A. L. Spek, *J. Organomet. Chem.* **1976**, *118*, 183.

[11] R. J. Corriu, G. F. Lanneau, C. Priou, *Angew. Chem., Int. Ed. Engl.* **1991**, *30*, 1130; R. J. Corriu, G. F. Lanneau, B. P. Chauhan, *Organometallics* **1993**, *12*, 2001.

[12] R. Probst, C. Leis, S. Gamper, E. Herdtweck, C. Zybill, N. Auner, *Angew. Chem., Int. Ed. Engl.* **1991**, *30*, 1132; H. Handwerker, C. Leis, R. Probst, P. Bissinger, A. Grohmann, P. Kiprof, E. Herdtweck, J. Blümel, N. Auner, C. Zybill, *Organometallics* **1993**, *12*, 2162.

[13] M. Weinmann, H. Lang, unpublished.

Novel Metallo-Silanols, -Silanediols, and -Silanetriols of the Iron and Chromium Group: Generation, Structural Characterization, and Transformation to Metallo-Siloxanes[1, 2]

Wolfgang Malisch*, Stephan Möller, Reiner Lankat, Joachim Reising,

Siegfried Schmitzer, OliverFey

Institut für Anorganische Chemie

Bayerische Julius-Maximilians-Universität Würzburg

Am Hubland, 97074 Würzburg, Germany

Summary: The novel ferrio-silanols $(Cp(OC)_2Fe-Si(OH)(Me)]_2O$ (**3**) and $C_5Me_5(OC)_2Fe-Si(H)(R)OH$ (R = Me (**4a**), OH (**4b**)) are obtained from the corresponding ferrio-chlorosilanes via hydrolysis. For the silanols, silanediols and silanetriols $L_nM-Si(OH)_nR_{3-n}$ (n = 1–3, R = alkyl, aryl) bearing a $C_5R_5(OC)_2Ru-$ (R = H, Me) or $C_5R_5(OC)_2(Me_3P)M-$ fragment (R = H, Me; M = Cr, Mo, W) (**8–12**) the oxofunctionalization of the corresponding metallo-hydridosilanes with dimethyldioxirane proves to be the most general synthetic procedure. The activation of Si–H bonds towards this oxygen insertion process by transition-metal fragments is clearly demonstrated by the controlled conversion of the pentahydridodisilanyl complexes $C_5Me_5(OC)_2(Me_3P)M-SiH_2-SiH_3$ (M = Mo (**17a**), W (**17b**)), obtained via Cl/H exchange with $LiAlH_4$ from $C_5Me_5(OC)_2(Me_3P)M-SiCl_2-SiCl_3$ (M = Mo (**16a**), W (**16b**)), to the disilanediols $C_5Me_5(OC)_2(Me_3P)M-Si(OH)_2-SiH_3$ (M = Mo (**19a**), W (**19b**)). The metallo-silanols can be transformed to the transition metal substituted disiloxanes $C_5Me_5(OC)_2Fe-Si(H)(R)OSiMe_3$ (R = Me (**13a**), $OSiMe_3$ (**13b**)), $Cp(OC)_2Ru-SiPh_2OSiMe_3$ (**13c**), $C_5Me_5(OC)_2Ru-Si(o-Tol)_2OSiMe_2H$ (**13d**), and disilane-disiloxanes $C_5Me_5(OC)_2(Me_3P)M-Si(OSiMe_2H)_2-SiH_3$ (M = Mo (**20a**), W (**20b**)) via condensation with $Me_2Si(R)Cl$ (R = H, Me). The structures of $C_5Me_5(OC)_2Ru-Si(o-Tol)_2OH$ (**8e**), $C_5Me_5(OC)_2(Me_3P)Mo-SiMe(OH)_2$ (**11a**) and $C_5Me_5(OC)_2(Me_3P)W-Si(OSiMe_2H)_2-SiH_3$ (**20b**), established by X-ray crystallography, exhibit interesting features arising in the case of **8e** and **11a** from the formation of dimers via one or two hydrogen bonds.

Silanols and siloxanes are compounds of particular interest in context with the synthesis of silicones [3], "inorganic" polymers used in nearly every field of modern life. Usually organosilanols $R_{4-n}Si(OH)_n$ (R = alkyl, aryl; n = 1–3) undergo rapid self-condensation, a tendency that rises with increasing number of OH-groups. Therefore, isolation of silanediols or -triols has been achieved preferentially with bulky organic ligands [4–6] or, more efficiently, by using the steric requirement and high electron-donor capacity of metal-fragments directly linked to silicon. In this context preparation of the ferrio-silanols $C_5R_5(OC)_2Fe-SiR^1{}_n(OH)_{3-n}$, has been realized via hydrolysis of the corresponding ferrio-chlorosilanes $C_5R_5(OC)_2Fe-SiR^1{}_nCl_{3-n}$ (R = H, Me; R^1 = alkyl, aryl; n = 0–2), showing interesting hydrogen-bonded structures [7]. In an analogous manner, the osmium silanetriol $(Ph_3P)_2(OC)(Cl)Os-Si(OH)_3$ is accessible, remarkably exhibiting no intermolecular interaction involving the OH-units [8]. We now report extension of the iron-silanol series by species of the Si–H-functionalized type $C_5R_5(OC)_2Fe-Si(H)(R^1)OH$ (R^1 = Me, OH), which can only be isolated in the case of the C_5Me_5-derivative. Moreover novel silanols bearing a ruthenium, chromium, molybdenum or tungsten fragment at the silicon, have been made available for the first time via oxygen insertion into Si–H-bonds using dimethyldioxirane.

1 Si–H-functionalized Ferrio-Silanols $C_5Me_5(OC)_2Fe-Si(H)(R)OH$ (R = Me (4a), OH (4b)) and the Bis(ferrio)disiloxanediol $[Cp(OC)_2Fe-Si(OH)(Me)]_2O$

As we have demonstrated recently the introduction of a $Cp(OC)_2Fe$-fragment offers access to Si–H-functionalized ferrio-silanols $Cp(OC)_2Fe-Si(H)(R)OH$, provided that R represents a reasonable bulky group (e.g., R = *p*-Tol, *i*Pr) [7]. For R = Me the preparation via the hydrolysis of $Cp(OC)_2Fe-Si(H)(Me)Cl$ (**1a**) fails due to immediate condensation yielding the bis(ferrio)disiloxane $[Cp(OC)_2Fe-Si(H)(Me)]_2O$. This siloxane is an interesting compound by itself, since it can be easily converted to the analogous chloro-species **2** by H/Cl-exchange with CCl_4 and further on to the flourine analogue with $AgBF_4$, which is characterized by a linear Si–O–Si skeleton [9]. On treatment with two equivalents of water and in the presence of triethylamine **2** is transformed to the 1,3-bis(ferrio)disiloxane-1,3-diol **3** in high yields (Eq. 1).

An organometallic approach to kinetically stabilized transition metal main-group element compounds of the half sandwich type involves formal substitution of the C_5H_5- against a C_5Me_5-unit at the metal promising both a better steric shielding of the coordination sphere and an increased electron donor capacity of the metal fragment [10].

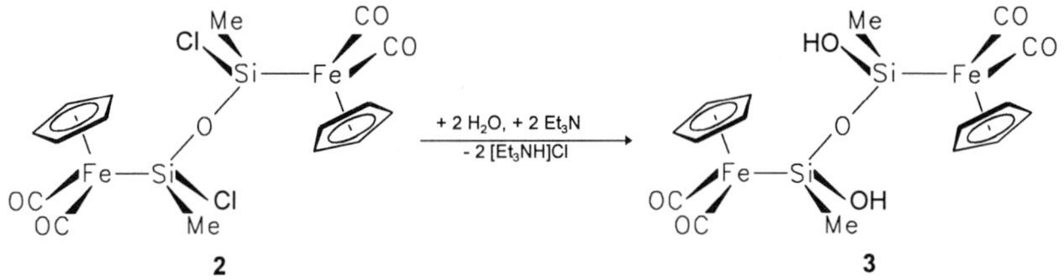

Eq. 1.

This fact guarantees controlled conversion of $C_5Me_5(OC)_2Fe–Si(H)(Me)Cl$ (**1b**) via reaction with H_2O to the corresponding ferrio-silanol **4a**, which shows the expected higher stability with respect to self-condensation. In the same manner, the ferrio-dichlorosilane $C_5Me_5(OC)_2Fe–Si(H)Cl_2$ (**1c**) reacts with two equivalents of water to produce the ferrio-silanediol **4b**, to our knowledge the first example of a hydridosilanediol.

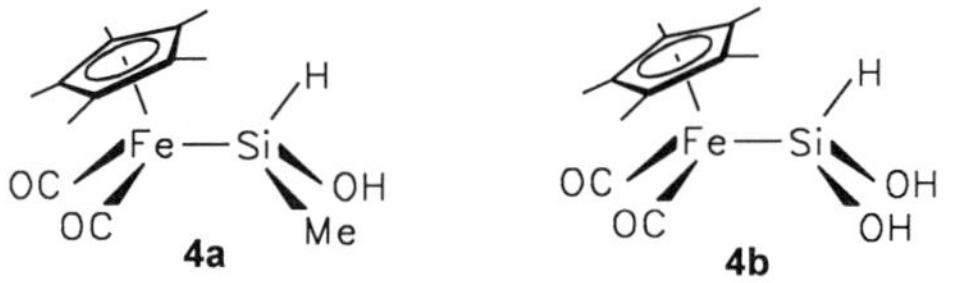

Fig. 1.

Compounds **4a**, **4b** are obtained as a pale yellow (**4a**) or colorless (**4b**) solid, which decompose at −30 °C within several weeks.

2 Ruthenio-Silanols $C_5R_5(OC)_2Ru–SiR^1{}_2OH$ (R = H, Me; R^1 = Me, Ph, o-Tol) (8a–8e) and the Ruthenio-Silanediol $Cp(OC)_2Ru–Si(p-Tol)(OH)_2$ (9)

While the hydrolysis route affords a series of ferrio-silanols, the ruthenio-silanols cannot be prepared analogously, since the ruthenio-chlorosilanes prove to be totally unreactive, concerning Cl/OH exchange at silicon. Therefore, oxofunctionalization of ruthenio-hydridosilanes provides the only synthetic access to this kind of silanols. In this context dimethyldioxirane, used as a solution in acetone, proves to be an excellent reagent [11]. Suitable starting materials for oxygen insertion, $C_5R_5(OC)_2Ru–SiR^1{}_2H$ (R = H,

R^1 = Me (**5a**), Ph (**5c**), *o*-Tol (**5d**); R = Me, R^1 = Me (**5b**), *o*-Tol (**5e**)) and Cp(OC)$_2$Ru–Si(*p*-Tol)H$_2$ (**7**), are obtained by the reaction of the sodium metalates M(Ru(CO)$_2$C$_5$R$_5$) (R = H, M = Na; R = Me, M = K) with the corresponding chlorosilanes [5] in cyclohexane followed by reduction with lithium-aluminiumhydride in the case of Cp(OC)$_2$Ru–Si(*p*-Tol)HCl (**6**) to give **7**.

Already at −78 °C, treatment of the ruthenio-hydridosilanes with stoichiometric amounts of dimethyldioxirane yields within 30–60 minutes the ruthenio-silanols as a light orange oil (**8a**, **8c**, **8d**) or pale yellow to light brown solid (**8b**, **8e**, **9**), respectively. In accordance to the experience with the iron-analogues the ruthenio-silanols show no tendency to undergo self-condensation under ordinary conditions. Compounds **8a**, **8c**, **8d**, however, are supposed to decompose due to β-H-abstraction, forming [Cp(OC)$_2$Ru]$_2$ (via Cp(OC)$_2$RuH) and polysiloxane.

8	a	b	c	d	e
R	H	Me	H	H	Me
R¹	Me	Me	Ph	*o*-Tol	*o*-Tol

Fig. 2.

The ruthenio-silanols are characterized by a strong downfield shift (30–40 ppm) of the ^{29}Si NMR resonance in comparison to that of the corresponding ruthenio-silanes (δ(^{29}Si) = 10.18/48.24 (**5b/8b**); 13.70/52.41 (**5d/8d**); −3.40/45.18 (**5e/8e**) ppm).

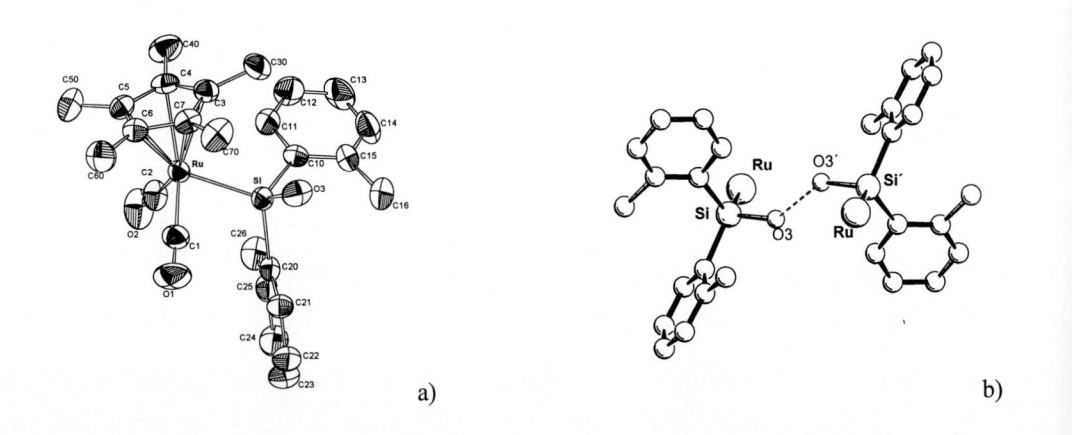

Fig. 3. Molecular structure of **8e**; in b) C$_5$Me$_5$ and CO are omitted for clarity; selected bond lengths [Å] and angles [°]: Ru–Si 2.411(2), Ru–C1 1.841(6), Ru–C2 1.863(6), Si–O3 1.647(4), Si–C10 1.891(5), C10–C15 1.408(7), Si–Ru–C1 83.4(2), C1–Ru–C2 93.1(3), Si–Ru–Z 124.79, C1–Ru–Z 124.93, Ru–C1–O1 174.9(7), Ru–Si–O3 105.8(2), Ru–Si–C10 116.2(2), O3–Si–C10 105.0(2), C10–Si–C20 107.9(3), C10–C15–C16 121.9(5) (Z = centroid of the C$_5$Me$_5$-ligand)

Referring to the ^{29}Si NMR data of organosilanols the presence of the transition metal fragment is expressed by a downfield shift of about 80 ppm [5]. In order to prove hydrogen-bonding for the ruthenio-silanols a single crystal X-ray diffraction analysis of **8e** has been performed (Fig. 3). It reveals formation of discrete dimeric units linked by one hydrogen bond (d(O3–O3′) = 2.82 Å). Coordination at the silicon is tetrahedral, the ligand arrangement at the ruthenium pseudotetrahedral.

3 Tungsten-, Molybdenum-, and Chromium-Silanols

Oxofunctionalization with dimethyldioxirane can be additionally applied for the synthesis of a vast number of chromium-, molybdenum-, and tungsten-silanols (**10a–10h**), -silanediols (**11a, 11b**), and -silanetriols (**12a, 12b**). These compounds are usually not available via the hydrolysis route due to the electronrichness of the metal fragments, for which the Me$_3$P- and C$_5$Me$_5$-ligand are responsible, creating a silicon unsusceptible towards nucleophilic attack.

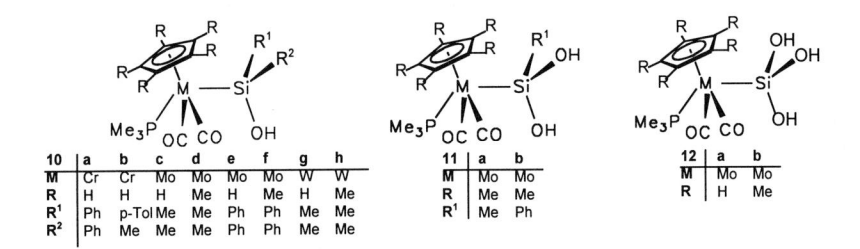

10	a	b	c	d	e	f	g	h
M	Cr	Cr	Mo	Mo	Mo	Mo	W	W
R	H	H	H	Me	H	Me	H	Me
R^1	Ph	p-Tol	Me	Me	Ph	Ph	Me	Me
R^2	Ph	Me	Me	Me	Ph	Ph	Me	Me

11	a	b
M	Mo	Mo
R	Me	Me
R^1	Me	Ph

12	a	b
M	Mo	Mo
R	H	Me

Fig. 4.

The thermal stability of the silanols depends on the nature of the metal fragment and increases in the series Cp(OC)$_2$(Me$_3$P)Cr < Cp(OC)$_2$(Me$_3$P)M < C$_5$Me$_5$(OC)$_2$(Me$_3$P)M (M = Mo, W). Moreover, electronegative or bulky substituents at the silicon raise the stability as a comparison between **10c** and **10e** clearly demonstrates. While the methyl-species **10c** decomposes in benzene at room temperature with formal loss of "SiMe$_2$O" to give Cp(OC)$_2$(Me$_3$P)Mo–H within 3 days, the phenylated species **10e** is stable for weeks under the same conditions.

The single crystal X-ray diffraction analysis of **11a** reveals that hydrogen bonding results in discrete cyclic dimers (d(O3–O4) = 2.84 Å). The formed six-membered Si$_2$O$_4$-ring shows chair conformation (Fig. 5), comparable to that observed for [(Me$_3$Si)$_3$C](Ph)Si(OH)$_2$ [6]. There is no hydrogen bonding between the dimers.

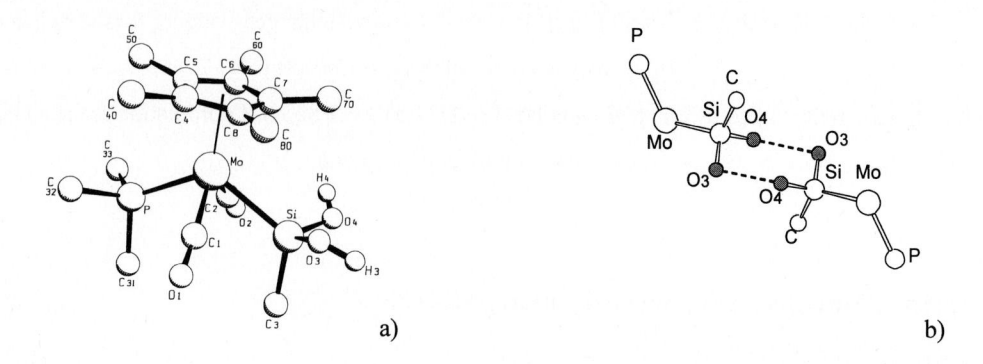

Fig. 5. Molecular structure of **11a**; in b) C$_5$Me$_5$, CO, and Me of Me$_3$P are omitted for clarity; selected bond lengths [Å] and bond angles in [°]: Mo–P 2.446(1), Mo–Si1 2.534(1), Si1–O3 1.673(2), Si1–O4 1.648(2), Si2–O4 1.613(2), Si2–O5 1.631(3), P–Mo–Si1 121.40(4), Si1–O4–Si2 145.8(2), Mo–Si1–O3 111.7(1), Mo–Si1–O4 112.61(9), O3–Si1–O4 104.1(1), O4–Si2–O5 111.5(1)

4 Transformation to Metallo-Siloxanes

The described metallo-silanols exhibit high stability concerning condensation, however, the OH groups display reasonable reactivity in context with chlorosilanes. These properties can be used to transform the ferrio-silanols **4a**, **4b** via reaction with trimethylchlorosilane to the ferrio-di- and -trisiloxanes **13a**, **13b** characterized by a hydrogen substituted α-silicon. In the case of **4b** the first siloxane unit is established after one day, the second takes five days (**13b**). The same reaction pattern can be applied for the generation of the ruthenio-disiloxanes **13c**, **13d**, resulting from the treatment of the ruthenio-silanols **8c**, **8e** with Me$_2$Si(R^2)Cl.

Fig. 6.

In a similar manner, the molybdenum-disiloxane **14** and -trisiloxane **15** are obtained from the molybdenum-silanediol **11a** on treatment with the stoichiometric amount of Ph$_2$Si(H)Cl or excess Me$_2$Si(H)Cl respectively. In context with **14** the presence of a metal-bound siloxy-unit, containing both an Si–H and an SiOH group is remarkable.

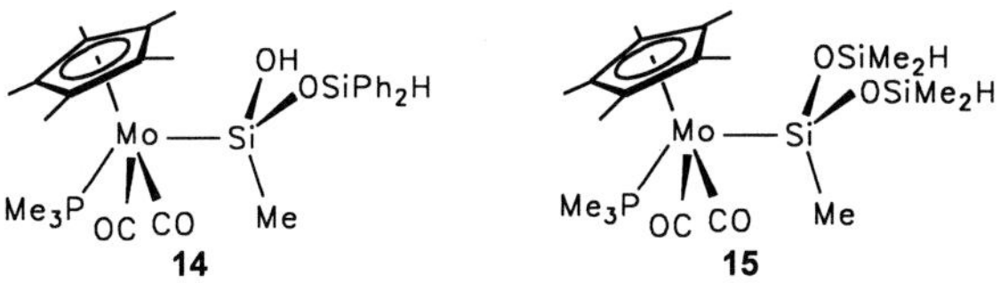

Fig. 7.

5 Tungsten- and Molybdenum-Disilanediols

An interesting topic in context with the oxofunctionalization of Si–H units concerns the application of this process to the pentahydridodisilanyl metal complexes since in this case two electronically different sorts of Si–H units are available. Appropriate complexes for this investigation can be obtained by a procedure worked out for the generation of H_3Si-molybdenum and -tungsten complexes [13]. It takes advantage of the fact, that the $C_5Me_5(Me_3P)(OC)_2M$-fragments guarantee M–Si bonds insensitive with respect to heterolytic cleavage by nucleophiles. The synthesis starts with the metallo-pentachlorodisilanes **16a**, **16b**, which are completely converted to **17a**, **17b** by Cl/H exchange with $LiAlH_4$. On treatment with carbon tetrachloride exclusive chlorination of the α-silicon takes place to yield **18a**, **18b** (Eq. 2). This fact clearly indicates that, due to the strong electron releasing effect of the metal fragment, the hydrogens at the α-silicon are preferentially activated.

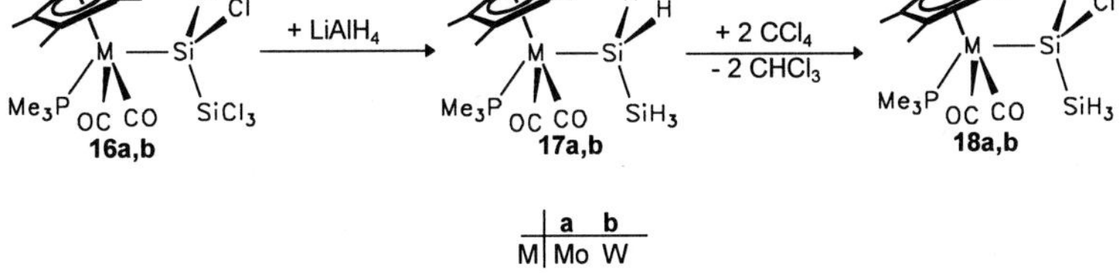

Eq. 2.

The same regioselectivity is observed for the oxygenation of **17a**, **17b** which is performed in a controlled manner only if two equivalents of dimethyldioxirane are used. Smooth insertion of oxygen into the α-Si–H bond takes place to give the 1-metallo-1,1-dihydroxydisilanes **19a**, **19b** (Eq. 3).

Eq. 3.

Attempts to convert additionally the SiH$_3$-unit into an Si(OH)$_3$-group by using an excess of dimethyldioxirane leads to uncontrolled decomposition. The complexes **19a**, **19b** suffer decomposition at room temperature [2], but the pronounced reactivity of the SiOH-units allows base assisted condensation with Me$_2$Si(H)Cl generating the rather unusual siloxanes **20a**, **20b** (Eq. 3). These polyfunctionalized silicon metal complexes, having an Si–Si unit incorporated into the siloxy fragment, promise interesting chemistry involving the different Si–H sites.

The single crystal X-ray diffraction analysis of **20b** [14] shows an Si1–Si2 distance of 2.371 (6) Å, similar to that found for metal-free disilanes, e.g., hexamethyldisilane (2.35 Å) [15]. The angles at the α-silicon Si1 prove a tetrahedral geometry with the Si2 being located in anti-position to the C$_5$Me$_5$-ligand at the metal (Fig. 3). The large angles Si1–O3–Si3 (157.0 (5)°) and Si1–O4–Si4 (152.0 (6)°) are indicative of a certain degree of π-interaction in the Si–O bond.

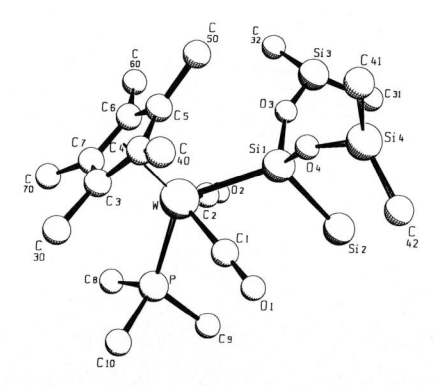

Fig. 3. Molecular structure of **20b**; selected bond lengths [Å] and bond angles [°]: W–P 2.431 (4), W–Si1 2.529 (3), Si1–Si2 2.371 (6), Si1–O3 1.633 (8), Si3–O3 1.574 (9), W–C1/C2 1.920 (1), P–W–Si1 123.1 (1), P–W–C1 78.0 (3), W–Si1–Si2 115.2 (2), Si1–O3–Si3 157.0 (5), Si1–O4–Si4 152.0 (6), Si2–Si1–O3 104.1 (3)

Acknowledgement: This work has been generously supported by the *Deutsche Forschungsgemeinschaft* and the *Fonds der Chemischen Industrie*.

References:

[1] Metallo-Silanols and Metallo-Siloxanes, Part 5. – Part 4: W. Malisch, K. Grün, N. Gunzelmann, S. Möller, R. Lankat, J. Reising, M. Neumayer, O. Fey, in: *Selective Reactions of Metal Activated Molecules* (Eds.: H. Werner, J. Sundermeyer), Vieweg Verlag, Braunschweig, in press; in addition Part 30 of the series *Synthesis and Reactivity of Silicon Transition Metal Complexes*.

[2] Some of these results appeared as abstracts: W. Malisch, S. Möller, K. Hindahl, *Xth FECHEM Conference on Organometallic Chemistry*, Crete, **1993**, Abstract of Papers, 235; W. Malisch, S. Schmitzer, G. Kaupp, K. Hindahl, *Xth FECHEM Conference on Organometallic Chemistry*, Crete, **1993**, Abstract of Papers, 59; W. Malisch, S. Schmitzer, G. Kaupp, K. Hindahl, *Xth International Symposium on Organosilicon Chemistry*, Posen, **1993**, Abstract of Papers, 80.

[3] W. Noll, *Chemie und Technologie der Silikone*, Verlag Chemie, Weinheim, **1968**.

[4] S. Schütte, U. Pieper, U. Klingebiel, D. Stalke, *J. Organomet. Chem.* **1993**, *446*, 45.

[5] N. Winkhofer, A. Voigt, H. Dorn, H. W. Roesky, A. Steiner, D. Stalke, A. Reller, *Angew. Chem.* **1994**, *106*, 1414; N. Winkhofer, H. W. Roesky, M. Noltemeyer, W. T. Robinson, *Angew. Chem.* **1992**, *104*, 670.

[6] Z. H. Aiube, N. H. Buttrus, C. Eaborn, P. B. Hitchcock, J. A. Zora, *J. Organomet. Chem.* **1985**, *292*, 177.

[7] W. Malisch, S. Schmitzer, G. Kaupp, K. Hindahl, H. Käb, U. Wachtler, in: *Organosilicon Chemistry – From Molecules to Materials* (Eds.: N. Auner, J. Weis), VCH, Weinheim, **1994**, p. 185.

[8] C. E. F. Rickard, W. R. Roper, D. M. Salter, L. J. Wright, *J. Am. Chem. Soc.* **1992**, *114*, 9682.

[9] W. Ries, T. Albright, J. Silvestre, I. Bernal, W. Malisch, Ch. Burschka, *Inorg. Chim. Acta* **1986**, *111*, 119.

[10] R. Maisch, W. Barth, W. Malisch, *J. Organomet. Chem.* **1983**, *247*, C47; W. Angerer, W. Malisch, M. Cowley, N. C. Norman, *Chem. Commun.* **1985**, 1811.

[11] W. Adam, U. Azzena, F. Prechtl, K. Hindahl, W. Malisch, *Chem. Ber.* **1992**, *125*, 1409.

[12] S. Schmitzer, W. Malisch, to be published.

[13] S. Schmitzer, U. Weis, H. Käb, W. Buchner, W. Malisch, T. Polzer, U. Posset, W. Kiefer, *Inorg. Chem.* **1993**, *32*, 302.

[14] The structure of **20b** shows a 50 %-disorder with respect to the methyl group C31 and the silicon-bound hydrogen at Si3.

[15] B. Beagley, J. J. Monaghan, *J. Mol. Struct.* **1971**, *8*, 401.

Oligosilanyl-Tungsten Compounds-Precursors for Tungsten-Silicide CVD?

Arno Zechmann, Edwin Hengge*
Institut für Anorganische Chemie
Technische Universität Graz
Strehmayrgasse 16, A-8010 Graz, Austria

Introduction

Improvements in the performance of integrated circuits and the trend towards VLSI-technology require the replacement of polycrystalline silicon by materials with a lower resistivity for use as gate electrodes. Transition metal silicides appear to be valuable possibilities for these applications. Tungsten-silicon compounds could be suitable precursors for the precipitation of tungsten-silicide thin films. Moreover tungsten-silicon compounds are nearly unknown and of scientific interest.

1 Syntheses

1.1 Hydrogenated Systems

By way of salt elimination, we were able to synthesize the first perhydrogenated α,ω-ditungsten oligosilanes. Using homogeneous conditions (mixture THF/$n$pentane or $n$heptane as solvent), yields were up to 55 %.

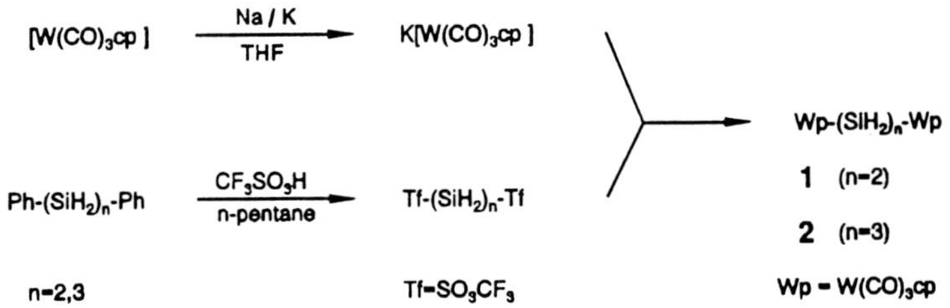

$$[W(CO)_3cp] \xrightarrow[THF]{Na/K} K[W(CO)_3cp]$$

$$Ph\text{-}(SiH_2)_n\text{-}Ph \xrightarrow[n\text{-pentane}]{CF_3SO_3H} Tf\text{-}(SiH_2)_n\text{-}Tf$$

$$\longrightarrow Wp\text{-}(SiH_2)_n\text{-}Wp$$

n=2,3

1 (n=2)
2 (n=3)

Tf=SO$_3$CF$_3$ Wp = W(CO)$_3$cp

Scheme 1.

Transmetallation to a dinuclear tungsten cluster could be avoided by using triflic acid as leaving group at the Si-part (no halogen-alkali exchange). The compounds which showed a surprisingly high stability (e.g., compounds are stable at room temperature for several hours and no evolution of CO could be detected) were purified by recrystallisation from $n$pentane [1]. Application of these compounds as precursors in CVD experiments will be examined in further studies.

1.2 Permethylated Systems

α,ω-Dihalosilanes do react very slowly (reaction times up to 4 weeks) with complex tungstenoates in nonpolar solvents (e.g., methylcyclohexane) giving monosubstituted compounds [2]. Even treatment with ultrasound does not shorten reaction times, but improves yields.

$$Cl\text{-}(SiMe_2)_n\text{-}Cl + K[W(CO)_3cp] \xrightarrow[RT / 14d]{\text{heterogenous} \atop \text{Methylcyclohexane}} Cl\text{-}(SiMe_2)_n\text{-}W(CO)_3cp$$

3 (n=3)
4 (n=4)

Eq. 1.

Further treatment of Wp–(SiMe$_2$)$_n$–Cl with Na[W(CO)$_3$Cp] leads to the transmetallation product only, with no difference in the solvent used (Methylcyclohexane or THF/$n$heptane respectively), according to the work of Malisch [2]. Attempted reduction with LiAlH$_4$ at –40 °C leads to cleavage of the Si–W bond.

Using DME/$n$heptane in the first reaction step, synthesis of ditungsten substituted oligosilanes is possible within 24 h. The bright yellow product can be separated by fractional recrystallisation from $n$pentane.

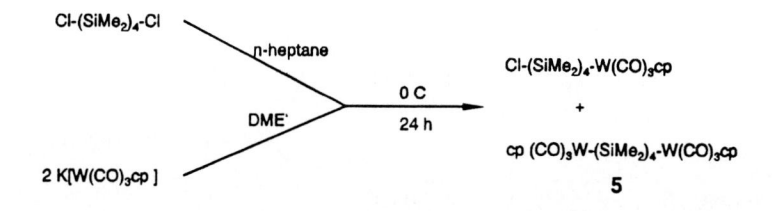

$$\begin{matrix} Cl\text{-}(SiMe_2)_4\text{-}Cl \\ + \\ 2\ K[W(CO)_3cp] \end{matrix} \xrightarrow[24\ h]{n\text{-heptane} \atop DME,\ 0\ C} \begin{matrix} Cl\text{-}(SiMe_2)_4\text{-}W(CO)_3cp \\ + \\ cp(CO)_3W\text{-}(SiMe_2)_4\text{-}W(CO)_3cp \\ \mathbf{5} \end{matrix}$$

Scheme 2.

2 ^{29}Si NMR Spectroscopic Results

All compounds show a distinct low-field shift in ^{29}Si NMR spectroscopy for Si$_{(1)}$. Coupling constants show the expected values (see Table 1).

Compound	δ [ppm] Si$_{(1)}$, Si$_{(2)}$, ...	J[Hz]
1	−65.79	$^1J_{Si\,W}$ = 20.0; $^1J_{Si\,H}$ = 184.5
2	−73.18; −94.88	$^1J_{Si\,W}$ = 21.0; $^1J_{Si\,H}$ = 185.0
3	−7.67; −34.53; 27.59	$^1J_{Si\,W}$ = 22.5
4	−5.26; −42.75; −32.79; 27.39	$^1J_{Si\,W}$ = 21.5
5	−4.06; −21.47	$^1J_{Si\,W}$ = 20.0

Table 1. ^{29}Si NMR data for oligosilanyl-tungsten compounds

References:

[1] B. Stadelmann, E. Hengge, M. Eibl, *Monatsh. Chem.* **1993**, *124*, 523.
[2] W. Malisch, *J. Organomet. Chem.* **1974**, *82*, 185.

Silicon Polymers: Formation and Application

Norbert Auner, Johann Weis

Materials based on silicon are found everywhere in nature in form of silica and silicates which count for more than ninety percent of the earth's crust.

A common feature of all these compounds is their tetrahedral structure at the silicon atom which is bound to four oxygen neighbors. A tremendous breakthrough in the history of silicon-based polymers has been achieved by the invention of the Direct Process by Müller and Rochow resulting in the industrial production of methyl chlorosilanes with hydrolytically stable Si–C bonds besides very reactive Si–Cl bonds which serve as building units for a wide variety of polydimethyl siloxanes including silicon fluids, resins, and elastomers.

The variation of organo-substituents at the silicon centers performed by Grignard reactions or by hydrosilylation of H-silanes or H-siloxanes with alkenes and alkines, respectively, gives even today materials with new industrially exploitable properties. Thus, besides the classical routes of condensation and addition curing mechanisms to build silicon network structures, new opportunities of photo-crosslinking have been put into practice. This counts as well for radical as for cationic systems.

The spectrum of silicon based polymers has been enriched most recently by particles with cores of precrosslinked silicones and shells of pure organic polymers thus overwhelming the incompatibility of the inorganic siloxane and the organic polymer phases. Those core-shell-structures have been established by using as well grafting to as grafting from techniques which offers a large potential for modifying organic polymers by silicones.

The variety of silicon based polymers is completed by high tech ceramics like silicon nitride and carbide. These materials are produced by pyrolysis of appropriate polymeric precursors such as polysilanes, -carbosilanes, and silazanes. Another important approach is realized by sol-gel processing.

Most of the contributions in Chapter IV deal with this area of research. Commercial interest is centered on the field of silicones with defined silicon substitution patterns, molecular weight, crosslinking degree and the study of structure-property correlations. In constrast the academic interest focusses on topics that may be recognized as "old" problems, such as "silyl modified surfaces", "synthesis and reactivity of silsesquioxanes", and "luminescent silicon". The reader will find answers to these problems in this chapter.

References:

[1] H.-H. Moretto, M. Schulze, G. Wagner, in: *Ullmannn's Encyclopedia of Industrial Chemistry, Vol. 24 A* (Eds.: B. Elvers, S. Hawkins, W. Russey, G. Schulz), VCH, Weinheim, **1993**.

[2] A. Tomanek, *Silicones and Industry*, Hanser, München, **1992**.

Silyl Modified Surfaces – New Answers to Old Problems

Joseph Grobe

Anorganisch-Chemisches Institut

Westfälische Wilhelms-Universität

Wilhelm-Klemm-Str. 8, D-48149 Münster, Germany

Summary: This contribution reports results of a cooperative research program of three groups at Münster University aimed at:

- The application of organosilyl esters to a variety of oxidic substrates, including sandstones of different origin, technical materials like concrete, glass, silica, alumina, or titania
- The preparation of protective organosilicon compounds and functional organoethoxy-silanes for the anchoring of biomolecules
- The elucidation of surface reactions (hydrolysis, condensation) by different analytical methods (IR, SIMS, and NMR)
- The immobilization and activity of glucose oxidase on the surface of controlled pore glass

Introduction

For about 30 years impregnation with organosilicon compounds is one of the most effective protective methods for monuments and buildings. The efficiency of the treatment depends on the building material, the protective agent and the choice of the application procedure. The necessity of such measures is obvious from the exponential increase of weathering decay of historical buildings (Fig.1), clearly demonstrating anthropogenic influences due to industrial development [1].

This enormous deterioration not only causes an irreparable loss of cultural identity but also implies a heavy financial load for the community budget. Therefore, closer inspection of either high performance results or failures of silyl modification is an important task for investigations with modern analytical methods. It is obvious that the increase in knowledge about surface reactions and the bonding processes not only enables the tailoring of more effective agents but will also provide a basis for the application of organofunctional silyl esters as immobilizing spacers for biomolecules. Both aspects will be discussed in this report, with particular emphasis on new results.

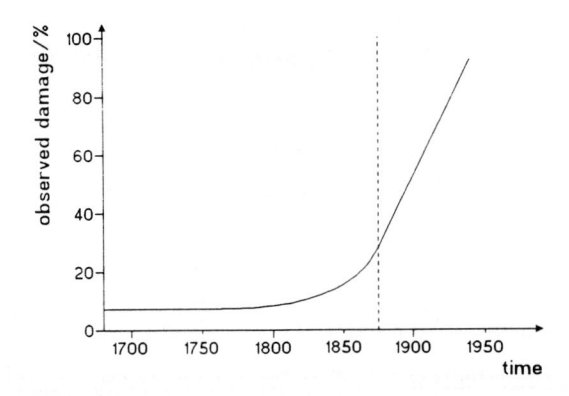

Fig. 1. Time profile of the observed weathering damage over the past 200 years

Protecting Coats for Baroque Angels
1 Mechanisms of Building Decay

In order to gain an understanding and insight into the possibilities of repairing damaged building materials and to avoid further deterioration a short overview of the main weathering effects is useful. Weathering comprises complex and synergetic combinations of physical, chemical, and biological processes generally linked to the presence of water on the surface and in the pore structure of the building material. Therefore, building protection essentially is protection against water. It is obvious that the simplest way to achieve this is to place valuable monuments in museums, churches, and other buildings, thus avoiding the entrance of water. Alternatively, outdoor monuments could be covered with plastic foils, a possible but impracticable option for buildings like Cologne cathedral.

Physical weathering generally is due to mechanical forces in the porous material. e.g. via freezing of water, swelling and shrinking of layer silicates or hydrostatic crystallization pressure caused by the formation of salt hydrates. In the *chemical deterioration* of building materials, moisture acts as a transport medium for aggressive gaseous compounds like SO_2, NO_x, or CO_2 and/or the corresponding salts (sulfates, nitrates, carbonates) formed by chemical reactions with the binder matrix. *Biological weathering* is caused by plant growth or microorganisms like nitro- or thio bacteria producing stone damaging nitrates and sulfates even without acidic gases. Note the important role of water for the transport of acids and salts as well as a nutrient for microorganisms.

The short discussion of the synergetic mechanisms shows that effective protection of building materials requires two different measures:

1) Consolidation of the damaged material by replacing or reinforcing the affected matrix with a synthetic binder
2) Reduction of the water uptake via hydrophobation of the stone

while maintaining the genuine appearance of the building material.

2 Protective Coatings on the Basis of Organosilicon Compounds – Quality Criteria

The two requirements indicate that effective and invisible coatings can be accomplished by using compounds that

- Reduce the spontaneous uptake of moisture
- Strengthen the porous material
- Form water repellant films on the inner surface of the porous system rather than covering the external surface
- Exclude negative changes of the natural physical and mechanical parameters (e.g., elasticity)

Organosilicon compounds belong to the first chemicals suggested and used for building protection. Already in 1861, A. W. Hofmann proposed the application of "silicon ether" for the restoration of the Houses of Parliament in London. However, extensive employment of these protectives had to wait until the industrial production of organochlorosilane precursors $X_n SiCl_{4-n}$ via Müller/Rochow synthesis was possible. From these precursors reaction with the corresponding alcohol ROH leads to compounds of the general type $X_n Si(OR)_{4-n}$ which are the basis of the main organosilicon preventatives.

Depending on the nature of X, the products obtained by hydrolysis and intermolecular condensation of the resulting silanols either show water repellant properties (X = alkyl) or allow consolidation of stones without significantly decreasing the spontaneous moisture uptake (X = OR). This is schematically shown in Fig. 2.

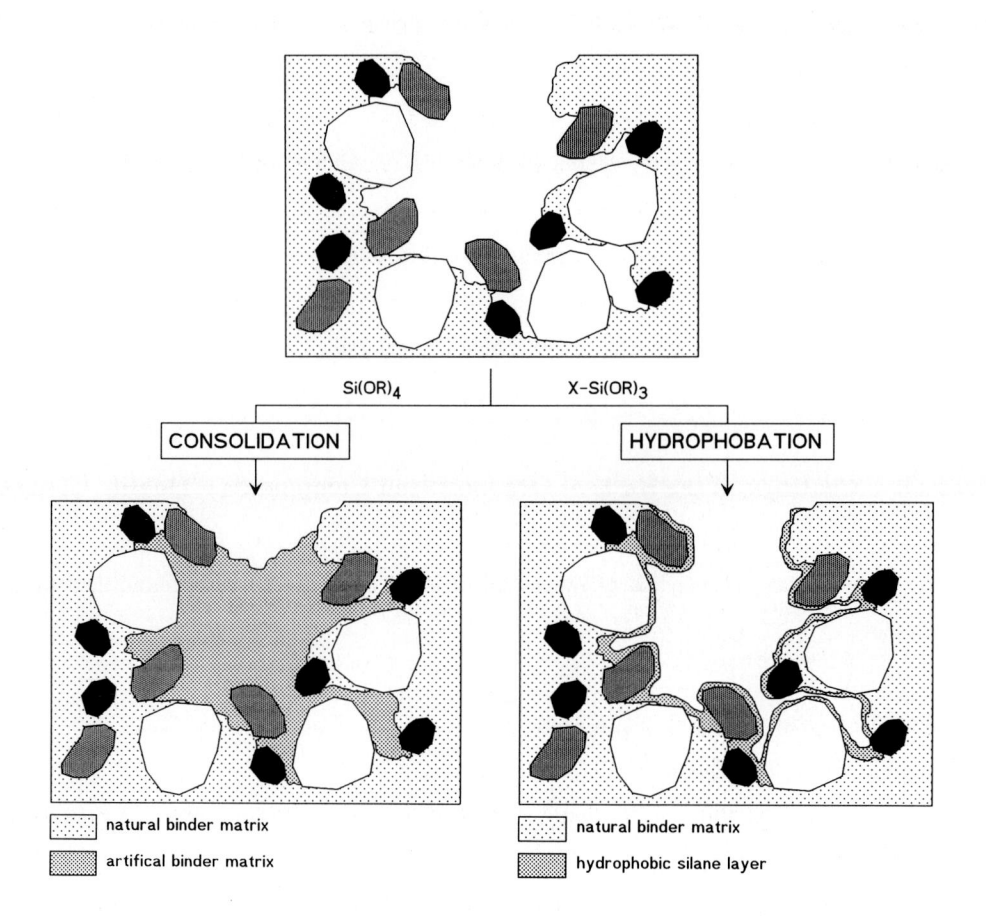

Fig. 2. Schematic description of the consolidation and hydrophobation of building materials with organosilicon compounds

In view of our aim to elucidate unexpected failures of impregnations with organosilanes the program of our research team at Münster University was directed to the use of simple model compounds and substrates. Instead of complicated oligomer mixtures containing various additives (e.g., catalysts) we usually applied $XSi(OR)_3$ monomers and besides natural sandstones included simple powdered oxides (SiO_2, Al_2O_3, TiO_2) and low surface area samples like glass and silicon wafers. This program allows a systematic variation of parameters and a step-to-step control of the changes on the solid surface by suitable physical methods.

2.1 Impregnation of Building Materials and Spontaneous Water Uptake-Criterion

In order to minimize water related damage, the primary purpose of protective agents is the reduction of the spontaneous uptake of moisture. With respect to the WTA recommendations for hydrophobations [2] a compound is considered to be effective if residual water uptakes < 30 % (relative to the untreated material) are obtained. As illustrated in Fig. 3 this criterion cannot be realized for all the combinations mentioned emphasizing that the application of a universal protective agent is not feasible.

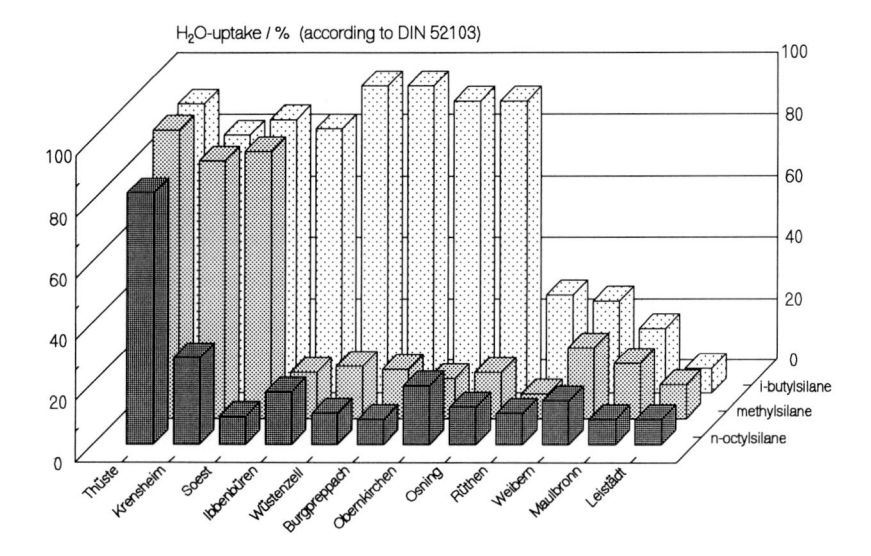

Fig. 3. Spontaneous water uptake of sandstones after impregnation with various ethoxysilanes (according to DIN 52103)

2.2 Spontaneous Water Loss of Impregnated Building Materials – A New Criterion

The measurement of the spontaneous uptake of moisture is one valuable criterion to evaluate the efficiency of the hydrophobation. However, this parameter cannot supply information about the dynamics regarding the water household and, therefore, underlines the importance to register the time profiles of both water uptake and water loss. According to our results obtained from more than 400 combinations of protective agents and building materials the rate of the water uptake generally decreases after the application of the organosilanes. In addition, impregnation with these applicates also affects the drying process; as indicated in Fig. 4, the hydrophobic properties of the organosilicon layer accelerate the water loss from the porous system.

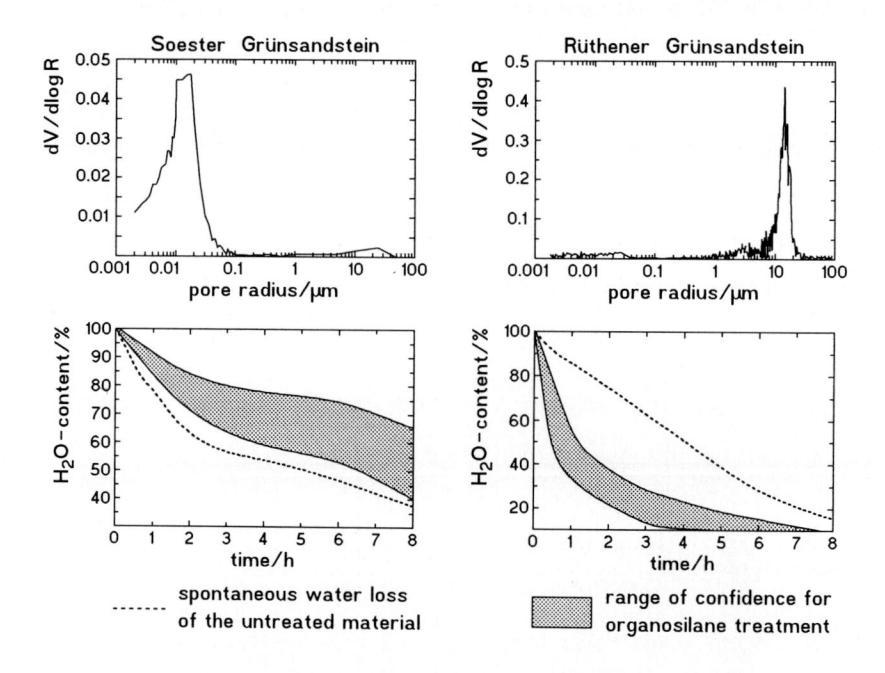

Fig. 4. Effect of the hydrophobation on the spontaneous water loss of building materials with various pore sizes

Nevertheless, this promising aspect does not apply to samples with smaller pore diameters. A comparison of the Soester and Rüthener Sandstein strongly suggests that for less porous building materials the moisture is retained inside the system. Evidently, the application of organosilanes here causes unexpected side effects, likely due to the narrowing of pores.

2.3 Penetration of Organosilicon Compounds in Natural Sandstones – Possible Routes to an Optimized Treatment

In order to avoid negative changes in the natural water household of building materials the applicates should be able to penetrate the pore structure sufficiently. Thus, depth profiling the organosilane after impregnation is important for the explanation of failures. A quantitative characterization of the depth profiles is possible by chemically altering the basis compounds $XSi(OR)_3$ to $Cl-XSi(OR)_3$ with the Cl atom acting as a tracer in the SEM experiment. Fig. 5 demonstrates the penetration properties depending on the kind of application procedure.

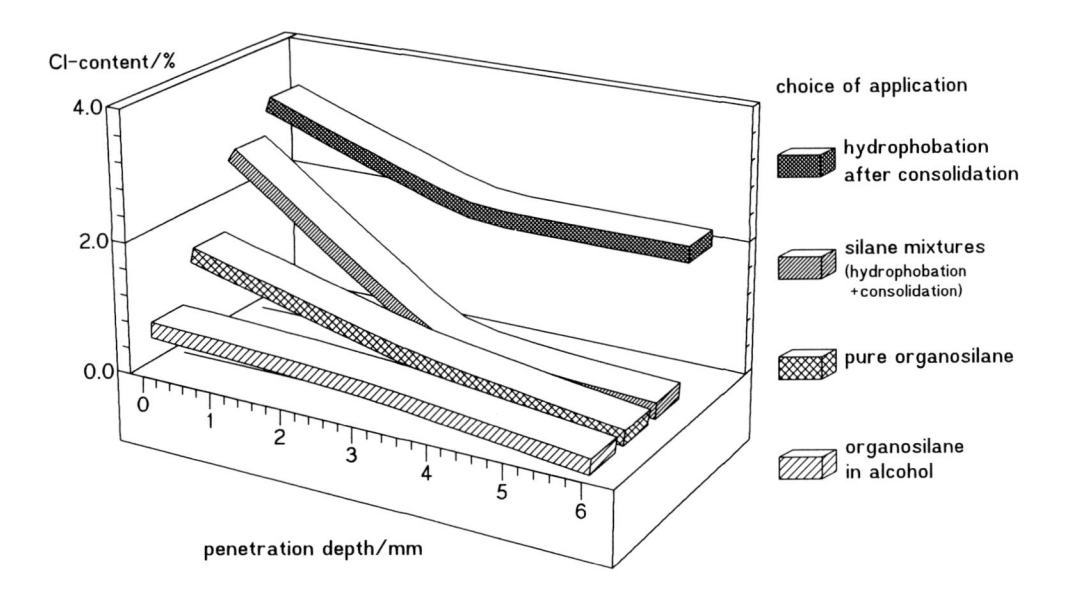

Fig. 5. SEM depth profiles of organosilanes as a function of the application procedure (material: Obernkirchener Sandstein)

Compared to the impregnation with the pure silane, consolidation prior to hydrophobation increases both, the silane uptake and the penetration depth. The positive contribution of stone strengthening compounds also becomes obvious when the material is treated with mixtures containing $XSi(OR)_3$ and $Si(OR)_4$. However, impregnation with silane/alcohol mixtures reveals surprisingly low values in terms of the silane loading; this behavior is due to the separation of the homogenous solution during migration through the pore structure (chromatographic effect).

2.4 Application of Mixtures $XSi(OR)_3/X'Si(OR)_3$

As mentioned before, the application of a universal organosilane is not realistic. The data presented in Fig. 3 indicate, however, that the hydrophobic properties of silanes with long alkyl groups (C4–C8) generally lead to lower water uptake compared to smaller applicates (e.g., C1). Due to higher manufacturing costs for C4–C8 organosilanes, the investigation of mixtures containing long chain as well as short chain silanes is necessary. Fig. 6 displays the water uptake of a Sander Schilfsandstein after impregnation with organosilane mixtures containing varying amounts of a C5 silane.

The WTA limit of 30 % can be obtained even for concentrations of 5 % C5 silane suggesting the impregnation with the mixture rather than with the cost prohibitive pure compound.

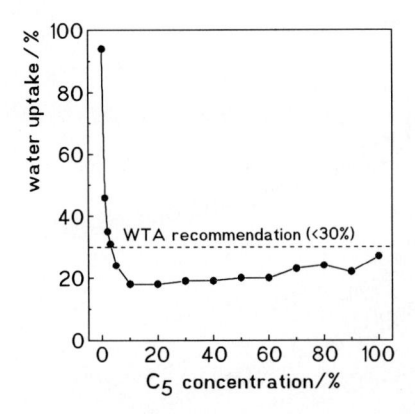

Fig. 6. Water uptake of a Sander Schilfsandstein after impregnation with mixtures of a C5 silane (normalized to the untreated stone = 100 %)

3 The Synthetic Program – Novel Compounds for Special Applications

Summarizing the various reasons for weathering indicates that a protection of the building material often requires more than simply the generation of water repellent layers. This aspect underlines the importance of organosilicon chemistry for developing new and promising compounds. The chemical structures given in Fig. 7 represent tailored organosilanes which have been synthesized in the Münster group.

Fig. 7 Principle synthesis paths for tailored organosilanes

Starting from the basic compound $XSi(OR)_3$, introducing alkyl- or additional OR groups leads to the well known hydrophobic (A) and consolidation agents (B), respectively. In type (C) two basic units are linked to each other via a flexible $(CH_2)_n$ chain of variable length to enhance the elasticity of the generated layer. The extent of reversible swelling and shrinking phenomena in layer silicates can be reduced with type (D) silanes. The terminal NR_3^+ groups in the structure act as molecular anchors to fix the layer distance in, e.g., the mica mineral.

4 Molecular Basis of Macroscopic Effects – Spectroscopic Investigations

The discussion of silyl ester applications to different stone materials demonstrates that both consolidation and hydrophobation are possible. However, the results of the treatment differ considerably depending on the type of substrate and/or silyl ester.

In order to elucidate the reasons for failures and to gain additional information about the bonding to the solid surface (covalent or physisorptive) and, furthermore, about the stability and durability of the hydrophobic layer a systematic cooperative study was undertaken using modern spectroscopic methods. Three powerful analytical techniques (IR, NMR, static TOF-SIMS) were used to elucidate the chemical structure of the condensation products resulting from various organosilyl esters $RSi(OR')_3$ applied to model substrates as well as natural sandstones. The different methods are supplementary with respect to depth and type of structural information. Whereas IR and NMR spectroscopy is well suited to detect characteristic groups and building units of the condensation products, static TOF-SIMS offers the possibility of analyzing the polymeric surface. The results to be discussed in this section exclusively refer to special examples.

4.1 IR Spectroscopic Investigations

The prominent goal of these studies was the detection of interactions between substrate and silyl esters with different organic groups R. The model substrates used include various kinds of silica, alumina, titania in addition to typical sandstones and silicon wafers. With respect to a contribution in the first conference report [3] one example may suffice here to demonstrate the procedure and the importance of the result obtained. The example concerns the treatment of a silicon wafer surface with 3-aminopropyltriethoxysilane (APTES) aimed at the detection of covalent bonding and possible interaction of the functional aminopropyl group with reactive surface centers. The method of choice is a combination of multiple internal reflection (MIR) FTIR with *time-of-flight*-SIMS measurements. The result of the IR investigation is shown in Fig. 8.

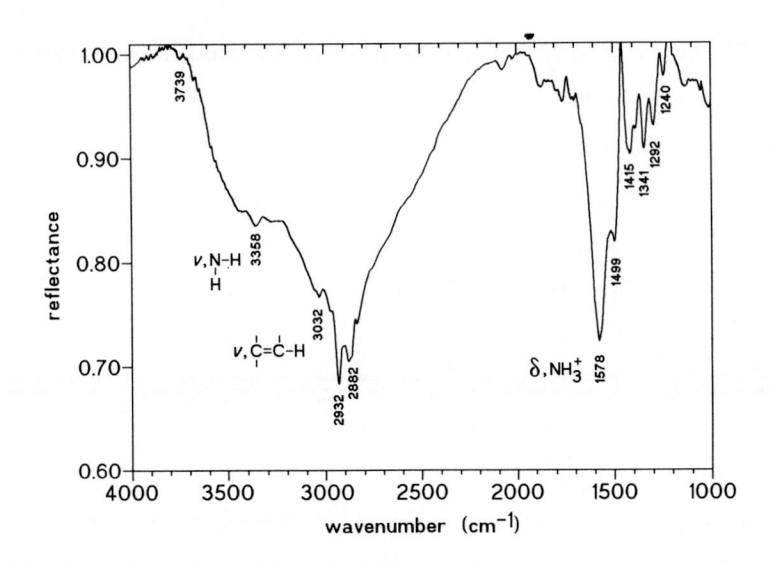

Fig. 8. MIR spectrum of a 3-aminopropyltriethoxysilane film on silicon

Absorption bands indicative of surface or surface catalyzed reactions result from alkylammonium and olefinic groups proving that hydrolysis and condensation of APTES on the wafer surface are accompanied by side reactions. Treatment of the condensate with water leads to the removal of the siloxane film which obviously is not covalently attached to the substrate. A surprisingly different result is observed for the condensate formed from 4-aminobutyltriethoxysilane (ABTES) with only one additional CH_2 group in the side chain. Bands due to $-CH=CH_2$ or $[-CH_2NH_3]^+$ are absent or appear in the IR spectrum with little intensities. Furthermore, the polycondensate is stable towards water in the pH range 2–11 over a period of 3 days.

4.2 High-Resolution Solid State NMR Studies of Polymeric Siloxanes

High resolution ^{29}Si MAS-NMR spectroscopy has proved to be a powerful tool for the structural investigation of oligomeric and polymeric organosiloxanes [4]. The informations obtained allow the detailed characterization of the different structural units as well as the determination of crosslinking within the polymeric network or interactions with the substrate surface [5]. Depending on the number of siloxane linkages $Si(OSi)_n$ to the silicon centre ^{29}Si resonances in typical ranges of the spectra are observed which can be assigned to different Q^n building units (Q^0 to Q^4) [4]. In case of alkyltriethoxysilanes $RSi(OEt)_3$, the extent of hydrolysis and polycondensation can easily be described using the characteristic resonances due to M- ($RSi(OEt)_2OSi$), D- ($Si(OEt)(OSi)_2$), and T- ($RSi(OSi)_3$) groups.

Fig. 9 shows the ^{29}Si MAS-NMR spectra of a silica model substrate and two different sandstones after silylation with methyltriethoxysilane.

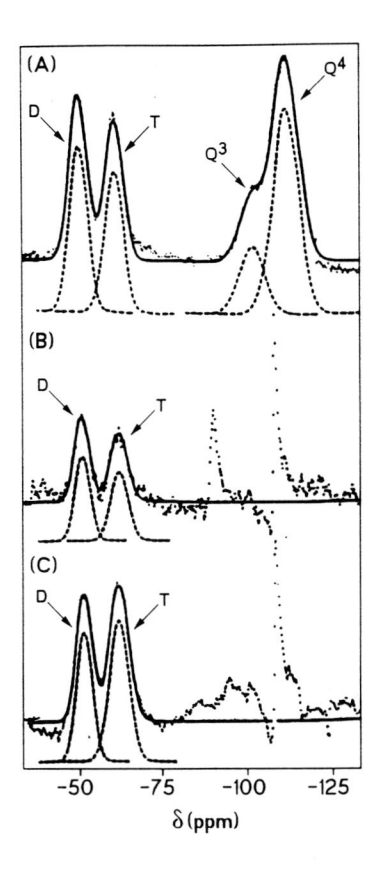

Fig. 9. ^{29}Si MAS-NMR spectra of silica (A), Burgpreppacher Sandstein (B), and Maulbronner Sandstein (C) after silylation with $CH_3Si(OEt)_3$

In case of the silica the substrate signals Q^2 (−91 ppm, $(HO)_2Si(OSi)_2$), Q^3 (−102 ppm, $HOSi(OSi)_3$), and Q^4 (−112 ppm, $Si(OSi)_4$) are affected by the formation of the methylpolysiloxane indicating changes in the surface structure of the material via reaction with $MeSi(OEt)_3$. The Q^2 resonance disappears completely and the simultaneous decrease of the Q^3 intensity proves the participation of the terminal (Q^3) and the geminal (Q^2) silanol groups in forming covalent bonds to the surface. The decrease of Q^2 and Q^3 intensities corresponds to the intensification of the Q^4 component in the spectrum of the treated substrate. The spectra of the polysiloxane on the different substrates are very similar. The observed differences are due to variable amounts of M, D, and T building units. The T/D ratio serves as a measure for the degree of crosslinking in the three-dimensional network. However, the presence of M and D groups indicates an incomplete condensation process, which additionally is underlined by the detection of residual Si–OEt components by means of ^{13}C MAS-NMR spectroscopy [6]. Thermal treatment simulating ageing processes successively converts M and D groups into T building units. Up to 500 °C the protective coating formed from MeSi(OEt)$_3$ reveals a surprising thermal stability, a result in accord with the IR spectroscopic investigations [3].

4.3 Secondary Ion Mass Spectrometry (SIMS) of Silyl Modified Surfaces

In addition to detailed information obtained by IR and NMR measurements static SIMS offers the important possibility of analyzing the surface of the polycondensate from organosilyl esters.

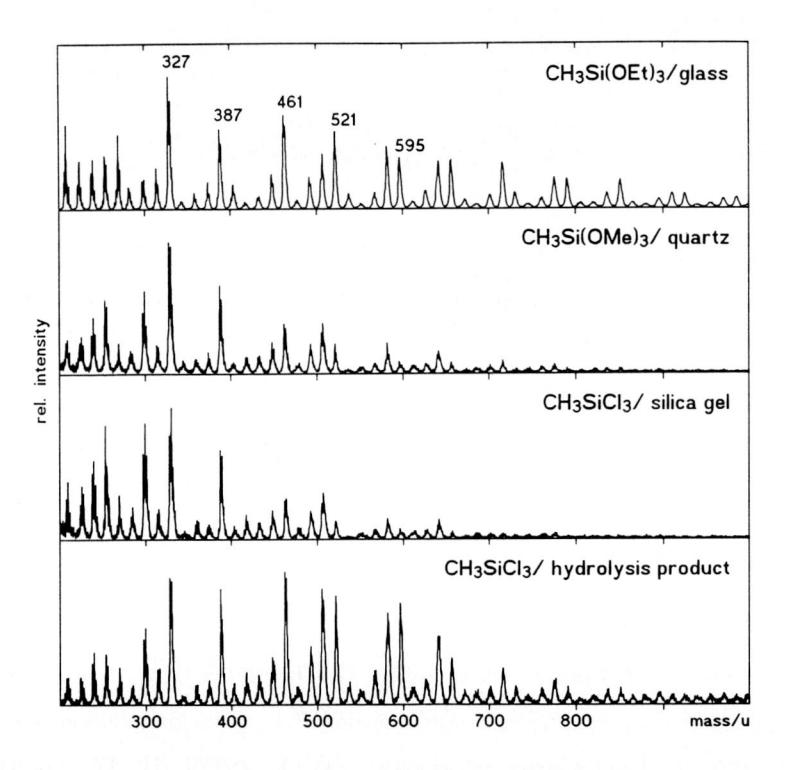

Fig. 10. Positive TOF-SIMS spectra of the polycondensates obtained from reactive methylsilanes CH$_3$SiX$_3$ (X = Cl, OMe, OEt)

In this respect, TOF-SIMS at present is a most efficient technique for obtaining molecular information from the uppermost monolayer [7]. Of particular value are the large mass range, the quasi-simultaneous ion detection and the high sensitivity of this method.

Fig. 10 demonstrates the special aptitude of SIMS measurements for investigating silyl modified surfaces prepared by the treatment of glass, quartz, or silica gel with different silanes CH_3SiX_3 (X = OEt, OMe, Cl).

All spectra, including that of the pure polycondensate from CH_3SiCl_3, show the same characteristic line pattern. Considering the low information depth of SIMS [7], this result indicates that the outmost monolayers consist of totally hydrolyzed polymeric siloxanes which do not contain the initial reactive group X. Similar results were obtained for the corresponding phenylsilanes $PhSiX_3$. Evaluation of the peak pattern gives evidence of the formation of silsesquioxanes $(CH_3SiO_{1.5})_n$ [8]. Comparison of the TOF-SIMS results with diffuse reflectance IR and NMR of these applicates shows that in near-surface layers and in the bulk material residual alkoxy groups are still detectable.

5 Functionalized Silyl Esters as Fetters or Anchors for Biomolecules

The interesting results obtained for silyl modified sandstones and model substrates via spectroscopic studies of surface reactions and bonding properties offered a good basis for the investigation of low surface area samples like silicon wafers. This material is used as one of the typical transducers for biosensors. The working principle of a sensor is schematically shown in Fig. 11.

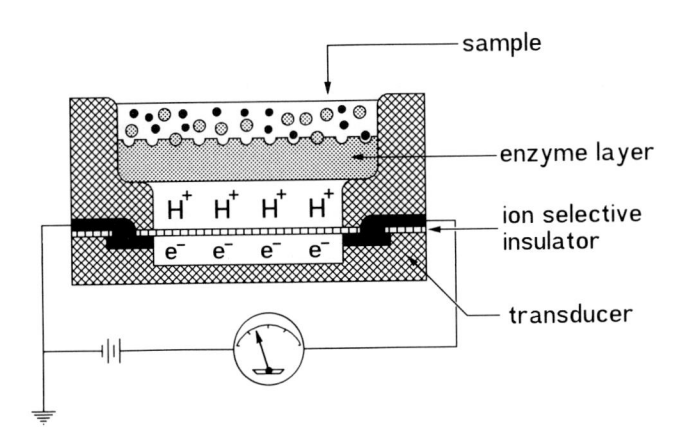

Fig. 11. Schematic build-up of a biosensor

Our contributions in this current area of research comprise the fixation of artifical or natural membranes and of detector molecules to the surface of a semiconductor, glass electrode, or optical conducting glass fibers using functionalized organosilyl esters of the general type $Me_n(EtO)_{3-n}Si(CH_2)_mX$ (X being a suitable group).

Spectroscopic results about the interaction of the traditional spacer compound 3-aminopropyltriethoxysilane (APTES) with a silicon surface have already been discussed in the first part of this paper leading to the conclusion that a considerable improvement regarding the stability and durability of the amino-functionalized siloxane film is possible by using the 4-aminobutyltriethoxysilane (ABTES) rather than the propyl compound.

The most popular techniques to immobilize biomolecules are based on the modification of a hydroxylated surface with APTES [9] followed by crosslinking with glutaraldehyde (GA) (Fig. 12A).

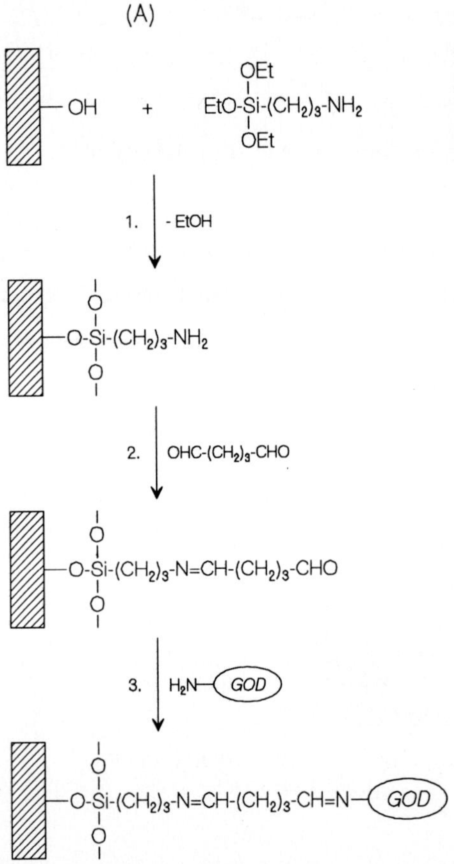

Fig. 12. Possible pathways for the immobilization of GOD on hydroxylated surfaces via APTES (A) or CHO-functionalized ethoxysilanes (B)

This procedure has a number of shortcomings:

- Crosslinking of amino groups of the spacer molecule by GA, thus reducing the number of aldehyde functions for the biomolecule interaction
- Undesirable polymer formation by GA [10]
- Poisoning of sensitive biological material (e.g. enzymes)

Our measures to overcome these problems include the preparation of aldehyde-functionalized silanes as a new class of spacer molecules for silica surfaces, the application of a series of such compounds with up to three ethoxy groups and different chain length of the anchor group to the surface of controlled pore glass (CPG) as a model substrate, the immobilization of glucose oxidase (GOD) and systematic activity studies over a period of four weeks.

Fig. 12 demonstrates that the novel technique allows direct reaction of the biomolecules with the silyl modified, aldehyde-functionalized surface to form Schiff base linkages between CHO and NH_2 groups (route B), thus eliminating the use of a crosslinking agent like GA. Omitting details of the experimental procedures [11] the results of three different immobilization methods are displayed in Fig. 13 presenting a comparison with the physisorptive attachment of GOD on native, underivatized CPG.

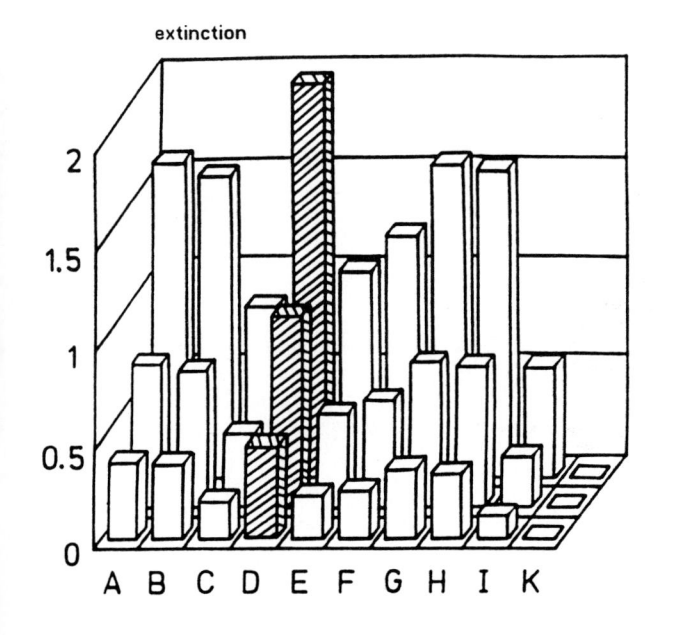

A: 3-Aminopropyl-triethoxysilane
B: 3-Aminopropyl-diethoxymethylsilane
C: Diethoxy-methyl-silyl-butyraldehyde
D: Ethoxy-dimethyl-silyl-butyraldehyde
E: Cyanuric chloride
F: Bis-(trichlorosilyl)-ethane
G: A + glutaraldehyde
H: B + glutaraldehyde
I: CPG native
K: blank

Fig. 13. Observed extinction values for three different GOD solutions (1:20, 1:10, 1:5) after immobilization on CPG via methods A–H.

After several hours all samples still show measurable enzymatic activities. In case of the physisorbed enzyme on native CPG the initial activity decreases almost to zero within two weeks. However, the covalently fixed samples exhibit a remarkable increase in activity by a factor of 8–11, which levels off after two weeks. The best results were obtained using aldehyde-functionalized ethoxysilanes (hatched bars in Fig. 13); here the very high activity is preserved even after storage of the samples at room temperature for more than four weeks. It is obvious that the new method described above represents a very promising alternative on high surface area materials to the traditional techniques using the combination of APTES/GA. Future investigations, therefore, have to focus on the reliability of aldehyde-functionalized silanes for the immobilization of biomolecules on low surface area samples like wafers, electrodes or glass fibers.

Acknowledgments: The studies reported in this lecture would have been impossible without the skilful collaboration of capable and enthusiastic coworkers. I would like to thank my graduate students and postdocs Markus Boos, Dr. Claudia Brüning, Roger Dietrich, Dr. Roland Fabis, Jutta Greb, Irmhild Krull, Dr. Karl Stoppek-Langner, and Manfred Wessels, who were involved in various projects of this program. I also gratefully acknowledge the fruitful cooperation with the research groups of Prof. A. Benninghoven and Prof. W. Müller-Warmuth at *Münster University* for supplying their expertise and equipment in TOF-SIMS and solid state NMR spectroscopy. Financial support came from the *Bundesminister für Forschung und Technologie*, from the *Institut für Chemo- und Biosensorik*, from the *Zollern-Institut* (Bochum), and from the *Fonds der Chemischen Industrie*. Valuable chemicals were made available by *Bayer AG*, *Hüls AG*, and *Wacker-Chemie GmbH*.

References:

[1]   E. M. Winkler, *Stone: Properties, Durability in Men's Enviroment, Appl. Mineral. 4*, Springer, **1985**.

[2]   *Wissenschaftlich-Technischer Arbeitskreis WTA*, Bautenschutz + Bausanierung **1978**, *3*, 72.

[3]   J. Grobe, K. Stoppek-Langner, A. Benninghoven, B. Hagenhoff, W. Müller-Warmuth, S. Thomas, in: *Organosilicon Chemistry – From Molecules to Materials* (Eds.: N. Auner, J. Weis), VCH, Weinheim, **1994**, p. 325.

[4] G. Engelhardt, D. Michel, *High Resolution Solid State NMR of Silicates and Zeolites*, Wiley, Chichester, **1987**; J. Kirkpatrick, R. A. Kinsey, K. A. Smith, D. M. Henderson, E. Oldfield, *Am. Mineral.* **1985**, *70*, 106.

[5] E. Lippmaa, M. Alla, T. J. Pehk, G. Engelhardt, *J. Am. Chem. Soc.* **1978**, *100*, 1929; E. Lippmaa, M. Mägi, A. Samoson, G. Engelhardt, A. R. Grimmer, *J. Am. Chem. Soc.* **1980**, *102*, 4889.

[6] S. Thomas, K. Meise-Gresch, W. Müller-Warmuth, K. Stoppek-Langner, J. Grobe, *Chemistry of Materials*, in press.

[7] A. Benninghoven, F. G. Rüdenauer, H. W. Werner, *Secondary Ion Mass Spectrometry*, Wiley, Chichester, **1987**.

[8] B. Hagenhoff, A. Benninghoven, K. Stoppek-Langner, J. Grobe, *Adv. Mater.* **1994**, *6*, *142*; J. Grobe, K. Stoppek-Langner, W. Müller-Warmuth, S. Thomas, A. Benninghoven, B. Hagenhoff, *Nachr. Chem. Tech. Lab.* **1993**, *41*, 1233.

[9] H. H. Weetall, R. D. Mason, *Biotechnol. Bioeng.* **1973**, *15*, 455.

[10] S. K. Bhatia, L. C. Shriver-Lake, K. J. Prior, J. H. Georger, J. M. Calvert, R. Bredehorst, F. S. Ligler, *Anal. Biochem.* **1989**, *178*, 408.

[11] C. Brüning, *Ph. D. Thesis*, Universität Münster, **1993**.

Alkenyl Ethoxysilanes for the Synthesis of Silylester Functionalized Copolymers and Surface Modification

Jutta Greb, Joseph Grobe*

Anorganisch-Chemisches Institut

Westfälische Wilhelms-Universität

Wilhelm-Klemm-Str. 8, D-48149 Münster, Germany

Summary: Silylester functionalized copolymers have been synthesized by free radical polymerization of alkenyl ethoxysilanes and vinyl acetate. The alkenyl ethoxysilane monomers and the preformed polymers have been applied to silica surfaces with hydroxy functions. Both treatments led to covalently modified samples of different properties as shown by DRIFT spectroscopy.

As bifunctional compounds, alkenyl ethoxysilanes can be used for the synthesis of silylester functionalized copolymers and for the modification of surfaces. With these products covalently fixed polymer films on surfaces can be produced, which are important for the construction of sensors and in coating technologies.

Silylester functionalized copolymers were synthesized by free radical polymerization of vinyl acetate with alkenyl ethoxysilanes in various molar ratios using AIBN (azo-bis-isobutyronitrile) as initiator. Eq. 1 shows the course of the reaction.

n = 1,4,6,8

Eq. 1.

^{1}H NMR spectroscopic studies have shown, that the obtained polymers contain the starting monomers vinyl acetate and ethoxysilane in the ratio of about 5:1. Molecular weight determination of the products using MALDI-MS (*m*atrix-*a*ssisted *l*aser *d*esoption *i*onization *m*ass *s*pectrometry) gave average values of 7 000–8 000 g mol^{-1}.

The alkenyl ethoxysilane monomers and silylester functionalized copolymers were used for the silanization of hydroxyl groups on silicate surfaces, thus producing silylester modified silicates or covalently fixed polymer films as demonstrated in Eq. 2 and Fig. 1.

$$\boxed{}Si-OH + R'O-SiR_3 \longrightarrow \boxed{}Si-O-SiR_3 + R'OH$$

Eq. 2.

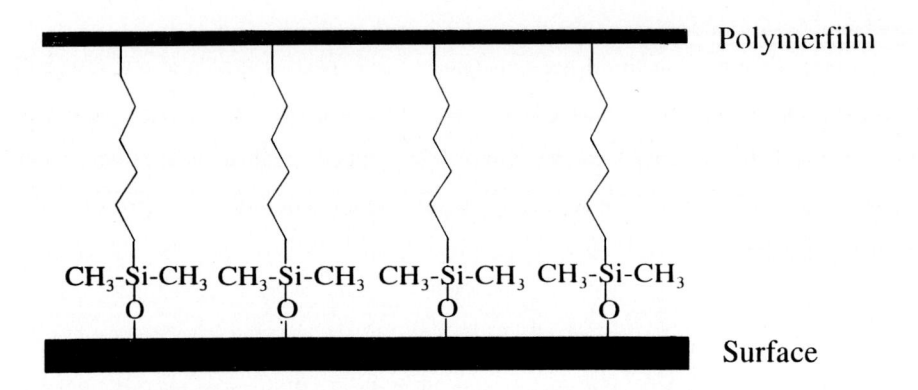

Fig. 1. Model of a fixed copolymer on a silicate surface

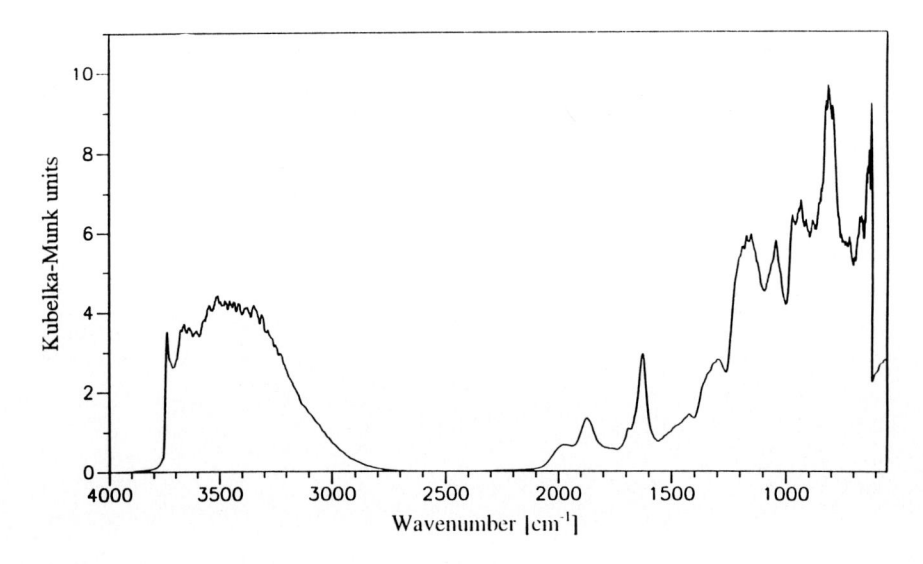

Fig. 2. DRIFT-spectrum of untreated silica gel

The silanized surfaces were characterized by DRIFT (diffuse reflectance infrared fourier transform) spectroscopy [1, 2]. Fig. 2–4 show the DRIFT-spectra of untreated silica gel and the analogous substrate after application of 1-octenyl-8-dimethylethoxysilane or of the copolymer.

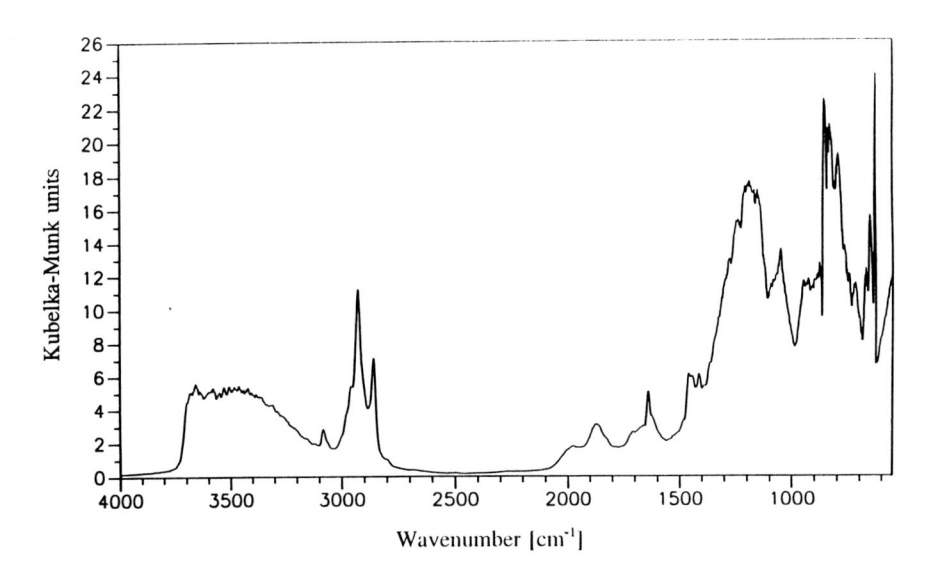

Fig. 3. DRIFT-spectrum of silica gel silanized by 1-octenyl-8-dimethylethoxysilane

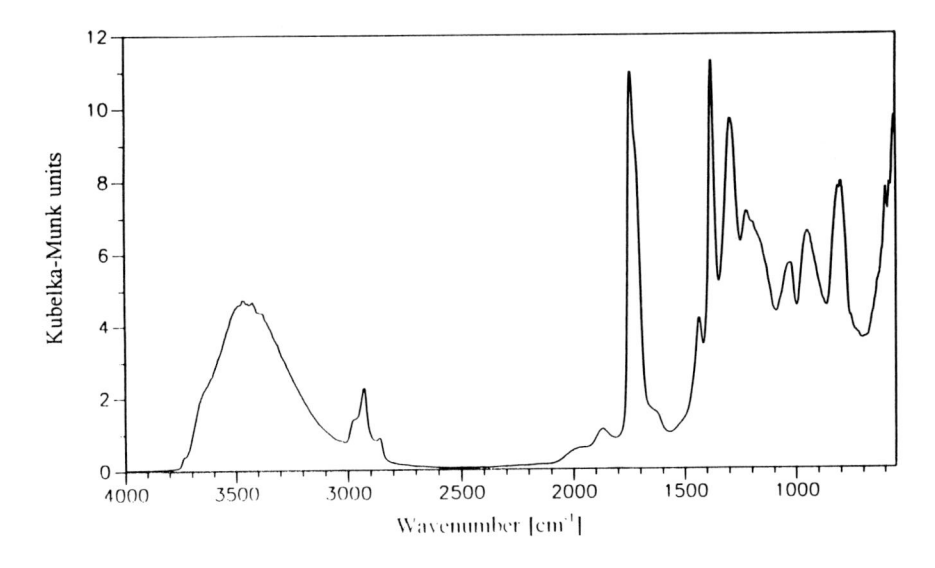

Fig. 4. DRIFT-spectrum of silica gel modified by the copolymer of 1-octenyl-8-dimethylethoxysilane and vinyl acetate

Fig. 3 presents the effective modification of silica gel by 1-octenyl-8-dimethylethoxysilane. The C–H valence bands at 2930 and 2870 cm^{-1} prove the presence of the alkenyl groups, the absorption at

3080 cm^{-1} is indicative of olefinic C–H groups. In contrast to Fig. 2, no peak at 3740 cm^{-1} caused by free silanol groups can be detected. The spectrum of silica gel treated with the copolymer shows the alkyl peak at 2930 cm^{-1} and the peak at 1740 cm^{-1} due to the C=O group of the vinyl acetate.

Systematic studies aimed at the formation of polymer films on silica surfaces by copolymerization of covalently immobilized alkenyl groups with suitable monomers are in progress.

References:

[1] M. Fuller, H. Ottenroth, *Labor Praxis* **1991**, 742.
[2] M. Fuller, P. R. Griffith, *Anal. Chem.* **1978**, *50*, 1906.

Silicone Surfactants –
Development of Hydrolytically Stable Wetting Agents

K.-D. Klein*, W. Knott, G. Koerner

Th. Goldschmidt AG

Goldschmidtstraße 100, D-45127 Essen, Germany

Summary: The hydrolysis of current organosilicone surfactants occurs at the Si–O–Si bonds. New silane surfactants free of Si–O–Si bonds have been developed recently. These materials show outstanding surfactant properties and are extremely stable in both acidic and alkaline conditions.

Organosilicones are favourable in a broad variety of industrial applications primarily because of their unique surface active properties. Especially in the case of aqueous applications the hydrophilically substituted trisiloxane derivatives are good wetting agents. Usually, the molecule consists of a lyophobic trisiloxane moiety attached directly to an alkyl spacer group via a silicon-carbon bond. This spacer group carries on the other side a nonionic or an ionic hydrophilic part.

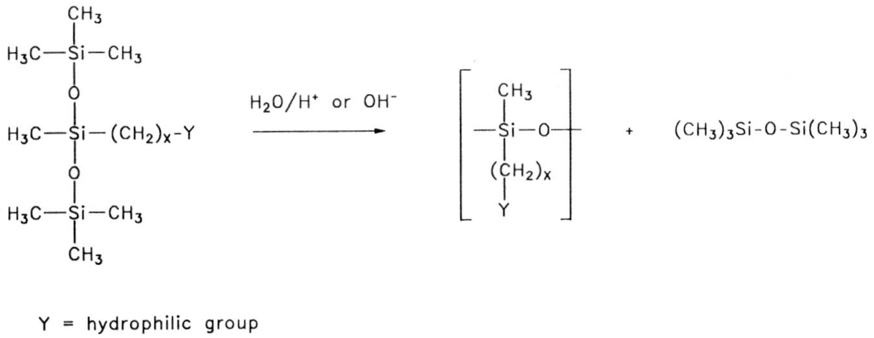

Y = hydrophilic group

Fig. 1. Hydrolysis of trisiloxane surfactants

Unfortunately the aqueous solutions of these surfactants are of limited stability due to a degradation process. The degradation of the trisiloxane surfactant may be explained by visualizing that the siloxane moiety undergoes a real reaction both with OH$^-$ and H$^+$ ions, respectively, thereby creating a poly-hydrophilic functionalized silicone oligomer. These oligomers exhibit very poor surfactant properties compared to the trisiloxane derivatives [1].

Reconsidering the past, many attempts have been made to enhance the hydrolytic stability of these products. The trials were based upon the variation of both the alkyl spacer part or similar hydrophobic moieties in order to protect the hydrolytically sensitive Si–O–Si substructure. These experiments lead to substances which show improved stability at neutral, slightly acid or alkaline pH. However, totally stable materials could not be obtained that way [2].

The only method to tackle that problem is to provide molecules which are simultaneously silicone-free and exhibit the unique surfactant properties of the trisiloxane derivatives.

Following that pathway, a first approach was made by Dow Corning [3]. Hydrogen carbosilanes were reacted with several α-olefins containing reactive moieties catalyzed by a platinum complex. Especially the resulting epoxy derivatives were useful starting materials directed to ionic organomodified carbosilanes after being reacted with nucleophilic reagents in a further step. The resulting ionic carbosilanes exhibited considerable surfactant properties combined with an outstanding stability in aqueous solutions over a wide pH range. However, the hydrogen carbosilane precursors were of limited availability because of a poor yield Grignard reaction.

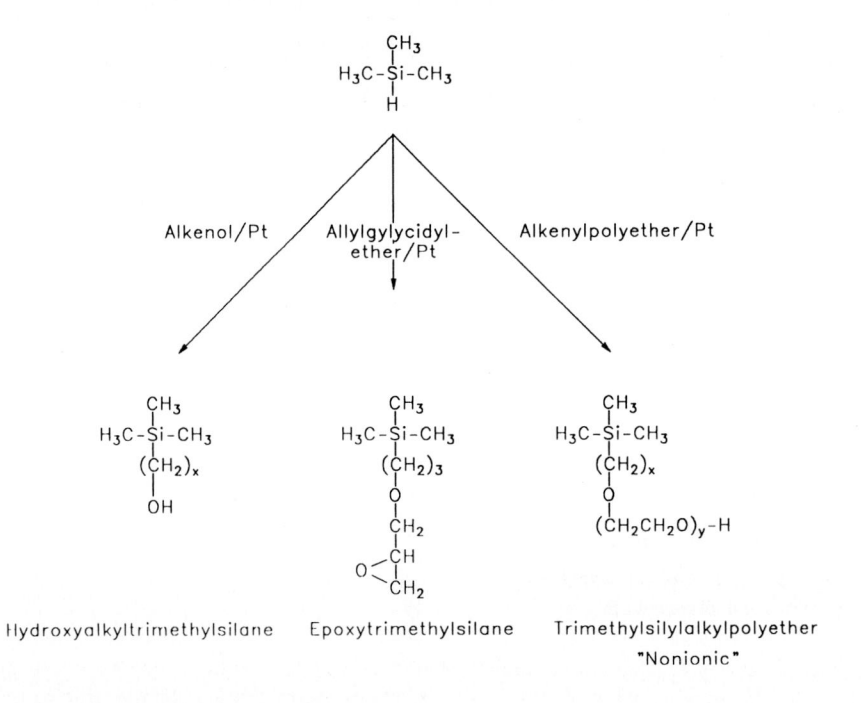

Fig. 2. Hydrosilylation reactions of trimethylsilane

Halogeno silanes can be reacted readily with hydrides, e.g. magnesium hydride, in the presence of an ethereal solvent to give the appropriate hydrogen silane [4]. Applying this method e.g. trimethylsilane was obtained in excellent yields. Trimethylsilane was reacted with α-alkenols, allylglycidyl ether, and various alkenyl polyethers whereby the latter directly leads to a nonionic silane surfactant. The easy availability of trimethyl silane and its derivatives opens up a very interesting route to new Si-surfactants from the view of economy, too.

Best surfactant behaviour was achieved in the case of the polyether trimethylsilane by introducing a hexyl spacer group in the molecule structure. For testing the wetting ability a droplet (50 µl) of an aqueous surfactant solution is applied by syringe to a clean sheet of polypropylene. Afterwards the increase of the diameter of the droplet is measured. The surface tension of the aqueous solution is usually determined by the well known ring method of du Noüy [5].

A 0.1 wt. % aqueous solution of the nonionic product shows a surface tension of 23 mN m^{-1} and a spreading ability of 80 mm – when applied as described above – in pH range 4.0–12.0 [6]. As we expected these solutions are stable over months at room temperature due to the fact that the hydrolytically unstable Si–O–Si substructure is omitted. Furthermore, this kind of substance is 88 % biodegradable at 28 days when tested according to OECD method 301 D.

The above mentioned hydroxy and epoxy functional silanes represent reactive intermediates which can be transformed into ionic trimethyl silane derivatives. To give an example, hydroxyhexyltrimethylsilane can be reacted with sulfamic acid in the presence of a polar aprotic solvent such as N-methyl pyrrolidone or dimethyl formamide. The transformation of the ammonium salt into the appropriate alkyl ammonium derivative is readily achieved by addition of amine under generation of ammonia gas.

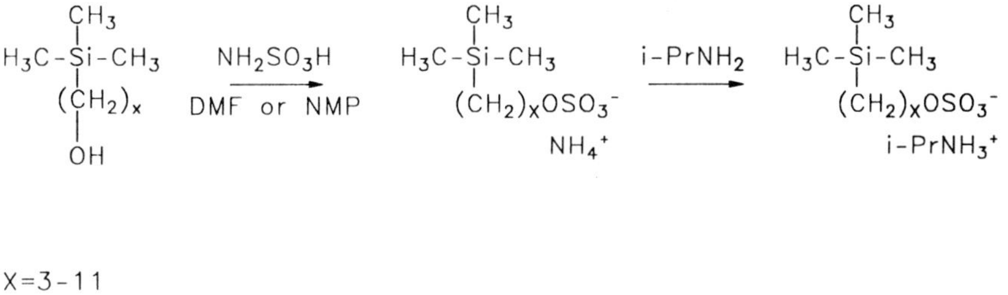

$$\underset{\underset{\displaystyle OH}{\displaystyle |}}{\overset{\overset{\displaystyle CH_3}{\displaystyle |}}{H_3C-\underset{\displaystyle |}{Si}-CH_3}} \quad \xrightarrow[\displaystyle DMF\ or\ NMP]{\displaystyle NH_2SO_3H} \quad \underset{\underset{\displaystyle NH_4^+}{\displaystyle }}{\overset{\overset{\displaystyle CH_3}{\displaystyle |}}{H_3C-\underset{\displaystyle |}{Si}-CH_3}} \quad \xrightarrow{\displaystyle i-PrNH_2} \quad \underset{\underset{\displaystyle i-PrNH_3^+}{\displaystyle }}{\overset{\overset{\displaystyle CH_3}{\displaystyle |}}{H_3C-\underset{\displaystyle |}{Si}-CH_3}}$$

with $(CH_2)_x$ chain where $X = 3-11$

Fig. 3. Sulfation of hydroxyalkyltrimethylsilanes

The products thereby obtained are anionic silane surfactants exhibiting excellent surfactant properties dependent on the ammonium counter ion. Some specimen are real foam boosters, too. For example an 1 wt. % aqueous solution of isopropylammonium sulfatohexyl trimetyl silane shows a surface tension of 21 mN m^{-1} and a spreading ability of 65 mm. As being familiar for the nonionics the aqueous solution of this anionic silane derivative is stable at pH 4–12 over months as well. This result can directly be compared with the trisiloxane analogue which loses its wetting ability completely after storage at room temperature and neutral pH for 14 d.

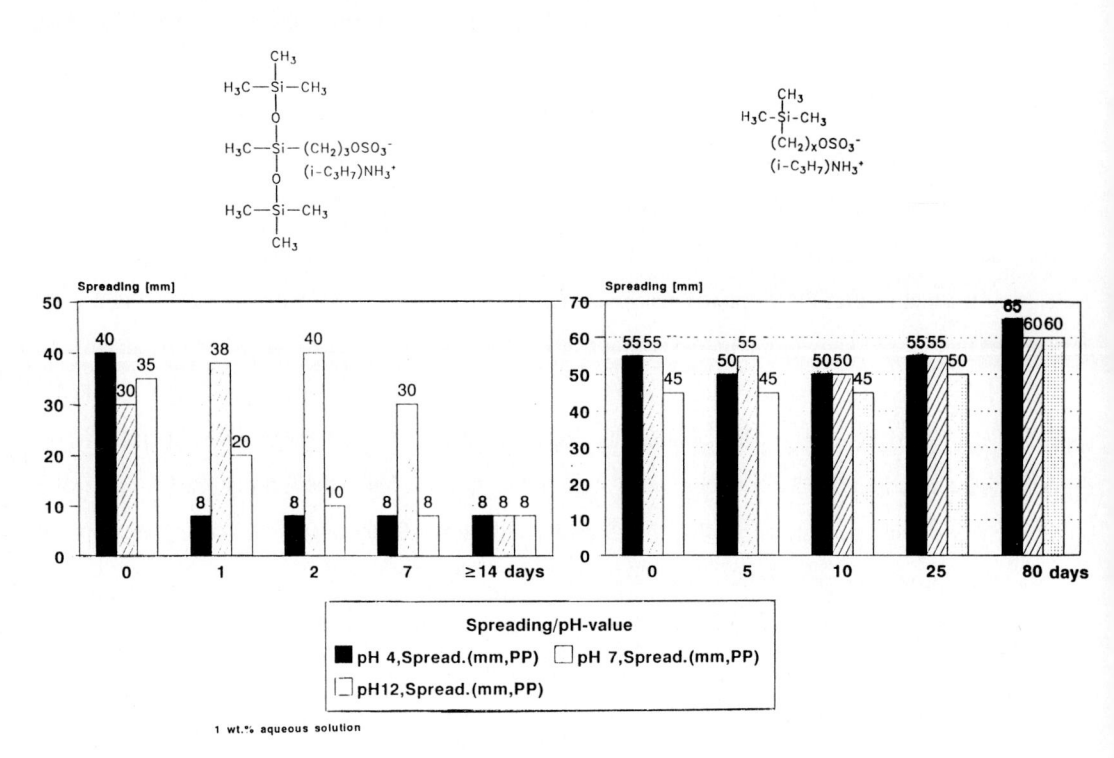

Fig. 4. Hydrolytic stability of isopropylammonium sulfatohexyltrimethylsilane compared with isopropylammonium-sulfatopropyltrisiloxane

The epoxy group is highly reactive towards nucleophilic reagents and can easily be transformed to various functions. Following that synthetic route quaternary ammonium, betaine as well as sulfonate groups were introduced into the silane structure. The reaction of one equivalent of trialkyl ammonium hydrogen sulfite and two equivalents of epoxy silane gives novel cation-anion complexes. The bundle of ionic silane compounds based upon the above epoxy silane showed reasonable surfactant properties and outstanding hydrolytic stability as well [7].

To sum it up we can stress that the substitution of the trisiloxane lyophobic part by a trimethyl silane moiety yields both nonionic and ionic silane surfactants. Aqueous solutions of these new surfactants are extremely stable in both alkaline and acidic conditions due to the fact that the siloxane substructure is left out. The surfactant properties of these substances are to a great extent comparable to the abilities of trisiloxane wetting agents.

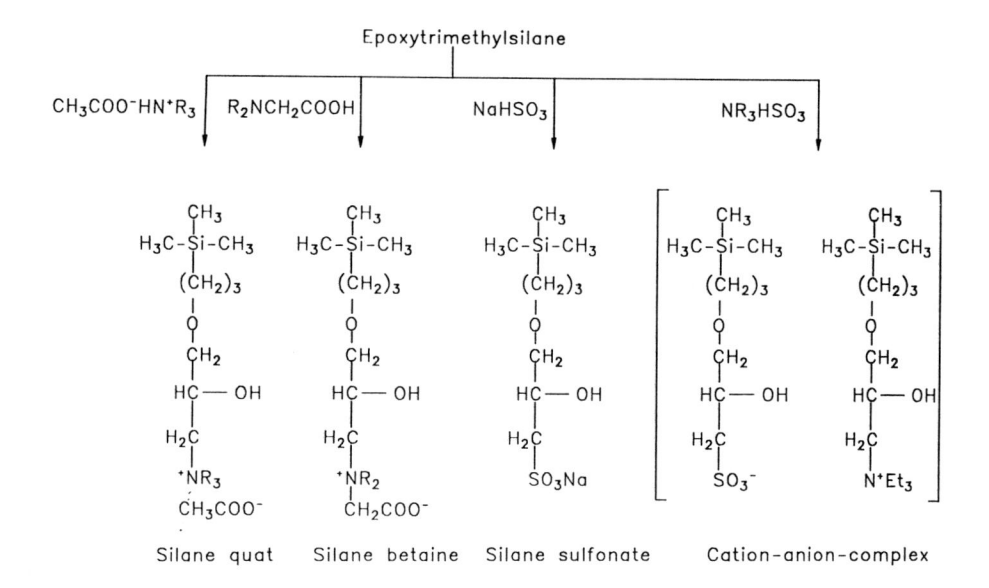

Fig. 5. Ionic silane surfactants based upon epoxysilanes

The combination of both the biodegradability and the stability towards hydrolysis is unique in the field of well known silicone surfactants and will ensure the widespread application of trimethylsilane based surfactants. Considering all these facts all applications for which trisiloxane surfactants are already used or at least recommended are opened up in general for the new silane surfactants. Furthermore, applications now can be taken into consideration for which the hydrolytically unstable trisiloxane derivatives failed in the past.

References:

[1] G. Koerner, *"Goldschmidt informiert..."* **1982**, 01/82, No. 56, p.5.
[2] K.-D. Klein, D. Schaefer, P. Lersch, *Tenside Surf. Det.* **1994**, *31*, 115; G. Feldmann-Krane, W. Höhner, D. Schaefer, S. Silber, *DE* App. 4 317 605.4, **1993**.
[3] A. R. L. Colas, F. A. D. Renauld, G. C. Sawicki, *GB* 8 819 567, **1988**; M. J. Owen, *GB* 1 520 421, **1974**.
[4] K.-D. Klein, W. Knott, G. Koerner; *DE* App. P 4 313 130.1, **1993**.
[5] A. W. Adamson, *Physical Chemistry of Surfaces*, Interscience, New York, **1968**.

[6] K.-D. Klein, W. Knott, G. Koerner, *DE* App. P 4 320 920.3, **1993**.
[7] K.-D. Klein, W. Knott, G. Koerner, *DE* App. P 4 330 059.6, **1993**.

Siloxanes Containing Vinyl Groups

Silke Kupfer, Irene Jansen, Klaus Rühlmann*

Institut für Anorganische Chemie

Technische Universität Dresden

Mommsenstr. 13, D-01062 Dresden, Germany

Summary: We studied the anionic ring-opening polymerization of D_3^{Vi} and D_4^{Vi} and the copolymerization of these compounds with D_3. The polymerizations were terminated by chlorosilanes with additional different functional groups or octyl groups. In this way we obtained siloxane polymers and copolymers with a larger number of vinyl groups at one end of the siloxane chain and additional silico- or carbofunctional groups or longer alkyl groups at the other end.

Introduction

For some years the subject of our work has been the syntheses of linear siloxanes with different silico- or carbofunctional groups terminating the siloxane chain [1–7].

At present we are engaged in the preparation of siloxane polymers and block copolymers with a larger number of functional groups attached to the chain. In the following we report on siloxanes containing vinyl groups.

Syntheses and Characterisation

The syntheses of siloxanes with vinyl groups at the silicon atoms of the chain was performed by anionic ring-opening polymerization of vinyl group-containing cyclotri- or cyclotetrasiloxanes (D_3^{Vi}, D_4^{Vi}) with alkyl lithium compounds as initiators. The polymerizations were terminated by chlorosilanes with additional silico- or carbofunctional groups or octyl groups. In the block copolymerizations we used D_3^{Vi} or D_4^{Vi} and hexamethylcyclotrisiloxane (D_3) (Scheme 1 and 2).

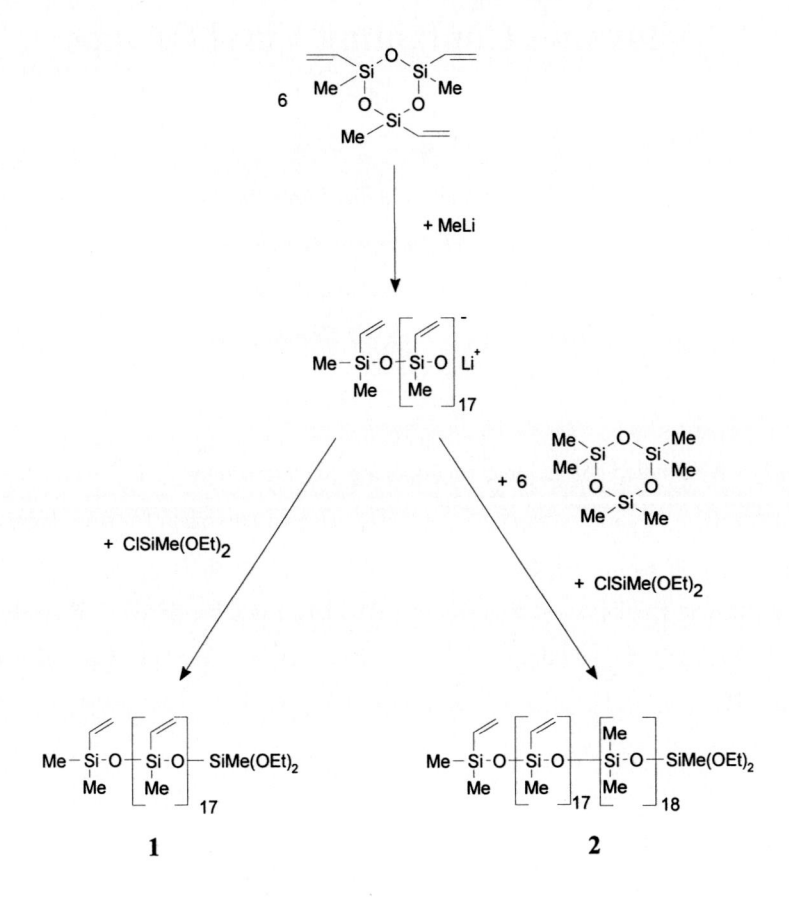

Scheme 1.

The consumption of cyclosiloxane during polymerization and copolymerization was monitored by GC. The conversion of D_3^{Vi} was nearly 100 % (Fig. 1). In contrast, the ring-opening polymerization of D_4^{Vi} yielded a conversion of about 60 % only (Fig. 2). During the copolymerization with D_3 the cyclic vinylsiloxanes showed a similar behavior as observed during homopolymerization.

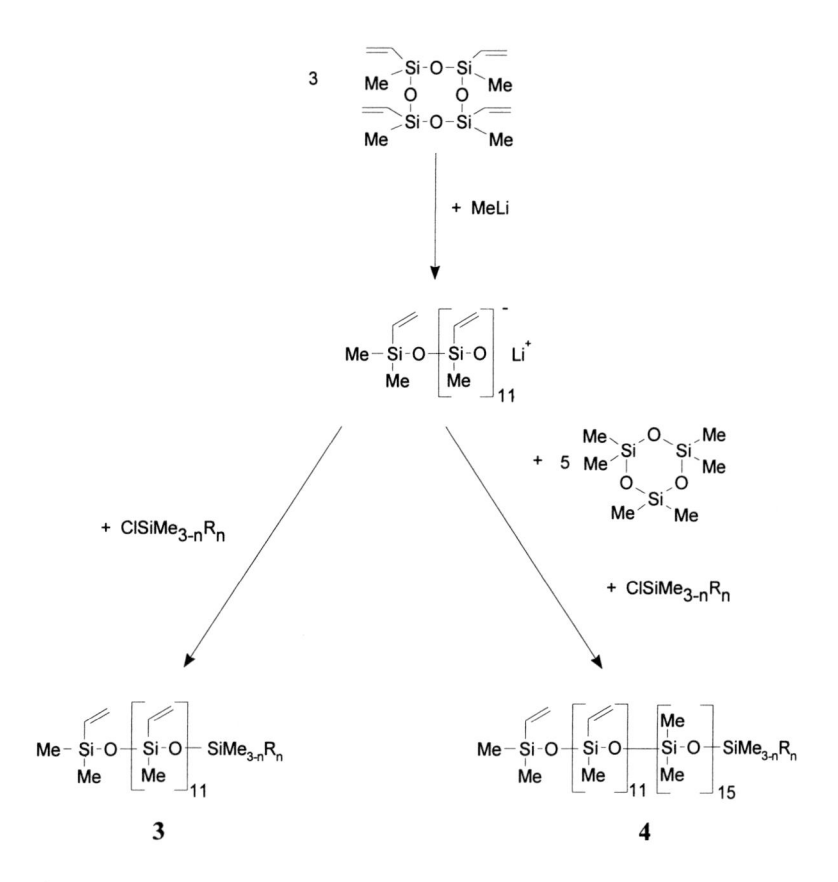

Scheme 2. 3a: $n = 0$; 3b: $n = 1$, $R = (CH_2)_3OSiMe_3$; 3c: $n = 1$, $R = n\text{Octyl}$
4a: $n = 0$; 4b: $n = 1$, $R = (CH_2)_3OSiMe_3$; 4c: $n = 1$, $R = n\text{Octyl}$

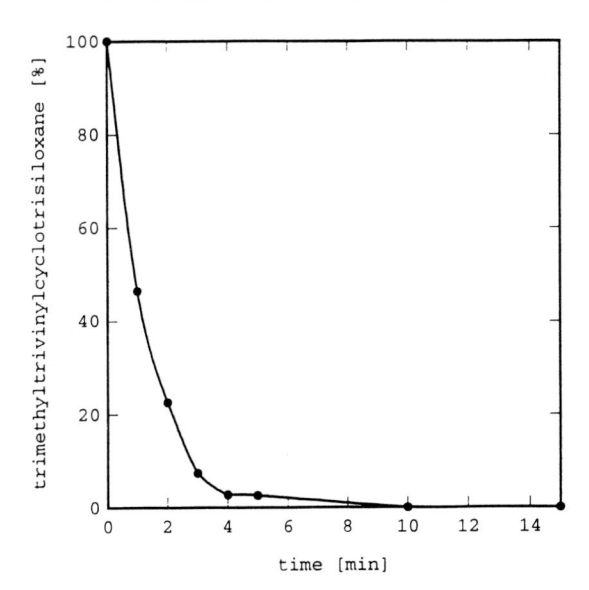

Fig. 1. Conversion of D_3^{Vi}

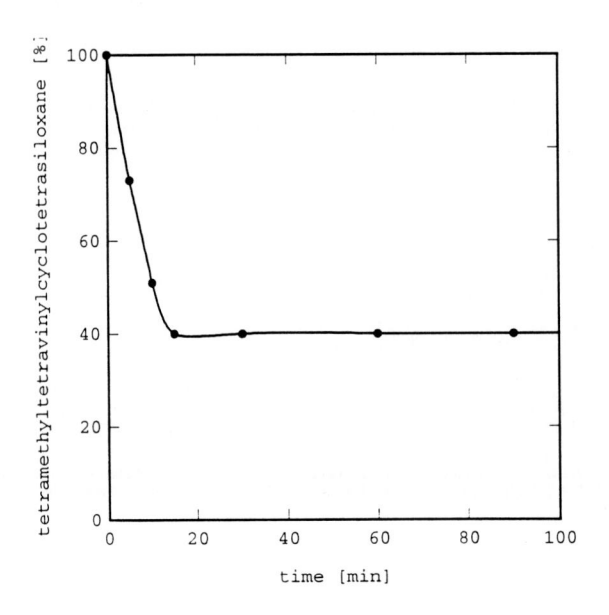

Fig. 2. Conversion of D_4^{Vi}

GPC studies proved the resulting siloxanes to have a narrow molecular weight distribution. The composition of the siloxanes was determined by ^{29}Si NMR spectroscopy (Table 1).

Product	Composition by ^{29}Si NMR	GPC			
		M(calc.) [g/mol]	M_w [g/mol]	M_n [g/mol]	M_w/M_n
1	$M^{Vi}D_{10}^{Vi}M_{0.7}^{OEt_2}$	1696	2003	1355	1.47
2	$M^{Vi}D_{16.80}^{Vi}D_{20.3}M_{0.7}^{OEt_2}$	3028	4246	2706	1.56
3a	$M^{Vi}D_{8.1}^{Vi}M$	1120	1290	990	1.30
3b	$M^{Vi}D_{3.9}^{Vi}M^{Pr\,OSiMe_3}$	1230	1035	843	1.22
3c	$M^{Vi}D_{5.5}^{Vi}M^{Octyl}$	1203	1003	827	1.21
4a	$M^{Vi}D_6^{Vi}D_{12}M_{1.5}$	2230	2994	2330	1.28
4b	$M^{Vi}D_{8.3}^{Vi}D_{17}M_{1.2}^{Pr\,OSiMe_3}$	2340	2863	1833	1.56
4c	$M^{Vi}D_{9.6}^{Vi}D_{21.4}M_{1.5}^{Octyl}$	2328	2891	1804	1.60

Table 1. Characterization of the products

References:

[1] I. Jansen, S. Kupfer, K. Rühlmann, X^{th} *International Symposium on Organosilicon Chemistry Poznan*, **1993**, p. 183.

[2] I. Jansen, H. Friedrich, K. Rühlmann, *DE* 42 34 898, **1994**.

[3] I. Jansen, H. Friedrich, K. Rühlmann *DE* 42 34 959, **1994**.

[4] H. Friedrich, I. Jansen, K. Rühlmann, *J. Inorg. Organomet. Polymers* **1991**, *1*, 397.

[5] H. Friedrich, I. Jansen, K. Rühlmann, *Polym. Degr. Stab.* **1993**, *42*, 127.

[6] H. Friedrich, I. Jansen, K. Dittmar, K. Rühlmann, *J. Inorg. Organomet. Polymers* **1994**, *4*, 155.

[7] H. Friedrich, I. Jansen, K. Rühlmann, *Polym. Degr. Stab*, in press.

Investigations of Chemical Heterogeneity in Polydimethylsiloxanes using SFC, HPLC, and MALDI-MS

U. Just*

Bundesinstitut für Materialforschung und -prüfung(BAM)
Unter den Eichen 87, D-12205 Berlin, Germany

R.-P. Krüger

Institut für Angewandte Chemie Berlin-Adlershof e.V. (ACA)
Rudower Chaussee 5, D-12489 Berlin, Germany

Summary: Supercritical fluid chromatography (SFC), high performance liquid chromatography (HPLC), and matrix-assisted laser desorption-ionisation mass spectrometry (MALDI-MS) were used for analysing polydimethylsiloxanes (PDMS) in the molar mass range 1.000–8.000 Da. The studied materials consisted of technical silicone oils with linear and cyclic components (the linear ones containing methyl or hydroxyl end groups), specially prepared standard mixtures and silicone rubber extracts. While SFC is successfully applicable in the low molecular mass range up to molar masses of 6.000 Da (M_2D_{80}), MALDI-MS facilitates proper differentiation in higher molar mass ranges; in the range below 1.000 Da the problem of discrimination arises when MALDI-MS is applied. Both techniques can provide results that are supplementary to each other. For quantitative separation of cyclic components from mixtures with linear diols HPLC was used, and collected fractions were characterized by MALDI-MS distinguishing chemical heterogeneity. Interpretation of molar mass distribution by MALDI-MS, however, has to be done with caution.

Introduction

Size exclusion chromatography (SEC) is normally used for determining molar mass distribution in polymer and oligomer homologeous mixtures. Values for average molecular masses can be obtained. But it is difficult to get these results of specimen with different chemical heterogeneity; SEC cannot provide the resolution required for determining different functional end groups in siloxane mixtures; distinguishing between cyclic and linear components, for example, is not possible [1].

Gas chromatography (GC) enables determination of low molecular linear and cyclic components. Even polycyclosiloxanes and complex mixtures such as pyrolysis products can be determined by coupling GC with mass spectrometry (MS) and cryo-Fourier transformation infrared spectroscopy (FTIR). However, substances of higher molar mass may stay on the column, or they may elute with great peak broadening when high temperature gas chromatography (HTGC) is used.

The range of oligomeric siloxanes can be significantly increased by the use of supercritical fluid chromatography. Coupling SFC with mass spectrometry is possible, too [2, 3]. It allows determination of hydroxyl end groups beside cyclic siloxanes if ammonia is used as reagent gas. But it is very difficult to identify smaller amounts of cyclosiloxanes with higher molar masses in linear diols, e. g., by SFC-/MS. Baseline separation of cyclic and linear components can only be achieved up to monomer unit 30, approximately. Mostly used MS quadrupole equipments are also limited to a mass range of 2.000 Da, respectively 4.000 Da.

A recently introduced technique for the separation of larger molecules is _m_atrix-_a_ssisted _l_aser _d_esorption-_i_onisation _m_ass _s_pectrometry (MALDI-MS). Developed by Karas et al. [4, 5] in 1988, it has been successfully used to determine the mass of biomolecules up to 500.000 Da. This method is based on the principle that the dissolved specimen is mixed with a matrix, and then crystallizes. After drying, a laser pulse is directed onto the solid matrix to photo-excite the matrix material, resulting in desorption and soft ionisation of the analyte. The molar mass is then determined by the _t_ime _of f_light (TOF).

This paper reports the analysis of polydimethylsiloxanes using MALDI-MS in comparison with SFC. MALDI-MS may be used as a supplementary method to SFC of siloxanes in a qualitative manner. Further, the paper discusses advantages and disadvantages for using MALDI-MS in determining molar masses. The MALDI-MS technique allows assignment of molecules with different functional end groups, too. The present paper also shows the use of MALDI-MS to characterize fractions of siloxane mixtures with cyclic and linear portions. The fractions are collected after HPLC or gradient elution procedures to quantify these portions with different chemical heterogeneity.

Specially prepared substances were gratefully provided by Dr. U. Scheim, Institute of Organic Chemistry, TU Dresden, Germany, and the diols from Chemiewerk Nünchritz.

The silicone rubber was laboratory tubing material.

Results and Discussion

Fig. 1 shows SFC chromatograms of PDMSs with methyl end groups (a), with hydroxyl end groups (b) and a sample consisting of cyclic siloxanes (c); the detector used was a flame ionisation detector (FID). The peak numbers refer to the degree of polymerization. The chromatograms demonstrate that SFC is a powerful tool in separating homologeous mixtures of PDMS up to more than 50 monomer units. But it is not possible to distinguish between functional end groups using a FID alone.

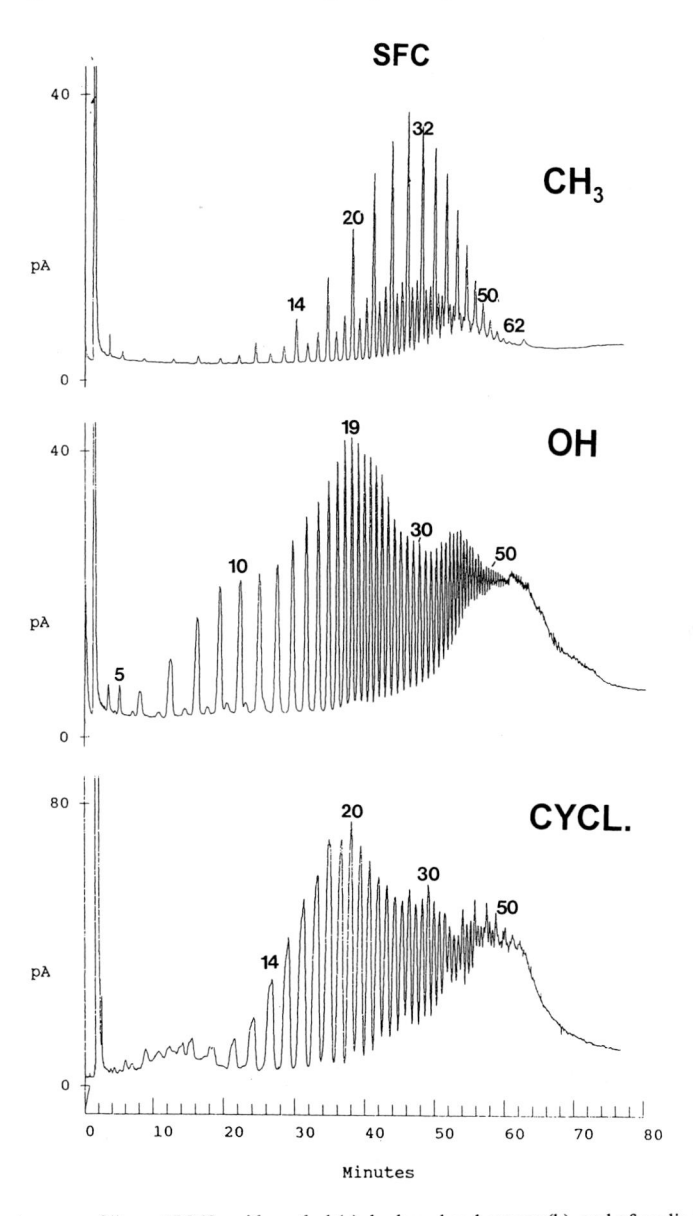

Fig. 1. SFC chromatograms of linear PDMSs with methyl (a), hydroxyl end groups (b), and of cyclic PDMSs (c)

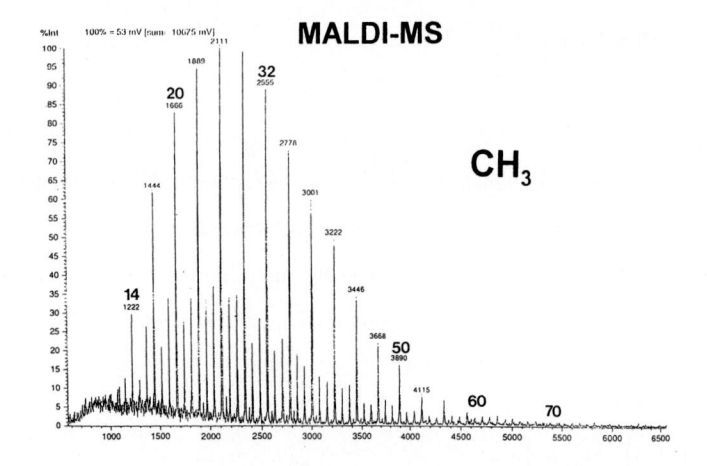

Fig. 2a. MALDI-MS spectrum of linear PDMSs with methyl end groups

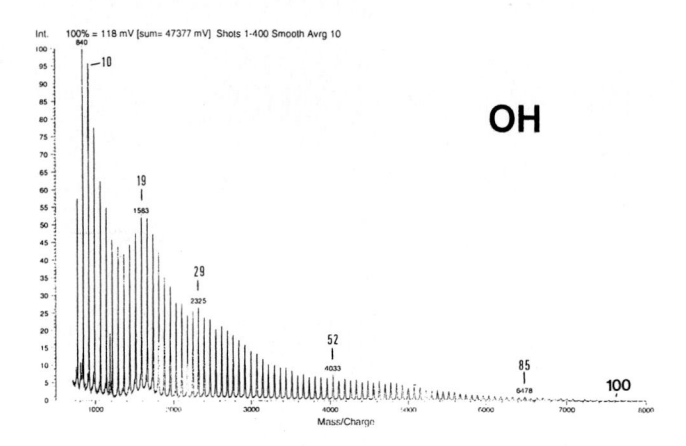

Fig. 2b. MALDI-MS spectrum of linear PDMSs with hydroxyl end groups

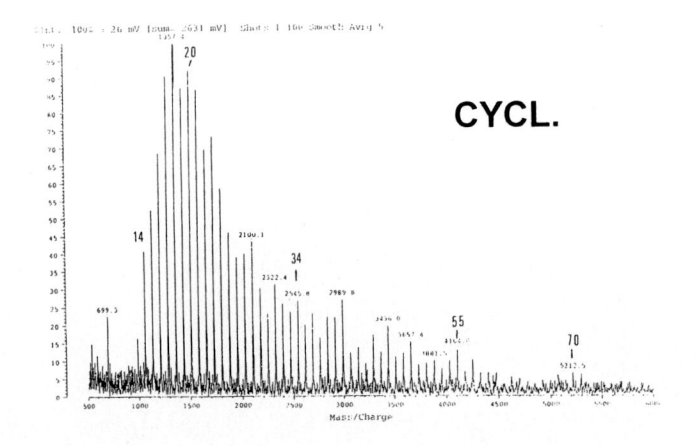

Fig. 2c. MALDI-MS spectrum of cyclic PDMSs

Fig. 2 shows MALDI-MS spectra of the same three samples. The spectra usually show the mole peaks plus the molar mass of Na (23 Da) and K (39 Da). Unless they have been specifically added, the sodium and potassium ions come from contaminants in the specimen and/or the matrix. Alternatively, another suitable salt may be added. The addition of lithium chloride suppresses sodium (and potassium) ions, so that the spectra can be more easily interpreted. Thus, the numbers in Fig. 2a refer to mole peaks M plus ^{23}Na (= M + 23 Da) and in Fig. 2b and 2c to M plus ^{7}Li (= M + 7 Da). As can be seen, the resolution in the higher molecular mass range is more pronounced than in SFC. Components with more than 100 monomer units can be separated in the diol sample.

Fig. 3 depicts a MALDI-MS spectrum (expanded plot) of linear siloxanes with methyl and hydroxyl end groups. It shows that linear siloxanes with methyl end groups may be differentiated from diols (the molar mass difference is 4 daltons) in the range 1.000–1.500 Da. Of course, there is no baseline separation between these specimen because of silicon and carbon isotope distribution in silicon-organic compounds. Smaller amounts of cyclosiloxanes, also present in the sample, are clearly separated from methyl end-capped components.

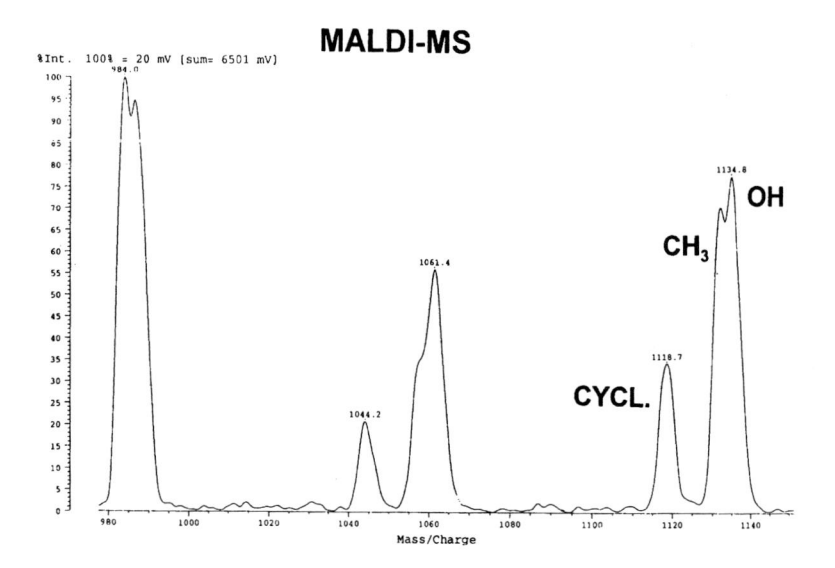

Fig. 3. MALDI-MS spectrum (expanded plot) of a mixture of methyl and hydroxyl end groups containing also smaller amounts of cyclosiloxanes

In the molar mass range below 1.000 Da SFC has advantages, because MALDI-MS cannot provide for indication of smaller molecules in this range. As the problem of discrimination remains in the lower molecular mass range, it is not yet possible to replace SEC of PDMS by MALDI-MS. But this technique is useful to get an idea about the order of magnitude of molar mass distribution. Calculations of weight average values of linear diols made by SEC and MALDI-MS have given nearly the same results. MALDI-MS may have advantages in that respect that separation problems as with SEC (e.g., different absorption depending on molar mass and functionality) do not occur. In order to calibrate SEC columns fractions may be collected, and the molar masses of the components may be determined by MALDI-MS.

We also used HPLC in connection with MALDI-MS. Our aim was to separate cyclic siloxanes, even larger ones not distinguishable by SFC, from linear ones quantitatively. Hexane and ethyl acetate were mixed in a ratio of 90 to 10 by volume and used as mobile phase in HPLC separations of mixtures of the cyclic siloxane sample with the diol sample. Cyclosiloxanes elute first, followed by the diols. Under these conditions only a small overlapping area is to be seen (Fig. 4a). The sample was fractionated and MALDI-MS spectra were taken from the fractions.

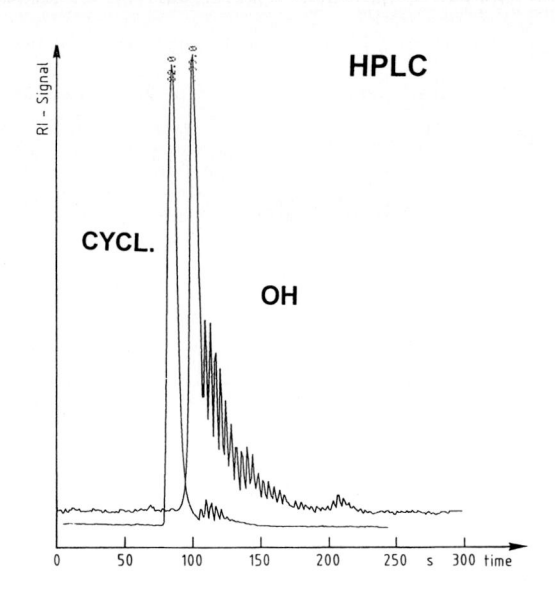

Fig. 4a. HPLC separation of silanols and cyclosiloxanes

Fig. 4b shows a MALDI-MS spectrum containing only linear diols which eluted at 100–150 s out of the column, Fig. 4c a cyclosiloxane fraction. Thus MALDI-MS may assist HPLC in developing separation methods; it enables characterizing cyclic and linear components in a qualitative manner. The short time required and its easy handling with MALDI-MS is noteworthy in comparison with SFC-MS techniques.

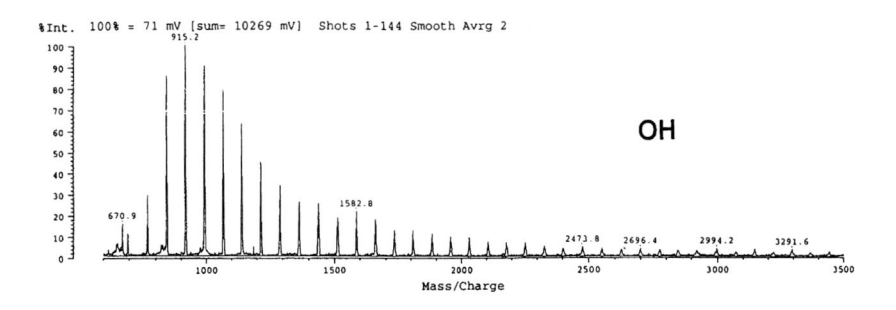

Fig. 4b. MALDI-MS spectrum of a silanol fraction

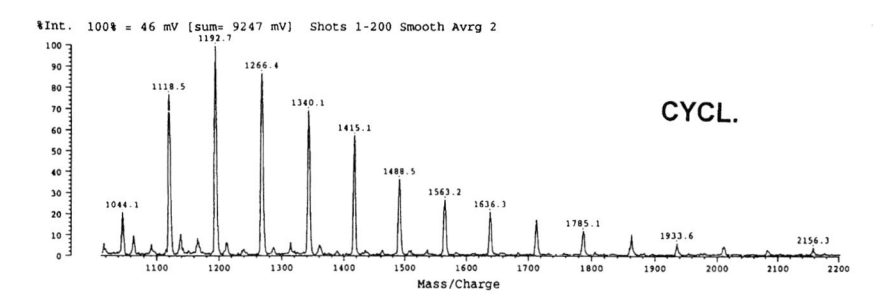

Fig. 4c. MALDI-MS spectrum of a cyclosiloxane fraction

By changing the composition of the mobile phase in HPLC it is possible to separate these components with different chemical heterogeneity completely and determine them quantitatively. An evaporative light scattering detector has advantages over a differential refractometer in developing a separation method by gradient elution. For example, we injected mixtures of PDMS with different chemical heterogeneity into a stream of hexane used as mobile phase for two minutes (250 x 4 mm silica column, 10 µm). Then the pump was programmed immediately changing to 100 % ethyl acetate.Thus cyclosiloxanes as well as unpolar PDMS with methyl end groups were separated from diols which only eluted when ethyl acetate was pumped through the silica column MALDI-MS may be used supplementary to HPLC in characterizing even traces of cyclosiloxanes with higher molecular masses.

Further work is done by us to separate completely cyclic and linear siloxanes with methyl endgroups in technical silicone oils as well as branched compounds by liquid adsorption chromatography at critical conditions (LACCC), where the species are separated not by molecular mass but by functionality and then analyse the separated compounds in a second dimension by MALDI-MS.

References:

[1] B. Hagenhoff, A. Benninghoven, H. Bathel, W. Zoller, *Anal. Chem.* **1991**, *63*, 2466.
[2] J. D. Pinkston, G. D.Owens, E. J. Petit, *Anal. Chem.* **1989**, *61*, 775.
[3] U. Just, F. Mellor, F. Keidel, *Fresenius J. Anal. Chem.* **1994**, *348*, 745.
[4] M. Karas, F. Hillenkamp, *Anal. Chem.* **1988**, *60*, 2299.
[5] U. Bahr, A. Deppe, M. Karas, F. Hillenkamp, U. Giessmann, *Anal. Chem.* **1992**, *64*, 2866.

Vulcanization Kinetics of Addition-cured Silicone Rubber

Dieter Wrobel

Bayer AG

D-51368 Leverkusen, Germany

Summary: A new type of measuring device enables the simultaneous investigation of the vulcanization and chemical conversion of addition-cured silicone rubbers. The crosslinking reaction is studied by means of an NIR analysis of the 1st harmonic of vinyl absorption and the vulcanization in terms of the storage modulus of a rheological measurement. Vinyl-terminated polydimethyl siloxanes react according to second order kinetics. The network produced does not have any recognizable effect on the chemical reaction. Polymers with lateral vinyl groups and very short chained polymers with vinyl end groups have an inhibitory effect and react approximately according to zero order kinetics. The dependence of the degree of conversion at gel point on the functionality of the crosslinker corresponds with the theoretical values.

1 Introduction

Kinetic tests on rubbers have up to now been limited primarily to the analysis of curemeter curves. In such investigations, the formation of a three-dimensional network is determined by means of its mechanical properties. The kinetics of the crosslinking reactions involved is virtually unknown. Detailed knowledge even of the chemical processes involved is frequently unavailable. This is, however, not the case for silicone rubbers, which cure as a result of the hydrosilylation reaction:

$$\equiv\text{Si–CH=CH}_2 + \text{H–Si}\equiv \xrightarrow{[\text{Pt}]} \equiv\text{Si–CH}_2\text{–CH}_2\text{–Si}\equiv$$

Eq. 1.

Silicone rubbers are therefore often used as model systems for fundamental investigations of elastomers [1–3]. The technological importance of these so-called addition systems is increasing due to their rapid vulcanization which does not produce any decomposition products. In this field, however, few investigations of the kinetics of addition reactions involving crosslinking have been published. In most cases, indirect measuring processes have been used. DSC measurements can be used for example to determine the degree of conversion of the crosslinking reaction in terms of heat development [4–8].

Swelling tests on the vulcanizates produced are an alternative option. The mean distance between two crosslinks can be applied to determine the concentration of crosslinks [9]. The reaction can only be studied once the gel point has been reached, since only then does a three-dimensional network form.

Only two studies have involved a direct determination of the SiH concentration by IR spectroscopy [5, 10]. The IR measurement of SiH absorption is, however, only suitable for analyzing the reaction in very thin layers or on surfaces. Supplementary to this analysis of the chemical conversion there are also many publications which describe the curing behaviour of industrial platinum-crosslinked silicone rubbers [11, 12] (Fig. 1).

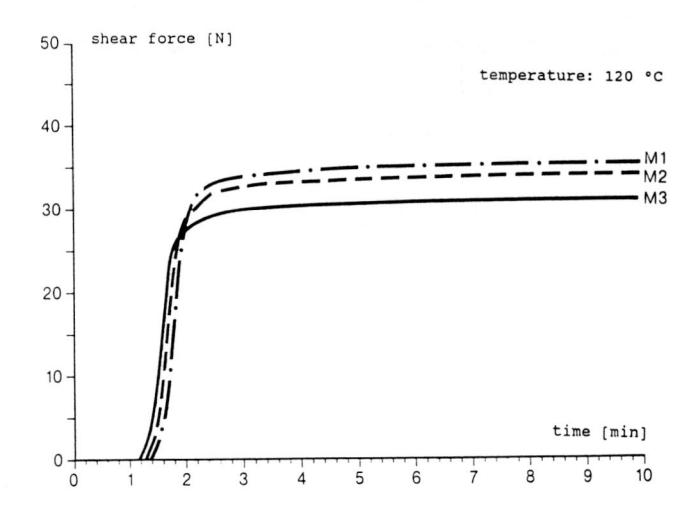

Fig. 1. Vulcanization test according to DIN 53 529

Up to now, the literature has contained no description of any method for studying the network structure and chemical reactions simultaneously.

2 Objective

The objective of this study is to record both network structure (curemeter curves) and chemical conversion simultaneously. The influence of the chemical reaction on the network formation is determined. The kinetics with respect to different polymers and crosslinkers will be also described.

3 Experimental Design

A special measuring cell with two double gaps and an NIR waveguide adapter is used for these investigations (Fig. 2).

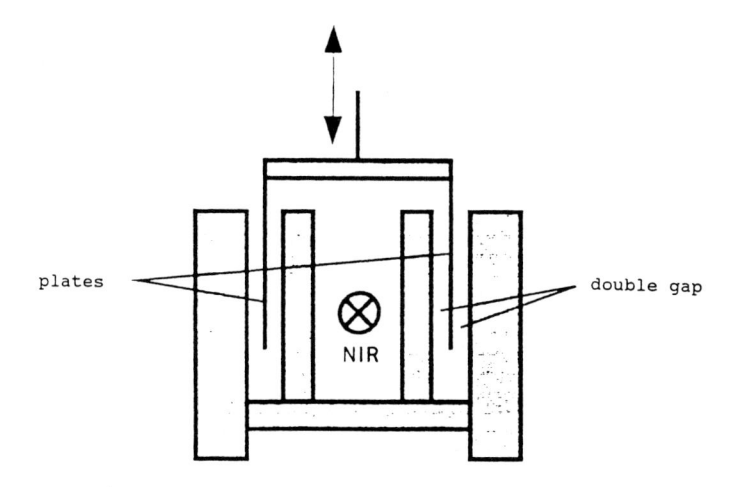

Fig. 2. Cross-section of the measuring cell

The reduction in vinyl concentration is determined from the first harmonic of the stretching vibration at a wave number of 6065 cm^{-1} (Fig. 3). At this frequency range, optical waveguides are sufficiently translucent for the measuring cell to be placed outside the NIR spectrometer. The first harmonic of SiH absorption at 4200–4600 cm^{-1} is outside the transmission range of normal optical waveguides. A further benefit is the high layer thickness (20 mm) through which transmission is possible. Calibration with vinyl siloxanes, which ensures a linear relationship between vinyl concentration and integral absorption, enables absolute concentrations of vinyl groups to be obtained.

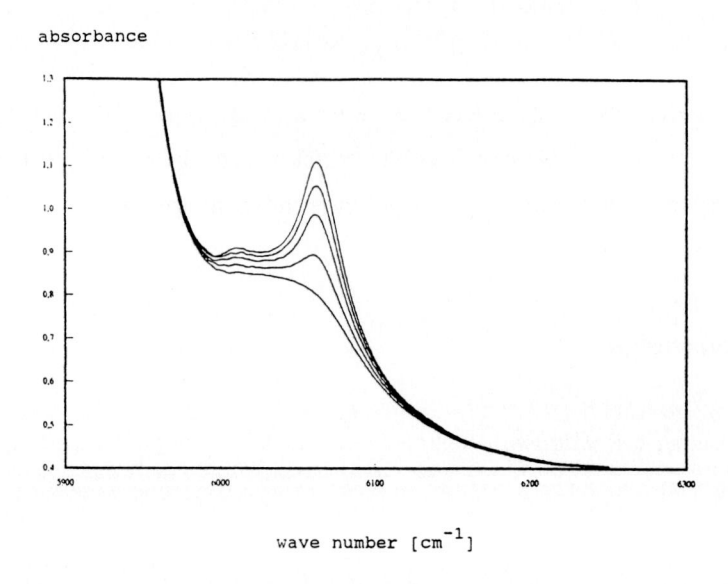

Fig. 3. NIR absorption of the vinyl groups at different conversions

Changes in mechanical properties are recorded by a piezoforce transducer which measures the power amplitude of the plates in the gaps (width = 1.45 mm) at a given frequency. The strain and frequency can be controlled in a wide range by a piezo translator. Depending on the frequency, storage modulus, loss modulus and complex viscosity can be calculated from the values of force obtained as a function of strain, frequency, and phase displacement (Eq. 2).

storage modulus $\qquad G' = \dfrac{\tau_0}{\gamma_0} \cos \delta$

loss modulus $\qquad G'' = \dfrac{\tau_0}{\gamma_0} \sin \delta$

complex viscosity $\qquad \eta = \dfrac{\sqrt{G'^2 + G''^2}}{\omega}$

Eq. 2. τ_0 = stress, γ = strain, δ = phase displacement, ω = angular velocity

A comparison of these values with measurements recorded by commercially available viscometers and rheometers shows that the values are similar.

4 Results and Discussion [13]

The tests were carried out on 1:1 mixtures of two components (A component = polymer + Pt catalyst; B component = polymer + crosslinker) which were prepared shortly before the test. The catalyst was a vinyl siloxane Pt^0 complex with tetramethyltetravinylcyclotetrasiloxane as a ligand. All the polymers and crosslinkers used were equilibrated and devolatized products with a molar mass distribution typical for silicone polymers.

Evaluation of the degree of conversion: On the basis of the NIR data and the calibration curve, the degree of conversion of the silane groups was evaluated using the following equation (Eq. 3)

$$\text{Conversion [\%]} = \frac{[Vi]}{[Vi]_0} \cdot 100$$

Eq. 3.

4.1 Basic system

The model selected was the crosslinking of a poly(dimethylsiloxane) with terminal vinyl groups using a crosslinker containing five SiH groups in its chain:

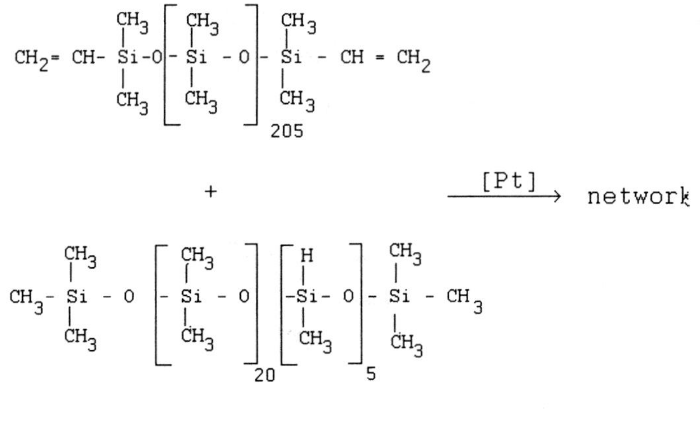

Scheme 1.

In addition, a two-fold molar excess of SiH groups was used in order to achieve an optimum crosslinking yield (see below). As Fig. 4 shows, the mixture achieves its final storage modulus value after 1.5 h at a reaction temperature of 30 °C with a platinum concentration of 1.5 ppm, i.e., after this time the network is fully formed.

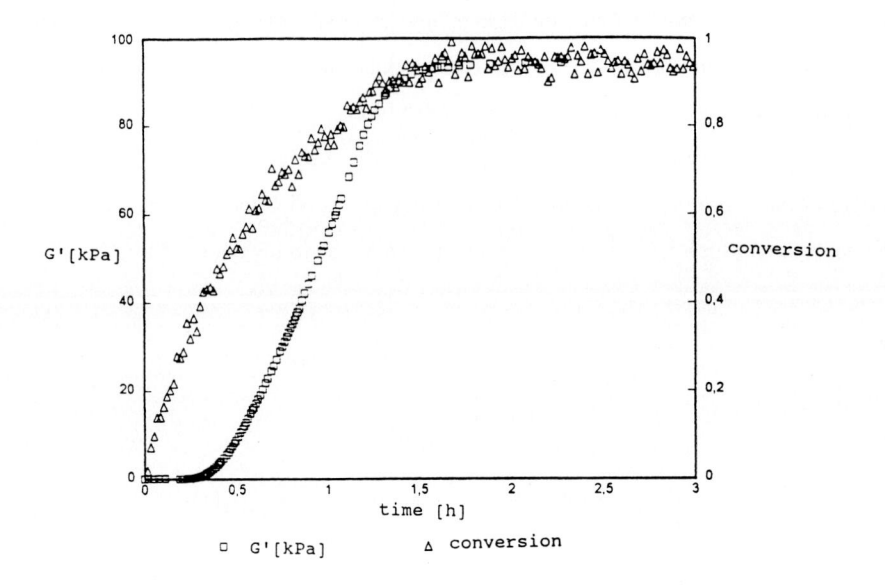

Fig. 4. Storage modulus and degree of conversion vs time ($T = 30$ °C, basic system)

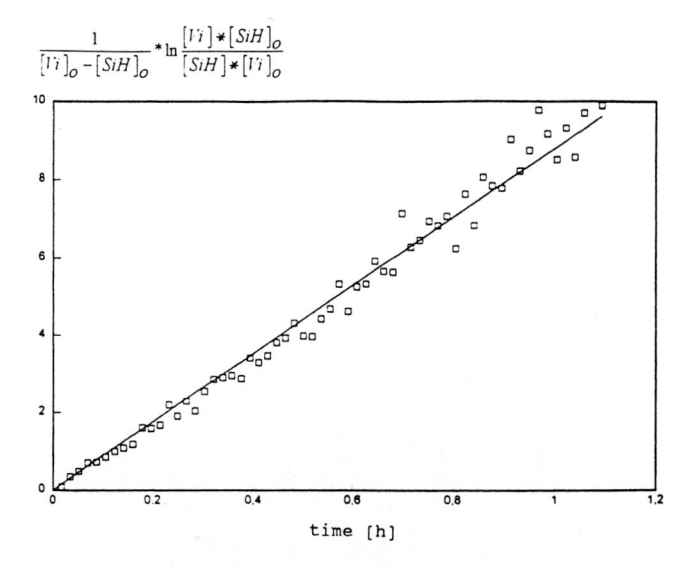

$$\frac{1}{[Vi]_o - [SiH]_o} * \ln \frac{[Vi] * [SiH]_o}{[SiH] * [Vi]_o}$$

Fig. 5. Reaction of the basic system at 30 °C (second order kinetic model)

The reaction begins without delay. The rate decreases as the degree of conversion increases. A graph showing vinyl concentration according to second order kinetics plotted against time gives a linear relationship (Fig. 5).

The same reaction order has also been noted in the literature for the conversion of a similar polymer using a tetrafunctional crosslinker with dimethyl hydrogen siloxy groups [5]. The gradient of the regression lines gives a rate constant k of 9.1 l mol^{-1} h^{-1}. As expected, the storage modulus, which is a measure of network formation, does not increase until the gel point is reached.

4.1.1 Temperature dependence

An Arrhenius plot of the rate constants obtained in this way at 30–45 °C gives a value for activation energy of 81 kJ mol^{-1} (Fig. 6).

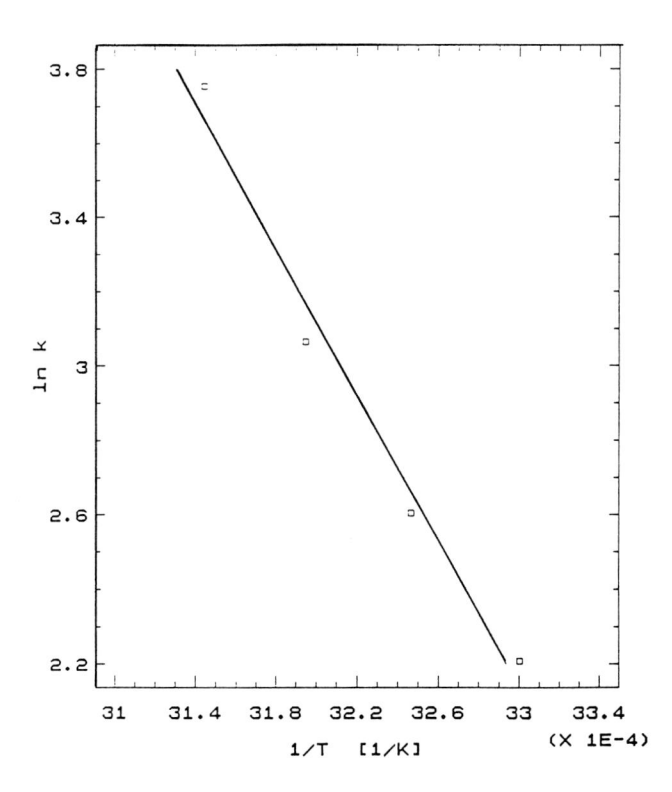

Fig. 6. Arrhenius plot for the second order reaction model (basic system)

This value is close to the value listed in the literature of 78 kJ mol^{-1} which was determined for the temperature range of 50–90 °C [9].

4.1.2 Molar Ratio of Reactants

Variations in the SiH/SiVi ratio of between 0.5 and 3 do not lead to any change in order of the kinetics. The storage modulus at the end of the crosslinking process is, however, at a maximum when the SiH/SiVi ratio is 2 (Table 1).

SiH/SiVi	Rate Constants [l mol^{-1} h^{-1}]	Storage Modulus G' [N/m^2]
0.5	9.5	3723
1.0	11.2	62200
2.0	9.3	94200
3.0	9.3	75700

Table 1. Variation of SiH/SiVi ratio (basic system, 30 °C)

4.1.3 Catalyst Concentration

The increase in the rate of reaction is linearly dependent on the platinum concentration (Fig. 7) in the range 1.5–2.5 ppm platinum (Eq. 4).

$$k_2 = -0.2 + 6.0 \, [Pt] \qquad [Pt] = \text{ppm Pt}$$

Eq. 4.

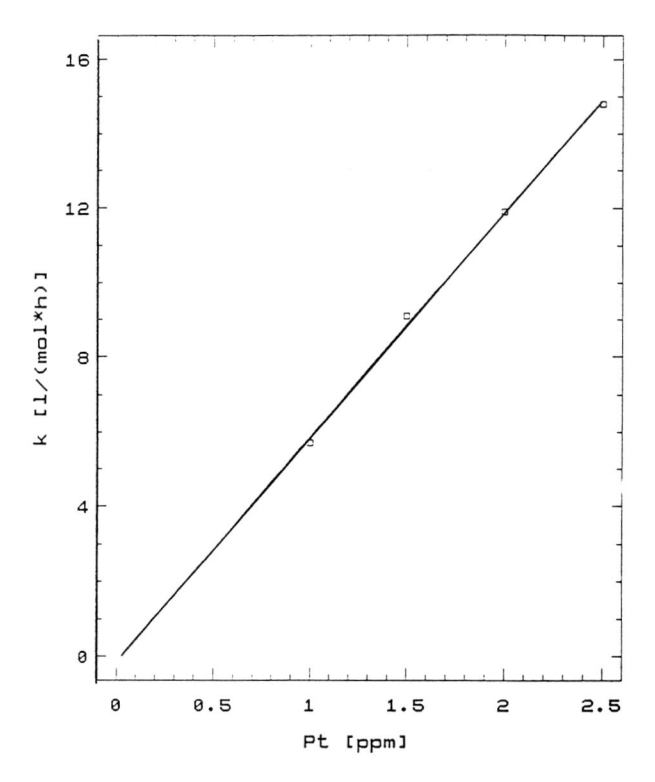

Fig. 7. Rate constant vs Pt concentration

4.1.4 Kinetic model

The reaction kinetics of the basic system can thus be described by an equation which also applies to monofunctional model compounds at high rates of conversion, i.e. low concentrations [14], (Eq. 5).

$$\text{rate} = k\,[\text{Pt}]\cdot[\text{SiH}]\cdot[\text{SiVi}]$$

Eq. 5.

Here, the network produced does not, therefore, affect the chemical reaction.

4.2 Crosslinker Functionality – Gel Point

An increase in the functionality of the crosslinker from 5 (basic system) to 20 does not alter the kinetics but it does affect the relative course of the curve indicating chemical conversion (SiH/SiVi = 2) and network formation (Fig. 8, 9).

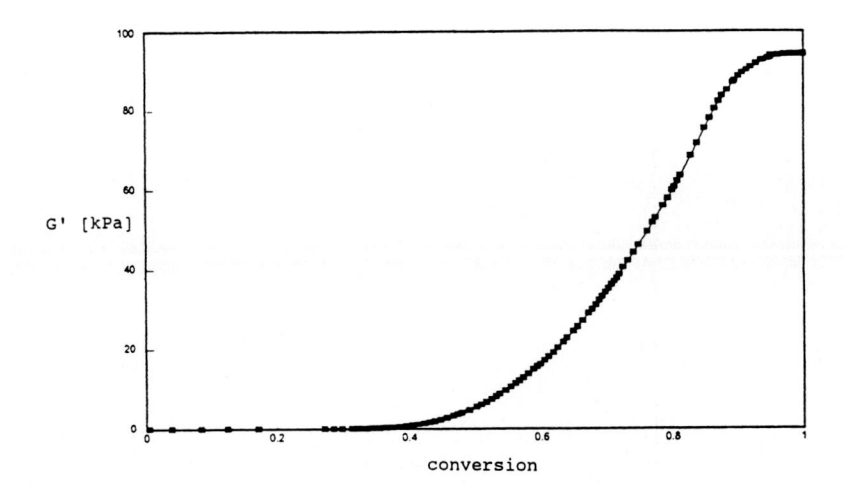

Fig. 8. Storage modulus vs conversion Crosslinker functionality $f = 5$

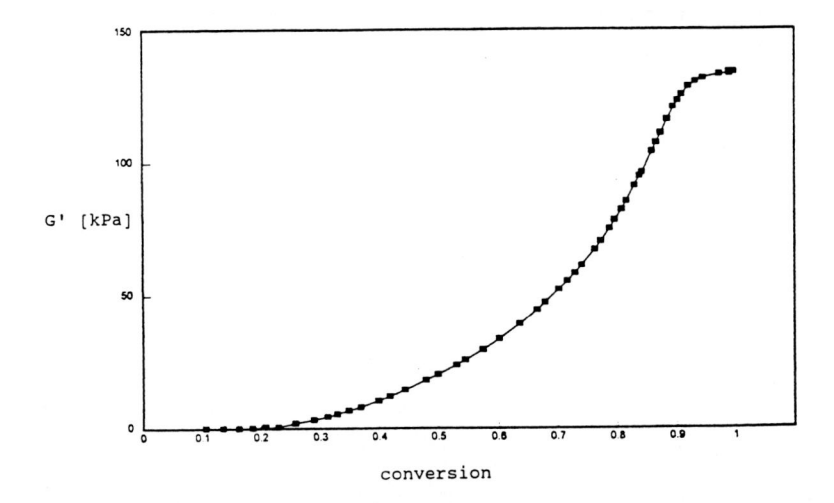

Fig. 9. Storage modulus vs conversion Crosslinker functionality $f = 20$

In the case of the crosslinker with higher functionality, the gel point occurs at a lower degree of conversion. This corresponds with the Flory-Stockmeyer theory in a more general form which describes the gel point as a function of stoichiometry and functionality [15] (Eq. 6):

$$r p^2 = \frac{1}{f-1}$$

Eq. 6. r = SiH/SiVi ratio, f = crosslinker functionality, p = conversion

When $r = 2$ and $f = 5$ or 20 and the gel point is 35 or 16 %, degrees of conversion can be calculated which correspond well with the values observed (Fig. 8, 9).

4.3 Reactivity of different polymers and polymer mixtures

Significant deviations from the kinetics described above are found in the case of other polymers. A shortening of the chain length of a vinyl-terminated polymer from 205 to 13 siloxy units (polymer **2**) results in a drastically reduced rate of reaction and completely different kinetics (Fig. 10).

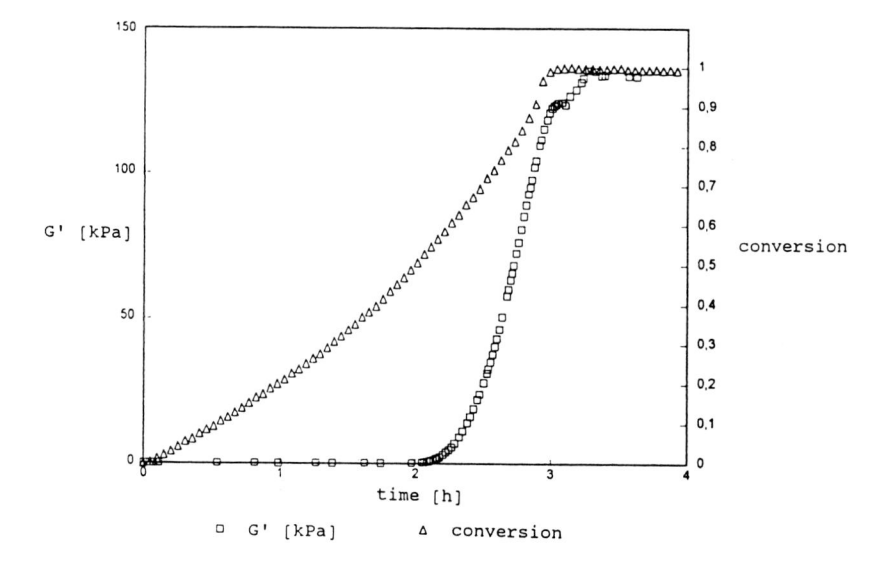

Fig. 10. Storage modulus and conversion vs time (polymer 2 SiH/SiVi = 2; 1.5 ppm Pt; 45 °C)

The rate of reaction increases in relation to the degree of conversion. An increase in chain length to 540 siloxy units does not change the order of reaction but only increases the rate constant ($k = 20.3$ $l\,mol^{-1}h^{-1}$) by two-fold. An even stronger inhibitory effect is caused by a high concentration of lateral vinyl groups (Fig. 12):

$$CH_3-\underset{\underset{CH_3}{|}}{\overset{\overset{CH_3}{|}}{Si}}-O\left[-\underset{\underset{CH_3}{|}}{\overset{\overset{CH_3}{|}}{Si}}-O\right]_m\left[-\underset{\underset{CH_3}{|}}{\overset{\overset{CH=CH_2}{|}}{Si}}-O\right]_n-\underset{\underset{CH_3}{|}}{\overset{\overset{CH_3}{|}}{Si}}-CH_3$$

Polymer 3

Fig. 11.

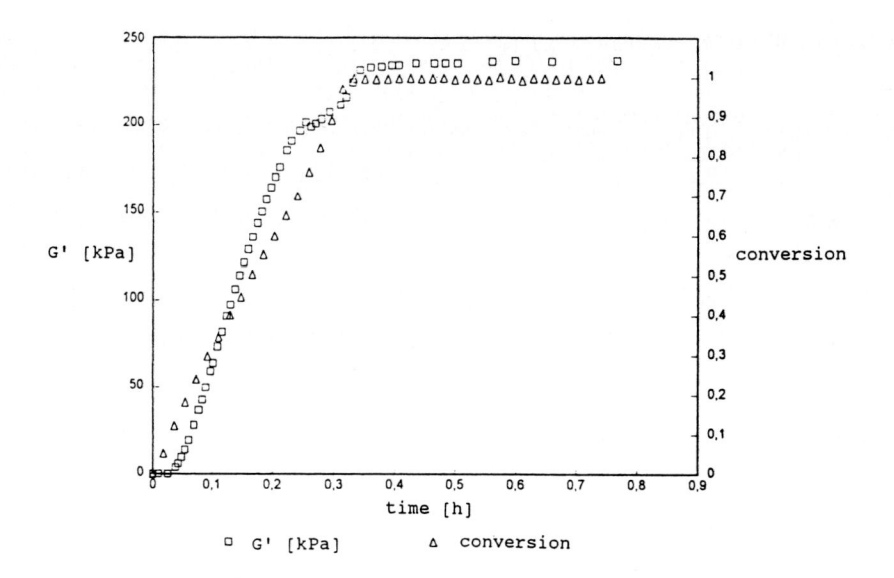

Fig. 12. Storage modules and conversion vs time (polymer 3; SiH/SiVi = 2; 200 ppm Pt; 65 °C)

The reaction takes place, giving high degrees of conversion, according to zero order kinetics. In the case of a SiH/SiVi ratio of 0.5, however, the mixture reacts according to second order kinetics (Fig. 13).

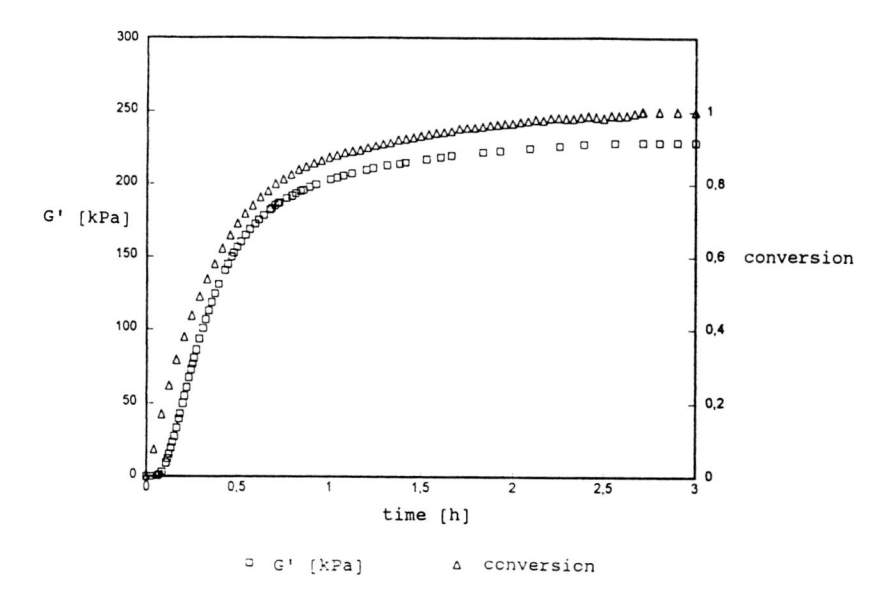

Fig. 13. Storage modulus and conversion vs time (polymer **1**; SiH/SiVi = 0.5; 200 ppm Pt; 65 °C)

If this is applied to the reaction conditions of the basic system, a reduction in rate constant of a factor of 10^4 is observed.

A similar inhibition of the reaction also occurs in the basic system when polymer **3** with lateral vinyl groups is added. A mixture containing equimolar amounts of vinyl groups from the basic system and polymer **3** reacts according to zero order kinetics at the same low rate of reaction as polymer **3** alone. This inhibitory effect of polymers with several lateral vinyl groups is also described in the patent literature [16]. The cause clearly does not lie in the varying reactivity of the different vinyl groups but in the inhibition of catalyst activity.

4.4 Inhibitor

Inhibitors such as 1-ethynylcyclohexanol are used to retard the reaction. They also cause a fall in the rate of reaction ($k = 3.5$ l mol^{-1} h^{-1}), as tests with 1.5 ppm of inhibitor show (Fig. 14).

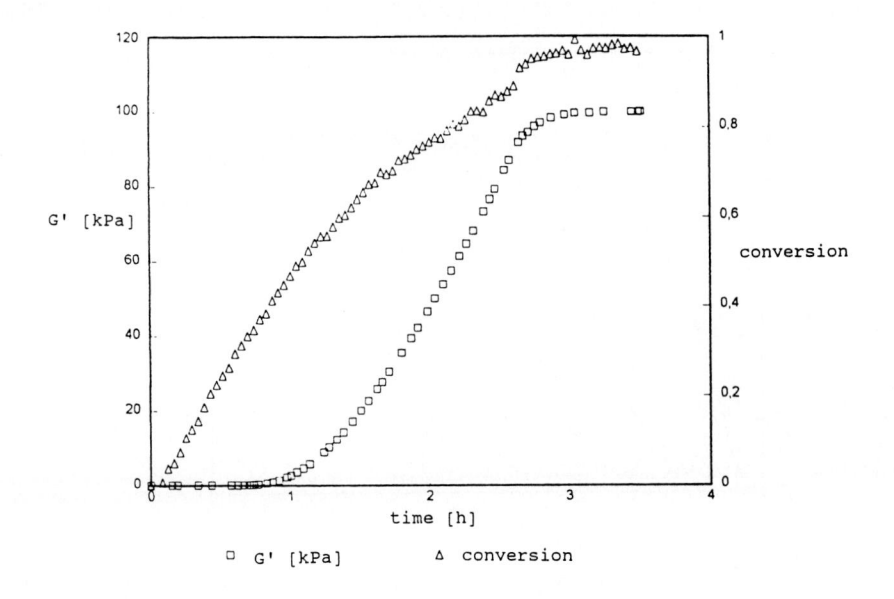

Fig. 14. Storage modulus and conversion vs time (basic system + 1.5 ppm inhibitor; 30 °C)

The conversion of the vinyl groups is also retarded since initially the triple bond of the inhibitor reacts. The order of reaction and the network structure remain unchanged, however.

5. Conclusions

The use of the measuring equipment described enables for the first time a simultaneous study of chemical reaction and vulcanisation. The crosslinking reaction can be easily investigated by measuring the 1st harmonic of the stretching vibration of vinyl absorption in NIR. In contrast to the IR region, layer thicknesses of 20 mm can be examined. A further advantage of this system is that the signal can be transmitted through optical waveguides which means that the measuring cell can be installed outside the spectrometer. This is important to enable the simultaneous testing of the course of vulcanisation with a rheometer in the same measuring cell.

Vinyl-terminated polydimethyl siloxanes react with hydrogen silanes according to second order kinetics. The same kinetics are found in model reactions on monofunctional reactants at a lower concentration range. Also, a direct proportionality between the rate of reaction and platinum concentration is observed in both cases. Network formation does not inhibit the reaction. Polymers with lateral vinyl groups and very short chain polymers with vinyl end groups have approximately zero order kinetics with a strongly reduced rate of reaction. The rearrangement (III) of the platinum complex by the "Chalk-Harrod-Mechanism" determines the kinetics [17] (Scheme 2).

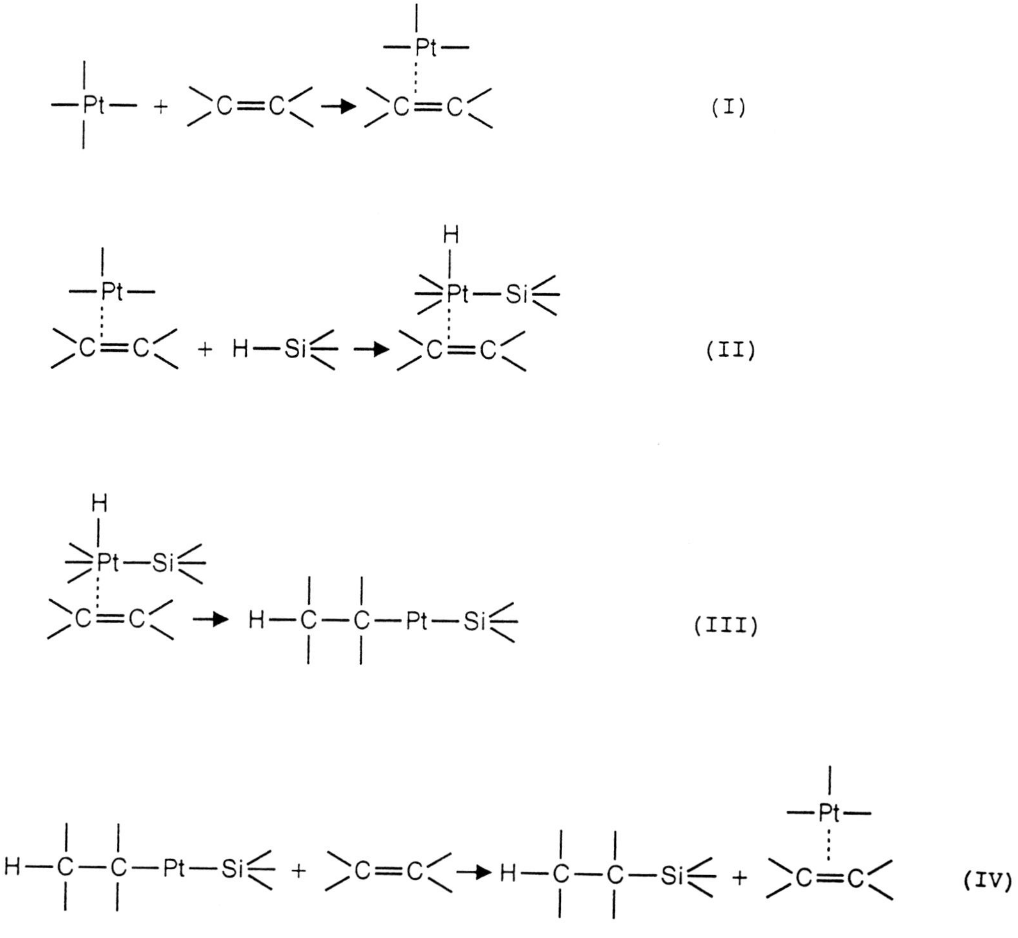

Scheme 2.

The rate of reaction is then determined by the activity of the platinum complex with different vinyl siloxane ligands.

Acknowledgements: The author wishes to express his thanks to Mr. Gerle and Dr. Schwindack.

References:

[1] M. A. Llorente, A. L. Andrady, J. E. Mark, *J. Polym. Sci. Polym. Phys. Ed.* **1980**, *18*, 2263.

[2] L. C. Yanyo, *Ph. D. Thesis*, University of Akron, **1987**.

[3] S. Venkatraman, *J. Appl. Polym. Sci.* **1993**, *48*, 1383.

[4] C. W. Macosko, L. J. Lee, *Rubber Chem. Technol.* **1985**, *58*, 436.

[5] J. Soltero, V. G. Gonzales-Romero, *Annual Technical Conf.* **1988**, *46*, 1057.

[6] R. H. Bogner, J.-C. Liu, Y. W. Chien, *J. Controlled Release* **1990**, *14*, 11.

[7] G. L. Beatch, C. W. Macosko, D. N. Kemp, *Rubber Chem.Technol.* **1990**, *64*, 218.

[8] M. I. Aranguren, *Ph. D. Thesis*, University of Minnesota, **1990**.

[9] K. G. Häusler, K. Hummel, K.-F. Arndt, H. Batz, *Angew. Makromol. Chem.* **1991**, *187*, 187.

[10] S. K. Venkataram, L. Coyne, F. Chambon, M. Gottlieb, H. Winter, *Polymer* **1989**, *30*, 2222.

[11] D. Wrobel, *Silicones – Chemistry and Technology*, Vulkan-Verlag, Essen, **1991**, 61.

[12] R. Tanton, *Meeting of the Rubber Division*, American Chemical Society, Orlando, Florida, **1993**.

[13] M. Gerle, *Diploma Thesis*, Universität Köln, **1993**.

[14] D. Brand, H.-H. Moretto, M. Schulze, D. Wrobel, *J. prakt. Chem.* **1994**, *336*, 218.

[15] C. W. Macosko, D. R. Miller, *Macromolecules* **1976**, *9*, 199.

[16] M. T. Maxson, *EP 257 970*, **1988**.

[17] A. J. Chalk, J. F. Harrod, *J. Am. Chem. Soc.* **1965**, *87*, 16.

Photo-crosslinked Polysiloxanes – Properties and Applications

Klaus Rose

Fraunhofer-Institut für Silicatforschung
Neunerplatz 2, D-97082 Würzburg, Germany

Summary: Organically modified polysiloxanes are synthesized, in a sol-gel process, from silicon alkoxide compounds bearing organic substituents which are stable against hydrolysis. Additional crosslinking is achieved by irradiation with UV-light, due to the presence of reactive organic moieties at the silicon. UV curable materials have been developed based on thiol/ene addition and on acrylate polymerization. These materials have applications as abrasion resistant coatings for PMMA, as flexible coatings and as adhesives for silica optical fibers. The incorporation of thioether moieties in the materials produces sensitivity to gaseous SO_2 due to adduct formation.

Introduction

UV curing of coatings, inks, and adhesives is well established as an industrial process. The main advantages of this technique are the capability for application on heat sensitive materials and the short curing times, which increase the rates of fabrication, as compared to thermal curing. Organically crosslinked polysiloxanes are currently used as coating materials for a variety of purposes, for example to provide abrasion resistance [1], corrosion resistance [2], or as an insulation layer in microelectronics [3]. These materials are prepared from silicon alkoxides bearing hydrolytically stable organic groups, $R'_n Si(OR)_{4-n}$ ($n = 1–3$). The Si–O–Si-backbone is formed by simple hydrolysis and condensation of the alkoxy groups (sol-gel process) [4]. Additional crosslinking is achieved by special reactive groups R' containing epoxy, vinyl, or acrylic moieties.

At the beginning of this materials' technology the curing reaction was mainly based on the thermally induced epoxide polymerization demanding high temperatures and long curing times [5, 6].

We developed polysiloxanes which can be cured or crosslinked by UV radiation at low temperatures in order to increase the curing rate and to open the field of application on heat sensitive substrates. In this paper some special properties and applications of photo-crosslinked polysiloxanes are described.

1 Crosslinking by Thiol/Ene Addition

A coating system, for the mechanical protection of PMMA, is obtained by polycondensation of $CH_2=CH-Si(OEt)_3$ and $HS(CH_2)_3Si(OEt)_3$, via sol-gel processing. As both components are present in equimolar amounts, crosslinking and curing is achieved by the UV induced thiol/ene addition. The curing is performed without the addition of photoinitiators and is completed after 20 s. The resulting layers (thickness around 5 μm) exhibit 2–3 Δ% haze after 100 cycles in the taber abraser test (ASTM D 1044 or DIN 52347). This mechanical stability is derived from the highly crosslinked, rigid inorganic Si–O–Si-backbone and the short spacer chain formed by thiol/ene addition (Fig. 1). This abrasion resistance is indeed remarkable, since it is achieved without the incorporation of inorganic and ceramic-like components derived from alkoxides of Ti, Zr, Al as described elsewhere [5].

Fig. 1. Photo-crosslinking of polysiloxanes by thiol/ene addition

2 Acrylate Polymerization in Optical Fiber Technology

An acrylate containing polysilicone resin has been synthesized as a coating material for silica optical fibers. UV curing of this material is completed within 0.1–0.5 s, due to the radical polymerization of acrylic groups. Thus, coating and curing can be performed while the fiber is drawn from a silica rod at a drawing speed of 10–15 m/s. With the incorporation of special spacers with different chain lengths and alterations in the degree of organic crosslinking through varying amounts of acrylate groups (Fig. 2) different Young's moduli are exhibited by the layers [7].

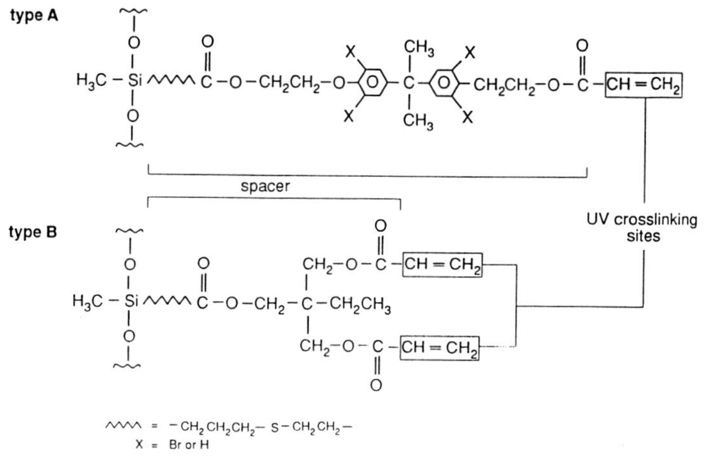

Fig. 2. Different spacers connected to the silicone chain bearing acrylate groups for photo-crosslinking

Siloxane type "A" (X = H) has a low modulus (10–20 MPa) and high flexibility. Therefore it is predetermined for use as the primary coating on the fiber, responsible for the absorption of mechanical stress. The higher degree of organic crosslinking in siloxane type "B" leads to a higher modulus (1400 MPa), an increase in fiber strength and good abrasion resistance. Due to these properties this siloxane type is very suitable for the secondary coating required to prevent mechanical damage to the fiber. The ability to tailor the mechanical properties of the material by mixing both components in various ratios is evident [7].

An additional very important aspect in telecommunications is the connection of silica optical fibers with each other or with integrated optical devices. The UV polymerizable acrylate siloxanes described are well suited for this special application because of their very fast curing rate, variable mechanical properties, and good adhesion to glass. They also possess high transparency and the transmittance is higher than 95 % between 700 and 1800 nm, the wavelength region commonly used in optical data transmission [8]. By varying the organic moieties in the spacer between the silicone chain and the acrylic network, it is possible to realize refractive indices between 1.52 and 1.60 (Table 1), which is very important for optoelectronics.

The properties described above and the thermal stability up to 150 °C, which is derived from the inorganic backbone, make these materials superior to common organic adhesives such as epoxy resins.

Organic Moiety	Refractive Index
linear aliphatic (Fig. 2, type B)	1.52
bisphenol A (Fig. 2, type A, X = H)	1.56
brominated bisphenol A (Fig. 2, type A, X = Br)	1.60

Table 1. Refractive index depending on different organic moieties in the spacer between inorganic and organic backbone

3 Chemical Sensitivity

On the treatment of crosslinked polysiloxane layers, containing thioether moieties (see Fig. 1 or Fig. 2), with gaseous SO_2, a UV absorption band was detected around 320 nm which is attributed to a SO_2/thioether complex. The signal is produced after treating the siloxane layers for 5 min with an atmosphere containing 0.1 % SO_2. The reverse reaction and corresponding loss of signal is complete after 30 min storage in ambient atmosphere at 30 °C. This special adduct formation is very sensitive to low SO_2 concentrations and the reversibility at ambient conditions, is superior to that of amine/ SO_2 adducts in polysiloxane matrix [9].

The application as optical sensor is possible based on sensitive films containing thioether moieties which exhibit changes in optical absorption when probed by guided light. Moreover, by combination with the fiber coating material described above (Fig. 2) the potential for fabrication of fiber optical sensors is evident.

4 Conclusion

Photo-crosslinking of polysiloxanes opens up new applications of inorganic organic hybrid materials not only in conventional applications such as hardcoatings for plastics but also in the field of high technology such as telecommunications or sensors for environmental monitoring. Due to the hybrid nature of these materials the combination of properties associated with both inorganic and organic materials is possible as well as special functionalization on a molecular level.

References:

[1] S. Amberg-Schwab, E. Arpac, W. Glaubitt, K. Rose, G. Schottner, U. Schubert, in: *High Performance Ceramic Films and Coatings* (Ed.: P. Vincenzini), Elsevier, Amsterdam, **1991**, 203.

[2] K. Greiwe, *Farbe Lack* **1991**, *97*, 369.

[3] H. Schmidt, H. Wolter, *J. Non-Cryst. Solids* **1990**, *121*, 428.

[4] C. Sanchez, J. Livage, *New J. Chem.* **1990**, *14*, 513.

[5] H. Schmidt, B. Seiferling, G. Philipp, K. Deichmann, in: *Ultrastructure Processing of Advanced Ceramics* (Eds.: J. D. Mackenzie, D. R. Ulrich), Wiley, Chichester, **1988**, 651.

[6] R. Naß, E. Arpac, W. Glaubitt, H. Schmidt, *J. Non-Cryst. Solids* **1990**, *121*, 370.

[7] K. Rose, H. Wolter, W. Glaubitt, *Mat. Res. Soc. Symp. Proc.* **1992**, *271*, 731.

[8] J. Labs, K. Rose, S. Werner, *Mikroelektronik* **1993**, 7.

[9] A. Brandenburg, R. Edelhäuser, F. Hutter, *Sens. Act.* **1993**, *11B*, 361.

Cationic Photo-crosslinking of α,ω-terminated Disiloxanes

A. Kunze, U. Müller*

Institut für Organische Chemie

Martin-Luther-Universität Halle-Wittenberg

Geusaer Str., D-06217 Merseburg, Germany

Ch. Decker

Laboratoire de Photochemie Generale

3rue Alfred Werner, F-68093 Mulhouse, France

J. Weis, Ch. Herzig

Wacker-Chemie GmbH

Geschäftsbereich S – Werk Burghausen

Johannes-Heß-Str. 24, D-84489 Burghausen, Germany

The photoinduced cationic crosslinking of α,ω-terminated disiloxanes containing vinyl ether (VE1 and VE2) and epoxy functional groups (Scheme 1) in dependence of the light intensity and the type of photosensitizers was investigated by means of *r*eal *t*ime *i*nfra*r*ed (RTIR) spectroscopy and calorimetric measurements.

The crosslinking of such types of disiloxanes can be described by means of a cationic chain process. The diphenyliodonium salt **1** was used as initiator system for the monomers.

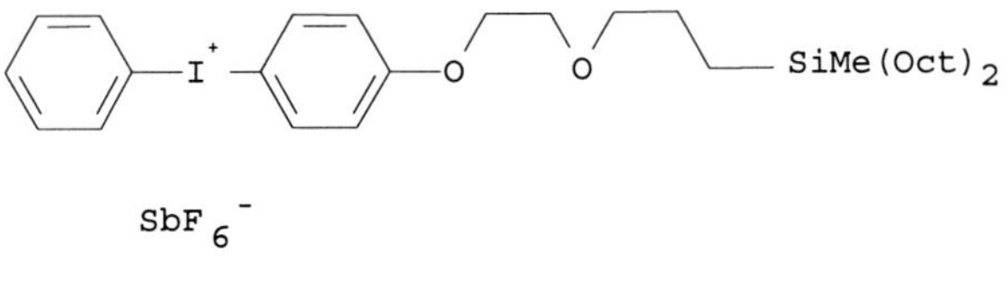

Fig. 1.   Diphenyliodonium salt **1**

By means of RTIR spectroscopy the change of the absorption of the functional group during the irradiation in dependence on the time has been recorded. From these measurements the polymerization rate at different times could be derived.

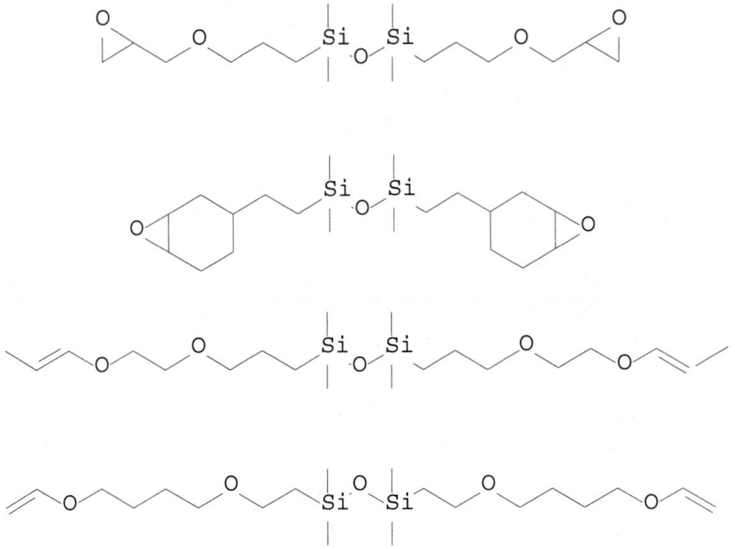

Scheme 1.

Differences in the rates and conversion degrees of epoxy- and vinyl ether systems have been found. The epoxide compounds polymerize significantly slower than the vinyl ether species. While the cyclohexenyloxide compound Ep2 exhibits, with a value of 0.4 $M \cdot s^{-1}$ the highest polymerization rate, the propylenoxide compounds exhibit only a tenth of this value. In the case of the vinyl ethers VE1 and VE2 a dependence of the polymerization rate on the type of their spacer group has been found. The methyl vinyl group in VE1 is less reactive than the vinyl ether group in VE2. Consequently, the vinyl ether system VE2 reaches a higher degree of conversion than the methylvinyl system VE1 due to a larger polymerization rate.

The photopolymerization of the polymers studied is influenced by the sensitizer as well as the intensity of light used. Aromatic hydrocarbons, like anthracene and tetracene, are used as singlet sensitizers whereas 1-chlorothioxanthone (CTX) is a triplet sensitizer. From the comparison of the polymerization rates of the systems studied as a function of sensitizer it follows that there is a dependence on the type of sensitizer. As seen from Fig. 2, CTX as triplet sensitizer has been found to be the best sensitizer for vinyl ether systems.

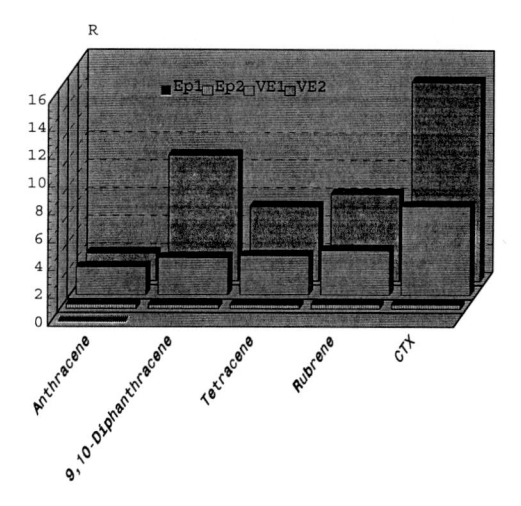

Fig. 2. Rate of polymerization (in M·s^{-1})

Both the conversion degree and the rate of polymerization exhibit in all cases a linear dependence on the light intensity (Fig. 3). The quantum yield (ϕ_P) of the photopolymerization reflects the effectivity of the reaction and corresponds to the number of polymerizable functional groups per absorbed photons. From the polymerization rate the following equation results:

$$\Phi_P = \frac{R_P}{I_{abs}}$$

Eq. 1. I_{abs}: absorbed light intensity; R_P: polymerization rate

The quantum yields of polymerization are shown in Table 1. The values for the epoxide systems are significantly lower than those of the vinyl ether systems. Because the yield is ca one for the Ep1-system, one can conclude that no chain reaction occurs.

Monomer	Ep 1	Ep 2	VE 1	VE 2
$1_1^+ SbF_6^-$	1.19	11.16	52.23	70.18

Table 1. Quantum yield of the different systems with anthracene as sensitizer ($6 \cdot 10^{-6}$ mol·g^{-1})

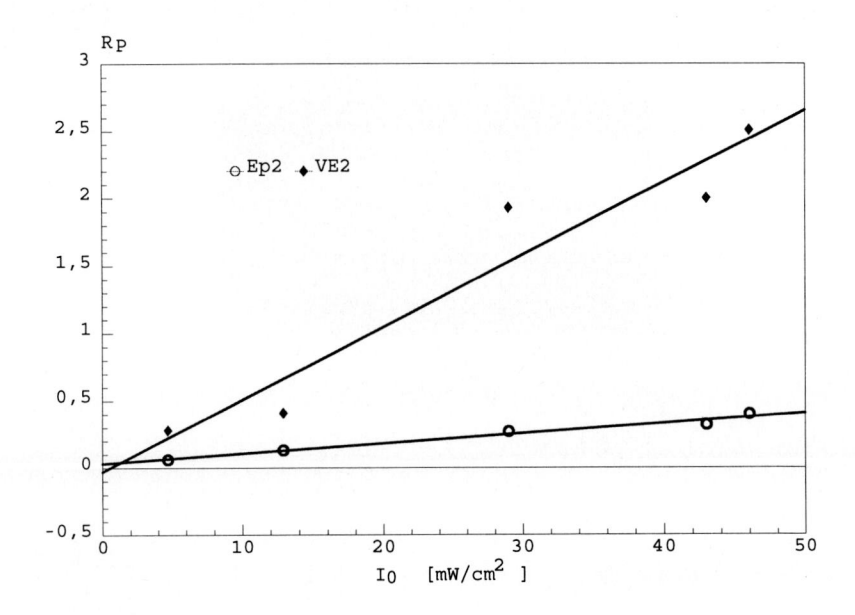

Fig. 3. Rate of polymerization as a function of the light intensity

In summary, the VE1-system exhibits the highest polymerization rate and conversion degree. For this, CTX and 9,10-diphenylanthracene have been proven as the best suited sensitizers. For the epoxide systems only the cyclohexenyloxide monomer reacts satisfactorily.

Azo- and Triazene Modified Organosilicones as Polymeric Initiators for Graft Copolymers

R. Kollefrath, B. Voit, O. Nuyken*

Institut für Technische Chemie, Lehrstuhl für Makromolekulare Chemie

Technische Universität München

Lichtenbergstr. 4, D-85747 Garching, Germany

J. Dauth, B. Deubzer, T. Hierstetter, J. Weis

Wacker-Chemie GmbH

Geschäftsbereich S – Werk Burghausen

Johannes-Heß-Str. 24, D-84489 Burghausen, Germany

Summary: Graft copolymers with poly(organosiloxane) backbone and thermoplastic side chains have been synthesized via the "grafting from" method based on azo- and triazene modified poly(organosiloxane)s. Initiation of free radical polymerization is possible from the polymeric azo and triazene initiators after thermal decomposition of the labile functions in solution. The graft products have been characterized by NMR, GPC, and DSC. Stable, free standing films can be cast from the graft copolymers.

Introduction

Combination of poly(organosiloxane)s and thermoplastics or elastomers by simply mixing the components is not possible due to high incompatibility of these polymers. Therefore alternatives have to be found to combine the interesting properties of silicones (e.g., low temperature flexibility, heat stability) with those of commodity polymers. Some interesting materials could already be synthesized by copolymerization of macromonomeric silicones [1], by blockcopolymer formation between functionalized poly(organosiloxane)s and polycarbonate prepolymers [1] or by interpenetrating network formation [2, 3]. Furthermore, EPDM could be combined with silicone using a compatibilizing low molecular weight organosilicone compound containing sulfur bonds [4].

Also compatibilizers can be applied for two-phase systems in order to reduce the phase size and improve the properties. This method has been successfully applied in an EPDM-silicone system using a silane grafted ethylene-propylene rubber (EPR) [5]. However, the combination possibilities of poly(organosiloxane)s with thermoplastics and elastomers are still limited to special systems.

In the following we would like to describe the synthesis of graft copolymers with silicone backbone based on polymeric initiators. These graft copolymers might find application as compatibilizers and, furthermore, the synthetic procedure allows broad combination possibilities.

Results and Discussion

The base for our polymeric initiators are amino functionalized poly(organosiloxane)s with different amino content and a molecular weight M_n of about 15.000 g mol^{-1}. These structures can be modified with azo or triazene groups according to Fig. 1 and 2.

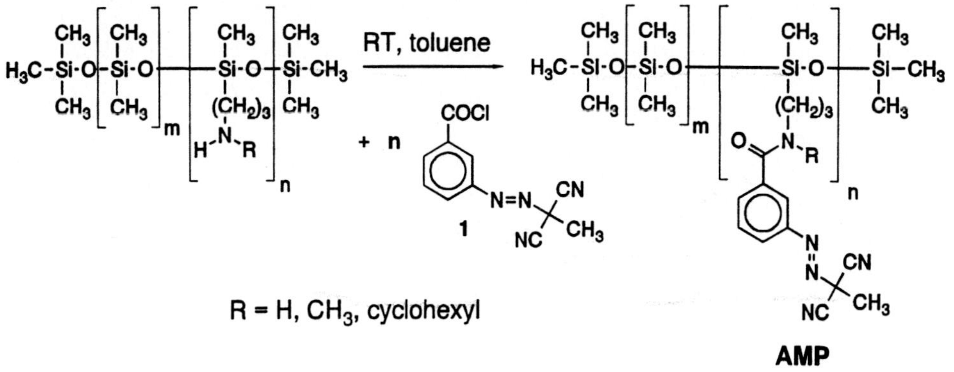

Fig. 1. Synthesis of azo modified poly(organosiloxane)s AMP

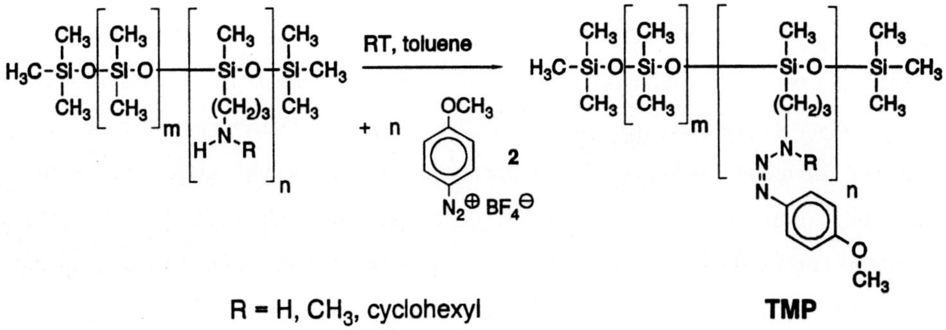

Fig. 2. Synthesis of triazene modified poly(organosiloxane)s TMP

The reaction of the secondary amines with the acid chloride 1 to AMP is quantitative. However, the azo compound 1 is not commercially available and has to be synthesized as described in Fig. 3.

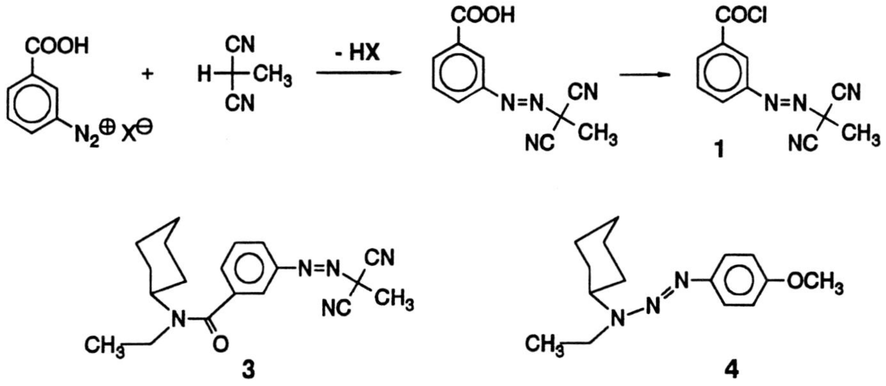

Fig. 3. Azo- and triazene compounds

In contrast, the synthesis of the initiator functionality and the modification of the poly(siloxane) is only one single step for TMP using an easily available diazonium compound 2 with a yield of about 60–80 %. Structure verification and determination of the initiator group content could be done by ^{1}H NMR spectroscopy. GPC analysis showed that the modified poly(organosiloxane)s retain their original molecular weights. About 4–8 initiating sites have been attached onto the silicone backbone. Due to the chromophores, AMP and TMP are yellow products.

Azo functions are well known to initiate free radical polymerization and have been used previously in polymeric initiators [6]. However, triazene functions are only rarely used in free radical polymerization [7], and not much is known about their thermal decomposition mechanism. The comparison of the thermal characterisitics of two model compounds, 3 and 4 (see Fig. 3), for AMP and TMP has shown that the azo function is much more thermally labile than the triazene group. The maximum of decomposition in bulk for 3 is at 135 °C and for 4 at 270 °C. Decomposition of 3 and 4 in DMSO solution at 80 °C followed by UV spectroscopy led to complete decomposition of 3 after 600 min whereas only about 5 % of the triazene functions are destroyed at this time.

Fig. 4 schematically describes the grafting reaction from AMP at 75 °C. The grafting reaction from TMP (carried out at 95 °C) follows a similar pathway but different radicals are formed during decomposition.

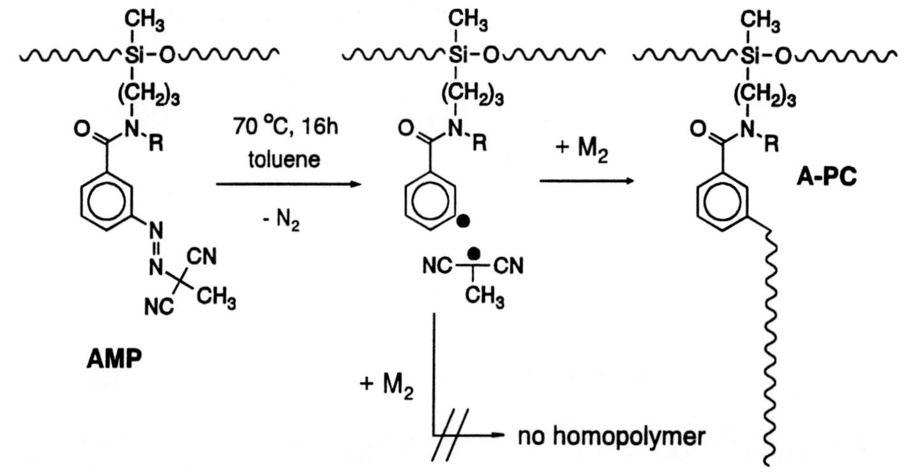

Fig. 4. Free radical grafting from azo modified poly(organosiloxane)s

The formation of graft copolymers was successful with both polymeric initiators as indicated by the increase of molecular weight after reaction. The characteristics of the graft products are given in Table 1 and 2. MMA and styrene have been used as graft monomers in various ratios toward the silicone backbone. As expected the molecular weight of the graft copolymers increases at a given initiator group concentration with increasing graft monomer/backbone ratio. The polymer yields obtained after grafting from TMP for 16 h are below 50 % due to the high thermal stability of the triazene functions. The graft products are colorless when based on AMP (A-PC) and slightly yellow when based on TMP (T-PC).

Polymer	Comonomer	m:n	Monomer/Silicone wt. Ratio	M_n	M_w
A-PC 1	MMA	≈ 25:1	1:1	19 100	30 300
A-PC 2	MMA	≈ 35:1	3.9:1	65 300	179 000
A-PC 3	MMA	≈ 25:1	5:1	16 100	23 800
A-PC 4	MMA	≈ 25:1	10:1	61 700	114 700
A-PC 5	MMA	≈ 50:1	5:1	77 200	137 000

Table 1. Composition and GPC results of the graft copolymers based on AMP (M_n = 15.000/mol)

Polymer	Comonomer	m:n	Monomer/Silicone wt. Ratio	M_n	M_w
T-PC 1	MMA	≈ 6:1	5:1	58 000	92 800
T-PC 2	MMA	≈ 50:1	1:1	33 900	65 200
T-PC 3	MMA	≈ 50:1	3:1	53 300	116 200
T-PC 4	MMA	≈ 50:1	10:1	130 900	296 000
T-PC 5	MMA	≈ 60:1	5:1	26100	41 600
T-PC 6	styrene	≈ 50:1	3:1	73 500	204 000

Table 2. Composition and GPC results of the graft copolymers based on TMP (M_n = 15.000/mol)

The azo initiator was chosen in a way that the low molecular dinitrile radical formed after decomposition does not initiate free radical formation. Therefore no homopolymer of the graft monomer in the graft products A-PC was found. However, in the case of the triazene initiator functions homopolymer formation can not be fully excluded in T-PC. Further studies on grafting efficiency, length and number of graft arms, and homopolymer formation, as well as on the initiator mechanism of the triazene functions are in progress.

DSC measurements on the graft products usually showed two separate T_g for the backbone and the graft arms indicating microphase separation. However, stable, free standing, slightly cloudy films could be prepared from all graft products.

Conclusions

Polymeric initiators based on azo or triazene modified poly(organosiloxane)s can be used to synthesize graft copolymers with silicone backbone and thermoplastic side chains. The azo functionality has some advantages such as lower thermal stability, known reaction mechanism, no homopolymer formation and cleaner graft products. However, the synthesis of the triazene polymeric initiators requires fewer synthetic steps. The graft products microseparate but form stable films.

References:

[1] B. Arkles, *Chemtech* **1983**, *13*, 542.
[2] H. Struckmeyer, *EP* 0322521 A1, **1988**; *Chem. Abstr.* **1988**, *111*, 234866.
[3] H. Block, M. Pyrlik, *Kunststoffe* **1988**, *78*, 1192.
[4] J. Mitchell, T. Wada, K. Itoh, *Gummi, Fasern, Kunststoffe* **1986**, *39*, 294.
[5] S. Kole, A. Bhattacharya, A. Bhowmick, *Plast. & Rub. Proc. Appl.* **1993**, *19*, 117.
[6] O. Nuyken, B. Voit, in: *Macromolecular Design: Concept and Practice* (Ed.: M. Mishra), Polymer Frontiers International, New York, **1994**, p. 313.
[7] L. Brodsky, M. Moskowitz, R. Moore (GAF Corporation), *US* 4 137 226, **1979**; *Chem. Abstr.* **1979**, *90*, 187621.

Core Shell Structures Based on Polyorgano Silicone Micronetworks Prepared in Microemulsion

F. Baumann, M. Schmidt*

Makromolekulare Chemie

Universität Bayreuth

Universitätsstr. 30, D-95440 Bayreuth, Germany

J. Weis, B. Deubzer, M. Geck, J. Dauth

Wacker-Chemie GmbH

Geschäftsbereich S – Werk Burghausen

Johannes-Heß-Str. 24, D-84489 Burghausen, Germany

Summary: The polycondensation of methyltrimethoxysilane in the presence of the surfactant benzethonium chloride shows the phenomenology of a polycondensation in microemulsion. These polyorganosiloxane micronetworks can be functionalized with azo groups which are capable of grafting reaction with vinylic monomers. The structure of the resulting core shell systems depends on the polarity of the organic solvent. In DMF molecularly dissolved star-like structures were observed.

Polyorgano-Silicone Micronetworks

Polymerization in microemulsion has developed into a powerful technique for the preparation of strictly spherical micronetworks [1, 2]. The final size of the polymerized particles is solely governed by the ratio of surfactant to monomer concentration, i.e., the fleet ratio S. To predict the final particle size at full conversion, two simple models for the polymerization in microemulsion have been proposed which differ only in some minor details. One of the models considers variable headgroup contributions to the particle radius [3]. This calculation finally arrives at Eq. 1.

$$S^{-1} \equiv \frac{m_{surf}}{m_{monomer}} = \frac{N_L \cdot \bar{\rho}}{3 M_{surf}} \cdot a_0 R - C$$

Eq. 1. m_{surf}, m_p: masses; M_{surf}: molar mass of the surfactant; a_0: headgroup; R: spherical radius; N_L: Avogadro number; ρ: particle density in solution; C: constant of the order of units depending on the headgroup contribution to the particle radius

We report here on the polycondensation of methyltrimethoxysilane [4–7] in microemulsion in the presence of the surfactant benzethonium chloride (1) [8].

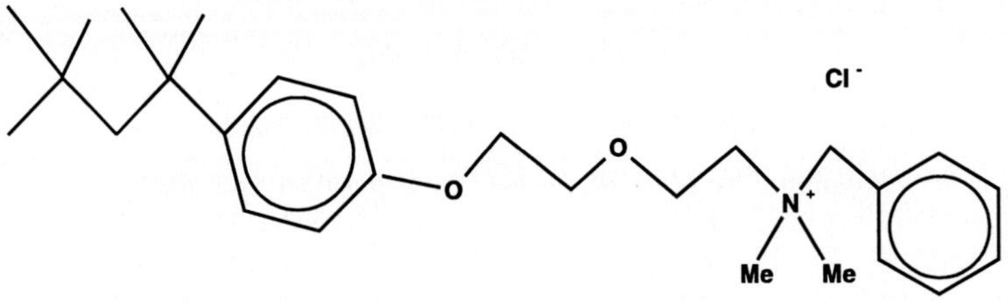

Fig. 1. Benzethonium chloride (1)

Results and Discussion

The results are summarized in Table 1 where the particle dimensions in terms of the radius of gyration, R_g, the hydrodynamic radius, R_h, and the particle molar masses, M_W, are given for different fleet ratios S. The molar mass and the radius of gyration are determined by static light scattering according to standard procedures and the hydrodynamic radius is evaluated from the diffusion coefficient measured by dynamic light scattering by formal application of the Stokes law (Eq. 2).

$$R_h = \frac{kT}{6\pi\eta_0 D}$$

Eq. 2. kT: thermal energy; η_0: solvent viscosity

A spherical particle shape is documented by the values of the ratio $\rho \equiv R_g/R_h \approx 0.775$. For the small particles, R_g can no longer be measured as the values drop below 10 nm.

S	1/S	R_h [nm]	R_g [nm]	ρ	$M_w \cdot 10^{-6}$ [g/mol][a]
0.06	16.67	19.5	14.5	0.75	4.2
0.07	14.44	16.7	13.3	0.80	3.64
0.075	13.33	14.7	11.3	0.77	2.34
0.09	11.11	13.4	< 10	< 0.75	1.87
0.12	8.33	12.7	10.5	0.82	1.65
0.15	6.67	11.3	< 10	< 0.88	1.32
0.2	5	11.0	< 10	< 0.90	1.28
0.25	4	10.6	< 10	< 0.94	1.03
0.3	3.33	10.5	< 10	< 0.95	0.82
0.4	2.5	13.6	16.8	1.24	1.55

Table 1. Characterization of the polyorganosilicone micronetworks; [a] particle molar mass including surfactant

In Fig. 2 the inverse ratio S^{-1} is plotted versus the hydrodynamic radius of the microspheres. A linear dependence is observed in agreement with Eq. 1. Thus, the described polycondensation of methyltrimethoxysilane resembles phenomenologically a polycondensation in microemulsion.

In Fig. 3 the particle radius is plotted versus the cube root of the volume of added monomer. A straight line is observed which extrapolates to $R_h \approx 0.5$ nm for no added monomer. This plot demonstrates that after a certain "induction" period the growth of the particles is homogeneous, i.e., the number of growing particles remains constant. Also, Fig. 3 gives a vague hint that the nucleation takes place in the water phase, i.e., at the early stage of polycondensation the surfactant does not form a micellarlike structure around the growing nuclei, because the curve in Fig. 3 almost hits the origin and does not yield the micellar radius of benzethonium chloride (**1**) at zero added volume, which is much larger than 0.5 nm.

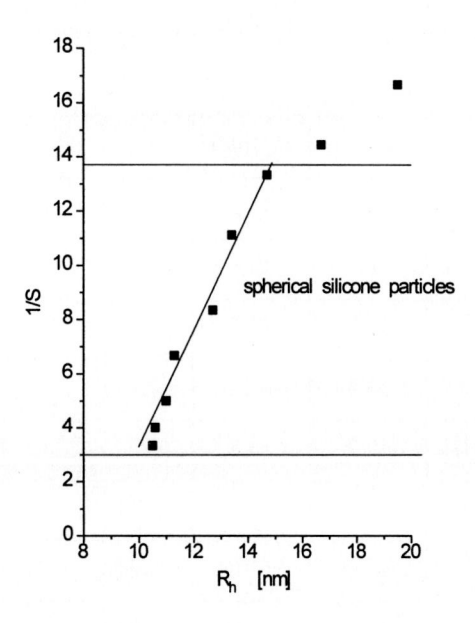

Fig 2. The particle radius R_h as a function of the inverse fleet ratio S^{-1} according to Eq. 1; the solid line represents the fit to the data in the linear regime marked by the horizontal lines

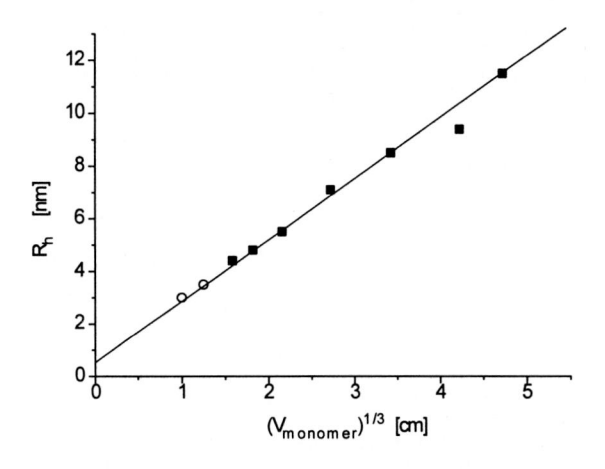

Fig. 3. The particle radius as a function of the cube root of added methyltrimethoxysilane volume; fleet ratio after all monomer was added: $S = 0.12$

Our conclusion that the nuclei do not start to grow within normal micelles is also supported by the surface tension measurements shown in Fig. 4, where the surface tension is plotted against the volume of added monomer. Below 12.5 mL added monomer the surface tension remains almost constant, whereas it increases with larger added volume of the monomer. Obviously, empty micelles are no longer present in the solution, as the growing particles now need more and more surfactant molecules for stabilization and eventually draw the surfactant molecules from the air/water interface, thus increasing the surface tension as displayed in Fig. 4.

The currently favoured working hypothesis for the stabilization mechanism of the particles is to claim an ampholytic surface structure, whereas the cationic surfactant only stabilizes the hydrophobic "patches" of the non-ionic methylsilane group, the remaining surface of the particles is stabilized by the anionic Si–O⁻ groups.

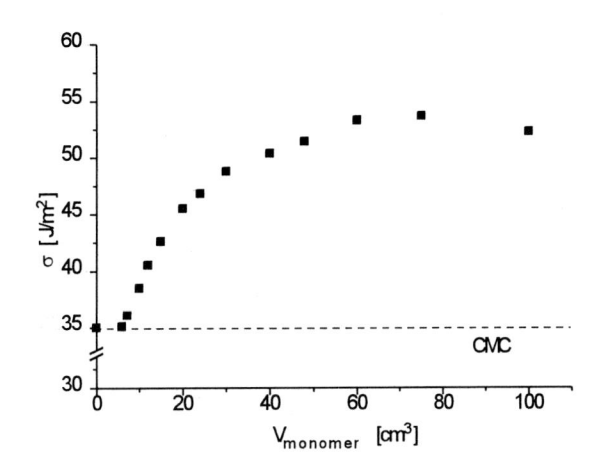

Fig. 4. Surface tension σ vs added methyltrimethoxysilane volume V; fleet ratio after all monomer has been added: $S = 0.09$; as a reference the surface tension of pure **1** in water above the cmc is given as the horizontal line

Core Shell Structures

Subsequent addition of the azomonomer (**2**) to the growing particle results in microspheres exhibiting azo groups on the surface, which, upon addition of methylmethacrylate and heating start a grafting reaction of linear polymer chains [9].

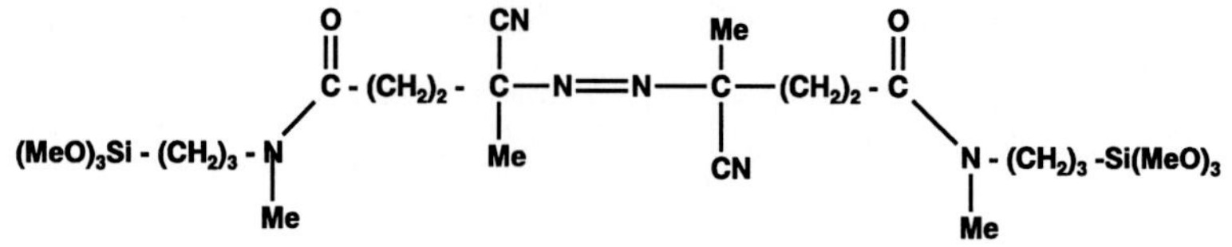

Fig. 5. Azomonomer (2)

Results and Discussion

The results are summarized in Table 2 and 3. Table 2 represents the results of the static and dynamic light scattering of the samples in aqueous suspension. The values of R_h, R_g, and M_w have increased from the azofunctionalized microsphere to the PMMA-copolymer, i.e., the grafting reaction was successful. The ρ value of 0.9 shows that the particles are not aggregated after the polymerization which should result in much larger values of ρ.

Sample	R_h [nm]	R_g [nm]	ρ	$M_w \cdot 10^{-6}$ [g/mole]
Functionalized microgel	9.7	< 10	< 1	0.65
Grafted microgel	19.9	17.0[a]	0.86[a]	9.0[a]

Table 2. Characterization of the functionalized and grafted microgels in water; [a] apparent values

Table 3 shows the results of the static- and dynamic light scattering in different organic solvents after removal of the surfactant. Utilizing the index matching technique the values of R_g and M_w in toluene (index matching of PMMA) indicate that the polar siloxane cores are aggregated. In DMF, however, (index matching of the siloxane core) the obtained value of ρ = 1.33 indicates a star-like structure. The deviation from the theoretical value for ρ = 1.2 [10] for a monodisperse, star-like structure in a Θ-solvent with high branching density and negligable core dimension, is most probably caused by the good solvent quality of DMF for PMMA and by the large dimension ($R_{h(core)}$ = 10 nm) of the centre of the star.

Sample	R_h [nm]	R_g [nm]	ρ	$M_w \cdot 10^{-6}$ [g/mole]
DMF	57	77	1.33	3.1
Toluene	215	101	0.47	17.2

Table 3. Characterization of the grafted sample in different organic solvents

Acknowledgement: Financial support by *Wacker-Chemie GmbH* and by the *Fonds der Chemischen Industrie* (M. S.) is gratefully acknowledged.

References:

[1] J. S. Guo, M. S. El-Aasser, J. W. Vanderhoff, *J. Polym. Sci., Polym. Chem.* **1989**, *27*, 691.
[2] W. Bremser, M. Antonietti, M. Schmidt, *Macromolecules* **1990**, *23*, 3796.
[3] C. Wu, *Macromolecules* **1994**, *27*, 298.
[4] J. Cekada, D. R. Weyenberg, *US* 3 433 780, **1964**.
[5] A. E. Bey, *US* 4 424 297, **1984**.
[6] M. Wolfgruber, B. Deubzer, V. Frey, *EP* 0 291 941, **1988**.
[7] D. R. Weyenberg, D. E. Findlay, J. Cekada, A. E. Bey, *J. Polymer Sci., Polym. Chem.* **1969**, *27*, 27.
[8] F. Baumann, M. Schmidt, B. Deubzer, J. Dauth, M. Geck, *Macromolecules* **1994**, *27*, 6102.
[9] M. Geck, J. Dauth, B. Deubzer, H. Oswaldbauer, M. Schmidt, F. Baumann, *DE* 4 338 421, **1993**.
[10] M. Schmidt, in: *Dynamic Light Scattering: Applications to Structures Analysis* (Ed.: W. Brown), Oxford University Press, Oxford, **1993**, 372.

Precrosslinked Poly(organosiloxane) Particles

Michael Geck*, Bernward Deubzer, Johann Weis

Wacker-Chemie GmbH

Geschäftsbereich S – Werk Burghausen

Johannes-Heß-Str. 24, D-84489 Burghausen, Germany

Gernot Pepperl

Vinnolit Kunststoff GmbH

Werk Burghausen

D-84489 Burghausen, Germany

Summary: Precrosslinked poly(organosiloxane) particles are synthesized by emulsion polycondensation/polymerization of alkoxysilanes and cyclic siloxanes; crosslink density and organic functionality can be varied widely. Graft copolymer particles are obtained by subsequent radical emulsion polymerization. Characterization by means of light scattering, electron microscopy, and thermal analysis reveals the properties of the disperse particles and of the polymer materials. The principal suitability of organopolymer grafted precrosslinked poly(organosiloxane) particles as silicone based modifiers for toughening thermoplastic polymers is verified using PMMA as a model (test) polymer for the large group of thermoplastics.

Introduction

"Precrosslinked" or "intramolecularly crosslinked" particles are micronetworks [1]. They represent structures intermediate between branched and macroscopically crosslinked systems. Their overall dimensions are still comparable with those of high molecular weight linear polymers, the internal structure of micronetworks (µ-gels), however, resembles a typical network [2]. Synthesis is performed either in dilute solution or in a restricted reaction volume, e.g., in the micelles of an emulsion. Particle size and particle size distribution can be controlled by reaction conditions. Functional groups can be incorporated into the microparticles and enable secondary reactions, such as grafting of monomers onto these microparticles.

Precrosslinked particles with low crosslink density exhibit elastic properties and can be applied for toughening thermoplastics or thermosets. The size of the elastic domains in blends consisting of elastic particles and a polymer matrix can be adjusted precisely, provided that the particles are dispersible. Via functional groups, microparticles can be covalently attached to a (thermoset) matrix. The grafting of polymer shells onto elastic microparticles improves the compatibility with the polymer matrix to be modified [3]. Thus, after processing of the polymer alloy discrete elastic particles can be observed as disperse phase in a continuous thermoplastic matrix.

Modification of thermoplastics or thermosets with precrosslinked poly(organosiloxane) elastic particles should impart impact strength as well as typical silicone properties, i.e., low temperature flexibility, stability against weathering, and improved chemical resistance [4].

Synthesis and Composition

Precrosslinked poly(organosiloxane) particles are composed of crosslinking trifunctional and linear difunctional siloxane units (T and D units, respectively) [5]. The molar ratios of D and T units can be varied without restrictions; thus, hard spheres (fillers) as well as soft, elastic silicone particles are accessible. In this study, the siloxane particles were synthesized in emulsion. The particle size was controlled by emulsifier concentration and crosslink density: highly crosslinked particles were obtained with particle diameters ranging from 20–50 nm; the size of elastic particles could be varied between 70 and 150 nm. The composition of precrosslinked poly(organosiloxane) particles is summarized in Scheme 1; further, organic radicals R which can be incorporated into the particles are listed [6, 7].

Poly(organosiloxane) microparticles containing functional groups are a suitable basis for graft copolymers with the organic polymer (e.g. polyacrylates and/or polystyrene) covalently bound to the crosslinked silicone core. Core shell structures can be realized by appropriate reaction conditions [7–9]. Organopolymer coated siloxane particles can be isolated from dispersion by spray–drying. They are compatible to polymers (especially to the polymers used for the shell) and can easily be dispersed in the organic matrix during processing of the compound.

Precrosslinked poly(organosiloxane) particles are synthesized (Scheme 2) by emulsion polycondensation/polymerization of (functionalized) alkoxysilanes and (optionally) cyclic siloxanes [6, 7].

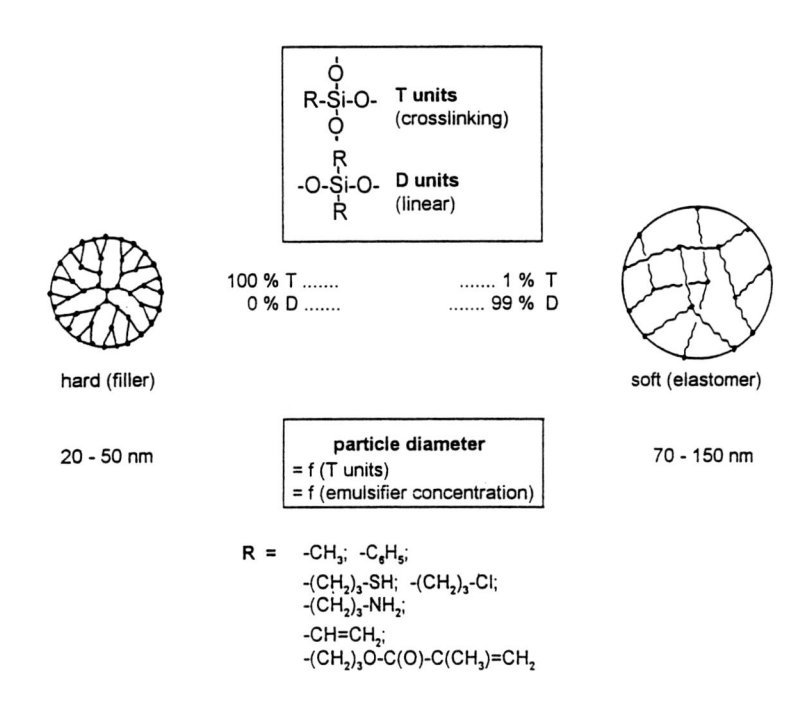

Scheme 1. Composition of precrosslinked poly(organosiloxane) particles

T units (crosslinking): R-Si(-O-)$_3$ with O above and below
D units (linear): -O-Si(R)(R)-O-

- 100 % T, 0 % D → hard (filler), 20 - 50 nm
- 1 % T, 99 % D → soft (elastomer), 70 - 150 nm

particle diameter
= f (T units)
= f (emulsifier concentration)

R = -CH$_3$; -C$_6$H$_5$;
-(CH$_2$)$_3$-SH; -(CH$_2$)$_3$-Cl;
-(CH$_2$)$_3$-NH$_2$;
-CH=CH$_2$;
-(CH$_2$)$_3$O-C(O)-C(CH$_3$)=CH$_2$

Synthesis and Characterization

$[R_mSi(OR')_{4-m}]$, $[R_2SiO]_n$

m = 1, 2
n = 4

R = (functionalized) Alkyl, Aryl
R' = Methyl, Ethyl

↓ emulsion polycondensation/-polymerisation

crosslinked poly(organosiloxane)

↓ graft polymerisation
H$_2$C=CH-C$_6$H$_5$;
H$_2$C=C(CH$_3$)-C(O)OCH$_3$

graft copolymer

Characterization
- in dispersion:
  - light scattering
- prepared from dispersion:
  - TEM
- after (thermoplastic) processing:
  - TEM
  - DSC
  - DMTA

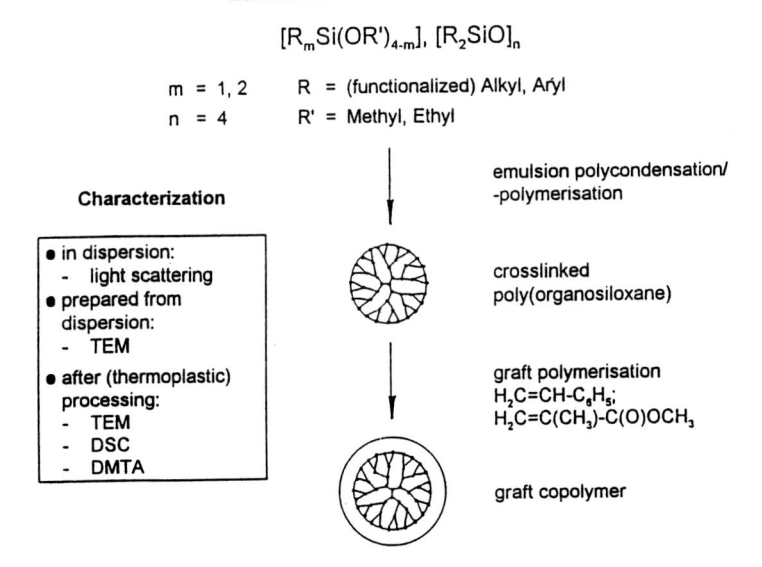

Scheme 2. Synthesis and characterization of precrosslinked poly(organosiloxane) particles

Graft copolymers are obtained by subsequent radical emulsion polymerisation of olefinic unsaturated monomers in the presence of functionalized siloxane particles. Scheme 3 illustrates that the graft polymerization can be accomplished by external radical initiation ("grafting onto") or, alternatively, by the use of siloxane radical macroinitiators ("grafting from") [7, 8].

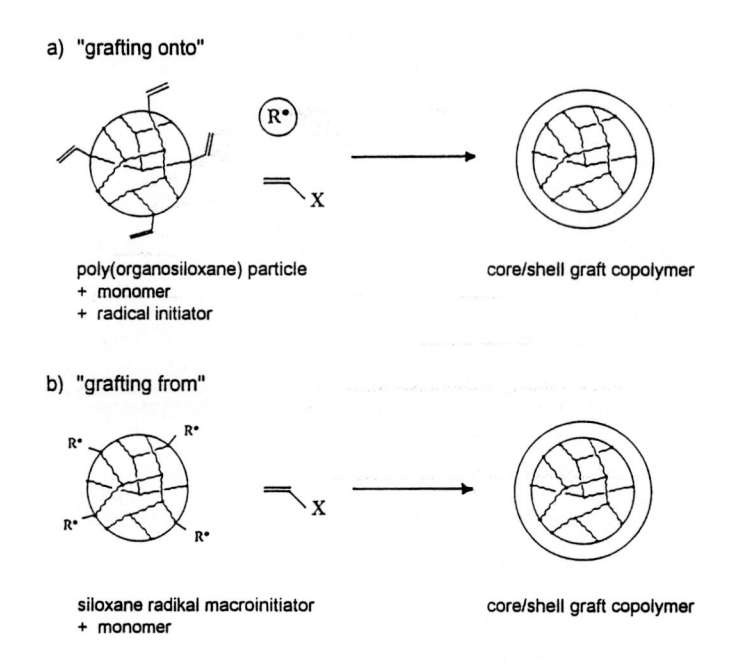

a) "grafting onto"

poly(organosiloxane) particle
+ monomer
+ radical initiator

core/shell graft copolymer

b) "grafting from"

siloxane radikal macroinitiator
+ monomer

core/shell graft copolymer

Scheme 3. Synthetic routes to graft copolymer microparticles

Characterization

The characterization of ungrafted and organopolymer grafted precrosslinked poly(organosiloxane) particles in dispersion by static and dynamic light scattering as well as by TEM (transmission electron microscopy) reveals information about particle structure, dispersity, molecular weight, and density. After particle isolation and processing, polymer properties can be investigated by means of DSC (differential scanning calorimetry) and DMTA (dynamic mechanical thermal analysis); information about morphology and microstructure is obtained by means of SEM (scanning electron microscopy) and TEM.

Using static and dynamic light scattering it could be proved that in dispersion spherical poly(organosiloxane) particles with monomodal particle size distribution are present. Fig. 1 illustrates the asymptotic decrease of the hydrodynamic particle radius R_h with increasing amounts of T units from approximate 50 nm to approximate 12 nm.

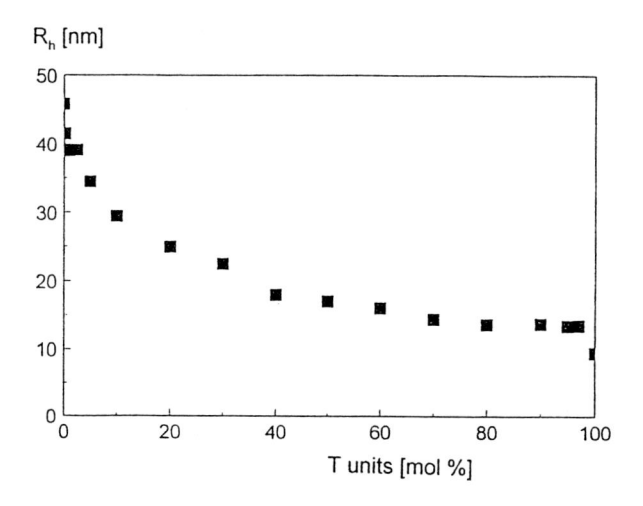

Fig. 1. The particle radius R_h as a function of the crosslink density (mol% T units) [9]

Simultaneously, the molecular weight of the particles is reduced from $M_w \sim 10^8$ g mol^{-1} to $M_w \sim 10^6$ g mol^{-1} [9]. Fig. 2a and 2b illustrate the monodispersity of precrosslinked poly(organo-siloxane) particles and of PMMA grafted core/shell particles, respectively.

The diameters of the particles, prepared from dispersion, are approximately 120 nm each. The powder morphology of spray-dried PMMA grafted siloxane particles is illustrated in Fig. 3a; the spherical agglomerates are approximately 1–20 µm in size. The microstructure of the graft copolymer after processing is shown in Fig. 3b.

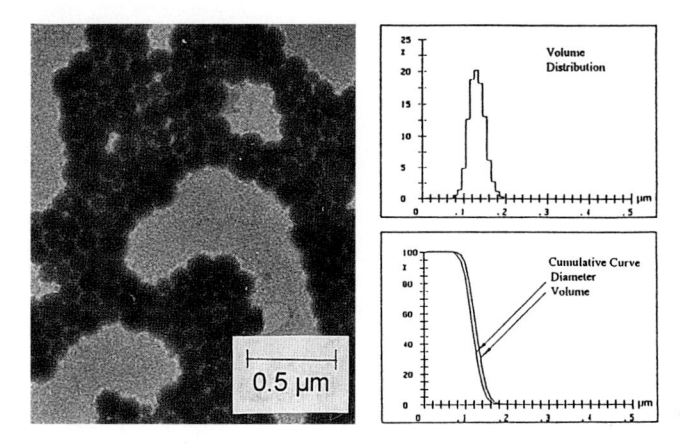

Fig. 2a. Precrosslinked poly(organosiloxane) particles (5 mol% T units) before grafting of PMMA (degree of grafting 50 %); transmission electron micrographs, volume distributions, cumulative curves

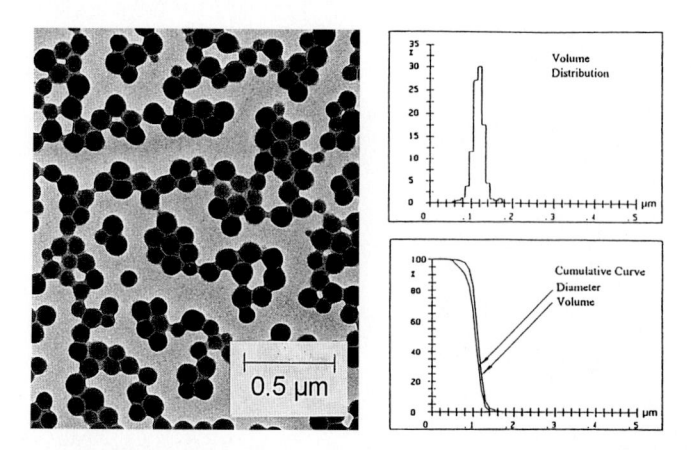

Fig. 2b. Precrosslinked poly(organosiloxane) particles (5 mol% T units) after grafting of PMMA (degree of grafting 50 %); transmission electron micrographs, volume distributions, cumulative curves

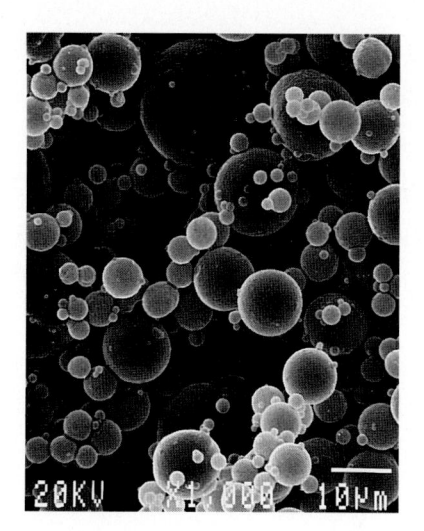

Fig. 3a. Powder morphology (scanning electron micrograph) of spray-dried PMMA grafted siloxane particles (obtained from the dispersion of Fig. 2b)

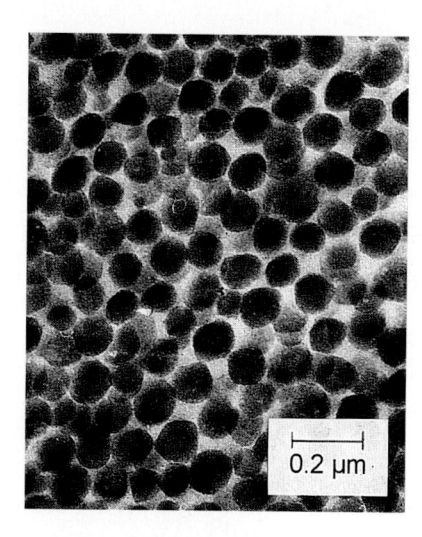

Fig. 3b. Graft copolymer microstructure (transmission electron micrograph, ultramicrotome section) after processing to a milled sheet

The dark spheres of the poly(organosiloxane) primary particles in a PMMA matrix are of approximate 120 nm in diameter; this is correlating with the particles size in dispersion (cf. Fig. 2a, 2b). Fig. 4a and 4b show micrographs of pressed sheets consisting of 10 (25) parts by weight of PMMA grafted elastic silicone particles and 90 (75) parts by weight of PMMA, illustrating the redispersion of the 120 nm microparticles as a result of processing.

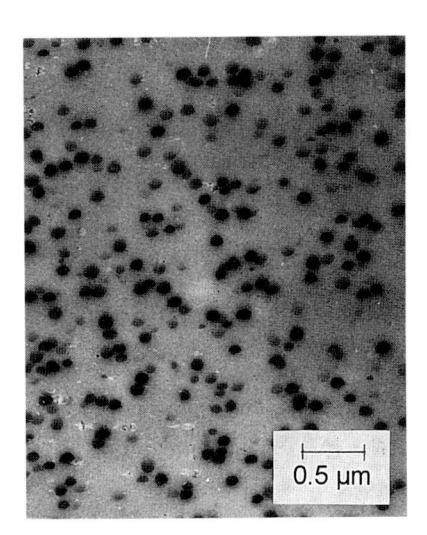

Fig. 4a. Pressed sheets consisting of PMMA and of PMMA grafted elastic siloxane particles (powder from Fig. 3a; transmission electron micrographs, ultramicrotome sections); 10 wt. % core/shell particles

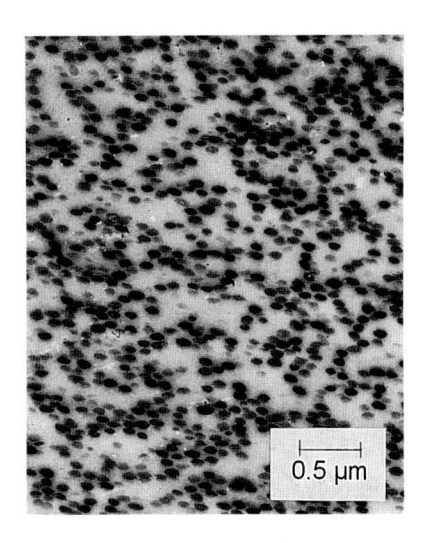

Fig. 4b. Pressed sheets consisting of PMMA and of PMMA grafted elastic siloxane particles (powder from Fig. 3a; transmission electron micrographs, ultramicrotome sections); 25 wt. % core/shell particles

As expected for two-phase polymer systems, two glass transitions are detected for thermoplast grafted elastic siloxane particles by means of thermal analysis (DSC, DMTA) [10]. In Table 1, glass transition temperatures of elastic siloxane graft copolymer particles with different crosslink densities (1–20 mol% T units) and distinct thermoplastic shells (PMMA, PS) are listed.

In addition, the glass transition temperatures of the corresponding poly(organosiloxane) particles and of PMMA and PS homopolymers are specified. In comparison to DSC data, DMTA glass transitions are shifted to higher temperatures according to the methods of measurement and to theory [11].

The glass transitions of the silicone elastomers at very low temperatures a priori are promising good low temperature properties of silicone elastomer modified organic polymers. In the case of PMMA, the low temperature glass transitions of thermoplast grafted siloxane particles are shifted about approximately 10 °C to even lower temperature, independent of the siloxane crosslink density. This implicates a modification of the interaction between the siloxane molecules in the presence of grafted PMMA molecules. For PS grafted siloxane particles, this effect is observed to a lower extent and only at very low siloxane crosslink density.

	DSC[a]			DMTA[b]	
Material[c]	Tg (core) [°C]	Tmelt [°C]	Tg (shell) [°C]	Tg (core) [°C]	Tg (shell) [°C]
E (1%)	−114	−45	−	−103	−
E (5%)	−115	−	−	−104	−
E (20%)	−105	−	−	−93	−
PMMA	−	−	112	−	131
PS	−	−	105	−	120
E (1 %)PMMA	−126	−54	118		135
E (5 %)PMMA	−128	−	117	−118	132
E (20%)PMMA	−116	−	115	−105	131
E (1%)PS	−119	−48	101	−111	113
E (5%)PS	−115	−	98	−104	111
E (20%)PS	−107	−	98	−95	112

Table 1. Thermal analysis glass transition temperatures and crystallization effects (melt temperatures of crystallites)
[a] DSC mettler TA 300 (temperature range −150 °C to 150 °C; heating rate 10 °C min^{-1});
[b] DMTA PL thermal sciences (temperature range −150 °C to 180 °C; heating rate 2 °C min^{-1}, frequency 1 Hz);
[c] nomenclature: E (x %) specifies the silicone elastomer crosslink density (in mol% T units); additionally, the shell polymer (PMMA, PS; degree of grafting 50 % each) is listed

Partially crystalline behavior at low temperature (endothermic melting peaks in DSC at approximate −50 °C) is observed for weakly crosslinked siloxane particles and the corresponding graft copolymers.

E' (modulus of elasticity) of PMMA and PS grafted silicone elastomer particles (crosslink density 1, 5, and 20 mol% T units; degree of grafting 50 % each) as a function of temperature is shown in Fig. 5a and 5b.

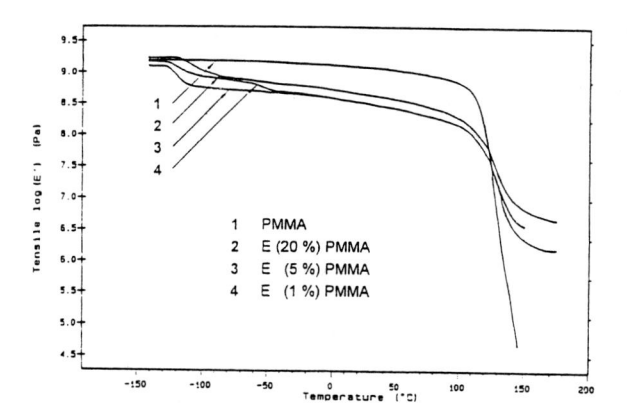

Fig. 5a. Dynamic mechanical thermal analysis: modulus of elasticity E' as a function of temperature of silicone PMMA graft copolymers, in comparison with E' of the corresponding homopolymers

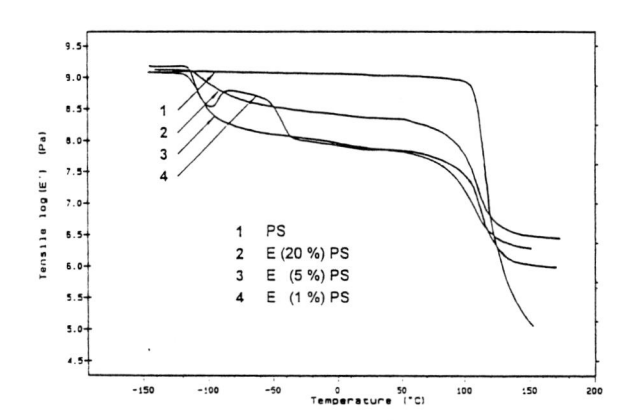

Fig. 5b. Dynamic mechanical thermal analysis: modulus of elasticity E' as a function of temperature of silicone PS graft copolymers, in comparison with E' of the corresponding homopolymers

In the temperature range between the low and the high temperature glass transition, E' moduli of the graft copolymers are at lower levels compared with E' of the corresponding hompolymers, i.e., the materials are more elastic. For graft copolymers with PMMA shells, the reduction of E' moduli is less distinctive than in the PS case. A reduction of silicone elastomer crosslink density from 5 to 1 mol% T units in PMMA and PS graft copolymers does not have any effect on further lowering the E' moduli. The partial crystallinity of the graft copolymers containing the weakly crosslinked (1 mol% T units) siloxane component is reflected in higher E' moduli at low temperatures (–50 °C to –100 °C), implicating disadvantageous low temperature properties.

Properties of Silicone Microparticle Modified Polymers

Thermal analysis as well as particle morphology and redispersion of precrosslinked poly(organosiloxane) microparticles proved to be promising with regard to an application as potential toughening agents. This caused us to investigate the principal suitability of silicone particles as modifiers for thermoplastic polymers.

Pressed sheets consisting of PMMA (Plexiglas 6 N, Röhm) and 0, 10, 25, and 100 parts by weight of PMMA grafted silicone elastomer particles (crosslink density 5 mol% T units, degree of grafting 50 %) were prepared by milling and compression moulding (DIN 16770). The following mechanical and thermal properties were investigated: Tensile test, and tensile modulus; hardness test (Shore D); Vicat B softening temperature; and Charpy test (impact bending test) [12]. The results are compiled in Table 2.

Sample	1	2	3	4
Modifier[a] [wt. %]	0	10	25	100
Yield Stress[b] [N/mm^2]	n.y.[c]	66.5	53.4	n.y.[c]
Yield Strain[b] [%]	n.y.[c]	4.3	3.9	n.y.[c]
Stress at Break[b] [N/mm^2]	55.7	60.7	47.4	13.2
Strain at Break[b] [%]	2	9	11.7	24.8
Tensile Modulus[d] [N/mm^2]	3260	2830	2230	391
Hardness Test[e]				
Shore D	87	83	81	48
Vicat B Softening Temperature [°C][f]	94	94	94.5	88
Charpy (Impact Bending) Test[g]				
a_n(−20 °C) [kJ/m^2]	6.6	10.6	22	32.4
a_n(0 °C) [kJ/m^2]	6.2	9.1	22.3	43.8
a_n(23 °C) [kJ/m^2]	6.1	10.7	22.1	37.2

Table 2. Mechanical and thermal properties of PMMA modified with PMMA grafted silicone elastomer microparticles
[a] PMMA grafted silicone elastomer (5 mol% T units, degree of grafting 50 %); [b] DIN 53455; [c] n.y.: no yield point; [d] DIN 53457; [e] DIN 53505; [f] DIN 53460; [g] DIN 53453

As a result of the tensile test, stress-strain diagrams of the samples investigated are shown in Fig. 6.

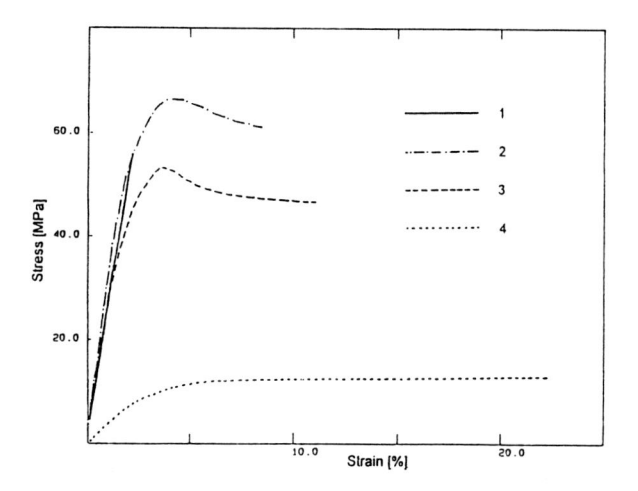

Fig. 6. Stress-strain diagrams of standard test samples consisting of PMMA and 0, 10, 25, and 100 wt. % (graphs 1–4) of PMMA grafted silicone elastomer particles (powder from Fig. 3a)

Unmodified PMMA is brittle, strain at break is low. When modified with 10 and 25 parts by weight of PMMA grafted silicone particles, PMMA becomes tough; yield points are observed in the stress-strain curves and strain at break is enhanced. Stress at break decreases with increasing modifier content. The tensile modulus E', determined by the slope at the origin of the stress-strain curve, proportionally decreases with increasing content of silicone modifier.

The results of the hardness test (Shore D) show that the hardness of the modified PMMA is reduced to a small extent only. The Vicat softening temperature, a measure for the temperature range of dimensional stability, remains constant up to a modifier content of 25 parts by weight. According to the Charpy test, the impact strength is increased about approximate 50 % (250 %) after addition of 10 (25) parts by weight of silicone modifier, independent of the temperature in the range investigated (−20 °C to 23 °C).

The results prove the principal suitability of organopolymer grafted precrosslinked poly(organosiloxane) particles as silicone toughening agents for thermoplastic polymers in the case of PMMA, which has been chosen as a model polymer out of the large group of thermoplastic polymers. Toughness is improved, while the Vicat softening temperature remains unchanged; hardness and modulus of elasticity of the thermoplastic polymers are reduced only to a small extent.

Acknowledgements: We thank Dr. M. Schmidt and F. Baumann, *University of Bayreuth*, for light scattering measurements, the *Central Analytical Department, Wacker-Chemie GmbH*, for electron microscopy, and H. Oswaldbauer and I. Wombacher, *Wacker-Chemie GmbH*, for their engaged work.

References:

[1] W. E. Funke, *J. Coatings Technol.* **1988**, *60*, 69.
[2] M. Antonietti, *Angew. Chem.* **1988**, *100*, 1813.
[3] R. Mühlhaupt, *Chimia* **1990**, *44*, 43; and references therein.
[4] G. Koerner, in: *Silicone – Chemie und Technologie*, Vulkan-Verlag, Essen, **1989**, p. 1.
[5] The term "functionality" of a siloxane unit is equivalent to the number of covalent Si–O bonds originating from the central Si atom; cf. *Chemistry and Technology of Silicones* (Ed.: W. Noll), VCH, Weinheim, **1968**, p. 3.
[6] M. Wolfgruber, B. Deubzer, V. Frey, *EP* 291 941, **1988**.
[7] K. Mautner, B. Deubzer, *DE* 4 040 986, **1992**.
[8] M. Geck, J. Dauth, B. Deubzer, H. Oswaldbauer, M. Schmidt, F. Baumann, DE 4 338 421, **1993**, patent pending.
[9] F. Baumann, *Ph. D. Thesis*, Universität Bayreuth, **1993**; M. Schmidt, F. Baumann, personal communication.
[10] H.-G. Elias, *Makromoleküle*, Hüthig & Wepf, Basel, **1981**, p. 953.
[11] H.-G. Elias, *Makromoleküle*, Hüthig & Wepf, Basel, **1981**, p. 352.
[12] See, for instance: H. Saechtling, *Kunststoff-Taschenbuch*, Hanser, München, Wien, **1986**, p. 500; H. Saechtling, *Kunststoff-Taschenbuch*, Hanser, München, Wien, **1986**, p. 512; H.-G. Elias, *Makromoleküle*, Hüthig & Wepf, Basel, **1990**, p. 953.

Silsesquioxanes of Mixed Functionality – Octa[(3-chloropropyl)-*n*propyl-silsesquioxanes] and Octa[(3-mercaptopropyl)-*n*propyl-silsesquioxanes] as Models of Organomodified Silica Surfaces

Benedikt J. Hendan, Heinrich C. Marsmann*

Anorganische und Analytische Chemie
Universität – Gesamthochschule – Paderborn
Warburger Str. 100, D-33098 Paderborn, Germany

Summary: Cohydrolysis of some organotrichloro or organotrimethoxy silanes yields cube shaped octa(organosilsesquioxanes) of mixed functionality with the general formula $[X(CH_2)_3]_n[H(CH_2)_3]_{8-n}[Si_8O_{12}]$ with X = Cl **1**, SH **2** and n = 0, 1, 2 (3 isomers), 3 (3 isomers). They can be separated by normal phase HPLC. The degree of substitution and the substitution pattern is recognizable by ^{29}Si 2D-INADEQUATE NMR spectroscopy. By the reaction of octa[mono(3-chloropropyl)-hepta(*n*propyl)-silsesquioxane] $[Cl(CH_2)_3][H(CH_2)_3]_7[Si_8O_{12}]$ **1.1** with potassium-diphenylphosphide octa{mono[3-(diphenylphosphino)propyl]-hepta[*n*propyl]-silsesquioxane} $\{[(C_6H_5)_2P](CH_2)_3\}[H(CH_2)_3]_7[Si_8O_{12}]$ **3.1** is obtained.

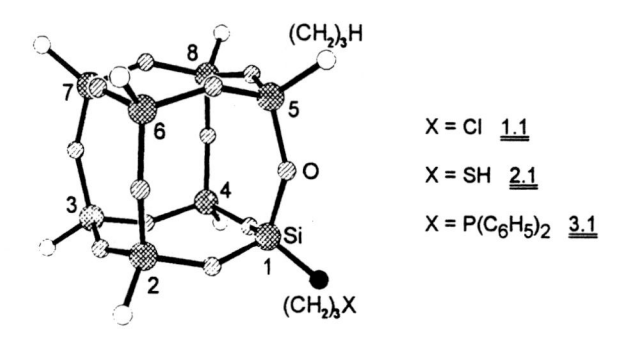

X = Cl **1.1**
X = SH **2.1**
X = P(C_6H_5)_2 **3.1**

This compound and octa[mono(3-mercaptopropyl)-hepta(*n*propyl)-silsesquioxane] $[HS(CH_2)_3][H(CH_2)_3]_7[Si_8O_{12}]$ **2.1** have been used as ligands in Rh(I) and Pt(II) transition metal complexes for modeling silica supported catalysts.

Introduction

Immobilising of transition metal catalysts on silica surfaces has been intensively studied during the last twenty years in order to combine the practical advantages of heterogeneous catalysts with the versatility of homogeneous transition metal catalysts. Despite many investigations, the nature of the catalytic species is generally unknown. So the synthesis of model compounds is neccessary. Silsesquioxanes resemble idealized parts of a silica surface. This similarity can be exploited in several ways. A short review of the use of partially condensed organosilsesquioxanes as model compounds is given by Masters et al. [1].

In the work presented here organofunctional silsesquioxanes are used as models for organomodified silica surfaces. The main advantage is that organosilsesquioxanes are soluble in organic solvents in most cases making them susceptible to a wide array of analytical methods not available for amorphous materials. Of the many systems possible and worthy of attention, this study is limited to octa-(organosilsesquioxanes).

Results

Cohydrolysis [2–4] of *n*propyltrichlorosilane and 3-chloropropyltrichlorosilane in the molar ratio 6.5:1.5 yields **1** and by cohydrolysis of *n*propyltrimethoxysilane and 3-mercaptopropyltrimethoxysilane in the molar ratio 7:1 **2** can be obtained [5]. The compounds can be separated in semipreparative amounts with normal phase HPLC on a Merck Hibar® RT 250-10 Si 60 LiChrospher® (5 µm) column with *n*hexane as eluent [5]. The retention times increase with the number of polar groups in the molecules (see Fig. 1).

Another parameter for the elution rate is the ability of the functional groups to have multidental interactions with the silanol groups of the silica gel. Thus, 1,2-di- and 1,2,3-trifunctionalized octa-(organosilsesquioxanes) have the longest retention times probably because of their local concentration of polar groups. ^{29}Si NMR spectroscopy is an appropriate method to investigate the substitution pattern of mixed substituted octa(organosilsesquioxanes) [5]. The connectivity of the silicon atoms of the cage can be monitored by ^{29}Si INADEQUATE NMR spectroscopy [6] (see Fig. 2).

Silsesquioxanes of Mixed Functionality 687

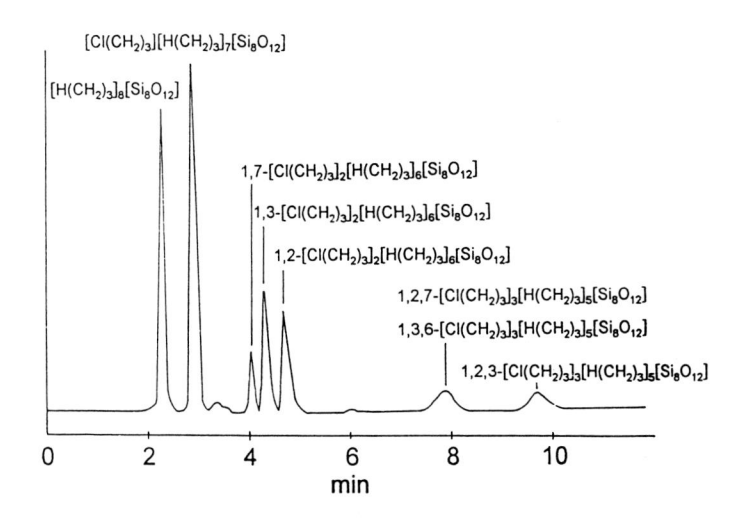

Fig. 1. HPLC chromatogram of **1**

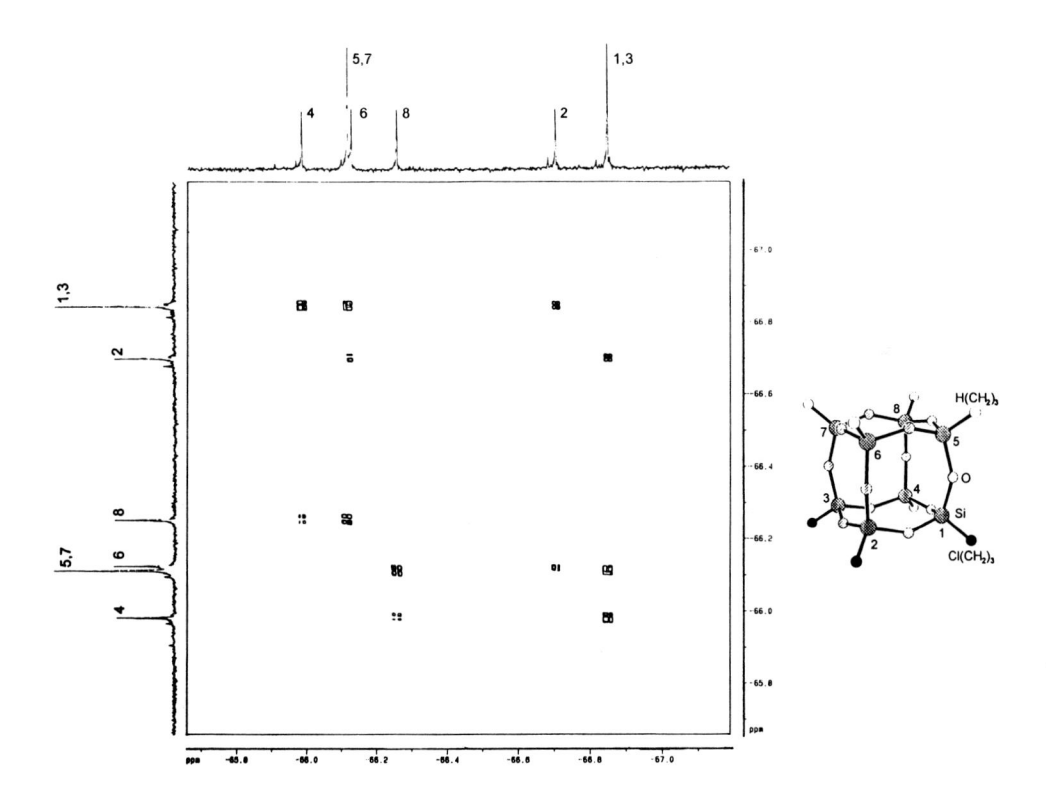

Fig. 2. ^{29}Si 2D-INADEQUATE NMR spectrum of 1,2,3-[Cl(CH$_2$)$_3$]$_3$[H(CH$_2$)$_3$]$_5$[Si$_8$O$_{12}$] (in C$_6$D$_6$)

Nucleophilic substitution of the chlorine atom of **1.1** with potassium diphenylphosphide ($KP(C_6H_5)_2$) gives **3.1** [7, 8]. The reaction of **3.1** with Rh(I) and Pt(II) transition metal complexes (Eq. 1–3) yields compounds which can be used as models for silica supported catalysts [8].

$$[Rh(CO)_2Cl]_2 \; + \; 4 \cdot \textbf{3.1} \; \longrightarrow \; 2\,[Rh(CO)(\textbf{3.1})_2Cl] \; + \; 2\,CO \quad (1)$$

$$[Rh(cod)Cl]_2 \; + \; 2 \cdot \textbf{3.1} \; \longrightarrow \; 2\,[Rh(cod)(\textbf{3.1})Cl] \quad (2)$$

$$[PtCl_2(cod)] \; + \; 2 \cdot \textbf{3.1} \; \longrightarrow \; [Pt(\textbf{3.1})_2Cl_2] \; + \; cod \quad (3)$$

For instance, $[Rh(CO)(\textbf{3.1})_2Cl]$ catalyses the hydrosilylation of 1-hexene with triethylsilane. For the preparation of transition metal complexes of **2.1** it was necessary to use the lead compound $Pb\{[S(CH_2)_3][H(CH_2)_3]_7[Si_8O_{12}]\}_2$, $Pb(\textbf{2.1a})_2$, as intermediate in order to avoid the formation of HCl which would destroy the siloxane cage [8] (Eq. 4).

$$(CH_3COO)_2Pb \; + \; 2 \cdot \textbf{2.1} \; \longrightarrow \; Pb(\textbf{2.1a})_2 \; + \; 2\,CH_3COOH \quad (4)$$

The reaction of $Pb(\textbf{2.1a})_2$ with chlorine containing salts yields insoluble lead chloride and the wanted complex [9] (Eq. 5, 6).

$$[Rh(CO)_2Cl]_2 \; + \; Pb(\textbf{2.1a})_2 \; \longrightarrow \; [Rh(CO)_2(\textbf{2.1a})]_2 \; + \; PbCl_2\downarrow \quad (5)$$

$$[Rh(cod)Cl]_2 \; + \; Pb(\textbf{2.1a})_2 \; \longrightarrow \; [Rh(cod)(\textbf{2.1a})]_2 \; + \; PbCl_2\downarrow \quad (6)$$

References:

[1] L. D. Field, T. W. Hambley, C. M. Lindall, T. Maschmeyer, A. F. Masters, A. K. Smith, *Chem. Aust.* **1993**, *60(7)*, 333.

[2] T. N. Martynova, T. I. Chupakhina, *J. Organomet. Chem.* **1988**, *345*, 11.

[3] T. I. Chupakhina, T. N. Martynova, P. P. Semyannikov, *Ser. Khim.* **1986**, *2*, 441.

[4] M. G. Voronkov, Y. V. Basikhin, A. N. Kanev, *Dokl. Akad. Nauk. SSSR* **1981**, *258(3)*, 642.

[5] B. J. Hendan, H. C. Marsmann, *J. Organomet. Chem.*, in press.

[6] D. L Turner, *J. Magn. Reson.* **1982**, *49*, 175.

[7] U. Dittmar, B. J. Hendan, U. Flörke, H. C. Marsmann, *J. Organomet. Chem.*, in press.

[8] B. J. Hendan, H. C. Marsmann, *J. Organomet. Chem.*, submitted.

[9] E. Delgado, E. Hernandez, *Polyhedron* **1992**, *11(24)*, 3135.

New Functionalized Silsesquioxanes by Substitutions and Cage Rearrangement of Octa[(3-chloropropyl)-silsesquioxane]

Heinrich C. Marsmann*, Uwe Dittmar, Eckhard Rikowski

Anorganische und Analytische Chemie

Universität – Gesamthochschule – Paderborn

Warburger Str. 100, D-33098 Paderborn, Germany

Summary: Functionalized polyhedral siloxanes, also called silsesquioxanes, present themselves as models for the investigation of surface modified silica gels. By nucleophilic substitution of the chlorine atoms at the cubic octa[(3-chloropropyl)-silsesquioxane], [Cl–(CH$_2$)$_3$]$_8$(SiO$_{1.5}$)$_8$ (**1**), functionalized octa[propyl-silsesquioxanes], [X–(CH$_2$)$_3$]$_8$(SiO$_{1.5}$)$_8$, with X= I (**2**), NCS (**3**), Ph$_2$P (**4**), and MeS (**5**) can be obtained. A new method for preparing deca- and dodeca-silsesquioxane cages is the partial rearrangement of octa[(3-chloropropyl) silsesquioxane], [Cl–(CH$_2$)$_3$]$_8$(SiO$_{1.5}$)$_8$, in deca[(3-chloropropyl)-silsesquioxane], [Cl–(CH$_2$)$_3$]$_{10}$(SiO$_{1.5}$)$_{10}$ (**6**), and dodeca[(3-chloropropyl)-silsesquioxane], D$_{2d}$-[Cl–(CH$_2$)$_3$]$_{12}$(SiO$_{1.5}$)$_{12}$ (**7**), catalyzed by sodium acetate or sodium cyanate. This mixture of silsesquioxane cages can be separated by NP-HPLC. In the presence of 18-crown-6 with potassium acetate or sodium cyanide in acetonitrile in addition to the partial rearrangement complete substitution of the chlorine atoms by acetoxy- or cyano groups is observed.

Introduction

Spherical siloxanes (RSiO$_{1.5}$)$_n$ with polyhedral frameworks, also called silsesquioxanes, can be used as defined oligomeric models for surface-modified silica gels or polysiloxanes. Whereas silsesquioxanes with alkyl-, aryl-, hydrido-, and trimethylsiloxy groups are known for a long time [1], functionalized octa-[propyl-silsesquioxanes], [X–(CH$_2$)$_3$]$_8$(SiO$_{1.5}$)$_8$, were synthesized for the first time in 1990 by Weidner, Zeller, Deubzer, and Frey [2]. Octa[(3-chloropropyl)-silsesquioxane], [Cl–(CH$_2$)$_3$]$_8$(SiO$_{1.5}$)$_8$, could be obtained by hydrolysis of (3-chloropropyl)-trichlorosilane.

Recently functionalized octa[(propyl)-silsesquioxanes] were obtained by hydrosilylation of $H_8(SiO_{1.5})_8$ [3] with functionalized allylic compounds [4–8]. In the following, the preparation of new functionalized propyl-silsesquioxanes by nucleophilic substitution or partial rearrangement of octa-[(3-chloropropyl)-silsesquioxane] is reported.

Results

1 Substitution Reactions at Octa[(3-chloropropyl)-silsesquioxane]

Octa[(3-iodopropyl)-silsesquioxane] can be prepared by the Finkelstein-reaction treating octa-[(3-chloropropyl)-silsesquioxane] with sodium iodide in dry acetone under reflux. The structure of $[I-(CH_2)_3]_8(SiO_{1.5})_8$ has been confirmed by X-ray crystallography [4, 5]. In the same way octa-[(3-thiocyanatopropyl)-silsesquioxane] was obtained.

$$1 + 8\ NaX \xrightarrow{acetone/\Delta T} [X-(CH_2)_3]_8(SiO_{1.5})_8 + 8\ NaCl\downarrow$$

Eq. 1.   X = I or NCS

Octa{[3-(diphenylphosphino)-propyl]-silsesquioxane} was prepared by treating $[Cl-(CH_2)_3]_8(SiO_{1.5})_8$ dissolved in THF with a solution of potassium diphenylphosphide in THF at room temperature:

$$1 + 8\ KPPh_2 \xrightarrow{THF} [Ph_2P-(CH_2)_3]_8(SiO_{1.5})_8 + 8\ KCl$$

Eq. 2.

The red color of potassium diphenylphosphide was helpful to control the reaction, because all other compounds were colorless.

The synthesis of octa{[3-(methylmercapto)-propyl]-silsesquioxane} was carried out by substitution of the chlorine atoms using sodium thiomethylate as nucleophilic agent. Because of the insolubility of NaSMe in toluene a phase transfer catalyst was used:

$$1 + 8\ NaSMe \xrightarrow{toluene/18-crown-6} [MeS-(CH_2)_3]_8(SiO_{1.5})_8 + 8\ NaCl$$

Eq. 3.

2 Rearrangement of Octa[(3-chloropropyl)-silsesquioxane]

Due to the weak nucleophilicity of the acetate and cyanate anions, no substitution of the chlorine atoms was observed. Instead, partial rearrangement to the greater cage molecules deca[(3-chloropropyl)-silsesquioxane] and dodeca[(3-chloropropyl)-silsesquioxane] was found.

$$1 \xrightarrow{\text{sodium acetate/acetone}} [Cl-(CH_2)_3]_n(SiO_{1.5})_n$$

Eq. 4. $n = 8, 10, 12$

Structure diagrams of the synthesized cage-types are shown in Fig. 1:

octa-silsesquioxane   deca-silsesquioxane   dodeca-silsesquioxane

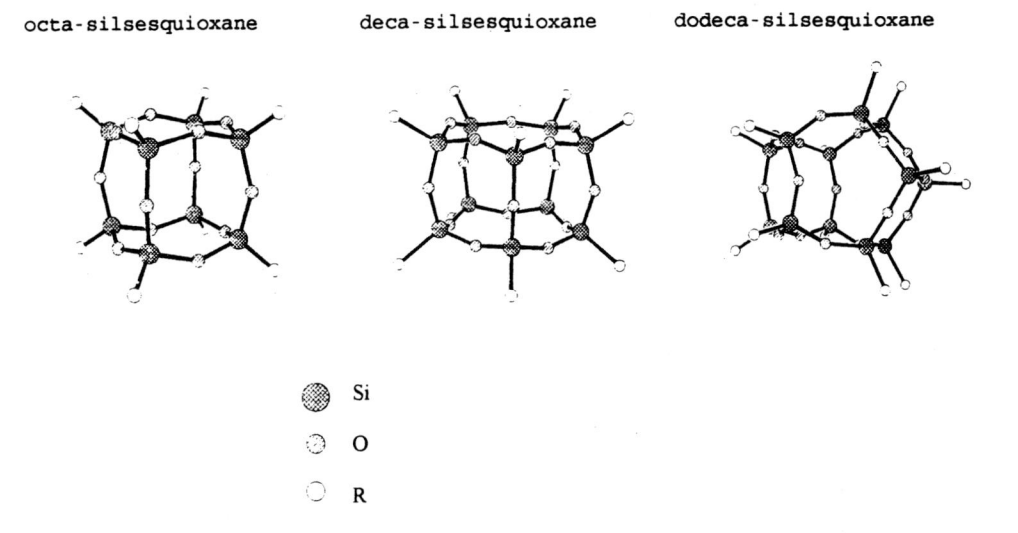

Si
O
R

Fig. 1. Octa-, deca-, and dodeca-silsesquioxane cage-types

Fig. 2 shows an ^{29}Si NMR spectrum after treatment of $[Cl-(CH_2)_3]_8(SiO_{1.5})_8$ with sodium acetate in acetone. The distribution between the cage sizes was 25.9 % (15.1 %) $[Cl-(CH_2)_3]_8(SiO_{1.5})_8$, 59.5 % (62.3 %) $[Cl-(CH_2)_3]_{10}(SiO_{1.5})_{10}$ and 14.6 % (22.6 %) $[Cl-(CH_2)_3]_{12}(SiO_{1.5})_{12}$ by catalysis with sodium acetate (sodium cyanate).

The compounds can be separated in semipreparative amounts with normal phase HPLC on a Merck Hibar® RT 250-10 Si 60 LiChrospher® (5 µm) column with a mixture of 50 vol % nhexane and 50 vol % $CHCl_3$ as eluent.

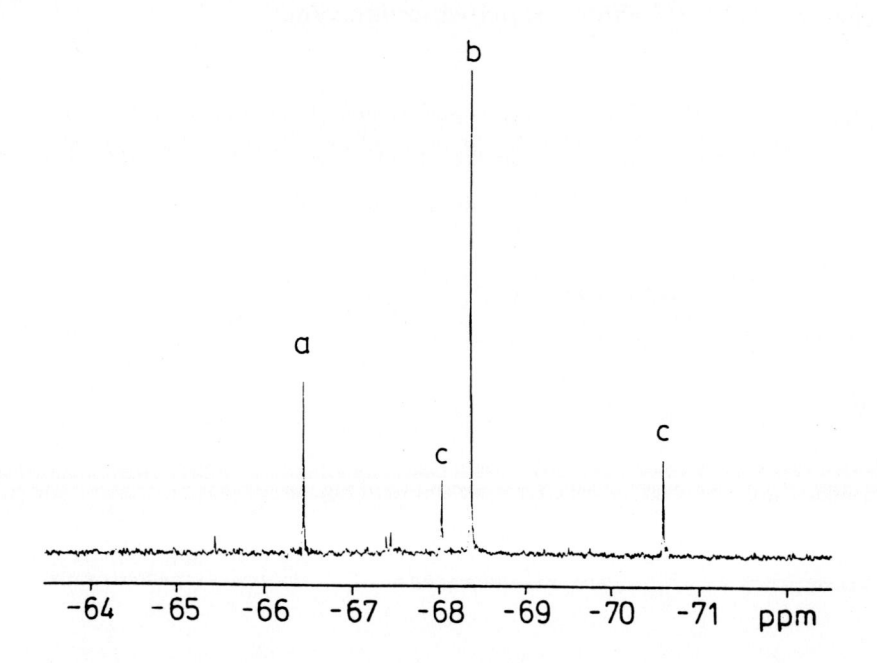

Fig. 2. ^{29}Si NMR spectrum in C_6D_6 of octa[(3-chloropropyl) silsesquioxane] (a), deca[(3-chloropropyl)-silsesquioxane] (b), and D_{2d}-dodeca[(3-chloropropyl)-silsesquioxane] (c)

The retention times are 8.2 min for $[Cl-(CH_2)_3]_8(SiO_{1.5})_8$, 16.0 min for $[Cl-(CH_2)_3]_{10}(SiO_{1.5})_{10}$ and 37.4 min for $[Cl-(CH_2)_3]_{12}(SiO_{1.5})_{12}$ at 7 mL min^{-1} flow rate. The substitution of the chlorine atoms at octa [(3-chloropropyl)-silsesquioxane] by acetoxy- or cyano groups is possible by using potassium acetate or sodium cyanide and 18-crown-6 as phase transfer catalyst in acetonitrile. In addition to the observed substitution, partial rearrangement to the greater deca- and dodeca-silsesquioxane cages, also took place.

$$1 \xrightarrow[\text{CH}_3\text{CN, 18-crown-6}]{+ \text{ potassium acetate or sodium cyanide, } - \text{KCl or NaCl}} [X-(CH_2)_3]_n(SiO_{1.5})_n$$

Eq. 5. X = acetoxy or CN; n = 8, 10, 12; 22.3 % $[Ac-(CH_2)_3]_8(SiO_{1.5})_8$, 60.2 % $[Ac-(CH_2)_3]_{10}(SiO_{1.5})_{10}$, 17.5 % $[Ac-(CH_2)_3]_{12}(SiO_{1.5})_{12}$ with Ac=acetoxy; 0.7 % $[NC-(CH_2)_3]_8(SiO_{1.5})_8$, 75.0 % $[NC-(CH_2)_3]_{10}(SiO_{1.5})_{10}$, 24.3 % $[NC-(CH_2)_3]_{12}(SiO_{1.5})_{12}$

References:

[1] M. G. Voronkov, V. I. Lavrent'yev, *Top. Curr. Chem.* **1982**, *102*, 199.
[2] R. Weidner, N. Zellner, B. Deubzer, V. Frey, *Chem. Abstr.* **1990**, 113, 116465m; *DE* 383797, **1990**, Wacker Chemie, Int. Cl 5: C 07 F 7/08:
[3] C. L. Frye, W. T. Collins, *J. Am. Chem. Soc.* **1970**, *92*, 5586.
[4] U. Dittmar, *Diploma Thesis*, Universität – Gesamthochschule – Paderborn, **1993**.
[5] U. Dittmar, B. J. Hendan, U. Flörke, H.-C. Marsmann, *J. Organomet Chem.*, in press.
[6] G. Calzaferri, D. Herren, R. Imhof, *Helv. Chim. Acta* **1991**, *74*, 1278.
[7] G. Calzaferri, R. Imhof, *J. Chem. Soc., Dalton Trans.* **1992**, 3391.
[8] A. R. Bassindale, T. E. Gentle, *J. Mater. Chem.* **1993**, *3*, 1319.

Investigations on the Thermal Behavior of Silicon Resins by Thermoanalysis, ^{29}Si NMR, and IR Spectroscopy

H. Jancke*, D. Schultze, H. Geissler

Bundesanstalt für Materialforschung und -prüfung (BAM)

Unter den Eichen 87, D-12489 Berlin, Germany

Summary: Thermoanalytical and spectroscopic (^{29}Si NMR and IR) methods were used to characterize the changes of structure of silicon resins when heated up to 700 °C. Building groups with unreacted Si–OH bonds perform a postcondensation to Si–O–Si bonds at 250 °C. At about 500 °C, Si–C bonds cleave by an oxidative conversion to Si–OH. At the end the silicon resin is converted to an aerosil like material.

Silicon resins, three-dimensional cross-linked silicon polymers, were synthesized by hydrolysis/condensation reactions from substituted chloro- or alkoxysilanes. Dependening on the silane components the chemical and technological properties may vary in wide limits for their use as protection material for circuit boards and electronic devices. ^{29}Si NMR spectroscopy has been proved to be a valuable tool to investigate the molecular structure of silicon organic compounds. That is demonstrated by spectroscopic investigations in the liquid state where the building groups of the polymer material were identified by signals in typical regions of the spectrum. In cases where the signals are narrow enough more subtle effects, e.g., microstructural or stereo chemical details of the SiOSi network, can be analyzed [1].

With high resolution ^{29}Si NMR in the solid state one can investigate even unresolvable material such as high temperature paints or moulding compounds, etc. In unresolved silicon resins the resonance signals are broad, of the order of 5 ppm. The spectroscopic resolution available is sufficient to identify the building blocks and allows to analyze the shift effects of small rings. Any further information of connectivity, stereoregularity and others is hidden under the broad resonance peaks measured [2].

Silicon resins are distributed commercially as solutions. After soaking, e.g., electrical parts and curing them at about 250 °C an insoluble and insulating protecting material is formed. The aim of this work is to reproduce the curing process and follow the fate of the building groups of silicon resins if heated to higher temperature. We use here thermoanalysis to check the gravimetric behavior throughout the process of heating the resins up to 700 °C. The change of molecular composition due to the curing process is investigated by ^{29}Si NMR spectroscopy. SiOH groups were identified best by IR spectroscopy. In the following, we will show the typical behavior of cured resins examplified on a methyl-phenylresin of the type NH 2400 from Chemiewerk Nünchritz.

The TG diagram from room temperature to 700 °C shows two main steps (Fig. 1). In the first step at about 200 °C, the region of technical application, the material looses 2.8 % of its weight. Here, the postcondensation of the remaining SiOH functions takes place what makes the resin no longer soluble.

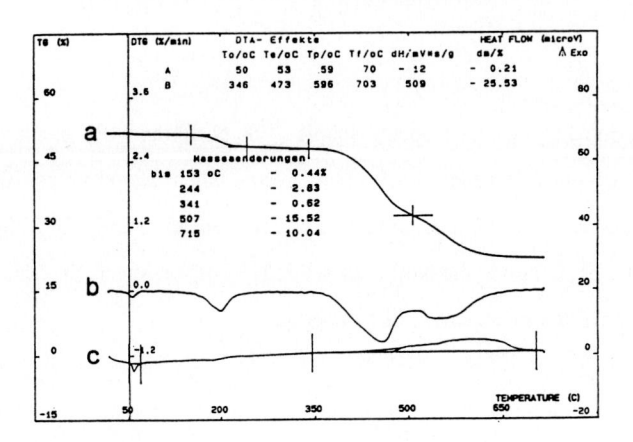

Fig. 1. TG (a), DTG (b), and DTA (c) diagrams for silicon resin NH 2400

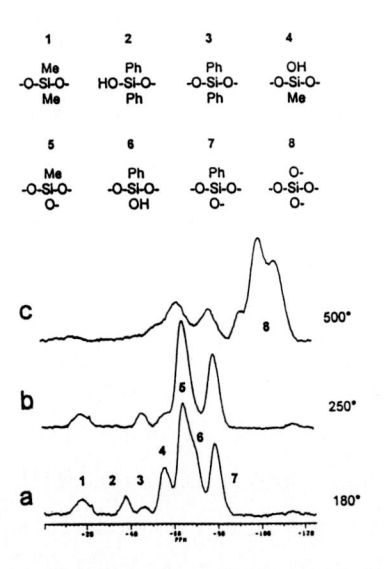

Fig. 2. ^{29}Si CP/MAS spectrum of NH 2400 (variable temperature)

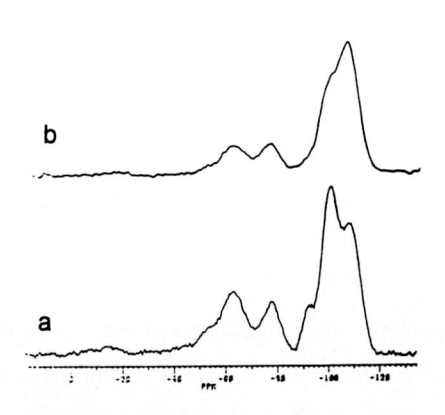

Fig. 3. ^{29}Si NMR spectrum of NH 2400 (500 °C) CP/ MAS (a), MAS (b)

In the second region 350–650 °C, another pyrolytic process causes a loss of weight of further 25 %, obviously in at least two separated steps. Both regions vary in position and height depending on the type of resin investigated. A pure methylresin (Wacker) is completely oxydized at 520 °C. A pure phenylresin (Type NH 670 Chemiewerk Nünchritz) begins its second oxidative conversion only at 480 °C and is not complete at 700 °C, the end of our experiment.

In the ^{29}Si NMR spectrum of several different resins one can identify methyl- and phenylsubstituted as well as alkoxyl- or hydroxyl containing siloxane units [2, 3]. In the samples heated to about 250 °C for two hours the SiOH functions disappeared. This clearly shows the result of the curing process. The groups containing SiOH functions perform a postcondensation to the respective SiOSi functions. On the other hand, this effect confirms our assignment of resonance signals to the building blocks only performed from substituent shift effects.

Fig. 2a shows the spectrum after two hours at 180 °C – technical samples of dissolved silicon resins are now only solvent-free – with the assignment of seven building groups mentioned on the top of the figure. The product cured for two hours at 250 °C (b) contains all the siloxane units without SiOH groups (**1**, **3**, **5**, and **7**). The groups **2**, **4**, and **6** containing SiOH functions disappeared and were transformed to the former groups. This is clearly seen at the transformation of MPh$_2$OH (**2** in Fig. 2) to the group **3**, i.e., D^{Ph_2}. Group **2** can only build group **3** by condensation and, consequently, the intensity of **3** increases.

After two hours at 500 °C in air a light brown powder is formed. In the NMR spectrum (Fig. 2c) the signals of former mentioned siloxane units in linear and crosslinked connections were found with diminished intensity. New intensive signals appear between −94 and −118 ppm. This spectral region contains the signals of silicate structures. Our finding of new signals here points to the building of crosslinked Q groups as the second step (and the end) of the pyrolytic process. The Si–C bonds to the organic substituents methyl and phenyl cleave, were oxidized to new SiOH bonds, and leave behind a silicate like material. Some remaining carbon from the pyrolysed organic groups may be the reason for the brown color. This process of transformation of siloxanes to silicates is similar to the investigations on the thermal stability of silylated inorganic surfaces, as studied by Grobe et al. [4].

The comparison of MAS and CP/MAS NMR spectra is the method of choice to differentiate among the three signal components in the Q-region of the spectrum. The CP/MAS spectra show a strong signal enhancement for the signals at −92 and −101 ppm in comparison to the intensity of the signal at −108 ppm (Fig. 3a). The enhanced signals suggest the vicinity of the Si nucleus under observation to polarizable nuclei, i.e., protons in a SiOH group. So, we identify the Q^2 group (−92 ppm) and the Q^3 group (−101 ppm) as may be assigned by chemical shifts [5]. The appearance of a Q^2 signal is caused obviously by the oxidative degradation of the chain groups **1**–**3** in the resin structure in contrast to the situation in [3], where only RSi(OEt)$_3$ comes to reaction.

Concerning the signal assignment of the building groups in polymeric siloxanes another detail in Fig. 2 is of interest. In the spectrum (b) all the SiOH bearing units are converted to siloxanes. A residue of signal intensity at the position of group **4** remains at –56 ppm. This is a further indication for the superposition of three-membered siloxane structure fragments and the D^{OH} group due to incomplete condensation [6]. The curing process converts D^{OH} groups to T groups and the cyclo-three siloxane fragments appear clearly.

The IR spectra of the materials under investigation confirm the change of the siloxane to the silicate structure (Fig. 4) and identifies the pyrolyzed powder as an aerosil like material with high content of SiOH groups (Fig. 4, b). The degradation of the organic substituents is seen by the nearly complete disappearence of any v(CH) bands at 2900–3100 cm^{-1} for the 500 °C sample. In contrast, the intensity of OH valence bands increased dramatically. We identify four distinct bands of 3640, 3540, 3475, and 3335 cm^{-1}, one of them belongs to adsorbed water. Moreover, a band is observed (not shown in Fig. 4) at 950 cm^{-1} that is characteristic for SiOH bonds.

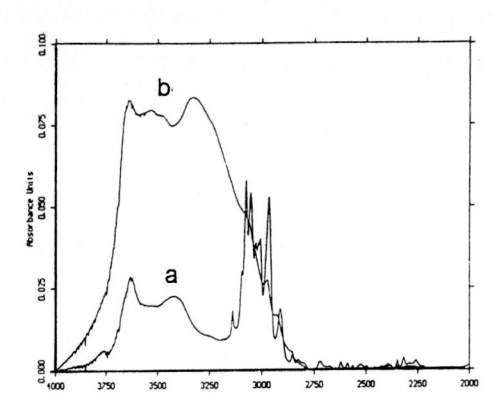

Fig. 4. IR spectra in the range 4000–2000 cm^{-1} for NH 2400; (a) 250 °C, (b) 500 °C

There is a straight way from silicon resin to an aerosil type silicic material with a high content of SiOH groups on the surface. Thermoanalytical and spectroscopic (NMR and IR) methods give deeper insight into the structure of the materials that are formed and to the reactions that proceed.

References:

[1] *The Analytical Chemistry of Silicones, Vol 112, Chemical Analysis* (Eds.: A. L. Smith, J. D. Winefordner), Wiley, Chichester, **1991**.

[2] H. Jancke, *Fresen. J. Anal. Chem.* **1992**, *342*, 846.

[3] G. L. Marshall, *Polym. J.* **1982**, *14*, 19.

[4] J. Grobe, K. Stoppek-Langner, A. Benninghoven, B. Hagenhoff, W. Müller-Warmuth, S. Thomas, in: *Organosilicon Chemistry – From Molecules to Materials* (Eds.: N. Auner, J. Weis), VCH, Weinheim, **1994**, p. 325.

[5] G. Engelhardt, D. Michel, *High-Resolution Solid-State of Silicates and Zeolites, Chapt. V*, Wiley, Chichester, **1987**.

[6] G. Engelhardt, H. Jancke, *Polym. Bull.* **1981**, *5*, 577.

Convenient Approach to Novel Organosilicon Polymers

Wolfram Uhlig

Laboratorium für Anorganische Chemie
Eidgenössische Technische Hochschule Zürich
ETH-Zentrum, CH-8092 Zürich, Switzerland

Summary: α,ω-Bis[(trifluoromethyl)sulfonyloxy]-substituted organosilicon compounds are prepared by protodesilylation of the corresponding phenyl- or aminosilanes with triflic acid. Reactions of these silyl triflates with organic dilithio reagents or di-Grignard compounds lead to the formation of new organosilicon polymers, which show a regular alternating arrangement of silylene or disilylene groups and π-electron-containing units in the polymer backbone. New organosilicon polymers are also prepared by the reductive coupling reaction of the silyl triflates with potassium-graphite (C_8K).

Introduction

In recent years, much attention has been directed to silicon-containing polymers as sources of novel materials [1, 2]. Numerous polymers have been prepared in which regular alternating arrangements of organo-silicon groups and π-electron containing units are found in the polymer backbone [3–5]. Such derivatives are of interest in order to explore the ability of a silicon atom to allow charge transport properties in a conjugated carbon backbone. Silyl- or polysilyl groups are known to be involved in conjugation with unsaturated groups either through $(d–p)_\pi$-overlap or a $(\sigma^*–p)_\pi$-type interaction. Therefore, the above polymer systems can be expected to be polymeric conductors.

The sodium condensation reaction of α,ω-bis(chlorosilyl)-substituted compounds and the coupling reaction of dilithio derivatives of compounds bearing π-electron systems with dichlorosilanes offer a convenient route to various silicon containing polymers. However, the polymers prepared by these methods always contain a small proportion of siloxy units in the polymer backbone, which would interrupt the electron delocalisation. Therefore, new synthetic routes to organosilicon polymers have been developed in which no alkali metal halide condensations are involved [6, 7]. We report syntheses of organosilicon polymers, based on silyl triflate derivatives, which are characterized by a high regioselectivity and excellent yields. Siloxy groups are not found in the resulting polymers.

Results and Discussion

In principle two synthetic routes to new organosilicon polymers based on triflate derivatives are realizable. Firstly, derivatizations can be carried out on finished polymers. Recent papers by Matyjaszewski and by our group have shown the feasibility of this route [8, 9]. We describe here the second synthetic route, which consists of the formation of the polymer chain by condensation of α,ω-bis[(trifluoromethyl)sulfonyloxy]-substituted organosilicon compounds with dinucleophiles (Eq. 1).

Eq. 1.

A large number of different α,ω-bis[(trifluoromethyl)sulfonyloxy]-substituted organosilicon compounds can be obtained by relatively simple methods from the corresponding amino-, allyl-, or phenylsilanes. Moreover, it is remarkable that these silyl triflate derivatives are often easily formed, when the synthesis of the corresponding chloro- or bromosilanes is difficult or does not appear to have been attempted. Eq. 2 and Eq. 3 show selected examples of this synthesis [10–12]. The products were prepared in high purities and yields. The resulting triflates should be used for the polycondensation without further purification, because they often cannot be destilled without decomposition.

$(Et_2N)Me_2Si$—⟨C$_6$H$_4$⟩—$SiMe_2(NEt_2)$ $TfOMe_2Si$—⟨C$_6$H$_4$⟩—$SiMe_2OTf$

$$\xrightarrow[-2\ Et_2NH_2OTf]{4\ TfOH}$$

$(Et_2N)Me_2Si$—$C{\equiv}C$—$SiMe_2(NEt_2)$ $TfOMe_2Si$—$C{\equiv}C$—$SiMe_2OTf$

Eq. 2.

$PhMe_2Si$—$\underset{Me_2}{C}$—$SiMe_2Ph$ $TfOMe_2Si$—$\underset{Me_2}{C}$—$SiMe_2OTf$

$PhMe_2Si$—$SiMe_2Ph$ $\xrightarrow[-2\ C_6H_6]{2\ TfOH}$ $TfOMe_2Si$—$SiMe_2OTf$

Me_3Si—$SiMePh_2$ Me_3Si—$SiMe(OTf)_2$

Eq. 3.

We prepared numerous new organosilicon polymers using the described α,ω-bis[(trifluoromethyl)-sulfonyloxy]-substituted organosilicon derivatives as electrophilic starting materials [13, 14]. The dinucleophilic reactants were either organic or organometallic compounds. The reactions of [phenylene-1,4-bis(dimethylsilanediyl)]bis(trifluoromethanesulfonate) with six dinucleophiles in Scheme 1 illustrate the potential of this method. We could confirm the formation of the polymers at low temperature, in short reaction times, and with high yields.

The structural characterization of the described polymers was mainly based on NMR spectroscopy. ^{29}Si NMR chemical shifts are particular useful for structural characterization. As expected, one observes the broad signals, which are typical for organosilicon polymers. However, the half-band widths of the ^{29}Si NMR signals, 1.5–4.0 ppm, are much narrower than those in the case of polysilanes or polycarbosilanes prepared by Wurtz reactions. The narrower signals indicate the regular alternating arrangements in the polymer backbones resulting from the fact that the condensation reactions are not accompanied by exchange processes.

Weight-average molecular weights in the range M_w = 2 000–7 000, relative to polystyrene standards, were found by gel permeation chromatography (GPC). They correspond to polymerisation degrees of n = 10–25. The polydispersities (M_w/M_n) were found in the range 1.3–2.1 and are consistent with purely linear structures for all polymers described.

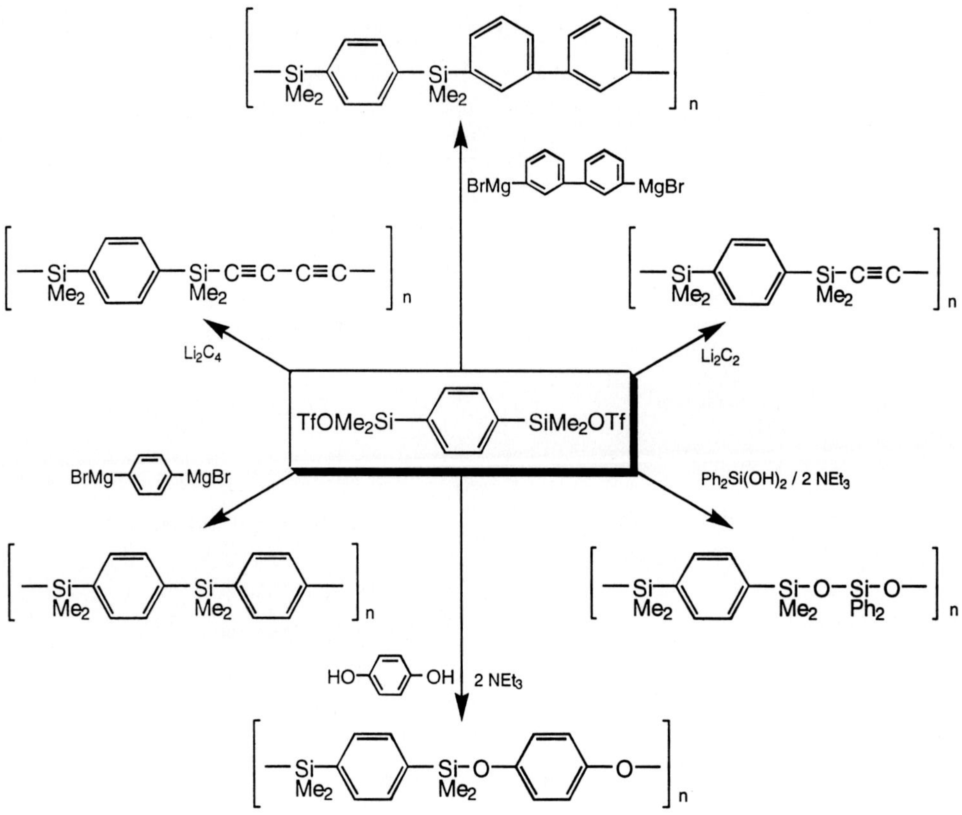

Scheme 1.

Normally, the polymers are yellow-brown solids. Defined melting points are not observed. The solids become highly viscous fluids, with contraction of volume, at temperatures between 70 and 200 °C. These conversions occur within temperature intervals of about 20 degrees. All polymers are not decomposed at temperatures to 250 °C. They are soluble in the usual organic solvents such as benzene, chloroform, and THF.

In the last time we also investigated the reductive coupling of the prepared α,ω-bis[(trifluoromethyl)sulfonyloxy]-substituted organosilicon derivatives with alkali metals. We found, that these coupling reactions did not occur completely. The isolated polymers always contained siloxy units after hydrolysis of the reaction mixture. Therefore we looked for a better reducing agent.

In recent years Fürstner [15, 16] described the very rapidly and completely conversion of chlorosilanes into disilanes by the help of potassium-graphite (C_8K). These coupling reactions were characterized by low temperatures (0–25 °C), short reaction times (5–60 min), and high yields (> 90 %). So we tried to reduce our silyltriflate derivatives with potassium-graphite and we succeeded in preparing new organo-silicon polymers [17]. The reductive coupling of 1,4-bis[dimethyl(trifluoromethyl-sulfonyloxy)silyl]butadiyne with C_8K illustrates the reaction priciple (Eq. 4). In Eq. 5 it is shown, that we were also sucessful in copolycondensations of different bis(silyltriflates).

$$n\ TfOMe_2Si-C\equiv C-C\equiv C-SiMe_2OTf \xrightarrow[\substack{-2n\ KOTf \\ -16n\ C}]{2n\ C_8K} \left[-\underset{Me_2}{Si}-C\equiv C-C\equiv C-\underset{Me_2}{Si}- \right]_n$$

20°C, DME, 30 min
73%, M_w=6400 g/mol

Eq. 4.

$$n\ TfOMe_2Si-C\equiv C-SiMe_2OTf\ +\ n\ TfOMe_2Si-\langle\ \rangle-SiMe_2OTf$$

$$\xrightarrow[\substack{-4n\ KOTf \\ -32n\ C}]{4n\ C_8K}$$

$$\left[-\underset{Me_2}{Si}-C\equiv C-\underset{Me_2}{Si}-\underset{Me_2}{Si}-\langle\ \rangle-\underset{Me_2}{Si}- \right]_n$$

Eq. 5.

Conclusions

Obviously, the polymer syntheses described here are too expensive for technical applications. However, the essential advantage of the method consists of the possibility of synthesizing, with relatively little efforts, small amounts of numerous differently structured organosilicon polymers for investigations in material science. Further work will be directed to the physical properties (thermal behavior, conductivity) of the new polymers.

Acknowledgement: This work was supported by *Wacker-Chemie GmbH*. Furthermore, the author acknowledges Prof. R. Nesper for support of this investigation.

References:

[1] *Silicon-Based Polymer Science; Advances in Chemistry, No. 224* (Eds.: J. M. Zeigler, F. W. G. Fearon), American Chemical Society, Washington, DC, **1990**.

[2] R. West, in: *The Chemistry of Organosilicon Compounds* (Eds.: S. Patai, Z. Rappoport), Wiley, Chichester, **1989**, 1207.

[3] K. Nate, M. Ishikawa, H. Ni, H. Watanabe, Y. Saheki, *Organometallics* **1987**, *6*, 1673.

[4] S. Ijadi-Maghsoodi, Y. Pang, T. J. Barton, *J. Polym. Sci., Polym. Chem.* **1990**, *28*, 955.

[5] J. L. Breford, R. Corriu, Ph. Gerbier, C. Guerin, B. Henner, A. Jean, T. Kuhlmann, *Organometallics* **1992**, *11*, 2500.

[6] M. Ishikawa, T. Hatano, Y. Hasegawa, T. Horio, A. Kunai, Y. Miyai, T. Ishida, T. Tsukihara, T. Yamanaka, T. Koike, J. Shioya, *Organometallics* **1992**, *11*, 1604.

[7] R. Corriu, W. Douglas, Z. Yang, *J. Organomet. Chem.* **1993**, *456*, 35.

[8] K. Matyjaszewski, Y. L. Chen, H. K. Kim, *ACS Symp. Ser.* **1988**, *360*, 78.

[9] W. Uhlig, *J. Organomet. Chem.* **1991**, *402*, C 45.

[10] W. Uhlig, *Helv. Chim. Acta* **1994**, *77*, 972.

[11] W. Uhlig, C. Tretner, *J. Organomet. Chem.* **1994**, *467*, 31.

[12] C. Tretner, B. Zobel, R. Hummeltenberg, W. Uhlig, *J. Organomet. Chem.* **1994**, *468*, 63.

[13] W. Uhlig, *Chem. Ber.* **1994**, *127*, 985.

[14] W. Uhlig, *Organometallics* **1994**, *13*, 2843.

[15] A. Fürstner, H. Weidmann, *J. Organomet. Chem.* **1988**, *354*, 15.

[16] A. Fürstner, *Angew. Chem.* **1993**, *105*, 171; *Angew. Chem., Int Ed. Engl.* **1993**, *32*, 164.

[17] W. Uhlig, *Z. Naturforsch.*, submitted.

Silicon Carbide and Carbonitride Precursors via the Silicon-Silicon Bond Formation: Chemical and Electrochemical New Perspectives

Michel Bordeau, Claude Biran, Françoise Spirau, Jean-Paul Pillot,
Marc Birot, Jacques Dunoguès

Laboratoire de Chimie Organique et Organométallique
Université Bordeaux
Université Bordeaux 1, F-33405 Talence, France

Summary: Polysilacarbosilanes and polysilasilazanes prepared according to a copolymer strategy offer an easy, coherent approach to polycarbosilanes and silazanes, precursors of SiC and SiCN-based materials with variable C/Si and C/Si/N ratios. In contrast with the polysilazane route which leads, upon pyrolysis, to carbon-containing silicon nitride, the synthesized polycarbosilazanes are finally converted into nitrogen-containing silicon carbide. The formation of silicon-silicon bonds constitutes the key step in these syntheses. To avoid the use of sodium, a simple, inexpensive, and practical electrochemical technique using an undivided cell, a sacrificial anode, and a constant current density has been developed allowing a facile synthesis of di-, tri-, or polysilanes including polydimethylsilane.

After about 20 years of intensive investigations for accessing to silicon carbide-based materials, the Yajima process [1] remains the reference, especially in the field of industrial obtaining of fibers (Eq.1).

$$n\ Me_2SiCl_2 \xrightarrow[\text{Toluene or xylene reflux}]{2n\ Na} (Me_2Si)_n \xrightarrow[\text{Autoclave}]{470\ °C} "(HMeSiCH_2)_n" \xrightarrow[\substack{\text{a) Spinning} \\ \text{b) Curing} \\ \text{c) Pyrolysis}}]{} "SiC"\ \text{fibers}$$
$$\text{PDMS} \qquad\qquad \text{PCS}$$

Eq. 1.

For the last six years our interest has been focused to a better understanding of the relationship between the composition and the structure of the precursor and the composition, the structure, and the properties of the inorganic material obtained upon pyrolysis. For that purpose we have synthesized the first, well defined, linear polysilapropylene [2]. Otherwise, because of the drastic conditions required to convert PDMS into PCS, especially due to the infusibility of the former, it was proposed to modify the regular structure of PDMS by using a copolymer strategy, consisting of the cocondensation of Me_2SiCl_2 with bis(chlorosilyl)methanes, mainly $(HClMeSi)_2CH_2$ [3] or silazanes, mainly $(HClMeSi)_2NH$, a novel compound [4], in order to synthesize polysilacarbosilanes (PSCS) or polysilasilazanes (PSSZ), expected to be more soluble and reactive than PDMS (Eq.2).

$$(1-x)\ Me_2SiCl_2 + x(ClHMeSi)_2Z \xrightarrow[\text{PhMe, reflux}]{2\ Na} -(Me_2Si)_{1-x}-(HMeSi-Z-SiMeH)_x-$$

Eq. 2.

In the case of PSCSs when $x > 0.1$, most of the products were found soluble and were converted into the corresponding PCS at 450 °C, under atmospheric pressure. These spinnable PCS were transformed into SiC-based materials with ceramic yields very close to those of Yajima (~ 60 %). PSCSs of formula $(MeRSi)_{1-x}-(HR'SiCH_2SiR'H)_x$ also were prepared in order to appreciate the influence of R and R' on the carbon content of the ceramic. Otherwise, multinuclear solid state NMR studies indicated that the network was first built around silicon atoms (formation of SiC_4) then, at higher temperature, around carbon atoms (formation of CSi_4).

In the case of PSSZ a more extensive study was carried out. In particular, it was shown that the nitrogen content of the ceramic is in good correlation with the value of x when $x \leq 0.75$. For $x \geq 0.5$ Si–Si–Si sequences were hardly detected. A PSSZ prepared with equimolar amounts of Me_2SiCl_2 and $(ClHMeSi)_2NH$ was converted into PCSZ upon thermolysis (temperature of the bath 425 °C, 8h), cross-linked by γ-irradiation, then pyrolyzed to afford SiCN-based material containing a very small amount of oxygen [4a]. In a preliminary study carried out at the Laboratoire des Composites Thermostructuraux, F–33600 Pessac, France (Prof. R. Naslain), SiCN fibers were obtained (tensile strength: 2 400 MPa; Young's modulus: 235 GPa).

Both approaches as well as the Yajima route require the creation of silicon–silicon bonds involving the polycondensation of (di)chlorosilanes by refluxing in toluene or xylene in the presence of sodium. Since metals different from the alkali ones are not reactive or require the use of special conditions, it was decided to investigate a simple and practical electrochemical way, involving the use of an undivided cell, a sacrificial anode, and a constant current density [5].

The general scheme is summarized in Eq. 3–6.

$$\text{Cathode:} \quad \equiv\text{SiCl} + 2e^- \longrightarrow \equiv\text{Si}^- + \text{Cl}^- \quad (3)$$

$$\text{in Solution:} \quad \equiv\text{Si}^- + \equiv\text{SiCl} \longrightarrow \equiv\text{Si}-\text{Si}\equiv + \text{Cl}^- \quad (4)$$

$$\text{Anode:} \quad 2/n\,M - 2/n\,e^- \longrightarrow 2/n\,M^{n+} \quad (5)$$

$$\text{overall reaktion:} \quad \equiv\text{SiCl} + 2/n\,M + \equiv\text{SiCl} \xrightarrow{\text{Electricity}} \equiv\text{Si}-\text{Si}\equiv + 2/n\,MCl_n \quad (6)$$

At the anode, the metal is oxidized in preference to $\equiv$Si$^-$ and Cl$^-$ anions, thereby avoiding secondary chlorination reactions (especially on aromatic rings) in an undivided cell. Normally, for the reaction to proceed, the reduction potentials of the different species present in solution must satisfy the order shown in Fig. 1a.

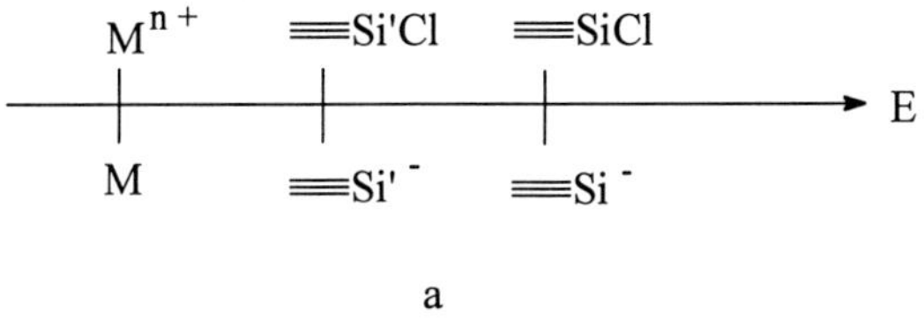

Fig. 1a.
a

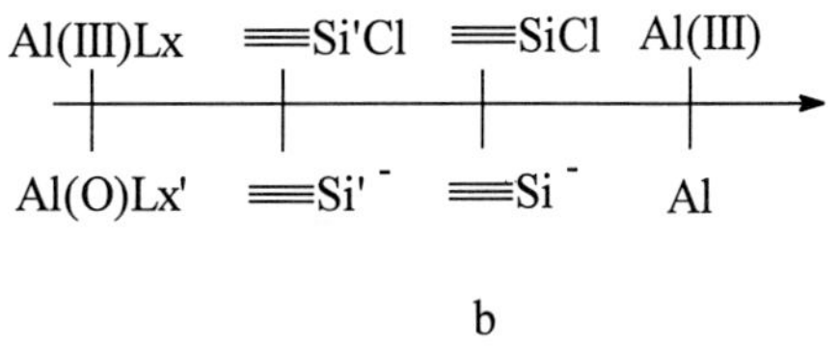

Fig. 1b.
b

When M^{n+} is Mg^{2+}, this order is always observed but with Al, Zn, or Cu cations this is not always the case, and a metal deposition can subsequently occur at the cathode. To avoid this undesired reaction, we added a small quantity of a complexing agent like HMPA or tris(3,6-dioxaheptyl)amine (TDA-1) to the aprotic solvent, to shift the reduction potential of the cation towards more cathodic values (Fig. 1b).

Since Me$_3$SiCl does not show any reduction wave at more positive potentials than -3.0 V vs SCE [6] in contrast with phenylchlorosilanes, it acts as the electrophile in the cross-coupling of phenyl(methyl)-chlorosilanes with Me$_3$SiCl. Eq. 7 and Table 1 summarize the results:

Substrate	Product						
	Electrosynthesis				Alkali Metal Route		
	Al Anode			Mg Anode			
	GC Yield %	Ref.	Isolated Yield %	Isolated Yield %	Intermediate	Yield %	Ref.
PhMe$_2$SiCl	90		80	83	≡SiLi	47	[8]
Ph$_2$MeSiCl	90	[7]	82	84	≡SiLi	74	[8]
Ph$_3$SiCl	90		71[a]	77[a]	≡SiLi/Na/K	75	[9]

Table 1. Electrochemical synthesis of unsymmetrical disilanes Ph$_n$Me$_{3-n}$Si–SiMe$_3$ – comparison with the classical route; [a] yield after recrystallization

$$\text{Ph}_n\text{Me}_{3-n}\text{SiCl} + \text{Me}_3\text{SiCl} \xrightarrow[\text{Al or Mg anode, THF + HMPA or TDA-1, room temperature, } n = 1-3]{2.2 \text{ F mol}^{-1}} \text{Ph}_n\text{Me}_{3-n}\text{Si–SiMe}_3$$
excess

Eq. 7.

After 1.2 F mol^{-1} of electricity had been passed through either Al or Mg anode, the symmetrical coupling of monochlorosilanes (65 mmol) led, as expected, to the symmetrical disilanes (Eq. 8) in yields comparable to those from the metal route (Table 2).

$$2 \text{ Ph}_n\text{Me}_{3-n}\text{SiCl} + \text{Me}_3\text{SiCl} \xrightarrow[\text{Al or Mg anode, THF + HMPA or TDA-1, } n = 0-3]{2.2 \text{ F mol}^{-1}} \text{Ph}_n\text{Me}_{3-n}\text{Si–SiMe}_{3-n}\text{Ph}$$

Eq. 8.

In this reaction, only one half of the engaged chlorosilane is reduced, the second half acting as the electrophile, so the theoretical total current is 1 F mol^{-1}. In spite of its very negative reduction potential, Me$_3$SiCl could also be reduced to Me$_6$Si$_2$ in the above conditions.

Substrate	Product	Electrosynthesis Isolated Yield [%]			Metal Route		
		Al Anode	Ref.	Mg Anode	Reaction Conditions	Yield %	Ref.
Me$_3$SiCl	(Me$_3$Si)$_2$	74	[7]	78	Mg/THF	40	[10]
					Li/THF	76	[8]
						84	[11]
					Na/K/xylene	70	[12]
						85	[13]
					MeMgCl/Et$_2$O	72[a]	[14]
PhMe$_2$SiCl	(PhMe$_2$Si)$_2$	80		83			
Ph$_2$MeSiCl	(Ph$_2$MeSi)$_2$	60		68			
Ph$_3$SiCl	(Ph$_3$Si)$_2$	70		71	Li/THF ultra sound	73	[15]

Table 2. Electrochemical synthesis of symmetrical disilanes (Ph$_n$Me$_{3-n}$Si)$_2$ – comparison with the classical route; [a] industrial disilane residue as the substrate

The coupling of dichlorosilanes (30 mmol) with excess of Me$_3$SiCl (10 mol. eq.) led, after the passage of 4.2 F mol^{-1}, to the respective trisilanes (Eq. 9) in isolated yields much higher than those obtained from chemical routes (Table 3). Both, Al and Mg anodes gave similar results.

$$Ph_nMe_{2-n}SiCl_2 + Me_3SiCl \text{ excess} \xrightarrow[\text{TDA-1}, n = 0-2]{4.2 \text{ F mol}^{-1} \text{ Al or Mg anode, THF + HMPA or}} Ph_nMe_{2-n}Si(SiMe_3)_2$$

Eq. 9.

This simple electrochemical method seems to be the most practical way to produce these trisilanes. Knowing that the reduction potentials of dichlorosilanes are much less cathodic than that of Me$_3$SiCl, we can assume that the dichlorosilanes are first reduced to monochlorosilyl anions which are then trapped by the Me$_3$SiCl, present in large excess, rather than by the dichlorosilane (Eq. 10).

	Product						
Substrate	**Electrosynthesis**				**Metal Route**		
	Al Anode			Mg Anode			
	GC Yield %	Isolated Yield %	Ref.	Isolated Yield %	Reaction Conditions	Yield %	Ref.
PhMeSiCl$_2$	98	90		92	Na/K/xylene reflux	32	[16]
Ph$_2$SiCl$_2$		77		79	Li/THF	14	[17]
Me$_2$SiCl$_2$		60	[7]	63	Li/THF	13	[11]
					3 steps[a]	42	[18]

Table 3. Electrochemical synthesis of trisilanes Ph$_n$Me$_{2-n}$Si(SiMe$_3$)$_2$ (4.2 F mol^{-1} passed) – comparison with the classical route; [a] industrial disilane residue as the substrate

$$\text{PhMeSiCl}_2 \xrightarrow[\text{excess Me}_3\text{SiCl,}]{2.1 \text{ F mol}^{-1}} \text{PhMeSiCl–SiMe}_3 \xrightarrow{2.1 \text{ F mol}^{-1}} \text{PhMeSi(SiMe}_3)_2$$

Al anode, THF + HMPA   83 % (G. C.)   98 % (G. C.)
                        63 % (isolated) 90 % (isolated)

Eq. 10.

Moreover, the kinetic curves (obtained from GC analysis) of the electrochemical reduction of PhMeSiCl$_2$ with an Al anode, as an example, show that the two silicon-chlorine bonds are reduced in two well-separated steps (Fig. 2):

This selectivity, not available by the chemical routes, allows a good synthesis of the 1-chloro-1-phenyltetramethyldisilane in 63 % isolated yield, but it must be pointed out that the optimization of the monosilylation procedure depends strongly on the total current passed because the formation of the trisilane begins just before 2 F mol^{-1}.

Similarly, tetrasilanes (Me$_3$SiSiMeR)$_2$ with R = Me, Ph were synthesized in 80–85 % yield and, from PhMeSiCl$_2$ and Me$_3$SiSiMe$_2$Cl, the expected pentasilane (Me$_3$SiSiMe$_2$)$_2$SiMePh was obtained in 85 % yield.

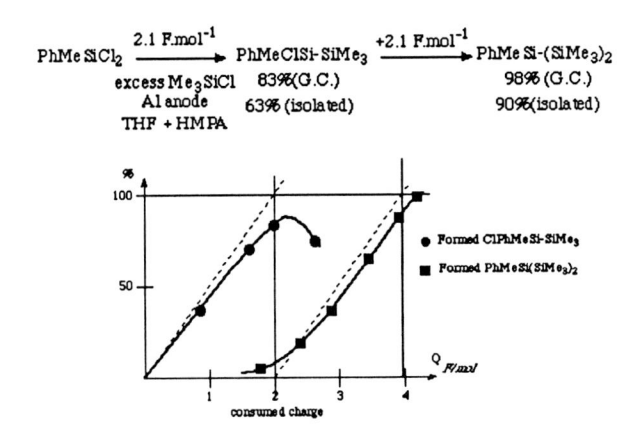

$$\text{PhMeSiCl}_2 \xrightarrow[\substack{\text{excess Me}_3\text{SiCl} \\ \text{Al anode} \\ \text{THF + HMPA}}]{2.1\ \text{F.mol}^{-1}} \text{PhMeClSi-SiMe}_3 \xrightarrow[]{+2.1\ \text{F.mol}^{-1}} \text{PhMeSi-(SiMe}_3)_2$$

$$83\% \text{(G.C.)} \qquad 98\% \text{(G.C.)}$$
$$63\% \text{(isolated)} \qquad 90\% \text{(isolated)}$$

Fig. 2. Kinetics of the electrochemical synthesis of ClPhMeSi–SiMe$_3$ and PhMeSi(SiMe$_3$)$_2$ with an Al anode

At last the electrochemical technique was used to obtain PDMS from dichlorodimethylsilane. Using various conditions such as THF as the solvent, with or without TDA-1 or HMPA as a co-solvent, aluminum as the anode and a stainless steel cathode, insoluble PDMS was formed in 17–30 % yield with a low current efficiency (16–27 %). PDMS was accompanied with oxygen-containing oligomers resulting from the THF ring-opening. Finally the reaction was optimized by performing the electrolysis "without solvent" i.e., using a large excess of dimethyldichlorosilane in the presence of a small amount of a complexing agent such as HMPA or TDA-1 to obtain PDMS in almost quantitative current efficiency (Eq. 11, Table 4).

$$n\ \text{Me}_2\text{SiCl}_2 \xrightarrow[\substack{\text{1. e}^- \text{(Al anode)} \\ \text{"without solvent"} \\ \text{2. Methanolysis}}]{} -(\text{Me}_2\text{Si})_n-$$

C.E. = 90; n > 25

Eq. 11.

A solution of Bu$_4$NBr (0.03 M, supporting electrolyte) in 70 mL of Me$_2$SiCl$_2$ could maintain a current intensity of 0.1 A for one day (run 1), but, in the absence of a complexing agent, the PDMS obtained in 72 % current efficiency was grey, due to the presence of metallic aluminum resulting from the reduction of AlCl$_3$. This was avoided by adding complexing agents in order to shift the reduction potential of AlCl$_3$ cathodically (runs 2 and 3). Moreover, sonication (35 kHz) resulted in a further increase of the yield of PDMS (90 %, run 4).

Electrochemically synthesized and Yajima PDMSs exhibit the same physicochemical properties and similar ability to be converted into comparable PCSs.

The sacrificial anode technique appears to be competitive with regard to the other electrochemical methods [19] to create the silicon–silicon bond and is very promising for the replacement of alkali metals in many syntheses, especially when high selectivity is required.

Run	Me_2SiCl_2 (Complexing Agent)	Vol [mL]	Charge Passed [F]	PDMS Current Efficiency [%]	$Al^{[a]}$ [%]
1	Me_2SiCl_2 alone	70	0.09	72 (grey)	>5
2	Me_2SiCl_2 (HMPA)	70 (13)	0.07	76	2.9
3	Me_2SiCl_2 (TDA-1, HMPA)	70 (15, 10)	0.18	84 (white)	0.1
4	Me_2SiCl_2 (TDA-1, HMPA)	70 (15, 10)	0.24	90 (white)	0.05

Table 4. Electroreductive polycondensation of Me_2SiCl_2 "without solvent"; [a] from elemental analysis

Acknowledgement: The authors are indebted to *CNRS*, *French Ministry of Defense* (DRET), *Rhône-Poulenc Co.*, *Société Européenne de Propulsion*, and *Conseil Régional d'Aquitaine* for their financial assistance.

References:

[1]   S. Yajima, *Am. Ceram. Soc. Bull.* **1983**, *62*, 195; and references therein.
[2]   E. Bacqué, J.-P. Pillot, M. Birot, J. Dunoguès, *Macromolecules* **1988**, *21*, 30; *Macromolecules* **1988**, *21*, 34.
[3]   J.-P. Pillot, E. Bacqué, J. Dunoguès, *F* 2 399 371, **1986**.
[4]   E. Bacqué, J.-P. Pillot, J. Dunoguès, C. Biran, P. Olry, *F* 2 599 037, **1986**.
[4a] C. Richard, *Thesis*, Bordeaux **1990**, n° 513; P. Roux, *Thesis*, Bordeaux **1993**, n° 948.
[5]   O. Sock, M. Troupel, J. Périchon, *Tetrahedron Lett.* **1985**, *26*, 1509.
[6]   T. Shono, Y. Matsumura, S. Katoh, N. Kise, *Chem. Lett.* **1985**, 463.

[7] C. Biran, M. Bordeau, P. Pons, M.-P. Léger, J. Dunoguès, *J. Organomet. Chem.* **1990**, *382*, C 7.
[8] H. Gilman, K. Shiina, D. Aoki, B.J. Gaj, D. Wittenberg, T. Brennan, *J. Organomet. Chem.* **1968**, *13*, 323.
[9] H. Gilman, T. C. Wu, *J. Am. Chem. Soc.* **1951**, *78*, 4031.
[10] O. W. Steudel, H. Gilman, *J. Am. Chem. Soc.* **1960**, *82*, 5129.
[11] G. Fritz, B. Grunert, *Z. Anorg. Allgem. Chem.* **1981**, *473*, 59.
[12] G. R. Wilson, A.G. Smith, *J. Org. Chem.* **1961**, *26*, 557.
[13] W. Sundmeyer, *Z. Anorg. Allgem. Chem.* **1961**, *310*, 50.
[14] A. Ronald, J. Maniscolo, *US.* 4 309 556, **1982**.
[15] P. Boudjouk, B. H. Han, *Tetrahedron Lett.* **1981**, *22*, 3813.
[16] M. Kumada, M. Ishikawa, S. Maeda, *J. Organomet. Chem.* **1964**, *2*, 478.
[17] H. Gilman, G. L. Schwebke, *J. Am. Chem. Soc.* **1964**, *86*, 2693.
[18] M. Kumada, M. Ishikawa, *J. Organomet. Chem.* **1963**, *1*, 153; H. Sakurai, K. Tominaga, T. Watanabe, M. Kumada, *Tetrahedron Lett.* **1966**, 5493.
[19] See, for instance, C. Jammegg, S. Graschy, E. Hengge, *Organometallics* **1994**, *13*, 2397; and references therein (namely from T. Shono *et al.*, M. Ishikawa *et al.*, and M. Umezawa *et al.*).

Synthesis of Spinnable Poly(silanes/-carbosilanes) and Their Conversion into SiC Fibers

Robin Richter*, Gerhard Roewer, Hans-Peter Martin, Erica Brendler,
Hans Krämer, Eberhard Müller

Institut für Anorganische Chemie
Technische Universität Bergakademie Freiberg
Leipziger Str. 29, D-09599 Freiberg, Germany

Summary: The heterogeneous catalytic redistribution reaction of methylchlorodisilanes provides spinnable poly(methylchlorosilanes/-carbosilanes). Especially copolymers like poly(methylchlorosilanes-*co*-styrenes) are suitable polymers for melt spinning. The high reactivity caused by Si–Cl bonds enables oxygen free curing methods of the melt spun polymer filaments with ammonia. The synthesis is achieved without the employment of highly reactive metals and any solvents. The thus produced SiC fibers exhibit oxygen contents lower than 1 wt. %.

Introduction

Silicon organic polymer derived SiC fibers were first developed by Yajima *et al.* [1]. Due to certain steps of the manufacturing process (introduction of oxygen) of such fibers their thermal stability is relatively low (1000–1200 °C).

In general, the following preparative methods of silicon organic polymer precursors have been developed giving spinnable polymers:

1) Dehalocoupling reactions of organohalogenosilanes
2) Catalytic ring opening polymerization of silacyclobutane
3) Catalytic polymerization of silicon containing vinyl monomers
4) Catalytic dehydrocoupling of organomonosilanes or organodisilanes
5) Catalytic redistribution reactions of substituted organodisilanes (disproportionation) [2, 3]

The disadvantages of the synthesis routes 1–4 are the application of highly reactive and expensive metals (Li, Na, K, Mg) and the enormous quantity of solvents. Particularly, as result of the dehalocoupling reactions, the polymers are unreactive at room temperature. To overcome these problems we synthesized spinnable reactive poly(silanes/-carbosilanes) via heterogeneous catalytic disproportionation of methylchlorodisilanes which have been wasted as a byproduct of the "Direct synthesis" of methylchlorosilanes so far.

Results

For example, in Scheme 1, the disproportionation of 1,1,2,2,-tetrachlorodimethyldisilane (**1**) to methyltrichlorosilane (**2**) and oligo(methylchlorosilane) (**3**) is shown.

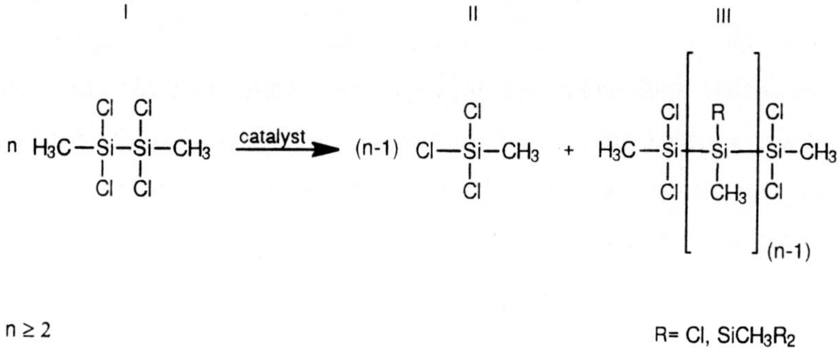

$$n\ H_3C-\underset{\underset{Cl}{|}}{\overset{\overset{Cl}{|}}{Si}}-\underset{\underset{Cl}{|}}{\overset{\overset{Cl}{|}}{Si}}-CH_3 \xrightarrow{catalyst} (n-1)\ Cl-\underset{\underset{Cl}{|}}{\overset{\overset{Cl}{|}}{Si}}-CH_3 + H_3C-\underset{\underset{Cl}{|}}{\overset{\overset{Cl}{|}}{Si}}\left[\underset{\underset{CH_3}{|}}{\overset{\overset{R}{|}}{Si}}\right]_{(n-1)}\underset{\underset{Cl}{|}}{\overset{\overset{Cl}{|}}{Si}}-CH_3$$

I II III

n ≥ 2   R= Cl, SiCH₃R₂

Scheme 1. Disproportionation of 1,1,2,2-tetrachlorodimethyldisilane (**1**) into methyltrichlorosilane (**2**) and oligo-(methylchlorosilane) (**3**)

The formation of tri- and especially tetrasilanes which are already branched (tertiary Si-units) as the first reaction products (described elsewhere [4]) suggests the appearance of intermediate silylene species which could enter in insertion reactions of Si–Si as well as Si–Cl bonds. The tri- and tetrasilanes undergo thermal crosslinking reactions at reaction temperatures of 165–250 °C. In addition dehydrochlorination reactions initiated by acid H-abstraction of methyl groups cause the formation of carbosilane (methylene) units in the polymer framework. Table 1 shows the gross compositions of poly(methylchlorosilanes) which are determined by the reaction temperature.

Table 1. Gross composition in dependence of the reaction temperature

Reaction temperature [°C]	Gross composition
Disilane fraction	$SiC_{1.2} H_4 Cl_{1.77}$
156	$SiC_{1.2} H_{3.9} Cl_{1.6}$
163	$SiC_{1.3} H_{3.6} Cl_{1.2}$
179	$SiC_{1.3} H_{3.7} Cl$
187	$SiC_{1.5} H_{3.8} Cl_{0.9}$
210	$SiC_{1.4} H_{3.6} Cl_{0.75}$
250	$SiC_{1.4} H_{3.3} Cl_{0.44}$

The polymers were melt spun under inert conditions into polymer filaments with diameters of about 25 µm. Spectroscopic investigations of these polymers are described in [5]. As shown in Scheme 2 the disproportionation of methylchlorodisilanes (**1**) in presence of olefins like styrene (**4**) gives poly(methylchlorosilanes-*co*-styrene) (**5**) with excellent spinnability.

Scheme 2. Disproportionation of 1,1,2,2-tetrachlorodimethyldisilane into methyltrichlorosilane and poly(methylchlorosilane-*co*-styrene)

Fig. 1 and Fig. 2 show the CP MAS NMR spectra of thus obtained poly(methylchlorosilane-*co*-styrene). The ^{29}Si NMR spectrum indicates a predominating share of branched Si structures (resonance at –65 ppm) beside $-SiCl_2CH_3$ and $-SiCH_3Cl$ groups (37 and 15–30 ppm, respectively). In the ^{13}C NMR spectrum the signals of polystyrene blocks (128 ppm, 146 ppm and spinning side bands, ethylene groups at 40 and 45 ppm) and polysilane blocks (10 to −15 ppm) can be seen, probably indicating the formation of a block polymer.

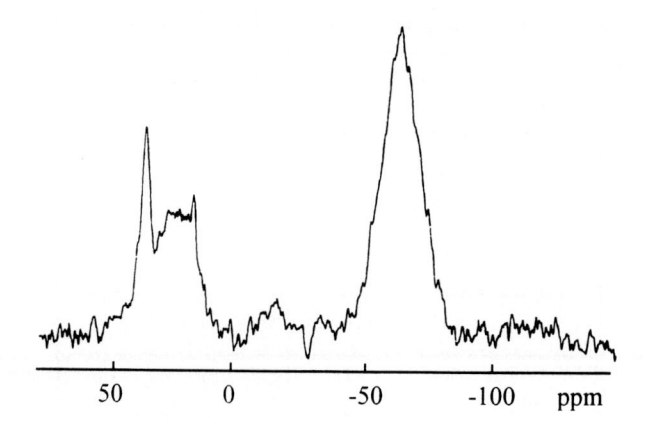

Fig. 1. ^{29}Si CP NMR spectra of poly(methylchlorosilane-*co*-styrene)

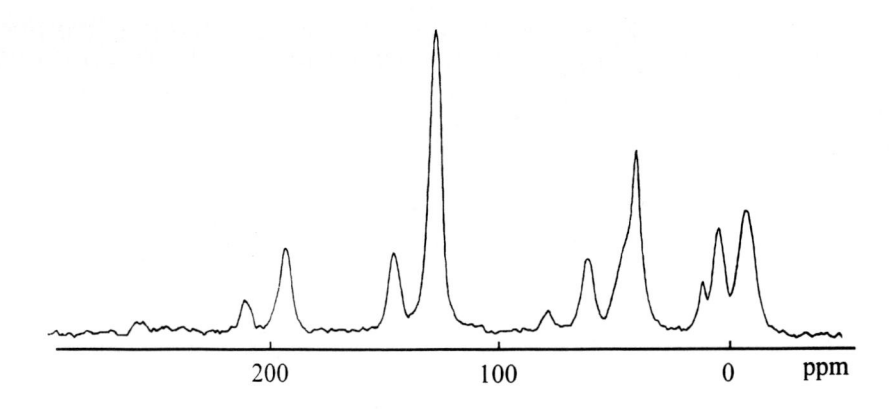

Fig. 2. ^{13}C CP NMR spectra of poly(methylchlorosilane-*co*-styrene)

Once spun, the shape of polymer fibers can be stabilized (cured) under argon mixed with ammonia. The reaction of SiCl groups with ammonia at the fiber surface gives Si–N bonds being the basis of superficial crosslinking resulting in unmeltability. The pyrolysis and conversion into silicon carbide is performed under pure argon at temperatures between 1000 and 1400 °C.

A completely amorphous structure was found by X-ray diffraction on fibers which were pyrolysed at temperatures up to 1300 °C. A crystallization starts around 1400 °C and nanocrystalline silicon carbide is formed with a crystallite size of about 2 nm. Compared to an uncured sample the crystallization is retarded. A significant crystallite growth occurs around 1500 °C connected with a decreasing of the fiber properties. The oxygen content of these SiC fibers is less than 1 wt. % found by neutron activation analysis. The current available data for tensile strength amounts to 1 GPa at 25 µm diameter (first measurements).

Acknowledgment: We are grateful to the *Deutsche Forschungsgemeinschaft* and the *Fonds der Chemischen Industrie* for financial support.

References:

[1] S. Yajima, *Chem. Lett.* **1975**, *9*, 1209.
[2] W. Kalchauer, in: *Organosilicon Chemistry – From Molecules to Materials* (Eds.: N. Auner, J. Weis), VCH, Weinheim, **1994**, p. 293.
[3] R. Calas, *J. Organomet. Chem.* **1982**, *225*, 117.
[4] E. Brendler, K. Leo, B. Thomas, R. Richter, G. Roewer, H. Krämer, in: *Organosilicon Chemistry II* (Eds.: N. Auner, J. Weis), VCH, Weinheim, **1995**, p. 69.
[5] R. Richter, *Freiberger Forschungshefte* **1993**, *A*, 832K.

Polymeric Silylcarbodiimides – Novel Route to Si–C–N Ceramics

Andreas Kienzle*, Kathi Wurm, Joachim Bill, Fritz Aldinger
Institut für Werkstoffwissenschaft, Pulvermetallurgisches Laboratorium (PML)
Max-Planck-Institut für Metallforschung
Heisenbergstr. 5, D-70569 Stuttgart, Germany

Ralf Riedel
Fachbereich Materialwissenschaft, Fachgebiet Disperse Feststoffe
Technische Hochschule Darmstadt
Hilpertstr. 31, D-64295 Darmstadt, Germany

Introduction

Ceramic fibers [1], coatings [2], and monolithic bodies [3] can be easily prepared by pyrolysis of element organic polymers at low temperatures. Silicon containing ceramics are mainly prepared by pyrolyzing polysilanes, -silazanes, -siloxanes, or -carbosilanes. We recently discovered oligomeric and polymeric silylcarbodiimides as promising precursors for Si–C–N ceramics [4]. These compounds are easily synthesized by reaction of dichlorosilanes with cyanamide in the presence of a base. Due to the high reactivity of the carbodiimide group, the ceramic yields are higher compared with that of methylated silazanes. Additionally, the carbodiimide group provides an easy introduction of, for example, boron. These boron containing precursors can be pyrolysed to Si–B–C–N ceramics, which exhibit high thermal stability.

Results and Discussion

Oligomeric and polymeric silylcarbodiimides are synthesised reacting dichlorodiorganylsilanes with cyanamide and pyridine in THF as solvent [4].

$$R^1-\underset{\underset{Cl}{|}}{\overset{\overset{CH_3}{|}}{Si}}-Cl \;+\; H_2N-C\equiv N \xrightarrow[\text{- pyridine*HCl}]{\text{pyridine}} \left[-\underset{\underset{R^1}{|}}{\overset{\overset{CH_3}{|}}{Si}}-N=C=N- \right]_n$$

Eq. 1. $R^1 = CH_3$ (1); $CH=CH_2$ (2); H (3)

Each reaction mixture contains different cyclic compounds of the hypothetical monomer [H$_3$C(R)Si=N–CN] (R=H, CH$_3$, CH=CH$_2$) as has been shown by NMR, IR, and mass spectroscopic investigations. Distillation of the solvent-free reaction mixture of **1** led to the isolation of crystals of the tetrameric product. Its structure has been investigated using X-ray diffraction. The tetramer consists of a slightly bent 16-membered ring. The silicon atoms are connected by carbodiimide-bridges. In contrast to the reaction of dichlorodimethyl- and dichloromethylvinylsilane the reaction of dichloromethylsilane with cyanamide results in the highly cross-linked polymer **3**. This polymer results from hydrosilylation reactions between the Si–H and the carbodiimide groups.

Pyrolysis of the silylcarbodiimides at 1 000 °C in argon atmosphere leads to ceramic materials in the ternary system Si–C–N. The ceramic yield strongly depends on the substituents at the silicon atoms. Reactive groups such as Si–H and Si–vinyl can be cross-linked during pyrolysis which results in higher ceramic yields (Si–H substituted silylcarbodiimides: 70 %; Si–vinyl substituted silylcarbodiimides 64 %). In contrast to these results cross-linking of the dimethyl substituted silylcarbodiimide occurs to a lower extent and therefore a lower ceramic yield (28 %) is obtained.

Highly cross-linked silylcarbodiimides can be easily obtained salt-free by reacting bis(trimethylsilyl)-carbodiimide and tetrachlorosilane with catalytic amounts of pyridine at room temperature:

$$Cl-\underset{\underset{Cl}{|}}{\overset{\overset{Cl}{|}}{Si}}-Cl \;+\; 2\,(H_3C)_3Si-N=C=N-Si(CH_3)_3 \xrightarrow{\text{toluene/(THF)}} \left[\underset{N=C=N-}{\overset{N=C=N-}{\diagdown\,Si\,\diagup}} \right]_n \;+\; 4\,(H_3C)_3SiCl$$

Eq. 2.

Depending on whether the reaction is carried out in toluene or in THF the product can be isolated as powder or as gel. The by-product chlorotrimethylsilane can be easily removed by distillation and reused for the synthesis of bis(trimethylsilyl)carbodiimide. The elemental analysis of the polymers depending on the solvent used for the synthesis are listed in the experimental part. As can be seen from the hydrogen content the polymers still contain trimethylsilyl groups. The product obtained in toluene as solvent exhibits a lower hydrogen content than that synthesized in THF.

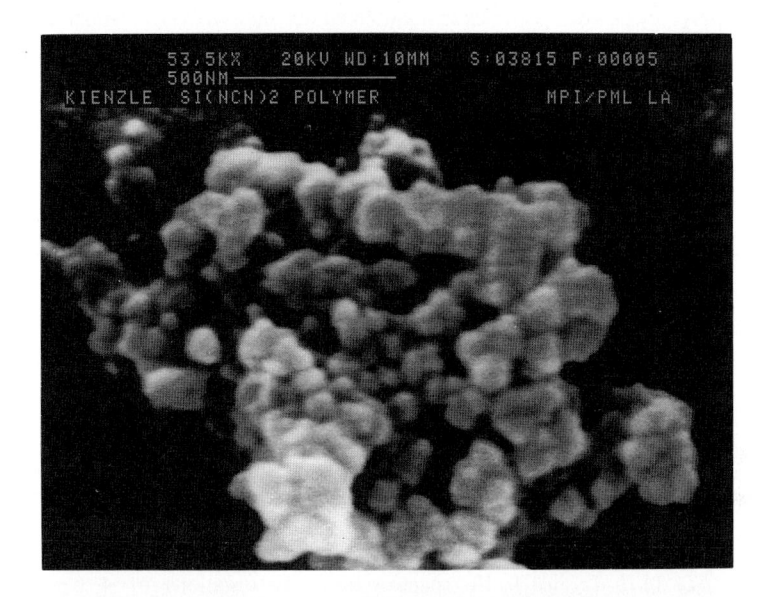

Fig. 1. SEM-micrographs of $[Si(N=C=N)_2]_n(SiMe_3)_o$ polymer powder before (top) and after (bottom) pyrolysis at 1100°C in argon atmosphere

This is referred to higher cross-linking than in case of THF. A low chlorine content (0.06 wt. %) is found for the reaction of tetrachlorosilane with bis(trimethylsilyl)carbodiimide which indicates that the reaction proceeds almost quantitatively. The pyrolysis of the highly cross-linked polymer leads to ceramics in the ternary system Si–C–N with 60 % yield. Due to further condensation of the polymer bis(trimethylsilyl)carbodiimide is formed between 100–500°C. Fig. 1 shows the polymer powder before and after pyrolysis at 1100°C in argon atmosphere.

The form of the polymer powder particles is maintained during pyrolysis. Nanosized Si–C–N-powder is obtained after pyrolysis. Furthermore, the high reactivity of the carbodiimide group provides an easy introduction of other elements, e.g., boron. As an example the hydroboration of bis(trimethylsilyl)carbodiimide (**4**) with dimethylsulfide-borane in toluene has been investigated:

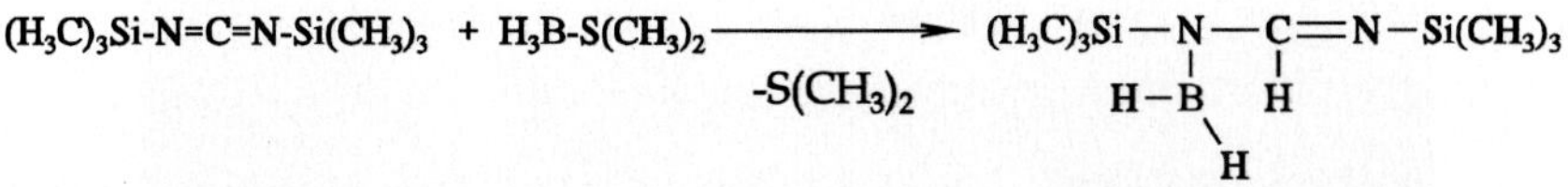
(H3C)3Si-N=C=N-Si(CH3)3 + H3B-S(CH3)2
-S(CH3)2
(H3C)3Si—N—C=N—Si(CH3)3
H—B H
H

Fig. 3.

The addition of two BH_3-molecules transforms the sp-hybridized carbon of the carbodiimide group into a sp^3-hybridized carbon of a methylenediamine group. At the same time, BN-bonds are formed. The resulting BH_2 groups react with **4** which results in the formation of cyclic oligomeric and polymeric compounds (GPC: M_w=1170 [g/mol]). IR spectroscopic investigations of **4** before and after the hydroboration with $H_3B–S(CH_3)_2$ (Fig. 2) show a decreasing intensity of the –N=C=N group ν_s vibration at 2205 cm^{-1}. New vibrational bands of B–N groups appear in the IR-spectra of the hydroborated product at 1599 (ν_{as} N–CH=N–) and 1429 cm^{-1} (ν_{as} BN_2).

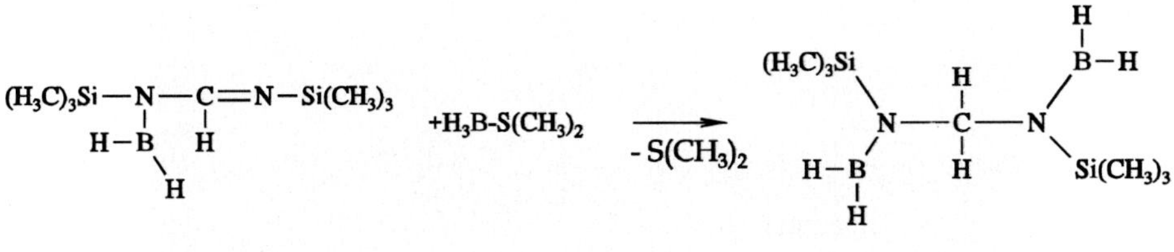
(H3C)3Si—N—C=N—Si(CH3)3
H—B H
H
+H3B-S(CH3)2
- S(CH3)2
(H3C)3Si
H
B—H
N
C
N
H—B
H
H
H
Si(CH3)3

Fig 4.

Polymeric silylcarbodiimides **1** and **2** react with dimethylsulfidborane forming a gel which leads to highly cross-linked, insoluble polymers **1HB** and **2HB**, respectively. Whereas **1** can only be cross-linked by the carbodiimide group, the addition of borane in **2** mainly takes place at the vinylic sites due to their higher reactivity. Pyrolysis of **1HB** and **2HB** at 1200 °C in argon atmosphere results in new SiBCN ceramics (ceramic yields: **1HB**: 61 %, **2HB**: 64 %).

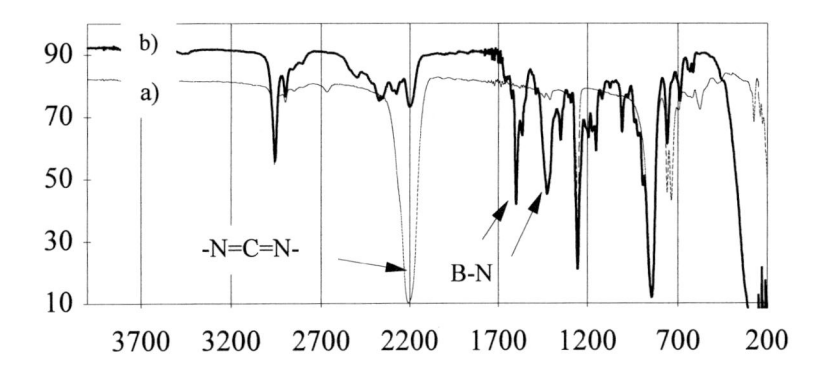

Fig. 2. IR spectra of bis(trimethylsilyl)carbodiimide before (a) and after (b) the reaction with $H_3B-S(CH_3)_2$ in toluene (film on KBr)

Ceramics From	C	N	Si Mass [%]	B	O	Chemical Formula
1HB	20.7	30.0	32.6	9.3	1.5	$Si_1N_{1.9}C_{1.5}B_{0.7}O_{0.08}$
2HB	27.3	26.7	29.0	6.1	5.0	$Si_1N_{1.9}C_{2.2}B_{0.5}O_{0.3}$

Table 1. Elemental analysis of the ceramic material obtained by pyrolysis of **1HB** and **2HB** in argon atmosphere

Experimental

All experiments were carried out in an atmosphere of dry argon. The solvents were dried according to common procedures. The experimental conditions for the synthesis of $[(H_3C)_2Si-N=C=N-]_n$ (**1**), $[(H_2C=CH)(H_3C)Si-N=C=N-]_n$ (**2**), and $[(H_2C=CH)(H)Si-N=C=N-]_n$ (**3**) are published in [4]. For the synthesis of bis(trimethylsilyl)carbodiimide see [5] (instead of chloroform we used THF as solvent).

Synthesis of highly cross-linked silylcarbodiimide $[Si(N=C=N)_2]_n(Si(CH_3)_3)_o$: A mixture of 10 mL (14.8 g, 0.09 mol) of tetrachlorosilane and 25 mL toluene is dropped to a solution of 39.6 mL (32.5g, 0,17 mol) bistrimethylsilylcarbodiimide, 50 mL toluene and catalytic amounts of pyridine (0.5 mL) while stirring. The highly cross-linked polymer is formed as white powder during the reaction. After stirring for four hours at room temperature, chlorotrimethylsilane and toluene are removed by distillation and 10 g of the cross-linked polycarbodiimide powder is obtained. A gel is formed if THF is used instead of toluene. Elemental analysis (wt. %): $[Si(N=C=N)_2]_n(Si(CH_3)_3)_o$ (THF as solvent) C 28.4, H 4.16, N 26.78,

Cl 0.06; $[Si(N=C=N)_2]_n(Si(CH_3)_3)_o$ (toluene as solvent): C 23.9, H 2.49, N 37.1, Cl 0,06; IR (nujol, KBr): ν [cm^{-1}] = 2185 (vs,b, ν-N=C=N–), 1258 (w, δ_{sy} Si–CH$_3$), 804 (vs, b), 578 (m, b).

Hydroboration of bis(trimethylsilyl)carbodiimide: 44 mL (1.21 g; 0.088 mol BH$_3$) of a 2 M solution of dimethylsulfide-borane in toluene are dropped slowly into a solution of 20 mL (16.4 g; 0.08 mol) of bis(trimethylsilyl)carbodiimide in 30 mL toluene that is stirred vigorously. The reaction mixture is heated to 110 °C for 10 h. After removing the solvent by distillation 12.6 g (71 % yield) of a viscous, colorless liquid and 4.2 g (23 % yield) of a yellow solid could be separated by distillation at temperatures up to 200°C/1·10^{-4} Torr.

^{1}H NMR(C$_6$D$_6$, 200 MHz, 25 °C): δ [ppm] = –0.01–0.32 (m, C–H, Si–CH$_3$), 2.1–2.53 (m, C–H, N–CH$_2$–N); 3.8–4.25 (m, B–H), 7.37–7.92 (m, C–H, N–CH=N).

^{13}C NMR (C$_6$D$_6$, 50 MHz, 25 °C): δ [ppm] = –2.6–3.29 (C, Si–CH$_3$), 53.78–67.45 (C, N–CH$_2$–N), 153.82–163,85 (C, N–CH=N).

^{29}Si NMR (C$_6$D$_6$, 39 MHz, 25 °C): δ [ppm] = –4.03–22.6 (Si, N–Si(CH$_3$)$_3$).

^{11}B NMR (C$_6$D$_6$, 64,21 MHz, 25 °C): δ [ppm] = –13,68 to –8,39 (B, H$_2$BN$_2$), 30 (B, HBN$_2$).

IR (film, KBr): ν [cm^{-1}] = 1599 (–HC=N–), 1423 (ν_{as} BN$_2$), 1260.0 (δ_{sy} Si–CH$_3$).

1HB Hydroboration of $[(H_3C)_2Si–N=C=N–]_n$ (1): 40 mL of a 2 M solution of H$_3$B–S(CH$_3$)$_2$ in toluene are added dropwise to a solution of 8.7 g of **1** in 20 mL toluene at 298 K. A solid, insoluble gel is formed after stirring 24 h at this temperature. The gel is dried (yield 9 g; 91 %), and then directly used for the pyrolysis (heating rate: 1.5 K min^{-1}) at 1 100°C in argon atmosphere (ceramic yield: 61 %).

2HB Hydroboration of $[(H_2C=CH)(H_3C)Si–N=C=N–]_n$ (2): The reaction has been carried out as previously described for the 1HB hydroboration of **1**. 10 g of **2** dissolved in 20 mL toluene; 45.3 mL H$_3$B–S(CH$_3$)$_2$ (2 molar in toluene). In this case the formation of the gel starts immediately after the addition of only a few drops of the H$_3$B–S(CH$_3$)$_2$ (ceramic yield 64 %).

Acknowledgement: The authors gratefully acknowledge the financial support of the *KKS (Keramik-Verbund Karlsruhe/Stuttgart)* and the *KSB-foundation*. We especially thank Prof. Dr. G. Becker for the possibility to carry out the chemical work at the *Institut für Anorganische Chemie* at the *Universität Stuttgart*.

References:

[1] S. Yajima, K. Okamura, J. Hayashi, *Chem. Lett.* **1975**, 1209.
[2] C. J. Chu, G. D. Soraru, F. Babonneau, J. D. Mackenzie, *Springer Proceedings in Physics* **1989**, *43*, 66.
[3] R. Riedel, G. Passing, H. Schönfelder, R. J. Brook, *Nature* **1992**, *355*, 714.
[4] A. Kienzle, A. Obermeyer, R. Riedel, F. Aldinger, A. Simon, *Chem. Ber.* **1993**, *126*, 2569.
[5] W. Einholz, *Ph. D. Thesis*, Universität Stuttgart, **1980**.

New Modified Polycarbosilanes

Wolfgang Habel, Werner Haeusler, Andreas Oelschläger, Peter Sartori*

Anorganische Chemie

Gerhard Mercator Universität – Gesamthochschule – Duisburg

Lotharstr. 1, D-47048 Duisburg, Germany

Summary: Starting with poly(dichlorosilylene-*co*-methylene) the synthesis of new poly(dialkylsilylene-*co*-methylenes), poly(dialkenylsilylene-*co*-methylenes), and poly(dialkinylsilylene-*co*-methylenes) was achieved by Grignard or organolithium reactions. Furthermore poly(dichlorosilylene-*co*-methylene) was photochlorinated under UV irradiation to poly-(dichlorosilylene-*co*-dichloromethylene).

Introduction

Polycarbosilanes are very useful as basic substances for synthesizing heat resistant ceramic materials and SiC fibers. Especially poly(diphenylsilylene-*co*-methylene), prepared from dibromomethane and dichlorodiphenylsilane with sodium metal by a Wurtz analogous reaction [1], was important as binding agent and sintering additive in the production of pressurelessly sintered SiC-ceramics [2].

In regard to substitution reactions of the functional phenyl groups, poly(diphenylsilylene-*co*-methylene) is a very interesting educt for the preparation of modified polycarbosilanes as precursors for SiC-based ceramic fibers.

Results and Discussion

The substitution of the phenyl groups by HCl under the catalytic influence of the Lewis acid $AlCl_3$ led to poly(dichlorosilylene-*co*-methylene) [3].

$$[Ph_2Si-CH_2]_x + 2x\ HCl \xrightarrow{AlCl_3} [Cl_2Si-CH_2-]_x + 2x\ PhH$$

Eq. 1.

The substitution of chlorine by different alkyl groups was achieved by a Grignard variant [4].

$$[Cl_2Si-CH_2]_x + 2x\ XMgR \longrightarrow [R_2Si-CH_2-]_x + 2x\ MgClX$$

Eq. 2. $R = CH_3, C_2H_5, C_3H_7, C_4H_9, C_5H_{11}, C_6H_{13}; X = Cl, Br, I$

The Grignard reaction could be transferred to organic groups with double and triple bonding.

$$[Cl_2Si-CH_2]_x + 2x\ XMgR^1 \longrightarrow [R^1{}_2Si-CH_2-]_x + 2x\ MgClX$$

Eq. 3. $R^1 = CH=CH_2, CH_2CH=CH_2, CH_2C(CH_3)=CH_2, (CH_2)_2CH=CH_2, (CH_2)_3CH=CH_2; X = Cl, Br$

$$[Cl_2Si-CH_2]_x + 2x\ XMgR^2 \longrightarrow [R^2{}_2Si-CH_2-]_x + 2x\ MgClX$$

Eq. 4. $R^2 = C\equiv CH, CH_2C\equiv CH, C\equiv C(CH_2)_2CH_3, C\equiv C(CH_2)_3CH_3, C\equiv C(CH_2)_4CH_3, C\equiv CPh; X = Cl, Br$

Furthermore the poly(dialkenylsilylene-*co*-methylenes) and poly(dialkinylsilylene-*co*-methylenes) could be synthesized by reactions with organolithium agents.

$$[Cl_2Si-CH_2]_x + 4x\ LiR^{1,2} \longrightarrow [R^{1,2}{}_2Si-CH_2-]_x + 4x\ LiCl$$

Eq. 5.

By these routes polymerizable poly(carbosilanes) could be obtained, useful as precursors, especially poly(diethenylsilylene-*co*-methylene) (**1**) and poly(diethinylsilylene-*co*-methylene) (**2**), for the production of SiC-based ceramic fibers. Green fibers spun from **1** or **2** could be rendered infusible prior to pyrolysis by treating them with daylight, UV-irradiation or a free radical generator such as dicumyl peroxide. The polymerized and stabilized fibers kept their shape when heated during the pyrolysis heating up to 1000 °C and 1700 °C. The resulting SiC-based ceramic fibers were oxygen-free. A further modification was achieved by the photochlorination of poly(dichlorosilylene-*co*-methylene) [5].

Eq. 6.
$$[Cl_2Si-CH_2]_x + x\ Cl_2 \xrightarrow{UV} [Cl_2Si-CCl_2-]_x + 2x\ HCl$$

Poly(dichlorosilylene-*co*-dichloromethylene) could be converted by reactions of organo-lithium compounds into peralkylated poly(carbosilanes) according to:

$$[Cl_2Si-CCl_2]_x + 4x\ LiR \longrightarrow [R_2Si-CR_2-]_x + 4x\ LiCl$$

Eq. 7.   R = $CH_3, C_2H_5, C_3H_7, C_4H_9, C_5H_{11}, C_6H_{13}$

Acknowledgement: We gratefully acknowledge financial support by *Deutsche Forschungsgemeinschaft* and *Fonds der Chemischen Industrie*.

References:

[1]   B. van Aefferden, W. Habel, P. Sartori, *EP* 0 375 994, **1990**; *Chem. Abstr.* **1991**, *114*, 63022b; B. van Aefferden, W. Habel, P. Sartori, *Chem. Ztg.* **1991**, *114*, 367.

[2]   P. Sartori, W. Habel, B. van Aefferden, A. M. Hurtado, H. R. Dose, Z. Alkan, *Eur. J. Solid state Inorg. Chem.* **1992**, *29*, 127.

[3]   W. Habel, L. Mayer, P. Sartori, *Chem. Ztg.* **1991**, *115*, 301.

[4]   P. Sartori, W. Habel, A. Oelschläger, *J. Organomet. Chem.* **1993**, *463*, 47.

[5]   W. Habel, W. Haeusler, P. Sartori, *J. Organomet. Chem.*, in press.

Heteropolysiloxanes by Sol-Gel Techniques: Composite Materials with Interesting Properties

Helmut K. Schmidt

Institut für Neue Materialien gem. GmbH

Im Stadtwald, Geb. 43 A, D-66123 Saarbrücken, Germany

1 Introduction

Chemical synthesis methods are not only used for the synthesis of chemical compounds, but also for materials. Examples are silicones, polymers, adhesives and ceramic powders [1]. But if one compares the potential of chemical synthesis with the number of materials produced by these routes, one has to say that the potential of chemical synthesis is only used to an extremely small extent. The focus of chemical research still is mainly directed to "new chemistry", generating new compounds. So far, sol-gel chemistry as an interesting route for material synthesis never has come to a real breakthrough in application and even is rarely used in silicon chemistry, although a large methodical overlap exists [2, 3].

The modification of silicon-type polymers by hetero atoms, for example, titanium, has been investigated by Adrian [4], but the expected increase in thermal stability could not be realized. Swelling of silicones with titanium alkoxides or silanes, as carried out by Mark [5], showed interesting effects on the mechanical properties of these systems, but did not lead to industrial materials. The use of siloxane precursors in sol-gel techniques has lead to the concept of inorganic-organic composites with a variety of interesting material developments [6–13]. The basic principles of these synthesis routes are the use of alkoxy silanes together with alkoxides or colloidal systems from oxides of other elements and the formation of inorganic molecular composite networks. An interesting feature in this connection is the question of the phase dimension of these composites.

Whilst in silicones the "inorganic" and "organic" phases are present within one molecule (intramolecular composite), the ormocer type of composite as desribed in [7, 8, 14] can be considered as a molecular level type of composite (intermolecular inorganic-organic composite). However, in this type of composite, the inorganic unit is not present as an inorganic phase, i. e., it does not maintain properties related to the inorganic solid state. The main function of the inorganic phases in the intramolecular composite is to act as a network modifier, introducing hardness or stiffness into the system, and several interesting properties resulting from these structural functions have been developed, such as hard coatings for plastics [14, 15]. It has been shown recently [16] that by controlling the synthesis parameters, it is possible to tailor the phase dimensions of the inorganic phase in a way that nanosized phases can be obtained and that this principle can be employed to obtain nanoscale metal, semiconductor and ceramic phases.

This leads to another interesting material development principle. In the present paper, several molecular composites as well as nanocomposites are presented and synthesis routes and properties will be discussed.

2 General Considerations

If one considers so-called molecular composites, one can distinguish between three basic types. As schematically indicated in Fig. 1, systems with inorganic backbones and organic groups directly linked to it, for example by covalent ($\equiv$Si–C$\equiv$) or chelate bonds, organic molecules dispersed in the inorganic backbone, for example, organic dyes dispersed in gels [17–19] or structures in which organic crosslinking units fixed to the inorganic backbone are present. It makes sense that the chemical nature of the link between the inorganic and the organic unit are of high importance for structure as well as for properties.

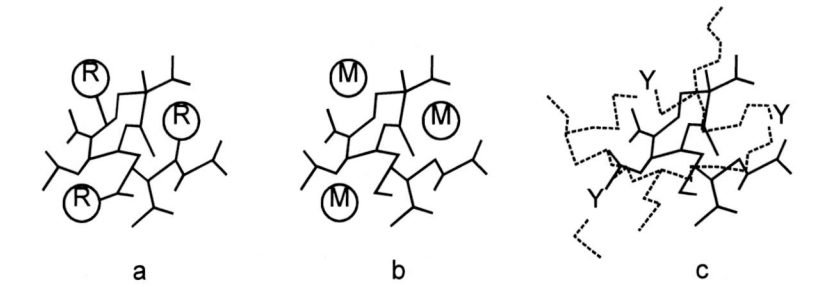

Fig. 1. Model of different types of inorganic-organic composites:
  a) organic groups modifying the inorganic backbone (R = organic or organofunctional groups like alkyls, aryl, acids, bases and others)
  b) organic molecules dispersed in inorganic networks (M = dyes, acids, bases, complexes and others)
  c) inorganic-organic composite with interpenetrating networks (Y = chemical link between the inorganic and organic backbone, for example chelates or covalent bonds); *broken line:* organic chains; *full line:* inorganic backbone)

In Fig. 2 some examples for chemical links between inorganic and organic groups are given. As one can see, there is a variety of possibilities, which, of course, could be extended.

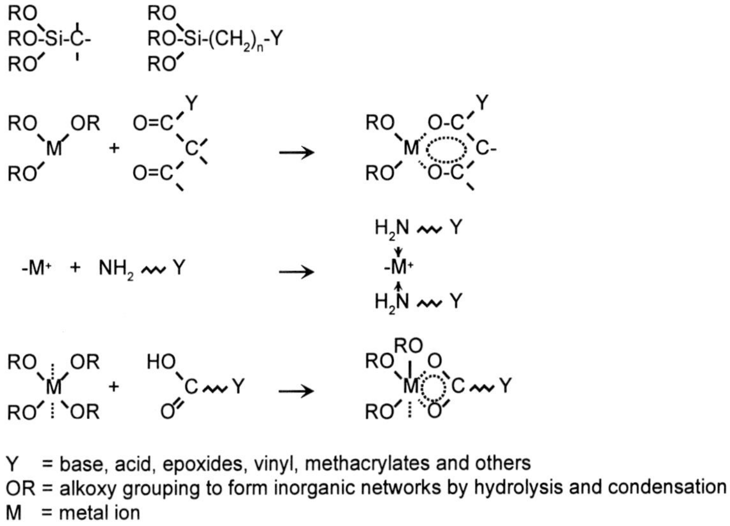

$$
\begin{array}{c}
RO \\
RO-Si-C- \\
RO'
\end{array}
\quad
\begin{array}{c}
RO \\
RO-Si-(CH_2)_n-Y \\
RO'
\end{array}
$$

$$
\begin{array}{c}
RO \\ \diagdown \\ RO'
\end{array} M \begin{array}{c} OR \\ \diagup \end{array} + \begin{array}{c} O=C \\ \diagdown \\ O=C \end{array} \begin{array}{c} Y \\ C \\ \diagup \end{array} \quad \rightarrow \quad \begin{array}{c} RO \\ \diagdown \\ RO' \end{array} M \begin{array}{c} O-C \\ \vdots \\ O-C \end{array} \begin{array}{c} Y \\ C- \\ \diagup \end{array}
$$

$$
-M^+ \; + \; NH_2 \sim Y \quad \rightarrow \quad \begin{array}{c} H_2N \sim Y \\ \downarrow \\ -M^+ \\ \uparrow \\ H_2N \sim Y \end{array}
$$

$$
\begin{array}{c} RO \\ \diagdown \\ RO' \end{array} \begin{array}{c} OR \\ M \\ \diagup \\ OR \end{array} + \begin{array}{c} HO \\ \diagdown \\ O \end{array} C \sim Y \quad \rightarrow \quad \begin{array}{c} RO \\ \diagdown \\ RO' \end{array} \begin{array}{c} O \\ M \\ \diagup \\ O \end{array} C \sim Y
$$

Y = base, acid, epoxides, vinyl, methacrylates and others
OR = alkoxy grouping to form inorganic networks by hydrolysis and condensation
M = metal ion

Fig. 2. Some examples for chemical links between inorganic and organic units

The given examples are well-known from chemistry, [1, 20] they are well investigated, but not used as principal routes for material synthesis. In the following, some examples are given how these principles can be used for making materials.

3 Composites

3.1 Molecular Composites

3.1.1 Transparent Coatings with a Tailored Index of Refraction

One of the advantages of the molecular type of composite is the high optical transparency of these materials. For this reason, they have been successfully applied as hard coatings for eye glass lenses from CR[39] as polymeric material [21]. Hard coating systems have been prepared from epoxy group containing silanes together with titanium alkoxide, leading to a refractive index between 1.52 and 1.55, depending on the amount of titania added to the system. The titania acts as a condensation catalyst, leading to a very dense, hard network, but can cause damage by UV light on extensive exposure due to its photocatalytic activity. Substitution of titania by alumina, which has been described elsewhere [22], leads to a coating system with similar mechanical properties and, if aluminum oxide nanoparticles are added, to properties superior to the cited system. However, the index of refraction is decreased substantially, which leads to problems due to interference if used as a hard coating for polymer substrates with a higher index of refraction, for example, 1.6, which is at present introduced into the market. For this reason, investigations have been carried out, the experimental details of which are described elsewhere [23]. The basic idea of the system to be developed was to use epoxy silanes as a basic network-forming unit and to substitute TiO_2 as a high refractive index component by organic groups such as phenyl or sulfide groups. In Fig. 3 the basic function of the used epoxy silane is shown. It can be polymerized by base catalysis to form a polyethylene oxide network. In addition to this, the alkoxide groups can be used for the formation of the inorganic backbone. Moreover, the epoxide group can also react with activated OH groups, for example, with phenols. If these phenols are used, an additional type of crosslinking between the epoxide units can be carried out.

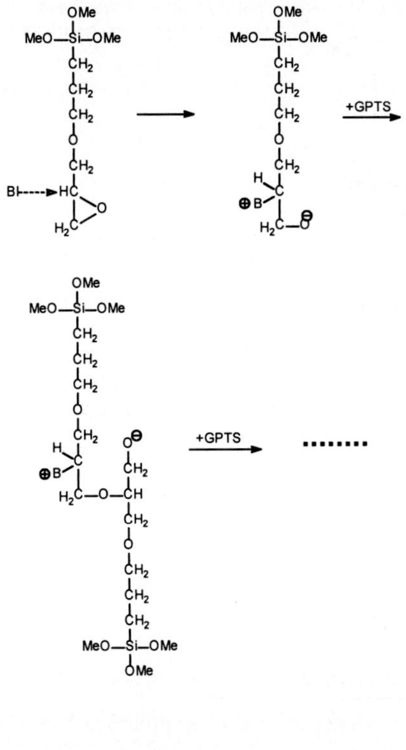

Fig. 3. Basic reaction of glycidyloxypropyl-trimethoxysilane (GPTS) to form polyethylene oxide chains

In this case, also bases like methyl imidazole can be used. In Fig. 4 a structural model of the reaction of GPTS with bisphenol A to form an inorganic-organic composite structure is shown.

Fig. 4. Model of a structure of a composite formed from GPTS bisphenol A and using methyl imidazole as a catalyst; R = glycidyloxypropyl group

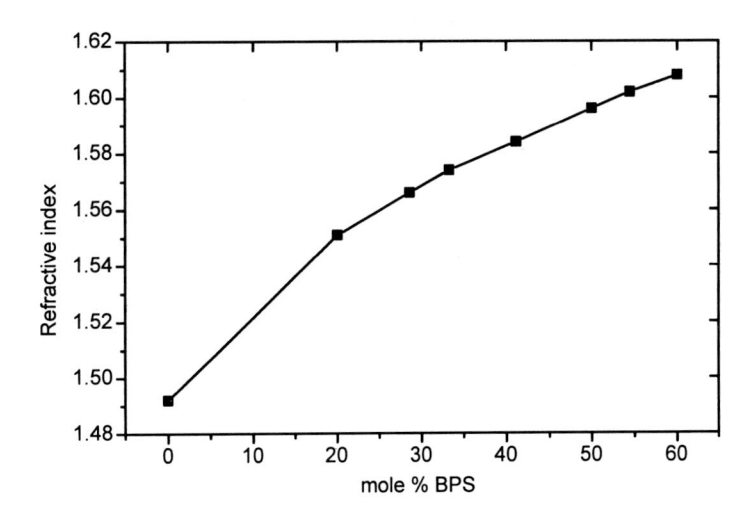

Fig. 5. Dependence of the refractive index of a bisphenol S composite on the BPS content, ref. [23]

In the same way, bisphenol S (HOC$_6$H$_4$SO$_2$C$_6$H$_4$OH, BPS) can be used as a crosslinking agent. Systematic investigations have been carried out to study the influence of BPS on the refractive index on the above described composites. The result is shown in Fig. 5.

As one can easily see, the refractive index can be varied from 1.49 up to 1.65 just by the variation of the percentage of BPS. The coating systems are prepared by hydrolyzing GPTS under acidic conditions to a viscous liquid and then adding the bisphenol including the catalyst. For the testing of the materials with respect to mechanical properties and abrasion resistance, a taber abrader test according to DIN 52347 (100 cycles, 500 g load) was carried out, and the haze was measured (% loss of transmission). In addition to this, a scratch test using a Vickers diamond scratching over the surface was used. The load in g causing the first visible scratch is registered as scratch resistance. Since the two-component system GPTS/BPS shows a rather poor performance with respect to abrasion and scratch resistance, tetraethyl orthosilicate and propyl triethoxy silane were used, acting as three-dimensional crosslinking agents. The results are shown in Fig. 6.

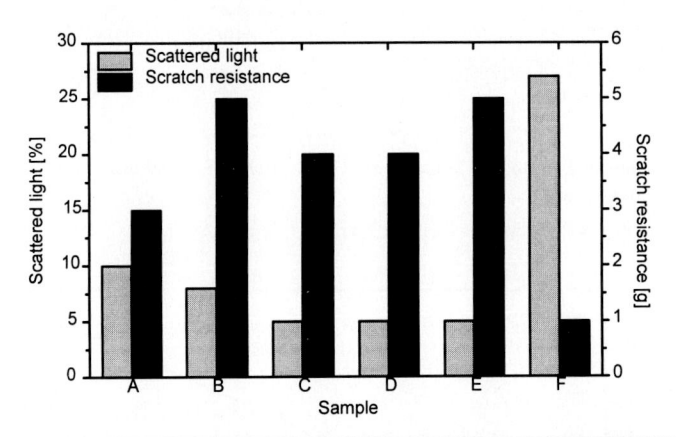

Sample	GPTS	BPS	TEOS	PTEOS
A	1	0.2	0	0
B	1	0.4	0	0
C	1	0.5	0	0
D	1	0.4	0.2	0
E	1	0.4	0	0.2
F	uncoated polycarbonate			

Fig. 6. Composition and scratch and abrasion resistance of various compositions

As one can clearly see, the composition E shows the best performance: low scattered light as a measure of the low abrasion and high scratch resistance at the same time. The refractive index of 1.56 of this composition is sufficient not to show disturbing interferences on CR^{39} as coating.

3.1.2 Molecular Composites with Tailored Adhesive Properties

As shown in the previous chapters, the chemical synthesis principles allow the introduction of a variety of functional groups into the composite materials and, by controlling the dimensionality of the network crosslinking tailoring of mechanical and thermal properties, too. As shown elsewhere [24], thermoplastic systems have been developed by using diphenyl silane diols, vinyl methyl silanes and EOS as starting materials. Systems containing more than 25 mol% of the diphenyl silane and more than 50 mol% of methyl vinyl silane show thermoplasticity and are stable against temperatures up to 250 °C, still having remarkable OH group contents after several hours of curing at these temperatures, but only if acids are used as catalysts. An infrared spectrum of the system containing 27.5 % of the phenyl (Ph) component, 70 % of the methyl vinyl (mevi) component and 2.5 % of the SiO_2 component is shown (Fig. 7).

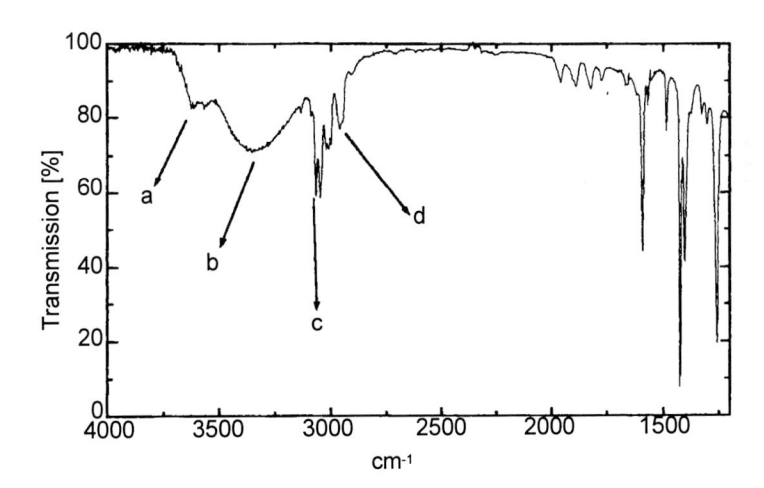

Fig. 7. IR spectrum of a condensate: 70 % mevi, 27.5 % Ph, 2.5 % SiO_2
a & b) free and bridged SiOH groups
c) aromatic and
d) aliphatic CH groups, ref. [25]

Investigations [25] have shown that these systems show good adhesion as a hot melt to all types of surfaces (glass, metals), if the free and bridged OH groups do not decrease below a certain level, determined as the ratio between the OH and the phenyl concentration, determined from the infrared peaks a–c (Fig. 8).

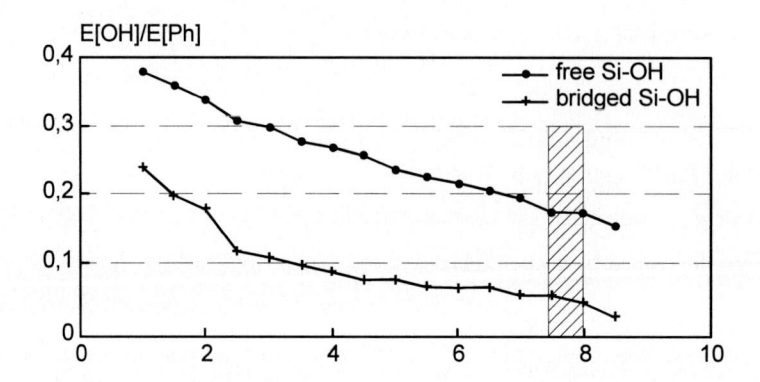

Fig. 8. Dependence of the OH group contents (free and bridged) as a function of curing time at 230 °C; hatched area: optimized adhesion, ref. [25]

It could be shown by the adhesion strength experiments that the best results are obtained if the ratio of free SiOH:phCH = 0.18 and bridged SiOH:phCH is less than 0.05. Higher and lower values lead to a decrease of adhesion, however, the overall adhesion is only in the range of about 3–3.5 N cm^{-1}, determined as peel strength. Investigations of the fracture mechanism show that the seal shows a brittle fracture behavior, as indicated in Fig. 9.

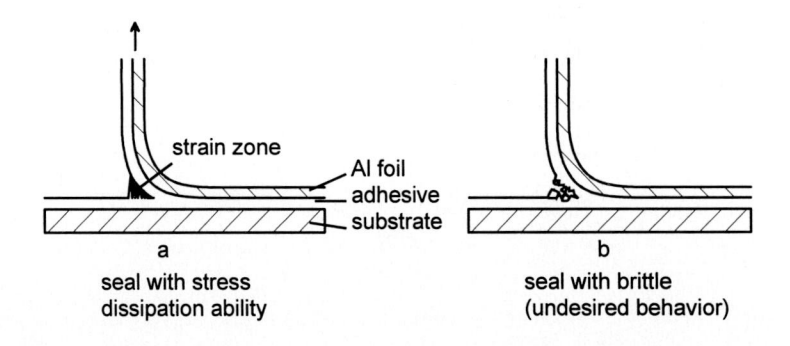

Fig. 9. Schematics of a peel strength experiment indicating good and undesired (brittle) behavior

It was found that after the curing time, three-dimensional crosslinking of the system has been increased due to a benzene elimination mechanism, as shown in Eq. 1.

$$\text{Si-OH} + \text{Ph-Si-} \longrightarrow \text{-Si-O-Si-} + \text{PhH} \uparrow$$

Eq. 1.

This could be proved by solid-state NMR analysis, showing that the T3 content increases with curing time at the expense of the diphenyl T1 content (index means the number of oxygen bridges to other silicon atoms of the considered silicon atom). It was of interest to investigate how far the stress dissipation ability of the seal can be improved without reducing the sealing properties. Moreover, it was of further interest for practical reasons how far these systems are able to be adapted to polyimide surfaces in connection with copper as a counterpart.

Preliminary tests of using this system as a seal between copper and polyimide showed that a seal strength up to 4.2 N cm^{-1} could be obtained with systems containing 70 mol% phenyl silane, 27.5 mol% methyl vinyl silane and 2.5 mol% EOS. Fracture investigations show that a cohesion rupture is also dominating in this case, confirming the above mentioned result that the brittleness of the seal is too high.

Several routes were investigated to improve the behavior. One route was to incorporate additional network modifiers in order to reduce the brittleness. For this reason, an additional silane (amino group containing silane, for example γ-aminopropyl triethoxy silane or γ-aminopropyl methyldiethoxy silane, AMDES) were introduced as a crosslinking agent for diepoxides acting as "flexible" chains between the highly crosslinked composite network.

At the same time it was assumed that interaction of the amino groups of the silane should act as additional adhesion promotor to the polyimide surface. The use of the γ-aminopropyl methyl triethoxy silane together with diepoxides (Araldite® GY 266) did not improve the brittleness remarkably. The use of AMDES, however, should lead to less brittle systems due to the only two-dimensional crosslinking ability of the diethoxy silane. The basic structure formation features are given in Fig. 10.

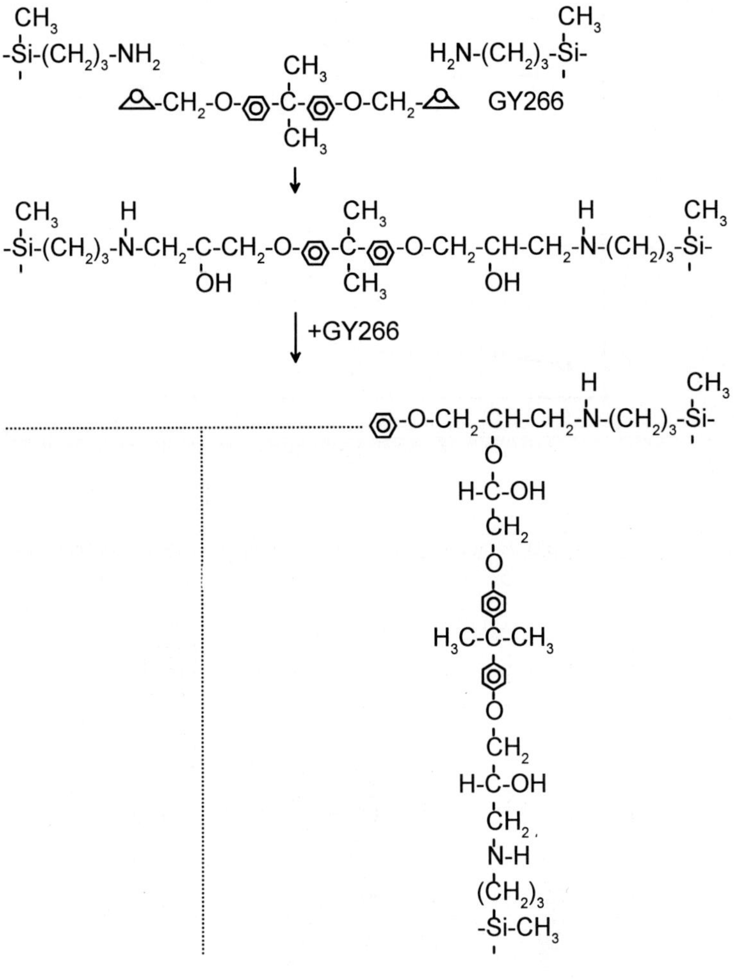

Fig. 10. Reaction model of amino silanes with diepoxides

In order to improve the condensation of the silane to the backbone of the basic composition, the curing time of the system was reduced in order to increase the OH group content. It could be shown that with ratios of the free SiOH:phCH of about 3.6 and of the bridged SiOH:phCH of 0.6 a crosslinking of the amino silane to the phenyl groups was possible. This was proven by ^{29}Si NMR spectroscopy [25]. If the amine was used in the ratio of 1:1 of the diepoxides (molar ratio), the peel strength was even decreased (2.5 N cm^{-1}), but when the ratio amine to epoxide was changed to 1:2, the maximum peel strength obtained as a function of the total content of the amine was above 10. The optimum seal strength was obtained by adding about 30 wt. % of the amine to the basic system. The results are shown in Fig. 11.

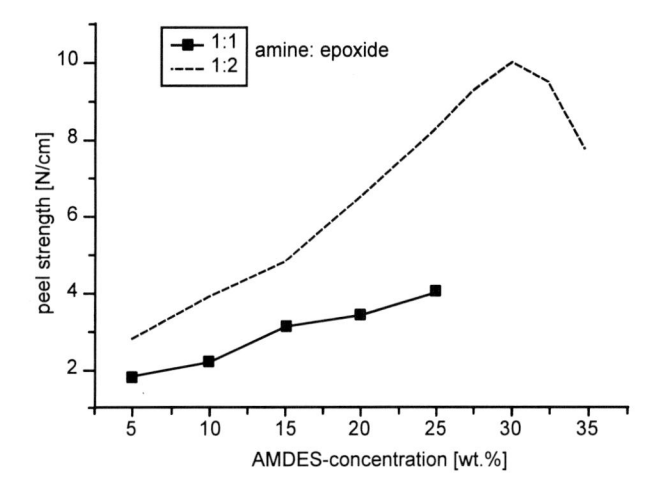

Fig. 11. Dependence of the peel strength on the amine content of the seal using two different amine-to-epoxide ratios

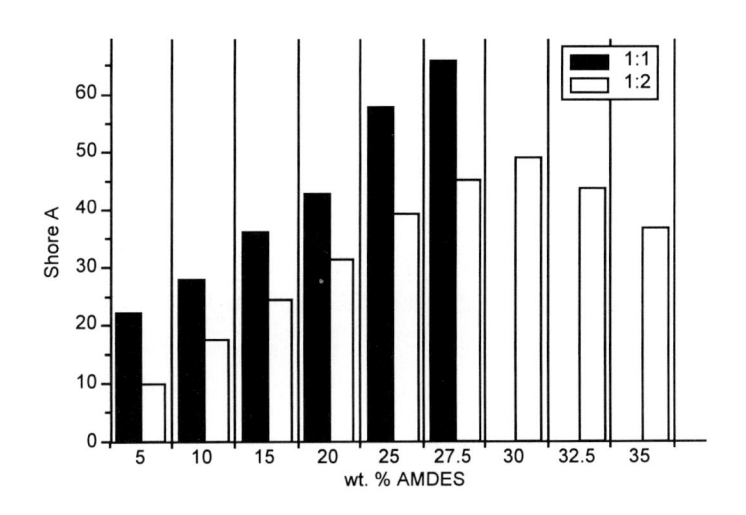

Fig. 12. Shore A hardness of composites with different AMDES/GY 266 contents

The investigation of the shore hardness of the different systems shows that the shore increases with the amino content, but the shore hardness of the 1:2 composition is still below that of the 1:1 composition. It is interesting that the shore hardness shows the same tendency as the adhesive strength (Fig. 12). An interpretation is that with increasing amount of amino groups the adhesive strength to copper as well as polyimide is increased due to the better adhesion to the different surfaces, but at the same time the shore hardness has to be kept low enough not to prevent stress dissipation. With too high amino contents the shore hardness decreases, but now the mechanical strength of the seal decreases, too (Fig. 12). The lower shore hardness of the 1:2 system indicates a better stress dissipation ability than the 1:1 system. The seal strength of 10 N cm^{-1} is an excellent value compared to the state of art (3–4 with silicones used in application).

The electrical properties of the system are shown in Table 1.

System	Resistivity ρ [Ω cm^{-1}]	Breakthrough Voltage D [V cm^{-1}]	DK 1 kHz	tan σ 1 kHz – 1 Mhz
17	1.1×10^9	10^5	3.9	2.8×10^{-2}

Table 1. Electrical properties of the optimized system

The values are within the limits for electronic application. The seal is thermally stable up to 180 °C. The investigations show that by tailoring chemical structures of sol-gel derived organically modified hetero polysiloxanes, mechanical and adhesive structures can be tailored for needs of application.

3.2 Nanocomposites

3.2.1 General Considerations

In the previous sections, two examples of the synthesis and material development based on molecular composite systems have been described. However, in these systems, due to the absence of inorganic extended networks (amorphous or crystalline), solid-state properties attributed to these extended networks cannot be introduced. These properties, however, are of interest for a variety of reasons, for example, mechanical properties, optical properties (linear and non-linear properties) or quantum size effects, if the phase dimension of these inorganic units can be kept in the nano range. Keeping the phase dimension in the nano range makes these composites interesting for optical purposes since Rayleigh scattering can be neglected. The combination of the nanophases with a molecular composite matrix as described above leads to a new type of nanocomposite based on hetero polysiloxanes. In Fig. 13, an overview over the potential of this type of nanocomposite materials is shown. The transparency T is a function of the coefficient of extinction in which Rayleigh scattering is included, and in this case the coefficient depends on the third power of the particle size. The interesting part is that passive as well as active functions can be generated by small particles in composites.

$T = \exp[(-\gamma_{ext} + \gamma_o)d]$

$\gamma_{ext} = f(c_v, \Delta n, R^3, \lambda^{-4})$

c_v = filler volume fraction,
Δn = difference of refractive index,
R = particle diameter,
λ = wavelength

large particle composites   nano composites

○ passive "functions": refractive index (n_E)
mechanical properties
interface properties
absorptive properties

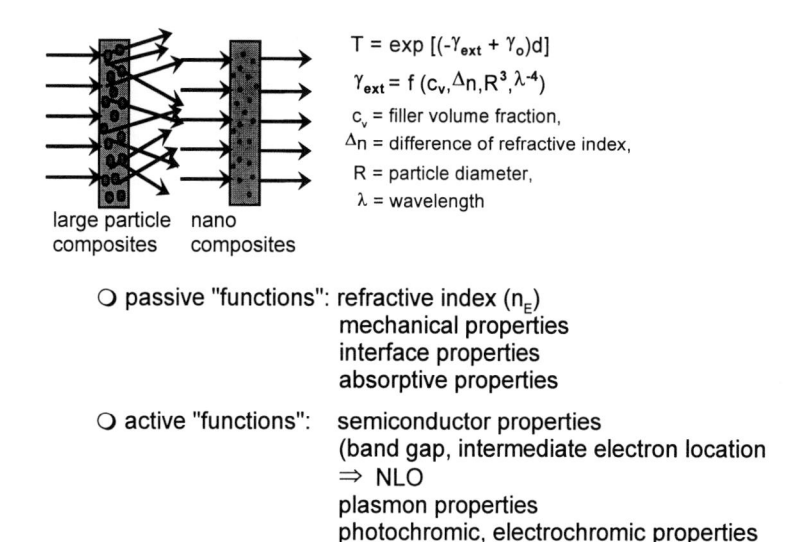

○ active "functions": semiconductor properties
(band gap, intermediate electron location
⇒ NLO
plasmon properties
photochromic, electrochromic properties

Fig. 13. Overview over some properties to be derived from inorganic-organic hetero polysiloxane nanocomposites and potential applications

For the synthesis of the nanocomposites, additionally to the reaction principles described above, nanoscale particles have to be generated within or added to the hetero polysiloxane composite matrix. As described elsewhere [26], the sol-gel process, is a suitable means. It can be considered as a growth process from solution where a controlled growth reaction takes place, that leads to a stabilized colloidal system and thus avoiding precipitation.

The sol-gel reactions have mainly been investigated in alcoholic solution, which is a reaction medium easily allows electrostatic stabilization by appropriate choice of the pH. This type of stabilization can only be used in a few cases as a means for incorporating the colloidal particle agglomerate-free into tailored matrices, since these matrices, as a rule, destroy the electrostatic "coating" around the particles. As a consequence, agglomeration takes place, and the high transparency required for optical application is lost.

For this reason, another type, the so-called short organic molecule stabilization by tailored surface modification for colloidal particles, has been developed, the principles of which in comparison to the electrostatic stabilization is schematically shown in Fig. 14.

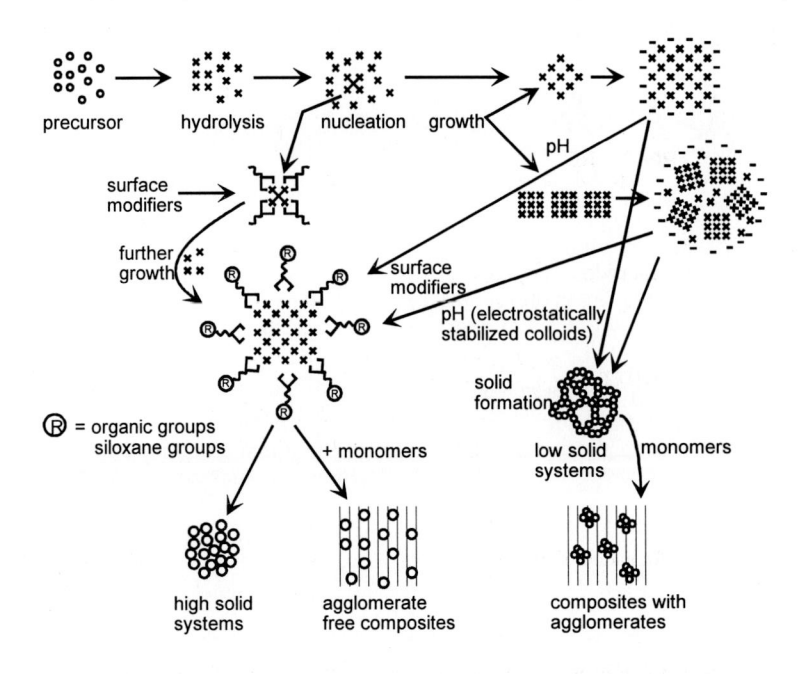

Fig. 14. Reaction paths for the stabilization and fabrication of nanocomposites using growth reactions controlled by surface modifications

The use of this route, of course, opens some interesting questions, for example, how far the surface-modifying molecules interfere with a growth reaction or how far bonds with the desired stability to the surface of the growing particles can be obtained. The interesting potential of this method is that by controlling the surface free energy of the growth process, a tailoring of particle size should be possible. Additionally to this, if it is further possible to tailor the bond strength to the surface, reactive colloids may be prepared by using so-called bifunctional molecules. In these molecules one function is used to link the molecule to the surface and the second function is used to create a desired activity like hydrolyzable and condensable groups of silanes, double bonds or epoxides for further polymerization or polyaddition reactions. These principles have been investigated in detail elsewhere [27–29].

3.2.2 ZrO$_2$ Containing Composites for Optics

Using zirconia as a model, it could be shown that if zirconium alkoxides are used as starting compounds in the presence of chelating agents like carboxylic acids [27] or β-diketones, [30] the precipitation of zirconia can be avoided through hydrolysis and condensation. In Fig. 15 the schematics of this reaction are shown.

Fig. 15. Reaction model of the formation of ZrO_2 colloids covered by methacrylic acid (ma) through a surface-controlled growth process

If, for example, methacrylic acid is used as a complexing agent, particles of about 2–5 nm in diameter can be generated by hydrolysis and condensation, and if, in addition to this, methacryloxy group containing hydrolyzable silanes are used in combination with the zirconia colloids, transparent composite systems have been synthesized.

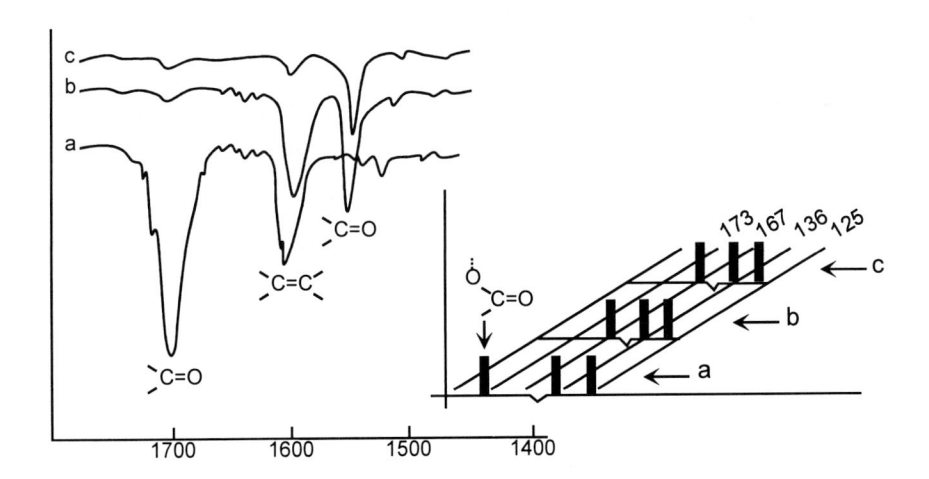

Fig. 16. Analysis of complexation and solidification of zirconia ma colloids by IR and ^{13}C NMR:
  a) methacrylic acid
  b) zirconia ma complex
  c) zirconia ma complex after hydrolysis, condensation and polymerization, before and after curing and exposure to water, according to [32]

As shown in [31], after curing at 120 °C these composites contain nanocrystalline monoclinic zirconia particles. IR and NMR analysis (as shown in Fig. 16) show that the carbon CO frequency obtained by complexation is still maintained after hydrolysis of the colloid and after polymerization of the double bonds together with methacryloxy silanes, no change in the carbonyl frequency can be observed. Even after using these systems as coatings, after two weeks exposure to water no change is observed. Similar results are also obtained by ^{13}C NMR analysis of the carboxylic carbon atom, as shown in Fig. 16.

In Fig. 17 a structural model of the composite before polymerization is given.

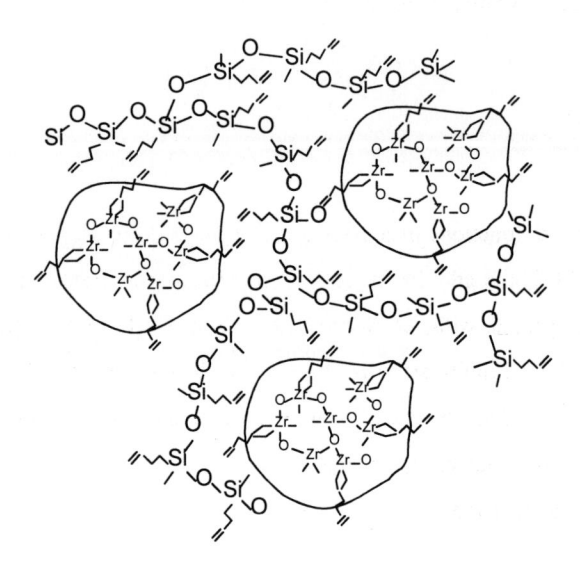

Fig. 17. Model for a colloidal organic-inorganic network interpenetrated by a siloxane network

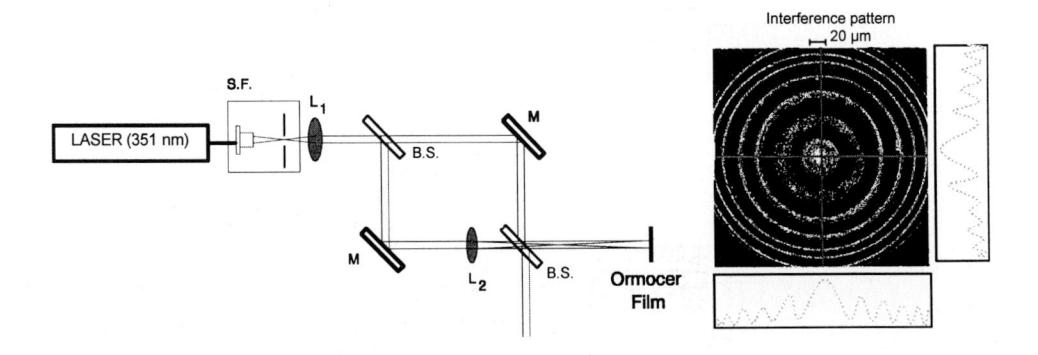

Fig. 18. Schematics of the fabrication of micro fresnel lenses by two-wave mixing, [35]

As shown by [31, 33, 34], these complexes can be polymerized to polymers by photo and thermal initiation, and at 120 °C conversions above 96 % have been obtained. By use of photoinitiators and two-wave mixing systems by a laser light source resulting in concentric interference patterns, these coatings can be used to create Fresnel lens micropatterns by partial polymerization and subsequent development by dissolving the unpolymerized parts by organic solvents. Suitable optical properties can be obtained by controlling the interference patterns and the refractive index by the zirconia content. An example of Fresnel lens micropattern fabrication is given in Fig. 18.

Using this principle, optical gratings and channel waveguides already have been developed [35].

3.2.3 Third-Function Coatings

The incorporation of other functions into these systems can be used for developing other interesting properties. As shown in [36], the addition of silanes carrying perfluorinated side chains as additional precursors leads to liquid systems with the fluorinated components homogeneously distributed through the material. As soon as these liquids are spread onto a surface in form of thin films (several μm in thickness), a phase separation takes place leading to a gradient material with the fluorinated side chains directed to the surface and the polar groups of the system directed to the substrate interface. In Fig. 19, the basic function of the system is shown [36].

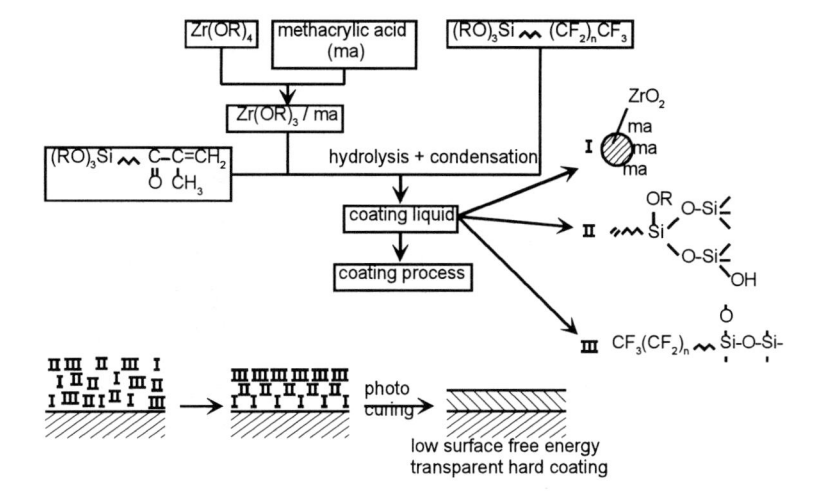

Fig. 19. Schematics of the thermodynamically driven alignment effect of the ZrO_2/perfluorinated silane composition

The alignment of the different components is driven by the reduction of the surface and interfacial free energy. This general principle can be extended to other colloidal systems as components, and by appropriate choice of composition, colloidal concentration and particle size, the gradient of the fluorinated system can be tailored in a wide range [37]. If SiO_2 is used instead of zirconia, surface-modified by methyl silanes, high-temperature systems up to 400 °C are obtained, also having the gradient effect. These systems have been developed for glass surfaces due to the low surface free energy. All these systems are dust- and soil-repellent, highly transparent and scratch-resistant compared to polymers.

In Fig. 20 the ESCA spectrum of the surface and of the bulk after two minutes argon sputtering is shown.

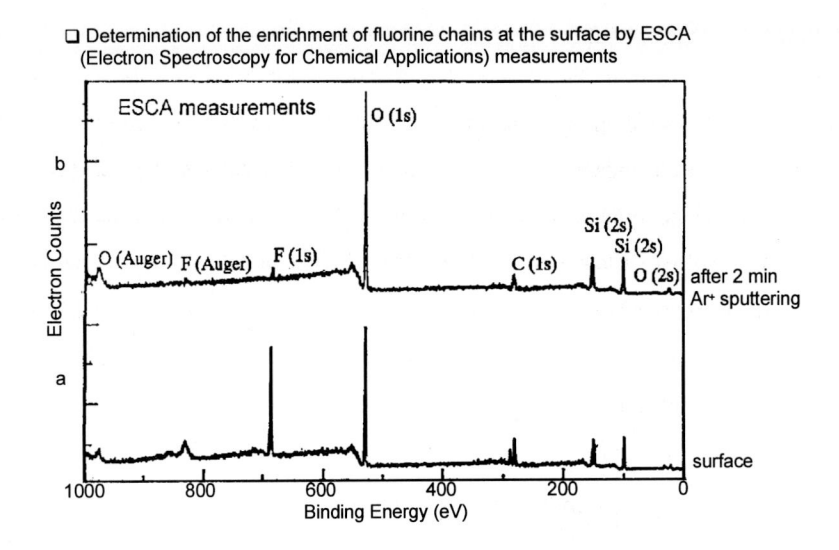

Fig. 20. ESCA spectrum of a coating of the ZrO_2/perfluorinated silane system
  a) surface
  b) bulk

As one can see, the concentration of fluorine is strongly enhanced on the surface. ESCA profiling shows that even in the ESCA machine the contamination content (adsorption of hydrocarbons) is extremely low compared with that of other samples (glasses, metals polymer surfaces). These systems already are successfully used in industry to avoid sticking to steel conveyor belts during production, for food containers as easy-to-cleaning inner coatings, for metal pipes in analytics to keep the carry-over with liquids near to zero and for the glass industry for keeping glass panes clean.

Summarizing, the in-situ growth of nanoscale particles in hetero polysiloxane type of matrix systems leads to an interesting type of nanocomposites, which is of special interest for application due to its possibility for photocuring in combination with high optical transparency (microoptical systems, waveguide systems).

The incorporation is of special interest for a variety of applications. One of them has been described in the previous sections, others are the incorporation of semiconductors [38] of metal colloids [39, 29] or silver halides [40].

3.2.4 Systems with Nanoparticles Added to a Matrix System

Another possibility mentioned above is the addition of nanoscale particles to a liquid matrix system where the nanoscale particles are grown outside of the system. Experiments have been carried out with boehmite in a matrix derived from $Si(OR)_4/Al(OR)_3$ and glycidyloxypropyl triethoxy silane (GTPS) [22]. Even the addition of 5 % by volume of γ-alumina or boehmite leads to systems which show a remarkably increased scratch-resistance compared to the unfilled material. The optical transparency is not influenced if the particle size of the boehmite is below 20 or 30 nm. In Fig. 21 the scratch resistance by the Vickers diamond test of the unfilled system is compared to the filled system and, as one can see, the scratch resistance is increased remarkably.

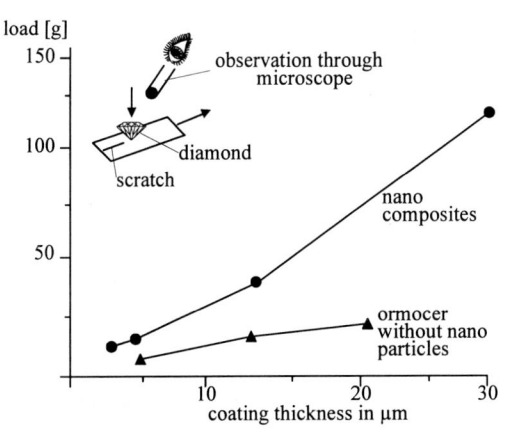

Fig. 21. Scratch resistance of the nanocomposite coating depending on the thickness in comparison to a conventional Ormocer coating, ref. [22]

This effect cannot be simply explained by the low degree of filling. Detailed investigations of the reaction mechanism of the composite formation by ^{27}Al, ^{29}Si, ^{13}C NMR spectroscopy, and NIR infrared spectroscopy show that in this system the boehmite acts as a condensation catalyst for the epoxide polymerization. At the same time, an ≡Si–O–Al= bond is formed by the epoxy silane to the boehmite particle surface, which also can be shown by NMR spectroscopy (Fig. 22).

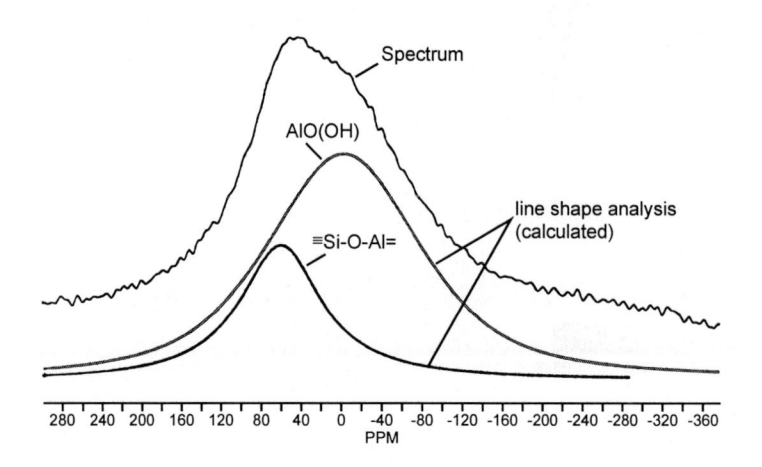

Fig. 22. ^{27}Al NMR spectrum of the ≡Si–O–Al= bond formation in the GPTS/- SiO$_2$/Al$_2$O$_3$/boehmite nanocomposite

For the described reason, a special structure is proposed, which is schematically shown in Fig. 23. In this figure, the structure is compared with a filled polymer structure (SiO$_2$ filled polysiloxanes, very often used as hard coatings for eye glass lenses). The figure also depicts the results of a so-called tumble test.
In this test the coated eye glass lens is tumbled in a tumbler filled with a special composition of abrasive materials. As one can see from the figure, the haze produced from coatings with a new nanocomposite is distinctively lower than that from other materials in use. This behavior is attributed to the special structures. NMR and IR data suggest that poly(ethylene oxide) chains are concentrated around the aluminum oxide particles, providing a flexible suspension of these particles in the stiff network and thus leading to a mechanical behavior showing a specific hardness on the one hand, but due to the suspension, the ability of not being broken out so easily by mechanical impact on the other hand. These systems have also been further developed to be used on other polycarbonates very successfully.

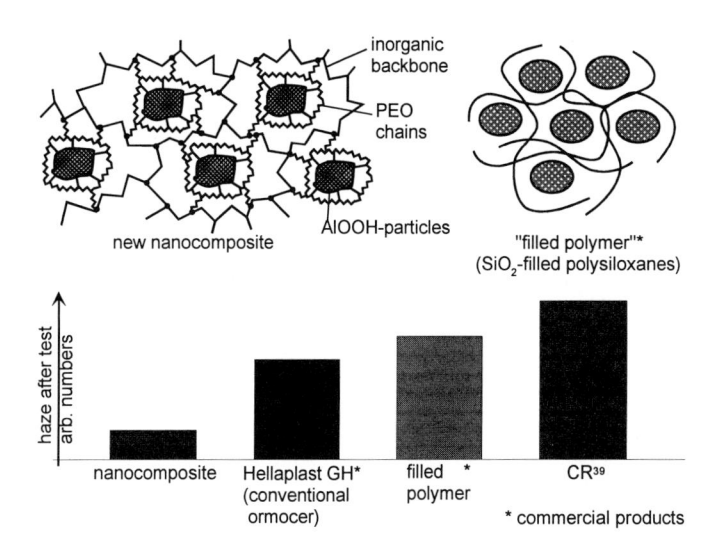

Fig. 23. Comparison of several coatings used on CR^{39} eye glass lenses by the tumble test and structural model of the nanocomposite coating

These few examples show how the use of nanocomposite systems with hetero polysiloxane type of matrices leads to interesting properties for applications. Further developments using these basic systems are transparent controlled release coatings for anti-fogging systems [41], anti-corrosive systems for metal protection [42], and nanocomposite optical bulk materials [43].

4 Conclusions

The development of materials based on hetero polysiloxanes in combination with or without nanoparticles is an interesting route to new materials with interesting aspects and applications. These materials, especially in the initial phase, cannot play the role as mass commodity materials to be produced in large quantities by chemical industry. This might be one of the reasons that the penetration of this type of material into industrial application is still at its infancy, but anyway, for creating special functions or special innovation in the field of materials users, this type of materials has already proven its usefulness for practice.

References:

[1] W. Noll, *Chemie und Technologie der Silicone*, Verlag Chemie, Weinheim, **1968**.

[2] J. D. Mackenzie, Y. J. Chung, Y. Hu, *J. Non-Cryst. Solids* **1992**, *147&148*, 271.

[3] J. D. Mackenzie, in: *Mat. Res. Soc. Symp. Proc.*, *Better Ceramics Through Chemistry II* (Eds.: C. J. Brinker, D. E. Clark, D. R. Ulrich), MRS, Pittsburgh, **1986**, p. 809.

[4] K. A. Adrianov, A. A. Zhdanov, *J. Polym. Sci.* **1958**, *32*, 513.

[5] L. Garrido, J. L. Ackerman, J. E. Mark, *Mat. Res. Soc. Symp. Proc.* **1990**, *171*, 65.

[6] H. Schmidt, H. Krug, in: *ACS Symposium Series*, *Vol. 572, Inorganic and Organometallic Polymers II: Advanced Materials and Intermediates* (Eds.: P. Wisian-Neilson, H. R. Allcok, K. J. Wynne), American Chemical Society, Washington, **1994**, p. 183.

[7] H. Schmidt, *J. Non-Cryst. Solids* **1994**, *178*, 302.

[8] H. Schmidt, in: *Sol-Gel Optics – Processing and Applications* (Ed.: L. Klein), Kluwer Academic Publishers, Boston/Dordrecht/London, **1994**, p. 451.

[9] H. Schmidt, in: *Ultrastructure Processing of Advanced Materials* (Eds.: D. R. Uhlmann, D. R. Ulrich), Wiley, New York, **1992**, p. 409.

[10] D. R. Uhlmann, G. P. Rajendran, in: *Ultrastructure Processing of Advanced Ceramics* (Eds.: J. D. Mackenzie, D. R. Ulrich), Wiley, New York, **1988**, p. 241.

[11] J. E. Mark, in: *Ultrastructure Processing of Advanced Ceramics* (Eds.: J. D. Mackenzie, D. R. Ulrich), Wiley, New York, **1988**, p. 623.

[12] G. L. Wilkes, B. Orler, H.-H. Huang, *Polym. Prepr.* **1985**, *26*, 300.

[13] C.-Y. Li, J. Y. Tseng, K. Morita, C. L. Lechner, Y. Hu, J. D. Mackenzie, in: *SPIE Sol–Gel Optics II*, *Vol. 1758*, **1992**, p. 410.

[14] H. Schmidt, B. Seiferling, in: *Mat. Res. Soc. Symp. Proc.*, *Vol. 73* (Ed.: Materials Research Society), **1986**, p. 739.

[15] R. Kasemann, H. Schmidt, *New Journal of Chemistry* **1994**, *18*, 1117.

[16] H. Schmidt, *Journal of Sol-Gel Science and Technology* **1994**, *1*, 217.

[17] J. McKiernan, J. I. Zink, B. S. Dunn, *SPIE Sol-Gel Optics II* **1992**, *1758*, 381.

[18] R. Reisfeld, in: *Sol-Gel Science and Technology* (Eds.: M. A. Aegerter, M. Jafelici Jr., D. F. Souza, E. D. Zanotto), World Scientific, Singapore, **1989**, 322.

[19] A. Makishima, K. Morita, H. Inoue, M. Uo, T. Hayakawa, M. Ikemoto, K. Horie, T. Tani, Y. Sakakibara, *SPIE Sol-Gel Optics II* **1992**, *1758*, 492.

[20] D. C. Bradley, R. G. Mehrotra, D. P. Gaur, in: *Metal Alkoxides*, Academic Press, London, **1978**.

[21] H. Schmidt, B. Seiferling, G. Philipp, K. Deichmann, in: *Ultrastructure Processing of Advanced Ceramics* (Eds.: J. D. Mackenzie, D. R. Ulrich), Wiley, New York, **1988**, p. 651.

[22] R. Kasemann, H. Schmidt, E. Wintrich, in: *Proceedings of MRS Spring Meeting*, San Francisco, Mat. Res. Soc. Symp. Proc. *346*, April **1994**, in print.

[23] V. Gerhard, H. Schirra, G. Wagner, H. Schmidt, in: *Extended Abstract of the Fourth Saar-Lor-Lux Meeting on Functional Advanced Materials* (Ed.: G. Kugel, CLOES-SUPELEC, Technopole de Metz/France), November **1994**.

[24] H. Schmidt, H. Scholze, G. Tünker, *J. Non-Cryst. Solids* **1986**, *80*, 557.

[25] T. Burkhart, *Ph. D. Thesis*, Universität des Saarlandes, Saarbrücken, **1993**.

[26] C. J. Brinker, G. W. Scherer, *Sol-Gel Science*, Academic Press Inc., Boston, **1990**.

[27] H. Schmidt, H. Krug, R. Kasemann, F. Tiefensee, *Development of Optical Waveguides by Sol-Gel Techniques for Laser Patterning*, in: *SPIE Proc.* **1991**, *1590*, 36.

[28] H. Schmidt, *Mat. Res. Soc. Symp. Proc.* **1984**, *32*, 327.

[29] M. Mennig, M. Schmitt, U. Becker, G. Jung, H. Schmidt, in: *SPIE Proceedings "Sol-Gel Optics III"*, *2288* (Ed.: J. D. Mackenzie), San Diego/USA, July **1994**, in print.

[30] J. Livage, M. Henry, C. Sanchez, *Progress in Solid State Chemistry* **1988**, *18*, 259.

[31] F. Tiefensee, *Ph. D. Thesis*, Universität des Saarlandes, Saarbrücken, **1993**.

[32] V. Gerhard, private communication.

[33] H. Krug, P. W. Oliveira, H. Schmidt, in: *Proc. 1994 Pacific Coast Regional Meeting* (Ed.: American Ceramic Society), October **1994**, in print.

[34] C. Becker, M. Zahnhausen, H. Krug, H. Schmidt, in: *Extended Abstracts of the Fourth Saar-Lor-Lux Meeting on Functional Advanced Materials* (Ed.: G. Kugel, CLOES-SUPELEC, Technopole de Metz/France), November **1994**.

[35] P. W. Oliveira, H. Krug, H. Schmidt, in: *Extended Abstracts of the Fourth Saar-Lor-Lux Meeting on Functional Advanced Materials* (Ed.: G. Kugel, CLOES-SUPELEC, Technopole de Metz/France), November **1994**.

[36] R. Kasemann, H. Schmidt, Oral presentation at "Fluorine in Coatings" Conference, Salford, Great Britain, September **1994**.

[37] S. Sepeur, *Master's Thesis*, Universität des Saarlandes, Saarbrücken, **1994**.

[38] L. Spanhel, M. Mennig, H. Schmidt, *Bol. Soc. Esp. Ceram. Vid. 31-C* **1992**, 7, 9.

[39] T. Burkhart, M. Mennig, H. Schmidt, A. Licciulli, in: *Proceedings 1994 MRS Spring Meeting, Symposium "Better Ceramics Through Chemistry VI"*, San Francisco/USA, Mat. Res. Soc. Symp. Proc. *346*, April **1994**, in print.

[40] M. Mennig, C. Fink-Straube, H. Schmidt, in: *Proc. 1994 Pacific Coast Regional Meeting* (Ed.: American Ceramic Society), October **1994**,in print).

[41] R. Kasemann, H. Schmidt, S. Brück, *Bol. Soc. Esp. Ceram. Vid. 31-C* **1992**, 7, 75.

[42] H. Schmidt, in: *Submicron Multiphase Materials*, *Vol. 274* (Eds.: R. H. Baney, L. R. Gilliom, S.-I. Hirano, H. Schmidt), Mat. Res. Soc. Symp. Proc., **1992**, p. 121.

[43] H. Krug, H. Schmidt, R. Naß, L. Spanhel, *Optische Elemente und Verfahren zu deren Herstellung*, Deutsche Offenlegungsschrift DE – 05 41 30 550 A1, 13.09.**1991**.

Fumed Silica – Production, Properties, and Applications

H. Barthel*, L. Rösch, J. Weis

Wacker-Chemie GmbH

Geschäftsbereich S – Werk Burghausen

Johannes-Heß-Str. 24, D-84489 Burghausen, Germany

Summary: Since its first production in the early forties of this century, fumed silica has found widespread use in industrial applications. Due to its pyrogenic manufacturing process by combustion of silicon tetrachloride in an oxygen-hydrogen flame, fumed silica offers a variety of fascinating properties. It consists of finely dispersed amorphous silicon dioxide and its surface is covered by highly reactive silanol groups which are available for chemical reactions. Additionally, fumed silica shows a space-filling particle structure related to a high surface area and lacking micropores. These characteristics enable fumed silica to act as a free flow additive in powder-like solids, as a thickener in various liquids and as a powerful reinforcing filler in elastomers. The aim of this paper is to combine results from surface chemistry and particle characterization in order to understand the basic mechanisms of fumed silica applications. Recent investigations in the field of surface characterization and interactions are reported.

Introduction

Fumed silica is a highly dispersed synthetic silicon dioxide product. Finely divided silicas may be classified according to their manufacturing process: natural products, byproducts and synthetic products. The product origin implies distinct differences of the properties of these silicas. Natural products as quartz powders or diatomaceous earth, and by-products that include, e.g. fused silica, silica fume, or fly ashes from metallurgy and power plants are mostly cristalline silica products of micron size particles or larger and have surface areas up to 1 $m^2 g^{-1}$, rarely 10 $m^2 g^{-1}$. In contrast, synthetic silicas from wet processes or thermal pyrogenic reactions are typically amorphous silica products with a high surface area of 100 $m^2 g^{-1}$ and larger. Wet processes, that are based on the reaction of soluble silicates with aqueous acids, lead to silica-gels or precipitated silica. By far the most important thermal pyrogenic path is the flame pyrolysis of silanes, which gives access to the product of interest – fumed silica. To better understand the properties and applications of this fascinating product, a detailed discussion of the production process will help.

Production

Fumed silica is produced by burning volatile silanes, such as silicon tetrachloride, in an oxygen-hydrogen flame. Ulrich [1] gave a description of the flame process based on the immediate formation of protoparticles directly related to the chemical reaction – rather than surface deposition. At high flame temperatures collision and coalescence of protoparticles lead to the formation of primary particles. The rate of coalescence depends on the viscosity of the molten oxide, which is exceedingly high for silicon dioxide at a flame temperature of about 1500 K. Therefore, the size d of the fumed silica primary particles, i.e., the surface area SA = $6/d*r_{silica}$ is strongly related to the flame temperature. At lower temperature, collision and sticking of primary particles only results in partial fusion and stable particle aggregates are formed. The silica aggregates leave the flame and cool, but they still collide. As their surfaces are now solid, agglomerates of aggregates are formed, that are held together by physico-chemical surface interactions (Fig.1).

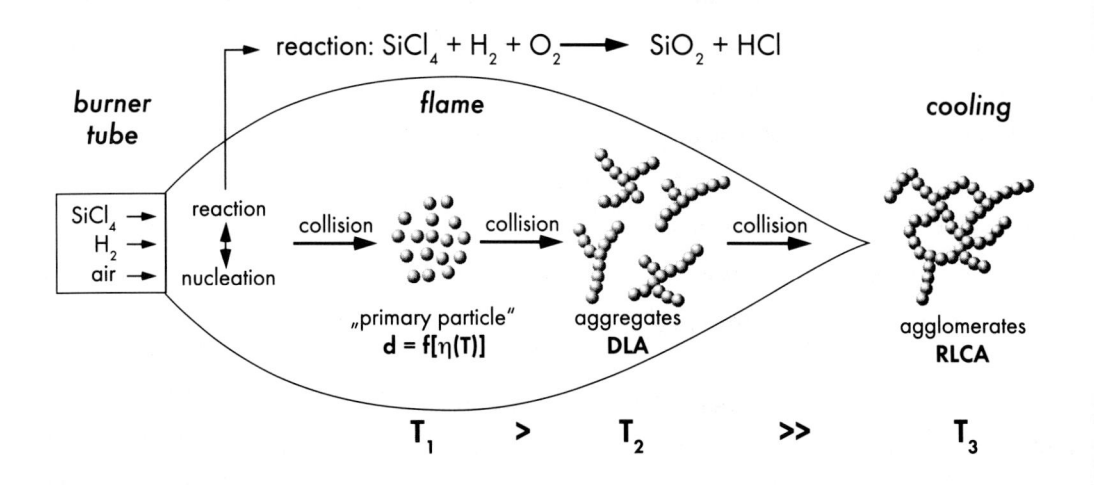

Fig. 1. Production of fumed silica in a flame process

A very helpful approach to understand the formation of particle clusters was given by Meakin [2], who uses the concept of mass fractal dimension D_m which relates the size d of a particle cluster to its mass M (Eq.1).

$$M \approx d^{Dm} \text{ with } 1 < D_m < 3$$

Eq. 1.

At high temperature the tacky primary particles adhere to each other to form aggregates at the moment they collide, i.e., the sticking coefficent is near 1. Thus, the formation of fumed silica aggregates in the flame is dominated by diffusion limited aggregation of primary particles (DLA), which gives an aggregate mass fractal dimension of $D_m^{aggregate} = 2.5$. Values close to this result from small angle neutron scattering (SANS), $D_m = 2.6$ [3], and small angle X-ray scattering (SAXS) or nitrogen adsorption at 78 K, with D_m from 2.6 to 2.7 [4]. At lower temperature the agglomeration of aggregates follows a reaction limited cluster aggregation (RLCA), as the aggregates may collide, and they will also rearrange their positions in order to optimize surface interactions. RLCA provides an agglomerate mass fractal dimension of $D_m^{agglomerate} = 2.1$, taking into account polydispersity of aggregate sizes.

Properties

Particle Structure

Fumed silica appears as a fluffy white powder characterized by an extremly low bulk density down to the range of about 20–50 g l^{-1}. In contrast, the submicron fumed silica particle consists of amorphous silicon dioxide and, hence, its true density is about 2200 g l^{-1}.

Any discussion of fumed silica particle structure has to take into account this enormous difference. The approach of mass fractal dimension may provide a rough but helpful estimation. A real mass fractal is limited by the size of the cluster as an upper limit and the size of the particles as a lower limit. Then, the density of the cluster $\rho_{cluster}$ may be calculated from the true density of the particle $\rho_{particle}$, the ratio of the cluster size $d_{cluster}$ to the particle size $d_{particle}$ and the mass fractal dimension D_m of the cluster (Eq. 2):

$$\rho_{cluster} = \rho_{particle} \cdot (d_{cluster} / d_{particle})^{Dm-3}$$

Eq. 2.

By electron microscopy the size of the primary particles in the aggregates is estimated to be about 10 nm. Particle size measurements using a nanosizer show the size of aggregates dispersed in a well wetting solvent to be in the range of 100 nm. Laser diffraction of fumed silica dispersed in air provides sizes of agglomerates larger than 5 μm.

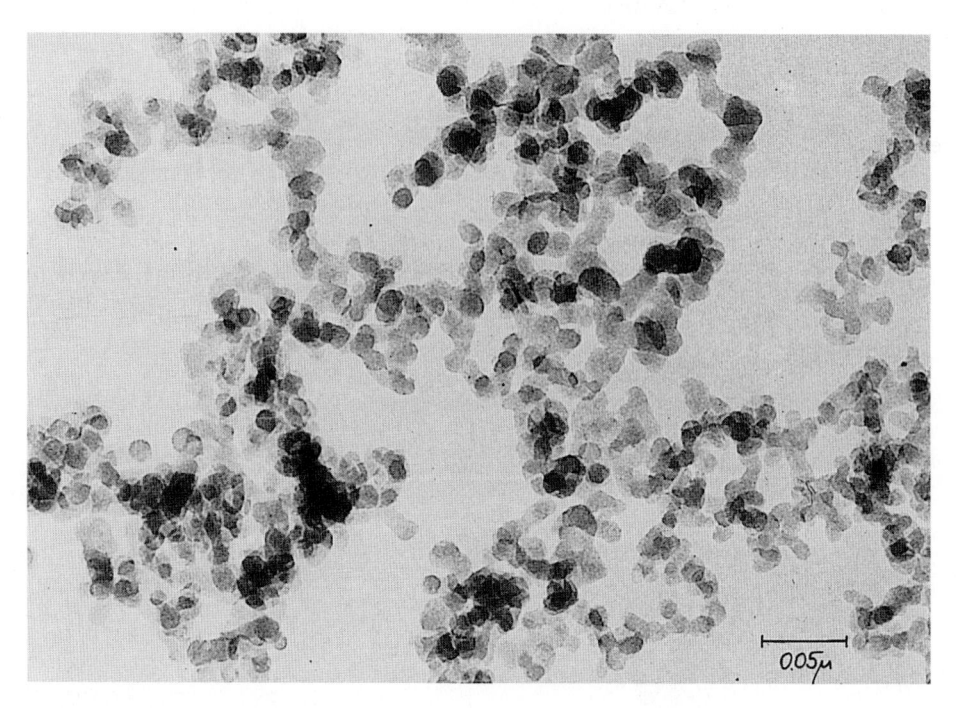

Fig. 2. Fumed silica (surface area 200 m^2 g^{-1}) – Transmission electron microscopy (TEM)

Fig. 3. Fumed silica (surface area 125 m^{2}g^{-1}) – Scanning electron microscopy (SEM)

By Eq. 2 and $D = 2.5$ from DLA we estimate the apparent density of the aggregate to be 700 g l^{-1}. This is the density found experimentally, when powdered fumed silica is pressed to a solid disc. $D_m = 2.1$ from RLCA then gives an agglomerate density of about 20 g l^{-1} – close to the bulk density of the freshly produced fluffy product.

From scanning electron microscopy (SEM) and transmission electron microscopy (TEM) (see Fig. 2 and 3) the predominant particle structures of fumed silica are aggregates, that consists of firmly attached and partially fused primary particles. Some characteristic features of these silica aggregates are summarized in Fig. 4.

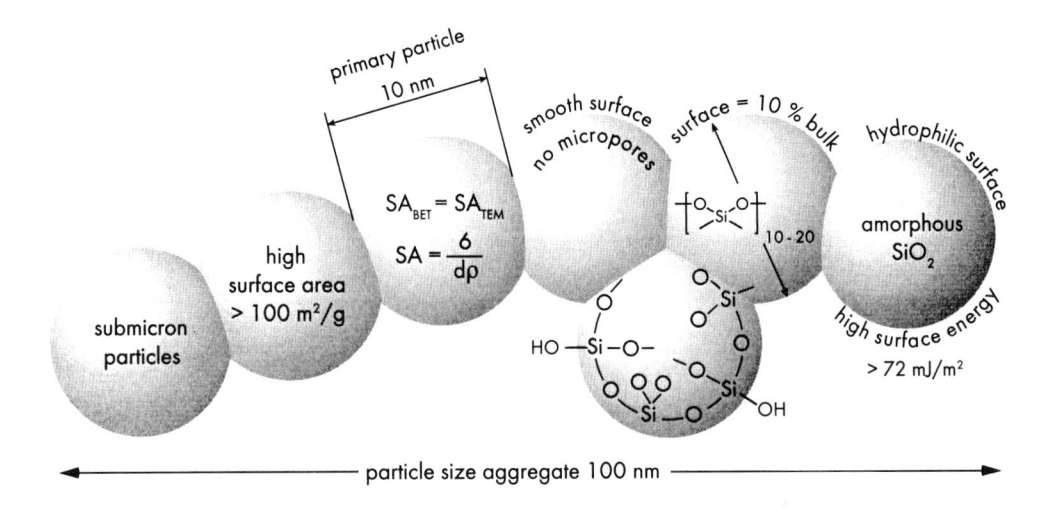

Fig.4. Characteristic features of fumed silica aggregates

Fumed silica aggregates are obviously linear and branched particle structures with a mean size of about 100 to 200 nm. By TEM we derive the size of the partially fused primary particles of about 10 nm. This very small particle size correlates well with the high surfaces area of fumed silica which usually is larger than 100 m^2 g^{-1} as determined by nitrogen adsorption at 78 K according to BET [5]. Adsorption techniques and electron microscopy provide very close values of surface areas. This indicates that fumed silica exhibits a smooth particle surface in the range of nanometers, apparently its surface is free of micropores.

This lack of micropores is further confirmed by SAXS and nitrogen adsorption data. Again a very helpful concept is fractality, in this case the surface fractal dimension D_s. A smooth, non-porous surface is described by a surface fractal dimension of $D_s = 2.0$, but a totally porous body will reach a D_s of up to nearly 3. SAXS and adsorption measurements on fumed silica result in a surface fractal dimension D_s very close to 2.0 [6]. The fact of a nonporous smooth surface of the fumed silica particles is a most important characteristic, as it will simplify the interpretation of chemical surface reactions [7]. Additionally, surface microporosity would markedly decrease the interaction efficiency of the silica surface [6].

Surface Chemistry

The chemistry of a finely divided solid with high surface area is at the borderline between molecular chemistry and physics of solid surfaces. The surface of a finely divided amorphous silica may be understood as a two-dimensional projection of the three-dimensionally linked silicon dioxide tetrahedra. By this the chemistry of the fumed silica surface is dominated by Si–O–Si units and, in particular, dangling surface Si–O bonds, which will create silanol groups under the influence of humidity at ambient temperature.

The density of the silica surface silanol groups strongly depends on the production process of silicon dioxide. There is no evidence that fumed silica contains internal silanols. In contrast to silica products from wet processes, fumed silica shows only a low surface density of surface silanols. Approximatively every second silicon atom on its surface bears a silanol group. A large part of these silanols are not hydrogen-bonded but isolated and give rise to a IR band at 3750 cm^{-1}. Additionally, surface silanols on fumed silica are statistically distributed over the surface [8]. For these reasons, the surface of fumed silica is highly reactive to chemical reactions. Due to its surface silanols but also from its oxide nature, fumed silica exhibits a high surface energy γ and is wettable by water – untreated silica therefore usually shows a hydrophilic character.

Many industrial applications of fumed silica basically depend on its high surface energy and the existence of surface silanol groups. But there are also a variety of systems where reactive silanol groups and water adsorbed on the hydrophilic surface will import serious disadvantages. In consequence, one of the most important reactions on fumed silica is the silylation of these surface silanol groups. The desactivation of the surface silanols groups will additionally lower the surface energy γ of the oxide.

Thus, the surface is rendered hydrophobic, as it will no be longer wettable by water. Most common silylating agents are alkylchlorosilanes, as dichlorodimethylsilane, or alkylsilazanes, as hexamethyl-disilazane (Eq. 3).

$$SiO_{4/2}-Si-OH + 1/2\ [(CH_3)_3Si]_2NH + H_2O \longrightarrow SiO_{4/2}-Si-O-Si(CH_3)_3 + 1/2\ NH_3$$

hydrophilic: $\gamma > 72$ mJ m^{-2} \qquad\qquad hydrophobic: $\gamma < 30$ mJ m^{-2}

Eq. 3.

Due to its high surface area, surface chemistry and physics dominate the properties of fumed silica. The O–Si–O being 0.3 to 0.4 nm let estimate about only 20 silicon dioxide units spanning the diameter of a primary particle of amorphous silica. Fumed silica therefore has an extremely high surface to bulk ratio up to about 10 %. This is why even bulk methods of chemical analysis are suitable to follow chemical reactions on its surface: elemental analysis, IR or NMR methods, etc.

The chemical reaction of the silica surface with hexamethyldisilazane leads to a coverage with trimethylsiloxy groups. The molecular area of a trimethylsiloxy group has been reported to be $a_{Me_3SiO} = 0{,}38$ nm^2 [9]. The limiting carbon content % C at full surface coverage will then be %C = 1.6 per 100 $m^2 g^{-1}$ of surface area – the completion of the reaction is easily followed by standard carbon analysis. Additionally, the disappearence of the band of isolated silanol groups at 3750 cm^{-1}, as monitored by DRIFT, is a convenient method to determine the yield of silanol conversion. Suitable methods for estimating the surface energy γ of hydrophilic or silylated silica are wetting tests or gas adsorption techniques.

Essential for silylation is the activation of the silica silanol groups. To accelerate the reaction of silica with hexamethyldisilazane molar amounts of water have to be added. But water may be omitted, without losing reaction activity, if a methyl group of the trimethylsilyl unit is substituted by a 3,3,3-trifluoropropyl group. DRIFT shows the isolated silanol groups at 3750 cm^{-1} to disappear in the course of silylation with trimethylsiloxy groups, but a broad band at 3650 cm^{-1} always remains. It belongs to hydrogen-bonded, associated and less active silanol groups being unavailable for reaction. On a 3,3,3-trifluoropropyldimethylsiloxy silylated silica surface the band of H-bonded, associated silanol groups is only shifted to 3690 cm^{-1} which indicates a weakening of the H-bonds. It is likely that the polar 3,3,3-trifluoropropyl group interacts with the less active, associated silanol groups in order to activate them.

Acid-base reactions are of particular analytical interest [10]. Titration of fumed silica with sodium hydroxide seems not to be a very exciting reaction. But the advantage of a acid-base reaction is that it distinguishes easily between the acidic silanols of silica and any other kind of monomeric, polymeric or resinous silanols. Titration may be a helpful tool to follow the degree of reaction of silica silanol groups with a variety of silylating agents.

$$SiO_{4/2}-Si-OH + NaOH \longleftrightarrow SiO_{4/2}-Si-O^- \, Na^+ + H_2O$$

Eq. 4.

Under suitable reaction conditions the acid-base reaction provides a hydroxyl capacity or silanol group density of fumed silica of about 2 SiOH per nm^2. This value falls well between the total amount of silanol groups on fumed silica of about 2.5 SiOH per nm^2 [11] and 1.7 SiOH per nm^2 as reported for the content of reactive isolated silanol groups [12]. Titration seems to give a good estimation about the content of chemically reactive and available silanol groups.

Surface Area

The surface area is one of the most important and most commonly used parameters to characterize a highly dispersed oxide. To quantify and to normalize chemical surface reaction data or physico-chemical properties usually the surface area rather than the oxide mass is used as a reference.

The most reliable technique to determine surface areas above 100 m^2 g^{-1} has turned out to be the adsorption of gases, e.g. nitrogen at 78 K, to calculate the BET surface area. This calculation bases on the monolayer capacity as evaluated from the adsorption isotherm and the molecular area of the adsorbate. When working with silylated silica samples this may lead to questionable results. Silylation and various other surface treatments of fumed silica clearly will change its surface properties, among them silanol content and surface energy. Koberstein [13] reported that even the so thought inert molecule of nitrogen may exhibit specific polar interactions due to its quadrupolmoment. Kiselev [14] pointed out that the apparent molecular area of the adsorbate may depend on its surface interactions with the adsorbent.

A more detailed study on silylated fumed silica that used nitrogen and argon at 78 K and also *neo*-pentane at 273 K as adsorbates revealed a distinct decrease of the BET surface area with growing carbon loading and decreasing content of surface silanol groups (Fig. 5) [15].

This decrease of the BET surface area is markedly larger than determined from TEM or as expected from the particle growth by the silylation layer. Obviously, the apparent molecular area of the adsorbate strongly depends on the surface energy of the solid as determined by wetting tests or the equilibrium spreading pressure from gas adsorption. On a high energy surface, as hydrophilic fumed silica, the molecular area of nitrogen $a_{N_2} = 0{,}162$ nm^2 (adopted from the liquid density) and, therefore, the related BET surface area from nitrogen adsorption, will provide reliable values. Whereas it seems that on a low energy surface, as for example silylated silica, no ordered monolayer is formed and therefore the monolayer capacity and consequently the BET surface area are evaluated markedly too low. In general it turns out that a good compromise is to use the nonsilylated hydrophilic surface area as a reference.

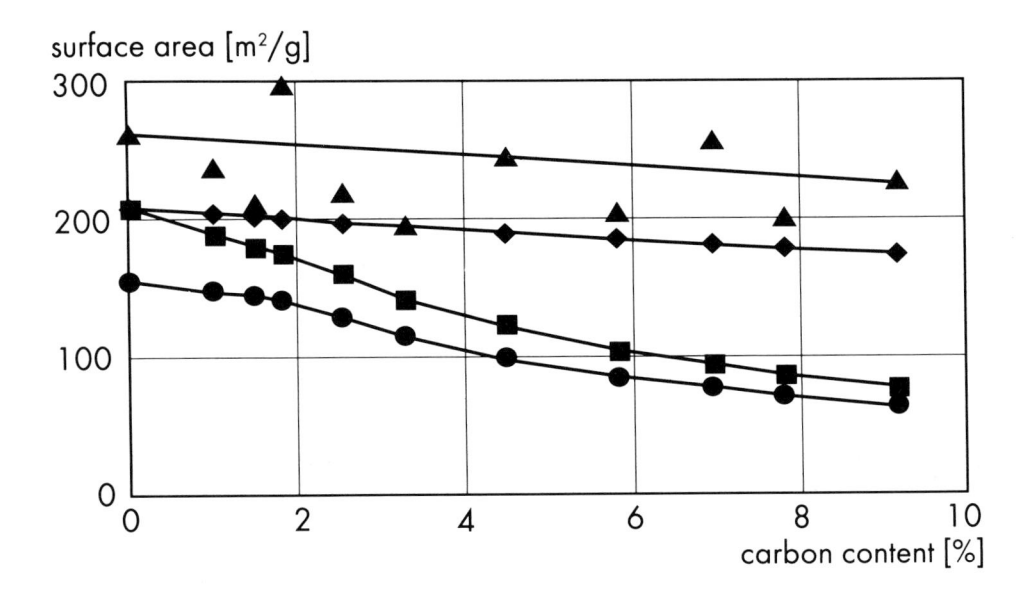

Fig. 5. Surface area of silylated silica: N$_2$-adsorption BET-surface area (■); Ar-adsorption BET-surface area (●); surface area from TEM analysis (◊); decrease of surface area according to mean thickness of silylation layer (▽) [15]

Applications

Fumed silica finds wide use in a variety of industrial applications. Most important are reinforcement of elastomers as active filler and thickening of liquids as a rheological additive, both applications cover more than two thirds of the market volume. Smaller volumes of fumed silica are used as free – flow additive in powder-like solids, for example, in toners for copiers and printers, in fire extinguishers, or even food. It also finds use in anti-foam agents, as anti-blocking, in cable insulation, catalysis, cosmetics, adsorbents, paper coating, pharmaceuticals, polishes etc. All these applications profit from the two basic properties of fumed silica that result from its pyrogenic origin: a structure of finely dispersed, aggregated particles and a large surface area of high activity. Exemplarily, this fascinating interplay of fumed silica particle structure and particle surface shall be discussed for the free flow of toners, the thickening of liquids, and the reinforcement of elastomers.

Free flow of toners

The ability of fumed silica to support and maintain the free flow of solid-like powders is directly related to the small particle size of its aggregates. The silica aggregates will cover the surface of the powder particle and thus prevent the powder particles to lump together and will additionally act as a ball bearing to let the powder flow (Fig. 6).

The free flow of a toner in a copy machine is essential to accurate toner flow and image quality. Without silica the free flow of the toner will diminish after multiple copies and the image quality breaks down. Highly hydrophobic silica as a free flow additive sustains the toner flowability. Additionally, any additive has to be compatible with the chargeability of the toner system. It turned out that the toner requires the addition of a silica type of same polarity and chargeability. The chargeability of the silica is controlled by surface chemistry – silylation and hydrophobization will retain the negative polarity of the silicon dioxide, but surface modification by amino or ammonium organosilicon compounds result in positive chargeability [16].

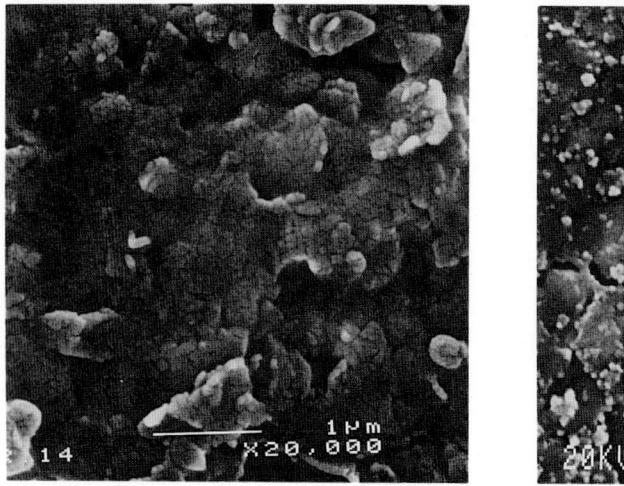

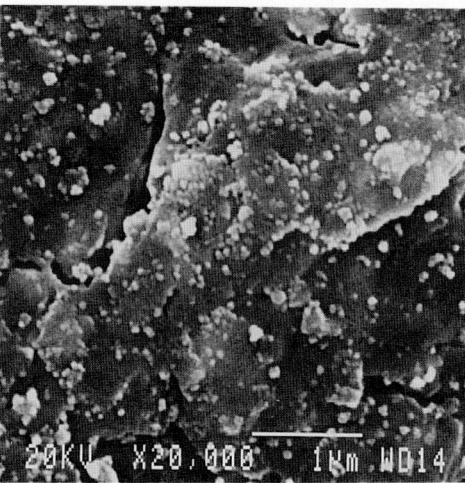

Fig. 6. Scanning electron microscopy of toner particles without silica (left) and with 0.4 % weight of fumed silica (right)

Thickening of liquids

Many industrial and manufacturing processes face the requirement to apply a liquid solution, resin or polymer onto vertical substrates as coatings, paints, lacquers, etc. During application, the liquid should be thin and easy to paint or spray but consecutively the applied liquid film should stay on the substrate the time needed to accomplish hardening.

Fumed silica bears the properties of an effective thickener which will not undergo swelling and exhibits chemically inertness. Additionally, thickening by fumed silica results in a non Newtonian system commonly accompanied by a yield point, shear thinning and thixotropy. Thus, fumed silica solves in an excellent manner the requirements of reversible shear thinning under high shear stress but a distinct yield point or high viscosity at lower shear stresses.

These rheological effects that fumed silica imposes on a liquid system may be understood by a fascinating interplay of particle structure and surface interactions, as depicted in Fig. 7. According to Eq. 1 a space-filling particle network will be possible at a loading of fumed silica at about 1 wt. % (with a primary particle size of 10 nm, an aggregate size of 100 nm and a size of agglomerates of 10 µm as measured by laser diffraction in liquids). Reality is not far from such a rough estimation, typical industrial loads of fumed silica as a thickener are in the range from 0.5 to about 2 wt. %.

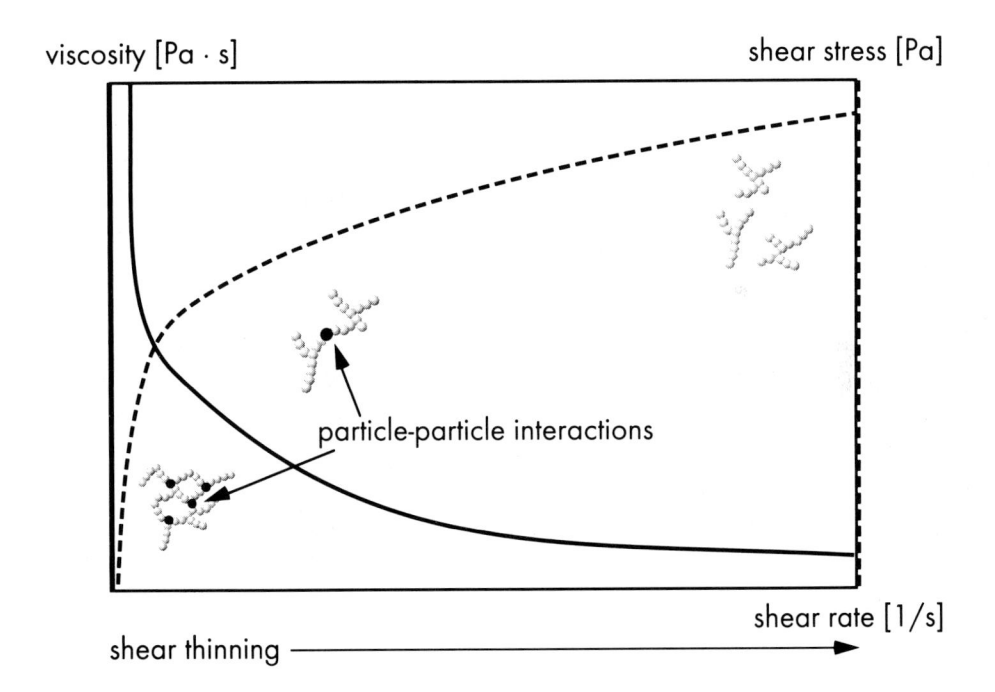

Fig. 7. Thickening of liquids by fumed silica

This space-filling network of percolating and interacting fumed silica particles may result in an enormous high viscosity or even a yield point. The thickened liquid gets a gel-like consistence and will resists shear stress until the shear stress overcomes the strength of the particle-particle interactions and starts to break the particle network and large agglomerates to smaller clusters of aggregates. With increasing shear rate more and more particle linkages are broken. By this the particle network is destroyed and the size of the agglomerates will diminish – the liquid is thinned under the stress of shear. At the moment the shear stress is decreased or stopped, then aggregates and small agglomerates will rearrange and restart to interact and grow to larger agglomerates – the liquid is thickened again by the silica, but as the rearrangement of particles requires time the system behaves thixotropic.

The origin of the linkage of silica aggregates to agglomerates and to particle networks are particle-particle surface interactions. Interparticle hydrogen-bonds have been discussed as a main force to bind fumed silica particles together [11]. The study of the thickening effect of hydrophilic and silylated hydrophobic fumed silica in liquids of different polarities revealed a more generalized picture of particle interactions and thickening ability. It turned out, that a similar system of silica surface and liquid medium will not show pronounced thickening whereas if the surface of the silica is markedly different from the liquid then in consequence a high thickening effect will occur (Table 1).

Hydrophilic fumed silica aggregates will very strongly interact in a nonpolar medium by hydrogen-bonds between surface silanol groups of neighbouring particles. For that, hydrophilic silica is an excellent thickener and rheological additiv for nonpolar liquids. Less than 5 wt. % of hydrophilic fumed silica will thicken a liquid alkane or silicone oil to a cuttable rubber-like gel.

The importance of surface silanol groups for linking silica particles and for thickening is demonstrated when fumed silica is silylated. Its thickening ability in a nonpolar liquid will sharply decrease when the silanol groups are blocked, the oxide surface is shielded and the silica is rendered hydrophobic. There seems to be no resulting energy of interaction for hydrophic particles in a nonpolar medium. A silicone oil may be loaded up to 30 wt. % with highly hydrophic fumed silica but always retains its liquid consistence.

liquid	silica	polar-hydrophilic	nonpolar-hydrophobic
polar		no interaction	hydrophobic interaction
		no thickening	high thickening
nonpolar		H-bonding	no interaction
		high thickening	no thickening

Table 1. Principles of particle-particle interactions

The thickening ability of hydrophilic fumed silica in a highly polar liquid is also rather low. In a polar medium the hydrophilic silica surface will be effectively wetted, the particles are shielded from each other and their interaction is prevented. Hydrogen-bonds and other polar interactions within the liquid, between surface and liquid molecules or between particle surfaces are of the same order of strength – no energy results from hydrophilic particles that interact in a polar medium and consequently no thickening is reached.

In contrast, if accurately dispersed highly hydrophobic fumed silica is in a polar liquid medium then strong thickening of the dispersion is achieved. The origin of the underlying particle interactions are so called hydrophobic interactions [17], which result from strong polar interactions in the liquid itself, for example intermolecular hydrogen-bonds. Polar liquids tend to build up rather stable supermolecular structures – water or aqueous solutions of alcohols are known to be well clustered and structured by hydrogen-bonding. A particle of hydrophobic surface will disturb this network. Therefore, the contact surface of the nonpolar hydrophobic particles towards the polar liquid will be minimized. This forces the hydrophobic particles to contact each other and this results in a rather strong particle-particle interaction.

By this indirect interaction the hydrophobic silica particle build up stable agglomerates and particle networks. Highly hydrophobic fumed silica thickens a liquid polar medium as excellently as a hydrophilic silica thickens a nonpolar one. It has to be mentioned that the interpretation of particle interactions in different media is not restricted to hydrogen-bonding. Qualitatively the interaction of particles on the basis of London dispersion forces in general also leads to particle interactions in a medium which differs in its Hamaker constant from that of the particle, but the interaction energy will tend to zero if the Hamaker constants of particles and medium are equal [18]. The rough principles of particle interactions as discussed above seem to be a general and helpful approach to understand thickening of liquid by fumed silica. It may depend on the particular system which types of interaction forces are quantitatively dominant, whether hydrogen-bonds, electrical charges or van-der-Waals interactions including London dispersion interactions of fluctuating dipoles, and Keesom forces from dipole-induced dipole or Debye forces from dipole-dipole interactions.

Reinforcement of Elastomers

From a rubber it is essentially demanded to elongate under stress, to withstand stress without breaking and to reversibly find back to its original shape after the stress ceases. An unfilled, cured polymer will rarely fulfill these requirements; this is a fact, which is true for most rubber-like systems. The industrial development of elastomers therefore is strongly related to the production of active reinforcing fillers.

The high mechanical strength of natural and organic rubbers as used in tires is due to the incorporation of pyrogenic carbon blacks as active fillers. Elastomers of a more polar polymer backbone, such as polyacrylates, polyurethanes or polysulphides, require fillers of higher polarity. In particular the performance of polydimethylsiloxane elastomers (silicone rubber) is basically related to the addition of fumed silica.

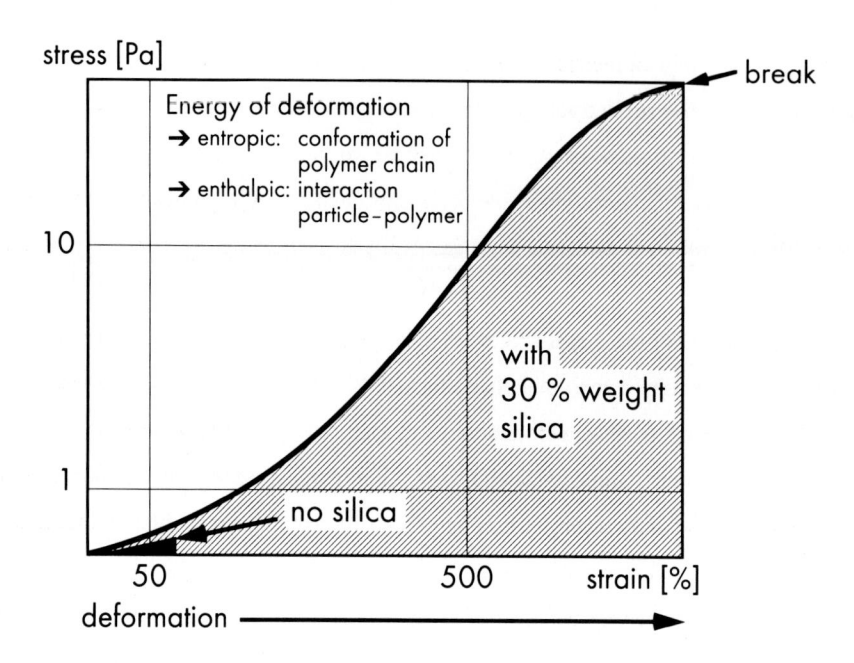

Fig. 8. Reinforcement of elastomers by fumed silica

The bare vulcanized silicone polymer, a polydimethylsiloxane network, shows only a low mechanical strength. Adding 30 wt. % of fumed silica by mixing and dispersing increases the elongation and stress at break by a factor larger than 10 (Fig. 8).

It is not the aim of this paper to discuss in detail theories of rubber elasticity, starting from the unfilled entropic polymer network of Kuhn and Flory [19] to the filled van der Waals network of Kilian [20]. Also, the experimental observation of hysteresis particular found in the first stress-strain cycles will not be followed further.

To understand elastic mechanical properties, the discussion of the storage of energy of deformation provides a powerful approach. Dynamic mechanical measurements at higher strain on filled silicone elastomers show that the energy of deformation may be related to an entropic and an enthalpic part. The entropic part is mainly due to the restriction of the conformational space of the polymer chain by the presence of the solid silica particles. Whereas the enthalpic part of the energy of deformation is related to the particle-polymer interactions at the silica surface. This enthalpic part of storage of deformational energy might be explained if the optimum number of polymer-on-particle surface interactions will decrease while stress is applied to the elastomer. The polymer chain may be drawn away from the particle surface or may loose its optimized position relative to the particle surface. When the strain is relaxed, the optimization and reorientation of the silicone chain to active adsorption sites of the particle surface will release interaction energy.

In order to achieve excellent mechanical properties of silicone elastomers the absolute amount of silica surface and additionally the surface activity of fumed silica should be as high as possible. To realize this aim high loadings of the filler are used which directly will lead to a large particle-polymer interface. A high quality silicone rubber requires a silica loading up to 30 wt.%. But the incorporation of highly active polar hydrophilic silica directly involves thickening of the unvulcanized rubber – crepe hardening will occur.

This is understandable by Eq. 2 that predicts that a filler loading of 30 wt. % establishes a percolating particle network of silica, even at a very small size of the silica agglomerates of about $d_{agglomerate} = 250$ nm, i.e., at an agglomerate size being only 2–3 times larger than the size of the aggregates $d_{aggregate} = 100$ nm. Going back to the thickening ability of silica, we know from Table 1, that a silica of high polarity exhibits strong particle-particle interaction in a nonpolar medium such as the silicone rubber.

In order to suppress thickening we are forced to silylate the silica surface and block surface silanol groups, to prevent the formation of hydrogen-bonds and interactions of the oxide surfaces between fumed silica particles. Consequently, as measured by wetting tests or gas adsorption this again will generally decrease the surface activity. As Litvinov [21] could clearly demonstrate by NMR, the strength of fumed silica particle-polymer interaction is distinctly decreased when silylating its surface. In order to optimize the compromise between reinforcement and thickening, it turns out that a key to further improvement is the investigation at the interactions of silicones on a fumed silica surface.

Adsorption measurements may provide helpful data in this field. In particular the adsorption of pure gases gives access to the interaction adsorbate – adsorbent without any interference from solute molecules. Fig. 9 shows the adsorption isotherms of hexamethyldisiloxane on a hydrophilic, a silylated but yet reinforcing and an over-silylated, no longer reinforcing silica at $T = 303$ K.

The total amounts of adsorbate adsorbed show no large differences near saturation pressure at $p_0 = 73.3$ hPa. All three samples base on fumed silica of surface area $SA = 300$ m^2 g^{-1} referring to a hydrophilic surface. In the high-pressure, multilayer and capillary condensation region the adsorption isotherm is dominated by particle size and aggregate interparticle voids in the mesopore range. But in the low-pressure submonolayer region of the isotherm at $p/p_0 < 0.07$, the adsorbate uptake is controlled by the strength of the adsorption sites. Compared to the hydrophilic silica sample the amount adsorbed of hexamethylsiloxane on both silylated silica samples is markedly diminished. The loss of adsorption capacity by silylation is also confirmed by adsorption of neo-pentane, argon and nitrogen [15]. As mentioned above, this reveals a sharp decrease of surface energy for both silylated silica samples – despite the differences of their reinforcing activity.

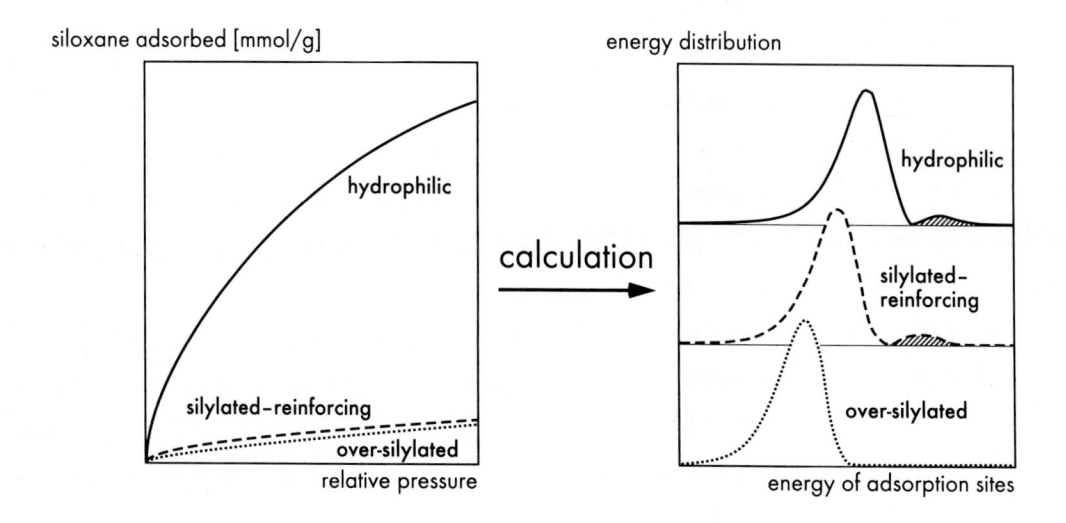

Fig. 9. Gas adsorption of hexamethyldisiloxane on hydrophilic, silylated reinforcing and over-silylated non-reinforcing fumed silica at $T = 303$ K, $0 < p/p_0 < 0.07$; $p_0 = 73.3$ hPa: adsorption isotherm (left) and calculated energy distribution (right)

Hines developed an analytical equation for an adsorption isotherm on a heterogeneous surface [22]. His analytical equation provides the advantage to more clearly interpret the shape of the adsorption isotherm by replotting it as a calculated energy distribution. By this helpful approach of high and low energetic sites of adsorption, we may at least qualitatively see differences between reinforcing and nonreinforcing silylated silica samples.

The hexamethyldisiloxane molecules probe adsorption sites of lower and higher energy on the hydrophilic fumed silica surface. When silica is silylated, the adsorption energy of the low-energy sites are consecutively shifted to lower values. Surprisingly the high-energy sites remain for the reinforcing silylated silica, but vanish for the non-reinforcing one.

For the moment we have no clear model to relate these low and high energetic adsorption sites to chemical groups such as silanols or the oxide backbone. Such an interpretation will need further investigations in order to achieve more quantitative data describing this picture of a heterogeneous silica surface.

Conclusion

Fumed silica is a highly dispersed silicon dioxide of large industrial importance and a wide spectrum of applications. Due to its production in a flame process fumed silica exhibits a smooth and nonporous particle surface. Additionally to its high surface area fumed silica bears isolated and statistically distributed surface silanol groups that render this product hydrophilic. A most important technical reaction, therefore, is the silylation and hydrophobization of the hydrophilic surface.

In the flame process firm aggregates are formed from primary particles by DLA followed by agglomeration of these aggregates by RLCA, according to mass fractal dimensions of $D_m^{aggregate} = 2.5$ and $D_m^{agglomerate} = 2.1$. The fractal approach to the spatial particle structure of fumed silica has been shown to explain the low bulk density of the product just as the loading needed to build up a percolating particle network to thicken liquid media. A model of particle-particle interactions has been evolved to interpret the high thickening ability of silanol group-free hydrophobic silica in polar media by hydrophobic forces.

Fumed silica acts as a highly reinforcing filler in silicone elastomers. Its activity results from its highly dispersed particle structure, high surface area and surface energy. To better understand the interplay of these properties first studies on gas adsorption of hexamethylsiloxane on hydrophilic and silylated silica have been conducted. The shape of the adsorption isotherm revels the existence of low- and high-energy adsorption sites, the latter qualitatively seem to be related to reinforcement of the silicone elastomer. Further quantitative studies in this field are needed.

Acknowledgements: The authors thank Dr. Achenbach and Dr. Heinemann, *Wacker-Chemie GmbH*, for providing elastomer and toner data and greatly appreciate the many interesting and helpful discussions.

References:

[1] G. D. Ulrich, J. W. Riehl, *J. Colloid Sci.* **1982**, *87(1)*, 257; G. D. Ulrich, *Chem. Eng. News*, **1984**, *8*, 22.

[2] P. Meakin, in: *The Fractal Approach to Heterogeneous Chemistry, Surfaces, Colloids, Polymers* (Ed.: D. Avnir), Wiley, New York, **1989**, p. 131.

[3] T. Freltoft, J. K. Kjems, S. K. Sinha, *Phys. Rev.* **1986**, *33B*, 269.

[4] H. Barthel, *unpublished results*.

[5] S. Brunnauer, P. H. Emmett, E. Teller, *J. Am. Chem. Soc.* **1938**, *60*, 309.

[6] A. J. Hurd, D. W. Schaefer, J. E. Martin, *Phys. Rev.* **1987**, *35(5)A*, 2362; H. Barthel, F. Achenbach, H. Maginot, *Proceedings of Mineral and Organic Functional Fillers in Polymers, International Symposium*, Namur, **1993**.

[7] D. Avnir, P. Pfeifer, *J. Chem. Phys.* **1983**, *79(7)*, 3558; D. Avnir, D. Farin, P. Pfeifer, *J. Chem. Phys.* **1983**, *79(7)*, 3566.

[8] E. Papirer, A. Vidal, M. Wang, J. B. Donnet, *Chromatogr.* **1987**, *23(4)*, 279; Legrand, *Adv. Colloid Int. Sci.* **1990**, *33*, 91.

[9] P. Larsen, O. Schou, *Chromatogr.* **1982**, *16*, 204.

[10] G. W. Sears, *Anal. Chem.* **1956**, *28(12)*, 1981.

[11] G. Michael, H. Ferch, *Schriftenreihe Pigmente*, Degussa AG, Nr. 11.

[12] B. Evans, T. E. White, *J. Catalysis* **1968**, *11*, 336; W. Hertl, *J. Phys. Chem.* **1968**, *72(12)*, 3993.

[13] E. Koberstein, M. Voll, *Phys. Chem.* **1970**, *71*, 275.

[14] A. V. Kiselev, A. Y. Korolev, R. S. Petrova, K. D. Shcherbakova, *Kolloid. Zh. 22*, **1960**, 671.

[15] H. Barthel, *Proceedings of the Fourth Symposium on Chemically Modified Surfaces, Chadds Ford*, (Eds.: H. A. Mottola, J. R. Steinmetz), Elsevier, **1992**, p. 243.

[16] M. G. Heinemann, R. H. Epping, *Japan Hardcopy*, Yokohama, **1993**.

[17] J. N. Israelachvili, *Chemica Scripta*, **1985**, *25*, 7.

[18] S. Ross, I. D. Morrison, *Colloid Systems and Interfaces*, Wiley, New York, **1988**, p. 205.

[19] W. Kuhn, F. Grün, *Kolloid-Z.* **1942**, *101*, 248; J. P. Flory, *Principles of Polymer Chemistry*, Cornell University Press, Ithaca, **1953**.

[20] G. Kilian, *Progr. Coll. Polym. Sci.* **1987**, *75*, 213.

[21] V. M. Litvinov, *Polymer Science U.S.S.R.* **1988**, *30(10)*, 2250; V. M. Litvinov, *Organosilicon Chemistry II* (Eds.: N. Auner, J. Weis), VCH, Weinheim, **1995**, p. 779.

[22] L. Hines, S.-L. Kuo, N. H. Dural, *Separation Sci. Techn.* **1990**, *25(7&8)*, 869.

Poly(dimethylsiloxane) Chains at a Silica Surface

V. M. Litvinov

DSM Research B.V., PAC–MC

P.O.Box 18, 6160 MD, The Netherlands

Summary: Solid state NMR studies of molecular motions and network structure in poly(dimethylsiloxane) (PDMS) filled with hydrophilic and hydrophobic Aerosil are reviewed and compared with the results provided by other methods. It is shown that two microphases with significantly different local chain mobility are observed in filled PDMS above the glass transition, namely immobilized chain units adsorbed at the filler surface and mobile chain units outside this adsorption layer. The thickness of the adsorption layer is in the range of one to two diameters of the monomer unit (~1 nm). Chain units in the adsorption layer are not rigidly linked to the surface of Aerosil. The chain motion in the adsorption layer depends significantly on temperature and on type of the filler surface. With increasing temperature, both the fraction of less mobile adsorbed chain units and the lifetime of the chain units in the adsorbed state decrease. The lifetime of chain units in the adsorbed state approaches zero at approximately 200 K and 500 K for PDMS chains at the surface of hydrophobic and hydrophilic Aerosil, respectively.

The mean average molecular mass of the network chains is determined for the elastomer matrix outside the adsorption layer. Contributions to the network structure from different types of junctions (chemical junctions, adsorption junctions, and topological hindrances due to confining of chains in the restricted geometry (entropy constraints or elastomer-filler entanglements) are estimated. The major contributions to the total network density are provided by the topological hindrances near the filler surface and by the adsorption junctions. The apparent number of the elementary chain units between the topological hindrances is estimated to be approximately 40–80 elementary chain units.

A molecular model is proposed for qualitative analysis of the stress-strain behavior for filled PDMS in a wide temperature range. The model emphasizes the importance of the following molecular characteristics for understanding stress-strain behavior of filled PDMS:

1) The dynamic origin of the adsorption layer which provides physical network junctions
2) The topological hindrances arising from the filler particles

Stress-strain properties for unfilled and filled silicon rubbers are studied in the temperature range 150–473 K. In this range, the increase of the modulus with temperature is significantly lower than predicted by the simple statistical theory of rubber elasticity. A moderate increase of the modulus with increasing temperature can be explained by the decrease of the number of adsorption junctions in the elastomer matrix as well as by the decrease of the ability of filler particles to share deformation caused by a weakening of PDMS-Aerosil interactions at higher temperatures.

Molecular mechanisms for stress-softening are also discussed. It is shown that this phenomenon is not related to the chain slippage or to a conversion of a "hard" adsorbed phase to a soft one. The obtained results assume that the stress-softening in silicon rubbers is caused by two possible reasons: changes in the positions of filler particles relative to the direction of stretching at the first deformation and by a re-distribution of the topological hindrances. It is shown that the tensile strength at break as a function of temperature is closely related to the chain dynamics at the elastomer-filler interface.

Introduction

Incorporation of active fillers into elastomers is of significant commercial importance since it causes improvement of mechanical properties of the final products. Despite numerous investigations on filled elastomers using different techniques, the molecular origin of the reinforcement effect is still under discussion. For investigation of this phenomenon, filled polysiloxanes are most suitable samples since these rubbers display the strongest improvement of mechanical properties among elastomers and their properties can be studied in a wide temperature range.

The presence of active fillers causes significant changes in the elastomer matrix adjacent to the filler surface due to:

1) Partial immobilization of elastomer chains as a result of chain adsorption at the adsorption sites on the filler surface
2) The restriction of mobility of chain fragments between adjacent adsorption junctions
3) Numerous topological hindrances (entropy constraints) for chain fragments neighboring filler particles

On one hand, it is generally believed that knowledge of chain dynamics at the elastomer-filler interface is of major importance for the molecular understanding of the reinforcement effect. On the other hand, some other reasons for the reinforcement should also be considered. They are:

1) Aggregation of filler particles and filler-filler interactions
2) An increase of the intrinsic chain deformation in the elastomer matrix compared with that of macroscopic strain
3) The density of chemical junctions in the elastomer matrix

Many experimental techniques are used to study the adsorption ability of fillers and the adsorption of polymer chains at surfaces. Different types of information are provided by these methods.

1) *The characterization of surface activity of fillers* is obtained by use of several analytical techniques [1]. Examples of them are: inverse gas chromatography [1, 2], the adsorption of a low molecular weight analog of elastomers [3], the adsorption of elastomer chains from dilute solutions [4], the wettability, viscosity of PDMS fluids in the boundary layer at the surface of solids [5], the determination of the specific surface area, and the analysis of surface groups [1]. It should, however, be mentioned that the results obtained by these methods do not provide direct information on the elastomer behavior at the interface, due to the use of small probe molecules or the presence of a solvent in the systems studied.

2) *The study of the elastomer-filler interactions at a macroscopic level* is carried out by different types of mechanical testing and by the determination of the content of bound (insoluble) rubber. These methods measure properties which are of interest for commercial applications of the materials and provide indirect information on elastomer-filler interactions.

3) *The characterization of the elastomer-filler interactions at a molecular level* may be carried out by spectroscopic techniques such as IR and NMR spectroscopy, X-ray and neutron scattering, dynamic mechanical and dielectric spectroscopy, and molecular dynamics simulations [6]. Up to now, the most comprehensive studies of silica filled PDMS [4, 7–22] and carbon black filled conventional rubbers [23] have been carried out by ^{1}H [4, 7–20, 23], ^{2}H [21], and ^{13}C NMR relaxation experiments [22]. Solid state NMR offers several advantages for the investigation of filled rubbers since molecular properties of elastomer chains can be measured selectively by NMR experiments. The method is very sensitive to the molecular scale heterogeneity in a sample. The network structure which is composed of chemical, physical and topological junctions can also be analyzed by NMR relaxation experiments [11, 12, 14, 15].

This paper is devoted to the study of a part of the complex phenomena of reinforcement, namely the behavior of the host elastomer in the presence of filler particles. The results of solid state NMR experiments and some other methods for filled PDMS are reviewed. The short-range dynamic phenomena that occur near the filler surface are discussed for PDMS samples filled with hydrophilic and hydrophobic Aerosils. This information is used for the characterization of adsorption interactions between siloxane chains and the Aerosil surface. Possible relations between mechanical properties of filled silicon rubbers on the one hand and the network structure and molecular motions at the PDMS-Aerosil interface on the other hand are discussed as well.

Samples Used for the Present Study

Two types of Aerosil, i.e., hydrophilic and hydrophobic Aerosils, with different surface activities, were used for the preparation of mixtures with PDMS. Two procedures were used to mix PDMS with Aerosil, namely *mechanically mixed blends* and *blends obtained from solution*. In the first case, PDMS was mixed on a laboratory mill with Aerosil. In the second case, Aerosil was added to a 0.5 wt% solution of PDMS in *n*pentane, and the suspension was kept for 2 days. Then, while stirring, the solvent was removed, and the resulting powder samples were dried to a constant weight.

Partially deuterated PDMS was synthesized for ^{2}H NMR studies, as was described previously [21]. A detailed description of the sample preparations and the NMR experiments are given in the original papers [7–15, 21].

Part I

Molecular Motions at PDMS-Aerosil Interface and Network Structure in the Elastomer Matrix

I.1 Motions of PDMS Chain Units at the Surface of Hydrophilic Aerosil

I.1.1 Solid State NMR Studies

I.1.1.1 ^1H T_2 NMR Relaxation Experiments

The proton T_2 NMR relaxation experiment is a very sensitive tool to study motional heterogeneity and the dynamics of elastomer chains. A typical T_2 relaxation decay (free induction decay (FID)) for filled PDMS is shown in Fig. 1 for two samples [12].

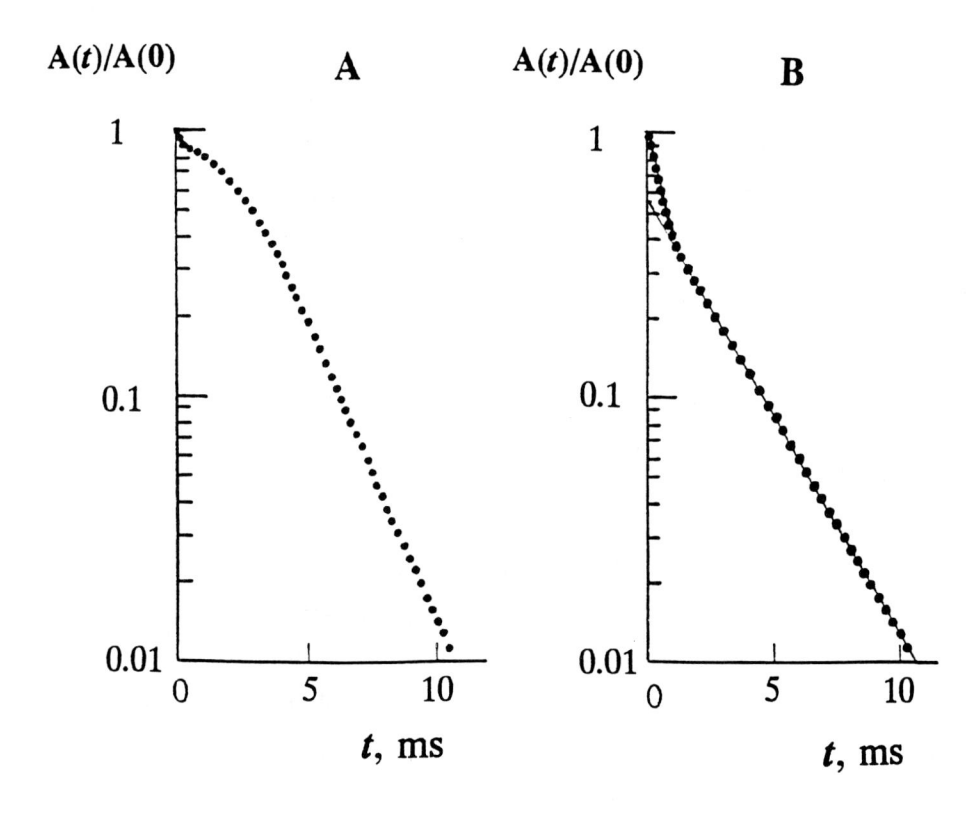

Fig. 1. ^1H free induction decay for (A) bound rubber containing 28 vol% of hydrophilic Aerosil (300 m^2 g^{-1}) (87 phr*), and (B) for the solution blend containing 60 vol% of the same Aerosil (336 phr) [12]; FID is measured by Hahn-echo pulse sequence
*here and below, the concentration of Aerosil in the rubber is represented by the number of weight parts of Aerosil per 100 parts of the rubber (phr)

The FID consists of two distinct components: a fast and a slowly decaying one. The time constant of these components is represented by the T_2 relaxation time. Like the proton relaxation, two relaxation components have also been detected in highly filled PDMS by a ^{13}C T_2 experiment [22]. The ^1H T_2 value for the fast decaying component is about 200 μs [7], whereas the ^1H T_2 value for polymer melts is usually of the order of milliseconds. The short T_2 relaxation time is in the range which is typical for chain molecules experiencing a strongly hindered mobility, like, for example, poly(diethylsiloxane) chains in a columnar mesophase [24]. Therefore, the fast decaying component (T_2^{ad}) is assigned to low mobile adsorbed chain units. These immobilized chain units form *the adsorption layer*. The slowly decaying component (T_2^m) corresponds to more mobile chain units outside the adsorption layer. The large difference in the decay time for these two components of the FID indicates that the local motion of adsorbed chain units is strongly hindered compared with chains outside the adsorption layer. Hence, the ^1H T_2 experiment supports a two-phase behavior for the elastomer matrix in filled rubbers [23]. The relative intensity of the component with the short T_2 provides the fraction of less mobile chain units in the adsorption layer. This fraction increases as the filler content increases which can be seen in Fig. 1.

The temperature dependence of the rate of T_2 relaxation, T_2^{-1}, is shown for chain units inside, $(T_2^{ad})^{-1}$, and outside, $(T_2^m)^{-1}$, the adsorption layer for solution blends in Fig. 2 [8, 10].

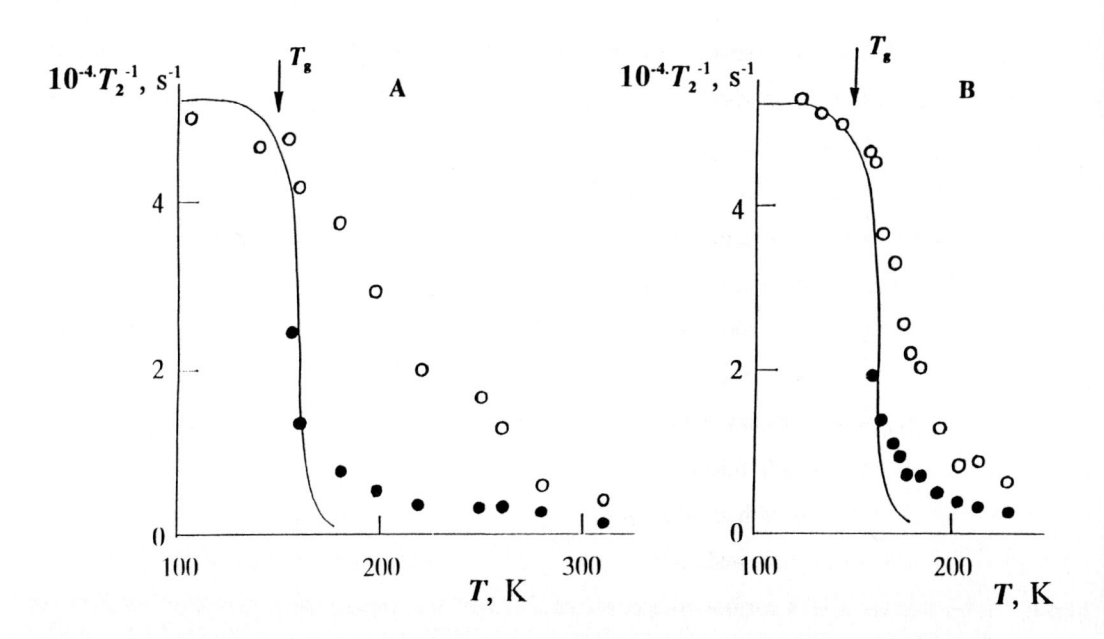

Fig. 2. The temperature dependence of a rate of the T_2 relaxation for chain units inside, $(T_2^{ad})^{-1}$ (○), and outside, $(T_2^m)^{-1}$ (●), the adsorption layer for a solution blend containing (A) 60 vol% of hydrophilic Aerosil (300 m² g⁻¹) (336 phr) [8], and (B) hydrophobic Aerosil (185 m² g⁻¹) [10];
the vertical arrow indicates the glass transition temperature T_g for pure PDMS. The solid line represents the temperature dependence of $(T_2)^{-1}$ for amorphisized pure PDMS.

With increasing temperature, a sharp increase of the $(T_2^m)^{-1}$ is observed. This change occurs in the vicinity of the glass transition temperature for PDMS (T_g = 150 K) and is caused by the local chain motion related to the glass transition for chain fragments outside the adsorption layer. The value of $(T_2^m)^{-1}$ in filled samples is close to that of $(T_2)^{-1}$ for purely amorphous PDMS (solid line in Fig. 2). Apparently, the local chain motion outside the adsorption layer in filled samples does not significantly differ from that for unfilled PDMS [7]. On the other hand, the $(T_2^{ad})^{-1}$ value for chain units adsorbed at the surface of hydrophilic Aerosil gradually decreases at temperatures well above the T_g as is shown in Fig. 2A. Therefore, the local chain motion in the adsorption layer is significantly lower than that for chain units outside the adsorption layer.

Thus, the motions of the adsorbed units are hindered compared with those of the non-adsorbed chain units. The fraction of the less mobile adsorbed chain units can be determined by the analysis of the T_2 relaxation decay. The fraction of the adsorbed chain units gradually decreases with increasing temperature [8].

I.1.1.2 ^{2}H NMR Solid-Echo Spectra

The influence of hydrophilic Aerosil on the chain dynamics in the adsorption layer becomes evident with the help of comparison of ^{2}H solid-echo spectra for a series of filled PDMS samples containing different amounts of Aerosil as shown in Fig. 3 [21].

At 143 K, below the T_g of PDMS, the spectra for all samples exhibit a Pake spectrum for the rotating methyl group with a fixed C_3-rotation axis. Above the glass transition, the line shape depends on the temperature as well as on the concentration of Aerosil. The most pronounced changes of the spectra occur in the temperature range 150 K–200 K as shown in Fig. 3a and 3b. These changes are due to the onset of chain mobility with frequencies between 10 kHz–1 MHz [25]. ^{2}H spectra for filled PDMS containing 40 and 60 vol% of Aerosil display both broad and narrow components. The central narrow resonance originates from the most mobile fraction of siloxane chain units, whereas the outer parts of the spectra correspond to the chain units with slow mobility (on the time scale of the experiment, i.e., 10^{-3} s). The line width of the motionally narrowed central region becomes smaller at higher temperatures, due to more intensive chain motion. This narrow component of the spectra corresponds to non-adsorbed siloxane chain units. Since highly filled PDMS do not crystallize [7, 8, 21], the broad component of spectra is assigned to low mobile chain units in the adsorption layer. The fraction of the broad component of ^{2}H spectra is proportional to the volume fraction of Aerosil up to approximately 90 vol%.

With increasing temperature, the shape of the broad component of the spectra is changed from a Pake spectrum (below 180 K) to a partially motionally averaged spectrum. This is clearly seen for filled PDMS containing 89 vol% of Aerosil, in which case the ^{2}H spectrum mainly consists of the broad component as seen in Fig. 3c. Apparently, the fraction of mobile non-adsorbed chain units is very low in this case. For this sample, a broad rectangular or triangular line shape, extending over almost the full spectral range between the singularities of the Pake spectrum, is observed in a wide temperature range above the T_g. The motional narrowing is not complete even at 433 K, nearly 300 K above the glass transition of PDMS.

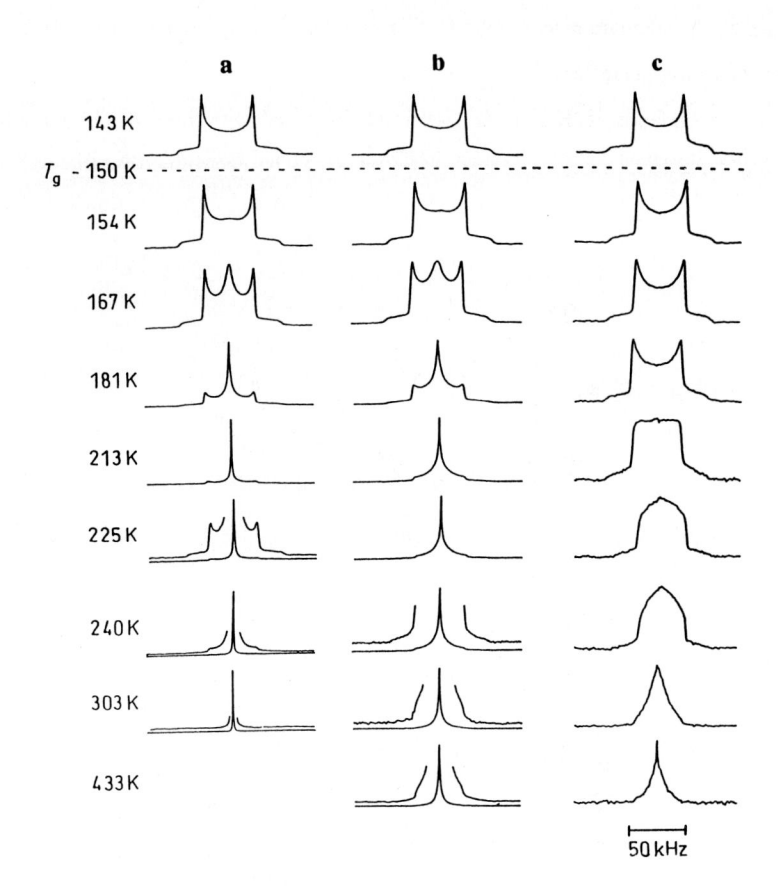

Fig. 3. Three fully relaxed ^{2}H solid-echo spectra for mixtures of PDMS containing (a) 40 (149 phr), (b) 60 (336 phr), and (c) 89 (1813 phr) vol% of hydrophilic Aerosil (380 m^2 g^{-1}) [21]; spectra containing a very weak broad component are also shown with the vertical scale increased by a factor of 20

Thus, the ^{1}H T_2 relaxation experiments and the analysis of the ^{2}H solid-echo spectra show that the adsorption of PDMS chain units on the surface of hydrophilic Aerosil significantly restricts the motion of chain units adjacent to the filler surface. However, chain units in the adsorption layer are not rigidly linked to the surface of Aerosil at temperatures well above the T_g. Motions of chain units in the adsorption layer become less restricted as the temperature increases.

I.1.1.3 ^{2}H Partially Relaxed NMR Spectra and ^{2}H T_1 Relaxation Times

Molecular motions not only affect the ^{2}H NMR line shape, but also determine the spin-lattice relaxation time T_1. Measurement of the T_1 relaxation provides information about fast motions with frequencies near the NMR frequency of deuterons, i.e., 46 MHz [25]. Moreover, ^{2}H T_1 relaxation experiments are very useful for detecting motional heterogeneity in polymers [25]. Motional effects in both relaxation and line shape studies are completely dominated by reorientations of C–D bonds. Therefore, in motionally heterogeneous polymers, different ^{2}H T_1 relaxation times can be related to chain units with different mobility as reflected in different line shapes.

The temperature dependence of ^{2}H T_1 relaxation times for filled PDMS containing 60 vol% of Aerosil is shown in Fig. 4 [21].

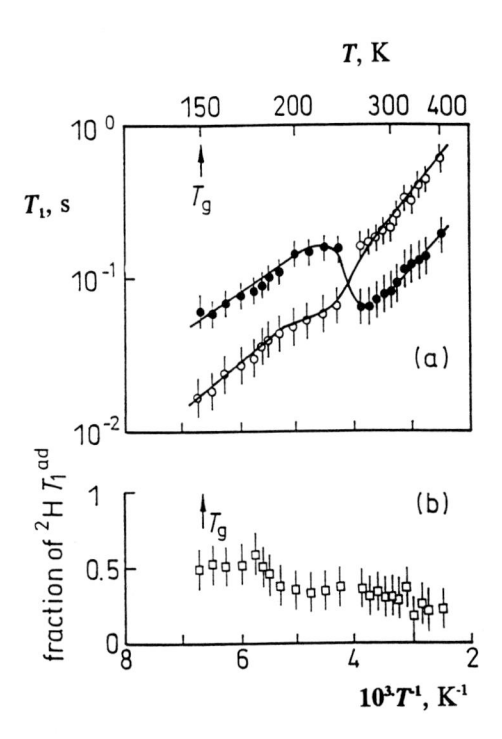

Fig. 4. a) Temperature dependence of ^{2}H T_1 relaxation times in filled PDMS containing 60 vol% of hydrophilic Aerosil (380 m^2 g^{-1}) (336 phr): T_1^{ad} (●) and T_1^m (○)
b) The fraction of the signal with T_1^{ad} relaxation time related to chain units in the adsorption layer [21]

The ^{2}H T_1 relaxation in this sample is characterized by two relaxation times, which indicate the dynamical heterogeneity of the sample. In order to assign the two ^{2}H T_1 components to the chain units inside and outside the adsorption layer, ^{2}H solid-echo spectra at different stages of T_1 relaxation have been measured. The evolution of the solid-echo spectrum at different times of the recovery to the equilibrium at 206 and 282 K is shown in Fig. 5.

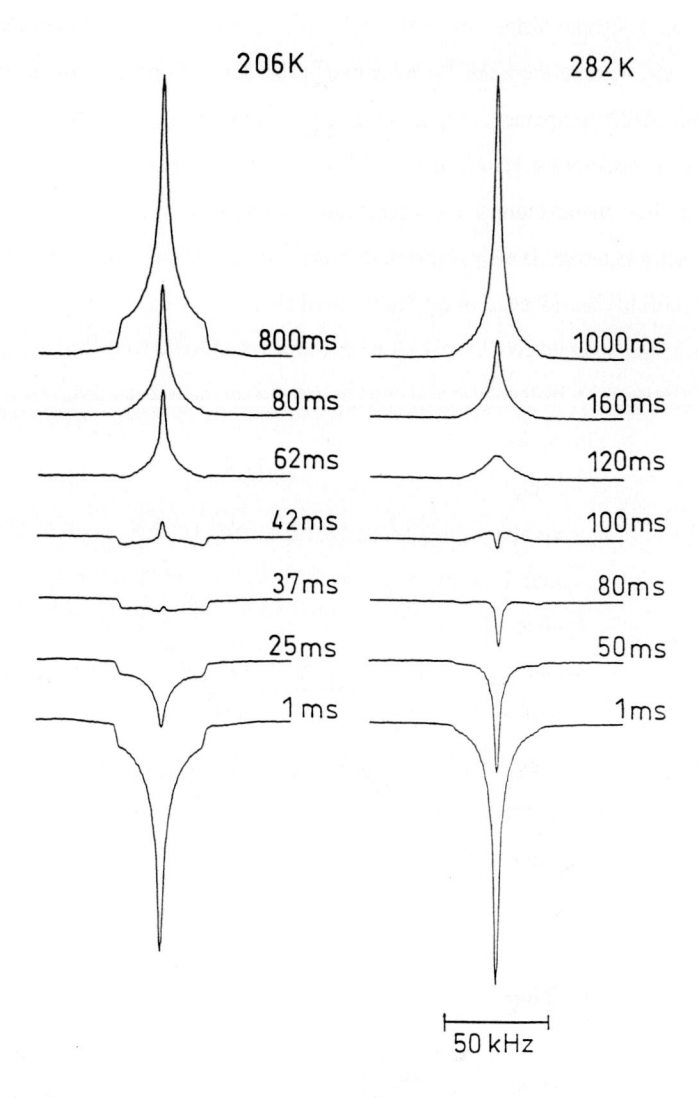

Fig. 5. Partially relaxed ^{2}H solid-echo spectra of filled PDMS containing 60 vol% of hydrophilic Aerosil (380 m^2 g^{-1}) (336 phr) at 206 K and 282 K [21]; spectra were recorded using an inversion-recovery pulse sequence; the time interval between 180° inversion pulse and the solid-echo pulses is indicated

It is clear that at both temperatures broad and narrow components of the spectra are characterized by different ^{2}H T_1 relaxation times. At 206 K, the recovery of the narrow component occurs faster than that of the broad component, whereas the opposite occurs at 283 K. This means that the chain units responsible for the different line shapes also reveal different dynamics for at least the time scale of the experiment, i.e., about 100 ms.

The analysis of the partially relaxed ^{2}H spectra allows a unique assignment of the ^{2}H T_1 values to the relaxation of chain units inside and outside the adsorption layer, as shown in Fig. 4 [21]. Two minima of the relaxation time ^{2}H T_1 can be distinguished on a plot of T_1 against $1/T$ as seen in Fig. 4. The frequency of motions at the temperature of the T_1 minima is close to the NMR frequency of deuterons, i.e., 46 MHz. The minimum at the lower temperature (at about 200 K) is attributed to the relaxation of more mobile chain units outside the adsorption layer. This minimum is caused by chain motion associated with the glass transition of PDMS (α-relaxation). The other minimum (in the vicinity of 270 K) is caused by chain motions in the adsorption layer. It is interesting to note that only one T_1 minimum at about 270 K is observed for highly filled PDMS containing 89 vol% of Aerosil [21]. According to the analysis of the ^{2}H solid-echo spectra, this sample contains mostly low mobile, adsorbed chain units.

The amount of chain units in the adsorption layer can be estimated from the weight fraction of the ^{2}H T_1^{ad} relaxation time. As shown in Fig. 4b, the fraction of adsorbed chain units decreases from 55 % at 148 K to 20 % at 400 K for highly filled PDMS. The amount of adsorbed chain units increases proportionally to the total surface area of Aerosil available for the adsorption [7, 8]. The weight fraction of adsorbed chain units is equal to only 3 % at 246 K for filled PDMS with the weight ratio PDMS/Aerosil (380 m^2 g^{-1}) equal to 100:25 [21].

It seems reasonable to consider the adsorption layer as a microphase consisting of relatively mobile chain units which are able to adsorb and to desorb on the Aerosil surface. The fraction of adsorbed PDMS chain units decreases upon heating, due to the shift of the adsorption-desorption equilibrium to chain desorption [9, 21]. Only a few percent of PDMS chain units form adsorption bonds with the surface of Aerosil in filled PDMS of practical use [7, 8, 12, 21].

I.1.1.4 ^{1}H T_1 NMR Relaxation Time

The proton T_1 relaxation time was determined as a function of temperature for samples with varying content of hydrophilic Aerosil (300 m^2 g^{-1}) [7]. Due to ^{1}H spin-diffusion, only a single ^{1}H T_1 relaxation time is usually measured in heterogeneous polymers [23]. The presence of Aerosil in PDMS suppresses the T_1 minimum at 195 K, ascribed to the chain motion (α-relaxation) in unfilled PDMS, and leads to the appearance of a minimum at higher temperature, in the vicinity of 280 K as shown in Fig. 6.

The minimum becomes more pronounced with increasing the total surface area of the filler particles in the mixture. These results are in good agreement with ^{2}H T_1 experiments, as shown in Fig. 4a, and with ^{13}C T_1 experiments [22]. The ^{1}H, ^{2}H, and ^{13}C T_1 minimum assigned to the adsorbed chain units is for highly filled PDMS at approximately the same temperature. In conclusion, the results of ^{1}H, ^{2}H and ^{13}C T_1 experiments support a two-phase model for PDMS matrix filled with Aerosil particles.

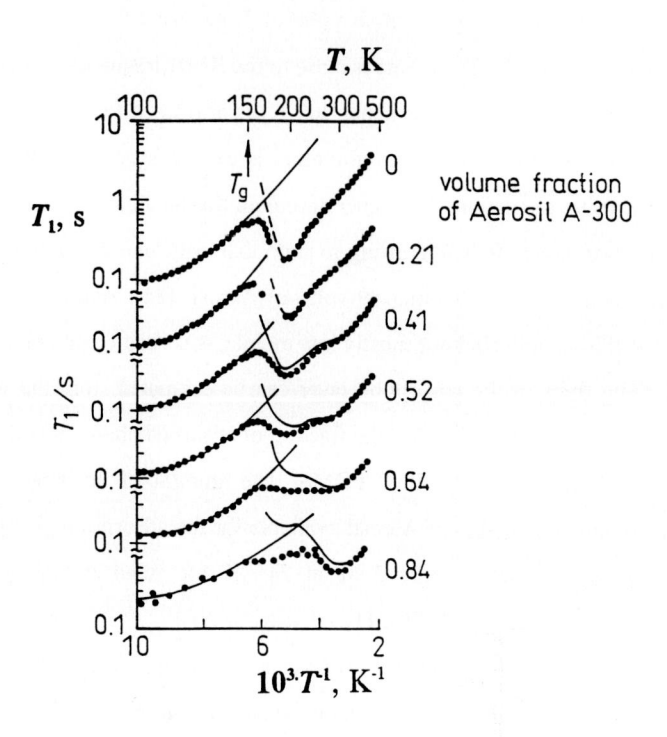

Fig. 6. Temperature dependence of ^{1}H T_1 of unfilled and filled PDMS [7]; experimental values are shown by points while the contribution to T_1 from C_3-rotation of the CH_3 group and from chain motions are shown by solid lines; the volume fraction of hydrophilic Aerosil (380 m^2 g^{-1}) in mixtures is indicated; below 230 K, measurements were performed for amorphisized samples [7]; the proton resonance frequency is 90 MHz

I.1.2 Dynamic Mechanical Experiments

The NMR data presented above reveal a dynamic heterogeneity of filled PDMS in the frequency range from about 10 kHz to 100 MHz. To determine whether the heterogeneity remains at lower frequencies, dynamic mechanical measurements are performed. The results for cured, unfilled silicon rubber are compared with those for filled samples containing different fraction of hydrophilic Aerosil (380 m^2 g^{-1}). For a more straightforward analysis of the mechanical experiments, a random poly(dimethyl/methylphenyl) siloxane copolymer containing approximately 90 mol% dimethyl- and 10 mol% methylphenylsiloxane units has been used for sample preparation. This copolymer is fully amorphous over the whole temperature range. The results of torsion experiments at a frequency of 1.6 Hz are shown as a function of temperature in Fig. 7.

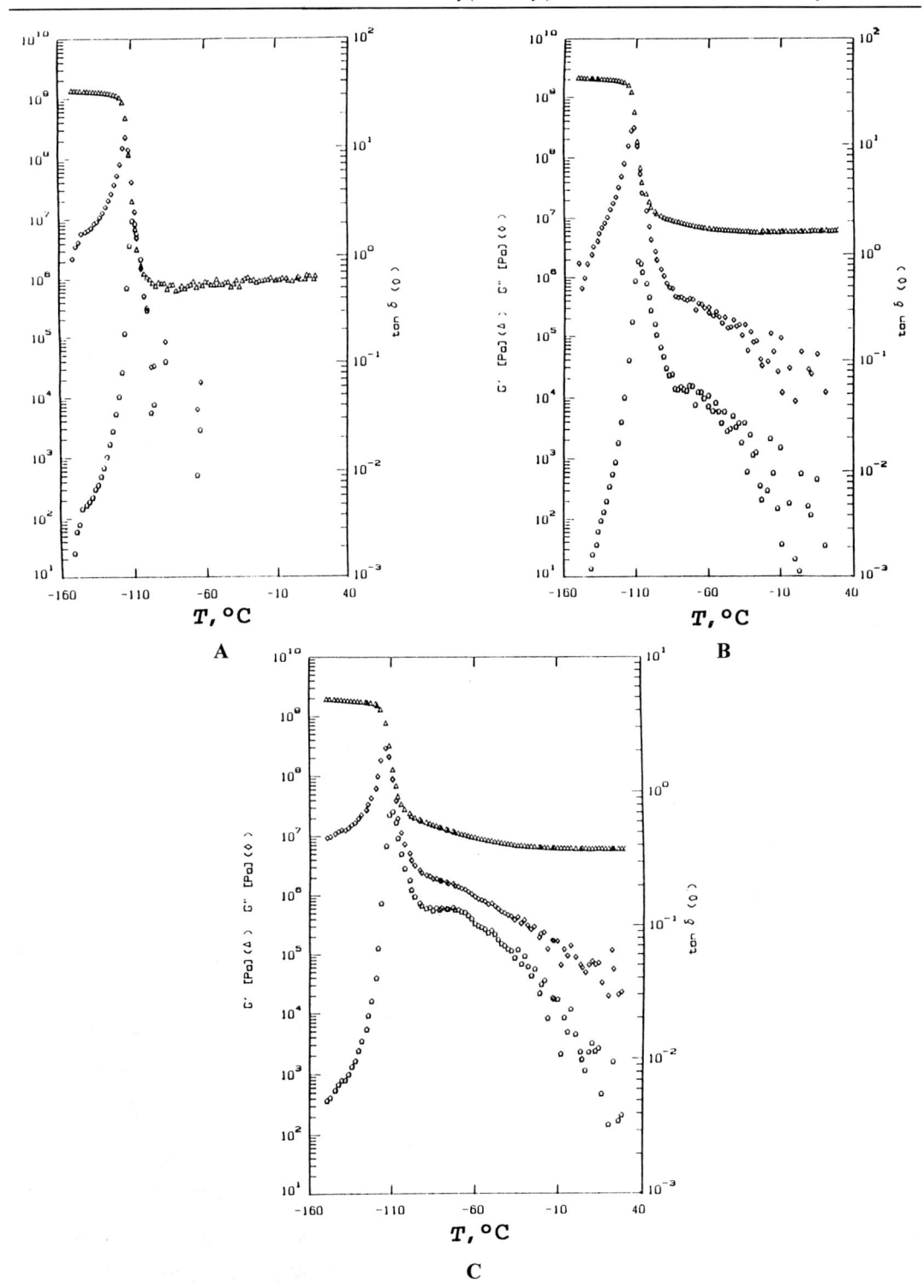

Fig. 7. Storage modulus (Δ), loss modulus (◊) and *tanδ* (○) as a function of temperature for unfilled (A) and filled silicon copolymer containing 12 vol% (30 phr) (B) and 17 vol% (46 phr) (C) of Aerosil (380 m^2 g^{-1}) [49]; the torsion with the amplitude of 0.05% was measured for a rectangular sample at 1. 6 Hz on Rheometrics, INC, RMS 800; samples were cured with 2 wt. % dicumyl peroxide

The rubber modulus increases with an increasing volume fraction of Aerosil. The modulus increase can be caused by the elastomer-filler and filler-filler interactions and by an increase of effective filler content. A very sharp peak for the *tan*δ is observed at 163 K for an unfilled crosslinked sample. This maximum corresponds to the glass transition of the rubber. Furthermore, it is observed that the T_g of the rubber does not change in the presence of filler. However, the second maximum of *tan*δ can be seen in the vicinity of 200 K for filled samples. The intensity of this maximum becomes more pronounced with increasing Aerosil content. This observation is in agreement with the results of the ^1H and ^2H T_1 relaxation study, as demonstrated in Fig. 4a and 6, respectively. Therefore, it seems reasonable to assign the maximum for *tan*δ at 200 K to the motion of adsorbed chain units. This maximum is observed at a lower temperature than the ^1H and ^2H T_1 minimum for the adsorbed chain units (at about 280 K) due to difference in frequency of these methods: 1.6 Hz and 46–90 MHz, respectively.

Finally, the NMR and the dynamic mechanical study show that two regions are present in filled silicone rubbers above the T_g, which differ significantly in local chain mobility: immobilized chain units adsorbed at the filler surface and mobile chain units outside the adsorption layer. The local chain motions outside the adsorption layer are similar to those for unfilled rubbers. Chain motions in the adsorption layer however are strongly restricted. The frequency of chain motions in the adsorption layer at 300 K is comparable to the frequency of chain motions in a crosslinked PDMS containing 3–4 elementary chain units between network junctions [26].

I.2 The Thickness of the Adsorption Layer

The thickness of the adsorption layer was estimated from the fraction of adsorbed chain units measured by means of ^1H T_1, T_2 and ^2H T_1 relaxation studies [7, 8, 10, 12]. From the known value of the specific surface of Aerosil, its volume fraction in mixtures and the fraction of low mobile chain units at the Aerosil surface, the thickness of the adsorption layer is estimated assuming uniform coverage of the filler particles by a PDMS layer of constant thickness. This calculation leads to a value of about 0.8 nm [7]. This value is increased by a factor 1.5–2, if a part of the filler surface will not be accessible for PDMS chains due to direct contacts between the primarily filler particles in aggregates [27]. Thus, the chain adsorption causes a significant restriction of local motions only in one or two monolayers adjacent to the filler surface. A similar estimation of the adsorption layer thickness has been obtained by other methods such as, e.g. dielectric experiment [27], adsorption study [3], the viscosity of the boundary layer for silicon liquids at the surface of a glass [5], molecular dynamics simulations [6], and ^{13}C NMR relaxation experiments [22].

It can be concluded that the direct energetic effect of the surface force field can be estimated to be of the order of one to two monolayers. Due to the local origin of the adsorption force and high flexibility of the siloxane chain, the local motions of PDMS chains at a distance more than 2 nm from the Aerosil surface are about the same as in unfilled PDMS [7, 8, 21].

I.3 Adsorption-Desorption of Dimethylsiloxane Chain Units at the Surface of Hydrophilic Aerosil

I.3.1 ^{1}H NMR study

Solid state NMR is a well established method to study the adsorption of low molecular weight molecules on the surface of solids [28]. However, the application of this method for a quantitative study of elastomer/adsorbent systems is complicated, due to the non-exponential T_2 relaxation for high molecular weight molecules and slowing down of chain diffusion in the direction perpendicular to the filler surface. Since the ^{1}H NMR relaxation behavior for siloxane oligomers at the Aerosil surface does not differ significantly from that for filled PDMS [9], a low molecular weight analog of PDMS has been used to study adsorption-desorption of dimethylsiloxane chain units on the surface of Aerosil. The oligomer molecule (ODMS) used to this aim has the structure $[(CH_3)_3SiO]\{Si(CH_3)_2O\}_3[Si(CH_3)_3]$.

Mixtures of hydrophilic Aerosil with the oligomer have been studied by ^{1}H T_1 and T_2 relaxation experiments in a wide temperature range [9]. A two-phase model was used to describe the relaxation of ODMS molecules at the surface of Aerosil. According to this model [29], molecules at the surface of a solid co-exist in adsorbed and non-adsorbed states. In a certain period of time, molecules perform jumps between these two phases. The lifetime of a molecule in adsorbed and desorbed states is detected from peculiarities of the temperature dependence of T_1 and T_2 NMR relaxation times [29]. It was shown that, already at 240 K, the lifetime of ODMS molecules in the adsorbed state is about 50 µs, i.e., the frequency of adsorption-desorption at this temperature is close to 3 kHz [9]. At about 500 K, nearly all ODMS molecules are desorbed.

Following the model suggested above, a schematic view of an adsorbed fragment of PDMS chain is shown in Fig. 8 [48].

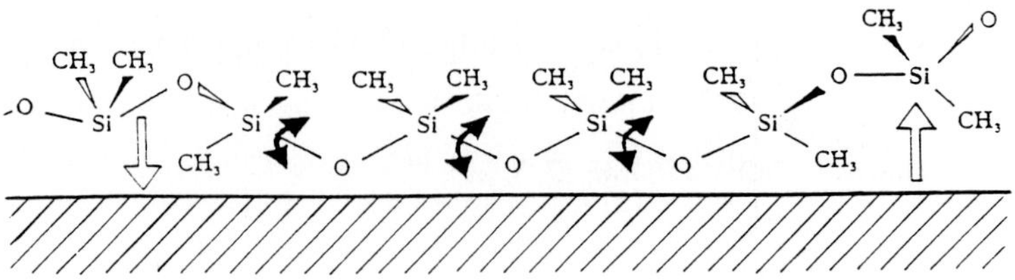

Fig. 8. Schematic view of an adsorbed fragment of PDMS chain; filled arrows denote anisotropic motions of chain units during the lifetime in the adsorbed state; open arrows denote adsorption-desorption of chain units; it is suggested that a length of adsorption-desorption jumps is of the order of 0.1–0.5 nm

According to this model, the temperature dependence of molecular motions for adsorbed and non-adsorbed chain units in filled PDMS containing hydrophilic Aerosil is shown in Fig. 9 [9]. The lowest temperature motion is a C_3-rotation of the CH_3 groups around the Si–C bond (line 1 in Fig. 9). The rate of the α-relaxation (points 2 in Fig. 9) in filled PDMS is close to that for unfilled sample (line 2 in Fig. 9). It has been proposed that independence of the mean average frequency of α-relaxation process on the filler content in filled PDMS is due to defects in the chain packing in the proximity of primarily filler particles [7]. Furthermore, the chain adsorption does not restrict significantly the local chain motion, which is due to high flexibility of the siloxane main chain as well as due to fast adsorption-desorption processes at temperatures well above T_g.

As shown in Fig. 9, the frequency of motion of adsorbed chain units is significantly lower than that of the chain motion outside the adsorption layer (line 3 and points 2 in Fig. 9, respectively). The amplitude of the motion of adsorbed chain units is hindered too, as was discussed in Section I.1.1.1. Besides the restricted motion of adsorbed chain units, the slower adsorption-desorption processes take place in the adsorption layer. It is suggested that the length of adsorption-desorption jumps is about 0.1–0.5 nm. The temperature-frequency domain for adsorption-desorption processes is shown by the dashed area in Fig. 9. The energy activation for the chain adsorption is about 8–12 kJ mol^{-1} per monomer unit [9].

Summarizing this section it can be stated that the adsorption bonds in filled PDMS have a dynamic origin. With increasing temperature, the frequency of adsorption-desorption processes in the adsorption layer increases and the adsorption-desorption equilibrium shifts to the chain desorption. At room temperature, the lifetime for the dimethylsiloxane chain units in the adsorption state is very short: chain units adhere to the filler surface only for tens of microseconds.

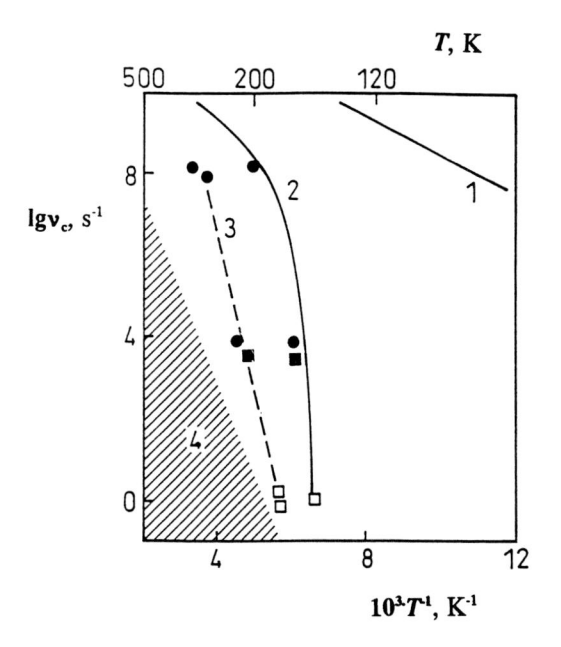

Fig. 9. Temperature dependence of the frequency of C_3-rotation of the CH_3 group around the Si–C bond (1), motion of chains outside the adsorption layer (2, points), anisotropic motion of adsorbed chain units (3) in filled PDMS containing hydrophilic Aerosil [9]; the temperature-frequency domain for adsorption-desorption processes is shown by the dashed area (4); solid line (2) shows the temperature dependence of chain motions (α-relaxation) in unfilled PDMS

I.3.2 Broad-Band Dielectric Spectroscopy Study

The broad-band dielectric study of highly filled PDMS is complementary to the NMR study of molecular motions in filled PDMS. The dielectric experiments were performed in the frequency range of 10^{-1}–10^6 Hz [27]. A combined analysis of the dielectric spectra both for filled PDMS and the pure components of the mixtures was used to assign the dielectric losses to motions of adsorbed and non-adsorbed PDMS chain units. As discussed above, the interpretation of the results is based on a two-phase model assuming the exchange of chain units at the surface of Aerosil between adsorbed and non-adsorbed states.

The dielectric spectra for a mixture of PDMS with hydrophilic Aerosil (380 m^2 g^{-1}) show three relaxation processes which are related to chain motions. The frequencies of these relaxations are shown as a function of temperature in Fig. 10.

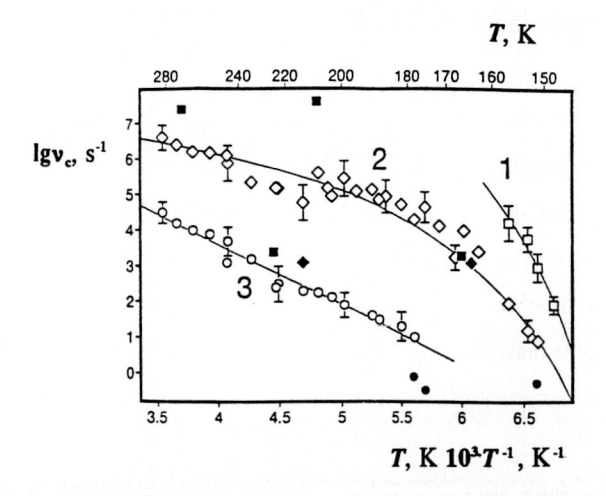

Fig. 10. Transition map for the mixture of hydrophilic Aerosil with PDMS [27]: the relaxation of chain units outside the adsorption layer is represented by symbol □, anisotropic motion of chain units inside the adsorption layer is shown by symbol ◊, the slowest chain motion related to adsorption-desorption processes in the adsorption layer is designated by symbol O; the data of the first two relaxation processes are fitted by the WLF function, the temperature dependence of the slowest relaxation shows the Arrhenius-like behavior; for comparison data from previous ^{1}H T_1 and T_2 NMR experiments ■, mechanical ●, and dielectric spectroscopy ♦ are given

Fast relaxation processes (□, ◊) show a Williams-Landel-Ferry (WLF) type temperature dependence which is typical for the dynamics of polymer chains in the glass transition range. In accordance with NMR results, which are shown in Fig. 9, these relaxations are assigned to motions of chain units inside and outside the adsorption layer (◊ and □, respectively). The slowest dielectric relaxation (O) shows an Arrhenius-type behavior. It appears that the frequency of this relaxation is close to 1–10 kHz at 240 K, which was also estimated for the adsorption-desorption process by NMR (Fig. 9) [9]. Therefore, the slowest relaxation process is assigned to the dielectric losses from chain motion related to the adsorption-desorption.

According to the dielectric experiments [27], an activation energy for adsorption-desorption processes at PDMS-hydrophilic Aerosil interface is equal to 32 kJ mol^{-1}. This value is about three times larger than that from NMR experiments [9]. This discrepancy may be due to the difference in the spatial scale of motions in the adsorption layer as detected by NMR and dielectric spectroscopy [54].

I.4 The Network Structure in Filled PDMS

The elastic properties of rubbers are primarily governed by the density of network junctions and their ability to fluctuate [35]. Therefore, knowledge of the network structure composed of chemical, adsorption and topological junctions in filled elastomers as well as their relative weight is of a great interest. The ^{1}H T_2 NMR relaxation experiment is a well established method for the quantitative determination of the network structure in the elastomer matrix outside the adsorption layer [14, 36]. The method is especially attractive for the analysis of the network structure in filled elastomers since filler particles are "invisible" in this experiment due to the low fraction of protons at the Aerosil surface as compared with those in the host matrix.

NMR analyses of the network structure in bound PDMS rubber containing different fractions of Aerosil (300 m^2 g^{-1}) were performed [11–14]. Several types of network junctions contribute to the total density of network junctions in filled elastomers, i.e., chemical and adsorption junctions, chain entanglements (about 300 monomer units between apparent entanglements) and topological hindrances for elastomer chains due to confining of chains in restricted geometry (elastomer-filler entanglements). The following procedure was used to determine the mean average molecular mass of PDMS chains between chemical, physical and topological junctions as well as the relative fraction of these junctions in the elastomer matrix outside the adsorption layer [13, 14]. It is assumed that the distribution of network junctions in the samples studied is rather uniform after 3 month extraction used for preparation of the bound rubber. The experiments have been carried out for bulk samples, swollen samples, and samples swollen in the presence of ammonium. The mean average molar mass of PDMS chains between all types of junctions was measured for bulk samples. It is assumed that topological junctions largely vanish in swollen samples. Low mobile adsorbed chain units are detected in swollen samples by the ^{1}H T_2 experiment [13]. Therefore, the network structure arising from chemical and adsorption junctions is determined for swollen samples. The molar mass of PDMS chains between chemical junctions is estimated for samples swollen in the presence of ammonium. Due to the high energy of adsorption of ammonium molecules, these cleave more weak adsorption bonds between PDMS and the Aerosil surface [37].

The apparent number of monomer units between different types of network junctions in bound rubber is shown as a function of the volume fraction of Aerosil in Fig. 11 [14].

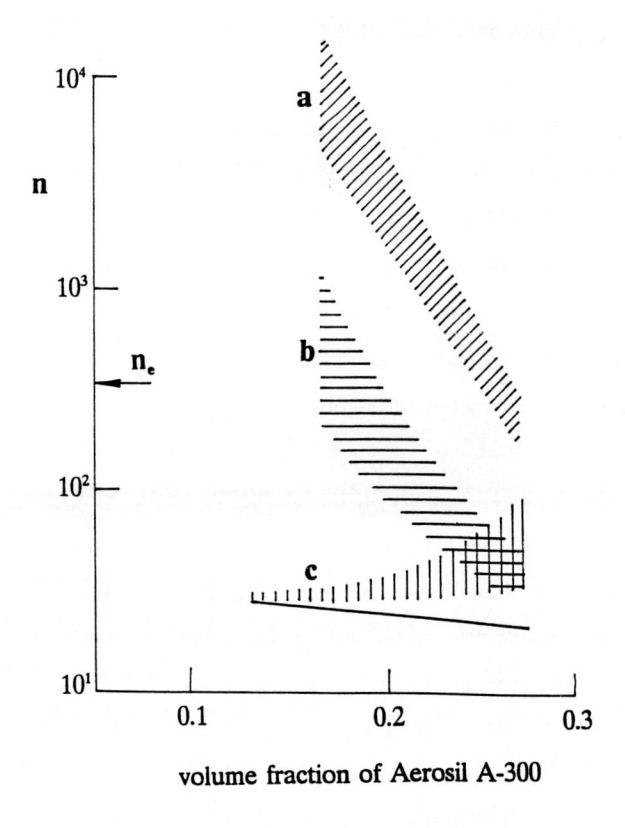

Fig. 11. The apparent number of elementary chain units (n) between network junctions (line) and contribution in it from chemical (a) and adsorption (b) junctions, and topological hindrances near the filler surface (c) as a function of the volume fraction of Aerosil (300 m^2 g^{-1}) in bound PDMS rubber [14]; the apparent number of elementary chain units between transient entanglements is shown by an arrow; the absolute error for the determination of the n value is shown by the dashed area

It follows from this Fig. that the amount of chemical junctions in silicon rubber increases with increasing fractions of Aerosil. The chemical junctions are apparently formed by scission of PDMS chains under the mechanical forces during milling. However, the fraction of these junctions is the lowest. The fraction of adsorption junctions increases proportionally to the filler content as shown in Fig. 11. The major contribution to the network structure is provided by topological hindrances near the filler surface as shown in Fig. 11.

With regard to the molecular origin of these hindrances, it should be mentioned that the linear dimension of Aerosil (300 m^2 g^{-1}) particles (about 7 nm) is comparable with the mean average distance between primarily filler particles in the PDMS matrix. Since the Gaussian chain statistics might be applied for PDMS chains in the filled rubbers [18], it is easy to show that the chain portions between the primarily filler particles should contain about 30–80 elementary units. This value is in the same range as the apparent number of elementary chain units between topological hindrances as measured by NMR

(Fig. 11). It is, therefore, highly probable that the bulky filler particles impose geometrical hindrances (entropy constraints) for the chain dynamics at the time scale of the T_2 NMR experiment (of the order of 1 ms). This effect may be compared with the effect of transient chain entanglements on chain dynamics in polymer melts. It should be remarked that the entanglements density estimated for PDMS melts by NMR is close to its value from mechanical experiments [38]. Therefore, it can be assumed that topological hindrances from the filler particles can also be of importance in the stress-strain behavior of filled elastomers.

I.5 Influence of the Chemical Nature of the Aerosil Surface on its Interaction with PDMS

Two types of Aerosil with different surface activity have been studied: hydrophilic and hydrophobic Aerosil. Hydrophilic Aerosil contains on its surface hydroxyl groups which are the sites of adsorption. The surface groups of hydrophilic Aerosil are able to interact with the PDMS chain both through permanent dipoles in the partially ionic siloxane bond via permanent dipole-dipole interactions and through even weaker van der Waals forces. In contrast to hydrophilic Aerosil, non-polar trimethylsilyl groups on the surface of hydrophobic Aerosil effectively decrease the dipole-dipole interaction and mainly weak van der Waals forces are formed between methyl groups of PDMS and trimethylsilyl surface groups of Aerosil.

In the present study, the surface activity of Aerosil is characterized by molecular motions of adsorbed chain units. Highly filled PDMS has been studied by ^{1}H T_2 NMR relaxation experiments and ^{2}H NMR spectra [8, 10, 21]. The ^{2}H NMR spectra are compared in Fig. 12 for highly filled samples containing hydrophilic and hydrophobic Aerosil [21].

According to the discussion in Section I.1.1.1, a broad Pake spectrum is observed for low mobile PDMS chain units, whereas a narrow line is recorded if the frequency of chain motions exceeds 10 kHz–1 MHz. The effect of hydrophilic and hydrophobic Aerosil on the chain motion is remarkably different. For PDMS filled with hydrophilic Aerosil, a broad ^{2}H NMR line is observed over the whole temperature range studied. The motional narrowing is not complete, even at 433 K, nearly 300 K above the T_g of PDMS. This means that mobility of PDMS chain units at the surface of hydrophilic Aerosil is hindered by adsorption interactions even at 433 K, although the strength of adsorption interactions decreases with increasing temperature. On the other hand, PDMS chains at the surface of hydrophobic Aerosil are already desorbed at about 200 K, since the NMR line is completely narrowed at this temperature as shown in Fig. 12.

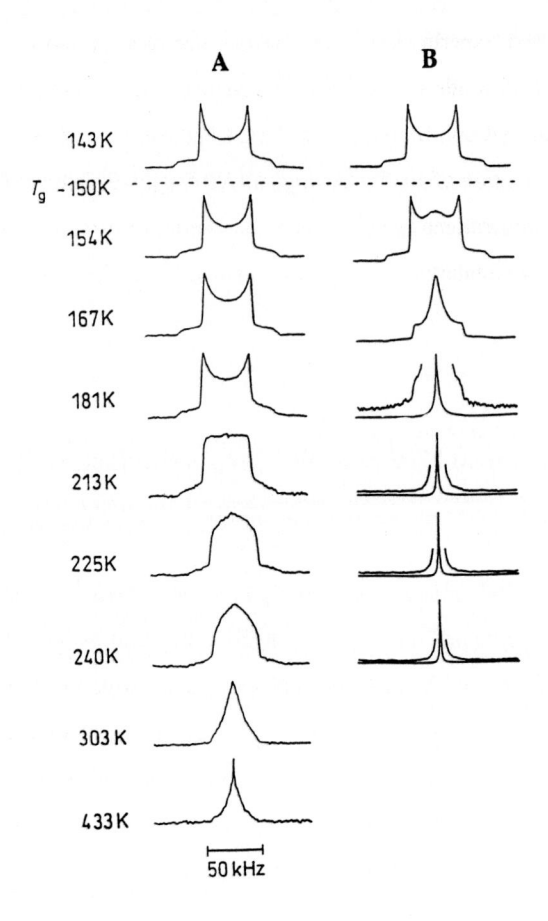

Fig. 12. Twelve fully relaxed ^{2}H solid-echo spectra for mixtures of PDMS with 89 vol% of hydrophilic Aerosil (380 m^2 g^{-1}) (1813 phr) (A) and hydrophobic Aerosil (180 m^2 g^{-1}) containing trimethylsilyl groups on its surface (B) [21]

The chain mobility at the surface of the Aerosil is also characterized by ^{1}H T_2 relaxation experiments [8, 10]. The relaxation rate, $(T_2^{ad})^{-1}$, for the chain units adsorbed by hydrophilic and hydrophobic Aerosil is compared in Fig. 2. A larger value for the relaxation rate corresponds to a lower mobility of the chain units. As shown in Fig. 2, the $(T_2^{ad})^{-1}$ value for chain units adsorbed by hydrophilic Aerosil decreases gradually with increase in temperature and remains larger than the T_2^m value for non-adsorbed chain units at temperatures well above T_g. In contrast, the $(T_2^{ad})^{-1}$ value for the adsorption layer at the surface of hydrophobic Aerosil decreases rapidly in the temperature range 150–200 K. Above this temperature, the T_2 relaxation experiment indicates the absence of any significant hindrances for the local chain motion caused by hydrophobic Aerosil [10]. In addition, the fraction of low mobile chain units in the adsorption layer is lower for PDMS filled by hydrophobic Aerosil than that for the same content of the hydrophilic Aerosil [10, 19].

Thus, the ^{1}H T_2 relaxation experiments and the analysis of the ^{2}H solid-echo spectra show that the strength of adsorption bonds depends strongly both on type of Aerosil surface and temperature. The chain adsorption at the surface of hydrophilic Aerosil significantly restricts motions of chain units adjacent to the filler surface. The local motion of chain units at the surface of hydrophobic Aerosil is not hindered by adsorption interactions at temperatures above 200–250 K. The lifetime of chain units in the adsorbed state approaches to zero at approximately ~250 K and ~500 K for hydrophobic and hydrophilic Aerosil, respectively [9].

Part II

Molecular Motions in the Adsorption Layer in Relation to Mechanical Properties

II.1 A Molecular Model Used for Qualitative Analysis of Stress-Strain Properties of Filled PDMS

It is generally believed that the nature of elastomer-filler interactions is of major importance for marked improvement in mechanical properties of the filled elastomers [39–44]. Adsorption of elastomer chains at the filler surface has a double effect on the enhancement of mechanical properties of filled elastomers. Firstly, *the ability of filler particles to share deformation* increases due to adsorption interactions between the filler particles and the host matrix. Secondly, these interactions provide significant amount of *adsorption and topological junctions in the elastomer matrix outside the adsorption layer*. It appears that less mobile chain units in the adsorption layer do not contribute directly to the rubber modulus, since the fraction of PDMS chain units in this layer is only a few percent of the Aerosil content used in commercial rubbers [7, 8, 12, 21].

A large number of macroscopic properties of elastomer networks are closely related to the density of network junctions and the extent of their fluctuations. *Qualitatively, any increase of network density causes an increase in stress, whereas fluctuations of network junctions leads to a decreasing stress.* It is generally believed that a formation of additional network junctions resulting from the presence of filler particles in the elastomer matrix is one of the reasons for the improvement of mechanical properties of filled elastomers. However, the application of macroscopic techniques does not provide reliable results for the network structure in filled elastomers. Furthermore, a lack of information exists on the dynamic behavior of adsorption junctions. The present study fills the gap of knowledge in this area.

It appears that the following peculiarities of the network structure in the elastomer matrix outside the adsorption layer are of importance for a molecular understanding of stress-strain behavior for these materials:

1) The main contribution to the network density in filled PDMS is provided by adsorption junctions and topological hindrances from the filler particles
2) The number of adsorption junctions and their strength depend significantly both on temperature and the type of Aerosil surface

Topological hindrances (entropy constraints) from filler particles provide the largest contribution to the total network density. The apparent density of these hindrances in bound rubber is significantly larger than that for chain entanglements in PDMS melts, i.e., about 40–80 [11–14], and 300 elementary chain units [38], respectively. It appears that these hindrances significantly restrict the large spatial scale fluctuation of PDMS chains and fluctuations of chemical and adsorption junctions, in a similar way as chain entanglements do. It seems that the amount of topological hindrance does not vary significantly with temperature due to the bulkiness of filler particles.

Adsorption junctions at the surface of active fillers are of importance due to the large total elastomer-filler interfacial area. The adsorption of chain units at the Aerosil surface causes a significant restriction of local chain motions in the first layer adjacent to the filler surface. The low mobile adsorbed chain units represents another type of network junction in filled elastomers. However, the adsorbed chain units are not rigidly linked to the surface of Aerosil above T_g. The lifetime of chain units in the adsorbed state is already very short at room temperature: chain units adhere to the filler surface for only tens of microseconds [9]. It was shown in Part I that the fraction of adsorbed chain units decreases on heating due to chain desorption. Therefore, the amount of adsorption junctions decreases with the increase of temperature as shown in Fig. 13.

$T \geq T_g$

$T > T_g$

$T \approx T_g + 350$ K
(at the surface of hydrophillic Aerosil)

$T \approx T_g + 50$ K
(at the surface of hydrophobic Aerosil)

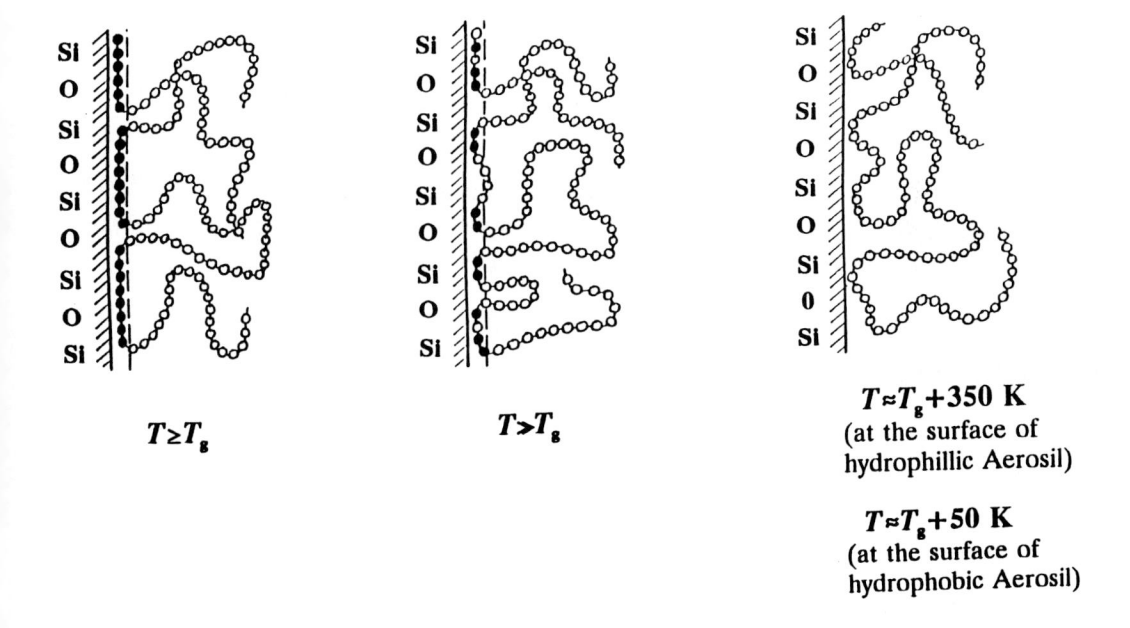

Fig. 13. Schematic view of adsorbed chain fragments at different temperatures; PDMS chains are completely desorbed at about 200 K and 500 K for hydrophobic and hydrophilic Aerosil, respectively; less mobile chain units in the adsorbed state are shown by filled circles; the dotted line corresponds to the estimated border for the adsorption layer

Moreover, motions of chain units in the adsorption layer become less hindered with increasing temperature. This means that the energy to overcome adsorption bonding and to break away chains from the filler surface in strained rubbers decreases at higher temperatures. Moreover, the ability of filler particles to share deformation decreases too, which is due to weaker PDMS-Aerosil interface as temperature increases. The strength of adsorption bonds strongly depends on the type of Aerosil surface. The chain adsorption at the surface of hydrophilic Aerosil significantly restricts motions of chain units adjacent to the filler surface, in contrast to hydrophobic Aerosil. The lifetime of chain units in the adsorbed state approaches zero at ~250 K and ~500 K for hydrophobic and hydrophilic Aerosil, respectively [9]. Finally, the response of filled elastomers to mechanical deformation is expected to be time-temperature dependent, due to the dynamic origin of the adsorption junctions.

II.2 Mechanical Properties of Filled PDMS in Relation to Network Structure and Chain Dynamics at the PDMS-Aerosil Interface

Considerable effort has been spent to explain the effect of reinforcement of elastomers by active fillers. Apparently, several factors contribute to the property improvements for filled elastomers such as, e.g., elastomer-filler and filler-filler interactions, aggregation of filler particles, network structure composed of different types of junctions, an increase of the intrinsic chain deformation in the elastomer matrix compared with that of macroscopic strain and some others factors [39–44]. The author does not pretend to provide a comprehensive explanation of the effect of reinforcement. One way of looking at the reinforcement phenomenon is given below. An attempt is made to find qualitative relations between some mechanical properties of filled PDMS on the one hand and properties of the host matrix, i.e., chain dynamics in the adsorption layer and network structure in the elastomer phase outside the adsorption layer, on the other hand. The influence of filler-filler interactions is also of importance for the improvement of mechanical properties of silicon rubbers (especially at low deformation), but is not included in the present paper.

II.2.1 Influence of the Type of Aerosil on Stress-Strain Curves

Stress-strain curves for PDMS, containing different types of Aerosil, are compared in Fig. 14. As can be seen, the total interfacial area between Aerosil particles and elastomer matrix and/or the amount of filler and its adsorption ability are of great importance for the improvement of stress-strain characteristics, i.e., modulus, energy of deformation, and tensile strength at break. However, these experiments do not provide straightforward information on the role of PDMS-Aerosil interactions present in the reinforced matrix. Apparently, the effect of chain adsorption is difficult to isolate from other aspects of reinforcement such as, e.g., aggregation of filler particles, filler-filler interactions and density of chemical junctions in the elastomer matrix.

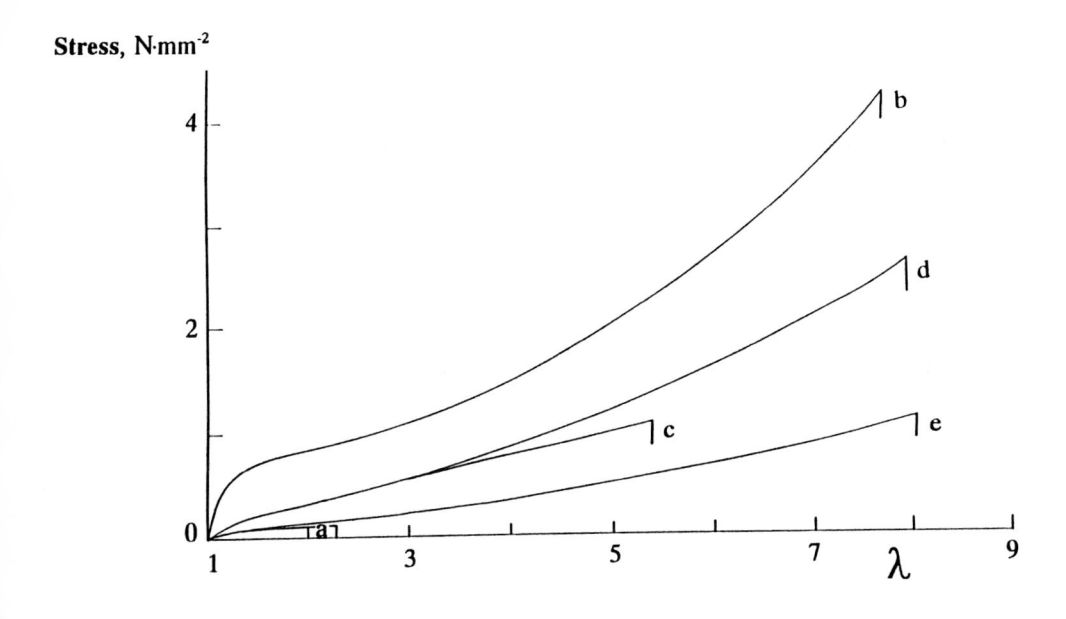

Fig. 14. Stress-strain curves for unfilled (a) and filled PDMS peroxide vulcanizates containing 9.4 vol% of hydrophilic Aerosil A 300 (300 m^2 g^{-1}) (23 phr) (b) and OX 50 (50 m^2 g^{-1}) (c), and hydrophobic Aerosil R 812 (260 m^2 g^{-1}) (d) and R 809 (45 m^2 g^{-1}) (e) [52]. T= 295 K

II.2.2 Stress-Strain Behavior at Different Temperatures: The Modulus and Energy of Deformation

It was shown above that the number of adsorption junctions as well as their strength strongly depends both on temperature and the type of Aerosil surface. Therefore, an analysis of the temperature dependence of the modulus and the energy of deformation can be of use for understanding the role of chain adsorption in these properties.

Modulus measurements have been performed for cured, unfilled and filled siloxane rubber which does not form a crystalline phase at low temperature. Therefore, the interpretation of stress-strain behavior for these samples is not complicated by crystallization. The temperature variation of the modulus is similar for 25 % and 100 % elongation of the samples. As an example, the temperature dependence of the modulus for 25 % elongation is compared for unfilled and filled samples in Fig. 15.

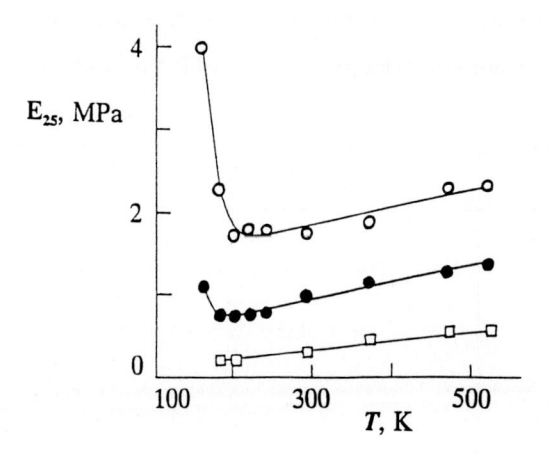

Fig. 15. Temperature dependence of modulus at 25 % elongation for unfilled (□) and filled (●, ○) cured, random poly(dimethylsiloxane/methylphenyl)siloxane copolymer containing approximately 90 mol% dimethyl-, 10 mol% methylphenyl-, and 0.3 mol% methylvinyl- units [45, 46]: (□) without filler, filled by (●) hydrophobic Aerosil (300 m^2 g^{-1}) and (○) hydrophilic Aerosil (60 m^2 g^{-1}); the weight ratio filler/elastomer is 30:100; the samples are cured at 400 K with 2 wt% of dicumyl peroxide in a presence of a platinum catalyst

The presence of filler in the rubber as well as the increase of the surface ability of the Aerosil surface causes an increase in the modulus. The temperature dependence of the modulus is often used to analyze the network density in cured elastomers. According to the simple statistical theory of rubber elasticity, the modulus should increase twice for the double increase of the absolute temperature [35]. This behavior is observed for a cured unfilled sample as shown in Fig. 15. However, for rubber filled with hydrophilic and hydrophobic Aerosil, the modulus increases by a factor of 1.3 and 1.6, respectively, as a function of temperature in the range of 225–450 K. It appears that less mobile chain units in the adsorption layer do not contribute directly to the rubber modulus, since the fraction of this layer is only a few percent [7, 8, 12, 21]. Since the influence of the secondary structure of fillers and filler-filler interaction is of importance only at moderate strain [43, 47], it is assumed that the change of the modulus with temperature is mainly caused by the properties of the elastomer matrix and the adsorption layer which cause the filler particles to share deformation. Therefore, the moderate decrease of the rubber modulus with increasing temperature, as compared to the value expected from the statistical theory, can be explained by the following reasons: a decrease of the density of adsorption junctions as well as their strength, and a decrease of the ability of filler particles to share deformation due to a decrease of elastomer-filler interactions.

This conclusion is supported by the analysis of the temperature dependence of the deformation energy. With increasing temperature from 203 K to about 300–400 K, a slight decrease of the deformation energy is observed for samples filled with both hydrophilic and hydrophobic Aerosil. Above 400 K, the deformation energy starts to increase, as shown in Table 1.

Sample		203 K	293 K	423 K	473 K
Siloxane/A300	A_1	10.50	9.90	9.20	11.50
	A_2	8.90	8.20	7.20	9.80
	ΔA	1.60	1.70	2.00	1.70
Siloxane/AM60	A_1	4.65	4.45	5.50	
	A_2	4.25	4.00	4.95	
	ΔA	0.40	0.45	0.55	
Unfilled	A_1	1.25	1.80	2.60	
Siloxane	A_2	1.25	1.80	2.60	
	ΔA	0.00	0.00	0.00	

Table 1. Energy for first (A_1) and second (A_2) deformation to the extension ratio $\lambda = 2$, and their difference (ΔA) as a function of temperature for cured, unfilled and filled silicon rubbers [45, 46, 56]; value for the deformation energy is given in kJ mol^{-1}

In contrast to the filled samples, the deformation energy for the unfilled ones increases proportionally to the increase in the absolute temperature according to the prediction of the simple statistical theory of rubber elasticity. Thus, it appears that the change of the modulus and the deformation energy with increasing temperature reveals a decrease of the density of adsorption junctions in the elastomer matrix, as well as a decrease of the ability of filler particles to share deformation, resulting from a weakening of elastomer-filler interactions.

II.2.3 Stress-Softening of Silicone Rubbers (Mullins Effect)

The data in Table 1 indicates that the deformation energy is lower for the second deformation compared with the first one. This effect is called the Mullins effect or stress-softening. Two explanations of this effect are given for the case of complete elastic recovery of a sample before its second deformation [39–43]:

1) A conversion of the "hard" phase adjacent to the filler particles to the soft phase
2) Progressive detachment of chain fragments adsorbed on the filler surface or their slippage along the filler surface; these explanations of stress-softening assume that adsorption junctions are low mobile at the time scale of deformation

It is remarkable that the stress-softening (ΔA) is constant within an experimental error in the temperature range 203–473 K for both filled samples as shown in Table 1. This is a very surprising result, since the widely accepted suggestions for the molecular origin of the stress-softening relate this effect to the properties of the adsorption layer. However, the present NMR study shows drastic changes in the dynamic behavior of the chain units at the Aerosil surface in the temperature range studied. Motions of chain units at the surface of hydrophilic Aerosil are strongly hindered at low temperatures (203 K), whereas PDMS chains are completely desorbed at about 473 K (see discussions in Sections I.1.1.2, I.3, I.5, and Fig. 9, 10) [9, 21]. Moreover, at about 240 K, the lifetime of siloxane chain units in the adsorbed state is about 50 µs in the case of PDMS filled with hydrophilic Aerosil [9]. It appears that this time is several orders of magnitude lower than the time that is required during stretching for pulling adsorbed chains over the filler surface over a distance comparable with the size of the elementary chain unit. If the stress-softening is related to a chain slippage or the conversion of hard phase to soft phase, a drastic difference in the mobility of adsorption chain units should cause a significant change in the stress-softening in the temperature range studied. However, the stress-softening does not depend on temperature. Hence, it is obvious that stress-softening has another molecular origin.

A possible interpretation for the stress-softening in silicon rubbers could be due to changes in the positions of filler particles relative to the direction of stretching at the first deformation. Since the filler particles provide numerous topological hindrances for elastomer chains (see Section I.4 and Fig. 11) [13, 14], the deformation energy for the second deformation could be lowered by redistribution of these topological hindrances caused by rearrangements of filler particles at the first deformation.

II.2.4 Tensile Strength at Break

The temperature dependence of the tensile strength at break (σ_r) of cured unfilled siloxane rubber is compared in Fig. 16 with that for two samples filled with hydrophilic and hydrophobic Aerosil [21, 46]. The difference in the tensile strength at break for the studied samples is determined only by effects of the filler, since the samples do not form a crystalline phase. With increasing temperature, the tensile strength at break for filled rubbers falls and approaches its value for the unfilled sample. This is observed at about 200–250 K and 500 K for rubbers containing hydrophobic and hydrophilic Aerosil, respectively. It is interesting to note that complete motional narrowing of ^{2}H solid-echo spectra for highly filled PDMS (Fig. 12) and the zero lifetime of polymer chain units in the adsorption state (see Section I.3) are observed at the same temperature. Therefore, it seems reasonable to assign the temperature change of σ_r to dynamic processes in the adsorption layer.

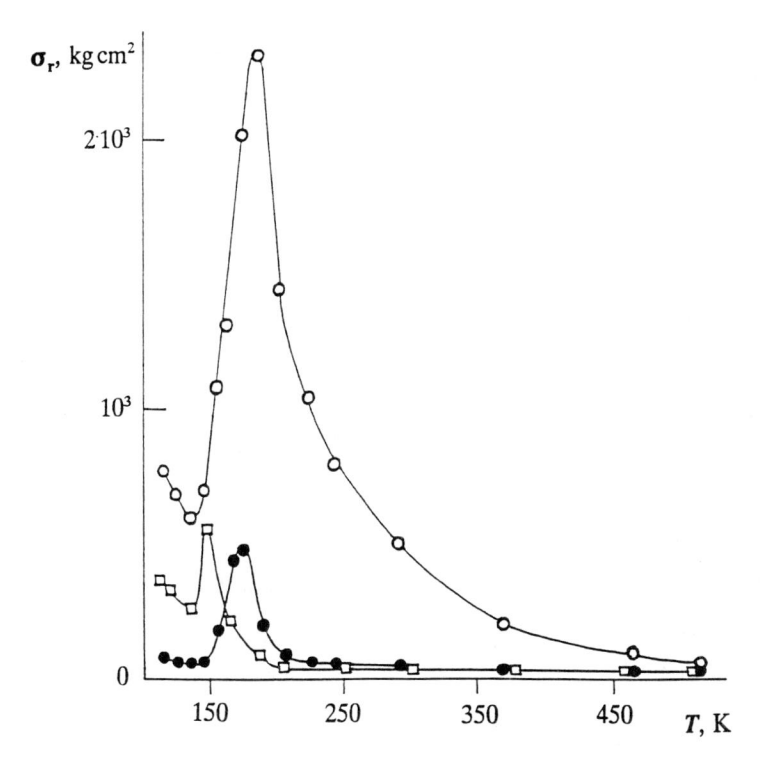

Fig. 16. Temperature dependence of the tensile strength at break referred to the actual cross sectional area (σ_r) for cured phenyl containing silicon rubber [21, 46]: (□) without filler, filled by (○) hydrophilic Aerosil and (●) hydrophobic Aerosil; the sample composition is described in the caption to Fig. 15

The NMR relaxation rate $(T_2^{ad})^{-1}$ characterizing the molecular mobility in the adsorption layer was related to the tensile strength at break for the studied sample [46]. As is seen in Fig. 2, a fast decrease of this relaxation rate is observed in the same temperature range as for σ_r, which is shown in Fig. 16.

The temperature dependence of the relative change of the tensile strength at break and the relative change of the $(T_2^{ad})^{-1}$ coincide as shown in Fig. 17.

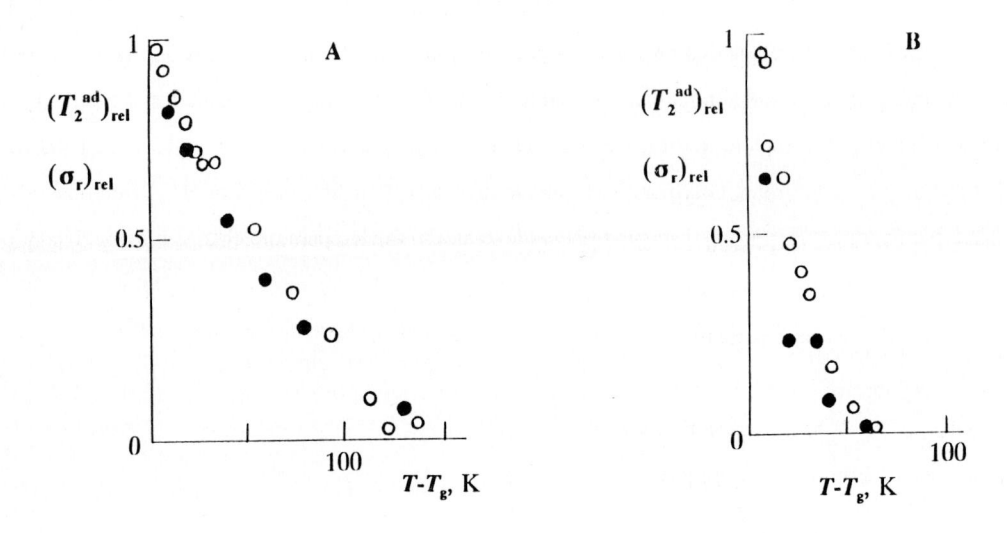

Fig. 17. Temperature dependence of the relative change of $(T_2^{ad})^{-1}$ (O) and the relative change of tensile strength at break (●) for silicon rubber filled by hydrophilic (A) and hydrophobic (B) Aerosil [46]; the value of $(T_2^{ad})^{-1}$ is measured for filled PDMS containing 60 vol% of hydrophilic Aerosil (300 m^2 g^{-1}) (336 phr) and hydrophobic Aerosil (60 m^2 g^{-1}); the sample composition is described in the caption to Fig. 15

This means that chain dynamics in the adsorption layer are closely related to the fracture behavior. The fracture of filled elastomers is a complicated phenomenon which is influenced by a number of parameters. Nevertheless, the results above suggest that the strength of elastomer-filler interactions is important in fracture behavior. The energy for the chain adsorption is estimated to be 8–12 kJ mol^{-1} per elementary unit [9]. Since a few chain units may form one adsorption junction, the energy of this junction would be higher. Moreover, it seems reasonable to suggest that the number of chain units forming an adsorption junction increases with sample elongation, due to the pulling out of short chain loops at the surface of the filler when strain becomes larger. Consequently, detaching of adsorption junctions may provide a temperature-dependent mechanism for high-energy losses during deformation. In addition, filler particles could effectively share the deformation and provide more uniform distribution of stress in the elastomer matrix, if elastomer chains are strongly bound to the filler particles by adsorption interactions. Therefore, it is assumed that the decrease of the tensile strength at break with increasing temperature is related to a decrease of the number of adsorption junctions as well as the strength of adsorption interactions at the elastomer-filler interface.

Conclusions

To sum it up it can be said, that solid state NMR is a very sensitive tool for the study of chain dynamics at the elastomer-filler interface as well as the network structure resulting from chemical junctions, adsorption junctions and topological hindrances from the filler particles. The method is of interest for establishing structure-property relations for filled elastomers.

A molecular model is proposed for the explanation of the temperature dependence of stress-strain characteristics such as, e.g., the modulus, the stress-softening and the tensile strength at break for filled PDMS. The model emphasizes the importance of the following molecular parameters:

1) Dynamical origin of the adsorption layer which provides the adsorption junctions
2) Topological hindrances (entropy constraints) arising from the filler particles

Acknowledgment: The experimental studies have been carried out in *Institute of Organoelement Compounds (Moscow, Russia)*, *MPI für Polymerforschung* (Mainz, Germany), and the *Institut für Makromolekulare Chemie* (Universiät Freiburg, Germany). The author is grateful to the many colleagues for the assistance. The fruitful collaboration with Prof. H. W. Spiess, Prof. V. S. Papkov, and Prof. W. Gronski is acknowledged. The generous fellowship support of the *Alexander von Humboldt Stiftung* and *SFB 60 (Deutsche Forschungsgemeinschaft)* is gratefully acknowledged. Comments on the manuscript of Dr. P. Steeman and Dr. W. Barendswaard are greatly appreciated.

References:

[1] M.-J. Wang, S. Wolff, *Rubber Chem. Techn.* **1991**, *64*, 559; J.-B. Donnet, *Kautschuk, Gummi, Kunststoffe* **1994**, *47*, 628.

[2] M.-J. Wang, S. Wolff, J.-B. Donnet, *Kautschuk, Gummi, Kunststoffe* **1992**, *45*, 11; S. Wolf, E. E. Tan, J.-B. Donnet, *Kautschuk Gummi Kunststoffe* **1994**, *47*, 485.

[3] A. Yim, R. S. Chahal, L. E. St. Pierre, *J. Colloid Interface Sci.* **1973**, *43*, 583.

[4]   T. Cosgrove, P. C. Griffiths, *Adv. in Colloid Interface Sci.* **1992**, *42*, 175; and references therein.

[5]   B. V. Deryagin, V. V. Karasev, I. A. Lavygin, I. I. Skorokhodov, E. N. Khomova, *Dokl. Akad. Nauk USSR* **1969**, *187*, 846.

[6]   I. Bitsanis, G. Hadziioannou, *J. Chem. Phys.* **1990**, *92*, 3827.

[7]   V. M. Litvinov, A. A. Zhdanov, *Dokl. Phys. Chem.* **1985**, *283*, 811; *Dokl. Phys. Chem* **1986**, *290*, 916.

[8]   V. M. Litvinov, A. A. Zhdanov, *Polym. Sci. USSR* **1987**, *29*, 1133.

[9]   V. M. Litvinov, *Polym. Sci. USSR* **1988**, *30*, 2250.

[10] V. M. Litvinov, M. Wobst, D. Reichert, H. Schneider, A. A. Zhdanov, *Acta Polymerica* **1988**, *39*, 243.

[11] V. M. Litvinov, A. A. Zhdanov, *Dokl. Phys. Chem.* **1986**, *289*, 759.

[12] V. M. Litvinov, A. A. Zhdanov, *Polym. Sci. USSR* **1988**, *30*, 1000.

[13] V. M. Litvinov, V. S. Papkov, A. A. Zhdanov, *Vysokomol. Soed.* **1988**, *B 30*, 343.

[14] V. M. Litvinov, V. G. Vasilev, *Polym. Sci. USSR* **1990**, *32*, 2231.

[15] V. M. Litvinov, *Int. Polym. Sci. Technol.* **1988**, *15*, T/28.

[16] O. Girard, J. P. Cohen-Addad, *Polymer* **1991**, *32*, 860.

[17] J. P. Cohen-Addad, P. Huchot, P. Jost, A. Pouchelon, *Polymer* **1989**, *30*, 143.

[18] J. P. Cohen-Addad, *Polymer* **1989**, *30*, 1820; *Polymer* **1992**, *33*, 2762.

[19] J. P. Cohen-Addad, R. Ebengou, *Polymer* **1992**, *33*, 379.

[20] J. P. Cohen-Addad, S. Touzet, *Polymer* **1993**, *34*, 3490.

[21] V. M. Litvinov, H. W. Spiess, *Makromol. Chem.* **1991**, *192*, 3005; *Makromol. Chem* **1992**, *193*, 1181.

[22] J. Van Alsten, *Macromolecules* **1991**, *24*, 5320; D. G. Rethwisch, J. Van Alsten, C. R. Dybowski, *Macromol. Symp.* **1994**, *86*, 171.

[23] V. J. McBrierty, J. C. Kenny, *Kautschuk Gummi Kunststoffe* **1994** *47*, 342; and refs. therein; N. K. Dutta, N. R. Choudhury, B. Haidar, A. Vidal, J. B. Donnet, L. Delmotte, J. M. Chezeau, *Polymer* **1994**, *35*, 4293; and refs. therein.

[24] V. M. Litvinov, B. D. Lavrukhin, V. S. Papkov, A. A. Zhdanov, *Doklady Phys. Chem.* **1983**, *271*, 543; *Polym. Sci. USSR* **1985**, *27*, 1715.

[25] H. W. Spiess, *Adv. Polym. Sci.* **1985**, *66*, 23.

[26] V. M. Litvinov, B. D. Lavrukhin, A. A. Zhdanov, K. A. Andrianov, *Polym. Sci. USSR* **1977**, *19*, 2330.

[27] K. U. Kirst, F. Kremer, V. M. Litvinov, *Macromolecules* **1993**, *26*, 975.

[28] H. Pfeifer, *NMR Basic Principles and Progress*, Springer, Berlin, **1972**, p. 53.

[29] J. R. Zimmerman, W. E. Brittin, *J. Phys. Chem.* **1957**, *61*, 1328.

[30] A. A. Kalachev, V. M. Litvinov, G. Wegner, *Makromol. Chem., Macromol. Symp.* **1991**, *46*, 365.

[31] V. M. Litvinov, unpublished results.

[32] M. Möller, *Adv. Polym. Sci.* **1985**, *66*, 59.

[33] A. E. Tonelli, *NMR Spectroscopy and Polymer Microstructure, The Conformational Connection*, VCH, New York, **1989**.

[34] S. Hayashi, T. Ueda, K. Hayamizy, E. Akiba, *J. Phys. Chem.* **1992**, *96*, 10922; and references therein.

[35] J. E. Mark, B. Erman, *Rubberlike Elasticity. A Molecular Primer*, Wiley, New York, **1988**.

[36] G. Simon, K. Baumann, W. Gronski, *Macromolecules* **1992**, *25*, 3624.

[37] K. E. Polmanteer, C. W. Lentz, *Rubber Chem. Techn.* **1975**, *48*, 795.

[38] A. Charlesby, R. Folland, J. H. Steven, *Proc. Roy. Soc. (London)*, **1977**, *A 335*, 189.

[39] G. Kraus, *Reinforcement of Elastomers*, Wiley, New York, **1965**.

[40] E. M. Danneberg, *Rubber Chem. Technol.* **1975**, *48*, 410.

[41] J. B. Donnet, A. Voet, *Carbon Black Physics, Chemistry, and Elastomer Reinforcement*, Marcel Dekker, New York, **1976**.

[42] Z. Rigbi, *Adv. Polym. Sci.* **1980**, *36*, 21.

[43] J. B. Donnet, A. Vidal, *Adv. Polym. Sci.* **1986**, *76*, 103.

[44] D. C. Edwards, *J. Mater. Sci.* **1990**, *25*, 4175.

[45] V. M. Litvinov, presented at Symposium *Measuring Techniques in Rubber Research*, Twente, The Netherlands, **1992**, p. 12.

[46] V. M. Litvinov, A. F. Bulkin, V. S. Papkov, to be published.

[47] M. G. Gerspacher, C. P. O'Farrell, *Kautschuk, Gummi, Kunststoffe* **1992**, *45*, 97; M. G. Gerspacher, C. P. O'Farrell, H. H. Yang, *Kautschuk, Gummi, Kunststoffe* **1994**, *47*, 349.

[48] The experiments were performed at the *MPI für Polymerforschung, Mainz, Germany*.

[49] It is suggested that adsorbed PDMS at the surface of Aerosil siloxane chain has preferably *trans-trans* conformation. This suggestion relies on the study of PDMS chain at the air/water interface [30], and analysis of ^{29}Si NMR MAS spectra for pure and filled PDMS containing 60 vol% of hydrophilic Aerosil [31]. Pure PDMS shows a single narrow resonance at –22.4 ppm in the ^{29}Si NMR spectrum. The NMR line for the filled sample is broadened and has a low field shoulder. The shift of the resonance to the lower field can be caused by an increase in *trans-trans* conformations [32, 33], and/or by an increase in the mean Si–O–Si angle due to flattening of adsorbed chain fragments [34].

[50] ^{1}H NMR parameters used depend mostly on amplitude and frequency of reorientations of the CH$_3$ protons due to the motion of the siloxane chain. A restricted reorientation of one monomer unit provides already an effective source for T_1 and T_2 relaxation. On the other hand, the dielectric relaxation is caused by the reorientations of the Si–O dipoles of the main chain. A few adjacent adsorbed units might be involved in the relaxation caused by adsorption-desorption.

[51] The experiments were performed at the *Institute für Makromolekulare Chemie, Universität Freiburg, Germany*.

[52] A random siloxane copolymer containing 90 mol% dimethyl-, 10 mol% methylphenyl-, and 0.3 mol% methylvinyl-chain units is used for the sample preparation. The copolymer is filled with hydrophilic Aerosil (300 m^2 g^{-1}) (A300) and hydrophobic Aerosil (60 m^2 g^{-1}) (AM60). The weight ratio filler/elastomer is 30:100.

2,5-Bis(*t*butyl)-2,5-diaza-1-germa-cyclopentane – A New Precursor for Amorphous Germanium Films

J. Prokop, R. Merica, F. Glatz, S. Veprek*

Institut für Chemie der Informationsaufzeichnung

Technische Universität München

Lichtenbergstr. 4, D-85747 Garching, Germany

F. R. Klingan, W. A. Herrmann

Anorganisch-Chemisches Institut

Technische Universität München

Lichtenbergstr. 4, D-85747 Garching, Germany

Summary: A novel organometallic precursor has been synthesized and used for the deposition of amorphous germanium by thermal OM CVD (*o*rganometallic *c*hemical *v*apour *d*eposition) at temperatures below 300 °C. The films are of high purity with an oxygen content below the detection limit of about $\leq 10^{18}$ cm^{-3}. OM CVD from a novel precursor 2,5-Bis(*t*butyl)-2,5-diaza-1-germa-cyclopentane is a promising alternative for the deposition of a-Ge at low temperature. Oxygen, nitrogen, and carbon impurities of the films are very low, but their electrical properties do not reach the quality of the best plasma CVD material. Work is in progress to solve these problems.

Introduction

Hydrogenated amorphous silicon and related alloys represent a special type of thin film semiconductor material that has attracted much attention in the last decade. The scientific and technological properties of these materials include a continuously adjustable band gap and the possibility of their n- or p-type doping during the deposition process. The films can be easily fabricated with low cost by plasma-induced decomposition of silane and germane. Amorphous germanium of electronic quality uses an intense plasma discharge at a moderate substrate temperature and a relatively high negative bias [1]. For three years no further improvement of the quality of the material could be achieved [2]. Nevertheless the recent progress brought some understanding of the reaction mechanism and of the discharge parameters which are needed for the preparation of films of a good electronic quality. Therefore, we have been trying to develop alternative deposition techniques on thermal OM CVD at low temperature.

As there are no suitable organometallic precursors commercially available, initial work dealt with the synthesis of such a precursor [4]. 2,5-Bis(*t*butyl)-2,5-diaza-1-germa-cyclopentane is a monomeric solid with a melting point of 45 °C and a sufficient vapour pressure of 0.40 mbar at 40 °C to allow its introduction into the CVD reactor. For the details about the synthesis and properties of this precursor we refer to a recent paper [4]. The present work deals with the investigation of the thermal decomposition of the precursor, the deposition of amorphous germanium (a-Ge) and the characterization of the deposited thin films. Finally some data should try to give some understanding about the deposition mechanism.

Experimental Set-Up

The deposition experiments were peformed in a CVD apparatus shown in Fig. 1. More details are described in [5]. The precursor was introduced from a thermostated bubbler (1) by adjusting its partial pressure via the temperature and if indicated via the flow rate of the carrier gas (10 sccm argon or 10 sccm hydrogen with 3 sccm argon).

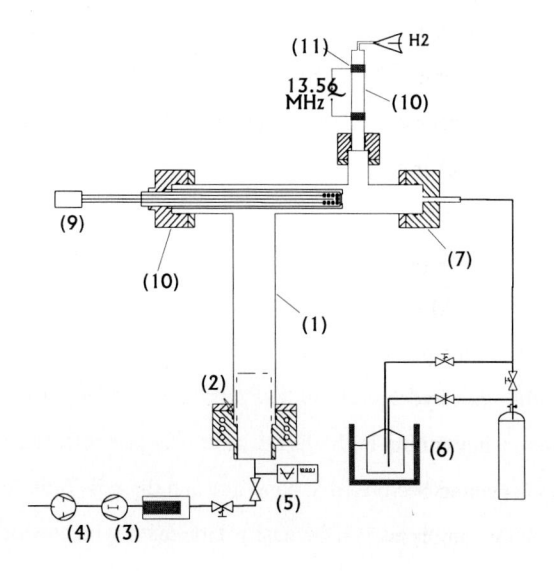

Fig. 1. Schematic diagram of the OM CVD apparatus used for the deposition study

The total leakage and desorption rate of the whole apparatus was below $1 \cdot 10^{-5}$ mbar $L \cdot s^{-1}$. The reactor consisted of a Pyrex glass tube (2). The substrates were mounted on the holder (3), the could be controlled by means of an external controller (4) and a thermocouple inserted into the copper block about 3 mm from the bottom of the substrate. The later described OM CVD experiments combined with a hydrogen afterglow were done in a similar apparatus, which included a side port (5) attached next to the substrate. In this tube a RF plasma (13.56 MHz) of hydrogen with external electrodes was used.

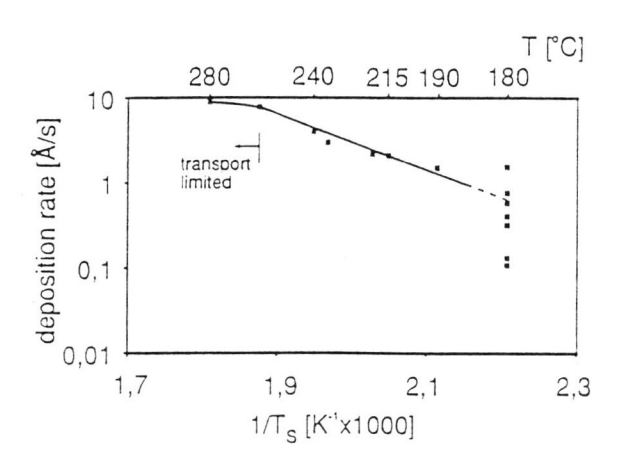

Fig. 2. Arrhenius plot of the deposition rate vs the reciprocal temperature; pressure and gas flow is shown in the graph

An Arrhenius plot of the deposition rate vs reciprocal absolute temperature is shown in Fig. 2. Depositions were made by indicated pressures with or without carrier gas. One notices in all cases that above 190 °C the deposition rate of several Å/s was found with an activation energy of about 50–60 kJ mol^{-1}. Below this temperature a strong decrease of the deposition rate was found. It did not matter whether the gas phase consisted of pure precursor or of a mixture of organometallic compound and argon carrier gas. Only the value of the deposition rate was varying with the different pressures which can be explained by the amount of precursor in the gas phase. Similar results (Fig. 3) were also obtained with in situ X-ray photoelectron spectroscopy (ESCA) studies, which indicate a sharp shift of the binding energy as an onset of the start of decomposition of the precusor at around 190 °C.

Another experiment with posthydrogenation of the deposited films by exposing the samples to a hydroen plasma did not bring any improvement of the optoelectronic properties.

The germanium films deposited between 190 and 300 °C were characterized for their structural properties by XRD. All films show amorphous scattering pattern with a detection limit of 1 vol % of crystalline component.

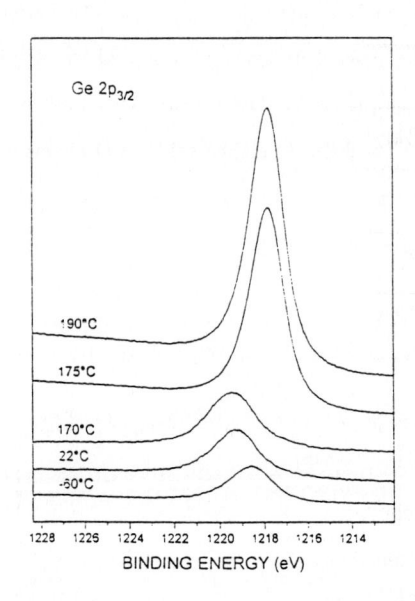

[Figure: ESCA spectra of Ge 2p$_{3/2}$ at temperatures 190°C, 175°C, 170°C, 22°C, −60°C, plotted against binding energy from 1228 to 1214 eV]

Fig. 3. ESCA study of the decomposition of the precursor at the given temperatures

XPS and IR absorption measurements indicated that the films were of a surprisingly high purity with carbon, nitrogen, and oxygen content below the detection limit for both methods. Using calibrated quantitative FT-IR absorption measurements [5] on ≥ 3.2 µm thick films deposited on a metallic substrate and measured in reflection (i.e., the effective optical thickness was ≥ 6.4 µm) we determined the detection limit of $3 \cdot 10^{18}$ oxygen atoms per cm^3. The oxygen content remained below this limit even after several months of their exposure to air at room temperature.

The electronic properties of the films must be compared like a standard with the data of good quality a-Ge:H material prepared by plasma CVD from germane [1, 2, 6]. The comparison is illustrated in Table 1. One notices a very low hydrogen content measured by IR absorption. This results in a high defect density, which effects a relatively high dark conductivity σ_D and very small activation energy E_A, which indicate pinning of the Fermi niveau close to the mobility edge of the conductance band. Also, a poor photo conductivity expressed by the $\eta\mu\tau$-product (quantum efficiency mobility lifetime) is measured. The value is five orders below that of plasma CVD deposited films. The high anticipations invoked by the high density and therefore low incorporation of oxygen on exposure to air were blighted by the low hydrogen content which is necessary for the relaxation of the amorphous network and which affects a low defect density giving good optoelectronic properties.

	Plasma CVD a-Ge:H	OM CVD a-Ge
[H] (at %)	~ 7	~ 0.8
σ_D (S cm^{-1})	$5 \cdot 10^{-5}$	$7 \cdot 10^{-3}$
E_A (eV)	0.45	0.2
$\eta\mu\tau$ (cm^2/V)	$2 \cdot 10^{-7}$	$1.2 \cdot 10^{-12}$

Table 1. Comparison of the electronic properties of good quality plasma CVD and the present OM CVD samples

In order to increase the hydrogen content of the films we have combined the thermal OM CVD with the afterglow of a hydrogen plasma. Thus the growing films should be exposed to atomic hydrogen radicals coming out of the plasma. Fig. 4 shows the effect of the afterglow on the deposition rate. As long as the RF power was < 5 W which affected that the discharge was limited to the side port, the deposition rate remained actually constant. However, when the discharge was extended into the reactor by increasing the power ≥ 35 W the deposition rate decreased to zero. A chemical etching of the germanium films by atomic hydrogen could be ruled out. More likely, the lone pair of the germylene is hydrogenated by the atomic hydrogen from the afterglow, forming the more stable germane.

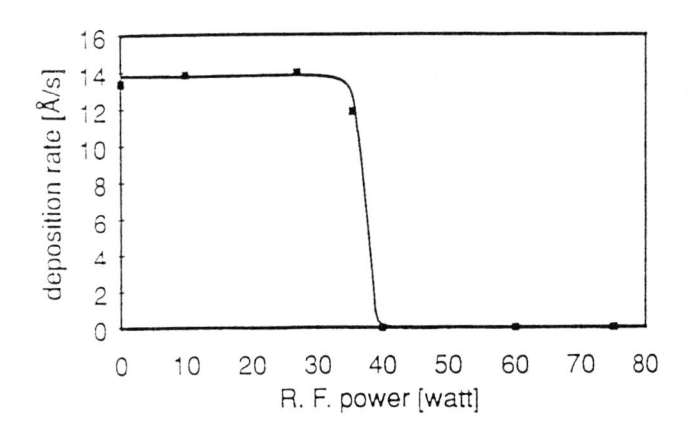

Fig. 4. ESCA study of the precursor decomposition at the given temperatures

Another experiment was done by exposing the deposited films to a hydrogen plasma for posthydrogenation. It did not bring any substantial improvement of the electronic properties. This can be associated with the dense nature of the material and the resulting slow diffusion of hydrogen into the bulk. The possible solutions of the problem are a better cleaning and activating process for the surface of substrate, a repetitive deposition of thin films and their subsequent hydrogenation and at last to give the problem back to synthetic chemists to design a modified precursor.

Acknowledgement: This work has been supported by the *Deutsche Forschungsgemeinschaft* und by the *Bundesministerium für Forschung und Technologie*. We would like to thank Dr. G. Ruhl for supporting one of the authors (R. M.) by the XPS measurements.

References:

[1]   F. H. Karg, B. Hirschauer, W. Kaspar, K. Pierz, *Solar Energy Mater.* **1991**, *22*, 169.

[2]   P. Wickboldt, S. J. Jones, W. A. Turner, F. C. Marques, D. Pang, A. E. Wetzel, W. Paul, J. H. Chen, *Phil. Mag.* **1991**, *B64*, 655.

[3]   F. Glatz, R. Konwitschny, M. G. J. Veprek-Heijman, S. Veprek, *Mat. Res. Soc. Symp. Proc.*, in press.

[4]   W. A. Herrmann, M. Denk, J. Behm, W. Scherer, F.-R. Klingan, H. Bock, B. Solouki, M. Wagner, *Angew. Chem., Int. Ed. Engl.* **1992**, *31*, 1485.

[5]   F. Glatz, J. Prokop, S. Veprek, F. R. Klingan, W. A. Herrmann, *Mat. Res. Soc. Symp. Proc.*, in press.

[6]   F. Glatz, *Ph. D. Thesis*, Technische Universität München, **1993**.

Getting Light from Silicon: From Organosilanes to Light Emitting Nanocrystalline Silicon

Stan Veprek

Institut für Chemie der Informationsaufzeichnung
Technische Universität München
Lichtenbergstr. 4, D-85747 Garching, Germany

Summary: Crystalline silicon crystal is a very inefficient luminescent material due to its indirect band gap. However, with decreasing grain size below 10 nm the increasing scattering of the wave functions at the grain boundaries leads to a progressive mixing of the electronic quantum states in the momentum space and increasing probability for radiative recombination of electron-hole pairs. Simultaneously, the band gap increases due to the localization of the electron and hole wave functions within the nanocrystal. Appropriate passivation of the surfaces of such nanocrystals is necessary to reduce the probability of non-radiative recombinations. Therefore, many organosilanes are very efficient photoluminescencing materials. Recent progress is summarized to make it understandable for readers who are not experts in the field.

1 Introduction

Silicon dominates microelectronics and there is no any other semiconductor to replace it in the foreseeable future. The only exception may be special applications, such as high temperature electronic devices operating above 200 °C, where silicon carbide represents a promising material which can be processed with a silicon-compatible technology.

There are several, relatively simple reasons for the dominance of silicon: The band gap of 1.1 eV is just suitable for devices operating at temperatures encountered in the most applications (−40 to +50 °C). From the chemical point of view, silicon is the only semiconductor which forms a stable oxide with excellent dielectric properties. In spite of the incommensurability of their crystal lattices, an almost perfect interface with a low defect density of $< 10 \text{ cm}^{-2}$ can be routinely prepared between silicon and silica by oxidation at high temperature and subsequent annealing in forming gas (4–5 mol% of hydrogen in nitrogen). This interface is the basis of metal oxide semiconductor (MOS) field effect transistor, storage capacitors in integrated circuits, charge coupled devices and other electronic elements. The refractory nature of silicon, together with the fact that it is an element, makes high temperature processing relatively easy because the likelihood of the formation of Frenkel defects is strongly reduced as compared, for example, with III–V compound semiconductors. Last but not least, the high mechanical stiffness of silicon makes the handling of the wafers in automatic production lines much easier because it reduces the danger of their damage and breaking.

The indirect band gap [1] of silicon, however, prevents one from building efficient optoelectronic devices from the crystalline material. Amorphous silicon, in which the momentum is not any good quantum number, shows efficient absorption as well as emission [4, 5]. Only a few years ago, Canham [6], Lehmann, and Gijsele [7] reported on an efficient photoluminescence in the visible spectral range ($h\nu$ ~1.6–1.8 eV) from "porous silicon" (PS) which is prepared by anodic etching of silicon wafers in an HF/water/ethanol solution. Although there are several earlier papers reporting on the electro- and photoluminescence (EL and PL) from small silicon nanoparticles (e.g., [8]) it were the reports of Canham, Lehman, and Gijsele which triggered an intense research in this field, resulting in more than 1200 papers published within the last three years.

Why does a material with indirect and relatively small band gap of 1.1 eV (corresponding to photons in the near infrared) emit efficiently visible light? The original explanation of Canham [6], which is nowadays called the quantum confinement model, is based on the theoretical prediction, according to which the band gap increases and becomes direct with decreasing crystallite size.

However, very soon it became clear that the situation is more complex (e.g. [9, 10]). The obvious problem arises with the fact that the red-yellow PL from PS is relatively slow with a decay time in the range of tens of microseconds, which, together with some further experimental observations [11] and theoretical calculations [9, 10, 12], is considered as strong evidence for an indirect band gap. However, as pointed out by Hybertsen [13], the electron and hole wave functions in small crystallites are spread in k space so that it is no longer meaningful to debate whether the gap is direct or indirect. Detailed calculations show that the phonon assisted transitions dominate in crystallites larger than about 1.5 nm, where an important part of the phonon contribution comes from scattering at the surface of the crystallites and a part from the bulk phonons.

The direct transitions dominate only in much smaller crystallites [13]. As the size of the crystallites covered in the current experiments is 2 nm or more, the porous and nanocrystalline light emitting silicon is "indirect" gap material. The calculated radiative life time decreases with decreasing crystallite size for both the phonon assisted and direct transitions. The predicted dependence [13] correlates reasonably with the measured dependence of the PL decay time on the photon energy, which is typically several tens of μs for the red and a few μs for the yellow PL [14].

An alternative explanation provides the "surface state mode 1" of Koch *et al.* [15]. Accordingly, the photogenerated electron-hole pair is trapped in surface states (which have an energy lower than the band gap of the nanocrystallite) and recombine, with some delay from such states. This model can explain various observed data, such as the polarization memory of the PL [17] and the dependence of the measured PL decay time on the phonon energy in terms of a distribution of the trap states as in a-Si [16]. In particular, the model can also explain the observation of the absence of any blue shift with decreasing crystallite size in compact nc-Si/SiO$_2$ films [18].

Future work may reveal that the difference between these models is not as large as it appears now. Much understanding has been achieved in the optical properties of small particles of compound semiconductors which resemble many phenomena observed in PS and nc-Si [21, 22]. In the limit of the small crystallite size of ~2 nm, many properties of the material, such as crystalline and electronic structure undergo significant changes [20] and approach those of molecules. The adequate theoretical description has to account for these changes, and even an adequate terminology should be applied. Is a particle of 1–2 nm size a small crystal or a large molecule? For these reasons it is interesting to compare the absorption and emission phenomena in PS and nc-Si with those observed in linear, two and three dimensional organosilanes.

2 Optical Properties of Organosilanes

The fundamental absorption in linear organosilanes of the general formula $R_1-(R_2-Si-R_3)_n-R_4$ corresponds to the allowed HOMO-LUMO transition (σ → σ*) [23, 24]. The corresponding sharp absorption at an energy around 3.5 eV in long (Si)$_n$ chains is typical for the density-of-states distribution in one-dimensional electronic systems. For a chain length of less than about 20–24 Si atoms, a blue shift of the absorption and a decrease of the extinction coefficient per Si atom are observed [24]. The maximum of the photoluminescence spectral distribution is only slightly red shifted to about 3.2 eV in the long chains, and it shows a high quantum yield. With decreasing chain length below 20 Si atoms the PL is also progressively blue shifted and the quantum yield decreases. These phenomena are due to the delocalization of the exciton over a length of about 20–24 Si-atoms in the long chain. Thus, the blue shift of the absorption and photoluminescence as well as the changes in the transition probabilities are due to localization of the exciton in a chain shorter than its extension in the long chains.

In analogy with this data one may expect similar localization effects in three dimensional particles when their linear dimension decreases below about 20 Si atoms, i.e. below about 4 to 5 nm, and becomes comparable or even smaller than the radius of the exciton. The Bohr radius of an exciton is:

$$a_B = (h^2 \varepsilon / e^2 4\pi^2)(1/m_e + 1/m_h)$$

Eq. 1. h: Planck's constant; ε: dielectric constant; m_e: electron mass; m_h: hole effective mass; this is indeed the case for isolated silicon particles [11] as well as for compact films of nc-Si [25]

The situation is somewhat more complicated in two-dimensional polysilanes, which have intermediate properties between the one-dimensional chain-like polysilanes and three-dimensional bulk silicon. The gap is of a quasi-direct nature as the indirect gap is only slightly smaller than the direct one [11]. However, the excitons strongly bind to the lattice which results in a large Stokes shift of the PL [26]. The observed blue shift of the absorption and photoluminescence with decreasing size of the polysilanes is considered to be due to confinement effects of the excitons [12, 26]. The strong coupling of the exciton to the lattice decreases somewhat the blue shifts as compared with the linear chains, and it results in a stronger localization of the exciton over a smaller number of Si atoms [12, 26].

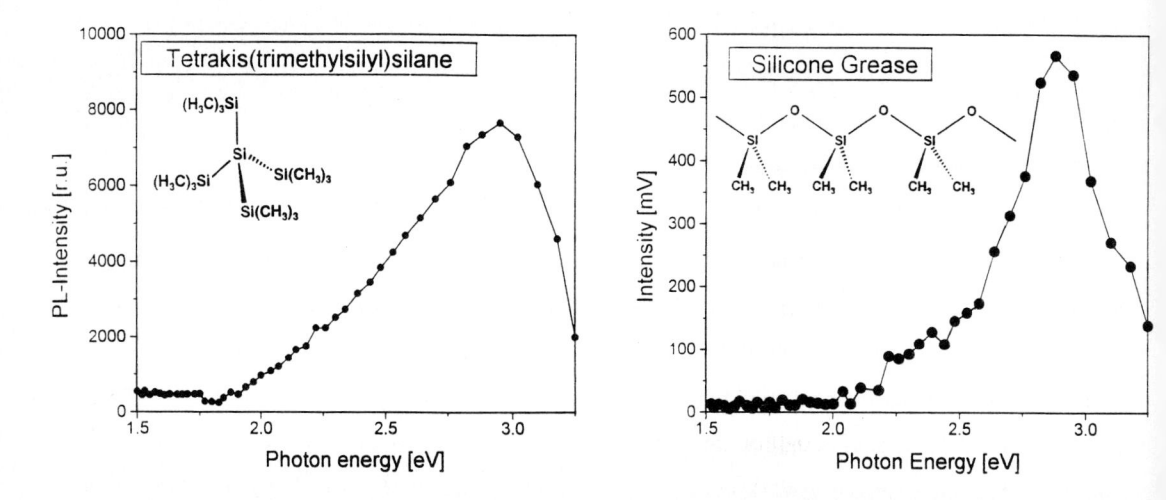

Fig. 1.   Photoluminescence from tetrakis(trimethylsilyl)silane and silicon vacuum grease [29]

It is interesting to consider the PL from small three-dimensional organosilanes. Furukava et al. [27] reported the PL from the α-form of *t*butyloctasilacubane with a maximum at about 1.65 eV and a quantum efficiency of 1. The molecule consists of a cube of eight silicon atoms in which each Si atom is bonded to three Si neighbors and its remaining valence is saturated by a *t*butyl ligand. The sensitivity of the electronic structure to small changes of the geometry of the Si-skeleton is illustrated by the fact that in a somewhat loosely bonded β-form the PL shifts to about 2.2 eV [28]. Fig. 1 shows for comparison the structure and the PL of tetrakis(trimethylsilyl)silane [29].

Here, the PL is shifted to the blue region where PL from heavily oxidized PS and nc-Si has been reported [19, 30–32]. For curiosity and to illustrate that photoluminescence is a very common phenomenon in organosilanes we show in Fig. 1 also the PL measured from silicone vacuum grease [29]. We shall return to this point later on.

In summary, I would like to emphasize, that many phenomena observed in nanometer small silicon particles are well known and understood in organosilanes. The example of the spectra of the α and β form of *t*butyloctasilacubane shows the large sensitivity of the PL spectra (and of the electronic structure) to small modifications of the silicon skeleton.

3 Optical Properties of Nanocrystalline Silicon

There are essentially three different ways how to prepare nanometer sized silicon particles. The porous silicon is, as already mentioned, prepared by anodic etching of silicon wafers in an HF/ethanol/water solution [6, 7]. The microporous silicon has typically a high porosity of 60–70 vol.%, and it consists of few nm thin wires which preserve the original orientation of the wafer. The thickness of the wires varies within the PS layer and the material is very brittle. Free standing PS films can be prepared by application of a high current density after the usual etching of the desired thickness of the PS.

Brus *et al.* prepared isolated silicon particles by high temperature pyrolysis of disilane with a subsequent passivation of the surface by oxidation [33]. The particles of various size are then processed by high-pressure, liquid-phase, size exclusion chromatography to separate sizes and obtain various fractions of monosize particles. Such particles represent an almost ideal model of silicon quantum dots.

For possible applications one may prefer compact nc-Si films. Such films are prepared by plasma CVD [34]. The best control of the crystallite size in the relevant range offers the chemical transport of silicon in a clean hydrogen plasma [20, 35–37]. After the deposition of nc-Si film with the desired crystallite size the sample is oxidized (typically at 700–800 °C) in pure oxygen and annealed in forming gas (about 4 mol.% of hydrogen in nitrogen). Fig. 2 shows the effect of the oxidation and FG annealing on the PL [25]. The advantage of this procedure is the possibility of preparing compact films with a relatively narrow distribution of nc-Si of a required size.

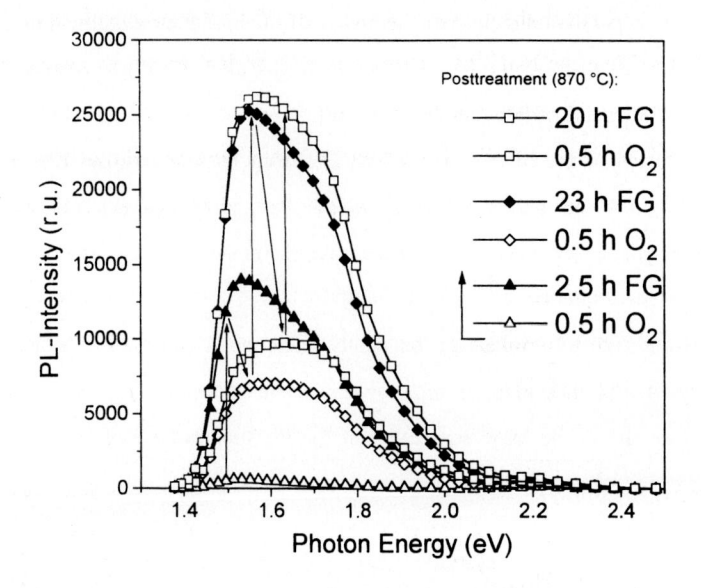

Fig. 2. Effect of a subsequent oxidation and annealing in forming gas on the intensity of the photoluminescence from nc-Si films [25]

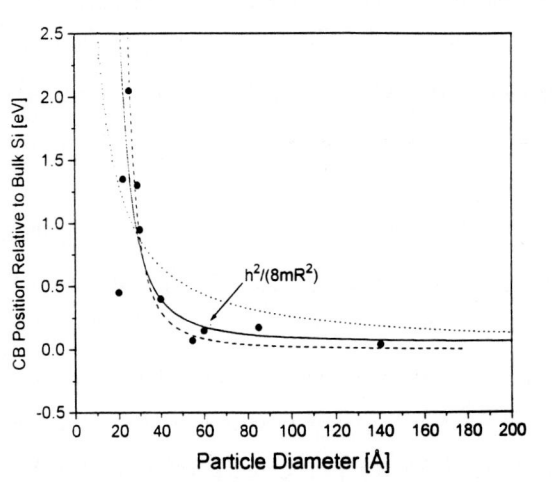

Fig. 3a. The dependence of the position of the conductance band on the size of small silicon nanocrystals determined by means of electron energy loss spectroscopy [38]:
solid line: effective-mass-like quantization
dotted line: $1/R$-dependence
dashed line: $1/R^3$-dependence

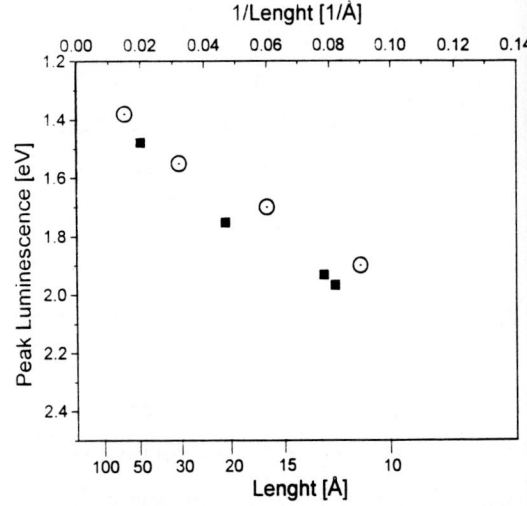

Fig. 3b. Correlation between the particle size and the position of the spectral maximum of the photoluminescence; the number of silicon atoms in the crystallites is indicated as well [39]:
⊙: Si nanocrystals
■: porous silicon

The dependence of the optical absorption in small isolated silicon nanocrystals on the photon energy shows a slow increase above the band gap energy which is typical for an indirect gap material [11]. A sharp increase is observed only when the photon energy reaches the value of the direct gap transition (about 3.4 eV) at the Γ-point ($k = 0$). With decreasing crystallite size below about 4 nm the band gap increases [38] and the PL shows a pronounced blue shift [39]. The results reproduced from ref. [38] and [39] are shown in Fig. 3.

The blue shift of the position of the conductance band measured by electron energy loss spectrometry [38] and of the PL [39] with decreasing crystallite size is clearly seen. In contrast, no such shift is found in bulk nc-Si/SiO$_2$ films (Fig. 4) [19].

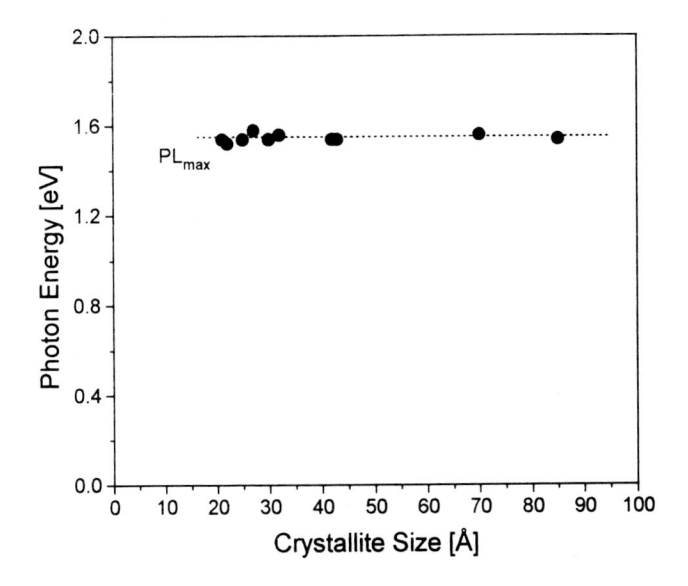

Fig. 4. Position of the PL maximum vs the average crystallite size in the plasma deposited nc-Si/SiO$_2$ films [19]

The radiative decay time of the red PL is of the order of several tens of microseconds in all cases. The lack of the blue shift in the compact films has been explained in terms of the surface state model [19]. The theory predicts an increase of the transition probability for absorption and PL with decreasing crystallite size [9, 10, 13]. Fig. 5 shows the measured dependence of the PL intensity on the average crystallite size in the compact ncSi/SiO$_2$ films [19, 25].

For comparison the theoretical values of the electron-hole recombination rate from several recent papers are included as well. One can see that below a crystallite size of about 4 nm the agreement between the measured data and the theory is reasonably good. The relatively high PL intensity from samples with an average crystallite size of > 5 nm is probably due to the presence of a small fraction of small crystallites in such films [19, 25].

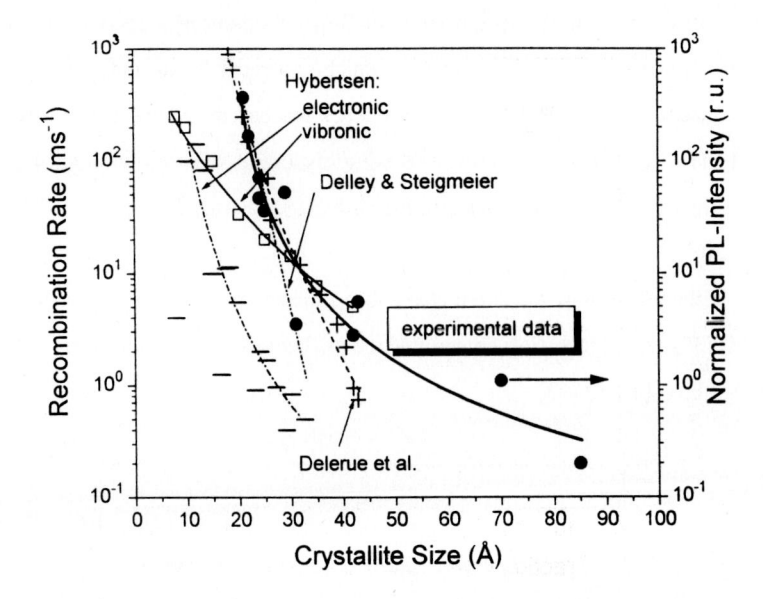

Fig. 5. Dependence of the peak intensity of the photoluminescence from Fig. 2 on the crystallite size [19, 25]; the theoretical calculations refer to the following papers: Delley and Steigmeier [9], Delerue et al. [10], and Hybertsen [13].

An important conclusion which can be drawn from these results is the strong increase of the PL intensity with decreasing particle size. The lower limit of crystallite size which can be currently prepared in the nc-Si/silica films is about 2 nm. For an efficient PL and EL one would like to prepare films with the maximum possible content of the smallest Si nanocrystals. For an efficient electroluminescence a thin silica interface between the crystallites is required in order to ensure good electric transport properties (tunneling of electrons and holes through the oxide). Therefore we have investigated the effect of the average thickness of the silica interface (or of the fraction of nc-Si in the films) on the intensity of the PL. The results are shown in Fig. 6 [40].

One can see that with decreasing average thickness of the silica grain boundaries below about 1 nm the intensity of the PL strongly decreases. The investigations being currently done in our laboratory should decide if this decrease is due to a delocalization of the electrons and holes. Such an effect has been described for CdS clusters trapped in zeolites [41]. To optimize the electroluminescence from nc-Si films will require to find the optimum compromise.

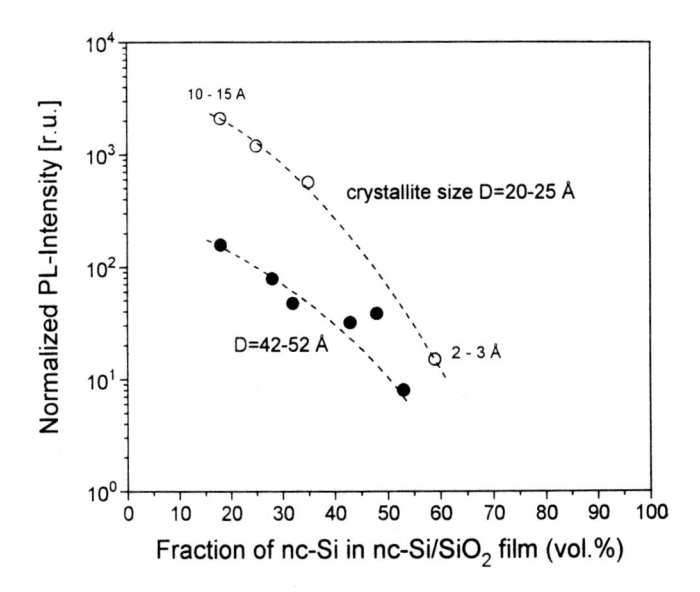

Fig. 6. Dependence of the photoluminescence intensity on the fraction of nc-Si in the nc-Si/SiO$_2$ films for a constant crystallite size; the approximate value of the thickness of the silica grain boundaries is indicated as well [40]

4 The Origin of the Green-Blue Photoluminescence

As mentioned above, in oxidized PS and nc-Si a green-blue PL is frequently found [19, 30–32]. Because of its fast decay of a few ns or even less, this PL is of great interest with respect to possible applications in fast optoelectronic devices, e.g, for the chip-to-chip communication in computers. The green-blue PL from oxidized PS, nc-Si and from spark-eroded silicon has been attributed to a blue shift in small Si crystallites and clusters which are formed by the oxidation.

In our recent papers, however, we have shown that this PL is associated with silanol groups in the silica [32, 42]. This conclusion is supported by many experimental findings:

- The green-blue PL does not arise due to a continuous blue shift of the red PL but it appears as a new feature which shows a different behavior than the red one.
- On boiling the sample in ultrapure water and subsequent annealing in forming gas the intensity of the green-blue PL increases, reaches a maximum at a temperature of about 450 °C where the silica surface is covered with isolated silanol groups and vanishes after further annealing at $\geq$ 800 °C where all silanol groups desorb. Alumina and several other metal oxides show the same or a very similar behavior and spectral distribution of the PL (Fig. 7).

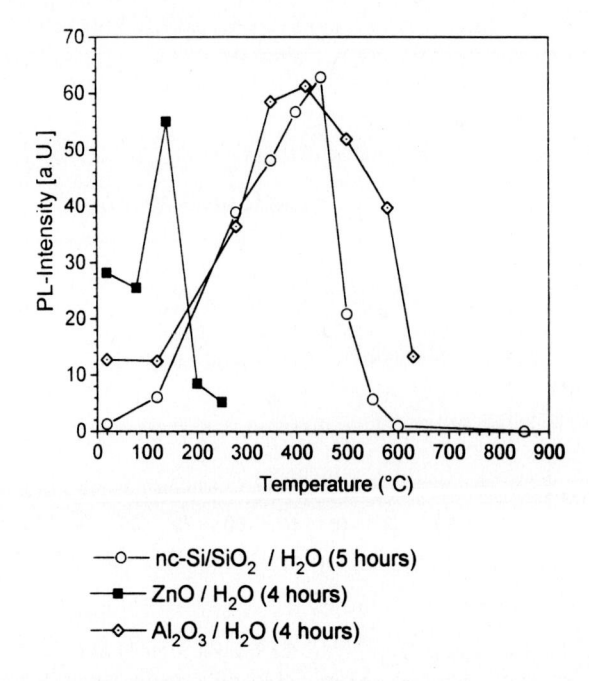

Fig. 7a. Dependence of the intensity of the green-blue photoluminescence from nc-Si/SiO$_2$ films, from alumina and from zinc oxide as a function of the temperature of isochronal annealing in forming gas after the samples have been immersed in boiling ultrapure water for a period of 5 hr [32, 42]

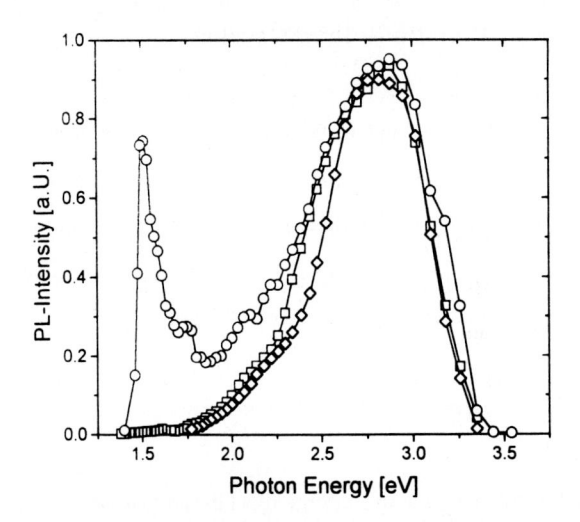

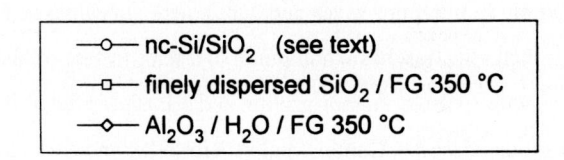

Fig. 7b. Spectral distribution of the photoluminescence from nc-Si/SiO$_2$ films, alumina, and spark eroded silicon

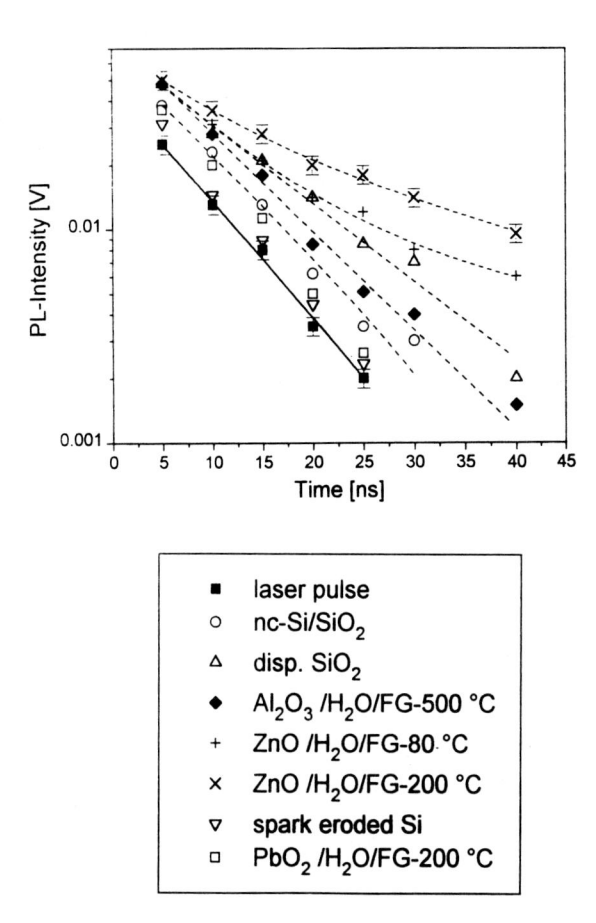

Fig. 7c. Decay of the green/blue photoluminescence from various samples [42]

Also the decay of the PL is equal within the resolution of our apparatus thus setting a few nanoseconds as the upper limit (Fig. 7c). All these results strongly indicate the molecular-like nature of the luminescence center. An important piece of evidence is shown in Fig. 8 where the degree of the polarization of the PL light, P, is plotted vs the angle between the axis z and the electric vector of the incident UV light [43].

$$P = \{I(z) - I(x)\}/\{I(z) - I(x)\}$$

Eq. 2. $I(z)$, $I(x)$: intensities of the polarized components of the PL with the electric vector in the direction of axis z and x;. the excitation UV light propagates along the axis x and the PL is observed along the axis y with the sample positioned in the origin of the coordinate system

One notices that the fast green-blue PL from nc-Si and alumina shows a strong polarization memory thus supporting the molecular nature of the absorption and luminescence center [44]. The red PL component from the same sample does not show any polarization memory. This can be understood in terms of the formation of the photogenerated exciton followed by the trapping of the electron-hole pair in the surface states from which the light emission occurs.

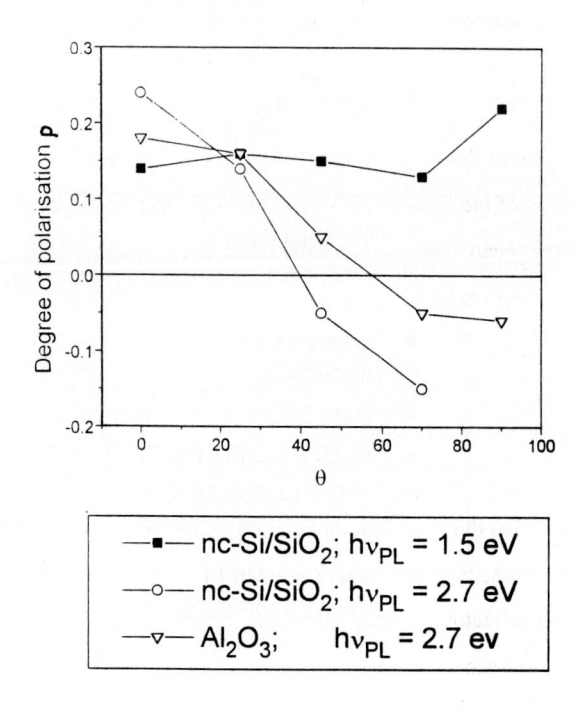

Fig. 8. Dependence of the degree of polarization of the PL on the polarization angle of the excitation UV light [43]

I refer to the original papers for further results [29, 42]. It is an open question wether this fast luminescence can be used for optoelectronic devices. The strong sensitivity of the PL to environmental humidity and temperature may be a limiting factor. Nevertheless, it is an interesting challenge to determine the microscopic mechanism that is common to this PL phenomena, which occurs in so many different systems.

5 Conclusions

The world of nano sized silicon opened up a large number of new phenomena associated with the quantum confinement of the excitons in the small crystallites. Many phenomena resemble those known from polysilanes. Some of the phenomena are still not unambiguously explained by appropriate theoretical models. In particular it is not quite clear why the blue shift which is theoretically predicted and experimentally observed in small isolated particles is absent in compact films. The suggested explanations in terms of the surface state model need further experimental verification.

The recent results clearly show that the green-blue PL from heavily oxidized nc-Si and PS is due to silanol groups. This rises the question if this very fast PL can be used in optoelectronic devices.

Acknowledgment: I would like to thank my co-workers for their enthusiastic collaboration, in particular to T. Wirschem for preparation of the figures used in this paper and my wife for critical comments on the manuscript. This work has been supported in part by the *Volkswagen Stiftung* and by the *Bundesministerium für Forschung und Technologie*.

References:

[1] Indirect band gap means that the minimum of the conduction band and the maximum of the valence band are at different positions in the Brillouin zone, i.e., the electrons in the corresponding quantum states have different wave vectors k. Because the momentum of a photon is very small the optical transitions (absorption, emission) of the electrons have to occur with $\Delta k = 0$. GaAs and many other compound semiconductors are efficient optoelectronic materials because of their direct band gap. In silicon, however, such transitions between the top of the valence and bottom of the conduction bands require assistance of a phonon which delivers the momentum needed [2, 3].

[2] C. Kittel, *Introduction to Solid State Physics*, Wiley, New York, **1971**.

[3] K. Seeger, *Semiconductor Physics*, Springer, Berlin, **1988**.

[4] *Amorphous Silicon and Related Materials*, Vol. 1 & 2 (Ed.: H. Fritzsche), World Scientific, Singapore, **1989**.

[5] *Amorphous Semiconductors* (Ed.: M. H. Brodsky), Springer, Berlin, **1979**.

[6] L. T. Canham, *Appl. Phys. Lett.* **1990**, *57*, 1046.

[7] V. Lehmann, U. Gijsele, *Appl. Phys. Lett.* **1991**, *58*, 856.

[8] D. J. DiMaria, J. R. Kirtley, E. J. Perkily, D. W. Dong, T. S. Kuan, F. L. Pesavento, N. Theis, J. A. Cutro, S. D. Brorson, *J. Appl. Phys.* **1984**, *56*, 401.

[9] B. Delley, E. F. Steigmeier, *Phys. Rev.* **1993**, *B47*, 1397.

[10] C. Delerue, M. Lanno, G. Allan, *J. Luminescence* **1993**, *57*, 249.

[11] L. Brus, *J. Phys. Chem.* **1994**, *98*, 3575.

[12] G. Allan, C. Deleure, M. Lannoo, *Phys. Rev.* **1993**, *B48*, 7951.

[13] M. S. Hybertsen, *Phys. Rev. Lett.* **1994**, *72*, 1514.

[14] The measured decay rate, $1/\tau$ (PL), is given by the sum of the rates of the radiative, $1/\tau$ (r) and non-radiative, $1/\tau$ (nr) electron-hole recombinations. Thus, the measured decay time of the PL, τ (PL) is shorter than τ (r), and only in the rare case of a negligible non-radiative recombination, τ (r) $\gg \tau$ (nr) it approaches the radiative time.

[15] F. Koch, V. Petrova-Koch, T. Muschik, A. Nikolov. V. Gavrilenko, *Mater. Res. Soc. Symp. Proc.* **1993**, *283*, 197.

[16] F. Koch, V. Petrova-Koch, T. Muschik, *J. Luminescence* **1993**, *57*, 271.

[17] A. V. Andrianov, D. I. Kovalev, I. D. Yaroshetskii, *Phys. Sol. State* **1993**, *35*, 1323.

[18] S. Veprek, M. Rückschloss, B. Landkammer, O. Ambacher, *Mater. Res. Soc. Symp. Proc.* **1993**, *298*, 117.

[19] M. Rückschloss, O. Ambacher, S. Veprek, *J. Luminescence* **1993**, *57*, 1.

[20] S. Veprek, Z. Iqbal, F.-A. Sarott, *Phil. Mag.* **1982**, *B45*, 137.

[21] L. Brus, *Appl. Phys.* **1991**, *A53*, 465.

[22] Y. Wang, N. Herron, *J. Phys. Chem.* **1991**, *95*, 525.

[23] R. D. Miller, J. Michl, *Chem. Rev.* **1989**, *89*, 1359.

[24] Y. Kanemitsu, K. Suzuki, Y. Nakayoshi, Y. Matsumoto, *Phys. Rev.* **1992**, *B46*, 3916.

[25] M. Rückschloss, B. Landkammer, S. Veprek, *Appl. Phys. Lett.* **1993**, *63*, 1474.

[26] Y. Kanemitsu, K. Suzuki, Y. Matsumoto, T. Komatsu, K. Sato, S. Kyushin, H. Matsumoto, *Solid St. Commun.* **1993**, *86*, 545.

[27] K. Furukawa, M. Fujino, N. Matsumoto, *Appl. Phys. Lett.* **1992**, *60*, 2744.

[28] H. Tashibana, M. Goto, M. Matsumoto, H. Kishida, Y. Tokura, *Appl. Phys. Lett.* **1994**, *64*, 2509.

[29] T. Wirschem, M. Rückschloss, S. Veprek, unpublished results.

[30] L. Tsybeskov, P. M. Fauchet, *Appl. Phys. Lett.* **1994**, *64*, 1993.

[31] L. Tsybeskov, Ju. V. Vandyshev, P. M. Fauchet, *Phys. Rev.* **1994**, *B 49*, 7821.

[32] H. Tamura, M. Rückschloss, T. Wirschem, S. Veprek, *Appl. Phys. Lett.*, in press.

[33] K. A. Littau, P. J. Szajowski, A. J. Muller, A. R. Kortan, L. E. Brus, *J. Phys. Chem.* **1993**, *97*, 1224.

[34] M. Rückschloss, B. Landkammer, O. Ambacher, S. Veprek, *Mater. Res. Soc. Symp. Proc.* **1993**, *283*, 65.

[35] S. Veprek, V. Marecek, *Solid St. Electron.* **1968**, *11*, 683.

[36] S. Veprek, *Pure Appl. Chem.* **1982**, *54*, 1197.

[37] S. Veprek, F.-A. Sarott, Z. Iqbal, *Phys. Rev.* **1987**, *B36*, 3344.

[38] P. E. Batson, J. R. Heath, *Phys. Rev. Lett.* **1993**, *71*, 911.

[39] S. Schluppler, *Phys. Rev. Lett.* **1994**, *72*, 2648.

[40] S. Veprek, *Mater. Res. Soc. Symp. Proc.*, in press.

[41] G. D. Stucky, J. E. MacDougall, *Science* **1990**, *247*, 669.

[42] M. Rückschloss, T. Wirschem, H. Tamura, J. Oswald, S. Veprek, *J. Luminescence*, submitted.

[43] T. Wirschem, S. Veprek, to be published.

[44] A. C. Albrecht, *J. Molec. Spectroscopy* **1961**, *6*, 84.

A Novel Liquid Xenon IR Cell Constructed from a Silicon Single Crystal

M. Tacke*, P. Sparrer, R. Teuber

Institut für Anorganische Chemie

Universität Fridericiana zu Karlsruhe (TH)

Engesserstr., Geb. 30.45, D-76128 Karlsruhe, Germany

H.-J. Stadter*, F. Schuster

Wacker-Chemitronic

Werk Burghausen

Johannes-Heß-Straße 24, D-84489 Burghausen, Germany

Single Crystal Silicon: An Excellent Material for Pressure Resistant IR Cells

Single crystal silicon possess two properties which are important for the design of a pressure resistant IR cell.

1) The IR transmission can be observed in the spectral range 4000–1300 cm^{-1}
2) The material can be tooled with diamond saws and ultrasonic drilling machines

For the manufacture a cylindrical silicon single crystal is first sawed to produce the outer windows. In a second step the probe room is drilled out. After final polishing of the inner and outer windows the cell can be fixed with *only one* lead gasket to the cryostat body. It is essential that the cell frame is made from Vacodil 36. This steel has nearly the same coefficient of expansion as silicon and allows the gas-tight connection between the cell and the cryostat body (see Fig. 1).

Liquid Xenon (LXe): A Special Solvent for IR Spectroscopy

Liquid Xenon is a low-temperature solvent for the temperature range between −112 °C (*mp, p* ~ 1 bar) and 17 °C (*scp, p* ~ 60 bar). The greatest advantage of this unusual solvent is its complete transparency over the IR spectral range. The limited solubility of substances in liquid xenon is often compensated by a longer optical pathway of the cell [1]. Because of the inertness of LXe and the weak interactions between LXe and the dissolved species the method allows a kind of "low temperature gas phase spectroscopy".

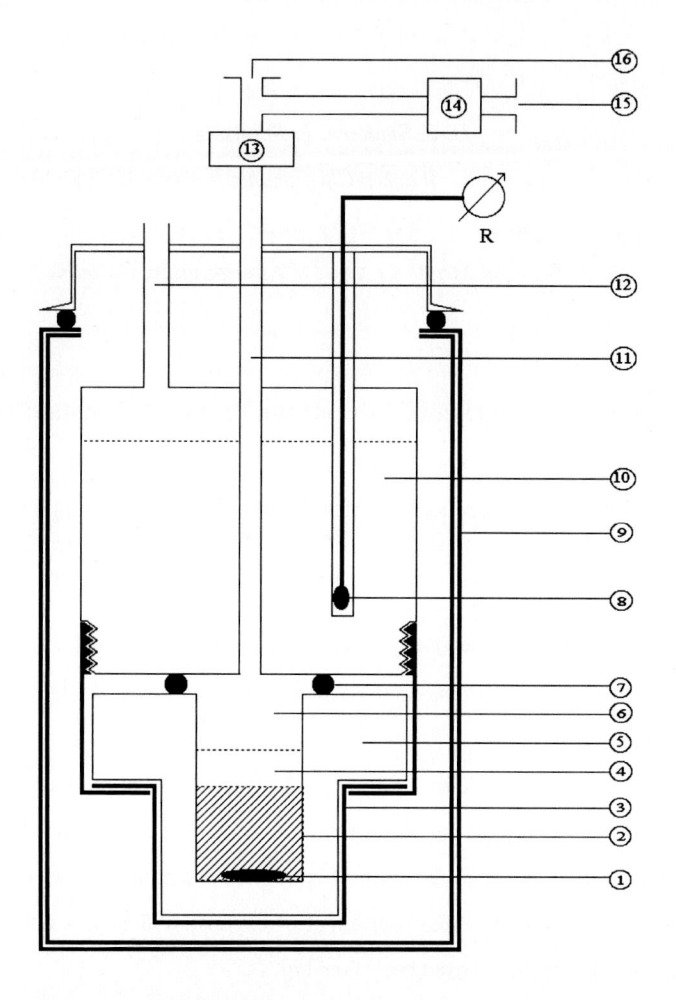

Fig. 1. The Liquid Xenon IR Cell (Schematic View):
(1) magnetic stirring bar; (2) the hatched area is part of the path of the IR beam; (3) cell frame made from Vacodil 36; (4) probe room filled with Xe(l); (5) silicon cell; (6) probe room filled with Xe(g); (7) lead gasket; (8) temperature control unit (Mo-1000 resistant); (9) cryostat housing; (10) cryostat body; (11) filling tube for the silicon cell; (12) filling tube for the cryostate; (13,14) high pressure valves; (15) flange connection to the vacuum system; (16) flange connection to the gas reservoir (Xe, CO etc.)

For an experiment a solution of the parent compound in pentane is injected into the Schlenk-type silicon cell with a syringe. After the solvent is removed in vacuo the cell is set under a pressure of Xe and cooled down. Then it is possible to react the parent compound with a gas like CO or N_2 [2].

Acknowledgement: The authors thank *Vakuumschmelze* (Hanau) for a gift of Vacodil 36, *Linde AG* (München) for high purity xenon and *Deutsche Forschungsgemeinschaft* for financial support.

References:

[1] R. R. Andrea, H. Luyten, M. A. Vuurman, A. Oskam, *Appl. Spectrosc.* **1986**, *40*, 1184.

[2] M. Tacke, Ch. Klein, D. J. Stufkens, A. Oskam, P. Jutzi, E. A. Bunte, *Z. Anorg. Allg. Chem.* **1993**, *619*, 865.

Author Index

A

Abele S. 511
Albrecht K. 321
Aldinger F. 725
Apeloig Y. 251, 263, 321
Auner N. 1, 41, 49, 249, 399, 467, 589

B

Backer M. 41
Barthel H. 761
Bassindale A. R. 411
Baumann F. 665
Baumann S. 295
Becker G. 161, 493, 511
Beckers H. 19
Beifuss U. 219, 225
Belzner J. 75, 459, 519
Bienlein F. 133
Bill J. 725
Biran C. 709
Birot M. 709
Bohmhammel K. 31
Bordeau M. 709
Brakmann S. 237
Braunstein P. 553
Brendler E. 69, 719
Brown S. S. D. 411
Brüning C. 243
Büchner M. 569
Bürger H. 19

C

Currao A. 469

D

Dannappel O. 427, 453
Dautel J. 511
Dauth J. 659, 665
Decker Ch. 655
Dehnert U. 519
Denk M. 251
Deubzer B. 659, 665, 673
Ditten G. 161
Dittmar U. 691
Dunoguès J. 709

E

Ehlend A. 141

F

Fey O. 575
Frey R. 525
Fuß M. 361

G

Geck M. 665, 673
Gehm H. 219
Gehrhus B. 289
Geissler H. 697
Glatz F. 815
Glynn S. G. 411

Graschy S. 89
Grasmann M. 41
Greb J. 609
Grobe J. 243, 317, 541, 591, 609
Gspaltl P. 101, 105, 109
Gross J. 231
Großkopf D. 127

H

Habel W. 733
Haeusler W. 733
Hassler K. 81, 95, 113, 203
Hausen H.-D. 141
Hayashi R. 251
Heckel M. 41
Heikenwälder C.-R. 399
Heinicke J. 289
Hemme I. 505
Hendan B. J. 685
Hengge E. 89, 101, 105, 109, 585
Herdtweck E. 41
Herrmann W. A. 815
Herzig Ch. 655
Herzog K. 63
Herzog U. 63
Hierstetter T. 659
Hiller W. 41
Hofmann J. 547
Hübler K. 161
Hübler U. 161

I

Ihmels H. 75

J

Jammegg Ch. 89
Jancke H. 697
Jansen I. 619
Jiang J. 411
Jones P. G. 231
Just U. 625

K

Kaim W. 141
Kapp J. 329
Karaghiosoff K. 195
Karni M. 251, 263
Karsch H. H. 133, 187
Kelling H. 215
Kienzle A. 725
Klein K.-D. 613
Klingan F.-R. 815
Klingebiel U. 127, 505
Klinkhammer K. W. 493
Knorr M. 553
Knott W. 613
Köll W. 81
Koerner G. 613
Kohlheim I. 215
Kollefrath R. 659
Krämer H. 69, 719
Krempner C. 389

Kroke E. 309
Krüger R. P. 625
Kunze A. 655
Kupfer S. 619

L

Lambrecht G. 231
Lang H. 569
Lange D. 215
Lankat R. 575
Lassacher P. 121
Ledderhose S. 225
Leistner S. 295
Leo K. 69
Lerner H.-W. 405
Linder U. 531
Linti G. 525
Litvinov V. M. 779
Lüdtke S. 499

M

Maas G. 149, 565
Maerker C. 329
Malisch W. 575
Maier G. 303
Marsmann H. C. 685, 691
Martin H.-P. 719
Marx G. 295
Mayer D. 565
Meinel S. 289
Merica R. 815

Merz K. 161
Merzweiler K. 531
Möller S. 575
Mörke W. 31
Moritz P. 19
Motz G. 511
Mühleisen M. 427, 447
Müller B. 361
Müller E. 719
Müller T. 263
Müller U. 655
Mutschler E. 231

N

Nesper R. 469
Niecke E. 209
Nieger M. 209
Niemeyer H.-H. 541
Niemeyer M. 161
Niesmann J. 127
Nöth H. 195
Nuyken O. 659

O

Oehme H. 389
Oelschläger A. 733

P

Pacl H. 303
Pätzold U. 55
Parker D. J. 411

Pauncz R. 251
Pepperl G. 673
Pillot J.-P. 709
Pinter E. 109
Pöschl U. 113
Polborn K. 525
Prokop J. 815

R

Reichel D. 231
Reinke H. 389
Reisenauer H. P. 303
Reising J. 575
Richter Rob. 69, 719
Richter Rol. 187
Riedel R. 725
Rikowski E. 691
Rösch L. 761
Roewer G. 31, 55, 63, 69, 719
Rose K. 649
Rühlmann K. 619

S

Schär D. 459
Sartori P. 733
Schenzel K. 95
Schierholt T. 317
Schleyer P. v. R. 329
Schmidbaur H. 3
Schmidt H. K. 737
Schmidt M. 665

Schmitzer W. 575
Schubert S. 547
Schultze D. 697
Schuster F. 837
Schwarz W. 493, 511
Schwenk H. 525
Seidler N. 161
Siegl H. 113
Siehl H.-U. 361
Sparrer P. 837
Spielberger A. 101, 105
Spirau F. 709
Stadter H.-J. 837
Stalke D. 519
Steinberger H.-U. 49
Strohmann C. 499
Stüger H. 121
Sünkel K. 547

T

Tacke M. 837
Tacke R. 231, 237, 427, 447, 453
Taylor P. G. 411
Teuber R. 837
Thomas B. 69
Tietze M. 219
Tsuji Y. 361
Turtle R. 411

U

Uhlig F. 109

Uhlig W. 703

V

Veprek S. 815, 821
Vogt F. 31
Voit B. 659

W

Waelbroeck M. 231
Wagner C. 41
Wagner S. A. 237
Walter H. 31
Walter O. 569
Waltz M. 209
Wehmschulte R. 541
Wengert S. 469

Weidenbruch M. 309
Weinmann M. 569
Weis J.   1, 249, 467, 589, 655, 659, 665, 673, 761
Westerhausen M. 161
Wessels M. 243
West R. 251
Wiberg N. 195, 367, 405
Will P. 309
Wörner A. 195
Wrobel D. 633
Wurm K. 725

Z

Zechmann A. 585
Zheng Z. 161
Ziche W. 41

Subject Index

A

α-elimination 289
α-hydroxycarbonylic acids 427
Ab initio calculations
    see calculations
acridones 225
adsorption layer 779
Aerosil 779
agostic interactions 511
alcoholysis 215
allyl cleavage 49
amidosilanes 511
aminofluorosilanes 127
aminosilanes 505
asymmetric silanes 231

B

β-silyl carbocations 361
β-silyl effect 361
β-silyl σ-hyperconjugation 361
Benkeser reaction 3, 133
bidentate ligands 427
biomolecules, immobilization 243, 591
bioorganosilicon chemistry 231, 237
biosensor 591
biotransformation 237

C

cage rearrangement 691
cages 203, 541
calculations 263, 303, 321, 329, 453, 469
carbene 149
carbene complexes 565
carbenoid 149
carbocations (β-silyl) 361
carbon oxysulfide 141
carbonyl ylides 149
catalyst support 685
catalysts 31
catalytic hydrogenation 55, 63
cations (organosilicon) 329
ceramics 709, 725, 733
chalkogen compounds 531
Chalk-Harrod mechanism 633
chlorination 101
chlorosilanes 55
clusters 209, 469
composite materials 737
copolymers 719
core shell structures 665
CP/MAS ^{29}Si NMR 697
crosslinking 649, 655, 719
CVD 295, 815, 821
CVD precursors 3, 815
[n+m] cycloaddition reactions
    see cycloaddition reactions
cycloaddition reactions ... 75, 289, 309, 367, 389, 399, 505
cyclohexasilanes
    see polysilanes, cyclopolysilanes

cyclopentasilanes
   see polysilanes, cyclopolysilanes
cyclopolysilanes 75, 101, 105, 109, 113, 519
cycloreversion reaction 367, 405
cyclotetrasilanes
   see polysilanes, cyclopolysilanes
cyclotrisilane
   see polysilanes, cyclopolysilanes

D

diastereoselectivity 225
diazo compounds 149, 565
dimerization of silaethenes 389
dimerization of silylenes 265
1,2-diphosphaallyl anion 161
diphosphenes 209
Direct Synthesis 31
disilanes 89, 195, 289
disilenes 263, 309
donor adducts 367
donor ligand 459

E

elastomers 659
electrochemical synthesis 709
electroluminescence 821
electron diffraction 95
electron transfer 499
electroreductive coupling 89
enantiomeric silanes 231
equilibrium constants 215

exchange
   see fluorine chlorine exchange
   see ligand exchange

F

filler 761, 779
fluorine chlorine exchange 127
fluoromethylsilanes 19
fumed silica 761

G

gallium silicon compounds 525
germanes 237
germanium, amorphous 815
germenes 367
glucose oxidase 243
graft copolymers 659

H

halogen lithium exchange 499
head to tail dimerization 389
heterobimetallic complexes 553
heterocumulenes 141
heterocycles 141, 219, 289
heteropolysiloxane 737
hexacoordinate silicon 459
hexasilanes
   see polysilanes, cyclopolysilanes
hydrazino ligand 459
hydrodehalogenation 31
hydrogenation, base catalyzed 55, 63

hydrogenation, mechanism 55
hydrogenation, partial 55
hydrolysis 101
hydrosilylation reaction 613, 633
hypercoordination 75, 329
hypersilyl group ((Me$_3$Si)$_3$Si–) 493, 525
hypervalent silicon 411

I

iminosilanes 127
insertion reactions 141, 553
IR cell 837
IR spectroscopy 299, 303
isomerism, rotational 95

K

kinetic studies 215, 411

L

λ^5Si-organosilicates 427, 447
ligand exchange 447
liquid xenon 837

M

MALDI-MS 625
matrix isolation spectroscopy 303
metal complexes 41, 127, 493, 531, 547, 553, 569, 585
metalla-sila-butadienes 569
metallo siloxane 575
metallosilatranes 541

microbial reduction 237
microemulsion 665
micronetworks 665
microparticles 673
MO diagrams 263

N

nanocrystalline silicon 821
network 779
nonclassical structures 329
nucleophilic substitution 411

O

oligosilanes
   see polysilanes
oligosilanylamines 121
oligosilylsulfanes 121
optical active silanes 237
optical sensors 650
organosilicates 453
oxofunctionalization 575

P

palladium complexes 553
pentacoordination 133, 411, 427, 447, 459
pentafulvenes 399
pentasilanes
   see polysilanes, cyclopolysilanes
Peterson reaction 389
phophaalkynes 161

phophorus-germanium clusters
  see clusters
phophorus-silicon clusters
  see clusters
phosphaalkenes 161
2-phosphaallyl anion 161
phosphine ligands 553
phosphinomethanides 187
phosphinoylide 187
phospholides 161
photoluminescence 821
photosensitizer 655
photolysis 46, 309
platinum complexes 553
poly(silyl)arenes 3
polycarbosilanes 733
polycarbosilazanes 709
polycondensation 703
polydimethylsiloxane 779
polymer films 609
polymeric initiators 659
polymerization kinetics 633
polymerization, radical 659
polyorganosiloxanes 659
polysilacarbosilanes 709
polysilanes 69, 81, 89, 95, 121, 519, 719
polysilasilazane 709
polysiloxane fibers 650
porous silicon 821
postcondensation 697

R

radical anions 105
radicals 105
rate constants 215
receptor affinities 235
reductive cleavage 499
reductive coupling 703
reinforcement 761
relaxation measurement 779
ring opening polymerization 619
ring systems 505

S

saccharomyces cerevisiae 237
sacrificial anode 709
salt elimination 367
SiC deposition 295
SiC fibers 719, 733
SiC precursors 709
silabutadienes 41
silacyclobutanes 399
silacyclobutenes 41
silacyclopropane 75
silacyclopropyne 303
silaethene 41, 49, 149, 367, 389, 399
silaheterocycles 187, 399
silaisonitrile 321
silanediol 575
silanediyl complexes 41, 251, 569
silaneimine 367, 405

silanetriol	575
silanitrile	321
silanol	215, 575
silanorbornene	49
silatetrazolines	405
silatranes	541
silenes	
see silaethenes	
silicates	447
silicenium ion	329, 459, 547
silicides	31, 469, 585
silicon antimony cages	
see cages	
silicon arsenic cages	
see cages	
silicon bismuth cages	
see cages	
silicon carbide	
see SiC	
silicon networks	469
silicon phosphorous cages	
see cages	
silicon polyanions	469
silicon radicals	105
silicon resins	697
silicon single crystal	837
silicone rubber	633
siliconium cation	459
siloxanes	3, 613, 619, 625, 673
siloxybenzothiopyrylium salts	219
silsesquioxane	685, 691
silyl amides	505
silyl amines	3
silyl hydrazines	3
silyl hydroxylamines	3
silyl metal complexes	41
silyl metal compounds	109, 519
silyl migration	553
silyl triflates	3, 81, 703
silylation	761
silylcarbodiimides	725
silylene	31, 133, 213, 251, 263, 289, 303, 309
silylene dimerization	263
silylenolether	219
silyloxyquinolinium salts	225
silylphosphanes	161
silylphosphanides	161
Si/Ni cages	
see cages	
^{29}Si NMR spectroscopy	69, 95, 113, 685, 698, 779
sol-gel chemistry	650, 737
spirocycles	453
stannenes	367
sulfur heterocycles	
see heterocycles	
supersilyl group (SitBu$_3$)	195, 367
surface chemistry	243, 591, 609, 761
surface tension measurement	668
surfactants	613, 665

T

tetrasilatetrahedranes 367
thermoanalysis 697
thermodynamic studies............................ 417
thermoplastics 659
thickening 761
thiochromanones 219
thiourethane 141
transmetallation 105
trimethylsilyl group 161
triphosphinogermane 209
triphosphinosilanes 209
trisilanes
 see polysilanes, cyclopolysilanes
tungsten silicide 585

U

Umpolung 133
UV curing 650
UV-Vis spectra 95, 263, 265

V

vinyl cations 361
vulcanization 633

W

wetting agents 613
Wolff rearrangement 149

Y

Yajima Process 709

Z

Zintl-Klemm concept 469
Zintl phases 469
zwitterions 427, 447, 453